Structural Geology of
Folded Rocks

Structural Geology of Folded Rocks

E. H. TIMOTHY WHITTEN
Professor of Geology
Northwestern University
Evanston, Illinois

RAND MᶜNALLY & COMPANY, CHICAGO

RAND McNALLY GEOLOGY SERIES

L. L. SLOSS, ADVISORY EDITOR

Preface

Within the past decade tremendous strides have been made in many branches of structural geology and an adequate treatment of the whole subject would now require a series of books. No attempt at such an encyclopaedic coverage is made here. Rather, this book is primarily concerned with current methods of studying the geometry of folded rocks and the associated minor structures. An attempt has been made to synthesize the scattered literature on these subjects into a single volume, and to present it in a format suitable for both senior undergraduate and graduate student teaching. This subject matter is not dealt with extensively in available textbooks, but the need for a book covering this field became apparent during the course of teaching at the University of London and Northwestern University. Until now many of the techniques and concepts currently used in field work have only been described in scattered journal literature. However, this book is no more than a guide, and those interested in embarking on research have no alternative but to read the original journal articles. For this reason every attempt has been made to refer to the original journal sources throughout the text.

It is considered that the approach outlined in this book can and should be introduced at an early stage of undergraduate work in structural geology. However, a well-rounded course of geologic training would naturally include instruction in some related subjects that have traditionally been considered part of structural geology, but which are not dealt with in this volume. For example, no attempt is made here to cover such subjects as the physical principles governing rock deformation, the development of mineral fabrics, the mechanical properties of rocks, regional structures and tectonics in relation to orogenesis, continental shields, or ocean basins, fault phenomena, or the structural characteristics of igneous rocks. The experimental approach to structural geology is extremely important, but it is axiomatic that the experimenter must have an accurate and correct picture of the quantitative nature and three-dimensional variability of natural features before the adequacy of a simulated example can be appraised.

The impression is often given that the nature of geological structures is well known; in fact, a search of the literature reveals an amazing dearth of published quantitative data on almost all features of folded rocks. A deliberate attempt is made here to base descriptions on actual described field relationships; this approach is intended to emphasize the limitations of present information and knowledge. A more generalized approach has been avoided because it tends to suggest that all initial problems have been solved, that elementary fold and fault relationships can always be detected easily in the field, and that all field examples conform to the categories defined in an elementary textbook. Many of the challenging and perplexing unsolved problems relate to the more obvious phenomena (e.g., the nature, three-dimensional variability, and origin of foliation in slates).

In practice, structures are commonly very variable and particular field relationships are rarely completely unequivocal. However, the student commonly finds it difficult to visualize the tremendous variability that exists in simple structures observed in the field. Teaching in London, England, and in Evanston, Illinois, emphasized that many students study structural geology without the benefit of *concurrent* observation of structural elements in the field. In this unhappy situation reliance has to be placed on occasional long field excursions to areas where structures can be seen. This creates a special need for accurate descriptions of actual field relationships that permit the reader to build correct mental pictures in advance of seeing the real things in the field. Unfortunately, published descriptions often do not permit the reader to erect a correct mental picture of the field relationships that were originally observed. It is the present thesis that field relationships must be the final arbiter, and that no amount of reading and laboratory study can be a complete substitute for observing the rocks themselves.

Another problem associated with students' reading of journal literature is terminology. The lack of a consistent terminology in structural geology makes it extraordinarily difficult to compare and contrast descriptions for different areas. It is often difficult to be sure whether two authors have described similar or different structural elements. An attempt is made to use a single consistent terminology in this book. Until rigorous operational definitions have been agreed upon for each structural element, there is little point in arguing about the best descriptive term, provided that unequivocal communication is established between all scientists. The attempt to constrain description within a single system of nomenclature has necessitated interpretation of the terminology used by others, and it is possible that erroneous "translations" have been introduced. Anyone who has struggled with the literature on cleavage and foliation will appreciate the impossible semantic problems involved.

For many years structural geology, like most of geology, was a qualitative art heavily dependent on the personal experience and intuition of individual geologists. This has precluded or seriously hindered direct comparison and correlation of work published by different people. At present geology is passing through a period of metamorphosis from the qualitative to the quantitative. Structural geology is rapidly growing into a quantitative science, and in the sequel an attempt is made to indicate some of the ways in which this change can affect the study of folded rocks. Rigorous operational definitions are being evolved slowly, a start has been made with investigating problems of probability sampling, and in a few cases structural elements are being described in terms that permit them to be handled by high speed computers. Although most of this work is still in its infancy, it is developing with most promising results. Because a large proportion of both undergraduates and graduate students has now had experience in matrix algebra and computer programming, these tools can be easily and appropriately introduced into the teaching of, and research on, structural geology. It would be unwise to preserve traditional qualitative approaches to structural geology when more appropriate and more powerful tools for advance become available. There is no question but that a decade hence geology will have emerged as a quantitative science, although, in some circles at the present time, there is militant resistance and scepticism about the initiation of quantitative approaches in any branch of geology, and about the gradual replacement of some traditional qualitative methods of research.

One of the more stimulating concepts introduced into geology within the past few years is that of the *model*. A large part of Chapter 14 is devoted to this topic. Structural geology, as a branch of physical science, is rapidly progressing to the stage in which observed elements can be viewed in the framework of a web of processes and responses (i.e., causes and effects) that will eventually be expressed mathematically (e.g., as differential equations) in terms of deterministic models. Any conceptual models based on experimental results must be controlled by reference to structures that actually occur in nature. Geologists are still some distance from erecting deterministic models on the bases of observed data, but the rapid growth of quantitative experimental and observational data is making the study of models an urgent necessity. The fact that models are discussed in Chapter 14 does not reflect relegation to last place — the topic should possibly have been treated within the first four chapters.

Unfortunately, there may appear to be a dichotomy between certain sections of this book because, in the present transitional state between the qualitative and quantitative study of folds, data are not available in many areas of the subject. At this time a wholly quantitative and integrated picture is impossible, and many chapters are, of necessity, purely descriptive. Marshalling the available descriptive information shows where information is lacking, and it is to be hoped that over the next few years appropriate quantitative data will be collected. Only then will it be possible for realistic statements to be made about the quantitative nature and three-dimensional variability of different structural elements. The first steps in the correct direction must lie in (a) obtaining an intimate knowledge of the rock structures and in identification of the variables involved, and (b) exposing the problems of sampling and of making quantitative syntheses of data so that structural geology field problems can be thought about within a quantitative framework. The former objective is sufficient justification for including the qualitative and descriptive material in this book.

Although many characteristics (attributes) of folded rocks have been studied for many years, the descriptions have often remained extraordinarily incomplete. Numerous additional variables are now being sought to provide more suitable data for quantitative analyses and syntheses. For many structural elements descriptions have only been published for one or two localities so that there is little information about their variability in space and time. It has been common in geology for world-wide generalizations to be based on detailed experience from one or two localities; two classic examples are the world-wide magma types and the Barrovian zones of regional metamorphism. At best this is a dubious procedure because extrapolation on the basis of field work within one or two small areas can very easily lead to erroneous concepts.

The problems associated with extrapolation and synthesis on the basis of small local observations are topics that have received extraordinarily little attention in the earth sciences. Fortunately, there is now a growing appreciation that the rigor of statistical sampling methods must be utilized in *all* geological work, not least in structural geology.

In the last ten years interest in the geometry of small-sized structural features has blossomed. This work has been vitally rewarding, and now it is beginning to be possible to "see the wood for the trees." It is not always easy to recognize the inherent problems associated with extrapolation from a described area to prediction about larger regions. For example, in several orogenic areas superposed folding has been recognized in regionally-metamorphosed rocks—events F_1, F_2, F_3, etc. Is each a local phase to be recognized and correlated over a square kilometer or two, or is each phase—F_1, F_2, etc.—characteristically of regional significance and to be correlated over several hundred square kilometers? What type of process or event gives rise to the fold phases in the metamorphic rocks, and what, if any, reflection of these events is to be expected at the Earth's surface and in the near-surface rocks? Is an event recognized as, say, F_2 strictly contemporaneous throughout the area in which it is mapped? The geologist studying structures in the field tends to be impressed by the relatively small-sized fold structures, but what criteria can be used to determine whether conclusions based on them are of regional significance, or whether folds of a different order of magnitude are intrinsically more significant? For example, folds seen in the field may simply serve to obscure a major nappe structure that is the dominating element of the region. Gravity tectonics and other equally-dramatic orogenic processes have been invoked to account for structural patterns, but it seems reasonable to insist that answers to some of these more limited geometrical questions must be forthcoming before major genetic hypotheses can be invoked for any particular terrane. Such questions are also of tremendous importance in any attempt to prescribe the range of physical conditions under which a set of rocks was deformed, and thus in defining the appropriate experimental conditions under which simulated structures should be studied.

Thus, this book is an attempt to portray the nature of the folded rocks and the various structures that are associated with them. Although only a part of "structural geology," this subject is at the core of the whole, and a thorough understanding of it appears to be a prerequisite to advance in most other segments of structural geology.

Structural geology is a live and rapidly developing subject that should not be relegated to the status of a discipline of nomenclature and classification only. An attempt is made to escape from an encyclopaedic catalog of structural forms. Although many of the concepts and techniques outlined in the sequel were developed for the study and analysis of metamorphic rocks, it is important to emphasize that this book is not about folds of metamorphosed rocks only. Unfortunately, many of the examples are drawn from metamorphic terrains; commonly this reflects the activity of geologists rather than that a technique is not appropriate in the study of folded unmetamorphosed rocks. It is hoped that this book will be useful to the students of the folds that occur in all types of rocks. Structural geology is a challenging and stimulating branch of the earth sciences that, with the aid of new tools, is on the threshold of tremendous advances.

E.H.T.W.

Evanston, Illinois
June, 1965

ACKNOWLEDGEMENT

Several geologists have kindly made original photographs available to me. This book would not have been possible without the liberal use of these and also of numerous illustrations and quotations published in the geological literature. Where such material has been used the original source is clearly indicated in the figure caption or in the reference accompanying each quotation. Grateful acknowledgement for permission to reproduce copyright quotations and illustrative material is made to:

Åbo Akademis Bibliotek, Finland.
Professor Sam L. Agron, U. S. A.
The Alberta Society of Petroleum Geologists, Canada.
The American Geophysical Union, U. S. A.
The American Journal of Science, U. S. A.
Edward Arnold (Publishers) Ltd., Great Britain.
Stephen Austin and Sons, Ltd. (*Geological Magazine*), Great Britain.
The British Association for the Advancement of Science, Great Britain.
Bundesanstalt für Bodenforschung, Germany.
Bureau de recherches géologiques et minières, France.
Canadian Institute of Mining and Metallurgy, Canada.
E. P. Dutton & Co., Inc., U. S. A.
The Edinburgh Geological Society, Great Britain.
W. H. Freeman & Co., U. S. A.
The Geological Journal, Great Britain.
The Geological Association of Canada, Canada.
The Geological Society of America, U. S. A.
The Geological Society of Australia, Australia.
The Geological Society of India, India.
The Geological Society of London, Great Britain.
The Geological Society of South Africa, South Africa.
Geologie en Mijnbouw, The Netherlands.
Geologische Rundschau, Germany.
The Geologists' Association, Great Britain.
The Geological Survey and Museum, Great Britain.
The Geological Survey of Canada, Canada.
Geologinen Tutkimuslaitos, Finland.
Her Britannic Majesty's Stationery Office, Great Britain.
Hessisches Landesamt für Bodenforschung, Germany.
Instituto Geológico y Minero de España.
The Johns Hopkins University (Geology Department), U. S. A.

Journal of Science of the Hiroshima University, Japan.
Macmillan & Co. Ltd., Great Britain.
Manchester University Press, Great Britain.
McGraw-Hill Book Company, Inc., U. S. A.
Meddelelser om Grønland, Denmark.
Methuen & Co., Ltd., Great Britain.
Professor Adolf Metzger, Finland.
The Mineralogical Society of London, Great Britain.
Nature, Great Britain.
New York State Museum and Science Service, U. S. A.
Norsk Geologisk Tidsskrift, Norway.
Oliver & Boyd, Ltd., Great Britain.
The Clarendon Press, Oxford, Great Britain.
Prentice-Hall, Inc., U. S. A.
Province of Saskatchewan, Department of Mineral Resources (Geological
    Sciences Branch), Canada.
Royal Irish Academy, Ireland.
Royal Society of Edinburgh, Great Britain.
E. Schweizerbartische Verlagsbuchhandlung, Germany.
Société Géologique de Belgique, Belgium.
Springer-Verlag, Austria.
University of California Press, U. S. A.
University of Chicago Press, U. S. A.
University of Groningen (Geology Department), The Netherlands.
University of Leiden (Geology Institute), The Netherlands.
University of Utrecht (Geology Institute), The Netherlands.
University of Upsala (Mineralogy-Geology Institute), Sweden.
Virginia Polytechnic Institute (Geological Sciences Department), U. S. A.
John Wiley & Sons, Inc., U. S. A.
Dr. Mary Vogt Woodland, U. S. A.
The Yorkshire Geological Society, Great Britain.

An early draft of Chapter 3 was read and criticized by Professors E. C.
Dapples, A. L. Howland, W. C. Krumbein, L. H. Nobles, and L. L. Sloss, and
most of my 1963 students at Northwestern University. Early ideas about the
model concept were read and debated by my colleagues in the Geology Department at Northwestern University; Chapter 14 was read by the same group and
subsequently revised; discussions with Dr. T. V. Loudon materially influenced
the presentation in Chapter 14.

Contents

Preface   v

Chapter
1. *Recording and presentation of structural relationships*   1
    Maps   1
    Stereographic and equal-area projections   10
       Wulff stereographic projection   10
       Schmidt equal-area projection   18
       Examples of problems involving stereographic manipulation   26
    Block diagrams   31
    Collection of oriented hand specimens   35
2. *Cylindroidal and noncylindroidal folds*   36
    Cylindroidal folds   37
    Noncylindroidal folds   63
    Orientation of certain folds   66
3. *Sampling and size in data collection*   70
    The concept of the population   71
    Patterns of variability of attributes   74
       Nondirectional attributes   76
       Directional attributes   78
    Size and patterns of variability of attributes   80
       Size and the population of objects   82
       Size and the population of attributes   85
    Size terminology   86
    Size and the study of folded rocks   88
    Statistical evaluation of structural data   90
4. *S-structures, tectonites, and axes*   96
5. *Symmetry concepts*   112
    Symmetry of fabric elements in observed rocks   115
    Symmetry of the movement picture   119
    Fabric elements and the movement picture   123
       Noncrystallographic fabric elements   123
       Crystallographic fabric elements   127
6. *Simple types of folds*   130
    The two principal varieties of folds   131
       Slip folds   133
       Flexural-slip folds   147
       Slip and flexural-slip folds contrasted   156
       Parasitic or "drag" folds   164
       Flexure folds   169
       Spatial relationships of slip and flexural-slip folds   170

Other varieties of folds   172
  Folds due to vertical movements   172
  Flow folds   173
  Folds in certain unconsolidated and near-surface rocks   175
  Fold terminology   178

7. *Disharmonic folds and transposition structures*   179
   Transposition structures   181
   Transposition structures: small examples   186
   Examples of large transposition structures   199

8. *Foliation*   216
   The terminology of foliation and rock cleavage   216
   Classification of foliation structures   220
     Axial-plane foliation   221
     Crenulation foliation   230
     Foliation parallel to bedding or lithologic layering   256
     "Fracture cleavage" — a variety of closely-spaced jointing   260
   Conclusion   262

9. *Linear structures*   264
   Important varieties of linear structure   267
     Lineations   267
     Other linear structures   293

10. *Superposed folds*   322
    Classification   323
      The classification of Sander (1948)   325
      The proposed classification   327
    The geometry of superposed flexural-slip folds   329
    The geometry of superposed slip folds   340
    Concluding remarks   357

11. *Examples of superposed folding*   358
    $B_1 \wedge B_2$-folds   358
      B_1 and B_2 approximately parallel   359
      B_1 and B_2 inclined   363
    $B_1 \perp B_2$-folding   472

12. *Microtextures in fabric analyses*   477
    Microtextural analysis   478
      Central Highlands of Scotland   484
      Central Pyrenees, Spain and France   492
      General discussion   501
    Microtextural evidence concerning porphyroblast growth   505
    Microtextural analyses in other regions   507
    Concluding remarks   511

13. *Sedimentary characteristics preserved in folded and metamorphosed
    rocks*   514
    Cross-stratification (cross-bedding) relicts   517
      Flexural-slip folding of cross-stratification   522
      Slip folding of cross-stratification   527
    Relicts of graded bedding   532
    Relicts of other sedimentary features   536
    Stratigraphic relationships   540

14. *Models and new methods for quantitative fold description*   545
    Process-response models   545
        When processes can be observed   549
        When actual processes cannot be observed   551
        Attributes and samples from structural complexes   554
        Applications of process-response models to structural geology   556
    Response characteristics and directional attributes for quantitative
    analyses   569
        Operations involving the use of direction cosines   575
        Some new uses of quantitative attributes of structural complexes   606
15. *Review of terminology used for folds*   613
    Types of fold and folding   616
16. *References cited*   640
Index   680

Chapter 1

Recording and Presentation of Structural Relationships

"Structural geology is handicapped by the difficulty of representing three-dimensional data by means of two-dimensional maps, plans, and sections" (Johnston and Nolan, 1937, p. 550). However, the analysis of folded rocks necessitates recording the three-dimensional features on two-dimensional paper. Several conventional methods are commonly employed, and some of these are described in this chapter so that they can be used as a basis for later descriptions.

The principal techniques in the study of folded rocks involve structural maps, block diagrams, and graphic projections on the basis of measurements and observations made in the field.

MAPS

Individual outcrops of each lithologic unit are plotted either on a standard large-scale contoured topographic map or on vertical aerial photographs. In reconnaissance work these base maps can be on a scale of, say, 1:60,000 (i.e., about 1 inch to 1 mile), but for detailed structural work the base map needs to be on a scale of at least 1:10,000; in the case of very complex areas a 1:2,500 scale (i.e., about 24 inches to the mile) is appropriate. In an area of folded sedimentary rocks the orientation of bedding planes must be carefully measured and plotted at the site of

1

observation on the map. The most effective method is to draw a line to represent the strike of the bed, and to indicate the dip by a small check mark on the strike line (Fig. 1). Commonly it is both convenient and timesaving to record the angular measurements in a notebook or on reference cards, and also to plot them on the map. The azimuth can be indicated by a three-figure number recording the degrees east of north (which thus range from 000 to 360), while the dip, which varies from 0 to 90 degrees, is shown by a two-figure number. Thus, in Figure 1A, $\frac{275}{62}$ represents a strike of 5° north of west, and a dip of 62° to the south. The intersection of the strike line and the check mark is plotted at the precise outcrop location on the map.

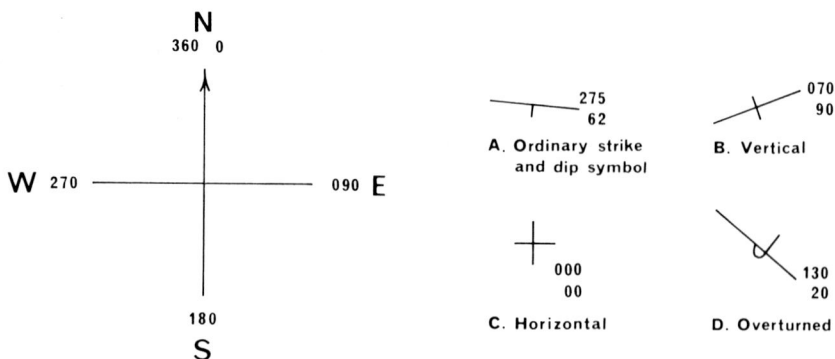

Figure 1. Compass bearings and map symbols to represent the dip and strike of a plane.

Commonly lithologic units, or the contacts between lithologic units, cannot be traced continuously across the countryside. Less complete maps may be adequate in reconnaissance or small-scale mapping projects, but in any detailed study the first step involves recording details of all the available outcrops on a map. It is important to be able to use the map to predict where outcrops should be expected, and where the mapped beds should be in the subsurface. Simple methods permit such predictions in restricted areas within which the structure is not too complex.

Figure 2 is a schematic map with 100-meter contour intervals. Suppose that strata have uniform dips throughout the map area and that at localities A, B, and C the junction between two well-defined lithological units is exposed.

On the basis of this information the geological map can be completed by extrapolation, although in an actual mapping project much more data would be required before an adequate map could be prepared. This example is only used to illustrate some of the cartographic techniques.

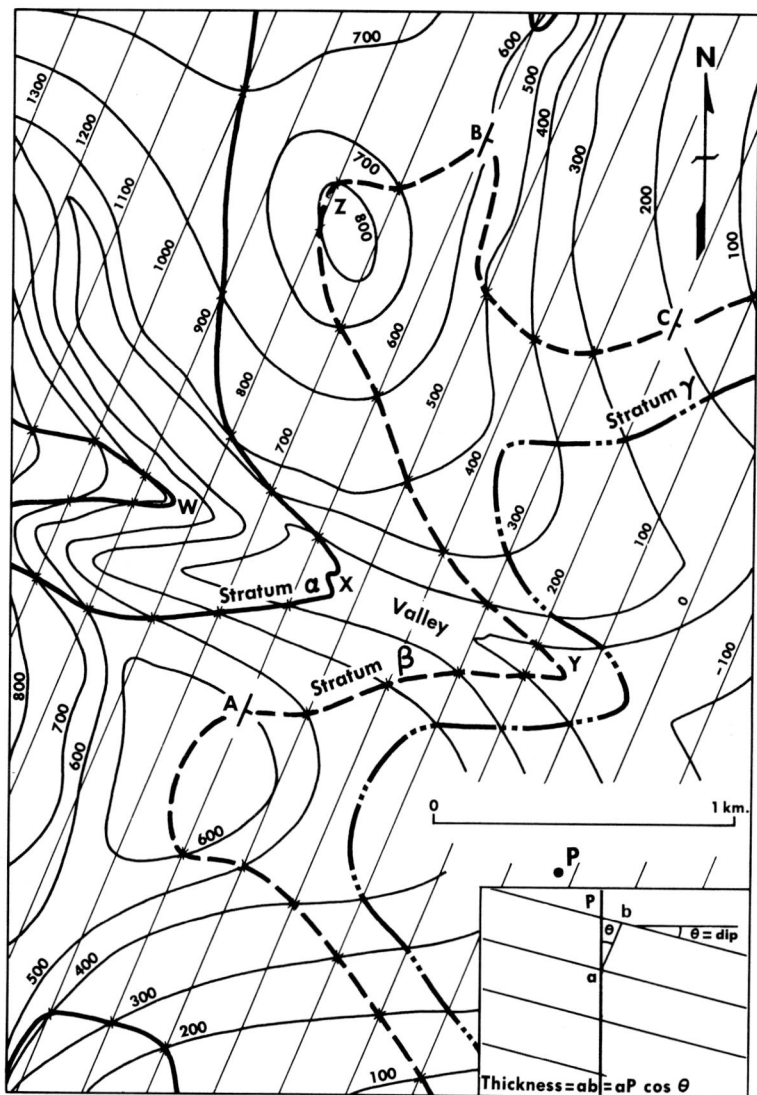

Figure 2. Schematic map with 100-meter contour interval to illustrate method of using stratum contours to construct an outcrop map; see text for explanation. *Inset*: Section across part of map (perpendicular to strike lines) to demonstrate how true and apparent bed thickness are related.

The elevation of localities A and B is 600 meters, so that a contour (stratum contour) drawn on the exposed stratum (stratum β) is represented by the line joining A to B. Because of the uniform dip the other stratum contours are parallel to AB, and, since C is at 200 meters, the 200-meter stratum contour passes through C. Keeping an even spacing between stratum contours the 300-, 400-, and 500-meter contours can be interpolated, together with contours for values above 600 meters and below 200 meters.

Commonly each stratum contour lies partly above and partly below the ground surface (indicated by the topographic contours). For example, the stratum contour for 600 meters between A to B lies well above the floor of the valley (Fig. 2). Theoretically, actual outcrops of the horizon (stratum β) observed at A, B, and C occur at every point where the stratum contours and the topographic contours intersect. If these intersections are marked on the map, the continuous line joining them represents the outcrop trace of the bed. It may be necessary to interpolate additional stratum contours or topographic contours at certain critical points; such points occur, for example, at Y and Z in Figure 2.

Suppose that a well is sunk at P (southeast corner Fig. 2) to determine the sequence of strata; if the boundary of the next unit of interest (stratum γ) is found in the well 100 meters above stratum β, the stratum contours can be relabelled with values 100 greater than previously (i.e., 100 greater than labelled in Fig. 2). The new intersections with the topographic contours permit this higher horizon to be traced on the map (stratum γ on Fig. 2). Similarly, by decreasing the original stratum-contour values by 300, the outcrop can be located for a unit (stratum α) recorded 300 meters below stratum β in the bore hole P. The outcrop of stratum α is shown in Figure 2; the usefulness of interpolated contours at W and X should be verified by the reader.

The depth below the ground surface of a particular stratum is found by subtracting the stratum contour value from the topographic contour value. However, the vertical distances between the lithological boundaries (used in the construction of Fig. 2) are smaller than the true stratigraphic thickness (measured perpendicular to the bedding). The inset to Figure 2 shows a vertical section through part of the map area. If Pa is the vertical distance between two units measured in a vertical bore hole, and θ is the angle of dip of the bedding planes, then the true thickness (ab) of the unit is ($Pa \cdot \cos \theta$). Similarly, if a unit has a thickness x and dip θ, the vertical distance between the stratum contours is $x/\cos \theta$.

Because A and B (Fig. 2) are at the same elevation, the strike of the stratum contours is defined immediately, but the strike and dip can be determined for a bed with uniform dip provided that any three points on the plane are known. For example, in Figure 3 suppose that actual outcrops of stratum β (of Fig. 2) are found at localities C, D, and E. Since the elevation of C is 200 meters and of D is 800 meters, the join CD can be divided into six equal units (Fig. 3); similarly, since E is at 500 meters, CE can be divided into three equal units to locate 300- and 400-meter points on the stratum surface. The stratum contours are drawn by joining

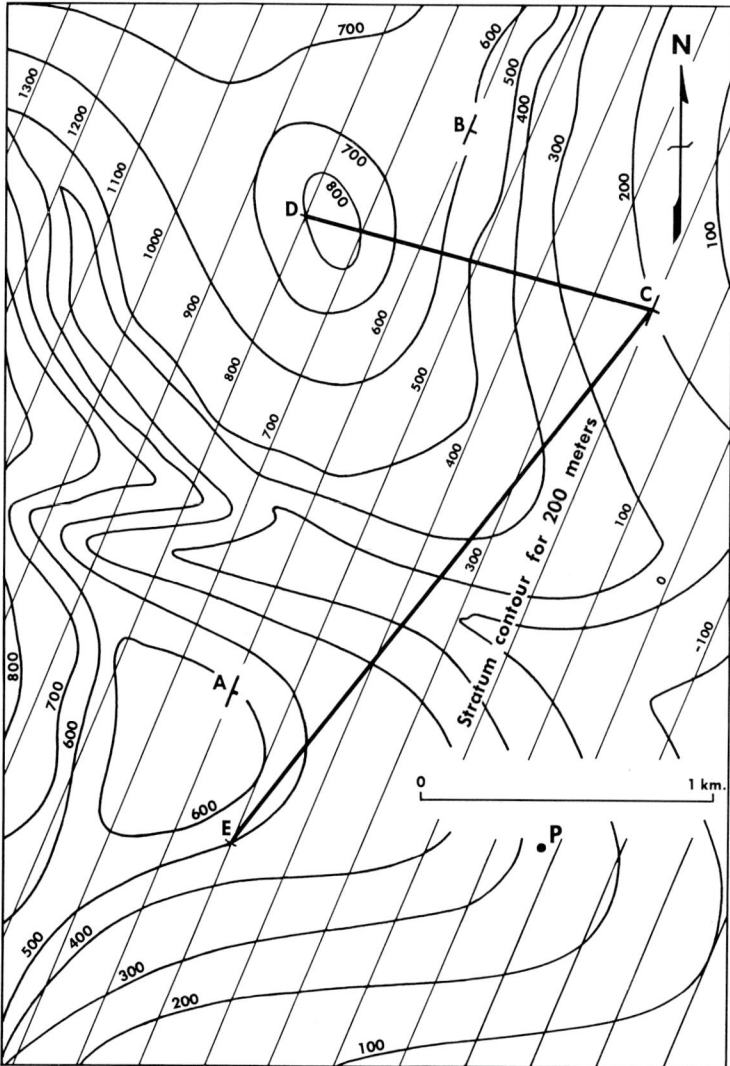

Figure 3. Schematic map (as Fig. 2) to demonstrate three-point method of constructing stratum contours; see text for explanation.

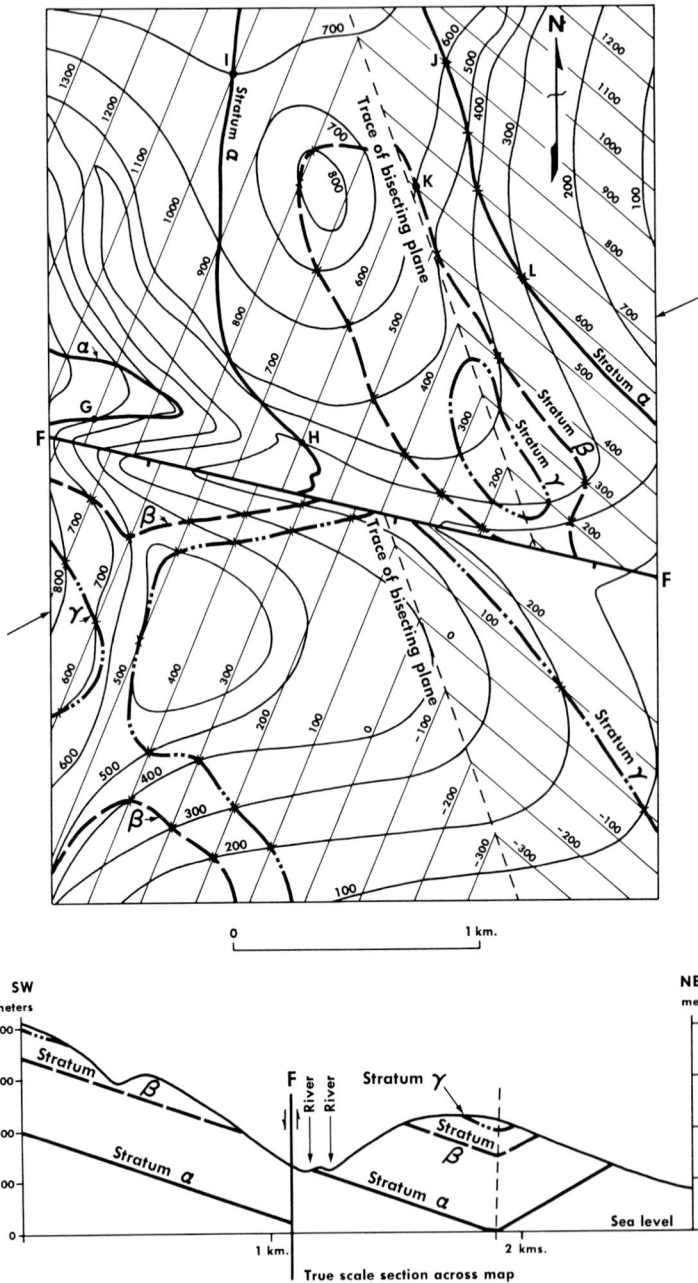

Figure 4. A: Schematic map (topography as in Fig. 2) to show use of stratum contours for a folded sequence of parallel strata; line F-F is a normal vertical fault with downthrow to south. B: True scale section along the line indicated by arrows at the margin of map; see text for explanation.

points of similar elevation on CD and CE; stratum contours can be constructed for the entire map by drawing additional parallel lines at equal spacing.

A more complex situation arises if the stratigraphic units are folded; in such a case stratum contours can be constructed with respect to each limb of the folded units. In Figure 4 the stratigraphic succession and the topography are that of Figure 2; suppose the uniform dip of Figure 2 is confirmed by exposures of the lowest mapped boundary (stratum α) at G, H, and I in the northwest corner of Figure 4. The same stratum (α) outcrops at J and L, and is found 300 meters down a well at K (Fig. 4); in these three localities the dip is to the southwest. Each strike line drawn with respect to J, K, and L is confluent with a contour of similar elevation drawn for G, H, and I. These stratum contours (Fig. 4) show that each lithologic unit in the stratigraphic succession is folded into a V-shaped structure with the base of the V tilted down (**plunging**) to the north.

The outcrop of the three folded units can be completed with the method used for Figure 2. The line F-F indicates a fault with a down-throw of 250 meters to the south. Provided the southern block is simply let down 250 meters the strike lines remain parallel on both sides of the fault, but the value of each stratum contour is decreased by 250 on passing southwards across the fault trace (Fig. 4). After relabelling the stratum contours south of the fault (Fig. 4) the outcrop pattern can be completed. South of the fault, stratum α does not rise to the present ground surface.

It is important to emphasize that Figures 2 through 4 are purely schematic; in practice individual beds are rarely perfectly planar, and folds do not tend to possess the geometric perfection shown in Figure 4. The fold in Figure 4 is not perfectly symmetrical because the strike lines for the southwesterly-dipping beds are closer together than those for the southeasterly-dipping beds; the former beds dip at about 33°, and the latter at about 26°.

The effect of topography on the shape of the outcrops is most marked. If the fold geometry of Figure 4 is maintained, but the ground surface is bevelled off to a uniform 100 meters above sea level, the map pattern changes radically (Fig. 5). The plane bisecting the angle between the fold limbs is the **bisecting plane**, and its trace on the ground surface is identified easily in Figure 5. This plane has the same position and orientation in Figure 4, although, because the topography controls the apparent shape of the fold, the relationships are less obvious.

In Figure 4 the fault F-F was assumed to be vertical, although fault planes are usually inclined to the vertical in nature. When a fault plane is less nearly vertical, it has a more sinuous trace. A low angle thrust fault could have an outcrop trace as sinuous as the outcrop traces in Figure 2. If the fault (of Fig. 4) is assumed to dip steeply to the south, strike lines can be drawn on the fault plane, and the resulting curved trace of the fault on the ground surface is shown in Figure 6.

These relationships illustrate a general rule. Any vertical plane (e.g., a dyke, vertical bedding plane, vertical foliation in schists, etc.) appears as a straight line on a map, no matter how rugged the topography. By contrast,

Figure 5. Schematic map showing the same geology as Figure 4A, but with the ground surface bevelled off to a horizontal plane 100 meters above sea-level.

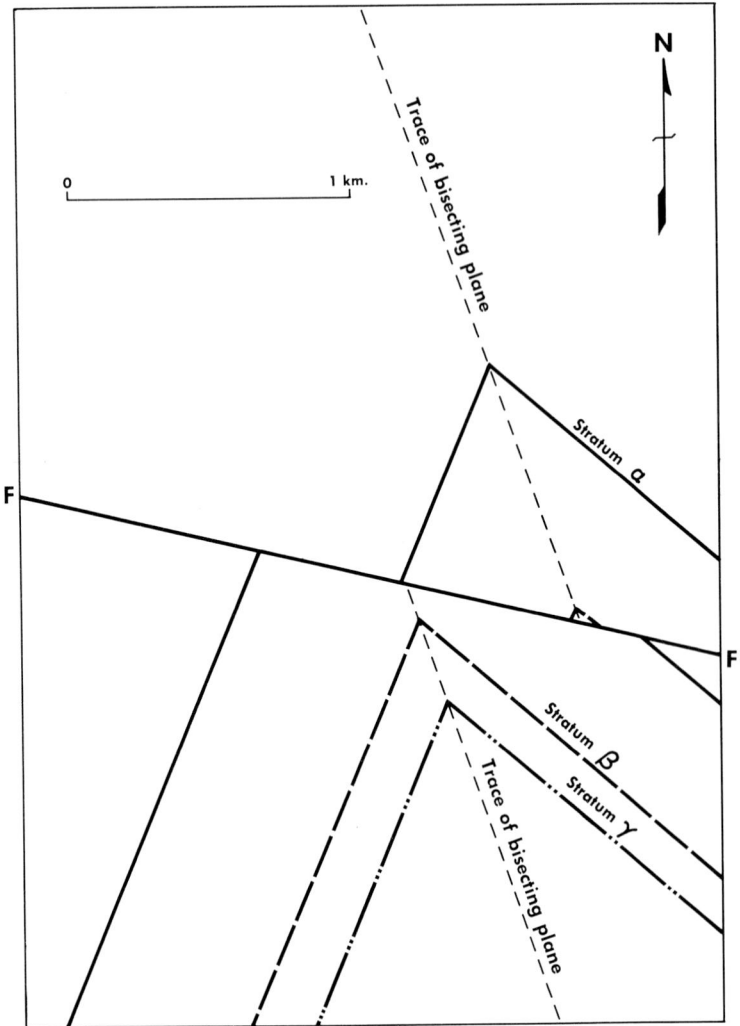

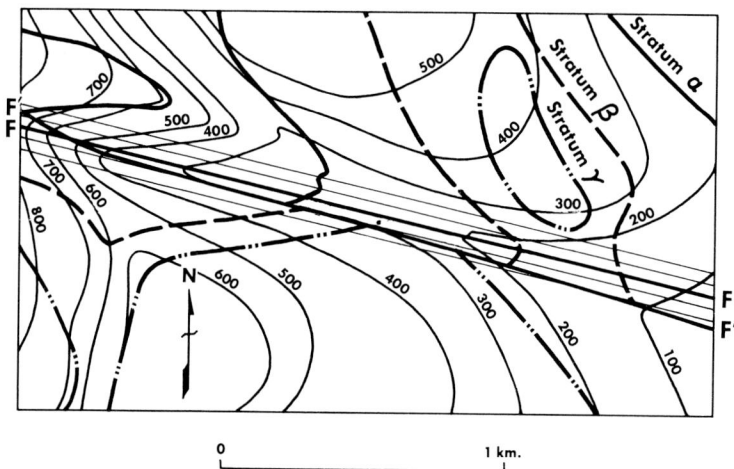

Figure 6. The schematic map in Figure 4A redrawn on the assumption that the fault plane dips southwards (instead of being vertical); stratum contours on the fault plane are shown (parallel to F-F) and F'-F' is the trace of the fault on the ground surface.

a horizontal stratum has a very sinuous outcrop pattern, especially in areas with deeply dissected topography, because the outcrop traces are parallel to the topographic contours. Naturally, every gradation between the straight outcrop of a vertical unit and the sinuous outcrop of a horizontal unit occurs in nature.

As an exercise these relationships can be established by modifying Figure 2. First, reproduce the topography shown in Figure 2 on a sheet of tracing paper. Now, complete the outcrop of the units between strata β and α, and between β and γ, on the assumption that stratum β outcrops at A and B, and that all the units are horizontal. For contrast, on another tracing, use the strike lines on Figure 2, but now make the dip to the northwest, instead of to the southeast. Complete the outcrop map for α, β, and γ on the assumption that β outcrops at A and B. Notice that the outcrops now V up the valleys and down the hill ridges, whereas in Figure 2 the outcrops V in the opposite direction. Finally, insert the outcrop of a vertical dyke which outcrops at A and B.

In field work, as well as in the study of maps, it is important to recognize the effects of varied topography on outcrop patterns.

In succeeding chapters maps of structural features are used extensively; the map is undoubtedly the basic method of representing structural relationships. However, other valuable methods of portraying three-dimensional relationships are in common use. Block diagrams commonly provide a very clear picture of the three-dimensional geometry of structures, and in recent years stereographic and equal-area projections have played an important role. These important projections are used extensively in this book, and their underlying principles need to be understood thoroughly.

STEREOGRAPHIC AND EQUAL-AREA PROJECTIONS

Many different methods of projection are used by cartographers in the preparation of maps of the curved surface of the Earth. However, distortion of some variables measured in three-dimensional space always arises when structural data are projected onto two-dimensional paper. For example, some map projections preserve areas correctly, some the angular relationships.

Two different varieties of projection are in common use in structural geology (e.g., Bucher, 1944; Phillips, 1960; Turner and Weiss, 1963). In many instances, as in crystallography, it is desirable to use a projection in which angular measurements are not distorted; in these circumstances the Wulff projection is used, which was named after the Russian crystallographer G. V. Wulff who published the first stereographic net to facilitate plotting observations in 1902, although the method was known in ancient Greece in the second century B.C. A slightly different projection, which preserves areas correctly, is of particular importance when any statistical appraisal of the density of structural elements in different orientations is required; Lambert invented a projection which meets these requirements, but Schmidt (1925) introduced its use to structural geology; in consequence, geologists usually refer to it as the Schmidt projection.

WULFF STEREOGRAPHIC PROJECTION

The principle of this projection is that points on the surface of a sphere are projected onto the equatorial plane by viewing each point through this plane from the north pole (Fig. 7A). Any circle drawn on the surface of the sphere is projected as a circle on the projection (Fig. 7B).

In structural work planes and lines (linear elements) are the principal geometric elements measured. Although measurements in the field are collected from different geographic sample sites, for projection purposes each observation is considered to be at the center of the projection sphere.

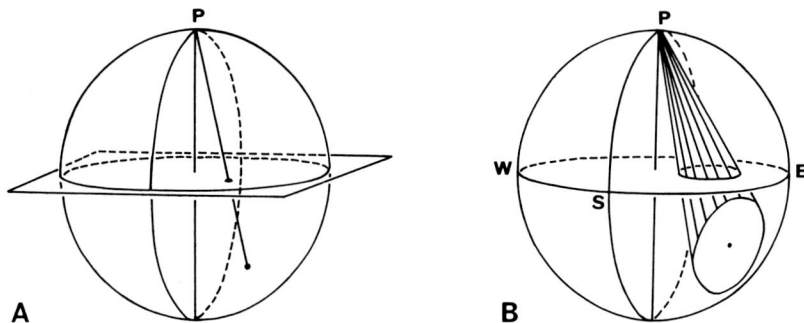

Figure 7. A: The principle of the Wulff stereographic projection. B: The projection of a small circle in stereographic projection (Based on Phillips, 1960, Figs. 31 and 36).

Thus, a bedding plane is moved parallel to itself until it passes through the center of the sphere; the plane then intersects the sphere along a **great circle.** Any plane passing through the center of a sphere intersects the sphere along a great circle, whereas any other plane intersects the sphere in a **small circle.**

The orientation of a plane (e.g., a bedding plane) can be defined uniquely by its normal; for projection purposes, this line at right angles to the plane is taken to pass through the center of the sphere. The normal intersects the sphere at a point.

In projection one dimension is eliminated. Thus, a plane is projected as a line, and a line as a point. To represent a bedding plane either the great circle, or the impingement of the normal on the sphere, can be projected. In Figure 8A the great circle defined by the bedding plane is shown together with the normal to the bedding. Only those structural elements which intersect the lower hemisphere, or which lie in the equatorial plane, are projected. In Figure 8B the equatorial plane—the plane of projection—has been isolated from the sphere, and in Figure 8C it is turned into the plane of the page. The standard convention is to use the lower hemisphere intersections only, although a few geologists have plotted projections of intersections with the upper hemisphere.

A considerable amount of computation and labor is avoided in plotting projections if the Wulff stereographic graticule is used (Fig. 9); considerable facility in the use of this three-dimensional protractor is essential. The graticule comprises two types of line: (a) a family of curves passing through a pair of opposite poles (these curves are projections of great circles) and (b) curves that do not pass through any common point and which represent the projections of small circles. Graticules with a 20

Bedding Plane within Sphere used for Stereographic Projection

A.

SPHERE

POLE to BEDDING PLANE

EQUATORIAL PLANE

N

W

E

S

θ = DIP of BEDDING PLANE

GREAT CIRCLE

INTERSECTION of POLE with SPHERE

BEDDING PLANE

Equatorial Plane of Projection Sphere with Projection of Great Circle Defined by Bedding Plane

B.

N

θ

W

E

θ = ANGLE of DIP

S

C.

N

PROJECTION of BEDDING PLANE

PROJECTION of NORMAL to BEDDING PLANE

W

E

θ

θ = ANGLE of DIP

POLE of BEDDING PLANE

Stereographic Projection

S

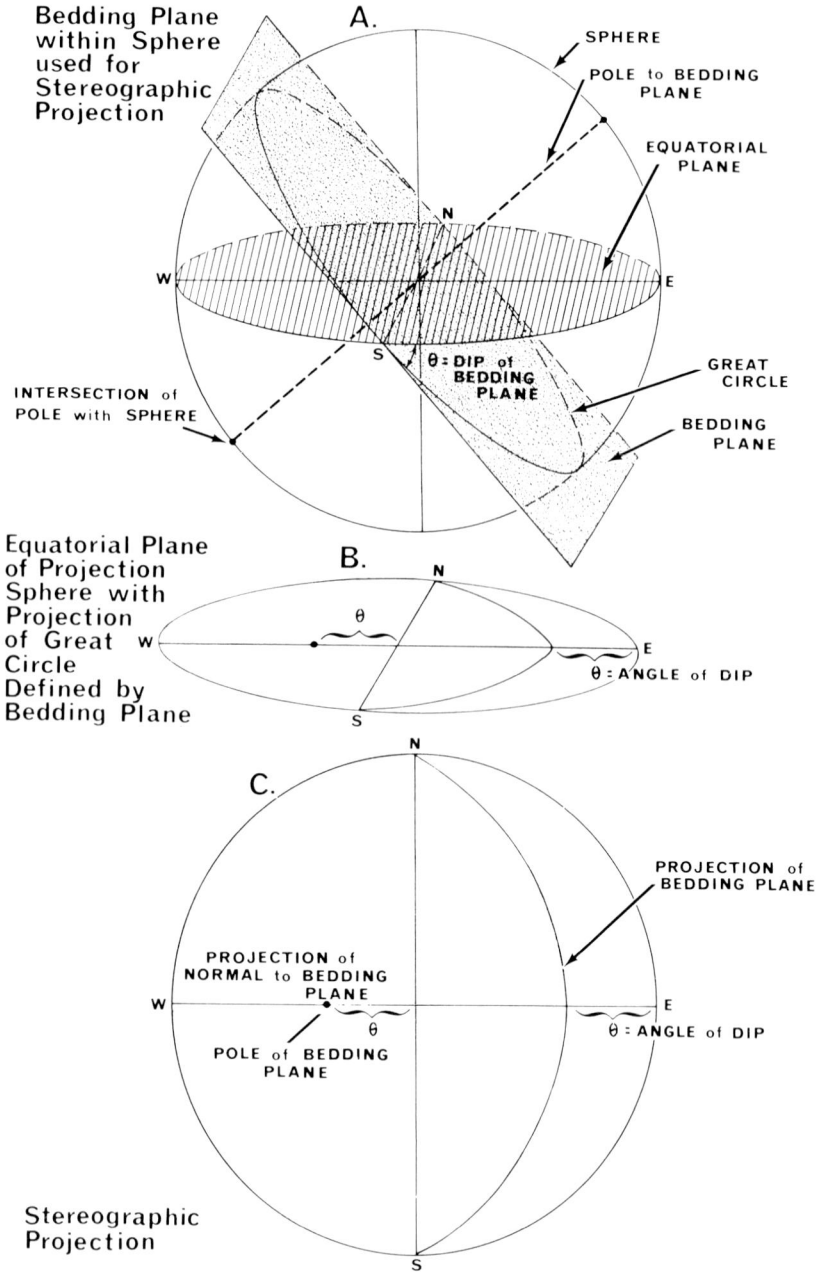

Figure 8. Construction of a stereographic projection: see text for full explanation.

cm. diameter are convenient[1]; they can be attached to a drawing table or a square piece of hard-board with a thumbtack (drawing pin) point projecting vertically upward through the center of the graticule. The bounding circle — the **primitive circle** — of the Wulff net should be boldly labelled at ten-degree intervals, commencing at the top and labelling counterclockwise (Fig. 9). Actual projections are plotted on sheets of tracing paper which can rotate freely over the graticule about the thumbtack point. A small piece of self-adhesive plastic tape attached to the center of both the graticule and the tracing paper avoids tearing during repeated rotation (Fig. 10). The primitive circle of the Wulff net must be copied on the

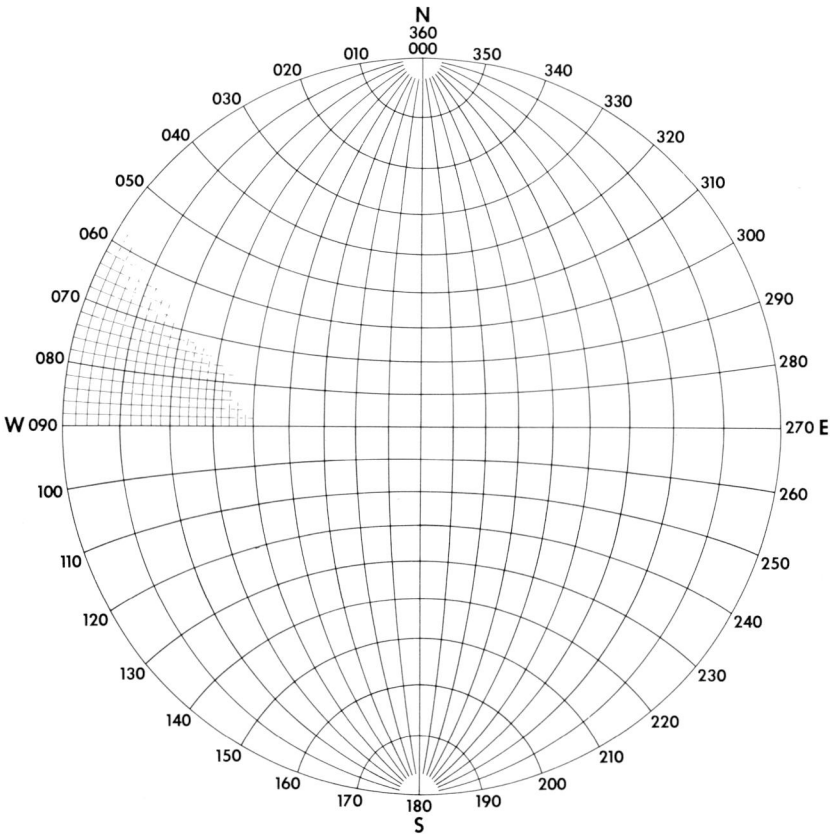

Figure 9. Wulff stereographic graticule labelled for plotting structural readings. Most graticules have great and small circle lines drawn at 2° intervals (as on part of this figure).

[1]Wulff and Schmidt graticules printed on stout paper are available commercially from several sources (e.g., Vickers Instruments, Inc.).

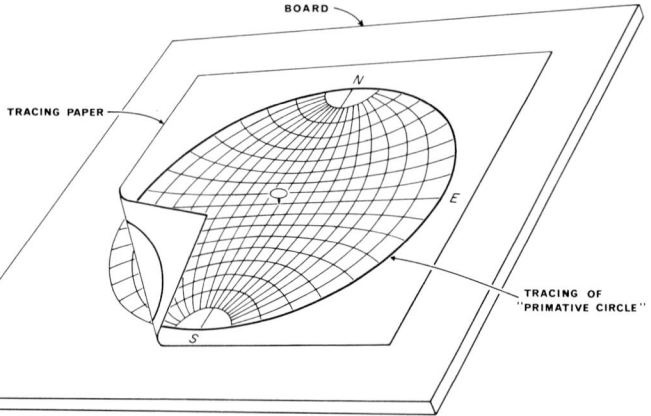

Figure 10. Tracing paper and board ready for plotting on a Wulff stereographic projection.

tracing paper, and N, E, S, and W points marked clearly before data are plotted. The technique of plotting data is illustrated by specific examples below.

In Figure 2 the strike of the bedding is 023 (i.e., 23° east of north) and the dip is about 26° to the southeast. To plot this plane the north-point (N) of the projection (the tracing paper) is rotated counterclockwise 23° according to the scale marked on the primitive circle of the Wulff net. The point now lying over the N-point of the Wulff net is marked as the strike of the bed. Keeping the tracing paper in this position, the projection of the bed is obtained by tracing the great circle that is 26° in from the right-hand side of the primitive; the pole (P) to the bedding plane is found by counting 26° out from the center along the east-west diameter of the Wulff net (see Fig. 11). The tracing paper can now be rotated 23° clockwise to bring the north-point of the projection back to coincide with N of the Wulff net.

In the northeast corner of Figure 4A the dip is about 33° to the southwest and the strike is 130. This limb of the fold can be plotted by rotating the N-point of the projection 130° counterclockwise, and then tracing the great circle which is 33° in from the right-hand side of the primitive circle (Fig. 11). The pole of this bedding plane (P') is located 33° from the center of the diagram, measured away from the projection of the plane along the east-west diameter of the Wulff net. When the N-point is rotated back to its initial position, the projection represents the folded bedding surfaces from Figure 4A. In Figure 11 the bedding plane traces intersect at F, which represents the line of intersection of the folded beds.

The orientation of the line represented by the point F is found by

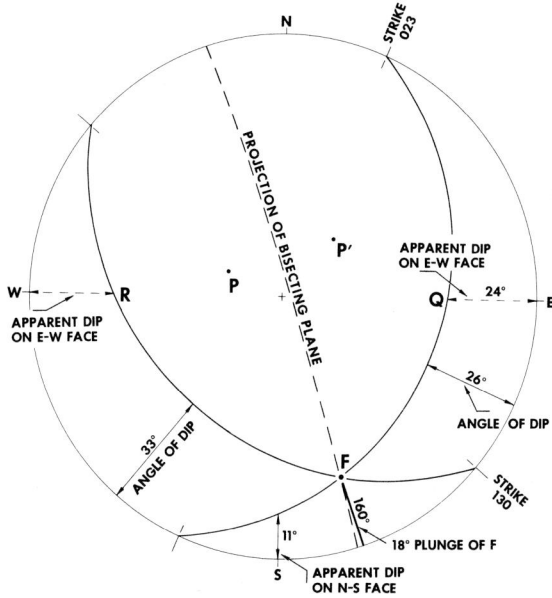

Figure 11. Stereographic projection of folded bedding surfaces from Figure 4A.

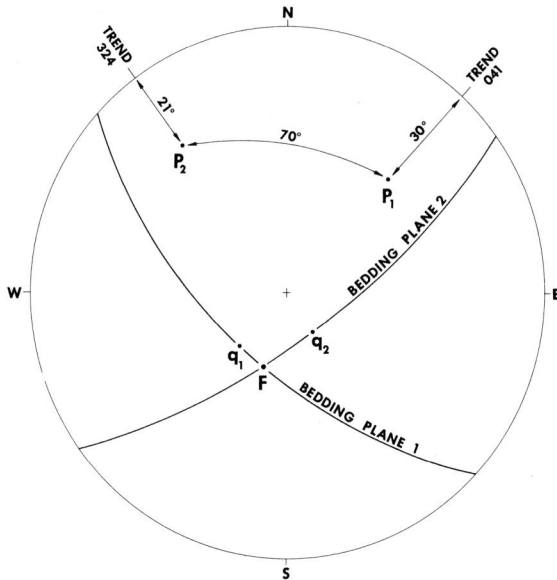

Figure 12. Stereographic projection of the lines P₁ and P₂, which are the poles to bedding planes 1 and 2, respectively.

rotating the tracing paper clockwise until F lies on the north-south diameter of the Wulff net. The south-point (S) of the projection now coincides with 160 on the Wulff net primitive circle, so F trends 160° east of north; in this orientation F lies 18° from the primitive circle, so the line of intersection is inclined down to the southeast at 18° below the horizontal. In the next chapter the line of intersection (F) is shown to be the fold axis which in this case has a **trend** of 160° and a **plunge** of 18°.

A line can be plotted directly on a stereographic projection. Suppose that a line (P_1) has a trend of 041 and plunges below the horizontal at 30°. The N-point of the projection is rotated counterclockwise to the 041 point on the perimeter of the Wulff net; P_1 is then located 30° in from the primitive down the north-south diagonal. Bring the N-points of the Wulff net and the projection together (Fig. 12).

A second line (P_2) has a trend of 324 and a plunge of 21°. To plot P_2 the N-point of the projection is rotated 324 degrees counterclockwise, and then P_2 is located 21° down the north-south diameter from N on the Wulff net.

The angular distance between P_1 and P_2 is found by rotating the tracing paper until both points lie on the same great circle of the Wulff net. Any two points can always be rotated to lie on a great circle, and the angular distance along the great circle is the required value – in this case 70°.

If the lines P_1 and P_2 are the normals to two planes, the planes can be constructed as follows. Rotate P_1 until it lies on the east-west diameter of the net; then from P_1 measure 90° along the east-west diameter and mark the point as q_1. Keeping the projection in the same position trace the great circle (from the Wulff net) which passes through q_1 – this is the trace of the required plane. Similarly, by moving P_2 to the east-west diameter, q_2 is located and the second bedding plane traced. These bedding planes intersect at F. If F is rotated to lie on the north-south diameter, the S-point of the projection lies at 196 on the primitive of the Wulff net, and F is 56° from the primitive. Hence, the line of intersection (F) trends at 196, and plunges at 56°.

The simplest method of determining the angular distance between two planes is to plot their poles, and to measure the distance along the great circle between these poles.

Notice that all the measurements referred to above are made along great circles (including the east-west diameter of the net, which is also a great circle). For some operations the small circles are used. For example, if it is necessary to view the stereogram in Figure 12 from the side the whole projection can be rotated 90° about the north-south diameter (so that the E-point becomes vertically upwards). With the projection and the Wulff net in the same orientation each pole is moved 90° along its small circle. Care must be taken to move each point in the correct direction. As the E-point (Fig. 12) moves to be vertically above the thumbtack (drawing pin), the W-point and P_2 move to the right across the projection to W' and P'_2 in Figure 13 (because Fig. 12 is a projection in the lower hemisphere). Similarly P_1 moves to the right until it reaches the primitive (i.e., until the line represented by P_1 becomes horizontal); P_1 then continues to move inward from the diagonally opposite point on the primitive along the small circle, as shown on Figure 13.

When the rotated poles P'_1 and P'_2 have been located (Fig. 13), they can be used to construct the planes to which they are normal (as in Fig. 12). These planes are the rotated equivalents of those projected in Figure 12, and they intersect at F' (Fig. 13); thus, F has also been rotated automatically along a small circle to F'.

Although a rotation through 90° has been illustrated, a rotation through any angle is possible by moving points on the stereogram the appropriate distance along small circles. Similarly, the axis of rota-

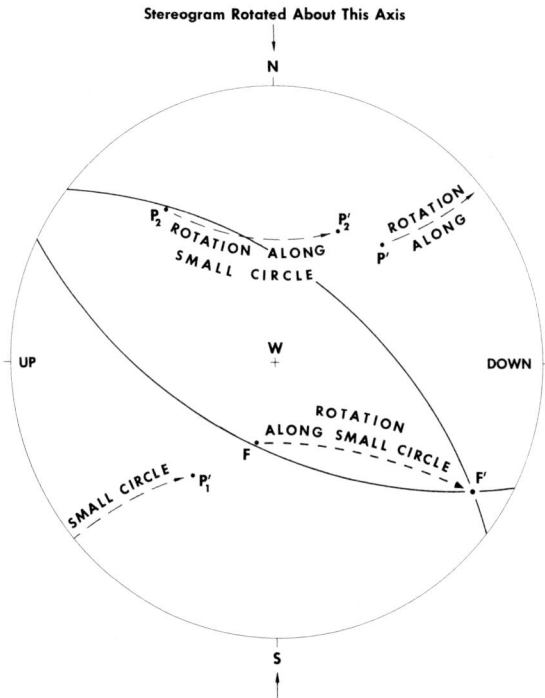

Figure 13.   90° rotation of the structural elements of Figure 12 about a horizontal north-south axis. P_1 and P_2 are poles to original bedding planes and F is the original fold axis.

tion can be varied. To rotate about an axis trending at 045, the 045 point on the projection primitive is brought to the N-point of the Wulff net first. Then each point is rotated about the small circles as in the previous example. When a plane is to be rotated, the easiest method is to plot and then rotate the pole; the rotated pole is used to construct the actual plane in its new orientation.

In the above examples the rotations have been about axes contained in the plane of the projection. Sometimes it is necessary to rotate data about an axis oblique to the projection plane. Suppose it is necessary to rotate the data in Figure 12 90° about an axis trending at 020, and plunging at 25°. This rotation axis is represented by A in Figure 14. It is easiest to proceed in three steps: (a) rotate the projection so that A is horizontal, (b) effect the 90° rotation about A, and (c) unroll the initial rotation (a). The stereogram is first rotated on the thumbtack until A lies on the east-west diameter. A is then moved 25° to A' on the primitive; concomitant with this rotation P_1, P_2, F, and the S- and W-points are rotated 25° along their respective small circles. For step (b) the whole projection is rotated about the thumbtack until A' is at the N-point of the Wulff net; now each point can be rotated 90° about the A'-axis (Fig. 14). Finally, for step (c), A' is returned to the E-point of the Wulff net, so that A' can be moved back 25° along its small circle to A; all the other points are rotated 25° along their small circles.

With these rotations complete, the planes represented by the rotated points P_1 and P_2 (P_1''' and P_2''' in Fig. 14) are traced from the Wulff net; these planes intersect at F''' (the rotated position of F). Notice that these rotations have carried the N and W direction lines down into the body of the stereogram while S and E do not appear because they are now in the upper hemisphere. The method will be understood more clearly if each rotation is followed through on an actual stereogram by the reader.

This type of rotation can be visualized in relation to a hand specimen. A specimen containing the two limbs of a small fold can be marked with the four compass directions and oriented correctly in relation to north. If the sample is now rotated 90° about an axis plunging 25° and trending 020, the type of rotation illustrated by Figure 14 will be effected.

SCHMIDT EQUAL-AREA PROJECTION

In the Wulff stereographic projection all angular relations remain true, but areas of equivalent size on the reference sphere are smaller near the center of the projection than in the peripheral areas. In the Schmidt equal-area projection equivalent areas on the reference sphere remain equal on the projection; however, this advantage is obtained at the expense of distorting angular relationships, particularly near the edge of the projection.

The Schmidt equal-area graticule is employed to facilitate plotting projections (Fig. 15). Data are plotted and rotated with this graticule in a manner similar to that described above for the stereographic projection.

Above, individual observations of bedding planes or fold axes have been considered. However, in much structural work a statistical approach has to be employed, because structural elements always show variability within an area. Commonly a large number of observations is made in the field and the data are plotted on an equal-area projection. This technique helps to identify preferred orientations within the data. In some slates the intersection of bedding and foliation (cleavage) defines a line (linear structure) whose orientation can be measured. Suppose that a large number of measurements is made within an area. Each measured line is now moved parallel to itself until it passes through the center of the reference sphere and impinges on the surface; the lines can now be plotted

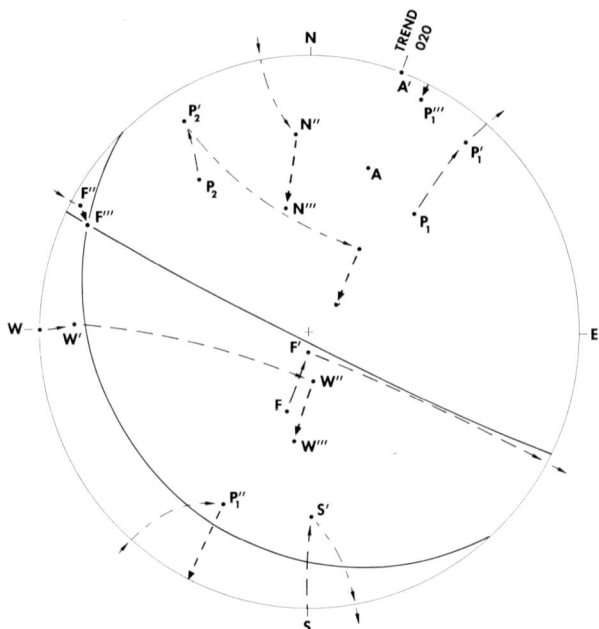

Figure 14.   90° rotation of the structural elements of Figure 12 about an axis (*A*) trending at 020 and plunging 25°. P_1 and P_2 are poles to original bedding planes and F is the original fold axis. *Step* 1: 25° rotation of these elements to P_1', P_2', and F', respectively; *Step* 2: 90° rotation to P_1'', P_2'', and F''; *Step* 3: 25° rotation to P_1''', P_2''', and F'''. The bedding planes (solid lines) now intersect at F'''.

on the projection. If these lines have a preferred three-dimensional orientation the points of impingement will be distinctly clustered on the surface of the reference sphere. Because the Schmidt equal-area projection insures that the number of impingements per unit area is preserved, it is a useful tool in analyzing the preferred orientation of any structural element. By contrast, the Wulff projection distorts (gives an incorrect picture of) the unit area distribution pattern.

When a large number of observations is plotted on a Schmidt diagram, a considerable amount of time is involved in rotating the tracing paper (the projection) over the graticule. Biemesderfer (1949) developed a helpful scale for use above the tracing paper (see Fig. 15). This counter rotates freely above the pivot pin, and is graduated in the same units as

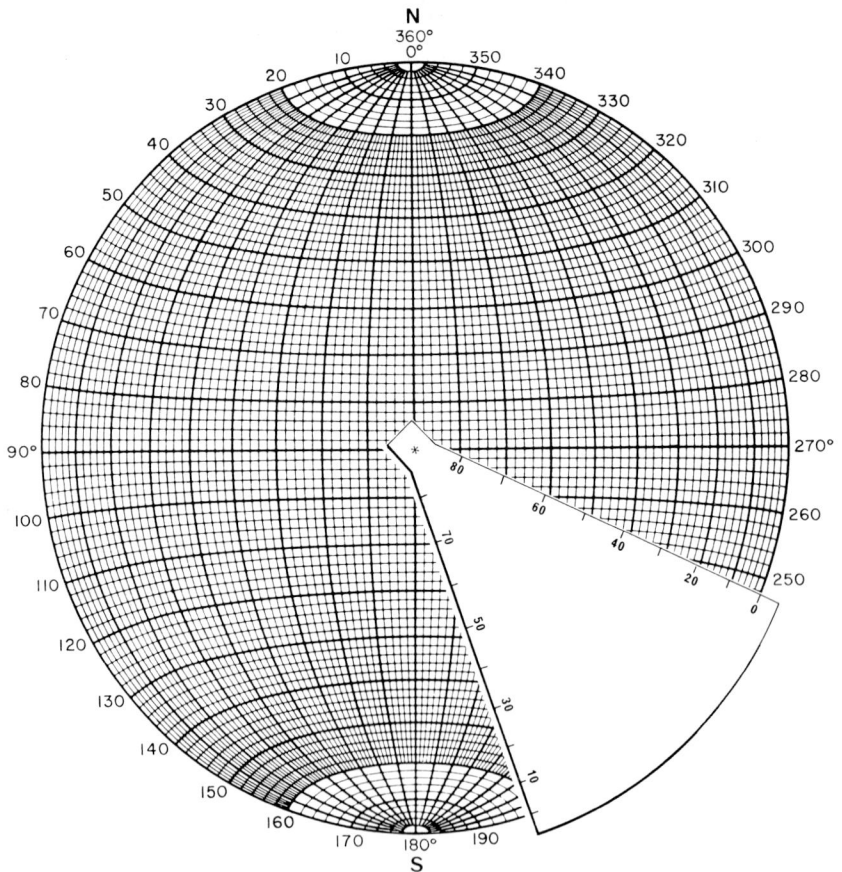

Figure 15. Schmidt equal-area projection with a Biemesderfer counter which is used to aid in plotting data.

the N-S and E-W diagonals of the net. With a little practice such a card-board, plastic, or steel scale is a most useful tool.

Commonly it is difficult to assess the geometrical pattern shown by an array of dots (projections of lines, for example) and when a Schmidt projection has been used it is most helpful to use contours; it is not legitimate to contour data on a Wulff stereographic projection because it lacks the equal-area property. The contour technique is widely used for both (a) directional and planar structural data collected in the field, and (b) petrofabric analysis data comprising measurements of the three-dimensional orientation of mineral grains in thin sections measured with a Federov universal microscope stage.

Contours are usually based on the number of data points within each 1 per cent unit area of the projection. Since standard Schmidt nets have a 20 cm. diameter, a circular counter of 2 cm. diameter has one-hundredth the projection area (Fig. 16). If a 1 cm. square grid is used below the projection, the counter can be moved systematically to each grid intersection (Fig. 16). The number of data points within successive 1 per cent unit areas is counted and written at the grid point as a percentage. For example, if four of a total 200 observations fall in a counter field, "2" is written at the grid point since $4/200 \times 100\% = 2\%$. A clean sheet of tracing paper secured above the projection is useful for recording the percentages.

A standard counter shown in Figures 16 and 17 is convenient in the middle of the projection. A special long counter (Fig. 17) is required near the primitive because points at opposite edges of the plot fall within the same 1 per cent unit areas. Such counters can be made in any machine

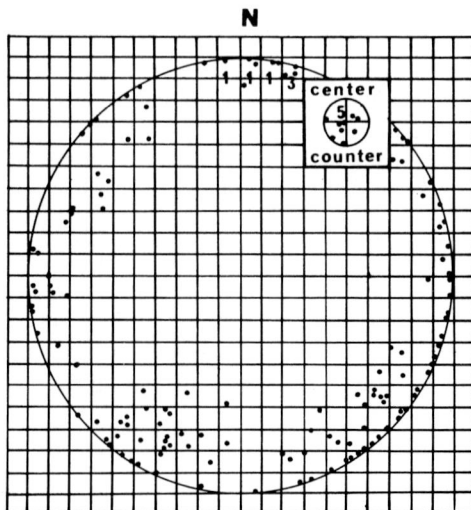

Figure 16. Counting as a preliminary for contouring a Schmidt projection; data points as in Figure 22 (Adapted from Phillips, 1960, Fig. 78).

shop from mild steel sheet. In use the long counter is pivoted about the thumbtack at the center of the projection. One of its counter circles is moved round the entire projection to lie successively over all the grid intersections near the primitive. Each count total is augmented to include all the points falling within the circle at the opposite end of the counter (Fig. 18).

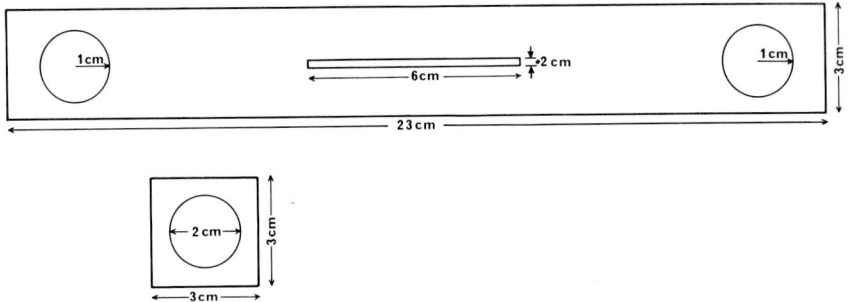

Figure 17. Counters used for contouring Schmidt projections.

Several additional counting and contouring methods have been introduced and examples will sometimes be found in the literature (cf. Turner and Weiss, 1963, pp. 61 ff.), but the system advocated above appears to be efficient in most situations. Minute differences of pattern resulting from dissimilar contouring methods are relatively unimportant since the plot is a statistical appraisal (whose significance should not be changed by the addition or subtraction of an individual point).

The projection shown in Figure 19 of 200 quartz c-crystallographic axes measured in a thin section of quartzite from Bloody Foreland, Donegal, Ireland, can be used to illustrate the successive steps. When the percentage counts are complete, contours are drawn on the sheet bearing these counts (Fig. 20). A fair copy of the contoured diagram can be traced on a clean sheet of paper; it is customary to shade the finished diagram (Fig. 21). The zero area and the first interval above zero are left unornamented; successive higher values are indicated by more intense shading. This scheme is useful because the empty areas, and the areas of especially high concentration, are usually the most significant. Any convenient contour values can be used, but there are advantages in (a) keeping the interval between successive contours equal, (b) using only whole numbers for contours, and (c) maintaining the same contour values in all diagrams used in a particular study.

In petrofabric analyses contour diagrams are usually based on a complete census of all the grains of the mineral of interest within a thin section area large enough to provide 100 to 300 measurements. With field data the nature of the sample population and of the sampling plan used in collecting the data are both of considerable importance (see Chapter 3). However, sampling problems are often severe in the field, and commonly neither a census nor a statistically-designed scheme is possible; this puts severe limitations on the interpretations that can be placed on the contoured diagrams. Again, with field data, only a very limited number of observations is sometimes available. Some published contoured diagrams

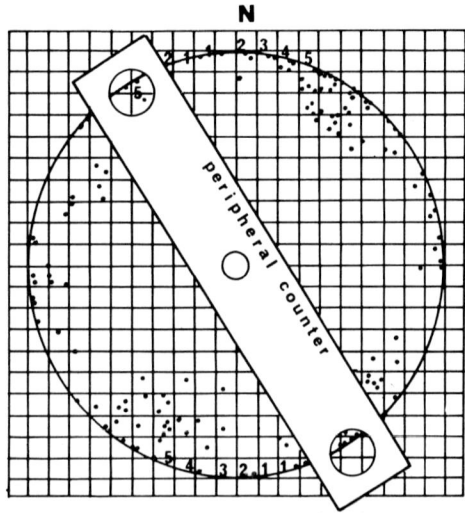

Figure 18. Counting in the peripheral area of a Schmidt projection as a preliminary to contouring; data points as in Figure 22 (Adapted from Phillips, 1960, Fig. 78).

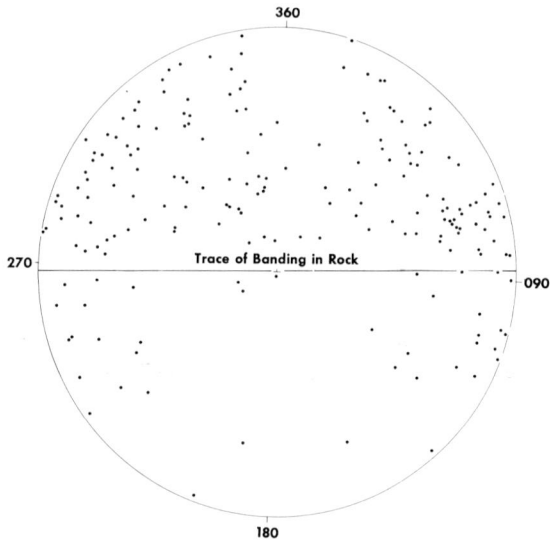

Figure 19. Equal-area projection for 200 quartz *c*-crystallographic axes from the Dalradian Errigal Quartzite, Bloody Foreland, Donegal, Ireland.

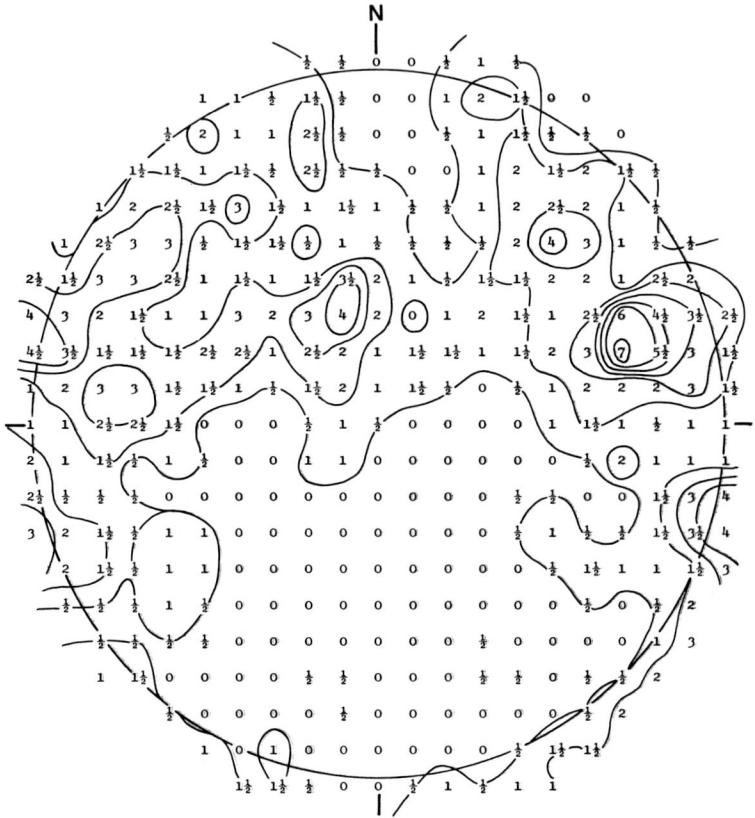

Figure 20. A completed count and sketched contours for the projected quartz *c*-crystallographic axes shown in Figure 19.

have been based on as few as 15 to 20 projected points. Such procedures appear to be unwise. The question of the minimum number of points that should be used for a contour diagram has not been examined rigorously. These questions deserve intensive investigation, but, in the interim, it would seem prudent not to base contour diagrams for field data on fewer than 50 observed values.

Figure 22 depicts the poles to 185 sea-eroded gullies along joints measured by Phillips (1960, pp. 58 ff.) in the mica schists at Start Point, SW England. The contoured diagram (Fig. 23) clearly illustrates the pattern possessed by the observed data; however, in such examples, the limitations of the sampling plan must be borne in mind constantly, because it is difficult (or impossible) to know whether the data are representative of *all* the measured and unmeasured joints.

It is often necessary to rotate equal-area projections, and the method outlined above for the Wulff stereographic projection can be used. If an equal-area projection has been contoured, each contour line can be rotated section by section; however, this is laborious and also introduces

Figure 21. Contoured equal-area projection for 200 quartz *c*-crystallographic axes from the Dalradian Errigal Quartzite, Bloody Foreland, Donegal, Ireland.

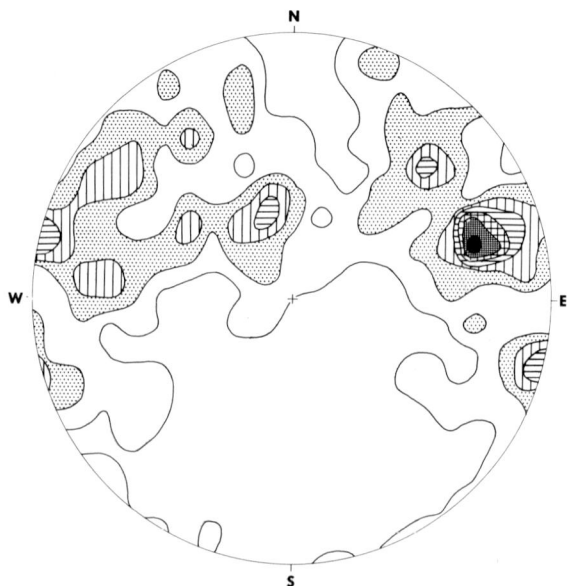

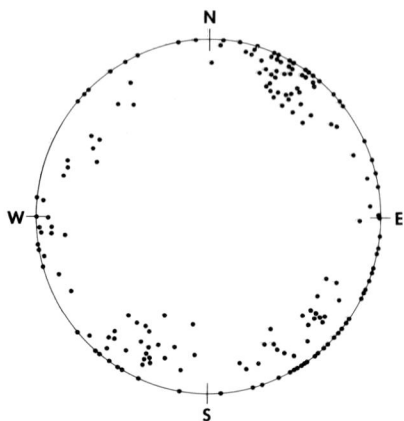

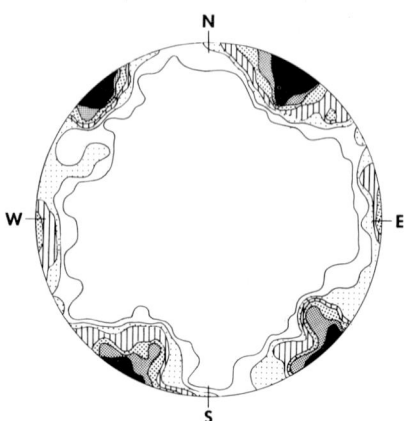

Figure 22. Equal-area projection of poles to 185 sea-eroded gullies along joint planes in mica schists, Start Point, S Devon, England (After Phillips, 1960, Fig. 77).

Figure 23. Contoured equal-area projection for the data illustrated in Figure 22 (After Phillips, 1960, Fig. 79).

some distortion. It is preferable to rotate each individual datum point and then to contour the rotated diagram.

When several hundred data points must be rotated, the operation should be handled in two steps. Those points that can be rotated along their respective small circles without going beyond the primitive are moved first. Points needing to go beyond the primitive—like point P_1 in Figure 13—are moved second. The first step is easily accomplished by aligning the required rotation axis with the N-S line of the equal-area graticule. Suppose each point is moved θ degrees. To simplify the second step, the sheet bearing the original data points is taken from its position for step one and turned upside-down (rotating the paper about an axis parallel to the E-W diameter of the graticule). The points to be rotated are now seen through the tracing paper and are moved along their small circles $180-\theta$ degrees in the opposite direction to that used for step one. During both steps of the rotation, it is helpful to insert half a Schmidt equal-area graticule (cut along the north-south diameter) as a mask beneath the new sheet of tracing paper receiving the rotated points; this eliminates considerable confusion during plotting by masking out those points on the original scatter diagram that are not being used in each step.

EXAMPLES OF PROBLEMS INVOLVING STEREOGRAPHIC MANIPULATION

Many useful constructions can be performed very efficiently with a stereogram (see, for examples, den Tex, 1954; Phillips, 1960). A few simple geometric examples follow. Readers with little experience of stereographic projections should plot these examples on a Wulff net. Additional useful examples were provided by Phillips (1960, pp. 79 ff.).

EXAMPLE 1

In a sedimentary sequence folded about a horizontal fold axis one bed has a strike of 130 and dips SW at 45°. Ripple marks are observed on this bedding plane; they **pitch** at 40° to the southeast (i.e., the angle between the strike line and the ripples is 40°). This fold limb is plotted in Figure 24; the linear ripple marks plot as a point (R), and, by turning R to lie on the north-south diameter, it is found that the ripples trend at 161 and plunge at 27° (i.e., the angle between the ripples and the horizontal is 27°).

What are the trend and plunge of the ripple marks on the following two limbs of the fold; (a) dip to the northeast at 30°, and (b) dip to the northeast at 80°?

These two limbs are included in Figure 24. On the assumption that the ripple marks were originally linear across the bed, the angle of pitch (i.e., angle between ripples and fold axis[2]) remains constant during folding. This can be demonstrated easily with a piece of lined notepaper creased about an axis oblique to the lines; the folded paper shows that the angle of pitch is between opposite ends of the fold axis for each flank of the fold. Thus, in Figure 24, the ripple marks are plotted (as R′ and R″) on the traces of the two new beds by counting 40° from the northwest end of the fold axis.

By rotating R′ onto the north-south diameter it is found to trend at 346 and plunge at 18°. Similarly R″ trends at 319 with a 40° plunge.

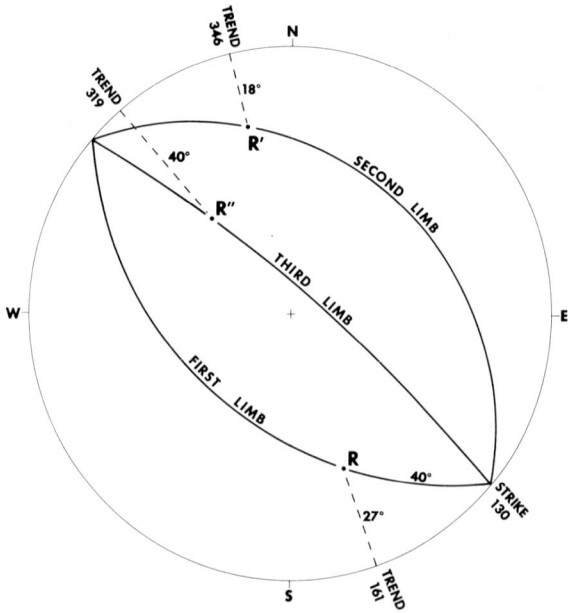

Figure 24. Stereo-graphic projection to illustrate Example 1.

EXAMPLE 2

On one bedding plane (strike 000, with dip west at 50°) ripple marks trend at 021. At a second outcrop of the same folded sequence a bedding plane (strike 110 with dip north at 30°) bears ripple marks which trend at 186. Is it possible that these ripple marks were a linear feature on the same folded bed, or do the ripple mark measurements suggest that the

---

[2]Note that it would be incorrect to call this angle between the ripples and the fold axis the pitch if the fold axis plunges; in Chapter 2, pitch is defined as the angle between a line and the strike of a plane containing the line. In this example the horizontal fold axis and the strike happen to coincide.

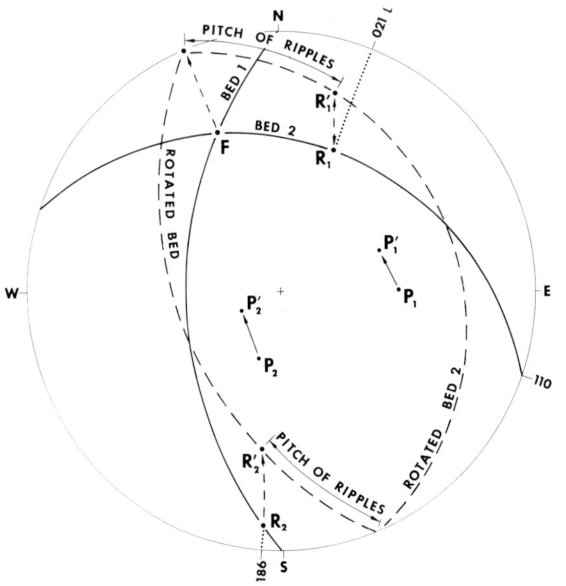

Figure 25. Stereographic projection to illustrate Example 2. R_1' and R_2' represent rotations of the ripple marks R_1 and R_2 that were originally observed on beds 1 and 2, respectively.

observations were made on different beds (i.e., that there were dissimilar ripple mark orientations on different beds)?

The data are plotted on Figure 25. The bedding planes intersect at F, which, for this problem, can be assumed to be the fold axis. If the fold axis is first rotated to horizontal, the technique outlined in Example 1 can be used.

If F is rotated to the east-west diameter of the Wulff net, then F, the poles to the two bedding planes (P_1 and P_2), and the projections of the ripple marks (R_1 and R_2) can all be rotated similar amounts until F reaches the primitive. The rotated bedding planes can be constructed by using the rotated bedding plane poles P_1' and P_2' (Fig. 25).

The pitch of the ripple marks can be determined directly by measurement along the traces of the rotated bedding planes. The similarity of these angles of pitch (Fig. 25) suggests that the problem relates to parallel ripple marks folded about F to their present orientation.

EXAMPLE 3

Suppose a folded sequence is overlain by an uncomformity that dips at 40° towards 17° east of north. Limbs of the fold in the underlying sequence include: (a) strike of 036, dip at 40° to the northwest, and (b) strike of 350, dip to east at 50°. Estimate the orientation of the fold axis in the underlying sequence prior to tilting of the unconformity.

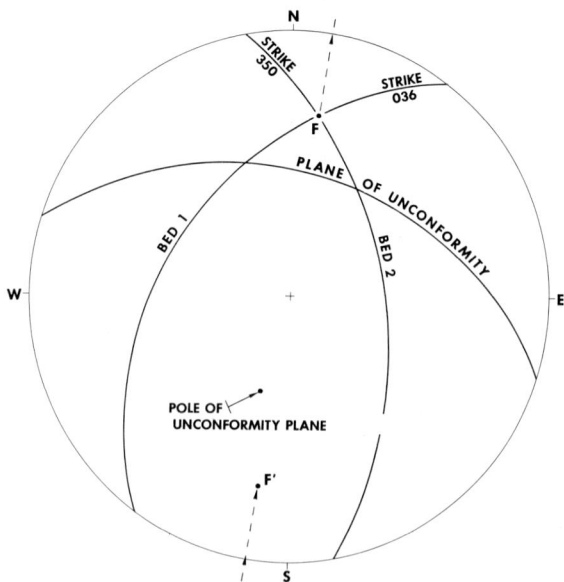

Figure 26. Stereographic projection to illustrate Example 3.

The two bedding planes intersect at F, which defines the fold axis (Fig. 26). To determine the earlier orientation of the fold axis (F), the plane of unconformity must be rotated to horizontal. So that the pole of the plane of unconformity can be moved along its small circle to the center of the projection, the strike of the plane is aligned parallel to the north-south diameter of the Wulff net. The pole to the unconformity is rotated through 40°, and F must be rotated a similar amount. F moves beyond the primitive to F′, which has a trend of 189 and a plunge of about 19°.

EXAMPLE 4

Suppose that the apparent dip is recorded on two quarry walls and that the true dip is required. For example, on a vertical NE-SW wall the apparent dip is 30° to the southwest. On a second wall, which has a strike of 110 and dip 70° to the SW, the apparent dip is 20° to the southeast. Apparent dip on a quarry wall is only a line. In Figure 27 the two quarry walls are plotted, and the apparent dips are points on these traces (plotted in a manner similar to that used for the ripple marks in earlier examples).

The unique plane through these two lines is the required bedding surface. This is found by rotating the stereogram until the projections of the apparent dips lie on one great circle. The strike is 091 and the true dip 38° to the south (Fig. 27).

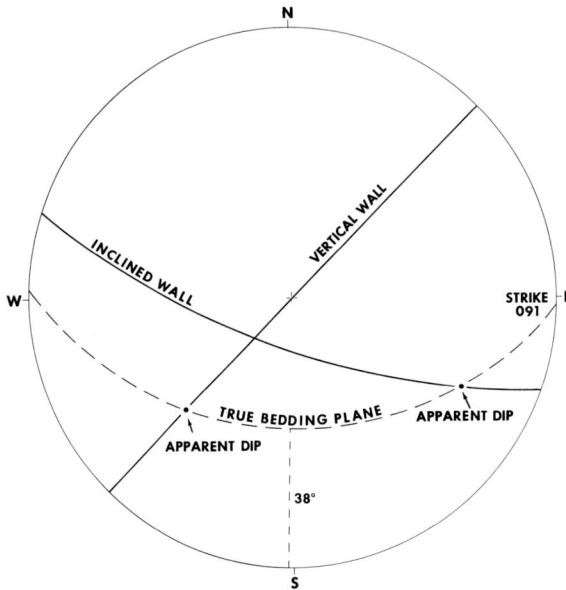

Figure 27. Stereographic projection to illustrate Example 4.

EXAMPLE 5

Suppose the intersection of bedding and foliation is seen to trend at 120 and to plunge at 37°; what range of orientations is possible if lineations can depart 20° from the observed value?

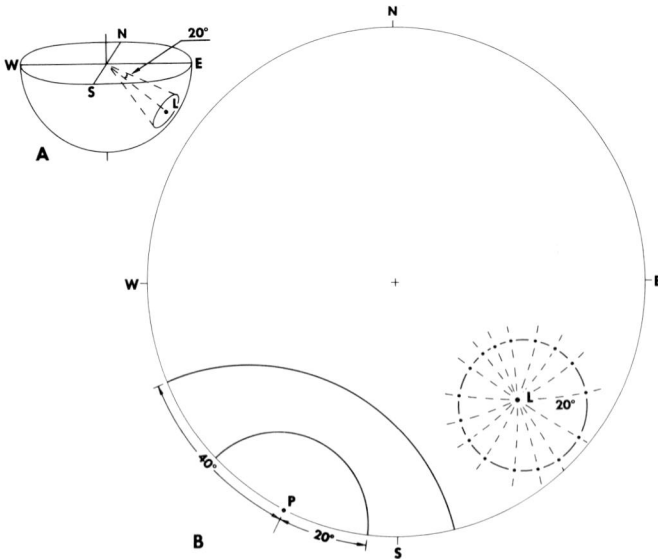

Possible lineations lie within a cone, with apical angle of 40°, about the observed lineation (L in Fig. 28A), and projections of these lines lie within a small circle about L on the stereogram. Hence the problem amounts to drawing a small circle about L. A small circle about a point on the primitive (e.g., P in Fig. 28B) can be traced directly from a small circle on the graticule. However, L lies within the projection. In this case the graticule must be turned so that successive great circles pass through L and points on them 20° from L are marked off. By joining these points the required small circle is drawn (Fig. 28B).

BLOCK DIAGRAMS

Block diagrams can often be useful for illustrating the three-dimensional structure of an area, although distortion makes them unsatisfactory for analyzing structural data or for solving structural problems. When the structure has been established firmly, a block diagram is commonly a more effective means of communication than either a map or a projection. The method is particularly effective for illustrating the relationships of geologic structure to topography. The block diagram of the overthrust fault of the Audibergue in the Alpes-Maritimes, France, drawn by Goguel (1936, p. 260) is a good example (see Fig. 29); Goguel (1962, pp. 117 ff.) briefly described the method of constructing such diagrams.

For rapid construction perspective block diagrams have sometimes been employed (e.g., Secrist, 1936). However, orthographic projections (Johnston and Nolan, 1937; McIntyre and Weiss, 1956) are preferable because three sides of a rectangular block are drawn in such a manner that each side is foreshortened equally. In the majority of cases isometric block diagrams are drawn; these are a special case of orthographic projection in which an equal amount of three faces of a cube are viewed. Commonly the horizontal and two perpendicular vertical planes are used for the edges of the projected block. The ready availability of isometric graph paper makes isometric block diagrams popular. Elementary techniques of drawing these block diagrams were discussed by, among others, Ives (1939) and Phillips (1960).

Figure 28. Stereographic projection to illustrate Example 5. A: Lower hemisphere showing observed lineation L and the cone within which possible lineations can occur. B: Stereographic projection showing two small circles constructed about P (representing cones with apical angles of 40° and 80°) and another about L representing the required cone with 40° apical angle.

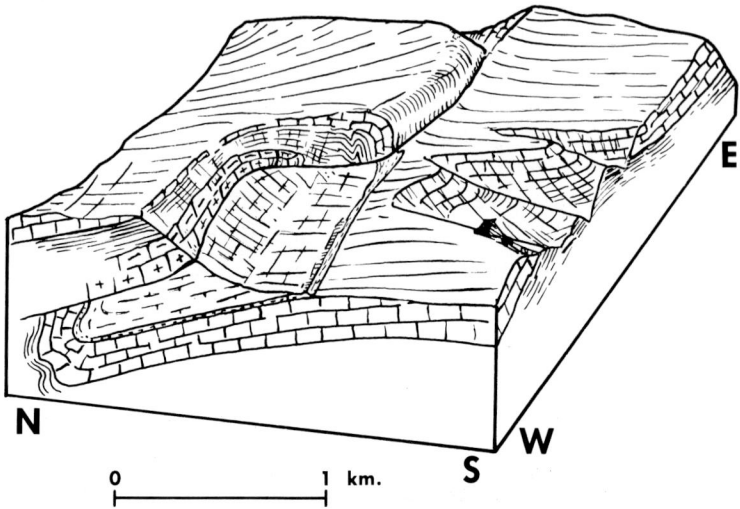

Figure 29. Block diagram of the eastern end of the Audibergue over-thrust, Alpes-Maritimes, France (After Goguel, 1936 and 1962, Fig. 83).

In an orthographic projection the rock body is projected along lines perpendicular to the plane of projection (the sheet of paper on which it is drawn). Hence, this projection is in perspective, but the vanishing point is at infinity so all parallel lines in the body are parallel on the projection; in an ordinary perspective drawing parallel lines in the rock body converge rapidly in the drawing because the vanishing point is not far behind the plane of the drawing. An observer on Earth views the Moon in orthographic projection.

An equal view of the two sides and the top of a cube in isometric projection is shown in Figure 30. The fold in the southern part of Figure 4A has been used to illustrate the method of constructing an isometric block diagram; each side of Figure 30 is taken to be two units of length. The trace of the bisecting plane in Figure 4A trends 70° north of west, and it is represented by PQ in Figure 30. The correct position of this plane on the block diagram is found from triangle PCQ; $\tan 70° = PC/CQ$, so, if $CQ = 1$, from tables $PC = 2.74$ and P can be plotted on the extension of CB. The join of P and Q represents the trace of the bisecting plane.

On the west of Figure 4A the strike is 023. If a bed is allowed to pass through C, then, in the triangle CBR, $\tan 23° = BR/2$, or $BR = 0.84$ (by use of tables). This value of BR permits R to be plotted, and CR represents the projection on the top of the model. Similarly, the beds at the east of Figure 4A strike at 140, so if a bed passes through D, $\tan 40° = TC/2$, or $TC = 1.68$. DT can represent the trace of this bed, but, since on any face of an orthographic projection all parallel lines project as parallel

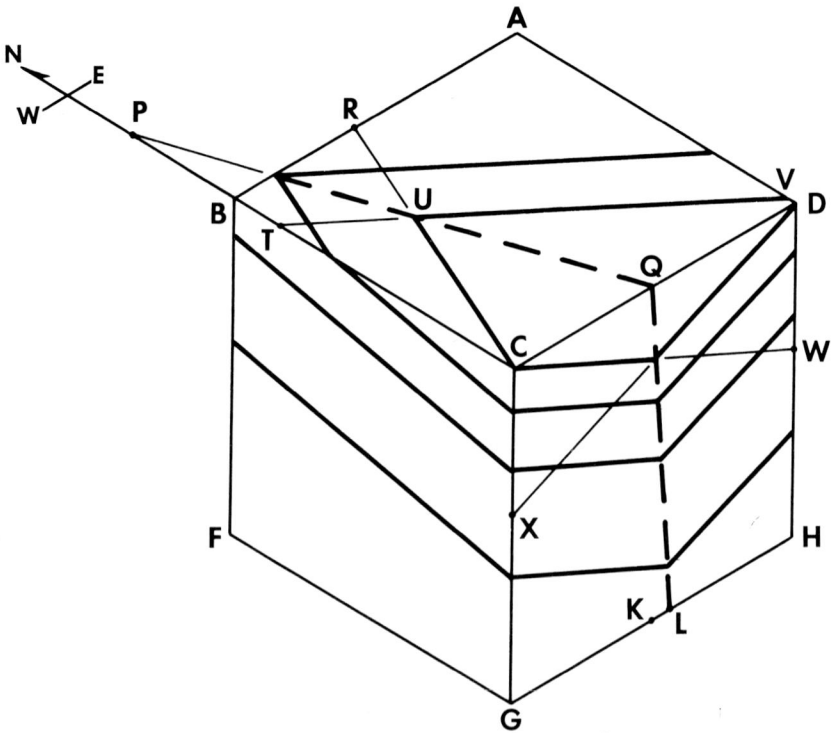

Figure 30. Isometric projection of the structure in the southern part of Figure 4A.

lines, DT can be moved parallel to itself until it intersects CR at U. CUV then represents the trace of a folded bed on the horizontal surface of the block. Other beds can be drawn parallel to CU and UV.

From the map (Fig. 4) it is found that the western flank of the fold dips at 26°, and the eastern at 33°. This information enables the east-west face to be completed, although the apparent dips on this face must first be determined on a stereogram. R and Q in Figure 11 are the intersections of the vertical east-west plane with the two bedding planes; projections of the 33° and 26° dips on the east-west plane are approximately 23° and 24°, respectively. Hence, in the triangle DCW on Figure 30, tan 24° = DW/2, or DW = 0.89; such values permit the east-west face to be completed.

The north-south face can be completed in a similar manner. Thus, the projection on the north-south vertical plane of the bed with 26° real dip has an apparent dip of 11° (Fig. 11).

The trace of the bisecting plane on the east-west vertical plane is also determined from Figure 11. On Figure 11 the bisecting plane passes through F (the fold axis) and bisects the angle between P and P′ (poles to the two bedding planes). The strike of the bisecting plane is found to be

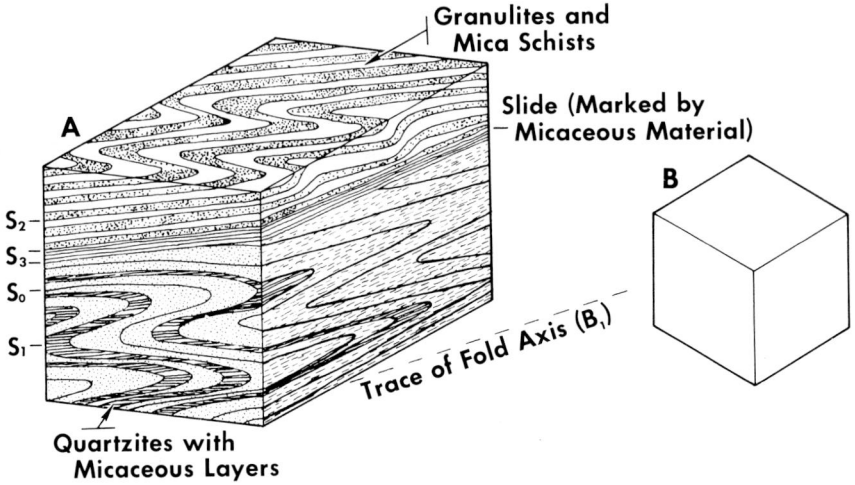

Figure 31. Block diagrams in orthographic projection. A: Complex structure within Dalradian rocks at Ord Ban, Central Highlands of Scotland; the projection is turned so that more is seen of one face of the block (After McIntyre and Weiss, 1956, Fig. 10). B: Isometric block diagram for comparison with Figure 31A.

160 and its dip 86° east. The intersection of the east-west vertical face and the bisecting plane is a line which plunges 86° east. Hence, on Figure 30 the angle DQL is 86°, so that in triangle QKL tan 4° = KL/2, and KL = 0.14.

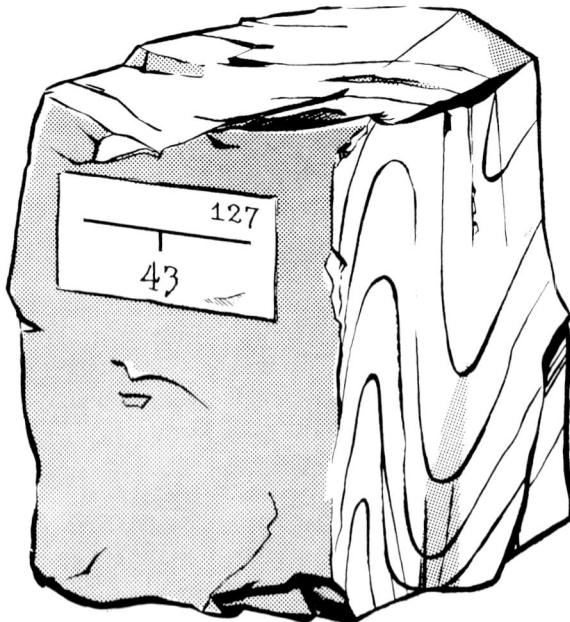

Figure 32. Labelling an oriented sample: adhesive plaster affixed to a smooth fracture surface is marked with the strike and dip of that surface.

Although the isometric block diagram is the most commonly used orthographic projection, McIntyre and Weiss (1956) demonstrated that other orientations can be drawn with relative ease. In the isometric case an equal amount of each of three faces is seen, but sometimes more interest attaches to one plane than to another. Then it is convenient to turn the block so that the interesting face is seen more fully; the block is projected in this preferred orientation. Detailed drafting instructions were given by McIntyre and Weiss (1956, pp. 147 ff.). Figure 31 was projected in this manner and represents structures described by Weiss, *et al.* (1955) from Ord Ban, mid-Strathspey, in the Central Highlands of Scotland. A series of granulites and mica schists above is separated from a body of quartzite below by a layer of micaceous material which marks the position of a slide, or tectonic break, between the two folded blocks.

COLLECTION OF ORIENTED HAND SPECIMENS

Most structural data are collected by measurement in the field. However, in many cases additional details can only be obtained by study of hand samples in the laboratory. For example, foliation and fold structures in pelitic rocks are sometimes too small for measurements and descriptions to be made easily in the field; sawed and varnished planes cut through samples often make minor structures much more clear. Reference was made above to petrofabric analyses whereby the three-dimensional orientation of individual mineral grains is measured under the microscope. In a later chapter reference is made to the value of studying textural features shown by thin sections for elucidating the structural and metamorphic history of a rock. For these, and many other purposes, oriented hand samples have to be taken to the laboratory; that is, the samples are labelled so that their three-dimensional orientation can be reconstructed unequivocally in the laboratory.

When an oriented sample is required a specimen is hammered free and trimmed before labelling. The sample is then fitted back into place and held securely; it is desirable to select a sample which can be labelled from above whenever possible. If a flat surface inclined to the horizontal is available on the sample (e.g., bedding plane, foliation plane, joint surface, etc.) the strike and dip of that surface should be marked clearly (Fig. 32). This information is sufficient to orient the sample unequivocally. The labels can be written on adhesive linen tape, or marked directly on the rock with a felt pen; as a safeguard, it is always desirable to record this information in a notebook too.

Chapter 2

Cylindroidal and Noncylindroidal Folds

In any quantitative study one must begin with precise operational definitions of the properties to be measured. In geologic work there are certain properties which it has been traditional to measure, such as dip and strike. From time to time additional variables have been measured quantitatively, and observations of new variables have frequently led to major "breakthroughs." For example, recognition of the importance of both graded bedding and the orientation of the axial surfaces of folds led to new strides in structural analysis and synthesis.

Features displayed by folded sedimentary and metasedimentary rocks will be considered first. Sedimentary rocks provide a relatively simple point of departure because initially they are approximately parallel flat lamellae. Even in the stable interior of a craton, gentle doming and

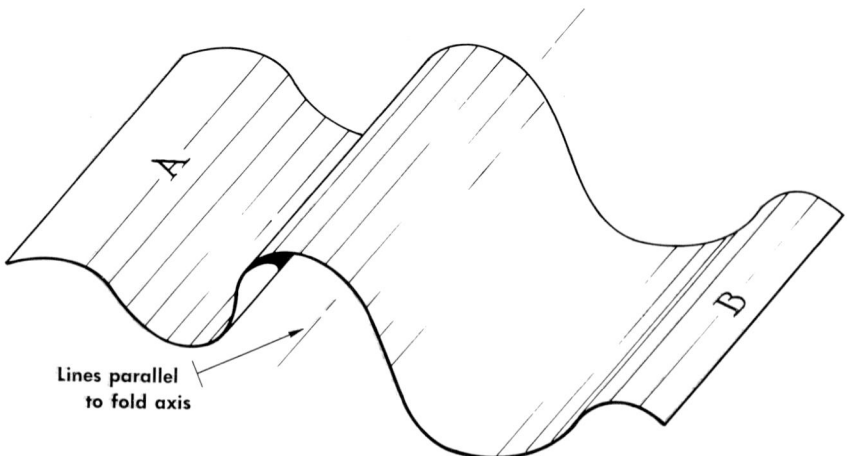

Figure 33. A cylindroidal fold produced from a sheet of lined paper; moving A towards B gives a fold axis parallel to the lines on the paper.

basining, coupled with compaction and diagenetic changes, cause sedimentary strata to be warped or folded. However, in orogenic zones more dramatic flexures occur.

In common with most geological phenomena it has been customary to name and classify folds. Innumerable names have been employed for the immense variety of slightly dissimilar geometric forms. Reference to a few structural geology textbooks shows that numerous different classifications of folds can be erected. This situation is similar to that in igneous petrology where multitudes of unnecessary rock names have been introduced for each slightly different rock "species." This proliferation of varietal names stems from a purely qualitative approach; however, structural geology is rapidly changing from a qualitative subject to a quantitative science, and this has necessitated a reexamination of the nature of folds and folding in quantitative terms. Nomenclature problems are examined in detail in Chapter 15.

CYLINDROIDAL FOLDS

To take the simplest possible example, a sheet of ruled paper can be flexed by slowly moving opposite ends of the paper towards one another. Suppose that "A" is written at the middle of the top line, and "B" at the middle of the bottom line of the paper. As A is moved towards B, the paper folds, a process which results in the paper between A and B being rotated about an axis perpendicular to A-B (Fig. 33). The axis of rotation is the **fold axis.** Each infinitely small part of the folded stratum is rotated about a line parallel to the fold axis. Hence, the fold axis is not a single line, but a direction parallel to an infinite number of parallel lines. With respect to the whole fold, the fold axis is a statistical concept.

So defined, the fold axis becomes the nearest approximation to the straight line which, when moved parallel to itself, generates the folded surface. In terms of the sheet of paper, the axis is the unique line which, when moved parallel to itself, remains contained in the sheet. Where A is moved towards B, the lines of the paper are parallel to the fold axis. The locus of the fold axis is the fold.

A line parallel to the fold axis that moves freely within a particular stratum draws a trace on the plane perpendicular to the fold axis. This trace is the **profile** of the folded surface. Together the profile and the orientation of the fold axis define uniquely the nature or **style** of a fold.

These definitions are based upon the usage of Wegmann (1929) and other great alpine tectonicians, such as Lugeon and Argand. McIntyre (1950A) drew attention to these formal definitions in describing the geometry of folded rocks in the Central Highlands, Scotland. Stockwell (1950) also used similar definitions and gave detailed instructions for constructing profiles from maps: he showed the fold axis can be found by taking the mean of intersections of numerous pairs of bedding planes, and used as an example the Sheila Lake fold, Sherritt Gordon mine area, Manitoba, Canada. Dahlstrom (1954) used a similar approach in 1951 for a detailed study of the Amisk Lake area, Saskatchewan, Canada.

Each fold axis is the statistical approach to a straight line that has a specific orientation in space. Suppose the line BC in Figure 34 is a fold axis; its orientation is uniquely defined by measuring its trend and plunge. If ADBE is horizontal, AB is vertically above BC and is the **trend** of BC. The angle ABC is the **plunge** of BC. Alternatively, if BC is a linear element within the plane BFCD it can be defined by the angle DBC – the **pitch** of BC – if the dip (angle EBF) and strike (BD) of the plane are stated. Although these terms are now standard (cf. Clark and McIntyre, 1951) the meaning of pitch and plunge has sometimes been reversed, as in much English literature prior to 1955, and some geologists still use pitch for the plunge defined here.

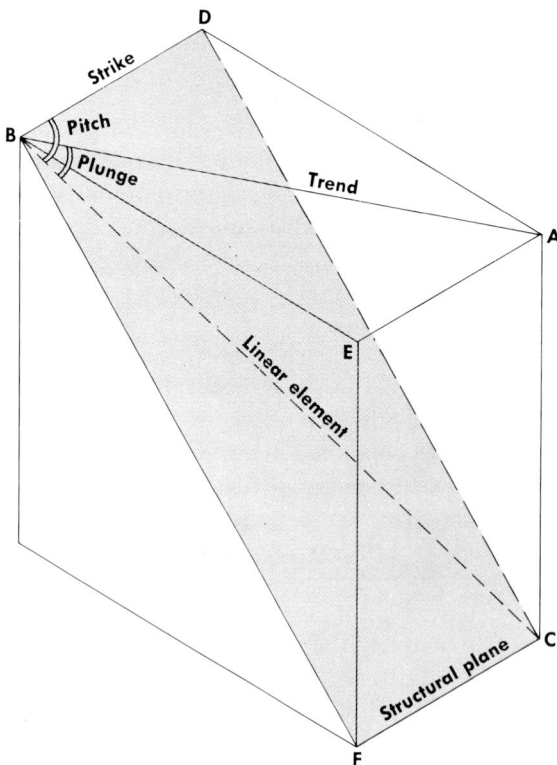

Figure 34. Plunge and pitch.

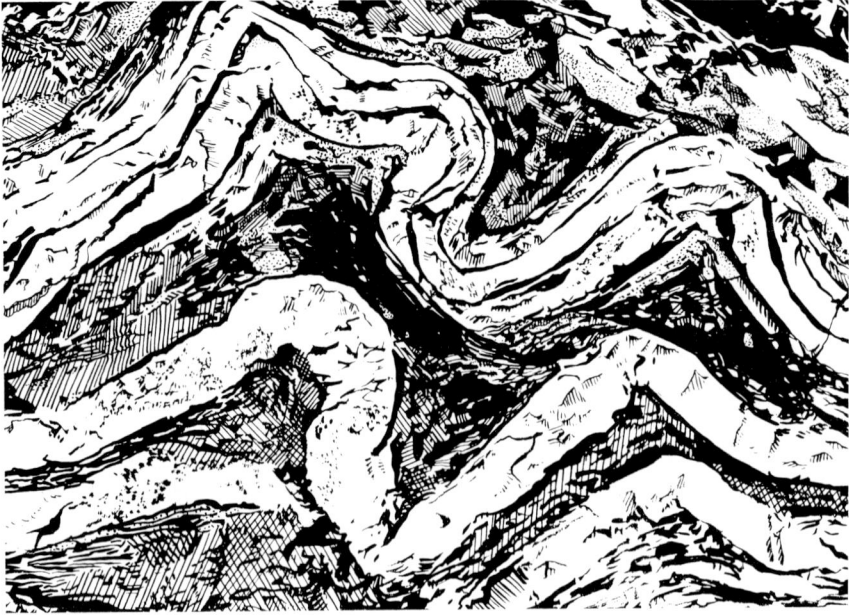

Figure 35. Disharmonic cylindroidal folds in limestone layers interbedded with shale, Whitehall, New York, U. S. A. Scale is 3 inches (Drawn from photograph in Donath and Parker, 1964, Plate 8, Fig. 1).

The definitions given above lead to the concept that such simple folds are cylindroidal. Reverting to the folded sheet of ruled paper, it can be made to rotate in such a way that the paper virtually forms a continuous cylinder with an axis parallel to the lines on the paper. Many natural folds

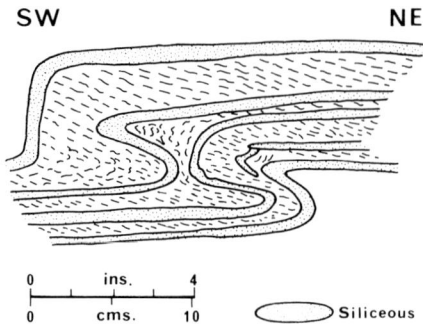

Figure 36. Profile of a disharmonic fold in Dalradian schists, Ord Ban, Central Highlands, Scotland (After McIntyre, 1951A, Fig. 1).

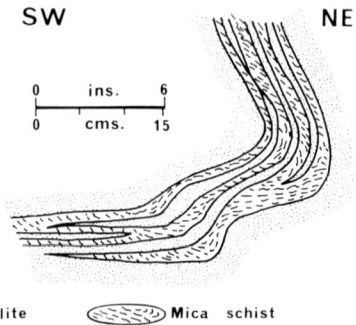

Figure 37. Profile of disharmonic isoclinal folds in Dalradian schists, Ord Ban, Central Highlands, Scotland (After McIntyre, 1951A, Fig. 2).

are also cylinder-like and are termed **cylindroidal** (or cylindrical) because, if enough points on the folded stratum are observed, some will be found to coincide with most parts of a cylinder. Most folds generated by compression of stratified sequences are cylindroidal and the definitions given above are appropriate; however, some noncylindroidal fold types also occur for which these definitions are inappropriate (see below, p. 63).

Competence is a relative term commonly used to describe the apparent behavior of a layer folded within a stratified sequence of dissimilar rock types; both composition and the thickness of individual units affect the competence of each layer. Members that tend to retain their stratigraphic thickness and to develop folds of large amplitude are said to be **competent;** those members that vary greatly in thickness and that accommodate to the space between competent units are **incompetent.** A relatively competent lithology in one environment may be relatively incompetent in another and vice versa (cf. Hills, 1963A, p. 219; Turner and Weiss, 1963, p. 471). Although competence expresses the differential rheological properties of a folded sequence it is a qualitative concept, and competence is affected by a large number of variables.

The varying competence of a stratified rock sequence often results in cylindroidal folds that are **disharmonic.** The folds in Figure 35 are disharmonic because, in passing from one lithological layer to another, the profile changes abruptly. Figures 36 and 37 show profiles of dramatically disharmonic cylindroidal folds from Ord Ban, Grampian Highlands of Scotland. In the Ord Ban folds thin interlayered quartz granulites and pelites had markedly dissimilar competence during the deformation, and the less competent pelitic rocks acted as a lubricant during the intense deformation of the granulites.

The description of a fold is not complete without reference to the **axial surface** — the planar or curved surface passing through the hinge lines, or regions of maximum curvature, of the successive folded lithological layers (Challinor, 1945, p. 82; Turner and Weiss, 1963, p. 108). If the surface is planar it is appropriately referred to as an **axial plane,** but commonly the surfaces are nonplanar. There has been some confusion concerning this nomenclature, and Billings (1954, p. 34) wrote that the axial surface divides a fold as symmetrically as possible; however, it is preferable to call such a symmetrical surface the **bisecting surface.** In the special case of a perfectly symmetrical fold the bisecting and axial surfaces coincide, but commonly these surfaces are not parallel and their inclination provides one measure of the asymmetry of the fold (Fig. 38). A bisecting surface may be planar and have any inclination to the horizontal; hence, it is viewed as a plane of symmetry with respect to the main fold limbs, although such a plane is difficult to define precisely because the

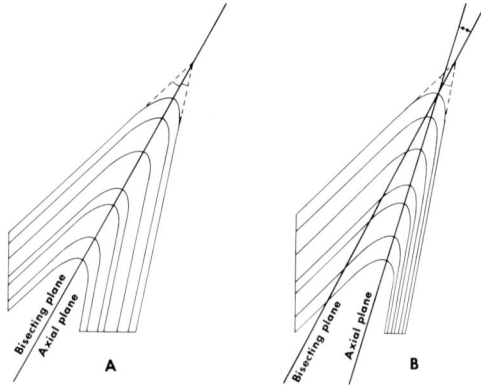

Figure 38. Bisecting and axial planes.

surface cannot be symmetrical with respect to all layers of an asymmetrical fold (see Fig. 38). This logical use of **symmetrical** and **asymmetrical fold** is strongly recommended, although it is contrary to the usage of some geologists (see Chapter 15).

Both the axial and bisecting surfaces contain the fold axis. Each planar segment of the axial surface has dip and strike, but the orientation is often difficult to define in the field without ideal three-dimensional exposures. Commonly these surfaces can be estimated more easily from a drawing or a photograph of the profile, on which the line through the

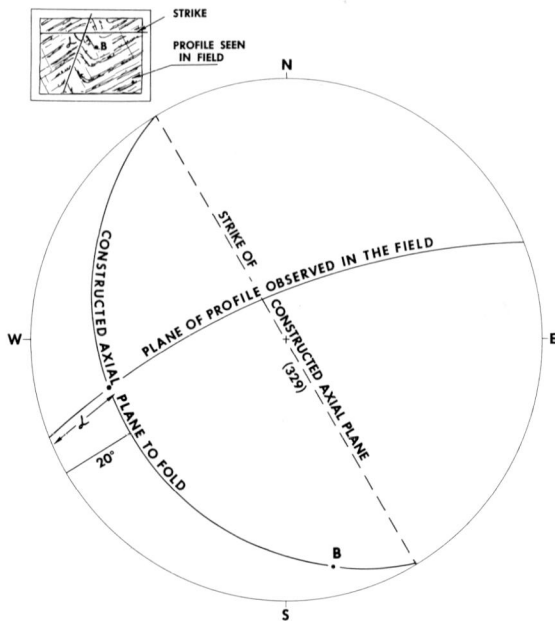

Figure 39. Stereographic construction of the fold axial plane from the observed fold axis (*B*) and the fold profile. *Inset*: A profile drawn in the field showing α, the pitch of the axial-plane trace. *Stereogram*: The profile plane is plotted normal to *B*, and the angle of pitch (α) is marked with a dot. The great circle passing through this dot and *B* defines the required fold axial plane; the strike of the plane is 329 and the dip 20° SW. Note that in the inset the axial plane is steeper.

points of maximum curvature of each layer is the trace of the axial surface. If the orientation of the fold axis can be determined independently, the orientation of the plane is readily calculated by stereographic techniques. In Figure 39 the fold axis, *B*, is plotted normal to the profile plane. From a photograph of the profile, the angle of pitch (α) of a planar section of the axial surface can be measured; α permits the trace of the axial surface (a line) to be plotted as a point in Figure 39. The great circle through this point and *B* defines the strike and dip of the required axial surface (Fig. 39).

Two precautions must be taken in collecting such data. First, the direction from which the profile was drawn must be recorded; although normal to the fold axis, mirror image profiles are obtained by viewing from opposite directions along the axis. For folds with approximately parallel axes, it is helpful to record their profiles with the same orientation (e.g., for approximately north-south fold axes, to record profiles looking north). Natural outcrops often do not permit this; for example, with north-south axes some profiles may be exposed on north-facing and some on south-facing outcrops. However, profiles for south-facing outcrops can be traced and viewed through the tracing paper to obtain a north-facing outcrop view, and vice versa.

Second, although normal to the fold axis, the profile plane can rotate about the fold axis. Hence, in order to orient the profile, a horizontal line on the outcrop should be included in the picture. This is adequate orientation except when the fold axis is subvertical; in such cases the projection of a compass direction must be drawn on the profile drawing.

In extremely simple cases, the axial surfaces may be parallel planes in successive anticlinal and synclinal elements (Fig. 40), whereas in disharmonic cylindroidal folds the axial surfaces can change orientation rapidly from one fold to the next (Fig. 41). When a fold axis is horizontal the axial surfaces have constant strike, even if they are strongly folded as

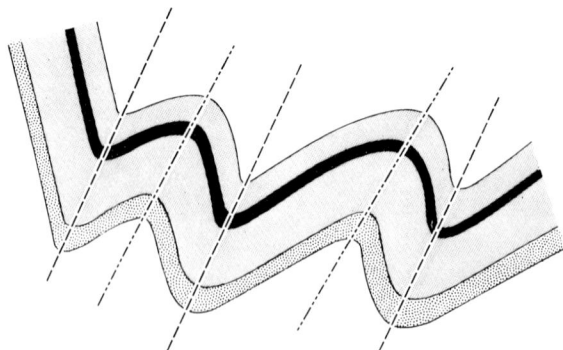

Figure 40. Diagrammatic profile of simple cylindroidal fold in which the axial planes are approximately parallel. The axial planes are shown by broken lines.

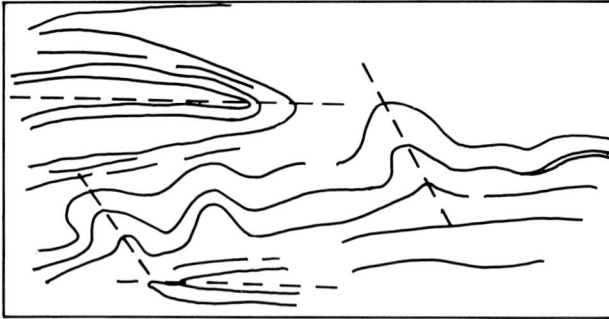

Figure 41. Profile of a disharmonic cylindroidal fold from Dalradian rocks, mid-Strathspey, Central Highlands, Scotland. Note the varied orientation of the bisecting planes (broken lines).

in Figure 37; however, the dips of the axial surfaces can vary widely – they may be in opposite directions – in neighboring flexures. Planar portions of axial surfaces have varying strikes if the fold axis plunges, but these surfaces are cylindroidal with respect to the fold axis.

Sometimes it can be assumed that the profile of a group of folds remains essentially constant; that is, if sections are cut normal to the fold axis, the profile is similar in each. In hand specimens this is commonly the case, although the profile often changes sufficiently rapidly for dissimilar profiles to be seen at each end of a small sample. Major folds usually show considerable variation in profile in areas of severe deformation. Such variability is a function of numerous interrelated factors including the physical and chemical properties of the rocks and the regional and local variations of stress. As Weiss (1959A) pointed out, the idealized concept of the fold axis as the rectilinear generator of the fold enunciated by Clark and McIntyre (1951) is, in actual practice, a statistical concept. Even the most regular of folds does not remain unchanged in profile indefinitely along the fold axis. If the profile changes along the fold axis, it is impossible for a single line to move parallel to itself so as to describe the dissimilar profiles. In such cases, the entire fold structure may be dissected into smaller geographical units within which a single axis can be considered the generator. While recognizing this dispersion of axes, it is commonly found that, when really large areas of folded strata are studied, the fold axes of individual sectors approximate closely to a single statistically-defined axis. For example, Atkinson (1960) used serial sections across the mountains of Prince Charles Foreland, Spitzbergen, to demonstrate that the folds are cylindroidal with axes which trend NNW to NW, and have plunges varying from zero to 5°, either NW or SE (see Figs. 42 and 43). Atkinson showed that there is continuity of the folds over many

N

Regional Trend of Fold Axes
Plunge 5° NW or SE

A

A

B

B

C

C

D

E

D

E

F

F

Thrust Sheets
Northern Block

SHEET 1
SHEET 2
SHEET 3
SHEET 4
SHEET 5

Thrust Sheets
Southern Block

SHEET 6
SHEET 7
SHEET 8

TERTIARY

Major Thrust

High-Angle Fault

| 0 | | 5 miles |
| 0 | 5 | kms. |

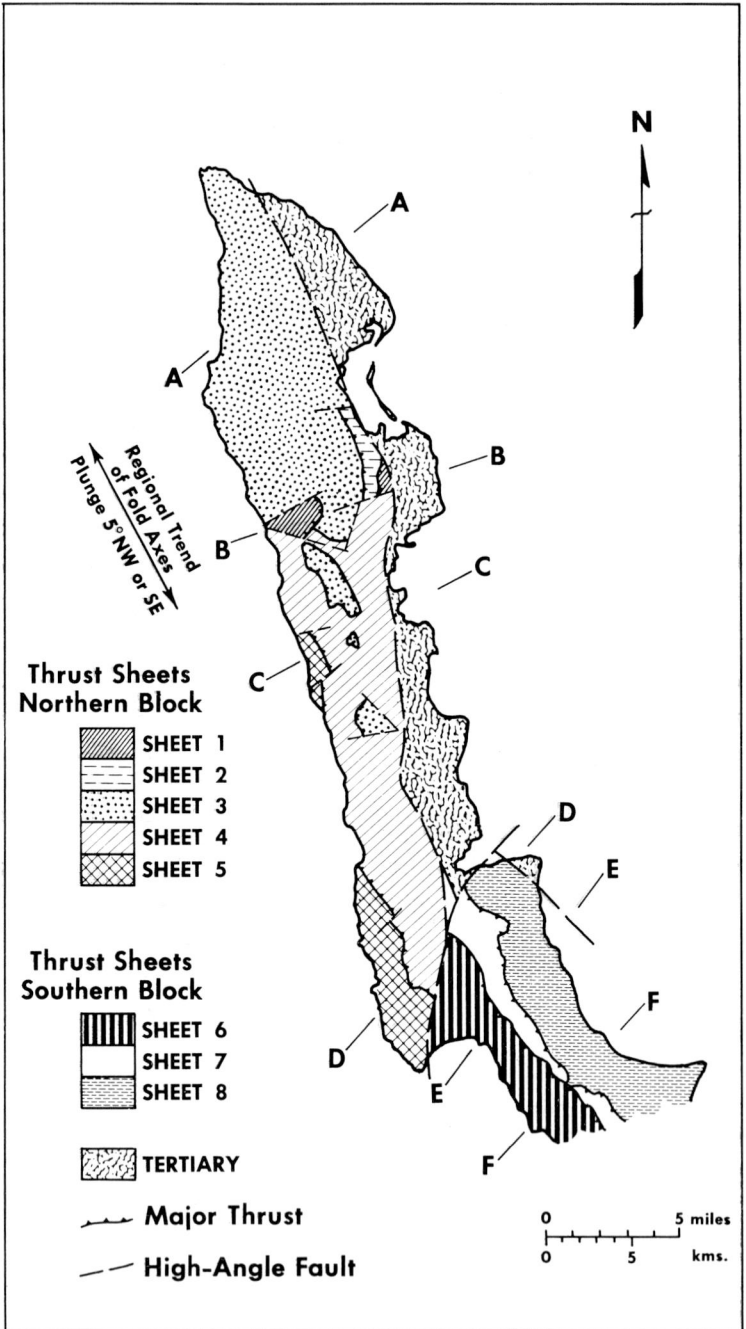

Figure 42.   Map showing distribution of thrust sheets, Prince Charles Foreland, Spitzbergen; A-A through F-F are section lines used in Figure 43 (After Atkinson, 1960, Fig. 1).

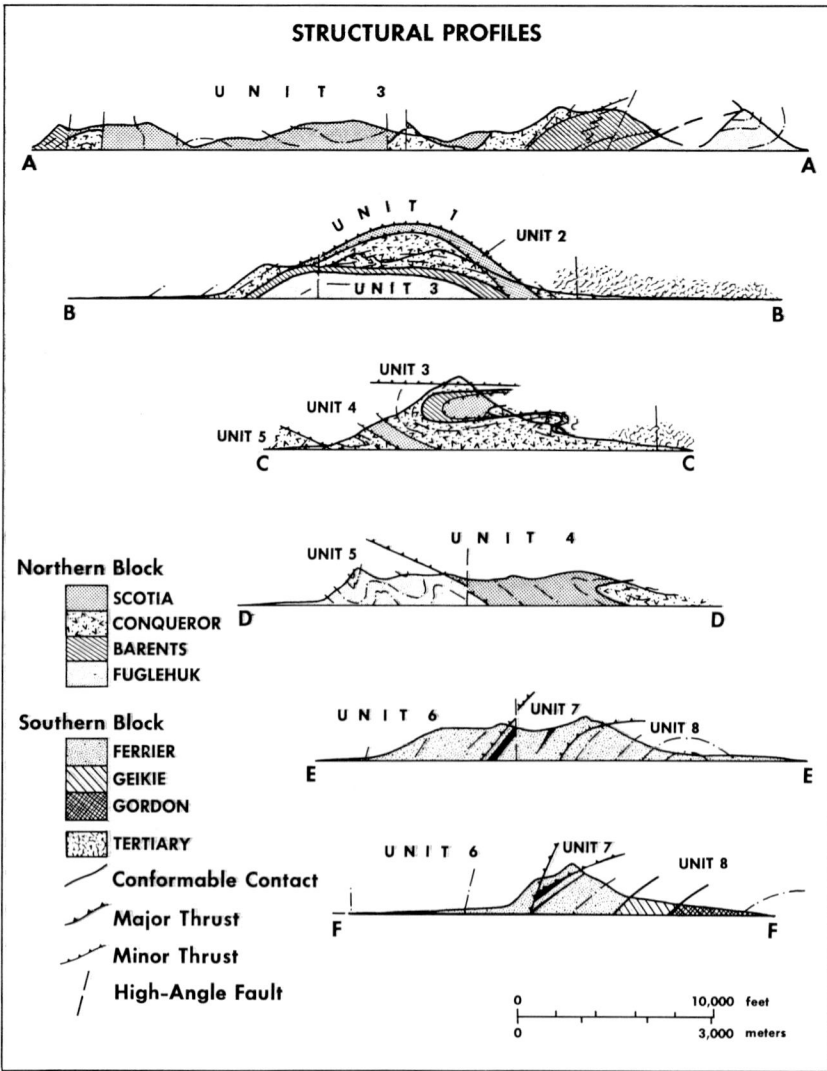

STRUCTURAL PROFILES

Figure 43. Serial profiles across the cylindroidal fold system of Prince Charles Foreland, Spitzbergen, drawn along the lines A-A through F-F of Figure 42 (After Atkinson, 1960, Figs. 2 and 3).

miles parallel to the fold axis, although over the same distance the profile slowly changes.

In the majority of areas affected by cylindroidal folding the folds of all different sizes are **homoaxial** within specific areas. That is, the fold axes are essentially parallel within an area, or, stated differently, the confidence band about the mean fold axial direction for the area is small.

Although the existence of homoaxial relationships must always be confirmed, such parallelism is of particular value in the study of areas of indifferent exposure. Dale (1896, p. 549) recognized this important point in studying the Green Mountain region, eastern New York, U.S.A.; he wrote:

> ...great movements which have produced what Sir William Dawson fitly calls "mountains of crumpling" are apt to leave their record on every cubic-foot of the region affected by them.... For this reason the characteristic features of such mountain-making movements are often quite as truly shown by single ledges, or even hand specimens, as by whole mountain sides.

One of the other well-known examples of the use of homoaxial relationships is the construction of a regional fold profile in the Grantown-Tomintoul area, Scottish Highlands, by McIntyre (1951B). The fold structure of this area is described in greater detail in Chapter 11.

These relationships imply that folds of all sizes can be homoaxial and can have similar fold styles at each scale of observation. This is a generally accepted principle that is widely used as a working hypothesis in structural analysis. However, some geologists seem to deny that such relationships can occur. For example, Goguel (1962, pp. 136 and 153) wrote:

> These minute secondary folds are very obvious, since they can be observed in outcrop, whereas the major folds can usually be identified only after laborious study of a whole mountain chain. For this reason, it would be to our advantage if we could use secondary folding in interpreting general structure. To this effect, some authors propose that the same mechanical principles can be used to explain the geometry of similar folds on different scales. However, this postulate has no rational basis.... Our efforts would be in vain if we were to seek a direct relation between the form of these minor folds and that of the major folds.

In fact, the correct answer probably lies between these extreme views. Successive phases of folding have occurred in most orogenic belts and the earliest very large folds are frequently overprinted by several generations of smaller folds not homoaxial with the older structures. One cannot expect all of the successive epochs affecting an area to be homoaxial.

Goguel used the illustration reproduced in Figure 44 to substantiate his opinion. In this diagram each individual fold appears to be cylindroidal, although, as Goguel implied, the whole structure departs significantly from homoaxiality. This is a question of size. In some regions homoaxial relationships pervade large volumes of rock, in others they exist for only very limited distances. Extrapolation from one scale of observation to

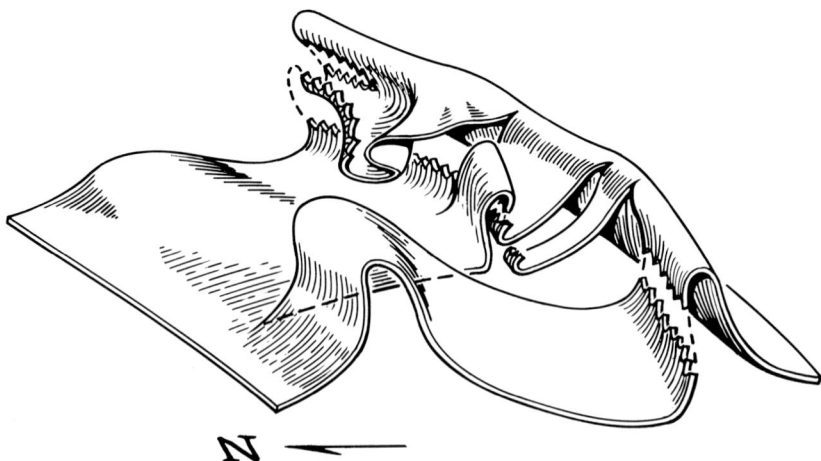

Figure 44. Folds in the Upper Jurassic, Vanige region, France, reconstructed by "removal" of younger and restoration of eroded rocks. Goguel thought minor folds have no direct relation to major folds, but, although the region is not homoaxial, each small part is approximately cylindroidal (After Goguel, 1962, Fig. 84).

another raises many fundamental problems in all geological studies; some of these problems are discussed in Chapter 3.

Thus it becomes clear that an important factor in defining the nature of folding is **size.** Particular attention is given to this in Chapter 3.

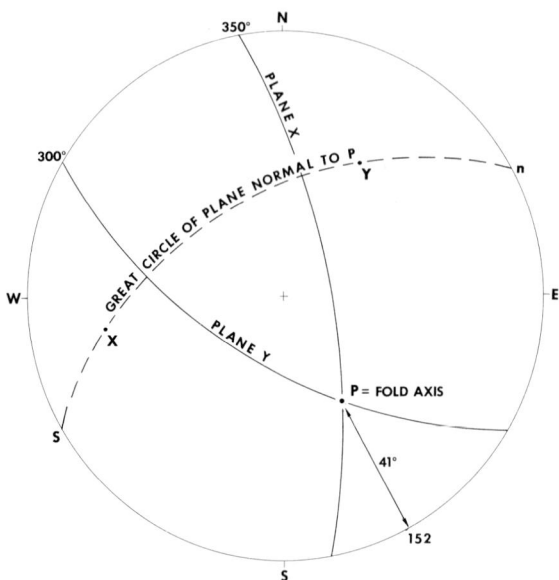

Figure 45. Stereogram representing the intersection of planes X and Y. See text for full explanation.

Many sets of folds tend to be homoaxial and this seems to have drawn attention away from the fact that the profiles and other structural features (e.g., axial surfaces, etc.) commonly vary greatly, even within quite small areas. However, brief study of any area of folded rocks quickly establishes this variability. As one example, some of the structures of the Monadhliath may be cited; Figures 183, 184, and 185 show a number of profiles from these mountains which lie north of Strathspey, Scotland.

The characteristics of cylindroidal folds can be represented on stereographic projections, and, as Dahlstrom (1954) emphasized, quantitative stereographic techniques clearly and impersonally state the bases on which many structural interpretations are made.

In Figure 45, the plane X dips east at 70°, and has a strike of 350, or 10° west of north, while plane Y dips to the southwest at 60° and strikes at 300, or 60° west of north. Each of these planes may be represented by either a line (great circle) or a point representing the normal to the plane. If these two planes are limbs of a fold, the line along which they intersect, P on the stereogram (Fig. 45), coincides with the fold axis. That is, plane X can be rotated about axis P to coincide with plane Y. The normals to these planes lie on a great circle, which represents the plane normal to P. If planes X and Y are part of a cylindroidal fold, all the other bedding planes measured on the fold have normals that fall on this great circle normal to P. An idealized cylindroidal fold is illustrated in Figure 46, and the normals to a number of "bedding planes" that comprise the cylinder lie close to the great circle perpendicular to the fold axis (P). In actual cases, folds are only approximately cylindroidal, so the normals to the bedding planes only lie close to a great circle. In cylindroidal folds the poles to axial and bisecting surfaces also fall on the great circle perpendicular to the fold axis.

This method was first developed by Wegmann (1929) and it is of great practical value. Even when exposures are poor and actual flexures are not seen in the field, it is commonly possible to deduce the nature of the folding. By measuring the orientation of all available bedding planes in a region of cylindroidal folding, and plotting their poles on a stereographic projection, the great circle most nearly passing through the poles can be found. The normal (perpendicular) to this great circle approximates the fold axis P, whose plunge is given by the distance between P and the primitive circle. In Figure 45 the plunge of P is 41°, and the trend 152.

In order to define the great circle adequately, as wide a variety of plane orientations as possible should be measured; often, in addition to all the dominant orientations within an area, the apparently unusual ones are very important. Vertical bedding planes are of particular significance because their strike is parallel to the trend of the fold axis in cylindroidal structures. Also, the plunge of the fold axis cannot be steeper than the shallowest dip of the folded surface. These relationships always apply, except when the fold axis is vertical; this is a relatively uncommon, though not impossible, situation. These properties should be confirmed by rolling a sheet of paper into a cylinder, and successively holding the cylinder axis horizontal and inclined.

In an area of indifferent exposure, measurement of all available dips and strikes of a folded unit permits a picture of the **modal fold** (Williams, 1959, p. 633) to be developed on a stereographic projection. The axial plane of a modal fold contains the axis and bisects the angle subtended by the **modal limbs.** If the inherent and severe sampling problems are ignored, one can follow Williams who said that the modal limbs represent the most commonly occurring orientation of fold limbs within an area, and that, in many cases, the modal limb can be identified as the maxima on stereographic plots of poles of folded bedding planes. This concept should not be confused with another referred to as the **average dip** (or **trend**) of a

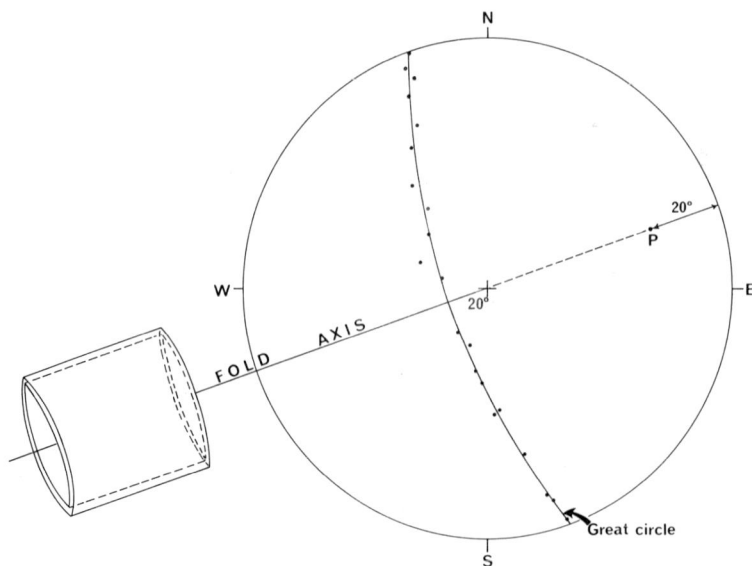

Figure 46. Idealized cylindroidal fold (left) plunging 070 (20); each dot on the stereogram is a pole to a "bedding plane" on the cylinder; the great circle is the approximation to these poles and P is the fold axis and the normal to the great circle. Note that the great circle is also 20° from the center of the stereogram.

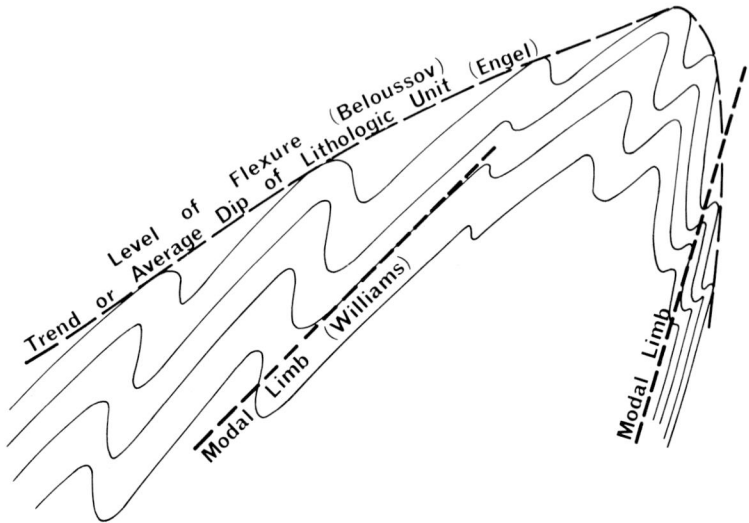

Figure 47. Fold terminology used by Beloussov, Engel, and Williams.

lithologic unit by Engel (1949, Fig. 3) and as the **level of the flexures** by Beloussov (1962, p. 113). Figure 47 shows the difference between these two terms. Both concepts should be used with extreme care; neither is used in this book.

In the foregoing pages the basic elements used in quantitative analysis of cylindroidal folds have been enumerated. The normals to bedding planes plotted upon a stereogram are S-poles on a π-diagram, and the axis defined by the great circle upon which the S-poles lie is designated β (Fig. 48).[1] Alternatively, the actual bedding planes can be plotted as shown in Figure 49 and these give a family of curves whose intersections approximate to the fold axis. Such a plot is a β-diagram, and the center of gravity of the intersections is called β. When an axis is definitely established as the fold axis, it is now customary to designate it B, and commonly $\beta = B$. Any family of S-structures can be plotted as a β-diagram, but in such cases β may not be equivalent to B.

It is generally accepted that in most cases a π-diagram is more accurate than a β-diagram for defining β unequivocally. On a β-diagram the $n(n - 1)/2$ intersections between the n measured S-surfaces are plotted, and their center of gravity defines β. Considerable labor is involved in locating all the intersections on a projection and in finding

[1]Sander (1942) originally designated this axis π, and there has been considerable confusion over the use of this term. β has been defined as in the next two sentences, so that, since π and β are in effect different symbols for the same fabric element, β is used exclusively in this book (cf. Turner and Weiss, 1963, p. 155).

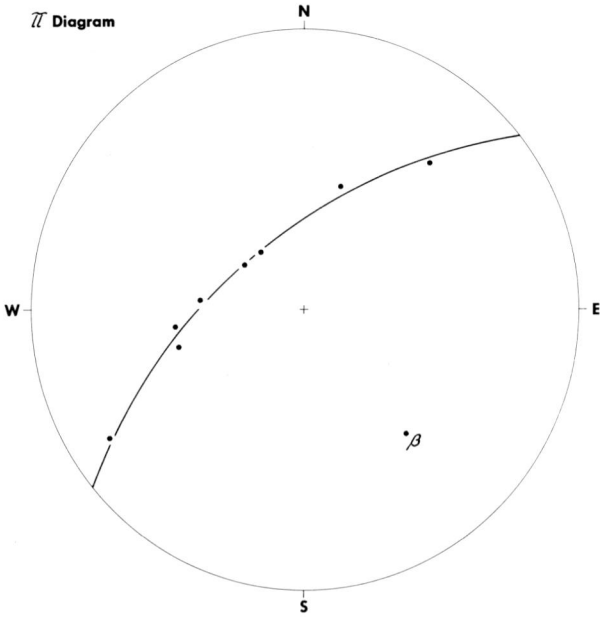

Figure 48. *S*-poles plotted on a π-diagram that defines an axis β.

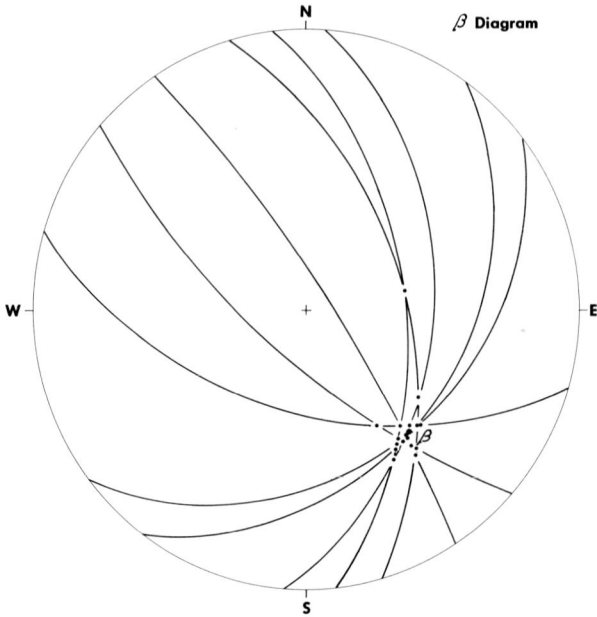

Figure 49. *S*-poles plotted on a β-diagram that defines an axis β.

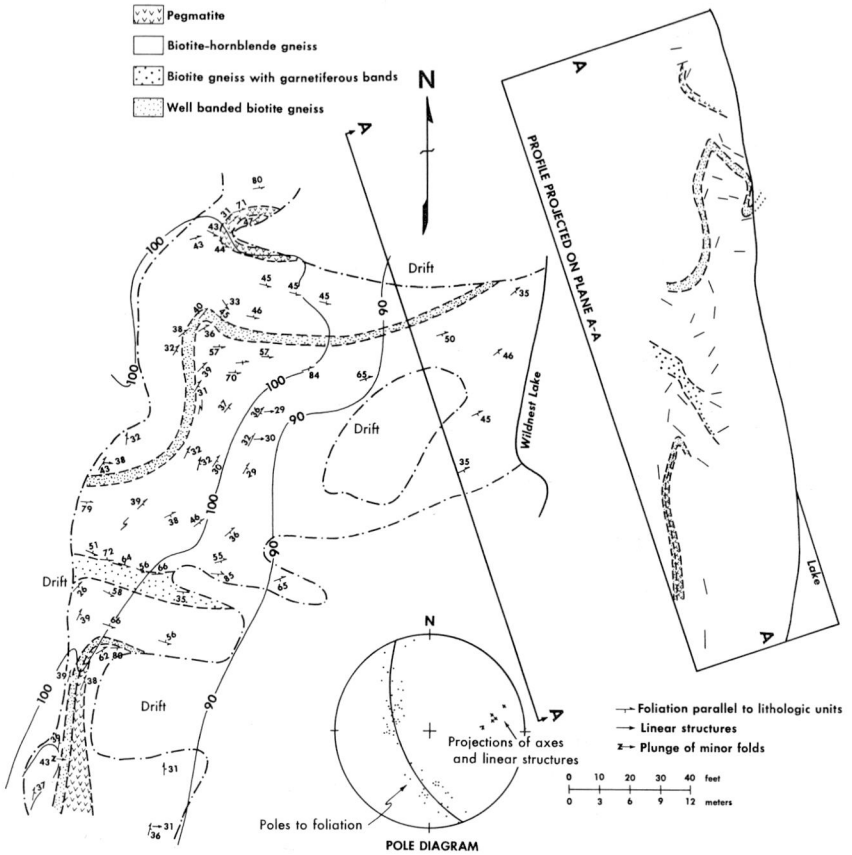

Figure 50. Structural geology at Wildnest Lake, Amisk Lake area, Saskatchewan, Canada. The map (left; contours in feet) shows the minor structural elements which are also plotted in stereographic projection (center bottom). The profile projected onto a plane A-A perpendicular to the fold axes is shown on the right (Geological and topographic mapping by J. E. Mullock; this Figure after Dahlstrom, 1954, Fig. 5).

their center of gravity, although at least one computer program has been prepared to handle the calculation algebraically (Noble and Eberly, 1964). By contrast a very large number of S-poles can be plotted on a π-diagram without difficulty and without introducing cumulative errors. More serious is the fact that many "spurious" intersections far removed from B tend to occur in β-diagrams when gently- or isoclinally-folded S-surfaces are involved, and where small errors occur in the field measurements (cf. Ramsay, 1964).

An actual example can be based on a small area at Wildnest Lake, Amisk Lake area, Saskatchewan, Canada (Fig. 50) mapped by Dahlstrom (1954). Here the foliation of the highly metamorphosed rocks is parallel to the boundaries of the lithologic units, and S-poles for these folded layers define a great circle. The normal to this great circle is essentially parallel to the B-axis defined by the minor fold axes in the area. Again, in the Gardner Lake area, Beartooth Mountains, Wyoming, U.S.A., Harris (1959) mapped a complex series of metasedimentary rocks which are now granitic gneisses. The orientations of foliation planes and fold axes collected on a regional basis were plotted on stereograms (Figs. 51 and 52); from these plots the B-axis was determined. Studies of this type are based on the assumption that individual small fold axes are homoaxial with the major regional fold axis or B-axis; although this relationship is common, more complex relationships can frequently occur.

For many years tectonicians have emphasized the necessity of viewing folded structures along the fold axis; Mackin (1950) called this the "down-structure method." The technique is useful because natural outcrops merely give random sections through fold structure, so that the shape of the folds and the thickness of the folded units are seriously distorted. Ordinary geological maps also distort fold structures seriously, because the map plane is a random oblique section through the flexures. To avoid such distortion, the structure can be projected onto a plane normal to the fold axis. This profile can be visualized in the field by viewing or photographing an outcrop along the fold axis. A similar effect can commonly be obtained by viewing a geological map "down the fold axis" rather than by looking at it obliquely from the south (as is customary with ordinary atlas maps, for example). Although the area is poorly exposed on the ground, the map of the Woodville area, Pennsylvania, U.S.A., is useful for illustrating this method. The simplified map published by Bailey and Mackin (1937) is reproduced in Figure 53; Mackin (1950, 1962) interpreted the structure as a fold system with nearly horizontal axial surfaces (recumbent fold system) that plunges southeast. Outcrops in the area confirm the homoaxial nature of the folds, and that the B-axis plunges southeast. Without this information the B-axis is difficult to determine from Figure 53 and the structure could be interpreted in a radically different manner (e.g., McKinstry, 1961). However, if Figure 53 is viewed systematically from all directions (as if it were a small horizontal outcrop), the fold axis can be "found"; by viewing the map obliquely from the northeast corner the true profile can be "seen" easily. Mackin (1962) prepared a revised profile (Fig. 54) across the area on the basis of McKinstry's (1961) detailed new map; it seems likely that the continuing mapping in this area will involve additional modifications to the geometry.

Unfortunately, the principle of viewing the profile is deceptively simple, and, in the field, because of the two-dimensional nature of most

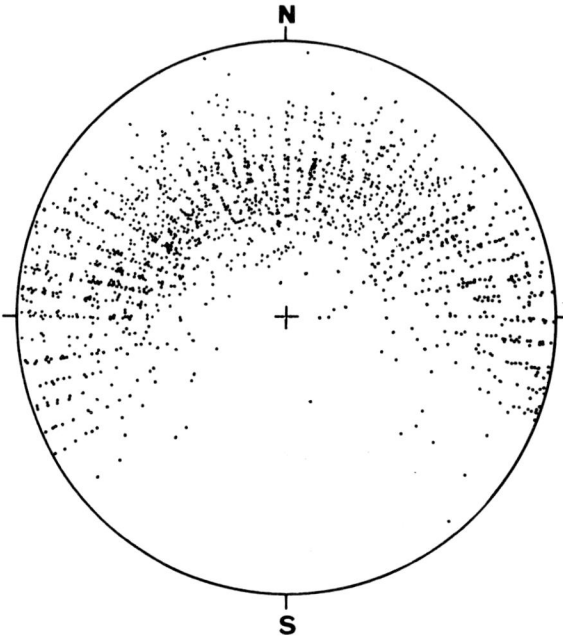

Figure 51. 1925 poles to foliation measured throughout the granitic gneiss in Gardner Lake area, Beartooth Mountains, Wyoming, U.S.A. Note that the distribution on the Schmidt projection reflects southward plunging (*ca.* 40°) structure that comprises one or more synforms or antiforms (After Harris, 1959, Fig. 5).

outcrops, considerable care is required to ensure that the true profile is being examined.

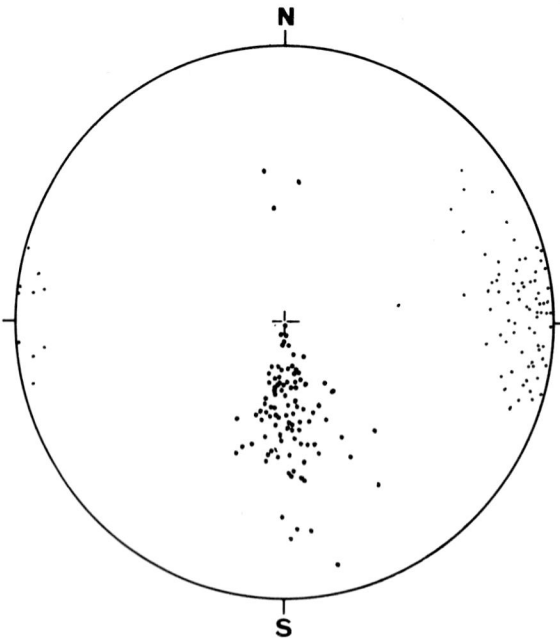

Figure 52. Structural elements of parasitic and small folds in highly contorted banded migmatite, west limb of Christmas Lake syncline, Gardner Lake area, Beartooth Mountains, Wyoming, U.S.A. Small dots: poles to axial planes; large dots: projections of fold axes (After Harris, 1959, Fig. 6).

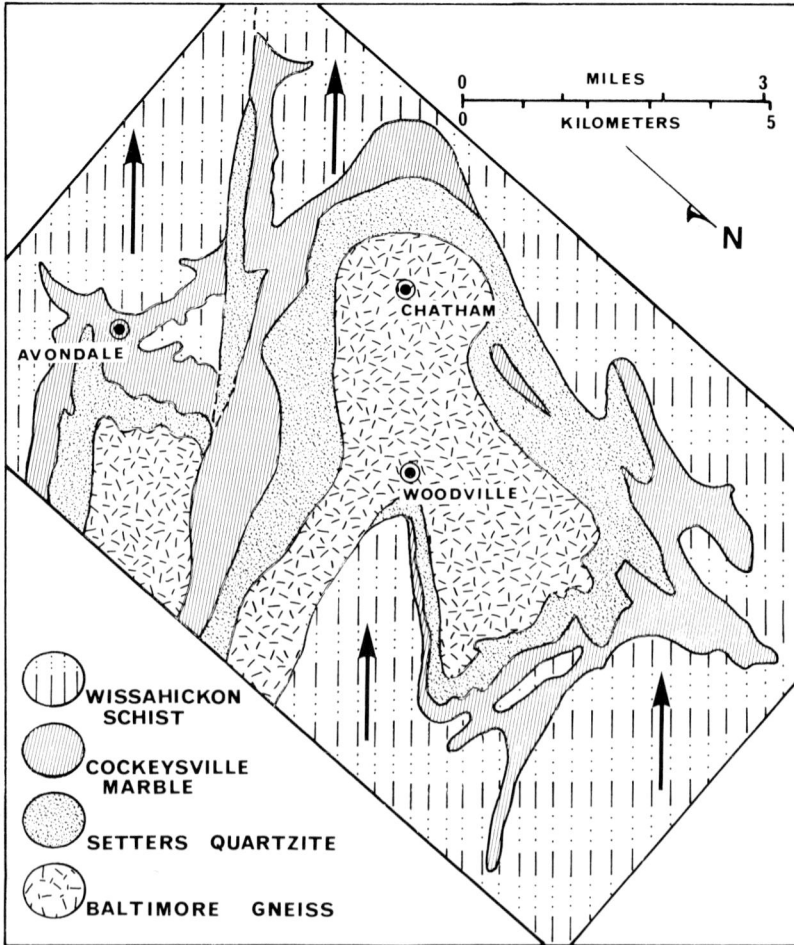

Figure 53. Map of bedrock geology, Woodville area, Pennsylvania, U. S. A. Large arrows indicate the trend and plunge direction of the inferred regional cylindroidal fold axes (After Bailey and Mackin, 1937, Fig. 5).

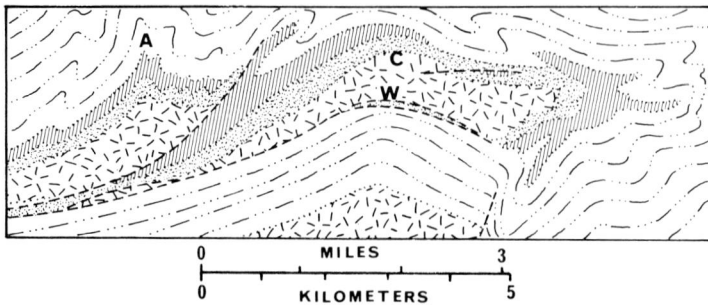

Figure 54. Hypothetical profile across the map (Fig. 53) of the Woodville area, Pennsylvania, U.S.A. A = Avondale, C = Chatham, W = Woodville (After Mackin, 1962, Fig. 1).

Figure 55. Map of hypothetical folds plunging north at 20° and (below) profile constructed from the map data. Levels 1 through 6 (from top to bottom at the margin of the section) correspond to east-west horizontal lines 1 through 6 from north to south in the map (After Mackin, 1950, Fig. 3).

If the *B*-axis has been identified, it is more realistic to construct cross-sections (profiles) normal to *B,* rather than in any other plane, because the profile gives the only undistorted picture of the fold geometry. Considerable distortion is introduced unavoidably when vertical sections are drawn through folded areas, unless the fold axes happen to be horizontal. The method of constructing a profile normal to *B* from a map is illustrated by the hypothetical example (Fig. 55) used by Mackin (1950). In Figure 55 the fold axis plunges N at 20°, the trace of the axial surface on the map has a northeasterly trend, and the inclination of the axial surface results in distortion of the bed thicknesses on the map.

The actual practical method of profile construction for an area of homoaxial folding is illustrated in Figure 56.

The preparation of profiles can be very helpful in studying a folded area. A profile across the Wildnest Lake structure mapped by Dahlstrom (1954) is included in Figure 50. Stockwell (1950, p. 117) drew a profile for the complex sequence of schists in the Sherritt Gordon mine area, Manitoba, Canada (Fig. 57); the trace of the axial surface trends approximately north-south although the fold axis plunges at 086 at 26°. Stockwell concluded that, because the foliation is parallel to the lithological boundaries and wraps around the mapped fold, the structure probably represents an isoclinally-folded series of rocks refolded subsequently to give the present major fold structure (see Chapter 8).

It is instructive to use differently-colored layers of modeling clay (plasticine) to make simple fold structures, and then to slice the models along various planes to simulate random outcrops. Figure 58 shows the appearance of a simple fold on the several faces of a block diagram.

Although the methodology associated with stereographic analysis of cylindroidal folds was essentially devised for unravelling the structural complexities of metamorphic rocks, the technique is very valuable in the study of folds in wholly-unmetamorphosed and simply-folded sedimentary rocks.

In what are believed to be the first applications of this type, Whitten (1957A) and Knowles and Middlemiss (1958) demonstrated the cylindroidal nature of folds in the Cretaceous rocks of the Isle of Wight and the northwest Weald, southern England, respectively. For the Isle of Wight

Figure 56. Construction of a profile for a mapped homoaxial fold system. A: Square grid pattern on map is shown projected along lines parallel to fold axis onto the projection plane (which is perpendicular to fold axis). B: Homoaxial fold on map is shown projected onto the projection (profile plane).

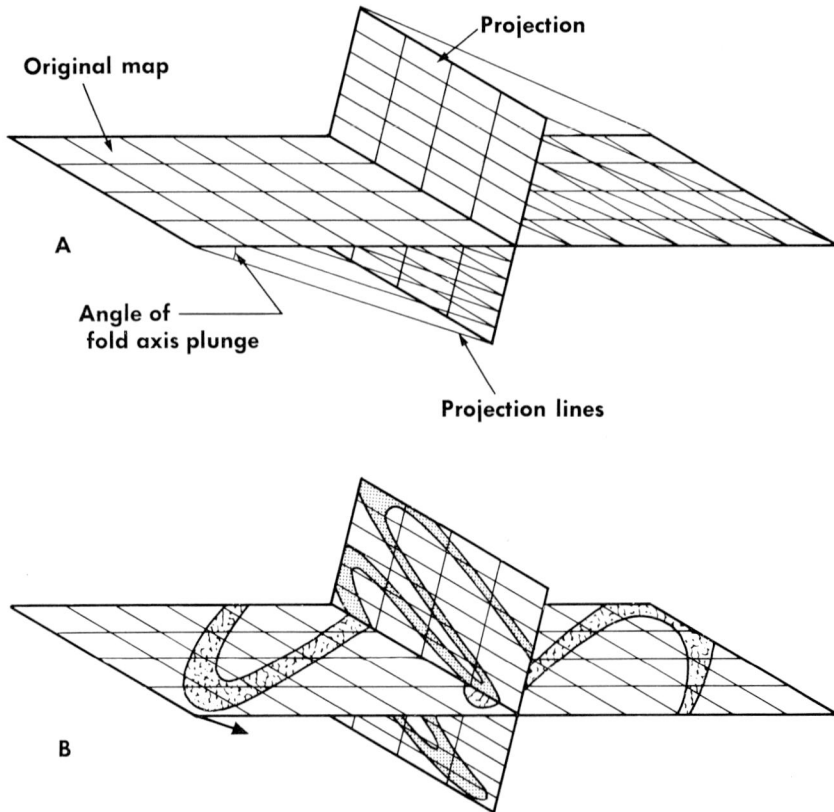

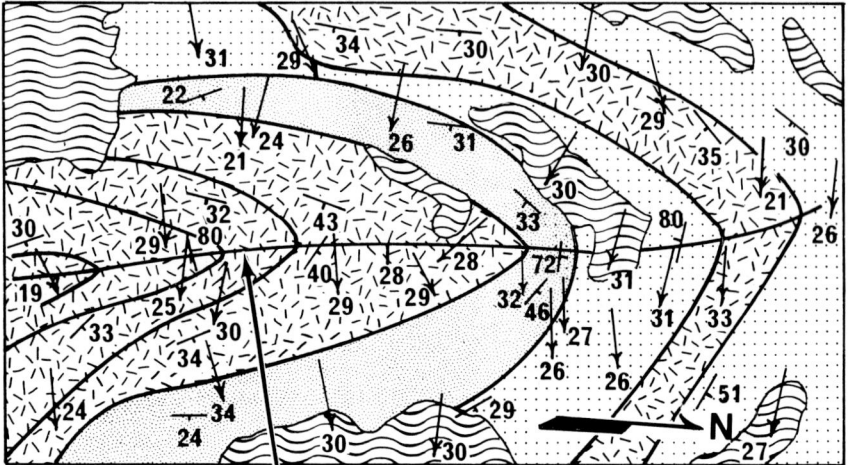

TRACE OF AXIAL SURFACE

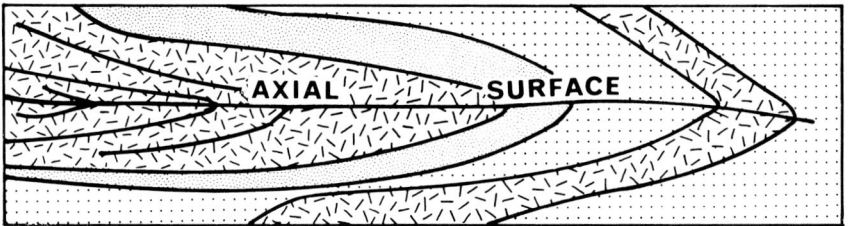

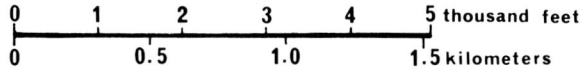

GNEISSIC QUARTZITE
HORNBLENDE PLAGIOCLASE GNEISS
STRATIFORM OLIGOCLASE QUARTZ GNEISS
LAKE
STRIKE & DIP OF S-STRUCTURE
ORIENTATION OF FOLD AXIS

Figure 57. The Sheila Lake fold that plunges eastward, Sherritt Gordon mine area, Manitoba, Canada. *Above*: Map. *Below*: Profile of the fold looking west along the fold axis (Mapping by J. D. Bateman; this illustration based on Stockwell, 1950, Fig. 12).

the data points and the outcrop pattern of the Chalk are shown in Figure 59. The π-diagram (Fig. 60) based on these observation points shows the fold axis could not have been located readily from the geological map, except that the strike of a few vertical Chalk bedding planes is parallel to the trend of the fold axes. Before the parallelism of fold axial trends and the strike of vertical beds can be assumed, however, the cylindroidal nature of the folding must be established.

In Figure 59 fold axes are shown, but, as indicated above, a cylindroidal fold axis is a statistical concept, and, although it has trend and plunge, it is no more appropriate to indicate a fold axis on a map in one place than in another. This requires emphasis because, in more recent studies of

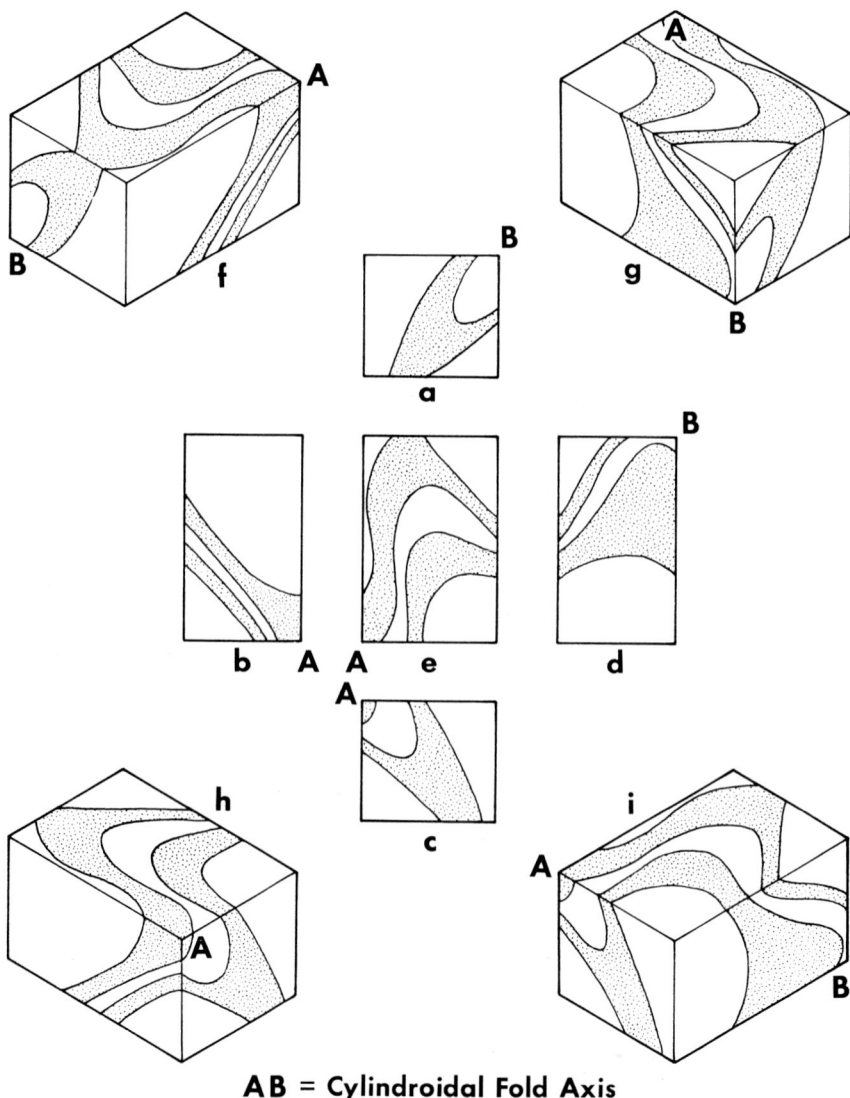

AB = Cylindroidal Fold Axis

Figure 58. Block diagram to illustrate a cylindroidal fold. The sides (a, b, c, d) and top (e) of the block are shown in exploded form at the center. The block is successively viewed from each top corner in the drawings (f, g, h, and i) around the margin (After Dahlstrom, 1954, Fig. 1).

eastern Isle of Wight by Falcon and Kent (1960; see also Falcon, 1961; Middlemiss, 1961; Whitten, 1961A), the statistical nature of the fold axes

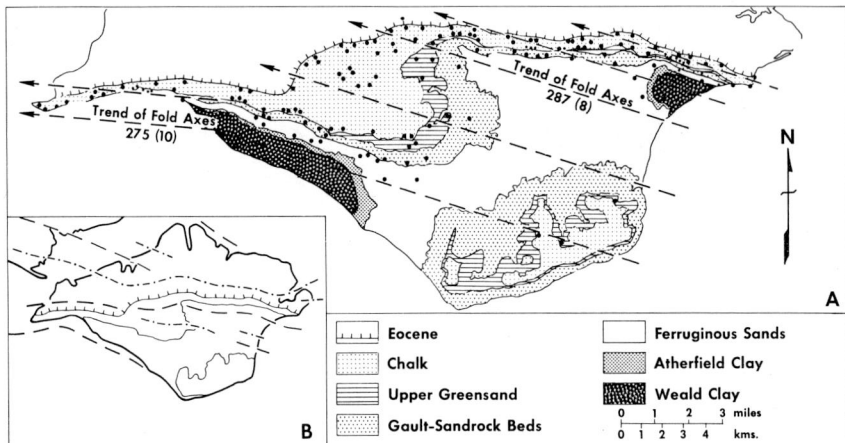

Figure 59. Cylindroidal fold axes of the Cretaceous rocks, Isle of Wight, England. A: Map showing the orientation of the constructed cylindroidal fold axes; dots show locations of dip readings used to define the axes. B: Trend of fold axes in the Isle of Wight according to White (1921); long dashes are anticlinal fold axes and dot-dash lines are synclinal fold axes (After Whitten, 1957, Fig. 3).

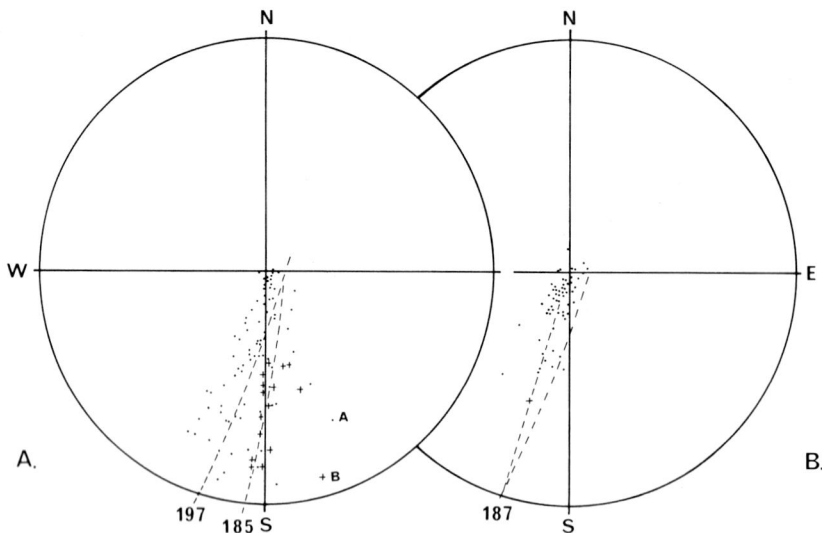

Figure 60. π-diagram for Cretaceous rocks of the Isle of Wight, England. Crosses relate to the western quarter of the map (see Fig. 59). Possible great circles for the π-poles are shown by broken lines. Data from Geological Survey of Great Britain 6 inch/mile field maps. A: Chalk (anomalous dips at A and B). B: sub-Chalk rocks (After Whitten, 1957A, Fig. 1).

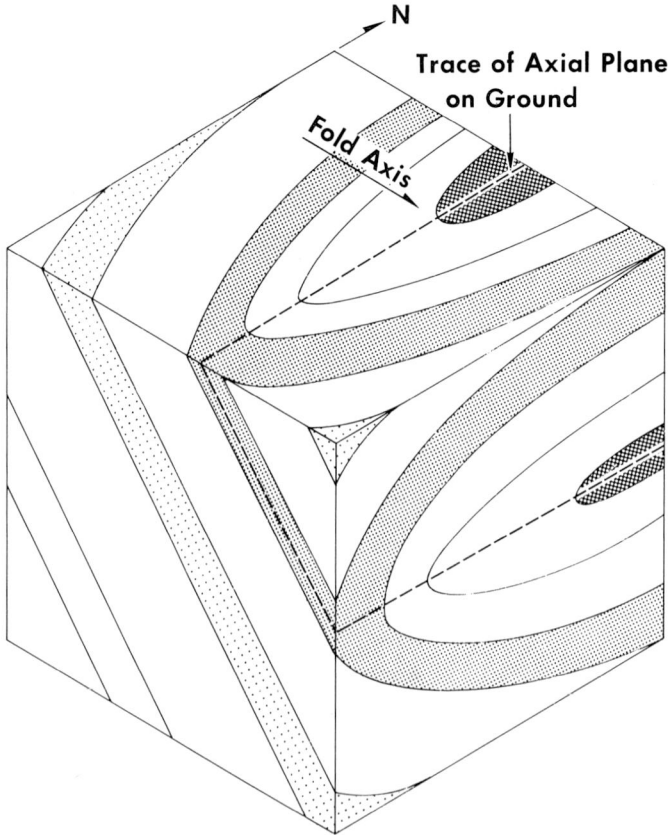

Figure 61. Block diagram showing a fold in which the trend of the fold axis is perpendicular to the trace of the axial plane on the ground surface.

seems to have been overlooked. The reason is not hard to find. It has long been common to draw the so-called fold axis on a map by indicating the central zone of the mapped fold system. The inset of Figure 59 shows such "axes" as mapped by White (1921). These latter "axes" approximate to the traces of the axial planes on the irregular ground surface, and in general they have little direct connection with the cylindroidal fold axial trend. It is only when the ground surface is horizontal and the axial plane vertical that the two dissimilar concepts give parallel trends. Although, for some purposes, it may be useful to plot the trace of the axial plane on the ground surface, it is entirely erroneous to designate such a trace as the fold axis. Figure 57 illustrates a case in which the cylindroidal fold axis has a trend inclined almost at right angles to the trace of the axial plane on the ground surface; realistic relationships are seen more easily in a block diagram (Fig. 61).

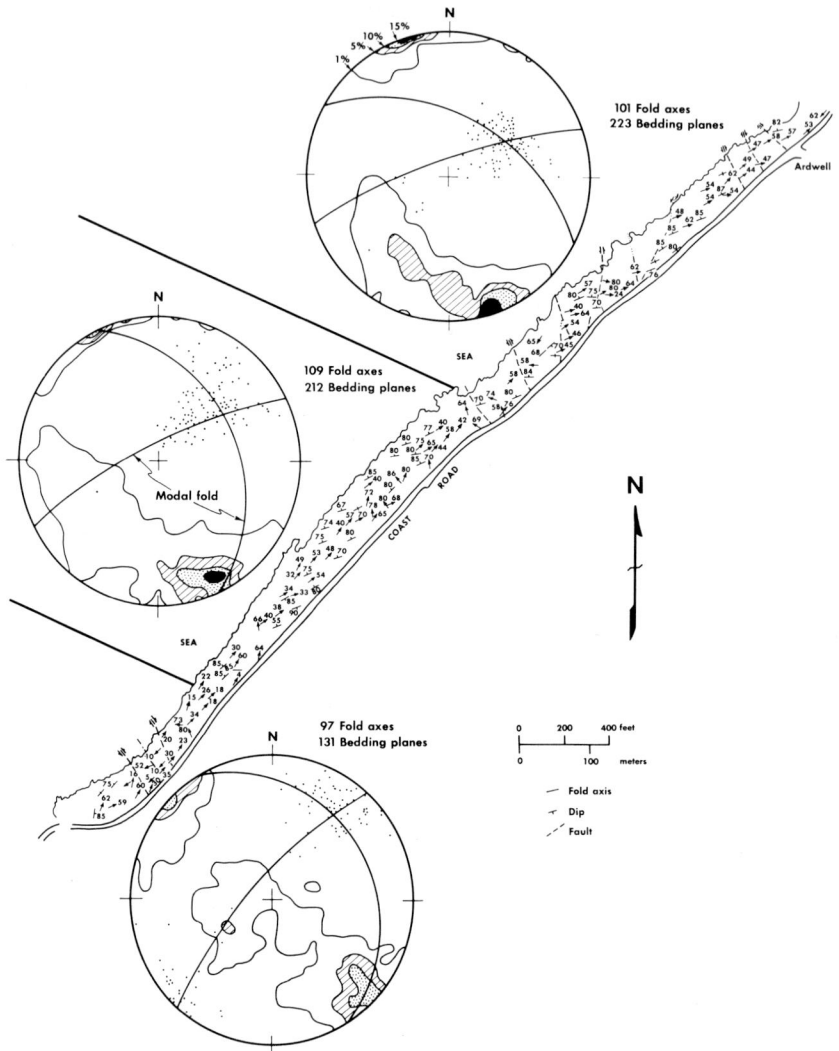

Figure 62. Map of sharply-folded Lower Paleozoic rocks exposed on the fore-shore between Kennedy's Pass and Ardwell Bay, near Girvan, Ayrshire, Scotland. The structural elements for the three subareas are shown in projection (fold axes represented by dots; poles to bedding planes contoured) (After Williams, 1959, Fig. 2).

Another noteworthy application of stereographic analysis to sharply-folded, but unmetamorphosed, sedimentary rocks was made by Williams (1959), in a detailed study of the Lower Paleozoic sequence of the Girvan district, Ayrshire, Scotland. Although he extended the study to the cylindroidal folds within a larger area, the excellent coastal exposures of Ardwell Flags permitted the most detailed study. Williams divided the coastal area into three subareas within which the fold axes are approximately parallel, and Figure 62 shows the orientations of fold axes and bedding planes.

NONCYLINDROIDAL FOLDS

Only cylindroidal folds have been discussed above, but some flexures are noncylindroidal. Both Stockwell (1950) and Dahlstrom (1954) recognized that, although cylindroidal folds seem to be the most common, conical folds also occur (see also Haman, 1961, Tischer, 1963, and Turner and Weiss, 1963). In a π-diagram for conical folds the S-poles fall on a small circle (rather than a great circle, as in the case of cylindroidal folds); Tischer (1963, p. 430) pointed out that cylindroidal folds can be considered a special case of conical folds.

Few examples of conical structures have been described in detail, although, on any scale of observation, fold systems may approach a conical form where the profile is changing along the B-axis. Clifford, *et al.* (1957) demonstrated that the Fionn Bheinn fold, Ross-shire (Northern Highlands, Scotland), shows these relationships. South of Loch Fannich limbs of the fold are almost parallel (isoclinal folding), but, when traced northwards, the structure becomes more open, and thus conical on a regional scale. On a map (Fig. 63) the linear structures have a fan-shaped arrangement, and converge southwards towards the vertex of the fold. Commonly, in the Moine Series of Scotland, and in most other folded areas, the trend and plunge of fold axes are variable; in consequence, conical, rather than cylindroidal, folds frequently occur locally (cf. Sutton, 1962, p. 84). Often, when a larger area is considered, such conical structures form part of an approximately cylindroidal regional system of folds.

Tischer (1962) mapped the Pégado anticline within the Wealden rocks of the western Iberian chain, Spain, and showed that the B-axis is subhorizontal with a northwest trend. The π-diagram for the northwest end of this anticline defines a small circle, which suggests a conical structure rather than a plunging cylindroidal fold (Fig. 64). Tischer (1963, p. 444) found it difficult to assume that these folds resulted from horizontal compression of parallel strata; he suggested that two phases of folding or rapid changes in thickness of the lithologic units might have produced the conical folding. Tischer (1963) described a second example of conical folds based on mapping by Habicht (1945) in Oligocene molasse between Ebnat and Krummenau, Switzerland. Symmetrical cylindroidal folds can be traced into a faulted area, but adjacent to a large fault the π-diagram

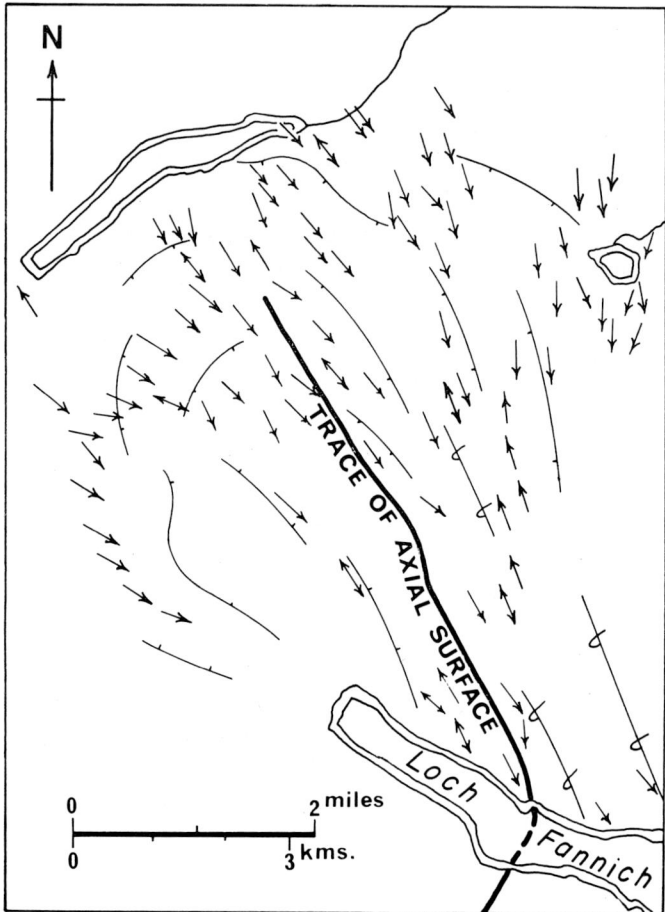

Figure 63.   Northern end of Fionn Bheinn conical synform, Ross-shire, Scotland. Trace of axial surface shown by heavy line; thin lines are generalized strike lines; arrows indicate minor folds and associated linear structures (After Clifford, *et al.,* 1957, Fig. 3; original mapping by Sutton and Watson).

defines a small circle (Fig. 65). Unfortunately, it is not certain whether the faulting pre- or postdates development of the conical folds (Tischer, 1963, p. 445).

Some of the most complete data for a specific set of conical folds were published by Evans (1963) for the Precambrian rocks of Charnwood Forest, Leicestershire, England. In each of four subareas π-diagrams define small circles (Fig. 66) appropriate to conical folds, and study of the foliation suggests that these folds probably do not represent refolding of earlier cylindroidal structures.

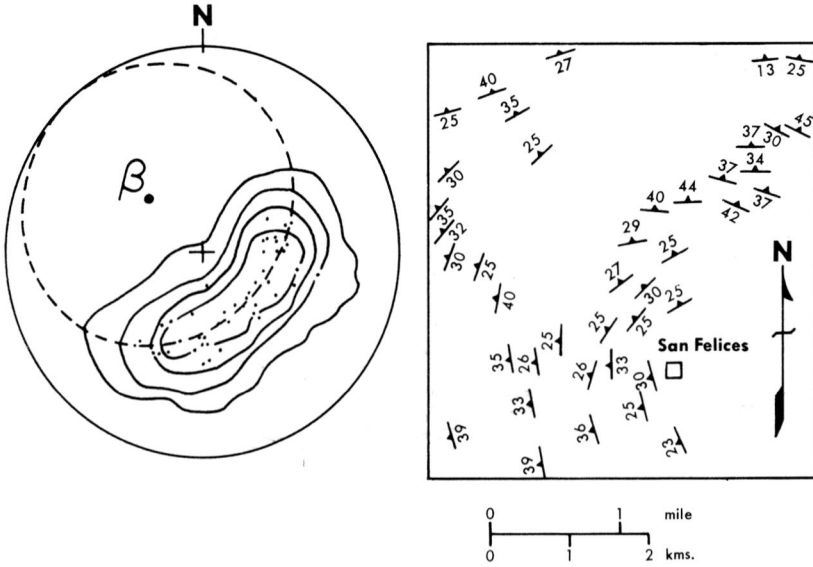

Figure 64. Conical fold: NW end of Pégado anticline in Wealden sedimentary rocks, western Iberian chain, Spain. Poles to 47 bedding planes are contoured at 0, 2, 4, and 6 per cent on the π-diagram, and the small circle is sketched in for the steeply plunging β-axis of the conical fold (After Tischer, 1963, Fig. 1).

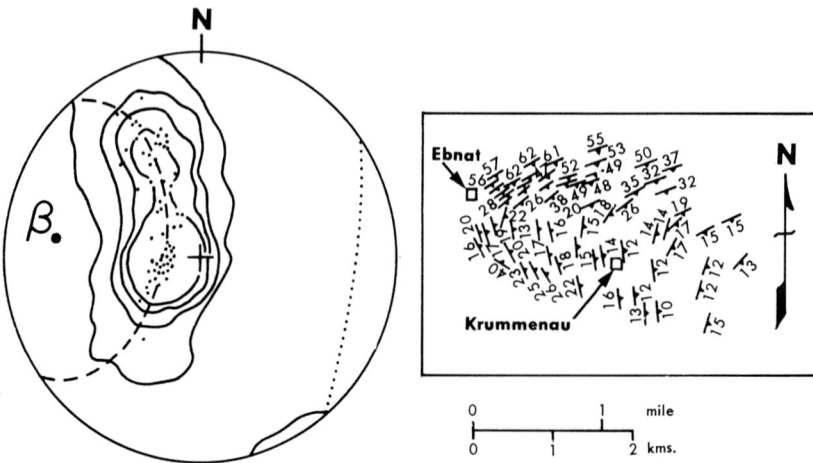

Figure 65. Conical syncline in Oligocene molasse, NW of the Santis Mountains, Switzerland. On the π-diagram poles to 64 bedding planes are contoured at 0, 2, 4, and 6 per cent. The broken line is a small circle about β, and the dotted line is the great circle normal to β (After Habicht, 1945, Plate 3, and Tischer, 1963, Fig. 2).

Ross and McGlynn (1963) described another possible example of conical folds from the metamorphosed graywackes and slates of the Yellowknife Group at Basler Lake, Northwest Territories, Canada. The Snare Group lies unconformably on the Yellowknife rocks. Ross and McGlynn reported that the Snare rocks are cylindroidally folded, and that this folding modified the already-metamorphosed and folded Yellowknife rocks, leaving them with conical fold structures.

In general it seems undesirable to include conical structures developed at the expense of cylindroidal folds with primary conical folds. However, discrimination between primary and secondary conical folds is liable to be difficult in many cases.

It seems certain that numerous examples of conical folds will be mapped in a wide variety of terranes in the future; they are likely to be as common in unmetamorphosed sedimentary rocks as in metamorphosed rocks.

Other examples of noncylindroidal folds include diapiric folds, domes, periclines, basins, etc. The nature of these structures is defined in Chapter 15.

ORIENTATION OF CERTAIN FOLDS

In a majority of cases the well-known structural terms **syncline** and **anticline** are adequate to describe folds. In more complex situations, and when the stratigraphic sequence is not well known, some ambiguity may be involved, which can be avoided in various ways.

In a mildly folded sedimentary sequence the term syncline implies fold limbs dipping downwards towards one another, so that successively younger strata occur towards the core of the fold, and, if exposed at the surface, towards the outcrop of the axial surface. In anticlines the fold limbs diverge downwards. Both syncline and anticline have come to imply that the relative age of the strata involved is known. Thus, in the syncline the youngest beds occur within the closure, and the older beds occur within the anticline closure.

If the structual complexity is such that the stratigraphic succession is upside down, but the relative age of the strata can be determined, completely-inverted synclines and anticlines occur. When the relative ages of the strata cannot be determined the terms **synform** and **antiform,** which were introduced by Bailey and McCallien (1937), are appropriate for downward-closing and upward-closing flexures (Fig. 67). These terms

Youngest

Oldest

Oldest

Youngest

A Syncline

SYNFORMS

B Inverted Anticline

Youngest

Oldest

Oldest

Youngest

C Anticline

ANTIFORMS

D Inverted Syncline

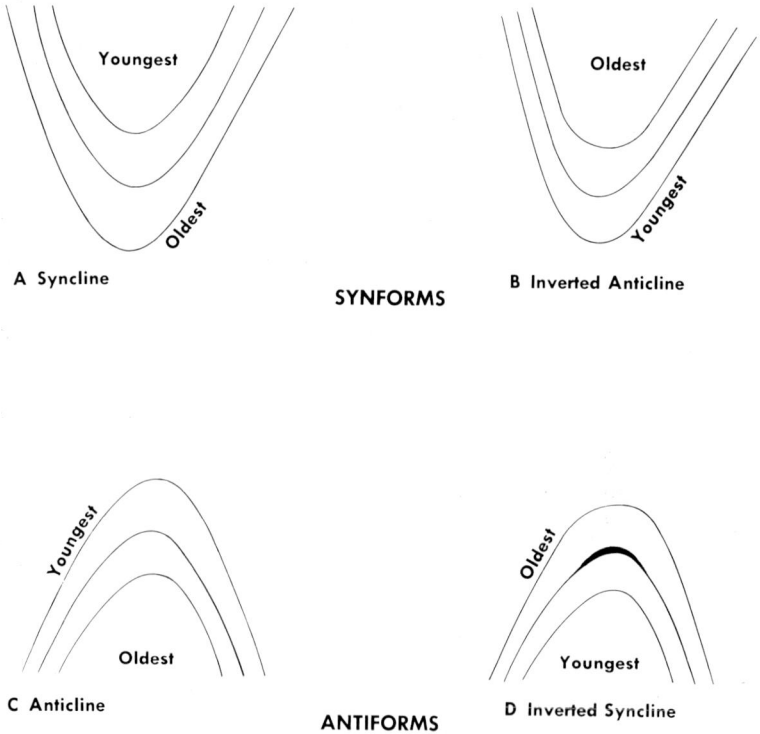

Figure 67. Synforms and antiforms.

involve purely geometric concepts, without any of the stratigraphic connotations attaching to syncline and anticline.[2]

Between the inverted and uninverted condition synclines and anticlines pass through a recumbent orientation in which the axial surface is subhorizontal. While **recumbent fold** is an appropriate name for folds in which the axial surface is subhorizontal, Naha (1959) correctly pointed

[2]McKinstry and Mikkola (1954, p. 3) used **structural syncline,** as opposed to stratigraphic syncline, for a "fold whose limbs dip toward each other regardless of the relative ages of the rocks." However, "synform" is to be preferred to "structural syncline."
Rather than use antiform and synform in conjunction with a statement of the direction in which the structure faces, Zen (1961, p. 313) introduced two new terms. In doing this he restricted anticline and syncline to imply purely geometric terms; this is contrary to common usage, and implied that anticline and syncline should be used in the manner in which antiform and synform are commonly used. Zen described a fold whose core contains relatively younger beds as a **topping fold.** If the core contains the relatively older beds the structure was called a **bottoming fold.** This terminology is likely to cause considerable confusion and should be abandoned; the new terms, and revised usage of anticline and syncline, do not have any advantage over the previously-accepted and adequate terminology.

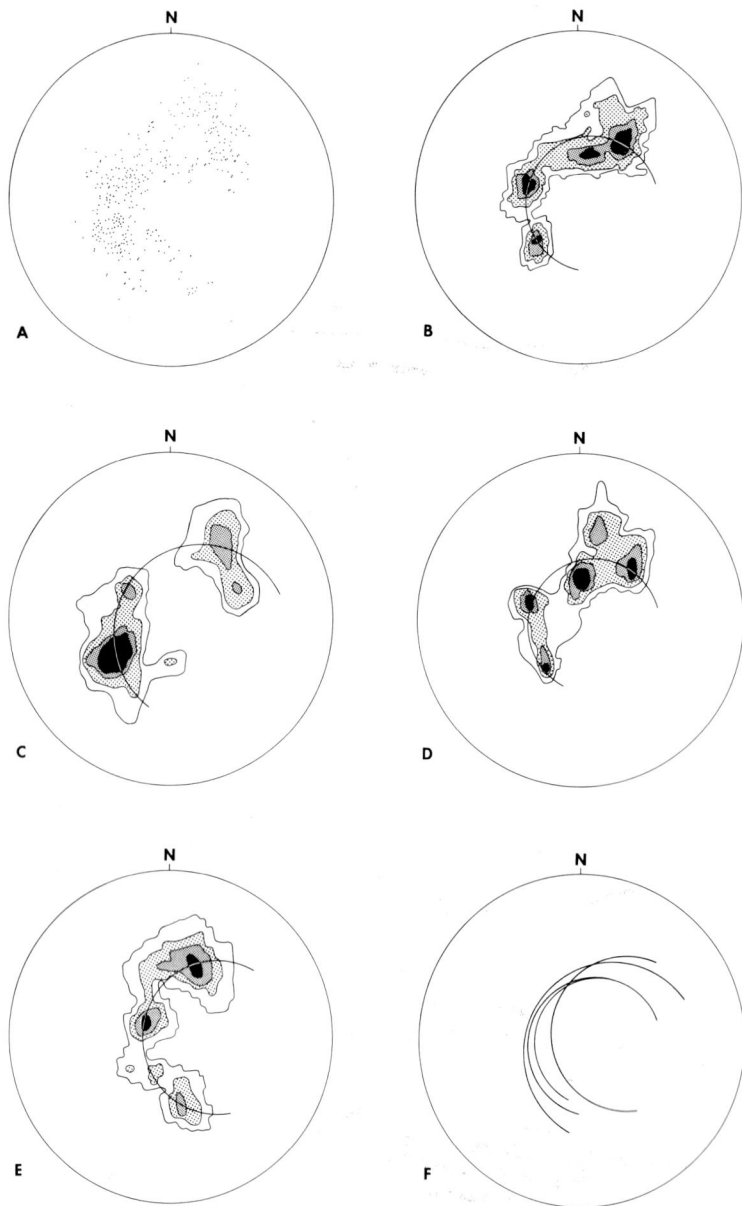

Figure 66. Projections for bedding plane data of Precambrian rocks, Charnwood Forest, Leicestershire, England. A: 350 π_{S_0} collected from all parts of the area. B: 120 π_{S_0} for subarea 1. C: 275 π_{S_0} for subarea 2. D: 100 π_{S_0} for subarea 3. E: 122 π_{S_0} for subarea 4. F: Synoptic diagram showing small circles for subareas 1 through 4. Contours at 2, 4, 8, and 12 per cent (After Evans, 1963, Fig. 1).

out that this definition is beset with difficulties if the fold axis is plunging. For such cases in which the axis plunges Sutton (1960A) used **reclined fold**; in these folds the axial surface strikes normal to the trend of the plunging fold axis. It follows that the axial surface has a subhorizontal trace on the profile plane of reclined folds.

In assessing structural data it has been common to refer to the direction in which the structures **young** or **face**. These useful concepts have been used for some years. "To young" was introduced by Bailey (1934) in his description of the area between Loch Leven and Glen Roy, western Scottish Highlands. Shrock (1948) described a bed or sedimentary succession as facing towards the side which was originally on top. The concepts of younging and facing are essentially the same. Cummins and Shackleton (1955), in describing the Ben Lui recumbent syncline, Scotland, extended Shrock's concept of facing to the orientation of folds. A fold is said to face in the direction, normal to its axis and in its axial surface, in which the younger beds are found. Beds intersected by foliation or schistosity young or face along this structure in the direction normal to the bedding/foliation intersection and towards the younger strata. It is often possible from a knowledge of local stratigraphy, by direct paleontological methods, or by study of sedimentation characteristics, to establish the sense of younging or facing, and thus to determine the nature of the flexures.

Chapter 3

Sampling and Size
in Data Collection

Although at first sight the contents of this chapter may not appear to have
direct bearing on the study of folded rocks, the concepts described below
are possibly more fundamental than those of any other chapter. Apart
from the routine description of a simple well-exposed structure, much of
structural geology is essentially a statistical study, because a quantitative
estimate of the three-dimensional variability of a folded series of rocks has
to be made on the basis of a limited number of observations. All geologists
are familiar with the acute problems that arise when "critical" outcrops
are concealed by superficial deposits, vegetation, or water. Because too
much uncertainty often attaches to results based on limited exposures
considerable funds are commonly expended in economic projects in order
to prove the details of a structure. Unfortunately, it is all too easy to base
conclusions on scanty data without seriously considering in a scientific
manner whether the information really permits extrapolation to the structures of interest.

In most structural studies the geologist does not have much control
over the number and location of the sampling sites; this places severe
restrictions on the interpretations that can be based on the observations.

If the data are demonstrably inadequate, it is inappropriate to justify
a conclusion on the grounds that funds will never be available to obtain
more outcrops or more observations. Thus, the importance of this chapter
lies in the fact that a structural geologist must be aware of the limitations
of the sampling plans used before he draws conclusions from his data.

The primary purpose of this chapter is to present a philosophical
approach or way of thinking; most of the conclusions have general
application to field data of all kinds. Petrographic examples are frequently
used because some people find it more difficult to think abstractly about
directional properties. It is the concepts that are important, and not the

means by which they are communicated. Once the concepts have been understood, there will be little difficulty in making the mental transfer to the structural geology of folds.

Commonly, it is very difficult to estimate what significance should be attached to **"typical"** samples and **observations.** Presumably, after thorough field study of an area, a geologist makes a subject-matter decision and selects "typical" specimens and makes structural readings to "represent" the area. These specimens and observations may be typical with respect to those properties which can be estimated readily, but atypical with respect to other less-easily-observed properties. Even if one property can be observed and measured easily, it is still necessary to establish the quantitative objective bases upon which several investigators can recognize the same phenomena as typical.

For decades debates have raged over the meanings of words and whether phenomena observed by one authority should be equated with structures seen by others elsewhere. In scientific disciplines it is important to ensure that observations are both correct and reproducible, and that different observers can make identical observations. This is possible if rigorous operational definitions are erected for each of the variables observed. It is unrealistic to reconstruct the geometry of an area without determining the nature and variability of the structural elements involved. Similarly, it is almost impossible to appraise the validity of any conclusions if details of the three-dimensional distribution and variability of the structural elements are not recorded.

THE CONCEPT OF THE POPULATION

In commencing a structural study it must be recognized that interest attaches to a **population of objects.** Each object has a variety of properties which can be measured, and these measurements provide **populations of attributes** (commonly expressed as scalars or vectors).

The population of objects of interest is different in each study. The members of the population have certain identifying characters in common; that is, by definition, they have approximately similar characteristics. In a beach-variability study a square-meter grid could be laid out and a population of objects defined as the exposed beach within each grid square. Again, a population of objects might be defined as 100 oriented samples collected according to a systematic plan from an area of folded schists, or all of the potential one-meter cubes of rock at the surface of a rectangular map area. In a major geological comparison, the population of objects of interest might be five continents or six mountain ranges; in a laboratory study the hand samples in a cabinet could be defined as the population of objects.

Having defined a population of objects, a number of attributes for each of the objects can be measured. For example, if a population of objects comprises 50 hand samples collected from a granite massif, the specific gravity of each object (sample) can be measured to provide a population of numerical attributes. Numerous different properties can be measured; for example, quartz content, color index, SiO_2 content, electrical resistivity, and many others. In this way an array of populations of attributes is built up. In a structural study directional attributes can be measured in addition to nondirectional properties (e.g., orientation of bedding, orientation of fold axis, etc.).

In most geologic work interest attaches to both the rocks which are readily examined and to materials which it is physically impractical to study (e.g., because of lack of exposure). As a consequence, it is convenient to subdivide the populations of objects and attributes on the basis of availability for study of the objects of interest; because the attributes relate to particular objects, this procedure automatically subdivides all the populations of attributes too.

The term **target population** has been assigned a technical meaning in statistical usage (e.g., Cochran, *et al.*, 1954), and it was introduced into geological literature by Rosenfeld (1954) and Krumbein (1960A). The target population is the population of objects (e.g., samples) of interest, about which the geologist wishes to make inferences or to draw conclusions. In a structural study of a highly deformed sandstone unit the target population could be defined as all potential $6 \times 6 \times 4$ cm. specimens comprising the whole unit. Alternatively, the target population might be defined to include only the uppermost layer of the sandstone unit that lies parallel to its eroded surface. In either case the population is a vast, but finite, number of specimens. The actual target population is chosen by the geologist on the basis of the objectives of his study, and in theory, at least, one or more attributes could be measured for each member of the population.

The target population is enormously important, although in the past it has commonly been treated as a purely academic question, represented vaguely in many geologists' thinking by "typical specimens" or the average of a few miscellaneous samples whose precise three-dimensional geographic locations are unimportant.

When a large number of specimens or structural readings is to be collected from a region, it might be assumed that standard statistical sampling techniques could be used. In an ideal situation sample locations on a grid superimposed on the map could be selected with the aid of a table of random numbers. Collection of samples or observations at such locations involves **probability sampling,** because every member of the population of objects has a prescribed chance of entering the gathered sample. In such a case, one of the standard statistical procedures could be used to analyze the data in a rigorous manner.

For a sampled population comprising the near-surface portion of a series of sand dunes, a grid could be laid out and sampling locations chosen with random-number tables; in this case probability samples can be collected from all designated localities. In structural geology problems, although grid squares (or intersections) can be located at random, specimens (objects) and structural observations (attributes) can be obtained from only a limited number of such grid intersections owing to the limitations of natural outcrop. Even in well-exposed terrains, large areas are totally unexposed, and other areas show scattered isolated outcrops only (cf. Krumbein, 1960A; Whitten, 1961B). Factors such as ease of collection of data, lack of weathering, etc., control the choice of the actual specimens or data collected from a prescribed locality. Sampling subject to such constraints is **upper semi-probability sampling** (cf. Cochran, *et al.,* 1954, pp. 294, 322).

A geologist might decide to define the target population of interest as the N potential hand samples which compose the eroded surface of a folded sandstone unit. However, superficial gravels, lakes, vegetation, etc., make part of this population inaccessible for all practical purposes. Therefore, the objects are drawn from the actual exposures, and the n potential hand samples composing the entire surface of the natural and artificial exposures constitute the **sampled population** of objects. Rosenfeld (1954) introduced "sampled population" into the geological literature in discussing petrographic variation in the Oriskany Sandstone, Pennsylvania, U.S.A. These concepts were discussed in some detail by Whitten (1961B; 1962) in relation to sampling granite masses.

In practice the sampled population of objects comprises the potential objects obtainable from all of the actual surface and subsurface exposures the geologist succeeds in visiting and sampling. He may not visit all outcrops, and the possible data locations at inaccessible exposures are not part of the sampled population.

Thus, there are severe restrictions on the direct probability sampling from the target population, but it is possible to take a probability sample from the hand samples composing each area of continuous outcrop (the sampled population of objects).[1]

---

[1]The widespread use of drilling apparatus in economic, and now in academic, research (cf. Link and Koch, 1962; Drever, 1964; and Baird, *et al.,* 1964) means that, in some cases, probability samples are becoming a practicable possibility.

Griffiths (in Milner, 1962, p. 606) defined an alternative series of terms, namely, the **hypothetical,** the **existent,** and the **available populations.** He noted that a fourth population is commonly introduced inadvertently as a result of inadequate sampling; this is the population that the collected samples represent. Thus, in effect, Griffiths divided the target population of Cochran, *et al.* (1954) into the hypothetical, existent, and available populations, while his available population equates with Rosenfeld's (1954) sampled population. The hypothetical population represented the hypothetical volume of rock about which a geologist wishes to draw conclusions despite the fact that part of this volume has been removed by erosion, etc., and is thus not available for study. The existent population concerns the body of rock which actually exists in the Earth's crust now and which may, as a result, be defined unequivocally both statistically and geologically, although it may not be possible to sample it in entirety.

In most structural problems to collect, or to make measurements of structural attributes of, *all* the *n* potential specimens which comprise the sampled (or available) population of objects is an impossibly large task. However, it is only by study of the sampled population that geological inferences can be made about the target population, which is the real subject of interest.

It follows from what has been written above that only in most favorable collecting circumstances do the limitations imposed by upper semi-probability sampling permit rigorous statistical statements to be made about the sampled population. The step from the sampled to the target population is still more severe. Even when ideal collecting conditions permit rigorous statements about the sampled population, only *inferences* about the target population can be made on the basis of geological reasoning and experience gained from the sampled population (Whitten, 1961B, p. 1335). It is a dictum that: "The step from sampled population to target population is based on subject-matter knowledge and skill, general information, and intuition — but not on statistical methodology" (Cochran, *et al.,* 1954, p. 19).

The sole exception to this rule occurs when the sampled and target populations are identical. In this possible, though not very common, situation in structural geology, statements about the sampled population apply directly to the target population.

PATTERNS OF VARIABILITY OF ATTRIBUTES

Commonly in geology it is necessary to investigate the pattern of areal or three-dimensional variability exhibited by a population of attributes. Each observation of an attribute relates to one object (e.g., a sample) that has a specific position in space; hence the variability of a population of

F (SYSTEMATIC, NO REGIONAL GRADIENT)

12	1	12	1	12	1
1	12	1	12	1	12
12	1	12	1	12	1
1	12	1	12	1	12
12	1	12	1	12	1
1	12	1	12	1	12

A (SYSTEMATIC, NO REGIONAL GRADIENT)

12	12	12	12	12	12
12	12	12	12	12	12
12	12	12	12	12	12
12	12	12	12	12	12
12	12	12	12	12	12
12	11	12	12	12	12

B (SYSTEMATIC, REGIONAL GRADIENT)

12	10	8	6	4	2
12	10	8	6	4	2
12	10	8	6	4	2
12	10	8	6	4	2
12	10	8	6	4	2
12	10	8	6	4	2

C (SUBSYSTEMATIC, NO REGIONAL GRADIENT)

17	12	5	12	19	12
12	8	12	15	12	23
17	12	2	12	18	12
12	1	12	20	12	7
9	12	11	12	1	12
12	10	12	1	12	7

D (SUBSYSTEMATIC, REGIONAL GRADIENT)

12	10	8	6	10	2
12	12	8	9	4	11
0	10	4	6	15	2
12	11	8	18	4	19
6	10	1	6	9	2
12	10	8	12	4	6

E (UNSYSTEMATIC)

24	4	19	5	2	8
5	4	19	14	14	5
3	17	0	2	5	17
6	7	17	13	23	5
2	9	19	11	2	6
17	17	6	14	3	5

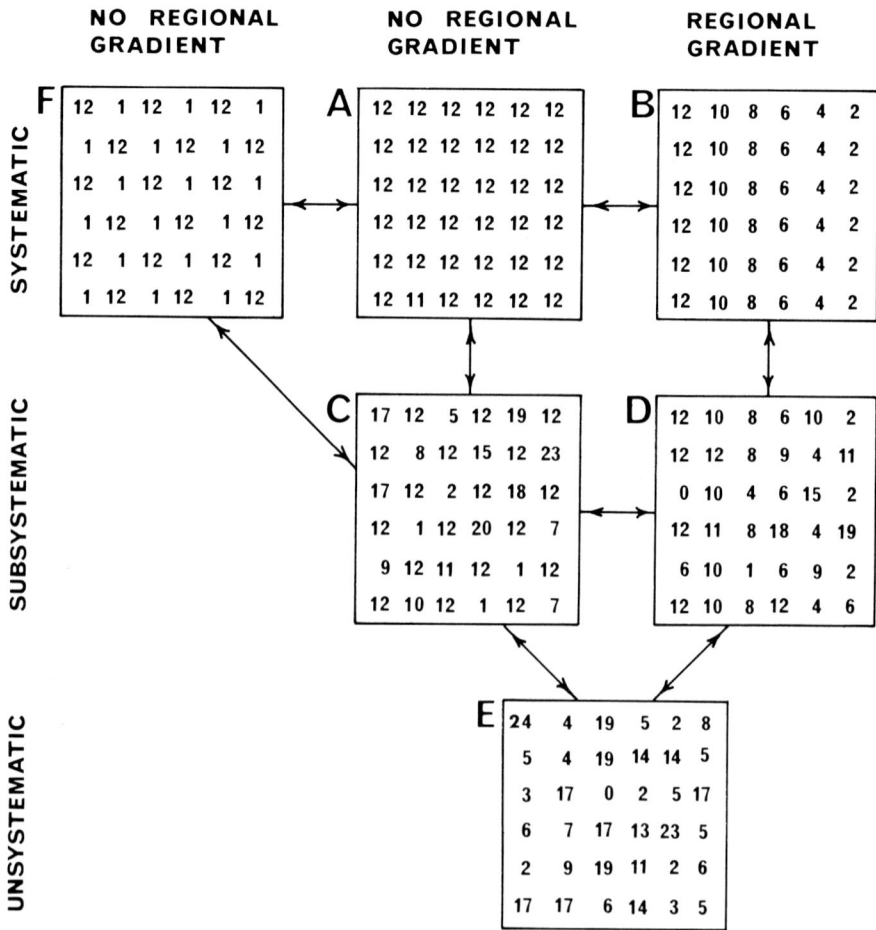

Figure 68. Schematic maps for populations of nondirectional attributes.

attributes is considered in relation to the U,V,W-location of each object, where U, V, and W are three orthogonal spatial coordinates. It is customary to define positions on the horizontal plane by U,V-coordinates, with the origin at the top left corner of the map, so that increasing positive values of U occur from top to bottom of the map, and increasing positive V-values from left to right across the map. W is the vertical axis and W-values increase upwards. It is convenient to keep the units of length equal on all three axes. U, V, and W are independent spatial variables, and the measured attributes are dependent variables, because they vary with respect to spatial location.

In structural geology most attributes are either planar or directional properties; these will be examined after a brief survey of the patterns shown by nondirectional variables such as those involved in analyzing the three-dimensional petrographic variability of a rock mass.

NONDIRECTIONAL ATTRIBUTES

Suppose each map in Figure 68 represents a population of non-directional attributes (i.e., a numerical value measured for a sample collected at each map point). The attribute might be the mean grain size of a sedimentary rock at each sampling site. In Figure 68 three main classes of areal distribution are illustrated, namely:

(a) Each individual in the population is numerically similar within the limits of experimental or observational error (Fig. 68A), so that there is no significant gradient across the area, and the confidence band about the mean value is small.[2] The grain size of samples from a dune sand area might yield such a pattern.

(b) The population of attributes has a strong gradient across the area, but the confidence band about the linear plane representing the gradient is small (Fig. 68B).[3] The grain size of samples collected across a bajada could yield a pattern of this type.

(c) The population is wholly random, in the sense that knowledge of several individual values does not permit predictions to be made about neighboring individuals (Fig. 68E); there is no significant gradient across the area, but the confidence band about the mean is large. A very poorly sorted glacial till might yield a map of grain size measurements similar to Figure 68E.

Systematic and **unsystematic patterns** are differentiated in Figure 68. Systematic is used when there is a small confidence band about the linear plane representing the variability in the map area; this applies to the horizontal linear plane in Figure 68A and the inclined plane of Figure 68B. Knowledge of several observed values of a systematic pattern enables unknown data to be predicted with reasonably high confidence.

---

[2]If there are n individuals in a population of attributes the mean is the sum of the n attribute values divided by n.

[3]The linear plane is assumed to be fitted to the observed data by the method of least squares. Suppose vertical wires are erected at each sample locality in Figure 68B, and that the length of each wire is proportional to the observed value; the linear plane approximately passing through the tips of all the wires can then be found by the method of least squares. With natural data all points will not fall exactly on any one linear plane, but the method of least squares ensures finding the plane which makes the sum of the squares of all departures from the plane a minimum. For Figure 68B a plane sloping down from left (values of 12) to right (values of 2) closely approximates the data; for Figure 68D departures from the plane are considerable. Computation methods were given by Krumbein (1959A) and Whitten (1963B).

When the appropriate plane is located, it can be inscribed above and below by a confidence band. These bands are calculated on the basis of a 95 per cent chance that any new datum point will lie within the confidence bands. For Figure 68B the confidence bands are narrow, and for Figure 68D they are much wider because the observed data have much greater variability about the linear plane. Computation of the curved surfaces that bound the confidence bands about a linear plane is quite complicated: the method was described by Krumbein (1963B).

The confidence band about the linear plane in Figure 68E is so wide that the pattern is unsystematic and predictions about unknown values would have little significance. Geological examples with every gradation between systematic and unsystematic occur (Fig. 68C). Systematic patterns with a strong gradient (Fig. 68B) grade into others, like those in Figures 68D and 68E, as the values of the attribute become less systematic and more random; the width of the confidence band about the inclined plane (Fig. 68B) gradually increases until the gradient becomes completely obscured by random values.

As suggested by Figures 68 and 69, there is also continuous gradation between the distributions with and without a gradient (i.e., between Figs. 68A and 68B, and between Figs. 68C and 68D).

For a particular population of geological objects many different attributes can be studied, and it is common for each attribute to have its own different distribution pattern. As a hypothetical example, consider a population of 25 hand specimens of granite collected from the intersections of a one kilometer U,V-grid. Quantitative measurements for these objects yield populations of attributes such as quartz percentage, plagioclase percentage, color index, SiO_2 percentage, specific gravity, etc. If quartz percentage has a random (unpredictable) variability with respect to (U,V), or only a very weak gradient, it may have a pattern resembling Figure 68D or 68E; in either case quartz values would be very variable from sample to sample. By contrast, the total feldspar percentage might have an almost identical value in each sample and virtually no gradient across the area; the pattern of distribution would then lie somewhere between those of Figures 68A and 68C. Color index might have a distinct gradient with respect to (U,V); this variable could be expected to have a systematic pattern and a relatively small confidence band as in Figure 68B.

The variability of most attributes is independent, but in some circumstances the regional variability of two or more attributes is interlocked. For example, Chayes (1960) drew attention to the strong interlock existing between variables drawn from closed tables (i.e., percentage data), whereby the variance of the major variables in the same closed table prescribes the correlations between the variables. Of the variables mentioned above, specific gravity appears to be an open variable not subject to the restrictions of closed-table data. Krumbein (1962A) showed that dependent relationships of this type extend to mapped percentage data, so that areal variation maps for closed-table data are not independent. Thus, in the example described above, quartz percentage, total feldspar percentage, and color index are subject to the restraints of closed-table data, and their areal variability, although different, is not independent but controlled by the variance of the variables.

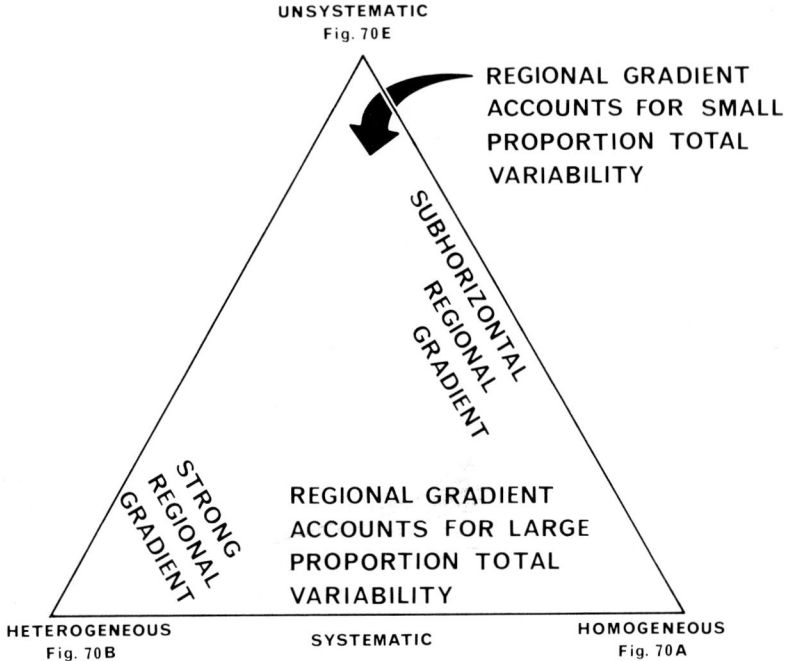

Figure 69. Patterns of variability of observed attributes.

In these examples the pattern of variability of each population was discussed in relation to one particular horizontal plane (two-dimensional U,V-section) through the rock unit. A three-dimensional population of objects can be studied statistically on the basis of a number of random two-dimensional sections. However, it is more efficient to consider the distribution patterns with respect to U,V,W-locations. In structural geology three-dimensional distribution is important, and the patterns illustrated in Figure 68 are readily extended to use with U,V,W-coordinates.

Another systematic pattern of distribution is illustrated in Figure 68F; there is no plane gradient across the map but the confidence band about the mean is large. Although possible, this type of distribution is uncommon in geology.

DIRECTIONAL ATTRIBUTES

The variability of directional properties can now be compared with the patterns for the nondirectional variables. Consider the linear structure defined by the intersection of bedding and schistosity in slate samples; suppose the orientation of this linear structure is measured as an attribute of a population of objects. The attribute may be considered in two parts,

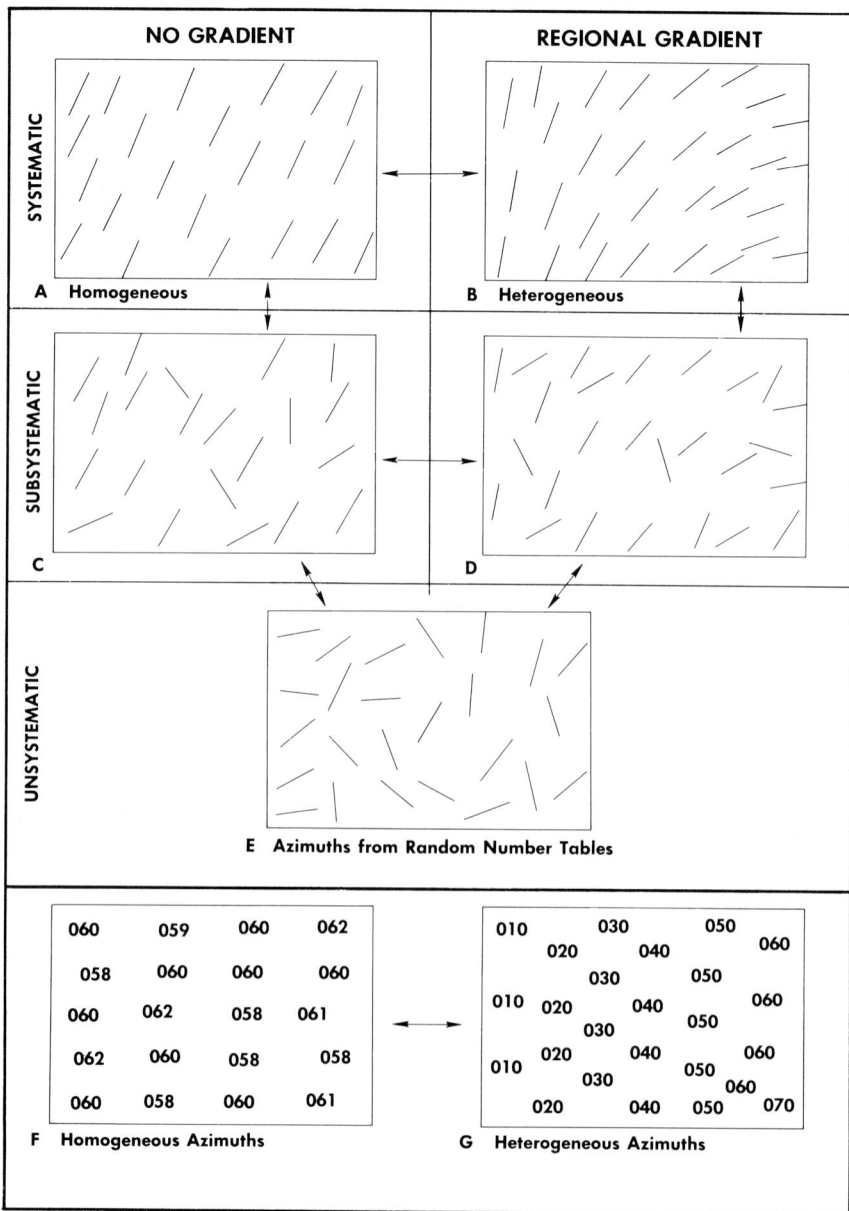

Figure 70. Patterns of variability of observed directional attributes (see text for explanation).

namely: (a) the trend of the linear elements expressed as a value ranging between 0 and 180 degrees east of north, and (b) the plunge ranging between +90 and −90 degrees (about the horizontal as zero).

In Figure 70 azimuths of a linear element are plotted on a series of maps; possible populations are shown in Figures 70F and 70G. Actual azimuths are plotted in Figures 70A through 70E, and their variability patterns correspond to those in Figures 68A through 68E, respectively. The plunge of a linear element can be independent of the trend, and a similar set of distribution maps for plunge should be considered in any complete study of the variability.

Instead of trend and plunge, each linear element could be expressed in terms of its direction cosines (i.e., the cosine of the angle between the element and each of the orthogonal axes U, V, and W). To investigate the areal variability of a single property the cosines with respect to any two of the axes must be mapped separately (cf. Chapter 14).

The areal pattern shown in Figure 70A is particularly important in structural geology and its characteristics should be noted carefully. There is only slight variability in trend, the confidence band about the mean is small, and there is no gradient across the map area. Such an attribute population will be called **homogeneous.** Specifically a **homogeneous attribute population** has a systematic pattern of variability, no gradient with respect to (U,V,W), and a small confidence band about the mean. Throughout the rest of this book homogeneous is used in this sense; unfortunately, the term has been used in several other specialized senses in structural geology and extreme care should be exercised in using the word. By the present definition homogeneous implies homoiousia, that is, essential likeness of the elements, but not their identity.

A pattern of the type illustrated in Figure 70A was referred to as *statistically homogeneous* by Paterson and Weiss (1961) and Turner and Weiss (1963). Although Turner and Weiss (1963, p. 16) traced the use of homogeneous to Kelvin and Tait (1883), who wrote that a "body is called 'homogeneous' when any two equal similar parts of it, with corresponding lines parallel and turned towards the same parts, are indistinguishable from one another by any difference in quality", a "homogeneous population" is not easily defined in statistical terms. Paterson and Weiss (1961, p. 854) said that a body of rock "is statistically homogeneous on a certain scale when the average of the internal configuration in any volume element is the same for all volume elements with dimensions not smaller than the scale of consideration." It would appear that "statistically homogeneous" was intended to imply "approximately homogeneous."

SIZE AND PATTERNS OF VARIABILITY OF ATTRIBUTES

The importance of the **size** of the volume of rock examined in relation to the pattern of variability is not a new concept, although the vital significance of the subject has often been overlooked. Long ago, Becker (1893, p. 15) wrote:

Orogeny can never be satisfactorily discussed until the dynamic significance of cleavages and cracks is clear. A necessary step toward this consists in the elucidation of those areas, great or small, throughout which the phenomena are uniform: for, however complex the conditions may be in any body of rock, they may be considered as uniform over a sufficiently small fraction of the whole mass.

Knopf (Knopf and Ingerson, 1938, pp. 15-6), in discussing structural geology, expounded much the same idea when she wrote:

> ... determination of isotropy or anisotropy in a rock may depend entirely upon the field that is under consideration. For example, an equigranular granite, if studied in a field several cubic meters in size, might be statistically isotropic. The same rock, if examined in a smaller field, say of a few cubic millimeters, may be anisotropic because it is made up of anisotropic components. Where anisotropic components such as quartz, feldspar, or mica are gathered together in an aggregate of random orientation, the rock is statistically isotropic.

Turner and Weiss (1963, p. 17) subscribed to a similar statement.

Hills (1953, p. 89) introduced the study of the mechanics of formation of individual folds by saying: "In order to understand the relationship of minor structures to folding, it is necessary to consider the genesis of individual folds rather than zones of folded rocks." Finally, Weiss (1959A, p. 7) succinctly stated the position as it applies to structural geology by enunciating the following rule:

> *For a given field, with respect to a given structural element, homogeneity of fabric increases inversely with scale.* Therefore, to obtain a greater degree of structural homogeneity within a given field, the field must be examined on a smaller scale.

The concepts underlying these varied statements require detailed scrutiny.

The pattern of variability within a series of rocks will not remain constant indefinitely when the field of interest is extended into adjacent areas. Hence, the size of the objects on which observations have been made must be stated. Both the upper and lower size limits of the objects studied, and of the volume of rocks about which inferences are drawn, are important; these questions are intimately bound up with the sampling techniques used in a particular study.

The size factor must be considered from at least two different aspects. First, the individuals comprising the population of *objects* (whose attributes are studied) can be of different sizes. Second, size considerations enter when portions of a population of *attributes* (for a particular population of objects) are examined.

SIZE AND THE POPULATION OF OBJECTS

Commonly, though not necessarily, a population of objects is defined in such a way that each individual is of approximately the same size. The size of the individual objects is important for two reasons:

First, the size of the objects limits the range of attributes that can be measured.

An attribute of interest might be jointing. In a population of objects each about 20 cubic meters in size, joints can be measured easily, but in thin sections this attribute could not be measured.

Again, a population of objects might be all the calcite crystals in a thin section of marble, and the attribute of interest the azimuth and plunge of the c-crystallographic axes. In a different population of objects, such as cubes of rock each one meter on a side, the individuals do not have a c-crystallographic axis, so the attribute no longer has meaning. Either a census, or a suitable sample, of the calcite crystals in each individual of this new population of objects could be examined, but these measurements would not represent the same attribute as that studied in the thin section.

If the attribute of interest had been specific gravity, it could have been measured with respect to both of the populations of objects considered above: one population of attributes would be specific gravity of the individual calcite grains, the other the specific gravity of the one cubic meter blocks.

Second, those attributes which can be measured in individuals of different size commonly show different patterns of variability in objects of different orders of size.

This situation can be illustrated with respect to a granite mass. First, suppose that mean modal quartz percentage could be measured for 25 adjacent one square kilometer areas of a granite massif (level A). Second, 25 very large hand samples, one from a randomly-chosen U,V-location in each of the square kilometer units, are collected (level B). Third, for level C one thin section is cut from each of 25 rock chips collected at randomly-chosen sites in each of the kilometer units (Fig. 71). For the thin sections (level C) the mean quartz percentage is likely to have an areal variability like that in Figure 68E, because, at the local level, this mineral tends to have a very erratic variability. However, at level B the attribute might have a subsystematic distribution pattern like Figure 68D, but with a considerable local variability about the regional gradient. Finally, at the regional level (level A), quartz values might have a systematic areal pattern like Figure 68B and comprise a population of attributes changing slowly on a regional scale (i.e., showing a gradient).

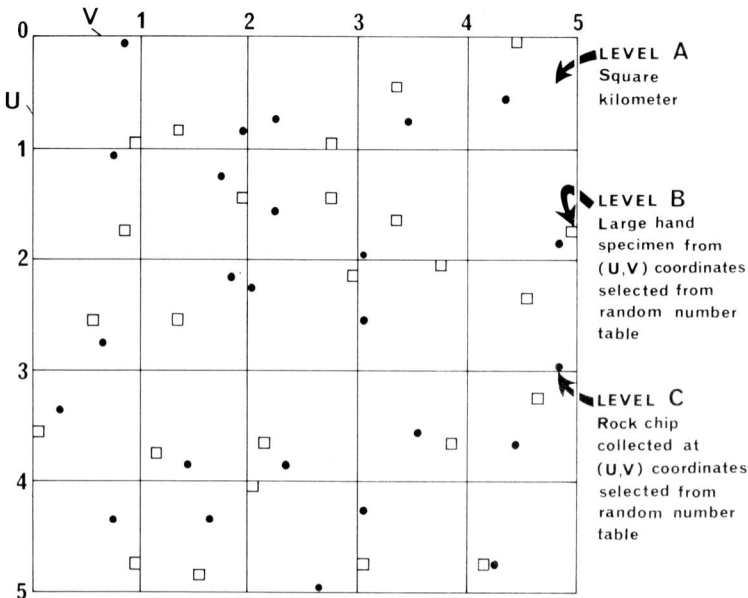

Figure 71. A sampling plan for part of a granite mass. See text for explanation.

Parallel sampling problems occur in the study of folded rocks. Suppose that the attribute of interest is the main fold axis in sample objects, and that populations of objects of different size are examined from the Appalachian fold belt between Newfoundland and Alabama, or from the Caledonian fold system which extends from western Ireland to northern Norway. Despite the regional ENE-trend of these fold belts that is obvious from a small-scale map, many smaller folds with strongly divergent trends have been found by detailed large-scale mapping (often 1:10,000). The type of problem is illustrated diagrammatically by Figure 72, where samples of 100 square kilometers each tend to have dominant N-S fold axes. The 25 square kilometer samples have dominant E-W fold axes, and, at a still smaller sample size, NW-SE fold axes might dominate the samples. Superposed folds will be considered in detail in later chapters, but it will be obvious that, if Figure 72 bears any resemblance to a major fold belt, the size of the sample units must play a significant role in determining the geometry deduced.

Now it is significant to enquire whether inferences can be drawn about the areal pattern of a variable in one population on the basis of observations based on a population of objects of a different size. In the examples considered above, each level of sampling is characterized by a

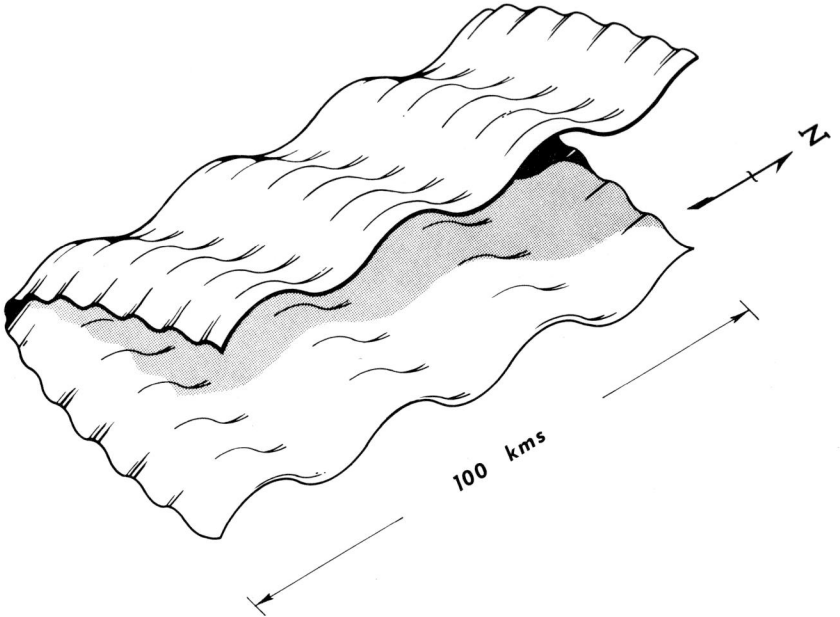

Figure 72. Diagram to illustrate the manner in which dissimilar fold trends can be dominant in samples of different size. See text for detailed explanation.

dissimilar pattern of areal variability that is likely to be independent of the type of variability at each of the other sampling levels. Information based on attributes of objects of one size does not commonly permit inferences to be drawn about the U,V-variability of attributes for objects of a different size; in a few instances the U,V,W-pattern of variability may be similar at each level of sampling. It has already been noted that: (a) particular attributes cannot always be measured for two populations of markedly dissimilar object size and (b) within each population of objects the pattern of U,V,W-variability of an attribute A can be either dependent or independent of the U,V,W-variability of attributes B, C, D, etc. These rules are important in considering structural observations made in the field, where A might be dip of bedding, B the plunge of a fold axis, and so on.

Characteristically, the sizes of the objects used in different levels of sampling in structural geology lie on an ordinal scale. The size of the objects of interest (quarries, outcrops, hand samples, thin sections, etc.) is usually dictated by convenience and custom, rather than by scientific principles. In consequence, the intervals between successive object sizes investigated are commonly dissimilar.

Little is known about the variance of different variables measured in objects of various sizes, or about the patterns of areal or three-dimensional variability shown by variables in populations of objects of different sizes. As already pointed out, however, the size of the objects appears to control the type of information that can be obtained. In consequence, sample (object) sizes different from those commonly studied might well be more desirable scientifically. Research is needed urgently to determine the optimum sampling plans and object sizes for different nondirectional and directional attributes studied in a *U,V,W*-system.

SIZE AND THE POPULATION OF ATTRIBUTES

Figures 68 and 70 show six simple distribution patterns which are of common occurrence in geology. In these examples there is either no gradient or a simple gradient that can be represented by an inclined plane. More complex patterns also occur in natural populations of attributes, but they can always be divided into **subpopulations** with simple patterns. Structural analysis is commonly aided by isolation of subpopulations with no areal gradient (Fig. 70A); this involves identification of **homogeneous subpopulations.**

Suppose a population of objects is divided into subpopulations each large enough to contain a significant number of objects. Sometimes an attribute has a similar pattern in all such subpopulations, as would be the case if Figures 68A and 70A were subdivided. Alternatively, subpopulations can have dissimilar patterns; a single plane cannot adequately represent the concentric pattern of Figure 73, but the area can be divided into subareas each with a dissimilar simple planar gradient.

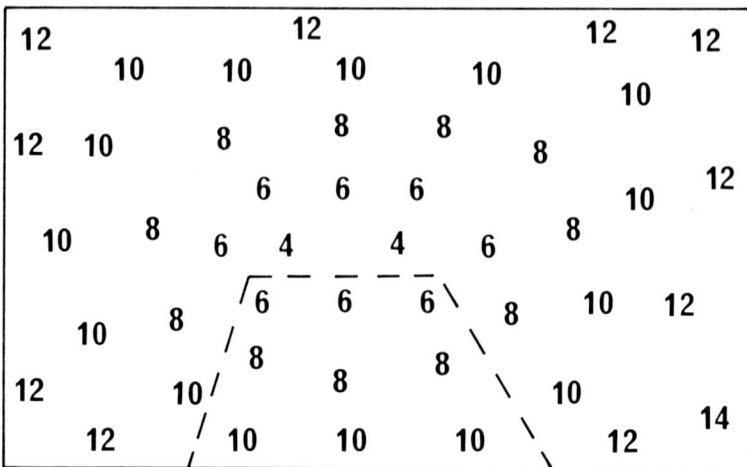

Figure 73. A population of attributes having a concentric pattern of variability: the subpopulation outlined by broken lines has a simple planar gradient.

Some examples involving directional properties are illustrated in Figure 74. The populations of attributes in Figures 74C through 74G are divided into subpopulations that do not have a gradient.

Figure 74E shows that, when a systematic pattern with a gradient occurs (e.g., Figs. 70B and 74B), subpopulations with no gradient can be identified. Such subpopulations are **homogeneous domains;** they correspond to the "statistically homogeneous domains" of Turner and Weiss (1963).

A subsystematic population (Fig. 74F) can be divided into subpopulations with a preferred orientation similar to that in Figure 74A, except that there is a rather large confidence band about the mean. If such subpopulations are further subdivided, smaller geographical domains that are homogeneous (i.e., similar to the systematic pattern of Fig. 74A) can be recognized. The boundaries between these smallest domains might be either gradational or sharp and represent actual structural breaks or discontinuities on this scale of observation (Figs. 74C and 74D).

Figure 74G shows that a simple folded sequence can be dissected into a large number of small homogeneous domains. It is most important to recognize that, for each different attribute of the population of objects studied, the boundaries of the homogeneous subpopulations may not coincide.

SIZE TERMINOLOGY

The structural geologist is concerned with the analysis of structures that range in size from individual mineral grains in a thin section to the total picture within an orogenic belt. The techniques involved at each level of sampling and at each level of size are essentially similar. It is not useful to erect an elaborate classification of size dimensions, because the size of units within which attributes are homogeneous varies enormously for each attribute and with each structural problem.

Weiss (Weiss and McIntyre, 1957; Weiss, 1959A) grouped the "scales" (i.e., sizes) at which structures can be examined into three classes: microscopic, mesoscopic, and macroscopic. If this type of classification is to be used, it is desirable to include submicroscopic, as was done by Turner and Weiss (1963, p. 15). The terms were defined as follows:

Submicroscopic scale: Fields of study in which the grain-size is so fine, or the specimens are so small, that X-ray analyses, etc., have to be employed for structural examination.

Microscopic scale: Fields of the dimension of a thin section which can be examined adequately with a microscope.

Mesoscopic scale: Fields ranging in size between a small hand specimen and a group of closely-related outcrops "... in which

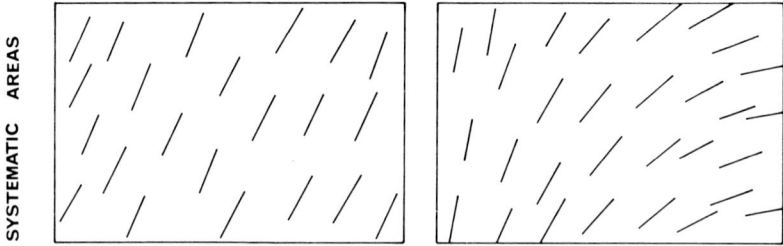

A. HOMOGENEOUS   B. HETEROGENEOUS

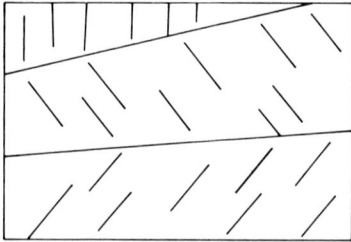
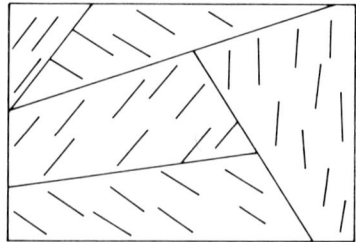

C. LEFT TO RIGHT TRA-   D. ALL TRAVERSES
   VERSES SYSTEMATIC   UNSYSTEMATIC

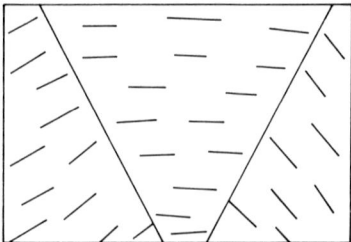
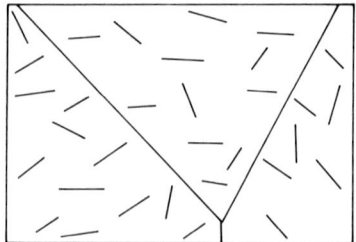

E. SUBPOPULATIONS   F. SUBPOPULATIONS
   HOMOGENEOUS   SUBSYSTEMATIC

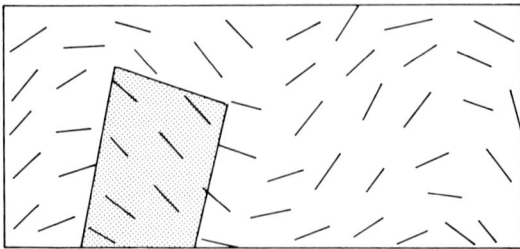

G. SYSTEMATIC VARIATION PATTERN:
   SHADED AREA IS HOMOGENEOUS

Figure 74. Various populations of directional attributes: see text for complete explanation.

data can be measured with sufficient accuracy and continuity to allow determination of its overall structural geometry" (Weiss, 1959A, p. 10).

Macroscopic scale: Fields of large size in which structures are discontinuous between outcrops, so that the structural pattern has to be inferred from quantitative geometric analyses.

Because of the entrenched usage of microscopic and macroscopic in igneous and metamorphic petrology and petrography, these new terms for size have led to confusion; this applies in particular to mesoscopic and macroscopic. For this reason, if a classification is indeed required, the three terms suggested by Rickard (1961, p. 325) for the same size categories are preferred. Adding submicroscopic to Rickard's terms, we have **submicroscopic, microscopic, minor,** and **major.**

These four categories lie on a purely ordinal scale and the limits between them are arbitrary. However, there is probably no advantage in defining boundaries more systematically.

The chief virtue of a classification is to emphasize the importance of recognizing the size of the individuals in a population. For this purpose, since the size of the objects is very variable and should not be prescribed by conventional arbitrary methods of sampling, Beloussov's (1962, p. 12) informal use of **small, medium,** and **large** for the sizes of structures is somewhat less restricting, and thus more useful, than a formal size (or "scale") terminology.

SIZE AND THE STUDY OF FOLDED ROCKS

Weiss (1959A) investigated the question of sample size with respect to the geometry of a simple fold. His concepts are modified in the following analyses of two attributes—bedding (S_0) orientation and fold-hinge-line (B) orientation. The scale of Figure 75 is not significant; in fact, it is important to appreciate that the whole diagram can represent a structure ranging in size from a small hand specimen to a fold system several kilometers long.

First, a population of m objects within the small volume I (Fig. 75) can be considered. Now, if the attribute "bedding plane (S_0) dip" is measured for each of the m objects, the confidence band about the mean is vanishingly small; this is a homogeneous attribute population because its systematic distribution pattern does not show a gradient (cf. Fig. 74A). The second attribute, the orientation of the hinge line (B), cannot be measured in the objects from domain I.

Next, suppose that a population of M objects represents a probability sample collected from domain II. The orientation of S_0 varies systematically in this population so this attribute has a gradient with respect to

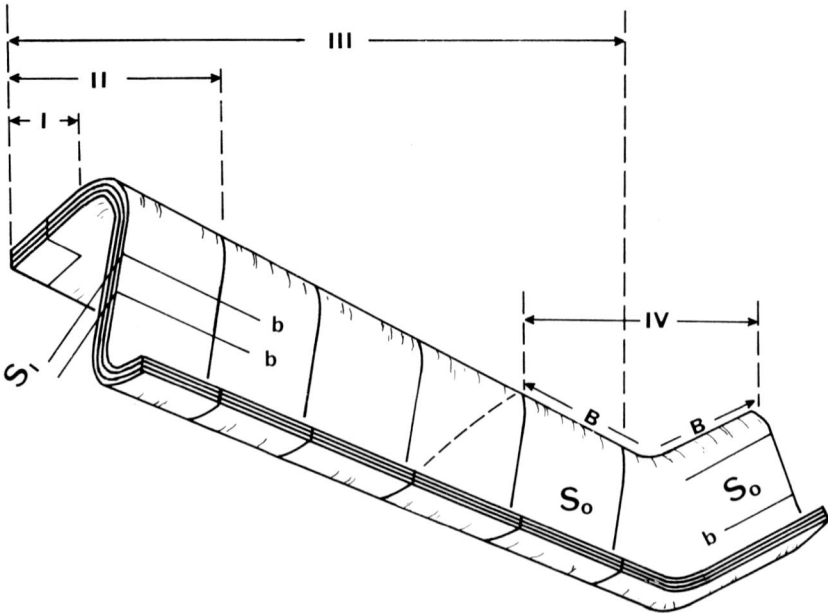

Figure 75. Different domains of homogeneity within a fold structure: see text for explanation.

(U,V,W); the population of attributes is heterogeneous, but it can be divided into several homogeneous subpopulations which lack a gradient. The hinge line (B) cannot be measured in all the M objects, although a few of the objects probably lie astride the fold closures and thus have a measurable attribute. Hence B is not an attribute of the population of M objects.

In a still larger domain (III) a population of n individuals each like the whole of domain II can be studied (Fig. 75). The attribute B can now be measured in each of the n individuals; it is homogeneous because it does not show any gradient with respect to the U,V,W-locations of the objects, and the confidence band about the mean orientation of B is very small. However, bedding plane dip now has no meaning as a single observable attribute of the n individuals.

It should be noticed that if axial-plane foliation (slaty cleavage) is present in domain II, the intersection of S_0 (bedding) and the foliation (S_1) can be defined as attribute b. The orientation of b is a measurable attribute of the population of M objects (domain II) and is approximately parallel in each sample. Hence, b is homogeneous in domain II.

The population of objects could be defined in different ways in domain IV. First, a probability sample of N small objects could be examined. In such a population the orientation of B (fold hinge) cannot be measured. However, b has a systematic pattern with a gradient with respect to (U,V,W); the population of b observations is heterogeneous, but it can be divided into homogeneous subpopulations in which the attribute has little variation about the mean. The dip of S_0 in the N objects has a complex pattern of variability, and several subpopulations are needed to ensure that the variance of S_0-dip is small in each. Second, a population of N' objects separated by equally-spaced planes perpendicular to B could be defined. In these N' objects both B and b can be measured, and they both show a simple gradient like that of b in the population N; however, the dip of S_0 is not a measurable attribute of the population N'.

If it is supposed that Figure 75 represents a single outcrop, it would be possible to examine a population of smaller objects than was examined in domain I; for example, a thin section cut perpendicular to B from within domain I provides a new size for study. If the sample comprises finely-bedded silts and shaly layers, the thin section will be compositionally anisotropic because of the lithological banding. During the process of folding some of the lithological layers are likely to have developed folds (Fig. 76) too small to affect the approximate parallelism of the bedding planes (S_0) in domain I. However, in the population of tiny objects comprising portions of the thin section, the parallelism of S_0 is disrupted by the small flexures. However, if these crenulations are widespread throughout the thin section, it may be possible to estimate the orientation of B (fold hinge) in each object of this population.

A statement about the pattern of geographic or U,V,W-variability of an attribute within a population is meaningless unless specific reference is made to the size of the objects on which the observations were made. The pattern of variability should be assessed with respect to all of the linear and planar structural elements that can be identified in a folded sequence of rocks. This can be accomplished only if a precise usable operational definition is used for each attribute.

STATISTICAL EVALUATION OF STRUCTURAL DATA

The main tools that have been used in statistical evaluation of structural data are the stereographic and equal-area projections; within the past decade these devices have been used extensively, permitting elucidation of many structures too complex for study by any other available method.

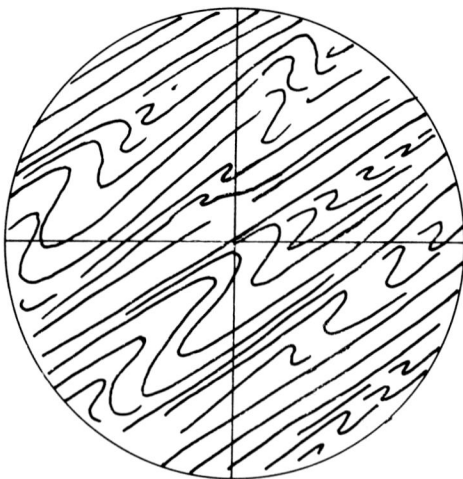

Figure 76. Diagrammatic view of a thin section that could have been cut perpendicular to *b* in domain I of Figure 75; numerous small folds of S_0 are shown.

The analysis of directional data measured in folded rocks falls into two categories:

(a) Stereographic analysis (Wulff graticule) of angular relationships between structural elements observed and measured in a single sample or a closely-related group of samples.

(b) Collation and evaluation of directional data from a geographic area using an equal-area (Schmidt) projection. Particular advantages accrue when the variable studied is homogeneous, i.e., has a small variability about the mean, and the population does not have a gradient with respect to the U,V,W-coordinates. If an attribute has a complex variability pattern in the population studied, a confused array of points will be plotted on the projection. One of the arts of analyzing complex fold structures is to divide the total mapped area into discrete domains, within which the subpopulations of attributes are homogeneous.

The following remarks apply to the use of equal-area projections rather than to stereographic projections.

Despite the unquestioned value of equal-area projection in regional analysis, the technique has some important limitations associated with both sampling and the three-dimensional variability of directional variables. Although work is proceeding with a view to circumventing these problems, complete answers are not yet available. It is important that both the limitations and the benefits of equal-area analysis be appreciated before the technique is used extensively.

When linear or planar elements are plotted on a projection, equal weight is inevitably given to each observation, although, commonly, no system of probability sampling guided collection of the individual readings (data). If the projection is only required as a qualitative and descriptive statement of the observations made, no difficulty arises. Problems enter when quantitative evaluation is introduced; for example, if the poles to joint planes are plotted and isopleths are drawn on the equal-area projection to represent the percentage of poles per unit area of the projection (Fig. 23).

Suppose that joints are studied within several square kilometers of a massive granitic complex, and that exposure is sufficiently good for probability sampling sites to be selected and used. The poles to joint surfaces measured at each site can be plotted on the equal-area projection. Suppose Figure 77 is a map of one of the outcrops; three different sets of joints are visible, but the character of each is dissimilar. The north-south set is characterized by many small, closely-spaced joints. The east-west set comprises a few widely-spaced joints, and each joint is opened widely. The northwest-southeast set is represented by isolated, widely-spaced, small joints. If each visible joint is measured and plotted on the projection as a single pole, considerably greater weight will be given to the north-south set when the results are contoured. If each set is represented by a single pole, the actual joints will be severely weighted in favor of the less abundant orientations. When the observations for numerous additional sampled locations are included in the plot, the vagaries of joint-spacing may eliminate the biases associated with each individual outcrop; on the other hand, the weighting associated with each outcrop might be compounded and result in a very strongly-biased final projection.

No adequate system seems to have been proposed for acquiring a statistically-adequate sample of joint surfaces.[4] However, until a definite sampling system, based on well-defined operational definitions of individual joint planes and/or joint sets, is developed, contoured projections of joint surfaces are apt to be misleading.

Similar considerations apply to all other linear and planar attributes of rocks represented by contoured equal-area projections. Such data, when collected to represent a geographic area, are wholly inadequate and misleading if the population of objects (whose attributes are measured) does not represent a close approach to a probability sample at the particular sampling level involved. Probability sampling is required in selection of the sampling sites within the whole complex, and also in

[4]Spencer (1959) and Wise (1964) made useful contributions to this subject with their quantitative data on the joints of Precambrian terranes in Wyoming and Montana, U. S. A., but the sampling problems remain acute.

selection of individual objects on which to measure the attributes of interest at each site. Unfortunately, a very large number of published contoured equal-area projections represent a selection of observations chosen subjectively as "typical" of the outcrops which a geologist happened to visit; such projections merely relate to Griffiths' (in Milner, 1962, p. 606) fourth population referred to earlier in this chapter. Thus, many projections, rather than being truly quantitative, are either essentially qualitative or even misleading. Apart from the exposure problems involved in collecting data at the required points, a statistically-valid method of sampling directional properties of a folded sequence does not appear to have been devised.

A second problem is connected with the areal and three-dimensional variability of attributes. When data are plotted in projection, each observation is divorced from its geographical location, and this involves loss of significant information. First, consider a petrographic example. In the study of granitic complexes it is common practice to determine the modal composition of a large number of samples, and to illustrate their variability on ternary diagrams (e.g., quartz-plagioclase-potash feldspar). Many granite masses possess marked internal compositional variability; although a ternary diagram provides one method of assessing the mean composition of the whole mass, such diagrams remove the modal compositions from their U,V,W-locations, so that the vital geographic element in the mineralogical variability is lost. The pattern of U,V,W-variability could be completely overlooked if the data were plotted on a ternary diagram only (Whitten, 1963A).

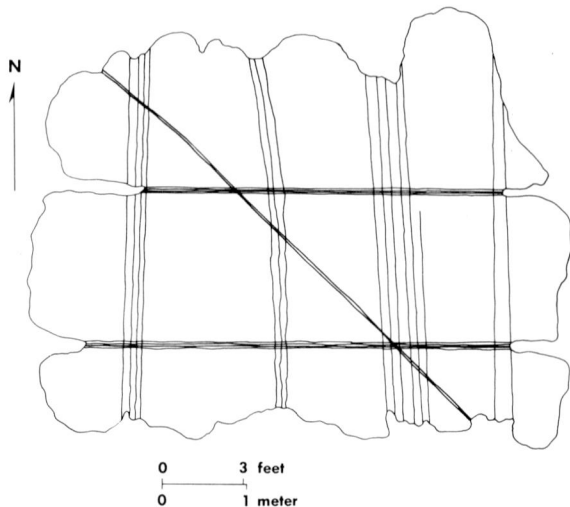

Figure 77. Map showing the joints in a hypothetical granite outcrop.

Second, it is easy to see that a simple equal-area projection can mask important information about a folded sequence. Suppose a great circle is defined by a π-diagram for data from several square kilometers of poorly exposed ground. The π-diagram may reflect a single simple flexure or alternatively a whole series of small folds; without preserving and analyzing the U,V,W-location of each datum point, these two possibilities could not be distinguished.

Cadigan (1962) developed a simple test based on the Chi-square index for determining whether mapped (areal) nondirectional quantitative geologic data have a random regional distribution. In the majority of studies, however, it is desirable to know what the pattern of areal variability is, rather than that the distribution is "nonrandom."

Current direction during deposition of a sandstone is a directional property that can be estimated on the basis of numerous sedimentary characteristics, such as sole marks on the under-side of bedding planes or cross bed orientations. Many geologists have estimated such current directions, and some attempt has been made to study their areal variability. Potter and Pryor (1961), for example, made a detailed study of the dispersal centers of Paleozoic and later clastic sedimentary rocks of the upper Mississippi Valley and adjacent areas on the basis of 9,141 cross-bedding observations made in rocks ranging from Pliocene through pre-Keweenawan age. Mean values of current direction were obtained for each of 26 stratigraphic units by using the Tukey test (see Harrison, 1957; Rusnak, 1957). Although Potter and Pryor divided their data on a time (stratigraphic) basis, the use of mean values within each unit divorced each observation from its essential geographic location. Steinmetz (1962, 1964; also Green, 1964) and Agterberg and Briggs (1963) also developed useful methods of handling current-bedding directional data.

One statistical method of representing regional and local elements of the areal variability of attributes measured for rock masses is **trend surface analysis.** The method involves computing a separate mathematical surface to approximate the spatial variability of each attribute measured at numerous map points. Contour-type pictures of the regional and local variability patterns of each attribute are then drawn. The arithmetic involved is tedious, but it can be handled by digital computer and several suitable computer programs are now available. The growing literature on trend surface analysis has only been concerned with nondirectional properties but it is relatively easy to extend the technique for analysis of the spatial variability of linear and planar directional properties of sedimentary and folded rocks (see Chapter 14).

Methods developed to seek areal correlations in mapped data and involving power spectra for map analysis have been explored by Cote, *et al.* (1960) and Horton, *et al.* (1962; 1964). Probably such methods could be usefully adapted for use with directional data of structural geology, but this has not been attempted to date.

Pending development of suitable methods of analyzing the regional variability of directional data, it is important to recognize that a composite equal-area projection of a simple directional property measured within a large geographic area amounts to the determination of a weighted average value. Agterberg (1959) in a preliminary study, showed that, when measured in a limited area, the scattering of minor fold axes around a mean direction represents a normal frequency distribution. However, when a population of mapped attributes has a systematic regional gradient within an area (cf. Fig. 70B), calculation of a mean value for the attribute, by seeking maxima on an equal-area projection, or by any other method, masks important characteristics of the data. These subjects are discussed more fully in Chapter 14.

Chapter 4

S-Structures, Tectonites, and Axes

In sedimentary rocks successive layers of detritus or precipitate have commonly accumulated in subparallel laminae or beds. The bedding planes define time planes in the rock. Bedding is a primary structural feature, which is clearly defined in those sedimentary rocks in which there is rapidly-changing lithology. However, within units which are more compositionally uniform (e.g., a massive quartzite), despite the fact that the sediment accumulated as successive layers, the individual bedding surfaces may not be visible.

In structural work, sets of parallel surfaces have been called S-surfaces (S-flachen: Sander, 1948, pp. 105-107). S-surfaces may be clearly visible as in the case of bedding or foliation planes. Alternatively, they may be more exoteric surfaces not readily visible in a hand specimen. For example, the weak planar preferred orientation of, say, biotite in a metamorphic rock defines an S-surface statistically. Thus S-surfaces may be either actual surfaces, or statistically-defined parallel or subparallel planes. In the course of progressive metamorphism and deformation new surfaces are frequently generated, and successive sets of surfaces are designated $S_0, S_1, S_2, S_3, \ldots S_n$.

Weiss (1955, p. 227) made a useful separation of the S-surfaces described by Sander into two categories, namely:

(a) **S-surface:** discrete, visible, more or less continuous surfaces such as bedding, foliation (rock cleavage), schistosity, and segregation banding.

(b) **S-planes:** not a visible feature in hand specimens, but defined statistically by the preferred orientation of one or more constituent minerals.

Weiss (1955, p. 227) claimed that:

> ... in general, plastic deformation ... favors development of S-planes, whereas discontinuous deformation favors appearance or accentuation of S-surfaces.... If deformation is continuous, that is, if each grain behaves as a deforming unit regardless of the behavior of the fabric as a whole, then no

inhomogeneities (*S*-surfaces) are induced by the movements. On the other hand, if movement is not continuous on the scale considered, then inhomogeneities appear transgressing the whole fabric.

In practice every gradation can occur between *S*-surfaces and *S*-planes. While the above distinction is useful, it is sometimes difficult to maintain in descriptive work; although a term to include both *S*-surfaces and *S*-planes has not been proposed previously, **S-structure** is used here. In metamorphic rocks an individual set of *S*-structures may be pre-, para-, or postmetamorphic, depending on whether its origin dates from before, during, or after a particular metamorphic event.

To comprehend the significance of *S*-planes some acquaintance with the petrofabric structure of rocks is necessary. The mineralogical constituents of a single specimen can be analyzed in two different ways. First, in terms of the mineralogical composition or petrography, and second, according to the geometrical relationships or petrofabric structure of the component grains. Until quite recently, the qualitative compositional petrography of samples has overwhelmed the parallel interest in the geometrical relationship of minerals to one another. In essence a petrofabric study involves measuring the three dimensional orientation of one or more directional attributes of the component grains of sedimentary, igneous, or metamorphic rocks; these measurements are most commonly made on the basis of an entire population of the grains of a particular mineral species in a small domain that is homogeneous with respect to a dominant *S*-surface. Petrofabrics is thus concerned with one aspect of rock texture.

In the English language, **texture** refers to the small features of the interrelationship of component grains as seen in the study of hand specimens and thin sections of rocks. By contrast, **structure** refers to the major large features. Some confusion has arisen, because this English use of texture and structure is opposite to German-language usage. The term **fabric** refers to the combination of both structural and textural elements. **Petrofabrics** is the descriptive aspect of fabric study, and is commonly, though not always, implemented by three-dimensional analysis with a universal (Federov) microscope stage. Professors Sander and Schmidt were the pioneers of petrofabric studies, but their pioneer works in German have only slowly become familiar to the English-speaking world.

As with *S*-structures, fabrics may be either primary or secondary. **Primary fabrics** develop when the rocks are initially formed by such processes as sedimentation, chemical precipitation, or magmatic flow of lavas. **Secondary fabrics** result from deformation of preexisting rocks during periods of stress, which are commonly closely associated in time with episodes of regional metamorphism, thrusting, etc.

The majority of **fabric elements** comprise planar or linear structures. Some elements, for example lineations defined by the intersection of *S*-surfaces, are clearly visible in rock outcrops. Other elements are very obvious on account of the large grain-size of the rocks, or because the constituent grains have a strong preferred orientation; in such cases the fabric elements can be measured easily in outcrops or in oriented hand specimens.

Some fabric elements (e.g., *S*-planes) may be obscure in hand samples because they are defined by a weak preferred orientation of the component grains. **Preferred orientation** is a technical term that implies that the grains of a mineral species depart from random (unpredictable) three-dimensional orientation on the scale of observation. It has been customary to make the estimate on the basis of a complete census[1] of all grains of that mineral within a small sample selected from a domain homogeneous with respect to the dominant *S*-surface. Thus, in a rock that is obviously folded, it is convenient to use a thin section from a fold limb. Sometimes the hinges of folds are analyzed, but special precautions are required: small homogeneous domains can be studied by sampling small bands perpendicular to the dominant *S*-surface. In order to define adequately the preferred orientation, or to establish that there is a random orientation, the positions of the crystallographic or optical properties of 100 to 300 grains of a particular mineral are measured with a universal microscope stage. Each of the minerals in a rock may show a preferred orientation, but each mineral species need not have the same orientation. In many rocks some minerals (e.g., micas) show a much more strongly-developed preferred orientation than other minerals (e.g., quartz).

Minerals which are abundant and whose orientation is easily determined are usually used for petrofabric studies. For example, biotite, quartz, and calcite are widely employed, although in special cases many other minerals, such as olivine, epidote, etc., have been studied. With quartz and calcite the positions of the *c*-crystallographic axes are measured. The orientation of each grain about its *c*-axis is more difficult to determine, especially in quartz where cleavage and twinning are not recognizable. With calcite, the cleavages and twin lamellae enable the three-dimensional orientation of each grain to be worked out in detail. The perpendicular to the dominant basal cleavage is measured in biotite and muscovite, and in detailed work with these biaxial minerals the orientation of the optic axial plane can be used.

The orientation of each measured grain is plotted on an equal-area projection. For example, the *c*-axes of quartz grains can be plotted, and a

---

[1]The relative merits of a complete census and a grid or random sample in petrofabric analysis need to be examined, because, with a universal stage, samples are easier to measure than censuses.

separate diagram is used for each mineral species studied. If a complete census is made of all the grains of one mineral within a specified portion of rock, it is legitimate to draw contours on the projection to represent the number of points per unit area (of the projection). For a population that yields a well-developed preferred orientation the contours tend to define either (a) a **pole diagram,** in which the plotted points are concentrated within one restricted area (Fig. 78), (b) a **girdle diagram,** in which the projected elements lie close to a single plane, and thus lie on a great circle (girdle) on the projection (Fig. 79), or (c) a well-defined but less-symmetric pattern (e.g., some of the triclinic girdle diagrams described in Chapter 5).

More diffuse or complicated patterns occur when the preferred orientation is less well developed, or when one structure is superposed on another.

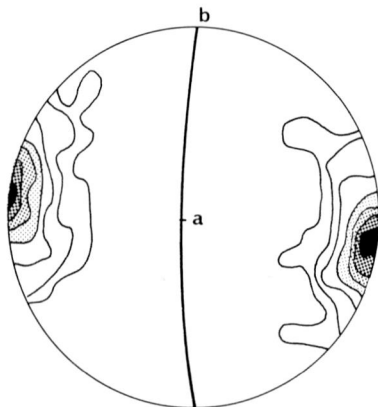

Figure 78. Pole diagram: equal-area projection of 200 poles to biotite cleavage in siliceous banded gneiss (Ryôké complex), north of Chausuyama, Yanai City, SW Japan. Contours at 1, 2, 4, 6, 8, 10, and 12 per cent; *a* and *b* are geometric fold axes (After Okamura, 1960, Fig. 22).

When a rock undergoes metamorphism all the major mineral components recrystallize. This process may involve (a) recrystallization of each particular phase (such as quartz or biotite) in response to the changing physical and chemical environment, and/or (b) the growth of some new mineral or group of minerals at the expense of some or all the preexisting ones. The term **neocrystallization** covers the concomitant recrystallization of existing phases and the growth of new mineral species.[2] For example, as a pelitic rock passes from the biotite zone of regional metamorphism into the almandine garnet zone, minerals like quartz and biotite may recrystal-

[2]Neocrystallization has been in use for several years (e.g., Fairbairn, 1942, p. 22), and as used now it includes both recrystallization and part of the process of **neomineralization.** The latter was defined by Knopf (Knopf and Ingerson, 1938, p. 107) as "... a transformation into minerals of new composition by reaction between minerals already present in a rock or by the introduction of new material from outside or by the loss of material already present."

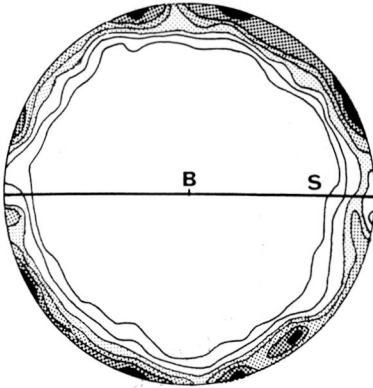

Figure 79. Girdle diagram: equal-area projection of 300 poles to muscovite in limb of a small fold in Ryôké metamorphic rocks, near Sugaki, SW Japan. Contours at 1, 2, 3, 4, 5, and 6 per cent; S is schistosity in sample and B the fold axis (After Nureki, 1960, Fig. 34).

lize while garnet crystals commonly grow as a new mineral phase at the expense of other phases.

Neocrystallization commences during the compaction and diagenesis of sediments: There has been some discussion (e.g., Greensmith, 1957, 1958; Stewart, 1958) as to whether metamorphic petrographic terms should be used to describe the textures produced at this stage. In practice diagenetic processes grade imperceptibly into those customarily classified as low grade metamorphic (e.g., the zeolite facies), and it is difficult or impossible to separate them; in consequence, their fabrics also show continuous gradation.

Although the fabric of each mineral in a rock can be studied, each does not show the same fabric; this is due in part to some phases responding to changing environments more rapidly than others. In a zone of continuing stress and metamorphism, some minerals reflect perfectly the last stress environment in which neocrystallization occurred, while others commonly reflect earlier phases and are mildly overprinted by the last epoch of deformation.

Neocrystallization may or may not develop a preferred orientation. For example, the thermal metamorphism associated with intrusion of dykes commonly results in hornfels texture in the wall rocks. Rather few petrofabric analyses have been published for such hornfels, but they are characterized by interlocking equigranular textures which probably reflect minimal directed stress, although the confining pressure may have been considerable. Aureoles adjacent to large igneous masses characteristically exhibit hornfelsic textures too, although locally schistosity is developed parallel to the contacts of an intrusive mass, and this probably reflects a weak radial stress pattern; such schistosity occurs, for example, at the margin of the Preissac-Lacorne granitic complex, Quebec, Canada (Dawson, 1945).

The major minerals of rocks affected by regional metamorphism almost invariably have a strong preferred orientation; this apparently reflects the influence of one or more periods of directed stress.

Although some progress has been made in relating fabric to the stress field, advances have been slow. Some of the most significant studies have been made by Griggs, Turner, and others (Griggs and Miller, 1951; Handin and Griggs, 1951; Turner and Ch'ih, 1951; Griggs, *et. al.,* 1951, 1953; Turner *et al.,* 1954, 1956), on the natural and experimental deformation of carbonate rocks (particularly the Yule marble, Marble, Colorado, U.S.A.). Despite this detailed work, the manner in which the preferred orientation or fabric is produced is still imperfectly understood. Fairbairn (1949) reviewed much of the research in this field and the related work on the deformation of metals. The symposium on rock deformation edited by Griggs and Handin (1960) presents the results of more recent experimental studies.

Much more work is required, but it is very difficult to utilize naturally-deformed rocks, because, in most cases, the rocks from major orogenic zones have been subjected to several successive but discrete phases of both neocrystallization and folding (see Chapter 12). In consequence, the stress field commonly was not constant throughout development of the fabric, and many rocks contain minerals which responded to neocrystallization during several phases of deformation. In more and more areas it has been recognized that neocrystallization occurred repeatedly during a single orogenic epoch, and that two, three or more phases of folding have been superimposed during a protracted cycle of deformation. Thus, it is difficult to determine at which stage in the deformation history a particular mineral species grew, and what the nature of the stress field was during its growth. With a mineral phase which was part of the mineral assemblage over a long period (which is often the case with, for example, biotite and quartz), its fabric is commonly a composite of several superimposed phases of neocrystallization.

The *S*-structures of rocks with secondary fabrics can be defined by at least two different types of orientation:

(a) **Habit orientation** (German: Form-regel): Preferred orientation of grains that have an elongate habit (shape); the texture gives the *superficial* appearance of elongated or flattened grains that have been bodily rotated into a preferred orientation, although other processes might have produced the texture.

(b) **"Lattice" orientation** (German: Gitter-regel): This lattice, or molecular-structural, orientation is commonly indicated by preferred orientation of crystallographic directions (indicated by optic properties, mineral cleavages, etc.). The orientation of optical properties may not be apparent from study of the external shape of the grains, as, for example, in a quartzite, in which the individual quartz grains might be subspherical.

It will be obvious that habit and "lattice" orientation are not mutually exclusive; rather, a fabric can be established by measuring grain shapes or "lattice" characteristics. Sometimes a rock may possess only shape or only "lattice" orientation.

Sander emphasized the necessity of studying the impact of deformative stresses on each individual grain in a rock. Where existing grains have recrystallized so as to assume a new orientation in response to the stress field, or a new phase has developed with a preferred orientation, the neocrystallization has been called **indirect componental movement.** Although it is commonly difficult to establish whether previously-formed crystals have been bodily rotated, the process—**direct componental movement**—is sometimes clearly demonstrated by the texture or fabric. In many regionally-metamorphosed and deformed rocks both direct and indirect componental movements appear to have been important simultaneously.

Sander's interest in "Teilbewegung," or the movement of the individual component parts of the rock, dates from about 1908-9, a time before the universal microscope stage was available. Even without the universal stage, Sander distinguished between the behavior of the rock unit as a whole, and that of each constituent mineral grain (Gefügeelemente = fabric element); he recognized that each mineral grain is individually (separately) and differently orientated.

Basing a definition on these concepts of Sander's, a **tectonite** is a rock in which componental movements were integrated to give a recognizable fabric. Suppose that in a rock comprising an aggregate of calcite grains, the *c*-crystallographic axes have a random (unpredictable) orientation in a small sample. If this rock were transformed into a tectonite, the grains would show a significant departure from the random orientation; that is, the *c*-axes and other measurable attributes would possess a preferred orientation. In actual rocks there may be some initial preferred orientation—resulting from sedimentation processes, for example—present before componental movements produced the tectonite fabric.

Sander recognized several different types of tectonite: *S*-tectonite, *B*-tectonite, and *R*-tectonite. However, because these types were defined in terms of genetic concepts, and we have only a very limited knowledge of how fabrics form, they have restricted usefulness. However, these terms are frequently referred to in the literature, so it is useful to review their definitions:

S-tectonite: tectonites in which the dominating structure is an *S*-plane or *S*-surface and the orientation of the components was produced by movement along (a) the most prominent *S*-surfaces — this variety was called a Scherungstektonite (*slip* tectonite); or (b) slip planes that intersect the megascopic *S* or *flattening* surface (cf. Knopf, in Knopf and Ingerson, 1938, p. 69). Sander thought that petrofabric diagrams for both of these genetically-different *S*-tectonites should be **pole diagrams** (Figs. 80 and 81).

B-tectonite: a tectonite with a fabric characterized by an axial direction, rather than a planar surface. Sander thought that sometimes the axial influence can be recognized only by the axis to a girdle on a petrofabric diagram; this axis is the normal to the girdle plane and is commonly designated the *b*-axis. Sander utilized *B* for *b* if the fabric *b*-axis is parallel to either (a) the intersection of two sets of slip planes, or (b) the axis of rotation about which slip planes have been flexed. These axes were then *B*-axes by definition, and were distinguished from a *b*-axis, which is simply the normal to the direction of slip in a single uncurved set of slip planes. It follows that petrofabric diagrams of *B*-tectonites would be characterized by **girdle diagrams** (Figs. 82 and 83).

R-tectonite: Sander distinguished as rotation tectonites (*R*-tectonites) those *B*-tectonites in which the *B*-axis is an axis of external rotation. Such tectonites comprise the second group (b) of *B*-tectonites mentioned above; because of the influence of external rotation the petrofabric diagrams are characterized by girdle diagrams. In common usage *R*-

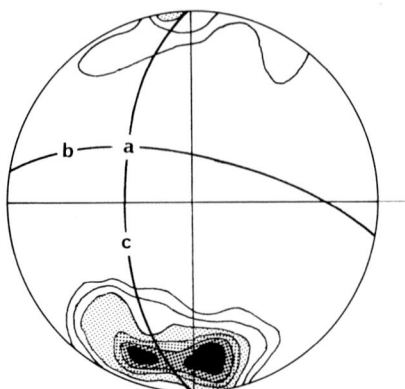

Figure 80. Fabric of an *S*-tectonite: 206 poles to calcite lamellae in marble, Griesscharte, Hochpfeiler, Tyrol, Austria; hand sample shows schistosity (*ab*-plane) but no lineation in this plane; *a b* and *c* define geometric planes in the fold (After Sander, 1930, Diag. 69).

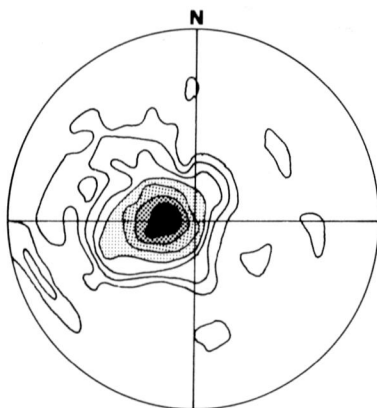

Figure 81. Fabric of an *S*-tectonite: 150 poles to biotite cleavage in Eilde Flags, A' Bhuidheanaich, Monadhliath, mid-Strathspey, Central Highlands, Scotland (After Whitten, 1959A, Fig. 7).

tectonites have frequently been referred to simply as *B*-tectonites (Figs. 84 and 85).

Because "componental movement" forms part of the definition, a tectonite is specifically a rock which has undergone deformation, and all other rocks possessing a fabric (e.g., micaceous siltstones) are excluded. Types of fabric other than those of tectonites are:

Depositional fabric: a preferred orientation produced by primary sedimentary or depositional features; this genetic term is frequently referred to as **apposition fabric** (e.g., Pettijohn, 1949, p. 59; den Tex, in Hills, 1963A, p. 413) or *Anlagerungsgefüge* in German.

Chemical precipitation fabric: examples include primary preferred orientation in unrecrystallized evaporite deposits, or in the ordered growth of crystals normal to the walls of some pegmatites; *Wachstumsgefüge* has been used for such fabrics in German.

Fabrics of certain igneous rocks: such fabrics are commonly transitional between depositional and true tectonite fabrics. Fluxion (flow) structures in lava flows provide a variety of apposition fabric because they are primary nondeformed features; **fusion tectonite** has been used (e.g., by den Tex, in Hills, 1963A, p. 413) for igneous rocks showing flow structure. In many large igneous massifs fluxion structures are partially modified by neocrystallization (e.g., autometamorphism), and Sander used *Schmelztektonite* for these complex mixed fabrics.

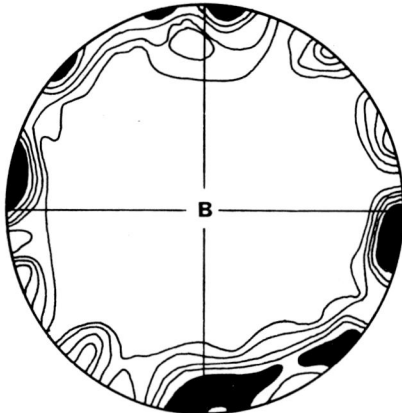

Figure 82. *B*-tectonite: 222 poles to muscovite cleavage planes, fibrous calcareous phyllonite, Brenner, Tyrol, Austria, measured in a thin section cut normal to axis of rotation in rock. Contours at 1, 2, 3, 4, and 5 per cent (After Knopf and Ingerson, 1938, Fig. 12).

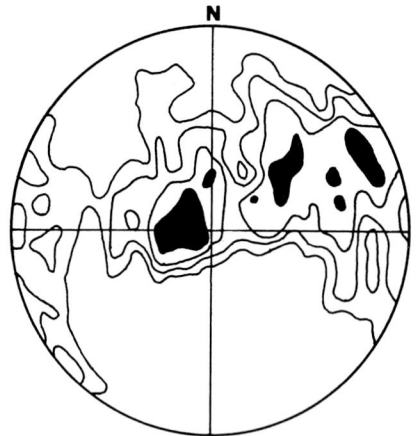

Figure 83. *B*-tectonite: poles to 150 muscovite cleavage planes, Dalradian pelitic schist, An Suidhe, Mid-Strathspey, Central Highlands, Scotland; contours at 0.5, 1, 2, 4, and 8 per cent (After Whitten, 1959A, Fig. 7).

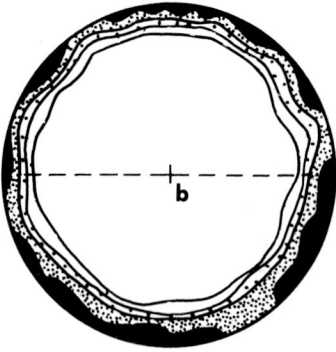

Figure 84. R-tectonite fabric: girdle in *ac*-plane of normals to muscovite cleavage, mica schist, Schimborn. Contours at 0.5, 1, 2, and 4 per cent (After de Sitter, 1964, Fig. 48D).

Figure 85. R-tectonite fabric: Collective orientation diagram (in *ac*-plane) of [0001] of 633 quartz grains measured in six quartz-albite domains of a phyllonitic schist, eastern Otago, New Zealand. Contours at 0.5, 1, 2, and 3 per cent (After Turner and Weiss, 1963, Fig. 6-15b).

Many of the rocks of the above three groups were designated **non-tectonites** by Billings (1954, p. 379), who defined a nontectonite as:

> ...a rock that results from the accumulation of many separate components, each of which moved into place independently of its neighbors; all undeformed sedimentary and igneous rocks belong in this category.

When a sequence of rocks is folded, the regional stress that produces the flexures also affects each constituent grain. Commonly a sequence of rocks is compositionally anisotropic, and this causes stresses to be slightly different in each local area during the regional deformation. For example, if a series of competent and incompetent metasedimentary rocks is folded together, local stress fields tend to develop small parasitic folds on the flanks of the main flexures, and to cause local adjustment of the whole fabric to the combined local and regional stress fields. To describe both the large and the small geometric features of folds, and to enable them to be related, **fabric axes** are commonly used. Unfortunately, premature genetic implications have commonly been assigned to structural and fabric axes. The use of the same terms for different types of axes by different strong "schools" of structural geologists has led to considerable confusion in the literature.

Sander (1926, p. 328) defined three orthogonal (mutually perpendicular) axes (*a*, *b*, and *c*) to describe the geometry of fold structures. Weiss (1959A, p. 25) and Rast (1964, p. 182), with considerable justification, drew attention to the fact that, until recently, the full philosophical content of Sander's axes has not been widely appreciated. Weiss (1959A, pp. 25-26) suggested that:

> To continue sprinkling papers with references to a- and b-axes — the significance of which is imperfectly understood by the writers themselves — can lead only to confusion and worthless argument. In Britain and Scandinavia, in particular, futile controversy on the subject of a- and b-lineations in the rocks of the Caledonides has occurred from time to time over a period of years. In the view of the writer this controversy arose out of a lack of appreciation on the part of the geologists involved of the importance of the symmetry of fabric in kinematic interpretation, and a confusion of fabric axes with kinematic axes.

The orthogonal a-, b-, and c-axes of Sander relate to the geometry of the fabric, and hence to the symmetry of the fabric elements. Orthogonal axes can also be erected with respect to the kinematic or movement picture that resulted in the geometry. In some cases fabric (or geometric) axes and kinematic axes have the same orientation.

The following definitions relate to the fabric or geometric axes and are appropriate where the fabric has monoclinic symmetry. The significance of monoclinic symmetry is discussed in Chapter 5; it is the most common variety of symmetry in cylindroidally-folded rocks, and it is marked by a single plane of symmetry perpendicular to the fold axis.

b-**axis:** is the normal (perpendicular) to the monoclinic plane of symmetry, and is thus parallel to the fold axis.

ab-**plane:** is the most prominent, or the principal, foliation (i.e., the fabric plane).

ac-**plane:** is the plane of symmetry of the monoclinic structure (or fabric) and is normal to the b-axis.

c-**axis:** is the normal to the ab-plane, but its orientation can vary because, although the tectonic axis (b-axis) is reasonably constant in orientation, the a-axis can vary from point to point within the fabric.

In the majority of fabrics, the ab-plane corresponds to the axial surface and/or the axial-plane foliation in cylindroidal folds. Figure 86 shows a diagrammatic fold in which the axial surface is slightly folded by rotation about b. Because the b-axis is commonly the only direction which is constant within a local unit of the fabric, it can be designated B. In Figures 86 and 87 the axial surfaces have varied dips, but they are cylindroidal with respect to B; that is, poles to the axial surfaces lie on a great circle whose axis coincides with B.

In some discussions of fold geometry the a-axis has been defined as the "direction of movement," the b-axis as the fold axis, and the c-axis as

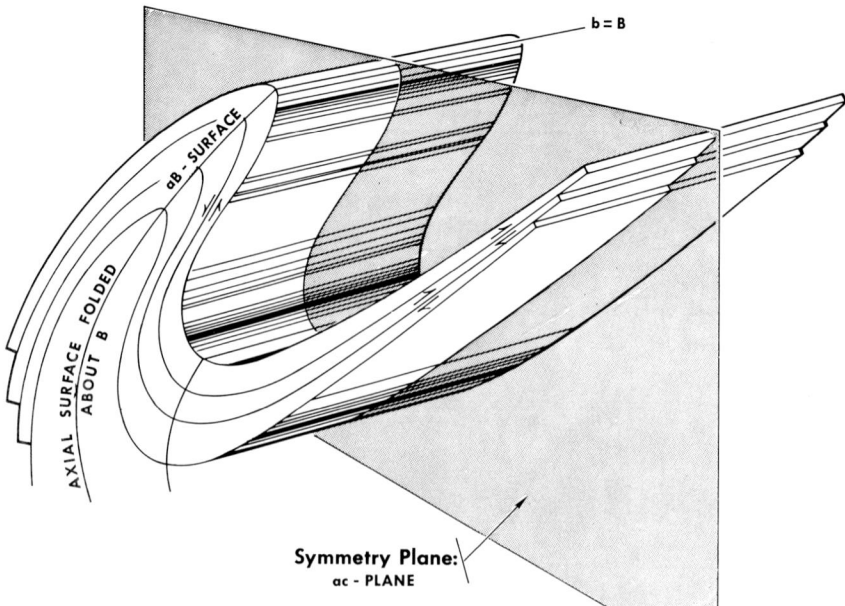

Figure 86. Diagrammatic monoclinic fold showing the geometric fold axes. B is perpendicular to the ac-plane.

the normal to the ab-plane. Apart from the difficulty and ambiguity of assigning the direction of movement in all of the many sizes of folds involved, such definitions confuse two dissimilar concepts. The "direction of movement" refers to kinematic concepts and not to a purely geometric description of the flexure. It is a genetic rather than a descriptive concept.

Kinematic axes can be assigned to simple cylindroidal folds (e.g., Fig. 86), in which development of the flexure involved slip between each successive bed. In Figure 88 the kinematic axes are shown for two artibrary points. The direction of slip is the a-axis, which has a different orientation at each point of the flexure; the orientation of the b-axis, the axis of external rotation, is homogeneous within the fold, and is thus designated B. Hence, the plane of slip is aB, while the deformation plane is perpendicular to the B-axis (i.e., the ac-plane).

It must be emphasized that the above definitions apply to movements with monoclinic symmetry. Figure 88 is based on the assumption that a sequence of strata, or lithological layers, with prominent S-surfaces has been subjected to external rotation, which caused slip along the S-surfaces. During such flexural slip, linear structures (like slickensides) sometimes develop on the S-surfaces parallel to the a-kinematic axis. If the metamorphic grade is high enough to cause extensive neocrystallization concomitant with the folding, a foliation (S_1) commonly develops normal to the ac-plane and containing the B-axis.

Figure 87. Cylindroidal folds in Precambrian gneiss of Idaho Springs Formation, near Bear Lake, Rocky Mountain National Park, Colorado, U. S. A. Arrows indicate some of the small isoclinal fold closures in thin quartzo-feldspathic units separated by biotitic schist.

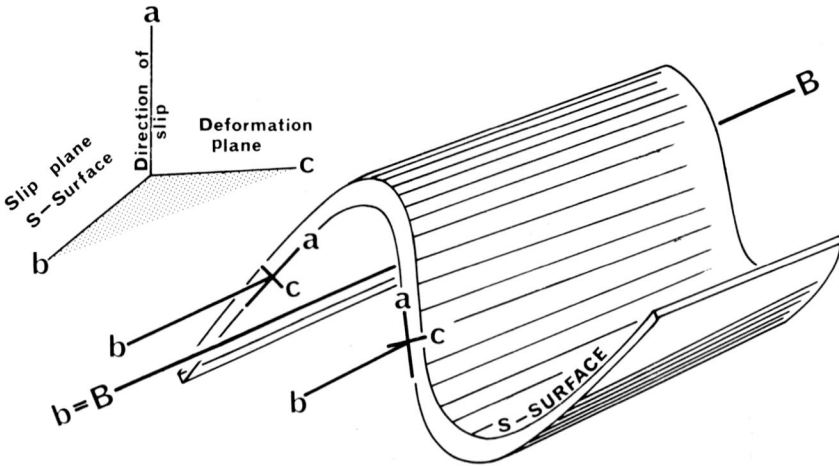

Figure 88. Kinematic fold axes for a monoclinic flexural-slip fold.

Emphasis has been placed on features of cylindroidal folds produced by flexural-slip folding (i.e., parallel folds). Other types of folding (e.g., slip folding) commonly occur in deformational belts and the concepts already outlined can be extended to meet these cases. The main types of simple fold structure are described in Chapter 6.

In some structural literature, especially some of that from Scandinavia and Scotland, the fold axes have been designated as a-axes. Anderson (1948) analyzed mathematically the development of monoclinic structures, and, by drawing analogies with canal flow in fluids, deduced that both the linear structures and fold axes are parallel to the direction of shear (i.e., parallel to a rather than to the B-axis as defined above). It appears that in a majority of cases in which the geometric fold axis has been equated with a, the kinematic a-axis has been "known" from the regional structure, and this "known" orientation has been forced on structures of a completely different (smaller) size. Unless otherwise expressly stated, the definitions enunciated above are used in this book.

The direction of movement in folded rocks is usually not unequivocal on all scales of observation. The size factor was emphasized above. The direction of movement (or a-kinematic axis) has a different orientation as units of different size are considered, and as different parts of the fabric are analyzed. In a section across a major fold belt it may be relatively easy to determine the gross direction of movement; however, the disposition of the kinematic a-axis could be completely different in a small parasitic fold. The direction of movement with respect to an individual sand grain or pebble undergoing strain may be different again (see Figs. 220 and 221).

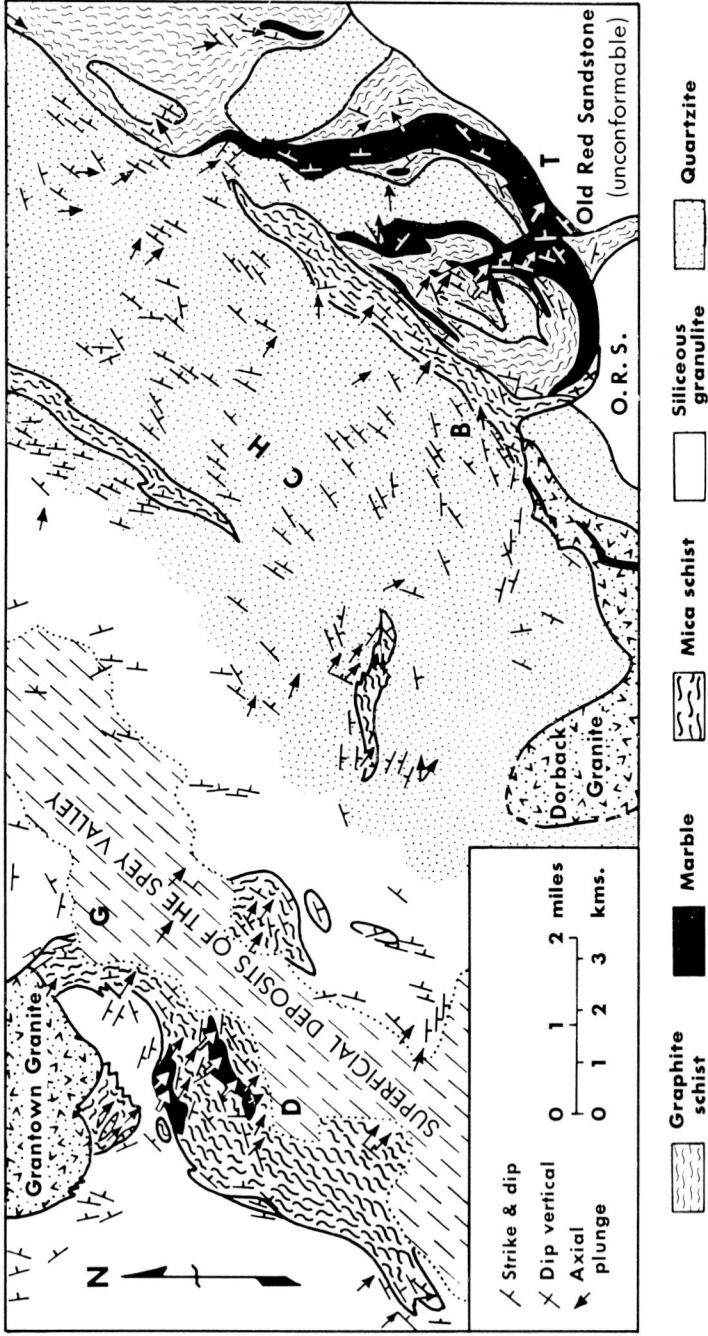

Figure 89. The Grantown-Tomintoul area of Moine and Dalradian rocks, Central Highlands, Scotland. B indicates Bridge of Brown; CH, Cromdale Hills; D, Dulnan Bridge; G, Grantown; T, Tomintoul; O.R.S., Old Red Sandstone (Devonian) (After McIntyre, 1951B, Fig. 2).

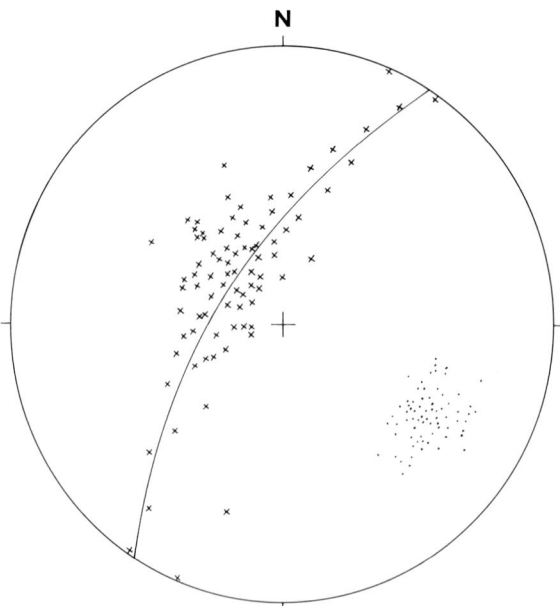

Figure 90. Equal-area projection of structural elements in the Dalradian Grantown Series, Grantown-Tomintoul area, Scotland (Fig. 89). Each plotted point represents the mean of a number of measurements at a particular locality; the great circle is drawn normal to the center of the lineation sheaf (After McIntyre, 1951B, Fig. 3).

· **Projections of Linear Elements**
× **Poles to S-Structures**

Without extremely detailed analyses the assignment of a-axes can be both subjective and liable to considerable error.

Within a regional fold system the fold axes are commonly not all parallel. However, as McIntyre (1951B) showed in a classic paper describing the Grantown-Tomintoul area of the Grampian Highlands, Scotland, extensive areas can be **homoaxial,** that is, have parallel fold axes (Figs. 89 and 90). Many areas have been subjected to two or more discrete or penecontemporaneous phases of penetrative deformation, and, in consequence, folding about new axes has been superimposed on earlier folding. In these cases the resulting fabric and kinematic pictures are composite, and the elements of each successive phase of deformation are more difficult to separate.

Thus, extreme care must be exercised in assigning a-, b-, or c-axes to folded rocks. Similarly, care must be used in interpreting published accounts, to ensure that the scale of observation and the meaning read into the assigned axes is the same as that visualized by the author. Weiss has done much to draw attention to the correct use of both fabric and kinematic axes, and the definitions above owe much to his statements (Weiss, 1955, p. 229; 1959A, p. 26).

Chapter 5

Symmetry Concepts

Knopf and Ingerson (1938) dedicated their memoir on structural petrology to Professor Bruno Sander in "... recognition of his pioneer service in the founding and establishment of structural petrology as a branch of geological science."

The ramifications of Sander's thinking (e.g., 1930, 1936, 1948, 1950) in relation to sedimentary, igneous, and metamorphic geology have been considerable. Although some of Sander's influence on structural geology has not been beneficial (cf. Rast, 1964, p. 182), his publications are particularly important in marking an early step in the continuing swing from purely qualitative petrography to a rigorous quantitative science.

At the turn of the last century there was growing awareness that the external forces producing a folded rock and the detailed internal texture are closely related. Curie (1894) wrote that "... certain causes produce certain effects, the symmetry elements of the causes must be found again in the effects produced [and that when] ... several phenomena of different nature are superposed in the same system, the dissymmetries are added. Therefore, there only remain as symmetry elements in the system those which are common to each phenomenon taken separately."[1]

However, it seems to have been Sander who first wrote extensively about the great significance of these relationships; he believed that the symmetry of the physical factors involved in the deformation is reflected in the geometric fabric of the resulting rock. Turner and Verhoogen (1960, p. 628) restated the dictum thus: "... [the] symmetry of a tectonite fabric reflects the symmetry of the movement plan of deformation."

Symmetry concepts are important for two major reasons. First, the symmetry of the observed rock, of the movements that produced the fabric of the rock, and of the forces that induced the movements can be analyzed separately, but these symmetries are intimately related. This is

---

[1]Translation by Paterson in Paterson and Weiss (1961, p. 858).

important in attempts to unravel the complete history of a rock, especially one devoid of fossils, ooliths, ripple marks, or other primary features whose original and distorted forms could be studied. Second, these three symmetries provide a basis for classification and discrimination between different folded rocks and fold processes.

Knopf (Knopf and Ingerson, 1938, pp. 42-62) gave a useful introduction to the symmetry of both movement and rock fabric. More recently Hoeppener (1961), Paterson and Weiss (1961), and Turner and Weiss (1963) reviewed this whole subject.

Before describing the actual types of symmetry that occur it is necessary to define some of the associated terminology.

Sander (1911; 1930, p. 1; 1948, p. 2) introduced the term *Gefüge,* which is now commonly translated as **fabric,** although Knopf (Knopf and Ingerson, 1938, p. 12) suggested that a reasonable translation of Sander's (1930, p. 1) own definition would be: "...the data that govern the arrangement in space of the component elements that make up any sort of external form...." The term fabric does not imply deformation, and the concept applies equally to all types of rocks — sedimentary, metamorphic, and igneous. Fabric (Gefüge) embraces both (a) *gestaltliche Gefüge:* the geometric arrangement of the elemental parts of a mineral aggregate (rock), and (b) *funktionale Gefüge:* the functional or behavioral directional physical properties associated with the mineral aggregate.

Paterson and Weiss (1961, p. 861) introduced **domain** as a technical term, and stated: "...a tectonite can be divided into parts or *domains* that are statistically homogeneous with respect to their structure on a given scale." This definition makes "domain" difficult to use because, as shown in Chapter 3, a population of objects can be homogeneous with respect to some attributes and heterogeneous with respect to others. "Domain" could be limited to refer to a volume of rock within which the population of a specified attribute is homogeneous; within such domains several attribute populations (for the same population of objects) might happen to be homogeneous. Alternatively, "domain" could be used more loosely for a body of rock within which the populations of the more obvious attributes are homogeneous. It must be remembered that, as the size of the objects (on which the attributes are measured) is changed, populations of different attributes may be homogeneous (cf. Chapter 3). In this book domain will not be limited to any one narrowly-defined usage; rather, the specific meaning intended will be indicated by the context.

The preferred orientation of an element may be either **penetrative** or **nonpenetrative.** For a specified domain size elements with a preferred orientation are penetrative if they are repeated with approximately the same orientation at imperceptible intervals. The size of the objects examined and the scale of observation are important. For example, shear planes in a hand specimen may be penetrative, but microscopic study of a thin section cut from the same rock may show individual anastomosing shear planes separated by slabs of wholly unsheared rock—such shear planes are nonpenetrative; some other elements may remain penetrative as the scale of observation is changed.

In a highly faulted area, the relative movement of adjacent slices of rock commonly generates slickensides (striations) on the fault planes. The slickensides are surficial nonpenetrative features because they do not effect neocrystallization of the whole rock.

For each size of object (and thus for every scale of observation) each fabric element may be penetrative or nonpenetrative and thus must be individually considered.

It is not easy to define **fabric element** because the nature of the penetrative geometric features involved varies widely according to domain size; that is, different features constitute fabric elements as the size of objects examined is changed. Unlike Sander (1948, p. 5) and Fairbairn (1949, p. 3), Paterson and Weiss (1961, p. 861) justifiably recognized two types of fabric element:

Crystallographic fabric elements which comprise lattice planes or lines within individual crystal grains. Mica {001} planes and [0001] directions of quartz and calcite are commonly studied quantitatively. Because such elements pervade each entire mineral grain they are penetrative. In practice, the orientation of such elements is often determined with the aid of nonpenetrative features of each grain (e.g., cleavages or twin lamellae).

Noncrystallographic fabric elements which comprise visible structural discontinuities within a rock. Linear elements include axes of small folds and the intersections of S-structures; planar elements include bedding, foliations, and schistosity.

The fabric of a rock is defined by the sum of *all* the penetrative fabric elements; nonpenetrative elements are not part of the fabric for the domain size concerned. In practice only the more easily measured elements have been recorded in most studies. The distribution of each individual fabric element is a **subfabric** (Paterson and Weiss, 1961, p. 863), and the measurements (the population of attributes) by which the orientation of an element is specified within a domain are the **fabric data.** The symmetry of a homogeneous population of attributes within a domain can be analyzed easily if the attributes (fabric data) are plotted on an equal-area projection. The purpose of the projection is to divorce the orientation of the fabric elements from their spatial location, as in the case of stereographic projection of crystallographic data. This enables the point symmetry of the fabric elements to be determined, following the methods of crystallography.

For this type of work a specific attempt is made to select domains within which the attribute population (subfabric) is homogeneous. If data are suspected of being heterogeneous within a body of rock studied, a record of the original spatial location of each fabric datum should be kept. The specialized technique of *Achsenverteilungsanalyse* (often referred to as AVA) was developed by Schmidt for studying heterogeneous fabrics in thin sections with a universal microscope stage (see Sander, 1950, pp. 39 ff.). On a photograph of the thin section, each grain is labelled according to the homogeneous subpopulation to which it belongs, so that a map of the several homogeneous subpopulations is built up. Although Achsenverteilungsanalyse was developed for thin section analyses, it could be extended with advantage to the study of larger units (e.g., hand samples, map areas).

SYMMETRY OF FABRIC ELEMENTS IN OBSERVED ROCKS

Paterson and Weiss (1961) showed that standard symmetry concepts used by crystallographers can be used to classify structural elements, and they reviewed the symmetry of natural fabrics in the light of the possible *space groups* and *point groups*; Paterson and Weiss (1961, p. 865) noted that only centrosymmetric fabric elements have been measured to date.[2] Theoretical analysis based on space group theory, combined with observations made so far, suggested that tectonite subfabrics fall into five symmetry classes, namely:

1. **Spherical symmetry:** the symmetry of a sphere, in which an infinite number of planes and axes of symmetry have every possible orientation. Rocks rarely exhibit this symmetry, which requires a random (unpredictable) orientation of the fabric elements. It might be anticipated that some sedimentary rocks (e.g., aeolian millet-seed sandstones) show such subfabrics, but examples do not seem to have been described (cf. Pettijohn,

---

[2]Most fabric elements are nonpolar; for example, the direction along the normal to a vertical schistosity plane is not differentiated and bearings of, say, 090 and 270 are equivalent. A few rock attributes are polar and can be considered vectors; for example, graded bedding, flame structures, and other common sedimentary structures define the original upward direction of bedding planes (S_0), and flute casts indicate transport directions. Such polar properties are noncentrosymmetric fabric elements, and, although they have not commonly been included in fabric studies, many can be penetrative and should probably be included in complete analyses.

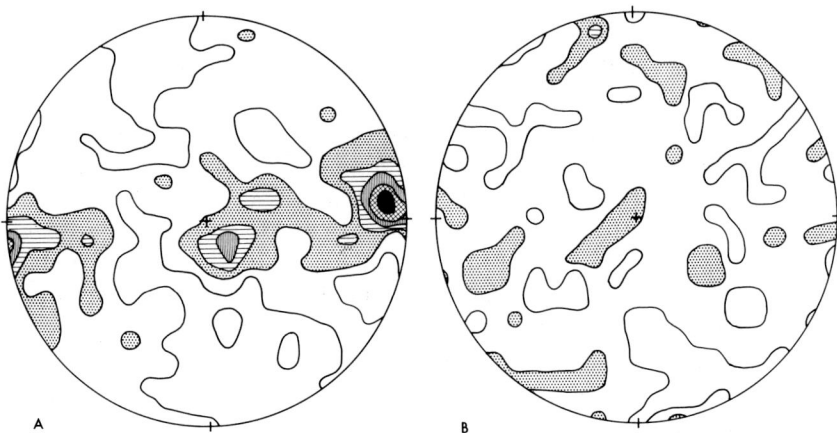

Figure 91. Spherical symmetry: Fabric diagrams with contours at 0.5, 1.5, 2.5, 3.5, 4.5, and 5.5 per cent. A: Vitreous hornfels formed from Quadrant Formation quartzite within the aureole of the Boulder Batholith, Montana, U.S.A.; hand sample has no visible bedding or foliation but diagram tends to monoclinic rather than spherical symmetry; based on 300 quartz c-axes. B: Spherical symmetry diagram produced by plotting 300 points located with aid of a random number table.

1957A, p. 77). Similarly, under the influence of simple thermal metamorphism neocrystallization might result in hornfels textures and subfabrics that approach spherical symmetry (cf. Paterson and Weiss, 1961, p. 866). However, this assumption may be false too because few examples have been recorded; within the southern aureole of the Boulder Batholith, Montana, U.S.A., the Quadrant Formation quartzite is a vitreous hornfels, but Figure 91 shows that one subfabric does not have spherical symmetry.

2. **Axial symmetry:** the symmetry of a cylinder, in which an infinite number of planes of symmetry intersect in a single axis, which is normal to another plane of symmetry. Subfabrics have pole diagrams in which the "pole" on the equal-area projection is circular. A rock with this symmetry is the deformed nepheline gneiss from Sørøy, northern Norway, described by Sturt (1961A); the c-crystallographic axes of the nepheline crystals have a strong preferred orientation which coincides with the b-geometric fabric axis (Fig. 92). Sturt suggested this fabric developed at the expense of a random orientation in the undeformed rock. Weiss (1954, p. 33) showed that calcite crystallographic axes in a marble from Barstow, California, U.S.A., have axial symmetry, while Paterson and Weiss (1961, p. 865) found that 171 linear structures measured in the field from quartzites at Ballachulish, Scottish Highlands, have a similar structure (Fig. 93).

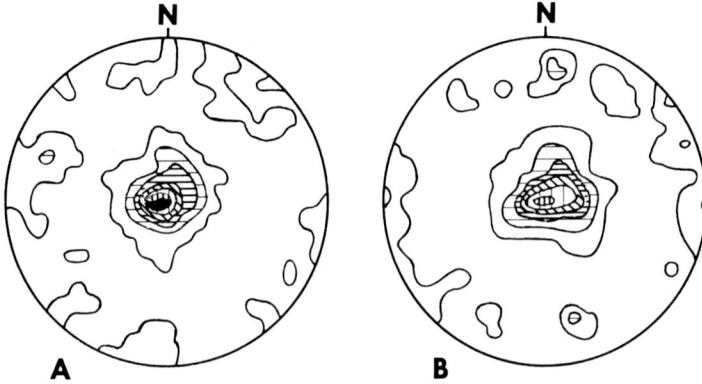

Figure 92. Axial symmetry shown by nepheline syenite gneisses from Sørøy, N Norway; each diagram is based on 300 nepheline grains. Contours in A at 1, 3, 5, 7, 9, 11, and 13 per cent, in B at 1, 3, 5, 7, 9, and 11 per cent (After Sturt, 1961A, Fig. 1).

3. **Orthorhombic symmetry:** this symmetry corresponds to that of the holosymmetric class of the orthorhombic system in crystallography, and is characterized by three mutually perpendicular planes of symmetry. Orthorhombic subfabrics have been thought of as uncommon, but an increasing number is being described. Examples include a plot of 1,000 foliation planes in gneisses from the Turoka area, Kenya, measured by L. E. Weiss (Fig. 94), and a mylonitized quartzite in the Moine Thrust Zone, Stack of Glencoul, Scottish Highlands, studied by J. M. Christie (Fig. 95).

4. **Monoclinic symmetry:** this symmetry has a single plane of symmetry with an axis of two-fold symmetry normal to it; the symmetry corresponds to the holosymmetric class of the monoclinic system of crystallography. This is a very common subfabric symmetry, and exam-

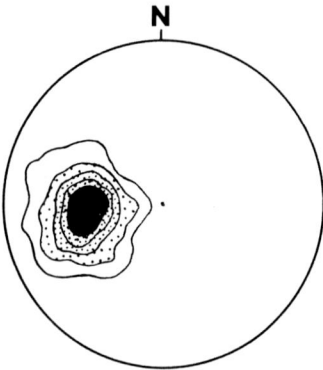

Figure 93. Axial symmetry shown by 171 lineations measured in quartzite from Ballachulish, Scottish Highlands; contours at 1, 4, 7, 10, and 13 per cent (After Paterson and Weiss, 1961, Fig. 7b).

Figure 94. Orthorhombic symmetry: poles to 1,000 foliation planes in gneiss, Turoka area, Kenya. Contours at 1, 3, 5, 7, and 9 per cent; measurements by L. E. Weiss (After Paterson and Weiss, 1961, Fig. 7d).

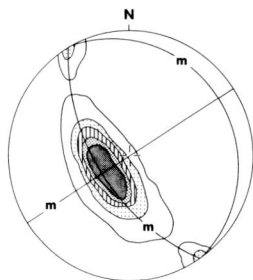

Figure 95. Orthorhombic symmetry: [0001] axes of 300 quartz grains in a mylonitized quartzite, Moine Thrust Zone, Stack of Glencoul, Scotland. Contours at 1, 3, 5, and 8 per cent; measurements by J. M. Christie (After Paterson and Weiss, 1961, Fig. 7c).

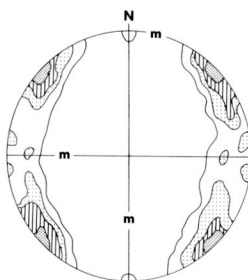

Figure 96. Monoclinic symmetry: [0001] axes of 255 quartz grains in a mylonitized quartzite, Barstow, California, U.S.A. Contours at 1, 3, 5, 7, and 9 per cent; measurements by L. E. Weiss (After Paterson and Weiss, 1961, Fig. 8a).

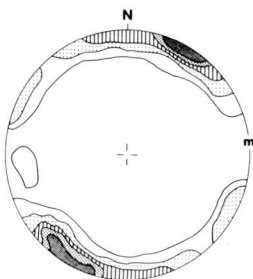

Figure 97. Monoclinic symmetry: poles to 2,000 foliation planes in schists and quartzites, Loch Leven area, Central Highlands, Scotland. Contours at 1, 2, 3, and 4 per cent; measurements by L. E. Weiss (After Paterson and Weiss, 1961, Fig. 8b).

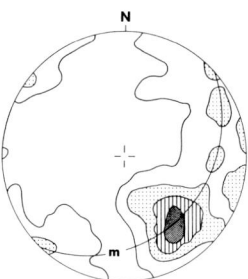

Figure 98. Triclinic symmetry: [0001] axes of 416 quartz grains in a deformed quartzite pebble, Panamint Range, California, U.S.A. Contours at 1, 2, 3, 4, and 5 per cent (After Paterson and Weiss, 1961, Fig. 8c).

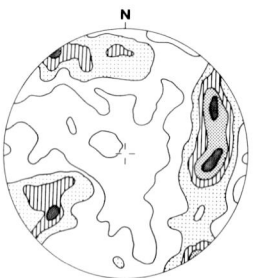

Figure 99. Triclinic symmetry: poles to 193 foliation planes in gneisses and schists, Lake O'Keefe area, Quebec, Canada. Contours at 1, 2, 3, 4, and 5 per cent; measurements by G. Gastil and L. E. Weiss (After Paterson and Weiss, 1961, Fig. 8d).

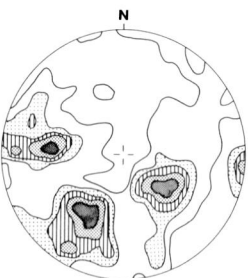

ples include the quartz subfabric of a mylonitized quartzite from Barstow, California, U.S.A. (Fig. 96), and the foliation of schists and quartzites, Loch Leven, Scottish Highlands (Fig. 97), studied by L. E. Weiss (Paterson and Weiss, 1961, p. 866).

5. **Triclinic symmetry:** this symmetry lacks planes of symmetry, and is characterized by one inversion monad-axis of symmetry ($\bar{1}$) only, as in the holosymmetric triclinic class of crystal symmetry. Triclinic symmetry is common and Paterson and Weiss (1961, p. 866) described examples from the quartz subfabric of a deformed pebble, Panamint Range, California, U.S.A., and from the foliations measured in the gneisses and schists near Lake O'Keefe, Quebec, Canada (Figs. 98 and 99).

Fabrics are commonly **homotactic** (Sander, 1930, p. 165); that is, the symmetries of all subfabrics are similar. When subfabrics have dissimilar symmetry the fabric is **heterotactic.** Prior to deformation each igneous, sedimentary, or metamorphic rock has an **initial fabric.** Inheritance of initial subfabrics to a varying extent by each of the fabric elements can contribute to the heterotacticity of the final fabric. When initial subfabrics have been overprinted, but are still discernible, the process was called *Überprägung* by Sander. *Umprägung* occurs when geometric features of the initial fabric have been completely destroyed.

An introduction to these symmetry groups (except the spherical) was given by Knopf (Knopf and Ingerson, 1938, pp. 42 ff.), who also included tetragonal symmetry as a separate group. As noted by Knopf, tetragonal symmetry is rarely found in nature, and it is a specialized limiting case of the orthorhombic. It is not desirable to maintain tetragonal as a separate additional symmetry group.

SYMMETRY OF THE MOVEMENT PICTURE

Kinematics is the study of motion (i.e., movement from one location to another), without reference to the forces causing the motion. The pattern of the constituent parts of a rock records a "frozen" picture of the movements of the individual particles which generated the fabric. It requires a separate study of the **dynamics** to determine the **kinetic forces** which produced the **movement picture** (i.e., kinematic picture or Bewegungsbild of Sander). The kinetics (or dynamics) of the fabric-producing processes are a matter of inference, and should not be confused with the kinematics which can be directly observed. The forces (kinetics) that produced the fabric have long since disappeared.

A movement or deformation is said to be **affine,** if straight lines within the original body are still straight lines after the movement; with affine

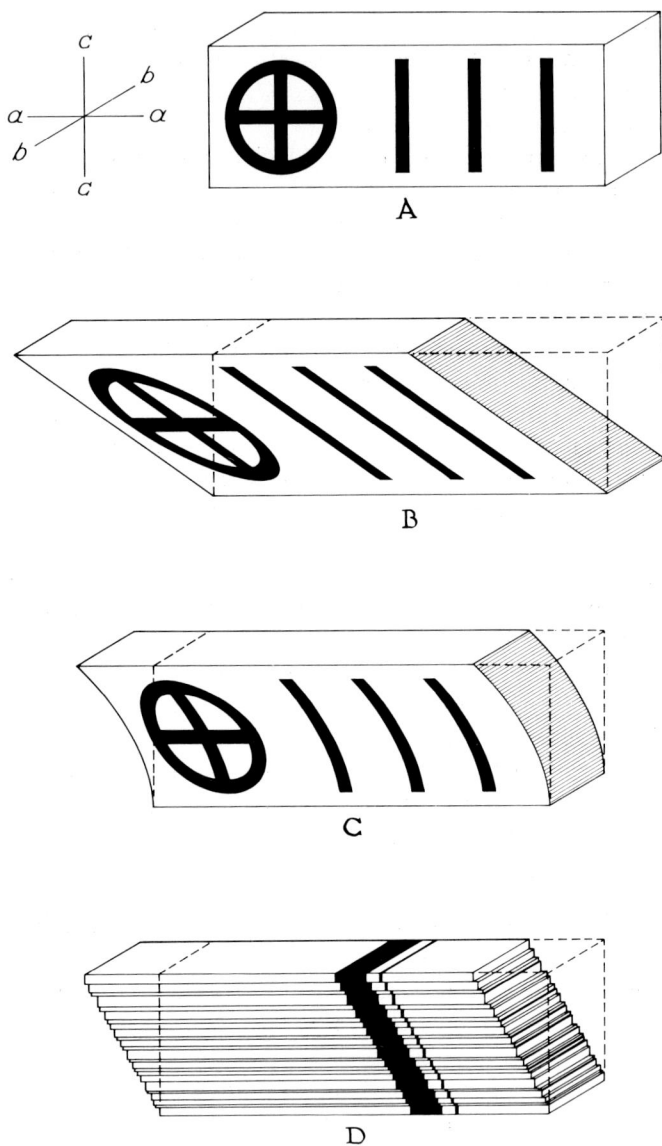

Figure 100. Affine and nonaffine movement. A: Patterns marked on a block prior to any movement. B: Block A affected by affine movement due to slip parallel to *a* along an infinite number of *ab*-planes. C: Block A affected by non-affine movement parallel to *a* along an infinite number of *ab*-planes. D: Enlarged portion of block B: at this enlargement, the block shows nonaffine movement, although, at smaller magnification, block B shows affine movement (A, B, and C based on Knopf and Ingerson, 1938, Fig. 8).

deformation, planes also remain planes. Figure 100 illustrates the difference between affine and **nonaffine** movement. Size is very important in this concept. At one scale of observation a domain may appear to have undergone affine deformation, but when examined in greater detail (e.g., higher magnification) deformations are commonly not completely penetrative and are thus nonaffine (Fig. 100D). Although affected by approximately affine deformation, local variations in both the rock composition and the stress pattern commonly result in local departures from a truly affine deformation.

In analyzing the movement picture of a sequence of folded rocks extensive use has been made of **stress** and **strain** theory, despite the fact that, because natural rocks are neither homogeneous nor isotropic, these concepts only apply in a general way. **Stress** refers to the forces acting within a rock body that result from a combination of (a) body forces acting at every point in the body (e.g., gravity) and (b) surface forces applied to the boundaries of the unit. The stress is expressed as gm/cm^2 (force per unit area) across any plane in the body.

Strain refers to both elastic and permanent deformations of the geometry of a body in response to stress. In studying folded rocks permanent changes of shape and volume are of primary concern — this is **finite strain.** Strain can only be assessed by comparing the changes in geometry before and after the stress. When the strain is affine the changes or transformations involve (a) movement (translation) of the rock body as a whole, (b) external rotation of the whole rock body, (c) volume change of the unit (dilation), and (d) distortion or change of shape of the unit (cf. Becker, 1893, p. 16). These four transformations can be expressed simply in terms of matrix algebra operations (cf. Sawyer, 1955, Chapter 8).

Of greatest importance in the development of a rock fabric are the relative displacements of the components *within* each domain; these involve (c) and (d) above, which are commonly referred to as **pure strain** and represented by a **strain ellipsoid.** Translation and external rotation — (a) and (b) above — only alter the orientation of the whole fabric with respect to the three-dimensional geographic coordinates.

Finite affine strain causes a sphere of the original rock to be transformed into an ellipsoid (cf. Becker, 1893, 1896), and the symmetry of a strain ellipsoid conveniently reflects symmetry elements of the kinematic or movement picture. **Pure strain** and **general strain** result in triaxial ellipsoids whose principal axes are either shorter or longer than the original sphere diameter (cf. Flinn, 1962; Turner and Weiss, 1963, p. 274). General strain is a combination of pure strain and external rotation. In a few special cases **plane strain** occurs in small domains; with plane strain, the intermediate triaxial ellipsoid axis remains equal to the length of the original sphere diameter, and the sphere and the ellipsoid are equal

Table 1.—*Symmetry and Geometric Elements of Strains Resulting in Tectonically-Significant Types of Finite Affine Strain*

	Type of Strain	Form of Strain Ellipsoid*	External Rotation	Symmetry System**
General cases	Pure strain	Triaxial ellipsoids	None	Orthorhombic
	General strain	A > D, B ≧ D, and C < D	Rotation about A, B, or C	Monoclinic
			Rotation about axis inclined to A, B, or C	Triclinic
Special cases	Pure shear	Triaxial ellipsoids A > D B = D C < D	None	Orthorhombic
	Simple shear		Rotation about B	Monoclinic

*A > B > C are the principal axes of the ellipsoid, and D is diameter of original sphere.
**Holosymmetric class of the system listed in each case.

in volume. Under controlled laboratory conditions additional forms of the strain ellipsoid are possible.

Further discussion of this complex theoretical topic would be out of place here; detailed reviews of the subject were given by Paterson and Weiss (1961) and Turner and Weiss (1963). Table 1 is based on Turner and Weiss (1963, p. 274) and summarizes the symmetry and geometric elements of those strains resulting from tectonically-significant types of finite affine strain.

The initial fabric of the rock and the strain responsible for the final fabric are independent. Curie's and Sander's principles (outlined at the beginning of this chapter) imply that the resultant symmetry of a domain comprises the symmetry elements common to the initial fabric and to the strain (referred to internal axes). Since the resultant symmetry must appear in the final observed fabric, any symmetry element absent in the observed fabric was absent in either (a) the initial fabric, or (b) some component of the strain. However, symmetry elements can be present in the final fabric that are not present in either the initial fabric or the strain components (cf. Paterson and Weiss, 1961, p. 879).

These arguments can be applied to both the complete fabric and to each of the subfabrics. The symmetry of fabrics that contain relicts of inherited subfabrics is commonly lower than the symmetry of the latest deformation, because of the influence of the initial fabric on the final fabric symmetry. The variety of tectonic forms which occur in nature is caused by the compositional anisotropism of the deformed rocks and by changes in the stress systems which affected the area in the course of time (Hoeppener, 1961, p. 695).

FABRIC ELEMENTS AND
THE MOVEMENT PICTURE

When all early structures have been obliterated it is more difficult to unravel the details of the deformation. Overprinted relicts of initial fabric carry a useful record of the strain. A folded rock can contain fabric elements that were (a) *inherited* in a transformed condition, (b) *imposed* during deformation, and/or (c) *combined* from inherited and imposed features. It is now appropriate to reconsider crystallographic and noncrystallographic fabric elements in terms of these categories.

NONCRYSTALLOGRAPHIC FABRIC ELEMENTS

These are structural discontinuities, visible at some scale of observation, which include:

(A) KINEMATICALLY-PASSIVE ELEMENTS

These are inherited elements that neither cause nor correspond to local structural discontinuities during the deformation (Paterson and Weiss, 1961, p. 878). Ooliths in a deformed limestone, or pebbles in a deformed sandstone, are possible examples; commonly such elements can be recognized after the deformation, so that they contain a complete record of the total deformation.

Cloos (1941, 1943, 1947A, 1953B) studied the deformation of ooliths in folded Appalachian carbonate rocks, Maryland, U. S. A., and demonstrated that the original subspherical ooliths are now ellipsoids elongated in the fold axial plane and perpendicular to the B geometric fold axis. Breddin (e.g., 1956A, 1956B, 1957, 1958A, 1958B) described comparable distortion of brachiopods, lamellibranchs, goniatites, etc., which suggested that the longest axis of the rock strain ellipsoid lies in the fold axial plane perpendicular to B. Breddin developed formulae for calculating the strain on the basis of the known shapes of the undistorted fossils (Figs. 101 and 102); for example, the angle between the median line and the hinge line of brachiopods becomes distorted if these lines are not initially parallel or normal to B (Fig. 103). Furtak and Hellermann (1963) made similar use of deformed fossil plant structures (Calamites, Lepidodendron, Sigillaria, etc.) recovered from coals. Deformed crinoid ossicles from Variscan folds of Middle Devonian sandstones, northern Rheinisches Schiefergebirge, Germany, were analyzed by Kurtman (1960). Initial fabrics—preferred orientation of fossil fragments resulting from original current action (cf. Schwarzacher, 1963; Nissen, 1964A) or diagenetic compaction (Breddin, 1956A)—can complicate analyses of this type.

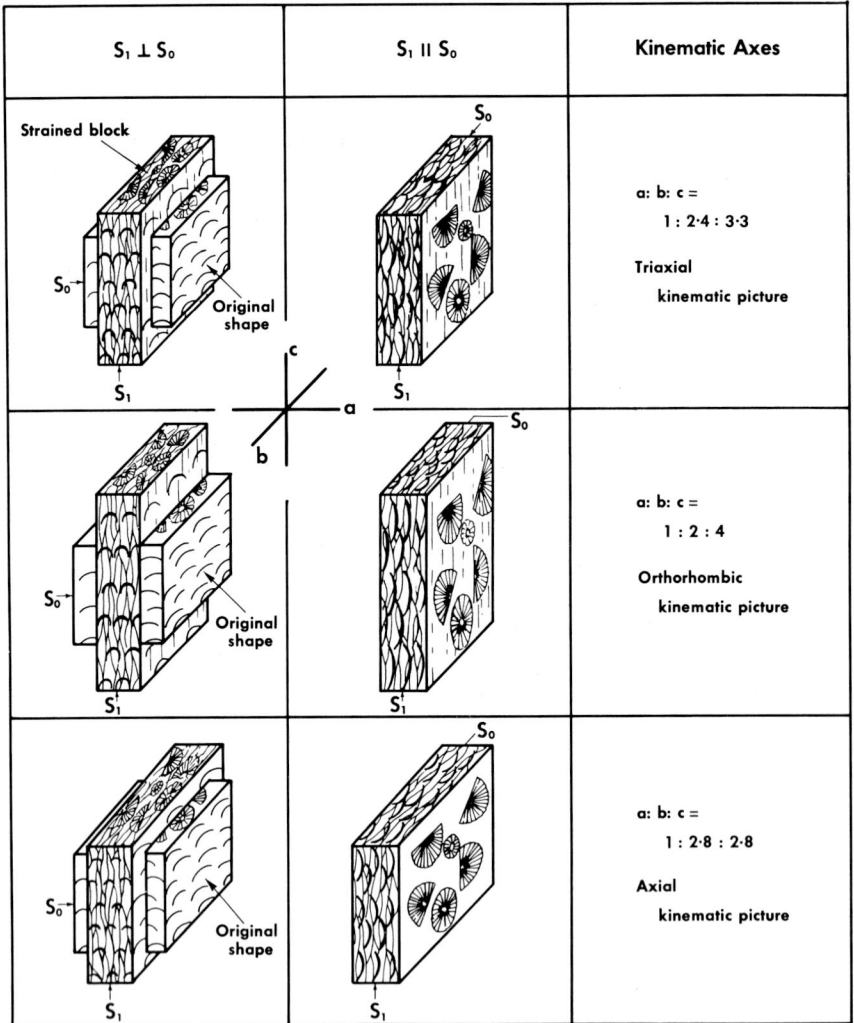

Figure 101. Distortion of fossils by different types of strain with S_0 either perpendicular to or parallel to S_1; $a:b:c$ ratios refer to lengths of axes of ellipsoid developed by strain of an initial sphere of rock (After Breddin, 1956A and Voll, 1960, Fig. 14).

It is not always possible to determine whether the strain shown by deformed ooliths, pebbles, fossils, etc., corresponds exactly with the

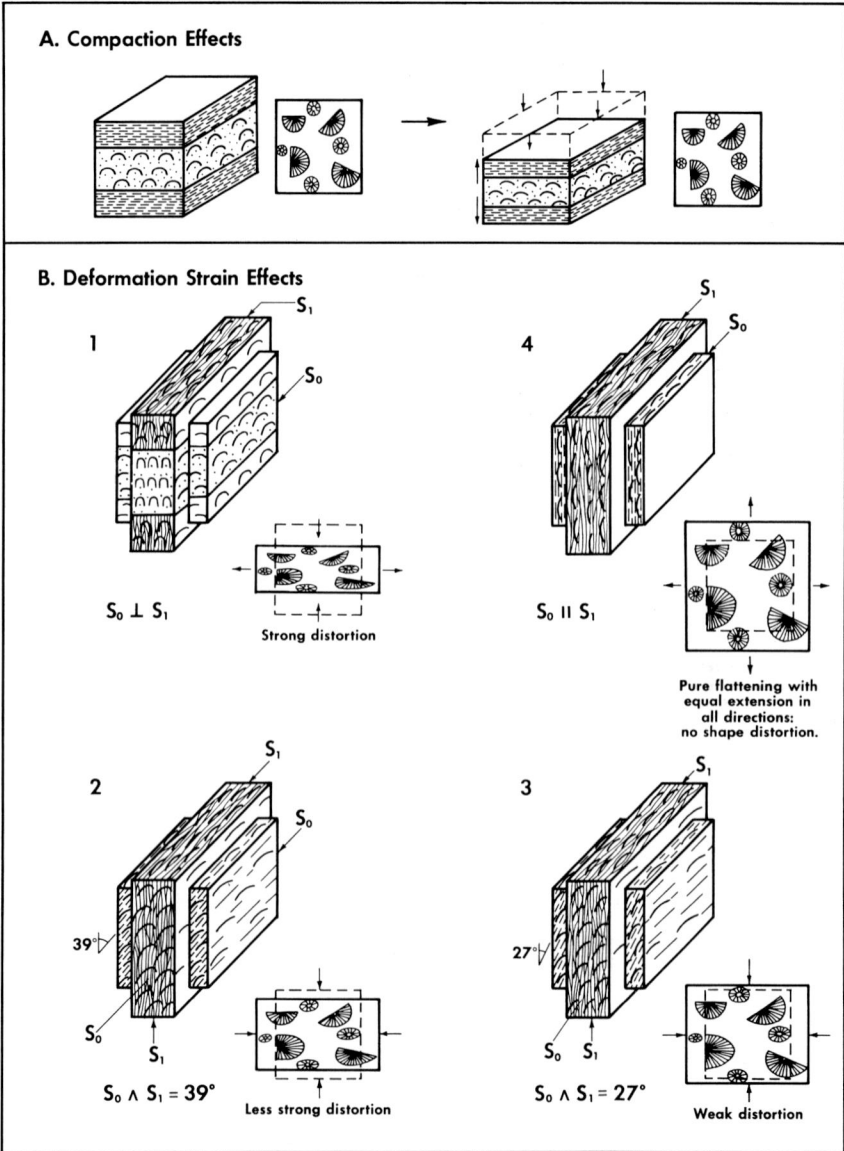

Figure 102. Distortion of fossils by compaction and deformation; distortion of the angle between hinge line and median line of fossils is shown in small inset beside each block diagram. A: The effect of compaction and diagenesis — fossil shape unchanged. B: Biaxial deformation for various original angles between S_0 and S_1 (After Breddin, 1956A and Voll, 1960, Figs. 13a and c).

w 90° w=90°
w=49°
w−39°   ψ−90° ψ−90°   w=49°
ψ−40.7   w−39°
w−49°   ψ−19.5°   ψ−40.7°   After strain
ψ−8.5° |STRAINED ——————— FORMS| ψ−19.5°
w−90°   B   ψ−8.5°   w=49°
| ORIGINAL | | ORIENTATION | w−90°
λ=22.5°   λ=45°   λ=67.5°   λ=67.5°   λ=45°   λ=22.5°   ψ=0°

ORIGINAL SHAPE

At onset of strain

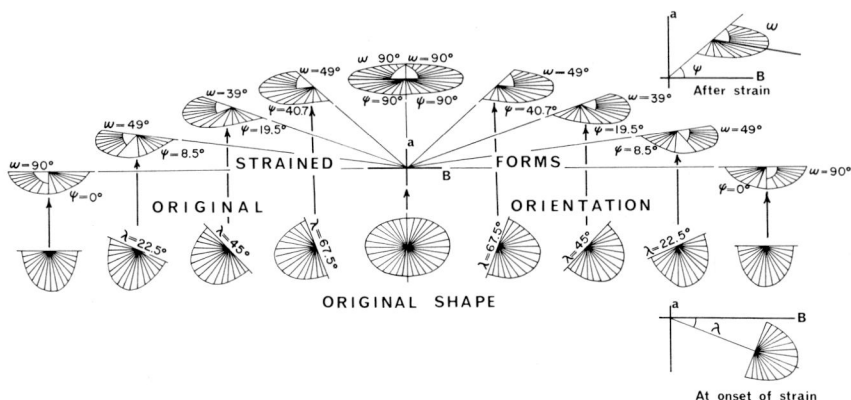

Figure 103. Deformation of brachiopods lying on a bedding plane for different angles (λ) between hinge line and the direction B before strain; after strain the angle between hinge line and median line of the brachiopod is ω, and the angle between hinge line and B is ψ (After Breddin, 1956A, and Voll, 1960, Fig. 15).

strain of a whole rock sample containing them. Commonly the whole rock and all of its components have not been subjected to the same strain because the mechanical properties of the fossils, etc., are dissimilar to those of the enclosing rock (cf. Voll, 1960). Cloos (1947A) showed that the deformed Ordovician Conococheague Limestone, Maryland, U. S. A., contains dolomitic spheres associated with ellipsoidal deformed calcitic ooliths. This topic is discussed further in Chapter 9.

(B) KINEMATICALLY-ACTIVE ELEMENTS

Kinematically-active elements may be either imposed or inherited fabric elements:

(1) *Inherited:* inherited fabric elements are loci of structural discontinuities present in the rock throughout the deformation process (Paterson and Weiss, 1961, p. 878). Unlike kinematically-passive elements, these active elements (e.g., bedding planes may be moved relative to one another; hence, unlike the ooliths discussed above, they need not remain as indicators of the mean strain. Bedding in Figure 104 is an inherited marker.

(2) *Imposed:* imposed fabric elements are always sites of unequal deformation that preserve a record of these local deformational inequalities. In polyphase deformation an element active at one time may be passive in a subsequent phase of deformation. The foliation in Figure 104 is an imposed fabric element, while the linear structure produced by intersection of bedding and foliation is a **composite fabric element,** because it is formed by the intersection of inherited and imposed fabric elements.

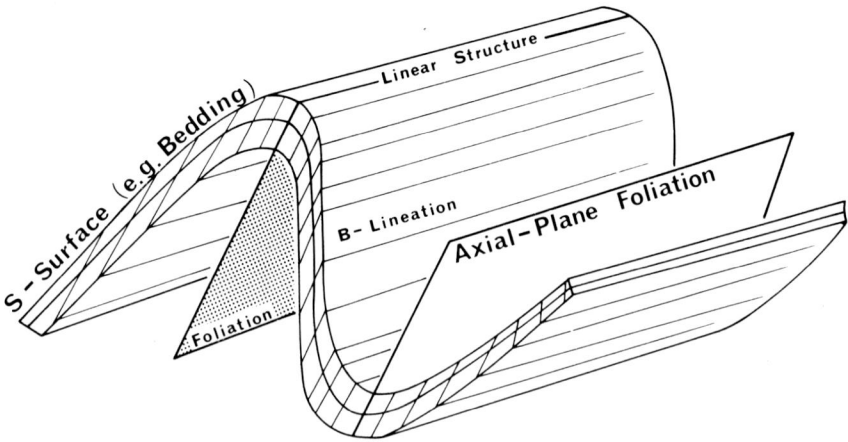

Figure 104. Examples of kinematically-active fabric elements: see text for discussion (Based on Paterson and Weiss, 1961, Fig. 16).

CRYSTALLOGRAPHIC FABRIC ELEMENTS

These elements are not directly visible in objects of any size; they include such features as c-crystallographic axes of quartz or calcite grains which are commonly observed with a universal microscope stage. A rock may, or may not, possess initial subfabrics defined by the preferred orientation of crystallographic elements. Commonly, it is difficult to determine the extent to which an observed subfabric consists of inherited, imposed, or combined components. Whether neocrystallization associated with the strain has completely or partially obliterated initial fabrics is often a debatable question.

Because the crystallographic fabric is reflected by the anisotropy of total-rock physical properties, several indirect methods of studying the strain ellipsoid have sometimes been used. The ellipsoid of anisotropic thermal conductivity was determined by Brinkmann, *et al.* (1961) at different points on a gently folded bed of no appreciable metamorphic grade. The data were obtained by coating thin sections of rock with wax and heating the rock at one point until an ellipse of melted wax formed; the major axes of the ellipses are proportional to the thermal conductivity (cf. Jannettaz, 1884). The longest major axis of the thermal conductance ellipsoid was found to be parallel to the fold B-geometric axis, and a second principal ellipsoid axis also lies in the fold axial plane. The major axis of the elliptical section in the ac-geometric plane of the fold is perpendicular to the axial plane trace at the crest of the fold, but it swings round to be perpendicular to this trace on the fold flanks too (Fig. 105).

Quartz grains in this lithological unit have been deformed into ellipsoids, and their longest axes have a preferred orientation almost parallel to the direction of maximum thermal conductivity. This is a habit orientation; the lattice orientation was not determined.

Magnetic susceptibility has also been used successfully, and Fuller (1964), for example, showed that the orientation of the susceptibility ellipsoid in some Welsh slates coincides with the orientation of the strain ellipsoid.

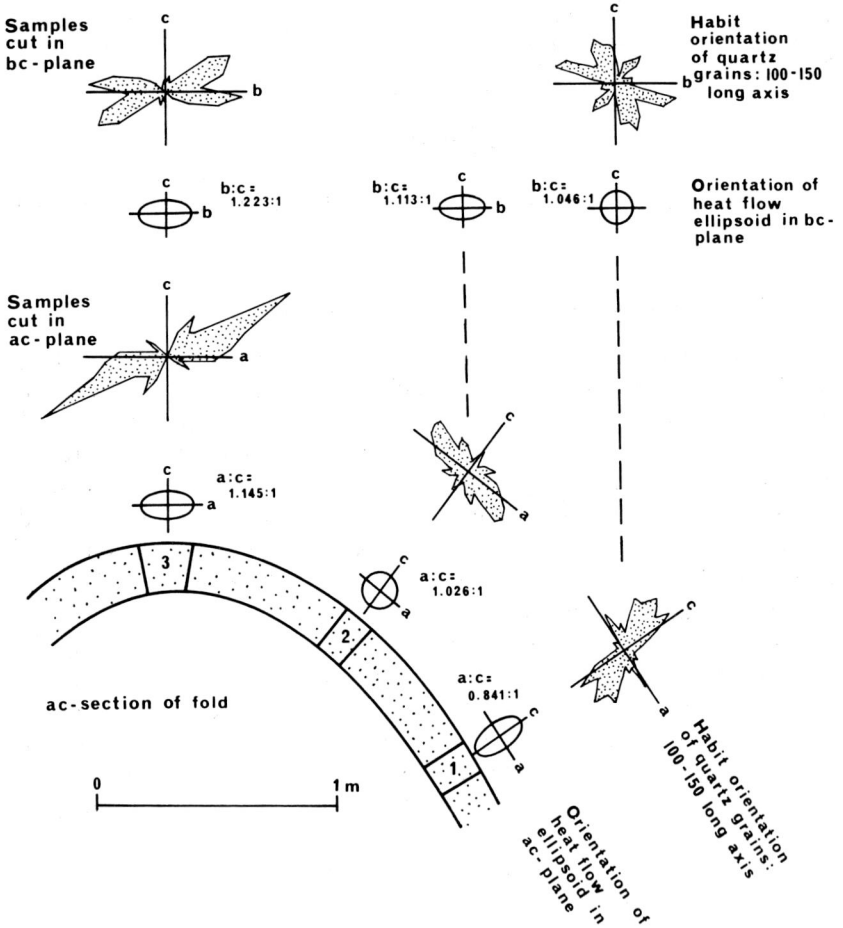

Figure 105. Thermal conductivity within samples of a gently folded sandstone: samples at three positions (1, 2, and 3 on *ac*-section at base of Figure) were studied in *ac*-section (center of Figure) and in *Bc*-section (top of Figure). Differential heat flow (see text for explanation) and habit orientation of the quartz grains are shown for both orientations (Based on data given by Brinkmann, *et al.*, 1961).

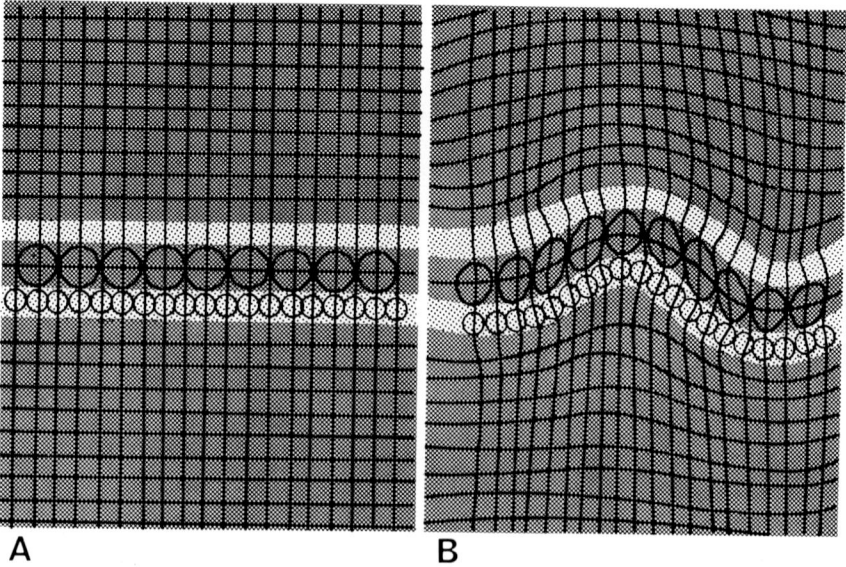

A   B

Figure 106. Variation of strain shown by distorted markers in an experimentally-made fold system. Light gray: stiff rubber layers (6.6 mm. thick) with modulus of rigidity $G = 51.6$ kg/cm^2; Dark gray: soft rubber ($G = 1.27$ kg/cm^2). A: Unstrained. B: Strained as a result of compression parallel to the layering (After Ramberg, 1963B, Fig. 12).

Little is known about the quantitative changes in strain during diagenesis or in the zeolite facies of regional metamorphism. However, it appears from the results of Brinkmann, *et al.* (1961) that definite fabrics begin to develop at a very early stage in the deformation and compaction of sedimentary rocks. In order to obtain an understanding of tectonite development it is imperative that intensive work on actual rocks be undertaken. Useful results can be obtained from experimental work. Ramberg (1963B), for example, used rubber sheets marked with circles to demonstrate the dissimilar strain in different parts of a flexure (Fig. 106). However, before such experiments can have direct application to rock structures, it is necessary to know whether the natural and synthetic strain patterns are similar or dissimilar.

Chapter 6

Simple Types
of Folds

In an earlier chapter some elementary quantitative geometric properties of folds were described. However, several dissimilar types of fold occur, and this chapter is devoted to description and discussion of their principal characteristics. Quantitative work on well-exposed natural examples is urgently needed, and the sequel serves to emphasize that considerable progress remains to be made along these lines.

It is important to recognize that both the geometry of folds and the kinematic picture (movement picture) can be observed and described, but that the kinetic or dynamic forces involved can be inferred only. With the imperfect state of our knowledge of earth processes, it is probable that inferences about the kinetics of a particular fold may be seriously in error. Thus, it is not desirable to use genetic terminology that implies that the kinetic picture is known; it would be preferable to give a geometric or kinematic description rather than to state, for example, that "*Flexure folding,* also known as *true folding,* may result from either compression or a couple" (Billings, 1954, p. 88).

In this chapter the principal varieties recognized are flexural-slip and slip folds — these are recognized on the basis of kinematic factors. Identification of these types is not dependent on assumptions about the kinetic or dynamic factors. The most precise definitions have been erected for folds in rocks that are obvious metamorphic tectonites, because such rocks have received the most quantitative study. However, "tectonite" simply involves componental movements and does not imply a particular grade of metamorphism. Since componental movements appear to occur whenever rocks are folded, and neocrystallization commences with diagenetic changes in newly-accumulated sedimentary rocks, most concepts developed for metamorphic tectonites apply to the majority of fold structures.

130

For simplicity of description, layered rocks with *S*-structures visible in a hand sample are considered initially. However, the concepts outlined below apply equally to compositionally-isotropic rocks (such as massive igneous rocks) and to layered rocks such as sedimentary sequences, gneisses, or flow-banded lavas.

THE TWO PRINCIPAL VARIETIES OF FOLDS

Deformation of layered rocks can be classified according to the relationship of the slip surfaces to the original *S*-structure. Although other types of fold occur (cf. Donath and Parker, 1964), the following are the dominant types in naturally-deformed rocks:

Flexural-slip folds: The dominant slip occurred along original S_0-surfaces. Analogous slip occurs between successive cards if a deck of cards is folded gently. From the definitions of the *a*-, *B*-, and *c*-kinematic axes given in Chapter 4, it follows that the *B*-kinematic axis is parallel to the fold axis $\beta = B$ (Fig. 107).

Slip folds: The dominant slip is on a new set of surfaces (S_1) oblique to the S_0-surfaces and along which movement occurs in preference to S_0.

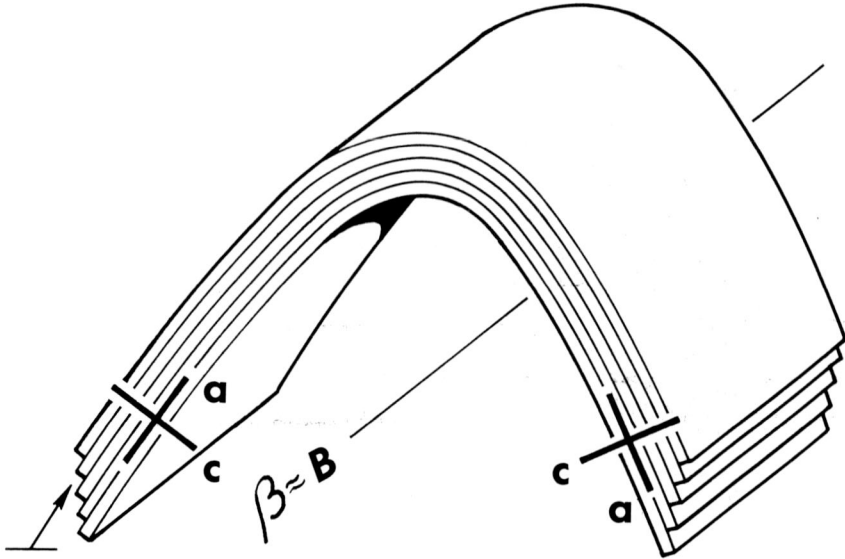

S-Structure along which slip occurs

Figure 107. Diagrammatic monoclinic flexural-slip fold showing the kinematic fold axes. *B* is perpendicular to the *ac*-plane.

Figure 108 illustrates a simple fold caused by slip along a single set of parallel S_1-structures.

Turner (1948, p. 174) suggested flexural-slip folds are the commonest type of fold in metamorphic rocks, although de Sitter (1956A, p. 594) thought that the slip fold is as important as the flexural-slip fold. In fact, flexural-slip folds are probably the commonest type of fold in all rocks, although it is difficult to base such a statement on firm quantitative foundations. Certainly flexural-slip folds are the most noticeable folds, but, as rocks are studied in more intimate detail, it becomes clear that slip folds are more important than was previously recognized. Slip folds commonly occur in compositionally-homogeneous rocks in which prominent S-surfaces are absent; without clearly-visible markers (such as S_0-surfaces) slip-folds can be overlooked easily.

These two types of fold develop distinctive geometric features as shown diagrammatically in Figure 109. Van Hise (1896) was the first to illustrate these geometric forms which he named **parallel folds** and **similar folds**. These descriptive terms refer to the geometry of the S-surfaces. Successive S-surfaces are parallel and concentric in simple parallel folds and flexural-slip folds (Fig. 109A). In similar and slip folds the successive

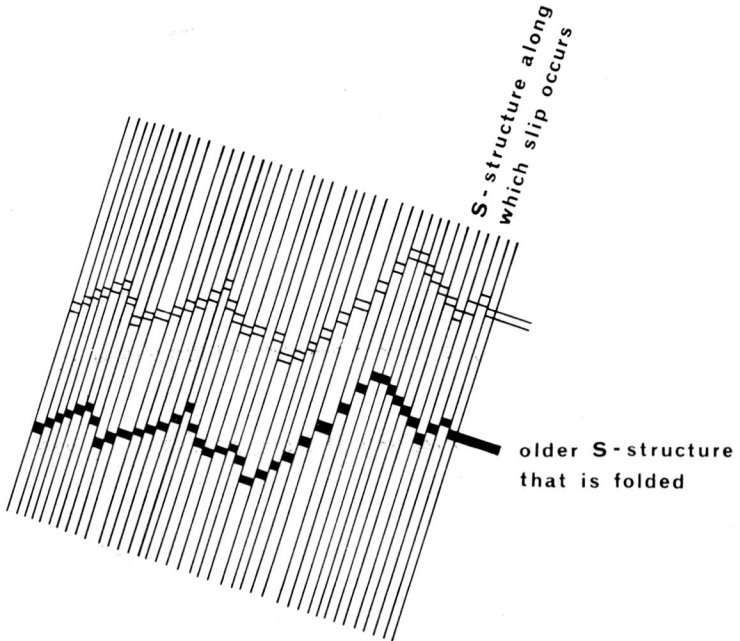

Figure 108. Diagrammatic sketch showing mode of formation of a slip fold.

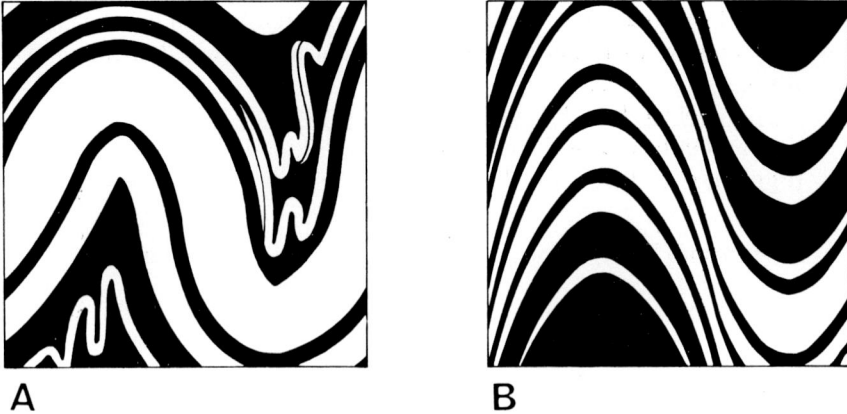

A   B

Figure 109. The geometry of idealized folds as seen in profile. A: Flexural-slip or parallel folds. B: Slip or similar folds (After Ramsay, 1962A, Fig. 1).

S-surfaces have similar profiles, so the fold profile remains essentially the same parallel to the trace of the axial plane. The terms parallel and similar folds are equally desirable to flexural-slip and slip folds, but for consistency the latter pair of terms is employed only.

Rast (1964, p. 177) pointed out that both Van Hise (1896) and Clough (in Gunn, *et al.*, 1897) were interested in the relationships of folds to the associated foliation, but that Van Hise was essentially concerned with fold morphology while Clough used minor folds and their foliations as markers of episodes of folding in deciphering the geological history. Becker (1882) used a mathematical and experimental approach that made him recognize the significance of differential movements along shearing planes during formation of folds. The type of folds that occurs within a particular sequence of rocks depends to a considerable extent on the cohesiveness, competence, and thickness of the successive layers (beds). A complex interplay of flexural-slip and slip folding commonly results when a varied sequence of more and less competent rocks is subjected to deformation.

SLIP FOLDS

For convenience it will be assumed initially that slip occurs on a single set of shear planes (S_1). In this case the fold axis (β) of the slip fold may be coincident with the *B*-kinematic axis so that $\beta = B$, but, with most orientations of S_1, the fold axis β cannot be parallel to *B*. These important relationships are somewhat difficult to comprehend at first. However, the geometry can be visualized easily by manipulation of a deck of cards (e.g.,

IBM cards). A single S_1-structure can be simulated by making the cards slip over one another. Three cases arise according to whether the intersection of S_0 and S_1 is:

(a) parallel to the a-kinematic axis (Fig. 110)
(b) parallel to the B-kinematic axis (Fig. 111)
(c) oblique to the a- and B-kinematic axes (Fig. 112)

A. Before Deformation   **B. After Deformation**

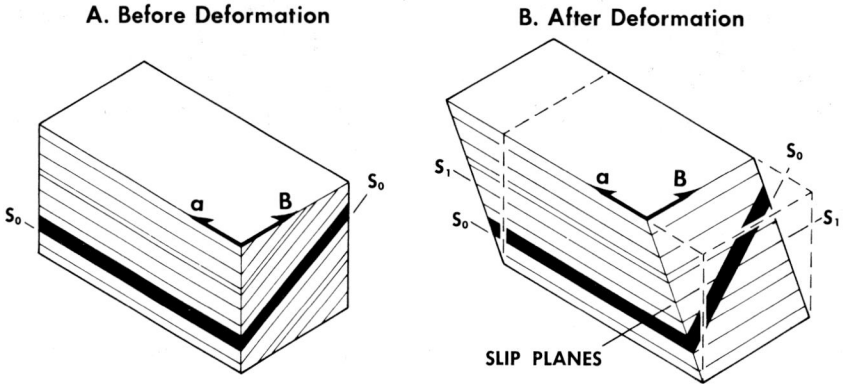

SLIP PLANES

Figure 110. The effect of slip on closely-spaced planes (S_1) when the intersection of S_1 and the bedding (S_0) is parallel to the a-kinematic axis.

Mastery of the solid geometry can be aided by marking S_0 across different decks of cards in these three possible orientations (Figs. 110A, 111A, and 112A). Nonaffine slip parallel to the length of the cards results in three fold relationships (Figs. 110B, 111B, and 112B); the significance of nonaffine was explained in Chapter 5. The geometric and kinematic pictures are shown in Figures 110, 111, and 112.

These relationships can be represented more easily on stereograms. Figure 113 represents a flexural-slip fold; the successive positions of S_0 (S_0^1, S_0^2, S_0^3, . . .) are shown as great circles together with the kinematic a-axis corresponding to each orientation. Because $\beta = B \perp a$, all of the a-axes lie on a great circle. However, in slip folds a has a constant orientation in each domain. In Figure 114 the three different cases referred to above are illustrated. When the intersection of S_0 and S_1 is parallel to a, no flexure of S_0 results (Fig. 114A). In Figure 114B, B is parallel to the intersection of S_0 and S_1 so the great circles representing the successive positions of S_0 (S_0^1, S_0^2, S_0^3, . . .) intersect at $\beta = B$. Finally, when the intersection of S_0 and S_1 is oblique to both a- and B-axes (Fig. 114C), the fold axis (β) does not coincide with the kinematic B-axis.

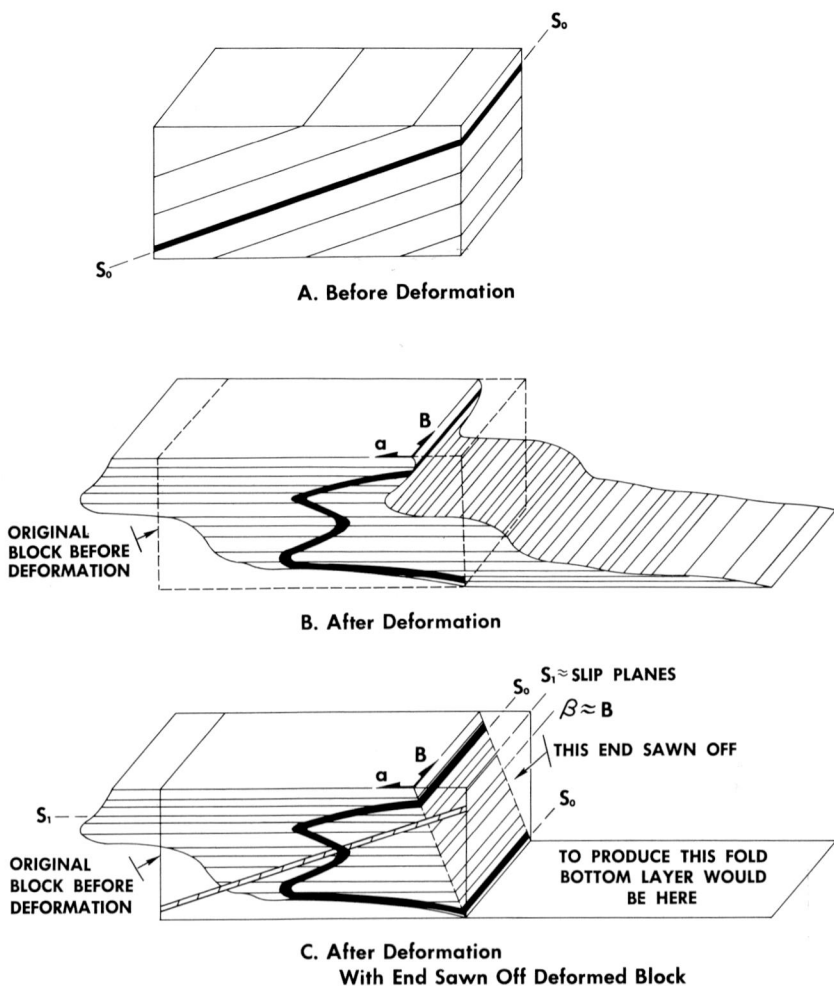

A. Before Deformation

B. After Deformation

C. After Deformation
With End Sawn Off Deformed Block

Figure 111. The geometry of a slip fold where the intersection of S_0 and S_1 is parallel to the B-kinematic axis, so that $\beta = B$. A: Before deformation; S_0 is bedding. B: After deformation by slip parallel to the a-kinematic axis. C: End of block B sawn off to show the geometry after deformation; S_1 are slip planes. The heavy black line (S_0) in C shows the post-deformation geometry of the original bedding plane indicated in block A; the position of the bedding plane (S_0) prior to slip is marked on the front of the original block in C.

Simple experiments with decks of cards show that numerous different types of fold profiles can be produced. Nonaffine transport is more common, because the amount of slip along each S_1-structure is usually unequal; in consequence, any original planar S_0 becomes folded (except

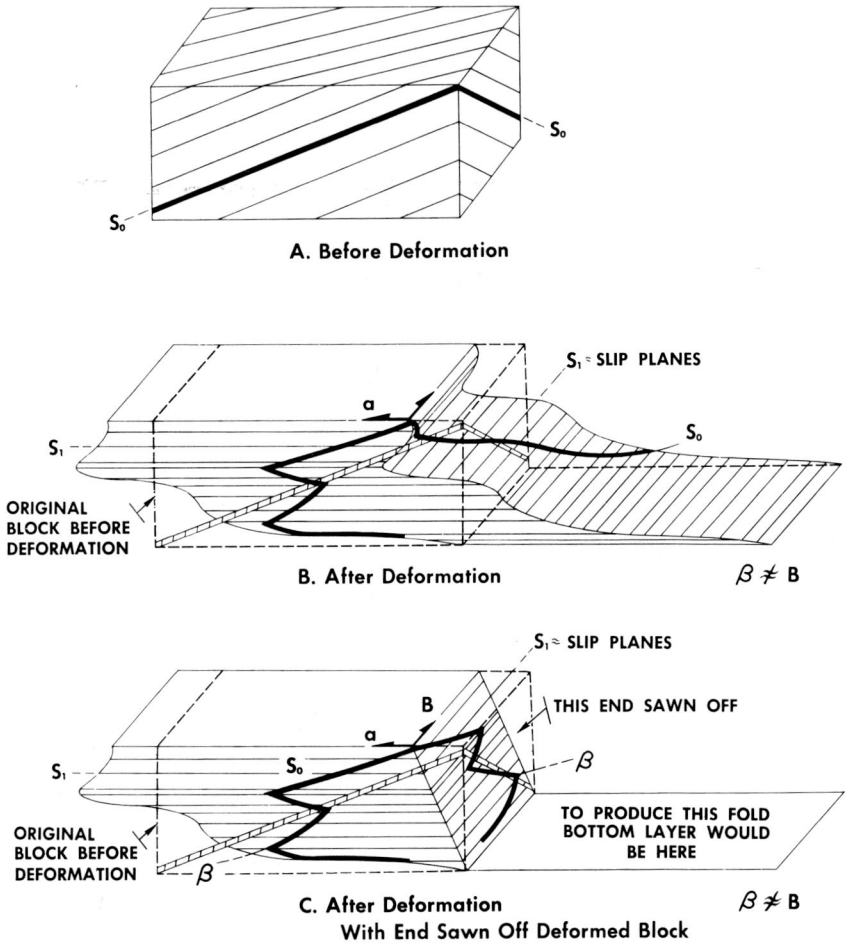

Figure 112. The geometry of a slip fold where the intersection of S_0 and S_1 is oblique to the a- and B-kinematic axes, so that $\beta \neq B$. A: Before deformation; S_0 is bedding. B: After deformation by slip parallel to the a-kinematic axis. C: End of block B sawn off to show the geometry after deformation; S_1 are slip planes. The heavy black line (S_0) in C shows the postdeformation geometry of the original bedding plane indicated in block A; the position of S_0 prior to slip is marked on the front of the original block in C.

when the intersection of S_0 and S_1 is parallel to a). Affine deformation results when each successive card is moved exactly the same distance (and with the same slip direction) with respect to its neighbors. Such movement is affine because an initial straight line marked across the deck of cards remains straight. The affine strain illustrated in Figure 100 should be compared with the nonaffine slip fold from the Stanley Shale, Arkan-

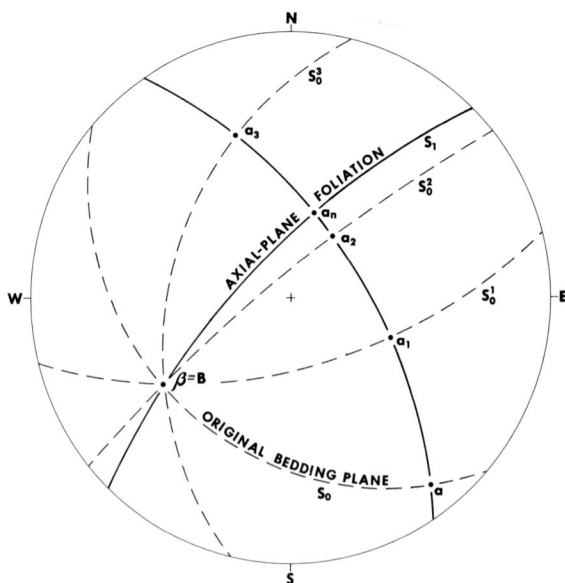

Figure 113. The geometry of a flexural-slip fold shown in stereographic projection.

sas, U.S.A. (Fig. 115), in which slight compositional and color changes define the original bedding (S_0).

Schmidt (1932, pp. 81-87) called flexures which result from nonaffine slip on S-surfaces **Gleitbrett folds;** he thought they develop by translation of relatively rigid undeformed layers along parallel slip surfaces. The slip planes are commonly irregularly-spaced within a domain. When successive S_1-planes are not close, the individual layers of rock parallel to S_1 are not necessarily subject to permanent deformation (i.e., penetrative componental movements did not occur within the layers). Development of Gleitbrett folds is illustrated in Figure 116 which is based on McTaggart's (1960) study of Keno Hill, Yukon Territory, Canada. McTaggart plotted 135 axes of overturned and recumbent folds (Fig. 117A), and showed that 150 intersections of S_0 and foliation (S_1) in Gleitbrett structures (Fig. 117B) have a similar distribution in this area. Slip planes developed in the axial part of the isoclinal folds, and foliation formed parallel to the limbs of the folds (Fig. 118).

During slip folding minor structural or compositional discontinuities (e.g., S_0) may become rotated (nonaffine deformation), although the rock as a whole is not rotated. Similarly, any parts of the rock caught between widely-spaced S_1-surfaces and not affected (deformed) by the penetrative deformation will glide along the slip surfaces without being rotated. By

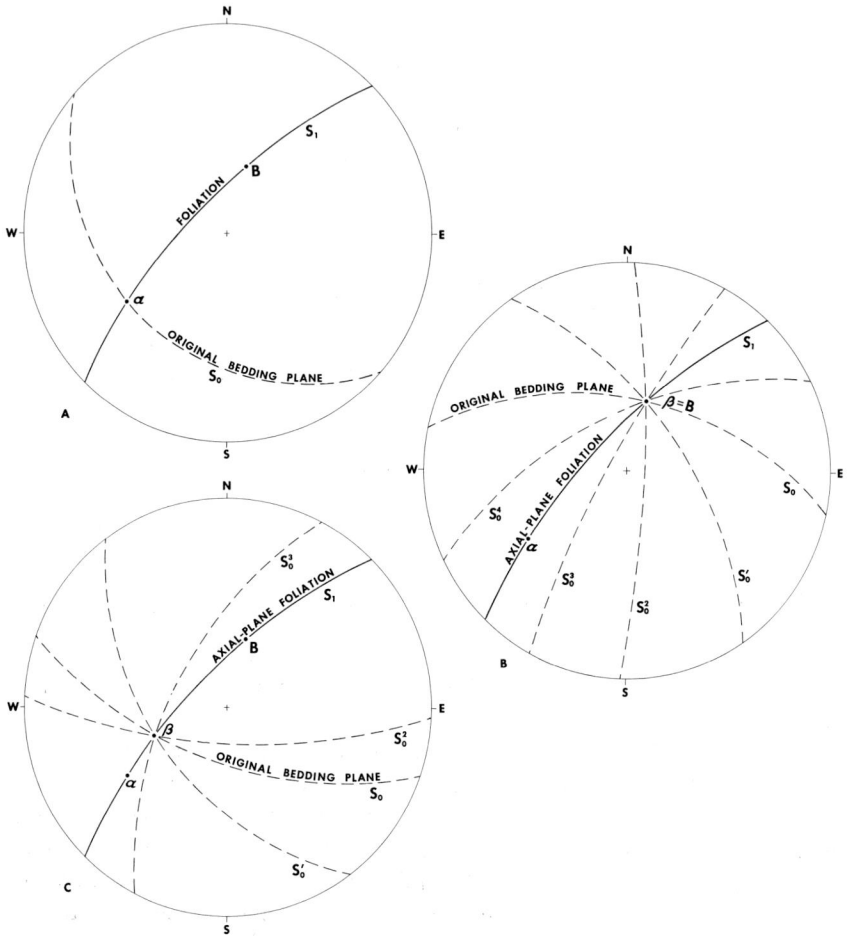

Figure 114. The geometry of slip folds shown in stereographic projection. A: The intersection of S_0 and S_1 is parallel to the a-kinematic axis. B: The intersection of S_0 and S_1 is parallel to the B-kinematic axis. C: The intersection of S_0 and S_1 is oblique to both a and B.

contrast, when flexural-slip folding occurs, every part of the fold is subject to both internal rotation and bodily rotation.

In some lithologies and environments, slip surfaces may be infinitely close together on the scale of observation, so that penetrative deformation results in foliation or schistosity coincident with the slip surfaces (S_1). The genetic term **shear schistosity** (*Scherung-S*) has been applied to such structures.

Above, consideration was limited to slip on a single set of shear planes, but the possibility of simultaneous slip on intersecting planes must

Figure 115. Slip fold in Stanley Shale (Mississippian), southern margin of Ouachita fold belt, Arkansas, U. S. A. Color changes define S_0, and the slip surfaces (S_1) are horizontal in this vertical section.

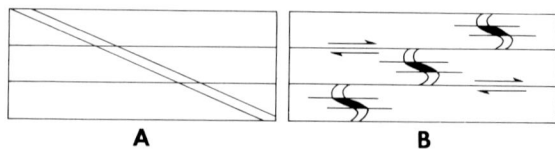

Figure 116. Development of Gleitbrett folds. A and B: Hypothetical examples in which a single bed (in A) is sliced into segments along closely-spaced fractures: in B the black portions have moved forward without apparent strain. C and D: Analogous structures from Keno Hill, Yukon Territory, Canada: the bedding is defined by graphitic layers and XY is the intersection of bedding and foliation in C (After McTaggart, 1960, Fig. 6).

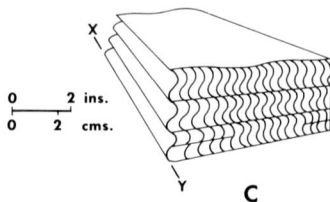

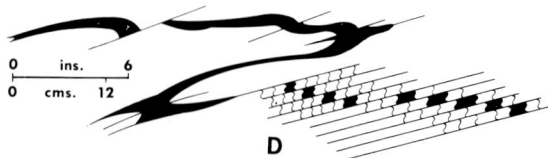

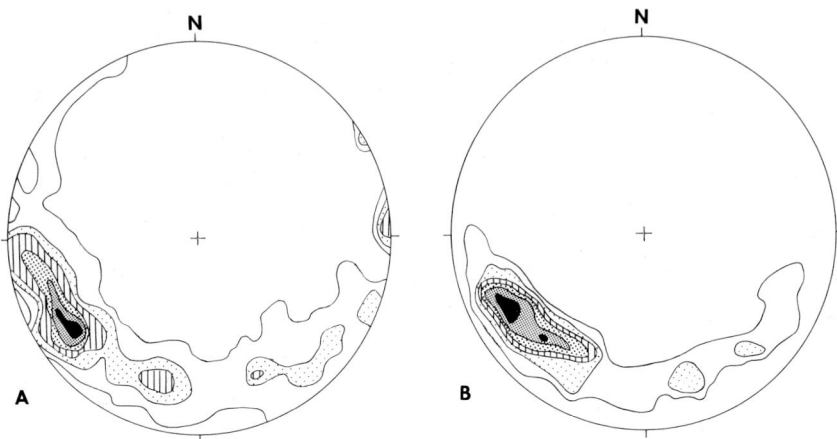

Figure 117. Structure at Keno Hill, Yukon Territory, Canada. A: 135 geometric axes of overturned and recumbent folds; contours at 3, 5, 8, 10, 12, and 15 per cent. B: 150 intersections of S_0 and S_1 in Gleitbrett structures; contours at 2, 4, 6, 8, 10, and 12 per cent (After McTaggart, 1960, Fig. 5).

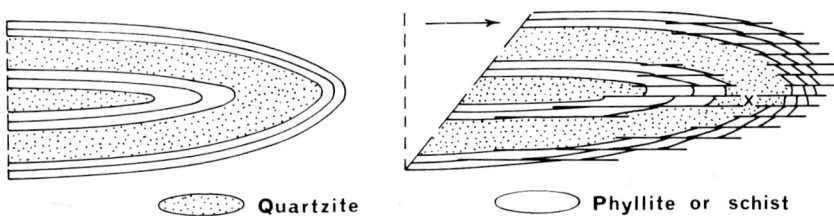

Quartzite   Phyllite or schist

Figure 118. Supposed mode of development of Gleitbrett structure in the axial zone of isoclinal folds, Keno Hill area, Yukon Territory, Canada. Continued deformation produced evenly-spaced foliation planes parallel to the limbs of the fold: each Gleitbrett advanced the same distance with respect to its neighbors. Some of the Gleitbrett (as at X) show folding produced during this movement (After McTaggart, 1960, Fig. 7).

also be borne in mind. Sander (1930, 1934) thought that slip on multiple planes is important in deep-seated metamorphic zones, and simultaneous symmetrical movement on two inclined sets of planar slip surfaces (Fig. 119) was referred to as "flattening" (Plattung). However, it is unlikely that large folds have resulted from deformation of this type.

Carey (1954) considered slip folds at length and placed special emphasis on the time factor in strain phenomena. It can be shown that the total strain within a domain is the sum of the elastic, plastic, firmoviscous, and viscous strain. In geological processes deformative stresses are commonly maintained for enormous periods. Carey claimed that with

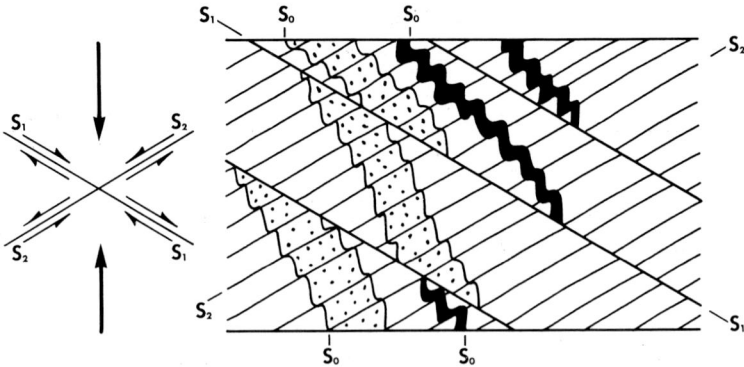

Figure 119. Movement picture for slip on S_2 followed by slip on S_1, in a Tyrolean mica schist: S_0 is bedding (After Turner, 1948, Fig. 41, based on Sander, 1934).

increasing time the elastic, plastic, and firmoviscous[1] components of strain may become unimportant, so that only the viscous property remains. Specifically, Carey thought that all strain except viscous can be neglected when deformation lasts longer than

$$\frac{\text{viscosity}}{\text{rigidity}} \times 10^3 \text{ seconds} = \textbf{rheidity} \text{ of the substance.}$$

Thus, a body of rock under stress for a time longer than the rheidity behaves essentially as a viscous liquid. Such a material was called a **rheid,** and Carey thought, possibly erroneously, that a rheid is a state of matter comparable to solid, liquid, or gas (Fig. 120). The recognition of rheid substances within the crust of the Earth implies that viscous behavior is compatible with crystallinity, and that substantial strength is compatible with widespread flow phenomena. Carey (1954, p. 110) gave compelling evidence to suggest that:

> ... the mantle must be regarded as an elastic solid for all phenomena of shorter duration than six months, and for such duration viscous flow phenomena in the mantle may be neglected. For phenomena lasting longer than 10,000 years the mantle behaves as a fluid except for a limited transitional field above 700 km. We may conclude then that for the time range of tectonic loads ... the mantle is a rheid ... and that, if it is treated mathematically as a fluid, no significant error will be made.

[1]Strain in an ideal **elastic** body is directly proportional to the stress, and it is completely and instantaneously reversible, in the sense that when the applied forces are removed the body immediately recovers its original shape and volume so that there is no permanent strain. When subjected to stress above a critical value an ideal **plastic** solid flows continuously under a constant stress, and the strain is permanent (irreversible). In a **firmoviscous** body elastic (reversible) strain is accomplished over a period of time, rather than instantaneously as in an ideal elastic body.

THE STATES OF MATTER

According to Carey (1954)

TEMPERATURE					
BOILING POINT	SOLID	Rheidity X 10^{-4}	Transitional State	GAS	COHESION ZERO ABOVE THIS LINE
MELTING POINT				LIQUID	STRENGTH ZERO ABOVE THIS LINE
				RHEID	FINITE STRENGTH

DURATION OF LOADING ⟶

| Deformation according to Elastic equation | Deformation according to Elastic and Fluid equations | Deformation according to Fluid equation |

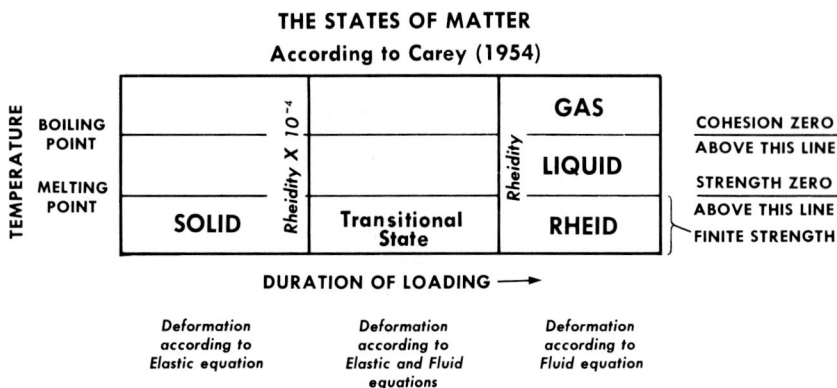

Figure 120. The states of matter according to Carey (After Carey, 1954, Fig. 1).

Carey (1954, p. 115) suggested that rheid behavior often extends to relatively shallow depths in the crust, and that migmatized and granitized zones temporarily have very low rheidity. Carey (1954, 1962A) considered slip folds to be characteristic of rocks that have been rheids, and he demonstrated the importance of nonaffine slip when halite, gypsum, serpentinites, mudstones, ice, and crystalline schists deform as rheids. Rocks that have been rheids are not restricted to orogenic zones, but the ubiquitous folds of Archean basement areas may reflect deep burial, and consequent high temperatures and lower rheidities that result in abundant "rheid folding" without appreciable crustal shortening. The theoretical geometry of the resulting folds is shown in Figure 121; this can be

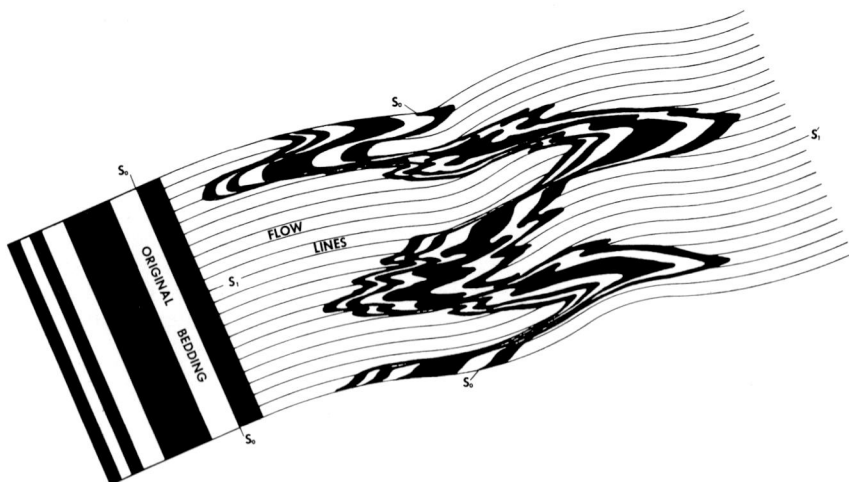

Figure 121. Geometry of theoretical example of rheid folding of bedding (S_0) (After Carey, 1954, Fig. 16).

Figure 122. Section through North Mine, Broken Hill, Australia, showing rheid folds (After Garretty, 1943, Fig. 1, and Carey, 1954, Fig. 17).

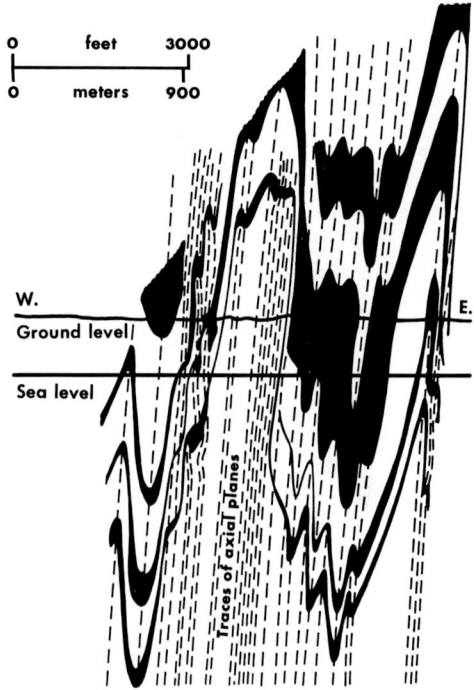

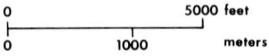

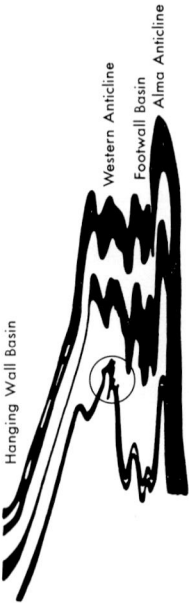

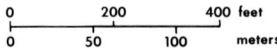

Figure 123. Rheid folds in crystalline schists, South Mine, Broken Hill, Australia; profile on right is an enlargement of area in circle of left-hand profile (After Gustafson, *et al.,* 1950, and Carey, 1954, Fig. 9).

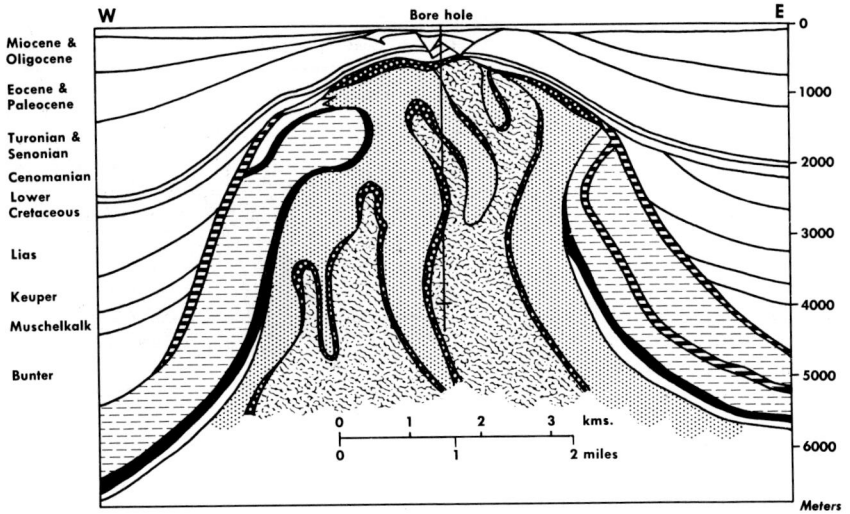

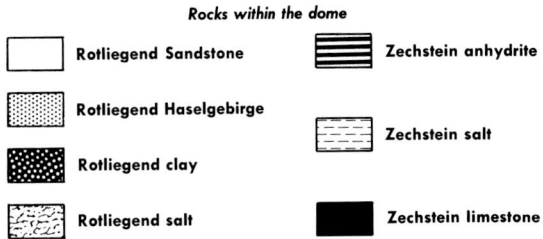

Figure 124. Schematic section showing the style of folding in a salt dome near Heide, NW Germany (After Bentz, 1949, Fig. 1).

compared with a cross section through the crystalline schists of North Mine, Broken Hill, Australia (Fig. 122). Carey (1954, pp. 93 ff.) pointed out that:

> ... this type of deformation involves no change in volume, no flow across the flow laminae, very little total displacement, and no horizontal shortening at all. ... It is a notable characteristic of rheid folding that even minor fold axes and inflection axes persist for very great distances.

He drew attention to the analogous folds (Fig. 123) in the crystalline schists of South Mine, Broken Hill, previously illustrated by Gustafson *et al.* (1950), and in a salt dome near Heide, NW Germany (Fig. 124) described by Bentz (1949).

Similar examples abound, and the rheid concept may well have wide application. An excellent example from the hornblende gneiss west of MacArtney Lake, Astrolabe Lake area, Saskatchewan, Canada (Fig. 125), was illustrated by F. J. Johnson (1961). Slip folds can commonly be seen in thin sections; Figure 126 shows a metamorphosed limestone described by de Sitter (1956A) in which the foliation (slip) planes have

Figure 125. Slip folds in hornblende gneisses, W of MacArtney Lake, Astrolabe Lake area, Saskatchewan, Canada (From photograph reproduced by F. J. Johnson, 1961, Plate 4B).

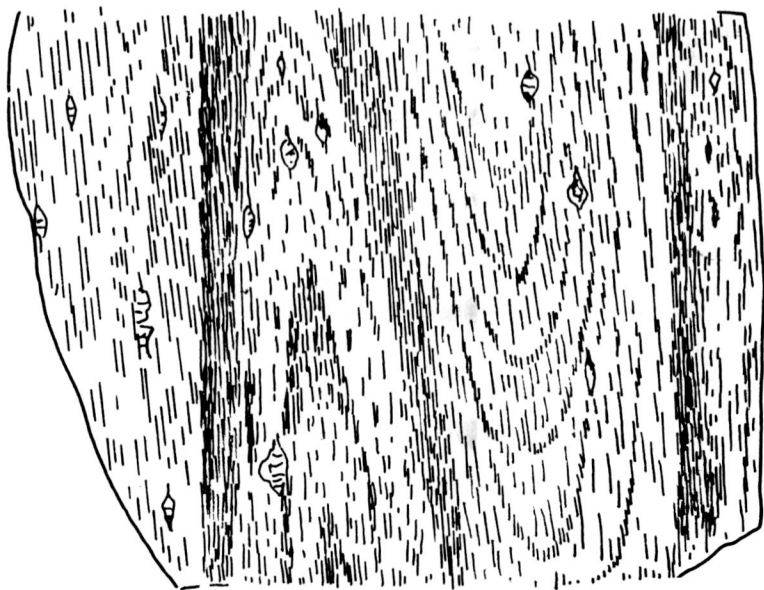

Figure 126. Slip fold in metamorphosed limestone: the slip planes have largely disappeared, although still emphasized by shadow zones around garnet porphyroblasts. Drawing of a thin section (After de Sitter, 1956A, Fig. 18).

disappeared, although the shadow zones at each end of the garnet porphyroblasts show that the movement was parallel to the axial plane.

Figure 127 shows the deformed medial moraines in the zone of viscous shear between two ice masses that advanced at different rates within a glacier; Carey pointed out that such folds are analogous to the slip folds of metamorphic rocks that deformed as rheids.

In slip folds developed in rheids it is supposed that different folds can advance at dissimilar rates, with a consequent change in the position of the main zone of shearing between the advancing folds (Carey, 1954, p. 95). Figure 128 illustrates this concept, and the complexities that could result from acceleration of the development of one lobe (A) at the expense of another (B). The possible reasons for the differential slip in actual rocks are discussed later in this chapter.

It would seem likely that there is a complete gradation between slip folds developed in rheids and Gleitbrett folds. However, although non-affine slip has geometric features similar to fluid laminar flow, it is inappropriate to refer to folds formed by such slip as "flow folds," where the analogy is with the flow of fluids. Flow in fluids may be laminar or turbulent. Bain (1931) described flow folds from the marbles between Phillipsburg, Quebec, Canada, and Pittsfield, Massachusetts, U.S.A., but his description of these complex structures suggests that they are slip folds.

Figure 127. Medial moraines of a glacier deformed in the zone of viscous shear between two ice masses that advanced at different rates: these structures have been interpreted as rheid folds of the glacier ice (After Carey, 1954, Fig. 13, who adapted a photograph by B. Washburn).

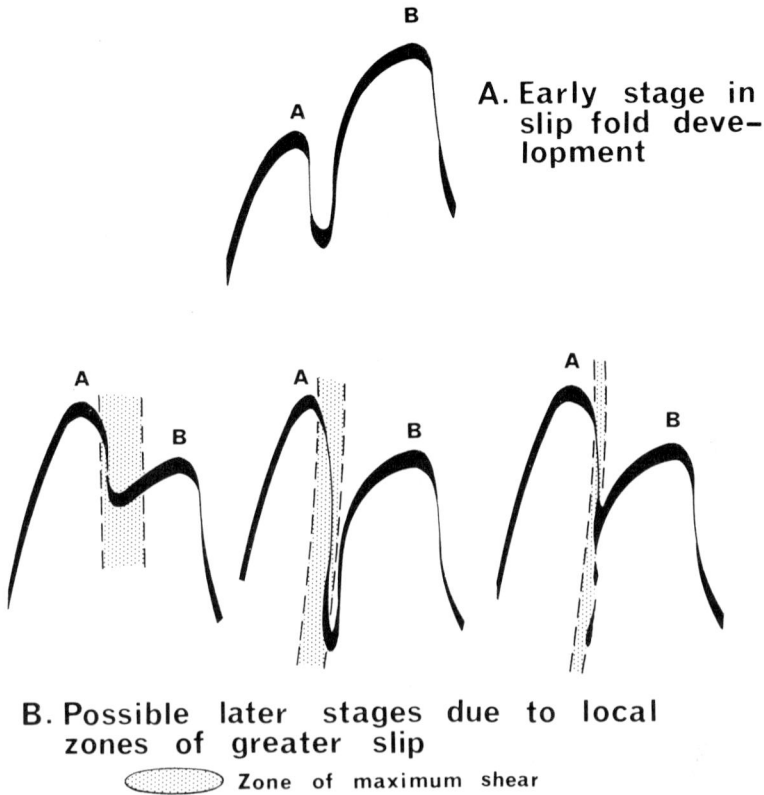

A. Early stage in slip fold development

B. Possible later stages due to local zones of greater slip

⬭ Zone of maximum shear

Figure 128. Slip folds developed in a rheid, showing the geometric effects of differential advance of two fold lobes (A and B). See text for additional explanation (After Carey, 1954, Fig. 20).

FLEXURAL-SLIP FOLDS

Most rocks which contain flexural-slip folds were not mechanically isotropic prior to deformation, so that progressive bending has promoted slip along the surfaces of discontinuity. These discontinuities (S-surfaces) rotate during the folding, and the essential geometry of flexural-slip folds was outlined in Chapter 2. Bedding planes, foliations, or schistosities are the commonest S-surfaces along which slip occurs. The magnitude of the slip is commonly not easy to measure because movement was parallel to the S-surfaces (Fig. 129B).

In a detailed study of an intimately-laminated series of pelitic, semipelitic, and psammitic metasedimentary rocks from Australia, Hills (1945) established the amount of bedding-plane slip by study of the shearing of

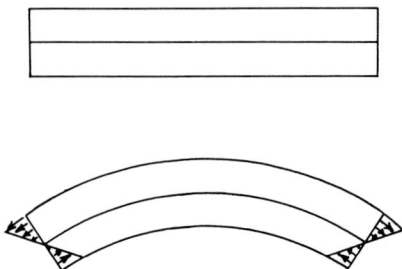

Figure 129. A laminated block before and after flexural-slip folding: slip occurs parallel to the laminae (*S*-structures) (After Stočes and White, 1935, Figs. 227a and b).

minute quartz veins that cross the bedding in the more sandy layers. The same quartz veinlets continue into the argillaceous layers (now slates), but there, although distorted, these veins maintain their continuity across the bedding planes. Thus, although mass movement occurred within the slate layers, no slip parallel to their stratification (S_0) took place. Faint color-banding parallel to bedding within the slates enabled slip folding to be seen clearly. Hence, bedding slip in the sandstones and mass flowage in the argillaceous rocks both accompanied the earlier stages of folding. At a later stage, slip on the bedding planes continued in the sandstones, while the incompetent rocks yielded by slip which produced slip folds in the slates (see Figs. 130 and 131).

Hills (1945, p. 54) suggested that the spacing of shear planes in each bed of an alternating sequence of pelitic and psammitic laminae is directly related to the internal friction within each rock unit. As a result, the thicker the sandy layer the wider the spacing of the planes of slip (Fig. 132); this is a relationship observed commonly (e.g., Balk, 1936, p. 707; de Sitter, 1954).

In banded rocks of this type, there is an obvious and intimate intermixing of deformation by slip and flexural-slip folding. Many other examples of this widespread phenomenon might be cited. Balk (1936, p. 709) showed that structures similar to those described by Scholtz (1930, 1931, 1932) and Kienow (1934) from the Variscan Mountains, Germany, occur in New York State, U.S.A., and that in both areas ". . . the plastic deformation of the slaty rocks may be accomplished by shearing rather than by folding, relatively competent strata continuing to bend during periods when less competent beds yield by minute shear." De Sitter (1954) also described in detail an intimately-interlayered shale and silt sequence in the Ordovician sericite schists near the Rio Lladorre, Spanish Pyrenees (Fig. 133). The pelitic rocks were deformed by slip folding,

Figure 130 (facing page). Flexural-slip fold in laminated sandstones and slates, Mitta Mitta River, Mitta Mitta, Victoria, Australia. Note the obvious slip planes parallel to the axial surface in the fine-grained member. The fold is cut by quartz veinlets. Scale is 1 cm. (From photograph reproduced by Hills, 1945, Fig. 3).

Figure 131. Portion of top right-hand corner of Figure 130 enlarged to show slip folds in laminated slate. Scale is 1 cm. (From photograph reproduced by Hills, 1945, Fig. 5).

and their foliation is parallel to the axial planes; the coarser-grained laminae show slip along bedding as a result of flexural-slip folding; the two mechanisms of deformation appear to have acted simultaneously and to have had the same origin (de Sitter, 1956A, p. 595). Knill (1960A) also described the interplay of these two types of folding in the Dalradian

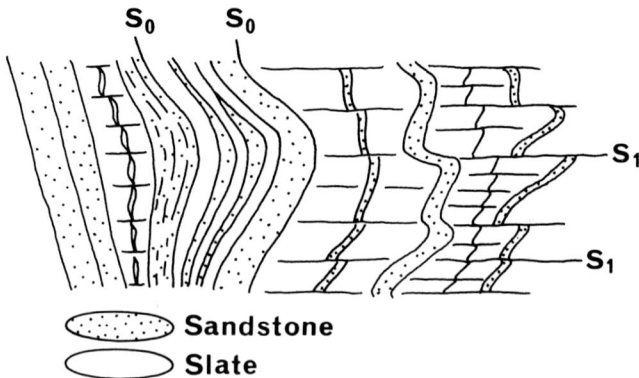

Figure 132. Variation of fold geometry and shear plane spacing in various lithologic units of a slate-sandstone sequence: S_0 = bedding, S_1 = shear planes (Based on Hills, 1945, Fig. 6).

metasedimentary rocks of the Craignish-Kilmelfort district, Argyllshire, Scotland; see also Mendelsohn (1959) and Evans (1963, p. 72).

When more competent and incompetent rocks alternate, as in many sedimentary sequences, folding is commonly accompanied by formation of miniature folds or crenulations within some of the units: these minor structures are called **parasitic** or **drag folds,** and although they are more common on the flanks of the folds, small folds are sometimes found around the closures too (Fig. 134). The gross fold pattern appears to be determined by the competent layers in sequences of this type (cf. Williams, 1961, p. 317).

Conjugate pairs of flexural-slip folds were described by Johnson (1956A) from the Moine Thrust Zone in the Lochcarron and Coulin Forest areas, Wester Ross, Scotland (Fig. 135); the intersection of the conjugate axial planes is parallel to each of the B-fold axes (cf. examples from Nova Scotia, Canada, described by Fyson, 1964).

Ramsay (1962B) maintained that conjugate folds commonly:

(1) are associated in space and time with fault and thrust planes (although also developed over several square kilometers in the surrounding area) and joints;

(2) are formed in thinly-bedded (i.e., less than 10 cm.) or laminated rocks during the later phases of orogenic deformation (Fig. 136); plastic deformation of competent rocks is involved (Hills, 1963B);

(3) have monoclinic, or even triclinic, symmetry because the axial plane intersection is oblique to the B-axes, which are themselves inclined (Fig. 137); the orthorhombic symmetry reported by Johnson (1956A) was stated to be a special case;

(4) appear to be produced by movements on one or more shear surfaces that are usually (but not invariably) inclined at 45° or less to the principal stress axis; and

Figure 133. Quartz-sericite schist near an anticlinal hinge in Ordovician rocks close to the Rio Lladorre, Spanish Pyrenees. Scales are both 3 cm. A: Polished surface of hand sample. B: Thin section showing enlargement of part of A (From photographs reproduced by de Sitter, 1954, Plate 1).

(5) have many similarities with the joint drags described by Flinn (1952) and Knill (1961) (Chapter 8).

Several types of secondary *S*-structures tend to form during development of flexural-slip folds. For example, shear fractures ("fracture cleavage") develop in the more competent rocks, and slight movement along each fracture can account for a considerable total displacement. Although subparallel to the axial surface of the fold, these fractures commonly diverge away from the fold closures (Fig. 138).

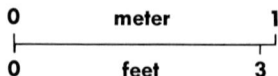

Figure 134. "Drag folds" in thin layer of quartzite embedded in mica schist sandwiched between folded strata of quartzo-feldspathic gneiss, Terningen quadrangle, Norway (After photograph in Ramberg, 1963A, Plate VII).

Planes of easy splitting develop in the less competent (e.g., more pelitic) members of the sequence; these planes of fissility comprise the so-called axial-plane foliation or "slaty cleavage." Although broadly parallel to the fold axial surfaces, these foliation planes tend to converge toward the fold closures (Figs. 138 and 139). Locally, other relationships may be found. For example, Fyson (1962, p. 216) noted that in Devonian argillites, from near Plymouth, SW England, axial-plane foliation converges gently away from the cores of folds. In associated silty or calcare-

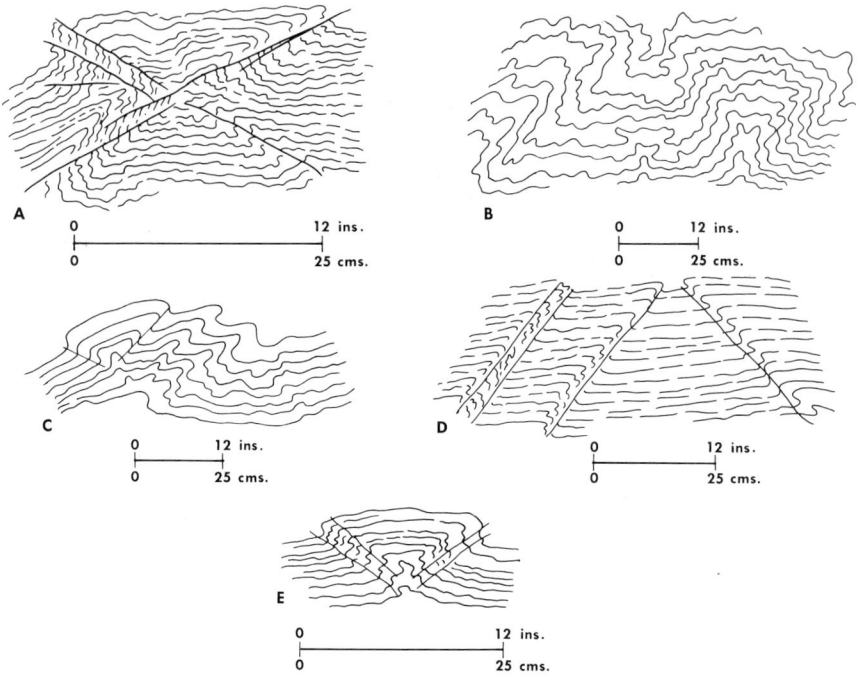

Figure 135. Profiles drawn from *ac*-joint faces of conjugate folds in the Moine Thrust Zone, Lochcarron and Coulin Forest, Wester Ross, Scotland. A and B: Lewisian mylonite, Lochcarron. C: Moine rocks, Coulin. D and E: Inverted Torridonian rocks, Lochcarron (After Johnson, 1956A, Fig. 1).

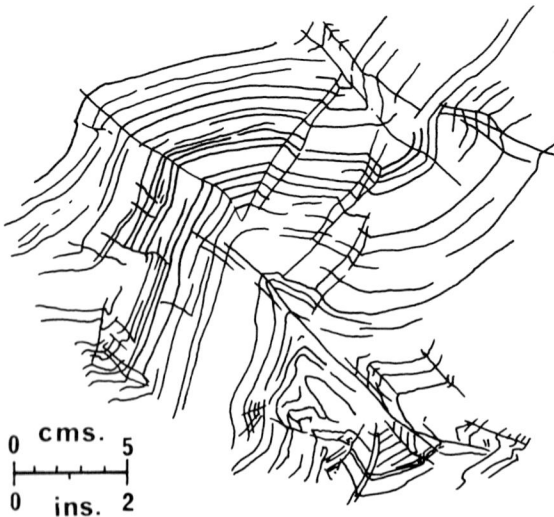

Figure 136. Conjugate folds in mylonitized quartzo-feldspathic Lewisian gneiss at base of the Tarskavaig nappe, Sleat, Isle of Skye, Scotland (After Ramsay, 1962B, Fig. 1).

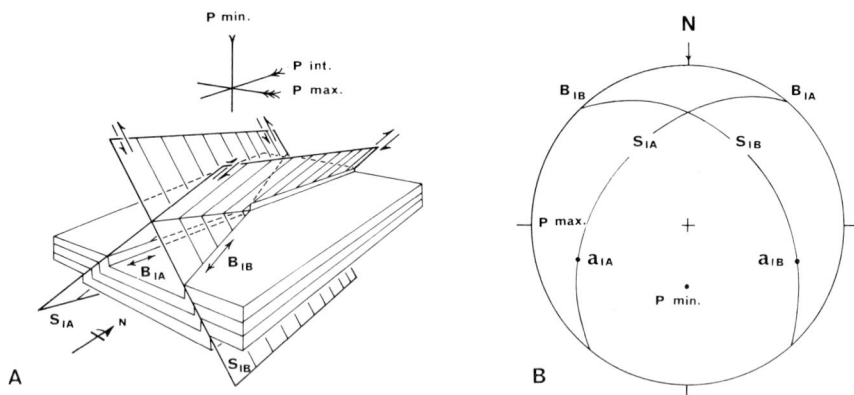

Figure 137. Geometry of a conjugate fold system with monoclinic symmetry that developed in an S_0 initially horizontal. The three stress components are $P_{max} > P_{int} > P_{min}$. Monoclinic structures develop when P_{max} is contained in the initial S_0, and P_{min} is not normal to S_0; structures with lower symmetry (triclinic) form when P_{max} is not contained in the initial S_0. A: Diagram showing the two shear surfaces (S_{1A} and S_{1B}) along which the rocks fail, and the two sets of fold axes B_{1A} and B_{1B}. B: Stereographic projection of the structures in A; P_{int} lies at the intersection of the great circles S_{1A} and S_{1B} (After Ramsay, 1962B, Figs. 4 and 5).

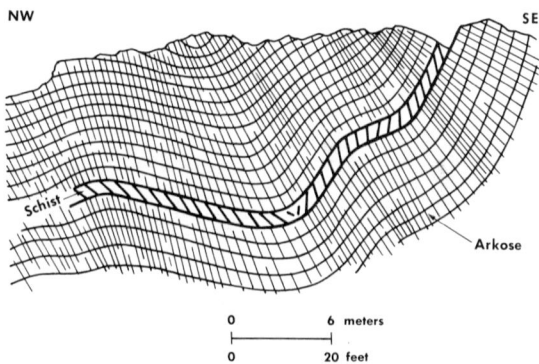

Figure 138. Fan of shear fractures in a syncline of Torridonian arkoses, Isle of Islay, W Scotland (After Wilson, 1961, Fig. 24).

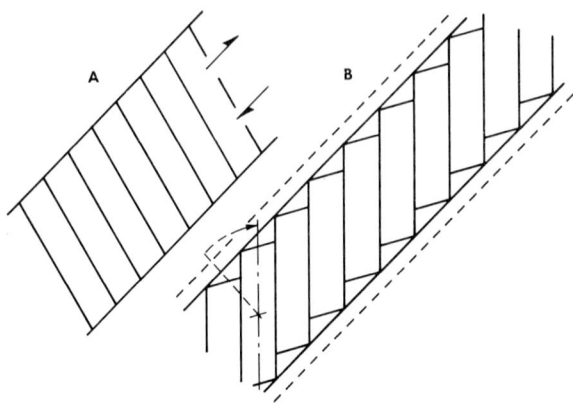

Figure 139. The spacing of shear fractures: thinning and extension of the fractured bed (from A to B) by rotation of the fracture slices during continuing folding (After Wilson, 1961, Fig. 25).

ous beds the foliation diverges away from the fold cores. J. L. M. Lambert (1959) noted similar relationships in Devonian rocks farther west in Cornwall.

When very thin incompetent layers (e.g., clay bands) are intercalated between thick competent beds, slip parallel to the bedding during flexural-slip folding commonly produces a foliation within the less competent rocks parallel to these slip planes (Fig. 140). This **bedding foliation** is quite widely developed in folded unmetamorphosed sedimentary rocks and in those metasedimentary rocks that have only been subjected to low-grade regional metamorphism. Despite the commonness of this structure few specific references have been made to it in the literature. Knill (1960A; 1960B) described bedding foliation in the Dalradian metasedimentary rocks of the Craignish-Kilmelfort district, Argyllshire, Scotland. D. C. Knill and J. L. Knill (1961) recorded bedding foliation in similar rocks from Rosguill, Donegal, Ireland.

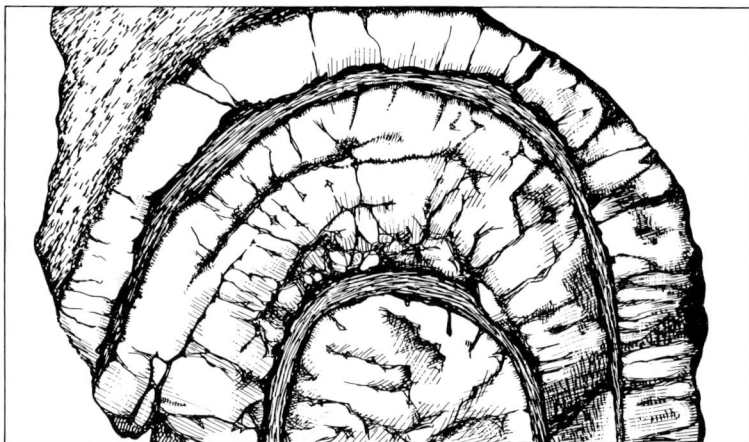

Figure 140. Flexural-slip fold in aplitic layers separated by graphitic schist, Passau, Germany (After photograph by Sander, 1930, Fig. 113).

SLIP AND FLEXURAL-SLIP FOLDS CONTRASTED

Slip and flexural-slip folds result from totally different movement pictures, and the folds themselves have distinctive features. Knopf (Knopf and Ingerson, 1938, pp. 160-161) listed several criteria for distinguishing between these flexures (see also Turner, 1948, p. 174); the three most significant criteria are:

(1) The distance between particular S_0-surfaces in slip folds remains constant when measured along the foliation (S_1) (Fig. 109B); in simple

flexural-slip folds the distance between particular S_0-surfaces is the same if measured perpendicular to the layers (S_0) (Fig. 109A).

(2) Slip folds occur only when all the beds are incompetent, so the folds show no relation to relative competence of the beds (layers). In flexural-slip folding competent layers develop folds of large amplitude and

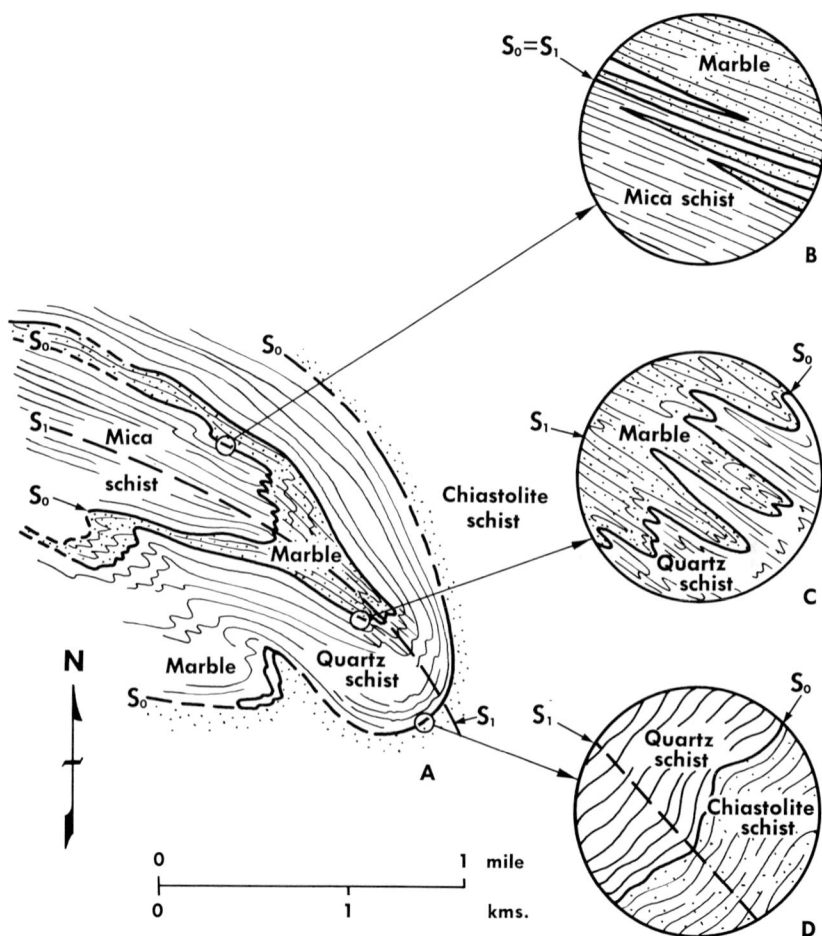

Figure 141. Transposition of lithological contacts in the Kernville Series, Lake Isabella, California, U. S. A.   A: Map of area. B: Detail along one fold limb lithological contact S_0 transposed into foliation S_1. C: Detail showing folding of S_0 with S_1 as axial-plane foliation. D: Detail showing weak folding of S_0 with S_1 as axial plane only (After Weiss in Turner and Weiss, 1963, Fig. 5-15).

wave length, while incompetent layers commonly develop smaller folds (Fig. 141).

(3) Knopf suggested that during flexural-slip folding arcuate "spaces" commonly develop between successive competent layers; as deformation proceeds such potential "spaces" become filled with incompetent material that migrates plastically, or with precipitates of the more mobile constituents (cf. sweat veins of quartz, calcite, etc.; see Fig. 142). This criterion is not often of use, although it is important in a few special cases. For example, Goguel (1962, p. 194) referred to the plastic behavior of coals, and to the thickening of coal beds in fold hinges which is often of importance in mining.

Folds tend to show subtle departures from such idealized models, so that these criteria require some qualifications in the light of recent studies and hypotheses.

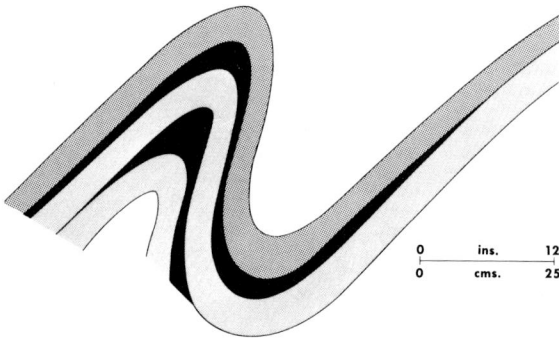

Figure 142. Diagrammatic flexural-slip fold profile showing potential "spaces" filled with incompetent material (solid black) between competent members (stippled).

Ramsay (1962A) carefully measured the thickness of individual layers in three flexural-slip fold sequences, and discovered that there is a systematic departure from uniform thickness. He measured the thickness normal to bedding (t) and parallel to the axial surface (T) on the fold profiles. As was to be expected, T shows minima at the fold closures and maxima in the fold limbs. However, it was somewhat surprising to find that t has reverse relationships with maximum values at the fold hinges. These results were based on a banded silicified siltstone (Fig. 143), a Cretaceous limestone and marl assemblage from the French Jura (Fig. 144), and a sample of Moine schist from Loch Hourn, western Scotland, comprising alternating bands of feldspathic quartzite, micaceous quartzite, and mica schist (Fig. 145) It is important to determine whether this pattern of t-variability is general in all flexural-slip folds, but measurements for additional folds are not available at this time.

Ramsay (1962A) suggested that these results reflect both thickening of the beds in the hinges and thinning in the fold limbs. Both de Sitter

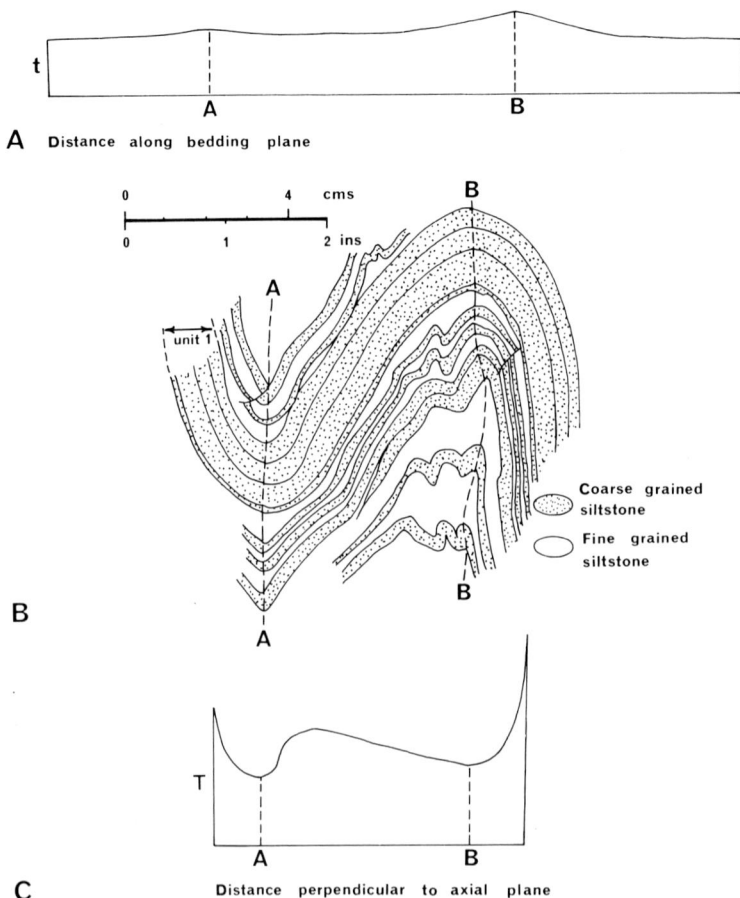

Figure 143. Flexural-slip fold in a siltstone; axial surfaces of the main folds (A and B) are shown by dashed lines. A: Thickness (*t*) of unit 1 measured normal to the bedding planes. B: Fold profiles. C: Thickness (*T*) of unit 1 measured parallel to the axial plane (After Ramsay, 1962A, Fig. 2).

(1958) and Ramsay (1962A) attributed such modifications of bed thickness to plastic deformation resulting from contraction of the rock parallel to the principal compressive stress and concomitant expansion in the plane normal to the compression. Ramsay followed de Sitter (1958, p. 283) in calling this process **flattening;** this term is not necessarily to be equated with the Plattung of Sander (1930) which has been translated as flattening (p. 140).

Several workers have considered that fold profiles are affected by flattening (in the de Sitter-Ramsay sense), although at the present time,

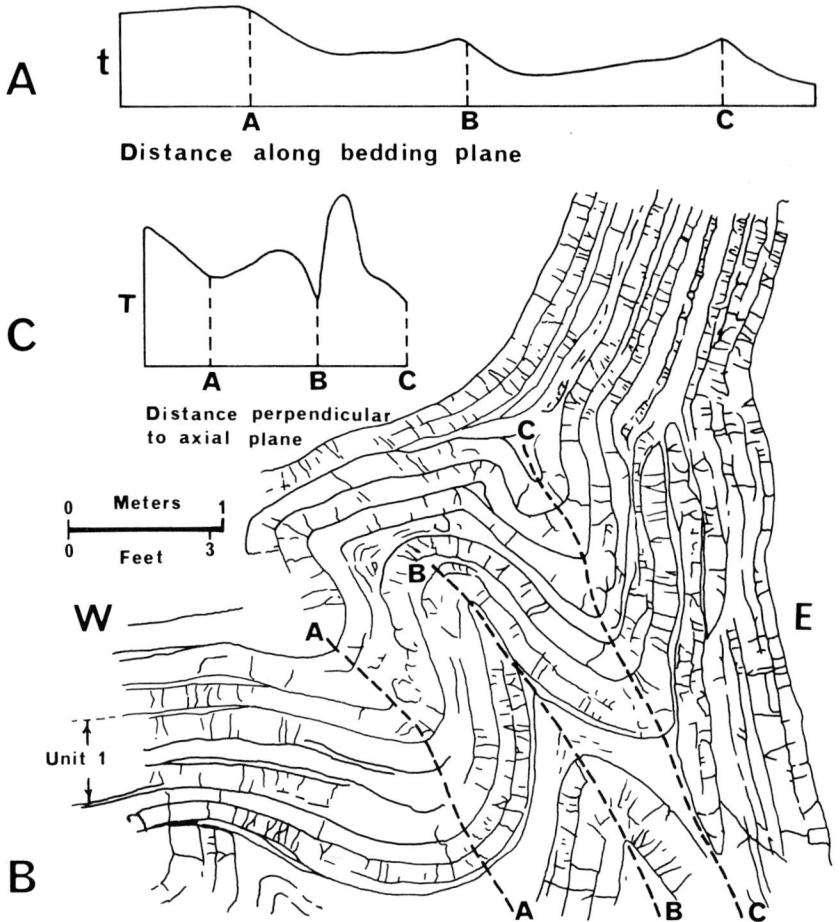

Figure 144. Flexural-slip folds in a Cretaceous limestone-marl assemblage, Val du Fier, French Jura. The fold profile (B) shows the limestone bed (Unit 1) measured as a basis for graphs A and C. A: Graph of thickness (t) of Unit 1 measured normal to the bedding planes. C: Graph of thickness (T) of Unit 1 measured parallel to the fold axial planes (Based on Ramsay, 1962A, Fig. 3).

there is disagreement about the sequence of events that commonly occurs. Voll (1953) thought the texture, density, and mechanical properties of the clays, marls, and arenaceous rocks of the Tertiary Molasse *Cyrena* beds of Bavaria, Germany, are not products of diagenetic processes, but of tectonic plastic deformation of the unconsolidated rocks; this strain was absorbed in a regional reduction of pore volume. Voll thought folding occurred later, but that because of the compression, constructive unfolding of the existing rocks cannot yield the original

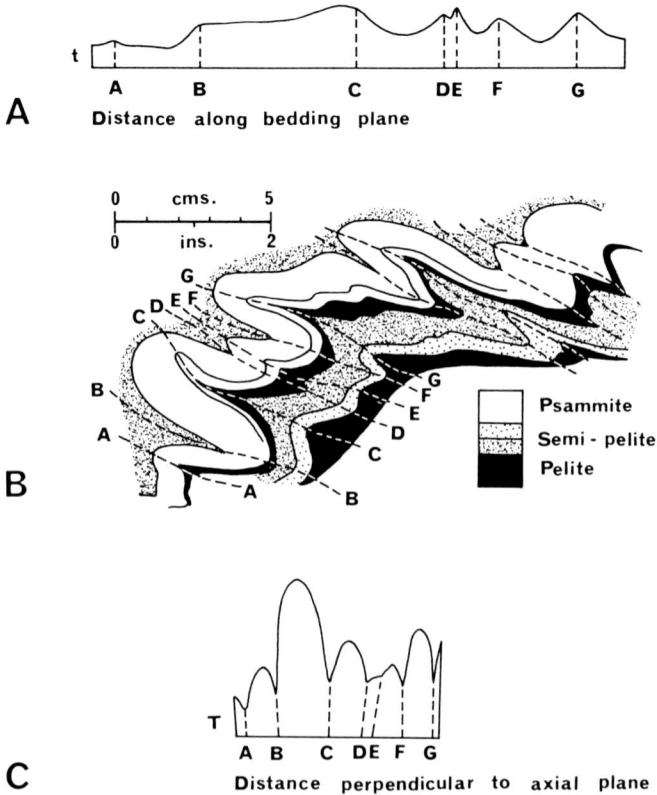

Figure 145. Flattened flexural-slip folds in Moine metasedimentary rocks, Loch Hourn, Northern Highlands, Scotland. A: Thickness (t) of the psammite unit measured normal to S_0. B: Fold profile with axial surfaces marked. C: Thickness (T) of the psammite measured parallel to the axial plane (After Ramsay, 1962A, Fig. 4).

length of the strata. For the Swiss Molasse Basin and for the Carboniferous rocks of the Ruhr, Germany, Breddin (1958A, 1958B) concluded that the plastic deformation was considerably greater than the compression caused by slightly earlier (older) folding. However, Voll (1960, pp. 540-1) claimed that in both of these areas the folding is younger than an early compressive event comparable to that studied by him in Bavaria.

In ideal slip folds (Fig. 109B) T should be a constant for each layer, but careful examination shows that this is not the case in the gneiss illustrated in Figure 146. Ramsay (1962A) measured T in several different lithologic bands of a Moine schist from the Scottish Highlands (Fig. 147) and demonstrated that both maximum and minimum values coincide with

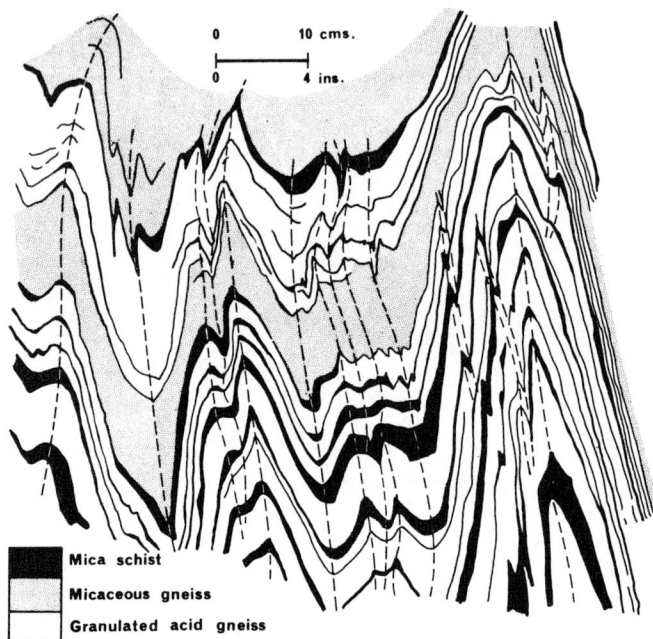

Figure 146. Profile of an almost ideal slip (similar) fold system in deformed gneiss, lower Pennine nappes, Switzerland (After Ramsay, 1962A, Fig. 12).

fold closures. This pattern is quite dissimilar to that associated with flexural-slip folds, and, although overall flattening may have affected these slip folds, this alone cannot account for both maximum and minimum T-values at the closures.

An explanation for the T-values for Figure 147 is not immediately available. This is, in part, due to the fact that neither experimental nor theoretical studies provide a convincing explanation of the mechanism of slip folding. Although intersecting shear planes have sometimes been invoked, it is generally believed that thin slices of rock (microlithons) slipped parallel to the axial-plane foliation in response to compressional stress normal to the axial-plane. This concept was supported by the studies of deformed ooliths, fossils, etc. by Cloos (1947A), de Sitter

Figure 147. Small slip (similar) folds in Moine schists, Scottish Highlands. A: Fold profiles with axial surfaces marked by broken lines. B: Thickness (T) variability of the units measured parallel to the axial surfaces (After Ramsay, 1962A, Figs. 10 and 11).

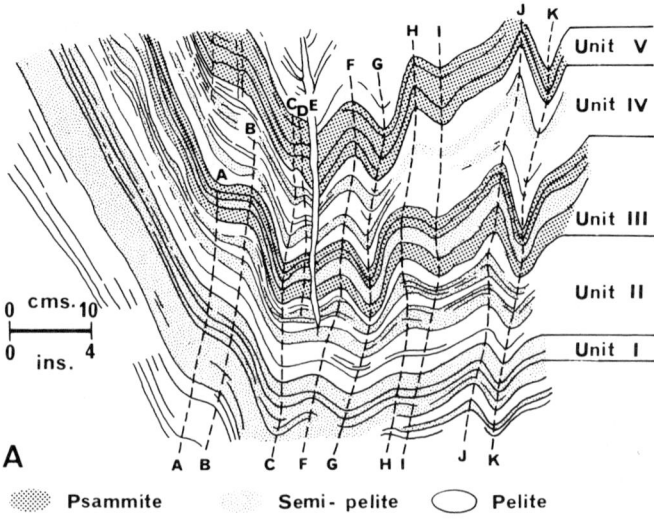

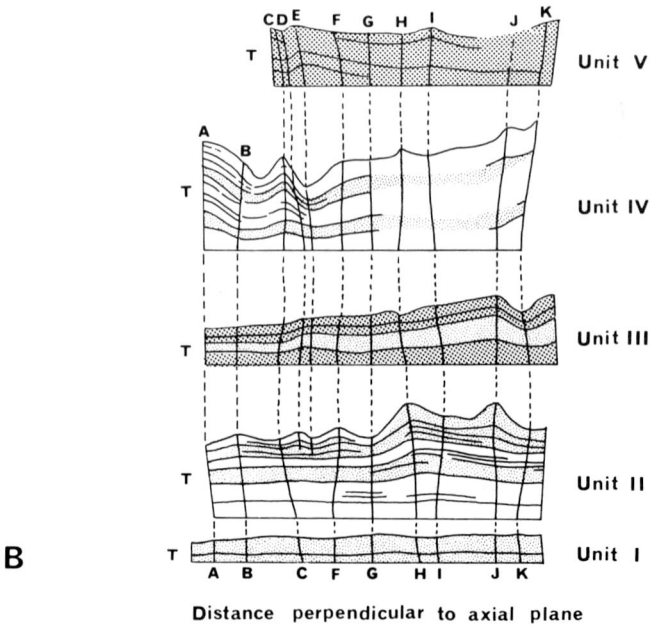

Distance perpendicular to axial plane

(1954), Breddin (1956A), and Voll (1960) and it is essentially the mechanism of viscous strain in rheids advocated by Carey (1954). Horizontal stress affecting a horizontally-bedded compositionally-isotropic sequence of sedimentary rocks could effect plastic flattening, but it is not immediately obvious that this would fold the bedding planes (S_0) or promote movement of microlithons at right angles to the compressive stress. Gonzalez-Bonorino (1960) vigorously attacked the inadequacy of current concepts used to explain the formation of slip folds.

Ramsay (1962A) proposed a possible mechanism of slip folding. Commonly the axial surfaces of slip folds are not strictly parallel (e.g., Fig. 147), and Ramsay supposed that this reflects unequal flattening in the rock. Uneven flattening would result in unequal extension in the plane normal to compression and, by hypothesis, the development of folds in the bedding surfaces of adjacent layers. If the unequal flattening is a three-dimensional phenomenon, neither the fold axes nor the axial surfaces of adjacent antiforms and synforms will be strictly parallel. The axial surfaces would be closest where the amount of flattening is greatest. Although plausible, this model must be tested with a considerable amount of exact quantitative data before it can be accepted fully. Ramsay drew some quantitative support for his hypothesis from the unequal flattening measured by Cloos (1947A) in deformed ooliths from different parts of the South Mountain structure, Maryland, U.S.A.

PARASITIC OR "DRAG" FOLDS

Van Hise and Leith (1911, Fig. 12, p. 123) apparently introduced the term **drag fold** for the Soudan Formation, Vermilion Iron district, Minnesota, U.S.A.; they stated:

> The pressure has been so great as to produce all varieties of folds, including isoclinal and fan-shaped. . . . A common type of

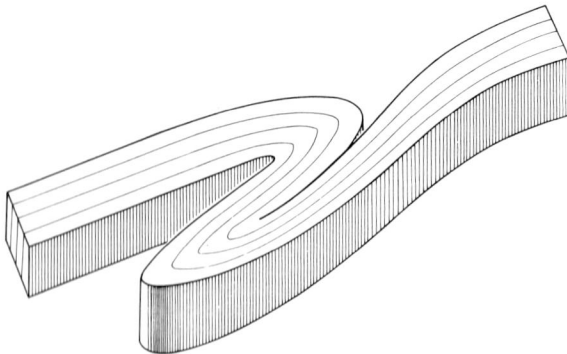

Figure 148. Diagram to illustrate folds of "drag" type in the Vermilion Ranges, Minnesota, U. S. A. (After Van Hise and Leith, 1911, Fig. 12).

fold is a drag fold . . . by which the formation becomes locally buckled along an axis lying in any direction in the plane of bedding. This type of folding, while leaving great local complexity, does not destroy the general attitude or trend of the bed.

This statement was illustrated by the drawing reproduced as Figure 148. In the same Lake Superior Region monograph Van Hise, Leith, and Mead portrayed minor troughs of drag type which they stated could occur in any portion of the iron-bearing member in the Menominee district of Michigan, U.S.A. (Fig. 149).

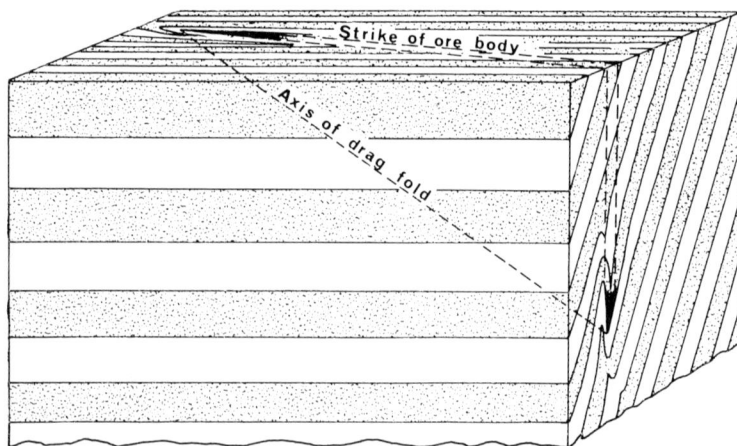

Figure 149. Sketch by Van Hise, Leith, and Mead to show the geometry of a drag fold within a monoclinal succession of the Menominee district, Michigan, U. S. A. The strike of the ore body (along closure of the synclinal fold) is slightly inclined to the strike of the strata (After Van Hise and Leith, 1911, Fig. 49).

On a regional scale any sequence of sedimentary rocks includes some more and some less competent units, and regional deformation tends to involve flexural-slip folding. Characteristically, when a region is folded multitudes of folds of every possible size develop. The thick competent units are thrown into broad synclines and anticlines, while the thinly-bedded and other less-competent units develop smaller folds. The relatively small folds have commonly been called *"drag folds"* because it was thought that slip of the competent members dragged the intervening less-competent units into small folds. Knopf (Knopf and Ingerson, 1938), de Sitter (1956B, p. 226; 1957, p. 57), Williams (1961), and others, recognized that the genetic implications of the term "drag fold" are open to question; **parasitic fold** (de Sitter, 1958) is the preferable term.

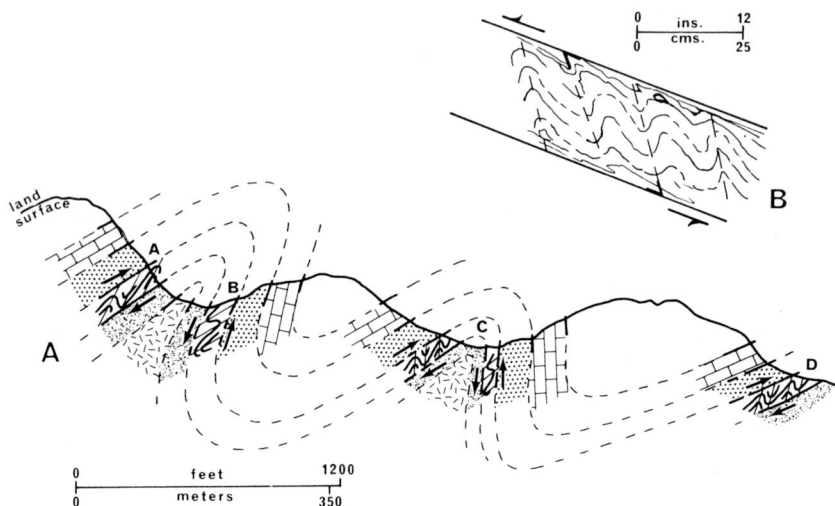

Figure 150. The geometry of parasitic folds shown in a diagrammatic profile of flexural-slip folds in a varied sequence of rocks. A: Regional profile showing how individual parasitic folds can be used to locate anticlines and synclines. B: Enlargement of a single parasitic fold.

Figure 150 represents diagrammatically the profile of a large series of folded metasedimentary rocks; the thinly-laminated units between the more massive competent layers abound with parasitic folds. Empirically it has commonly been found that isolated outcrops of parasitic folds in poorly-exposed country permit the major structures to be identified accurately, because the axial surfaces of parasitic folds are almost invariably subparallel to the axial surfaces of the major folds. This relationship is most useful in field work, but it would be misleading to assume that it invariably holds. The validity of the underlying assumption must be verified in each case; in areas of polyphase folding more complex relationships often obtain, but these may not be obvious in poorly exposed country (Chapters 10 and 11).

In practice, the axial surfaces of the small folds tend to diverge away from the major fold closures; they make an increasingly large angle with the axial surface of the main fold as the distance from the hinge line increases. The acute angle between the axial surfaces of parasitic folds and the slip surface between the major lithologic units always points in the direction of an arrow indicating the local direction of the slip (Fig. 150).

The folded layers on the concave side of a bedding slip surface move away from the hinge line. Consequently, in the majority of cases, it is easy to identify the location of antiformal and synformal structures. The varying orientation of the axial surfaces of parasitic folds demonstrates that the stress did not have a uniform orientation throughout the entire fold system.

Many parasitic folds, like those illustrated in Figure 151, are slip, rather than flexural-slip, folds; the parasitic slip folds in Figure 151 are details from folds described by de Sitter (1954) and illustrated in Figure 133. Williams (1961, p. 317) noted that such tiny slip folds simulate the folds in a rheid described by Carey (1954). Such parasitic structures cannot be the product of drag by competent beds on an incompetent unit (contrast Billings, 1954, Fig. 30, p. 45), and they do not closely resemble the "drag folds" originally described by Van Hise, Leith, and Mead (Van Hise and Leith, 1911).

The planes of slip associated with parasitic slip folds may be penetrative or nonpenetrative when seen in a hand sample. In the latter case the folds have almost undeformed Gleitbrett (cf. Schmidt, 1932, pp. 89-92) or microlithons (de Sitter, 1954, p. 433), separated by planes of slip. When the deformation is penetrative on the scale of the parasitic fold, neocrystallization is penetrative, axial-plane foliation is prominent, and the original bedding may be virtually obliterated.

Although it has often been stated that "drag folds" form in incompetent layers sandwiched between competent beds, parasitic folds commonly, though not exclusively, occur in relatively thin competent layers in

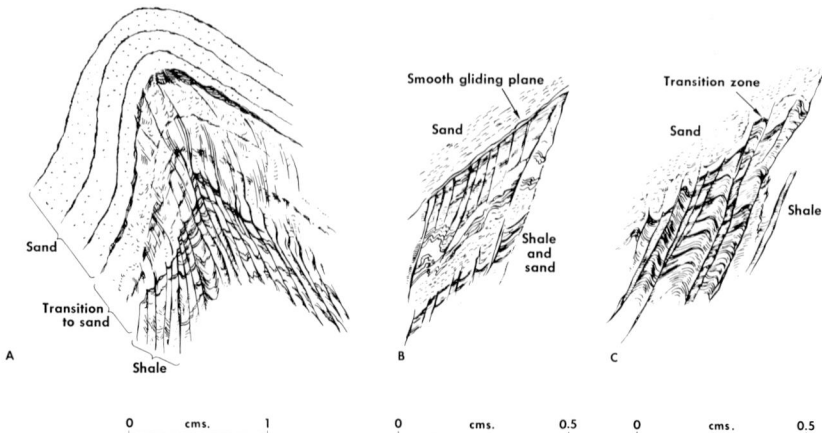

Figure 151. Details from small folds in alternating sandstone and slate sequence, Rio Lladorre, Spanish Pyrenees, illustrated in Figure 133. A: Flexural-slip fold in sandstone and crenulation foliation in slate members. When sandstone-slate interface is sharp, there is a shear along boundary (B), but with transitional interface, the crenulation foliation continues into the sandstone (C) (After de Sitter, 1954, Figs. 2 and 3).

the flanks of larger folds. Characteristically, such thin folded layers are enclosed between incompetent beds, which in turn are interlayered with thicker competent beds.

The fact that parasitic folds often do not die out towards the axial zone of major folds (cf. Figs. 134 and 152) militates against the drag hypothesis of formation. De Sitter (1958) and Ramberg (1963A) discussed possible alternative modes of formation, but additional quantitative information about the nature, variability, and spatial distribution in different environments is required before any final conclusions can be reached.

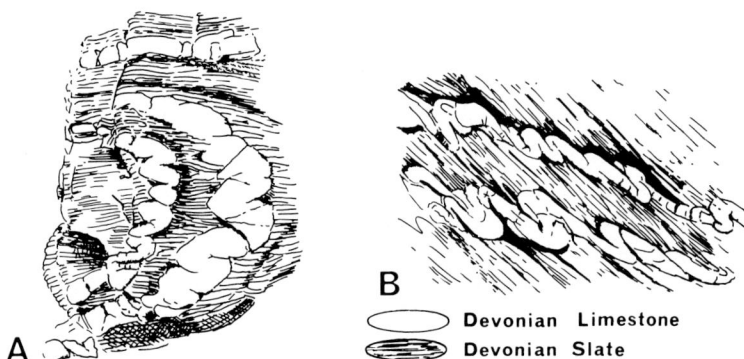

Figure 152. Parasitic folds of limestone members in Devonian slates, N Devonshire, England, drawn from photographs. A: Parasitic folds around an anticlinal fold closure. B: Strongly compressed parasitic folds (After de Sitter, 1958, Figs. 7 and 9).

Although "parasitic fold" is preferable to "drag fold," it would probably be more appropriate to use **minor slip fold** and **minor flexural-slip fold** for these structures.[2] Billings (1954, p. 82) distinguished between minor folds and drag folds in the following words:

> In many regions all the strata have essentially the same competency . . . large masses of rock move past one another, and the strata are thrown into *minor folds*. The minor folds usually bear the same relation to the major folds as do drag folds, and they are even called drag folds by many geologists.

However, it seems to be difficult and unnecessary to justify this distinction.

---

[2] Stočes and White (1935, p. 151) used drag fold for a completely different type of structure from those described here.

FLEXURE FOLDS

In addition to slip folds and flexural-slip folds most authorities recognize **flexure folds.** In laboratory experiments flexure folding has importance when the unit being flexed is compositionally isotropic. Such folding appears to be of limited significance in nature because most rocks subjected to folding are compositionally and mechanically anisotropic (cf. Turner, 1948, p. 172).

Stočes and White (1935, pp. 141 ff.) gave a useful introduction to the problems of tension and compression during flexure folding. When a plastic body, such as a stick of plasticine (modeling clay), is bent, the convex side is subjected to tension and is lengthened; the concave side is shortened and the intense accumulation of material often results in formation of small secondary folds (Fig. 153A). When a stick of plasticine is flexed in this manner the stress in the fold closure tends to be relieved by expansion and contraction parallel to the fold axis, but release

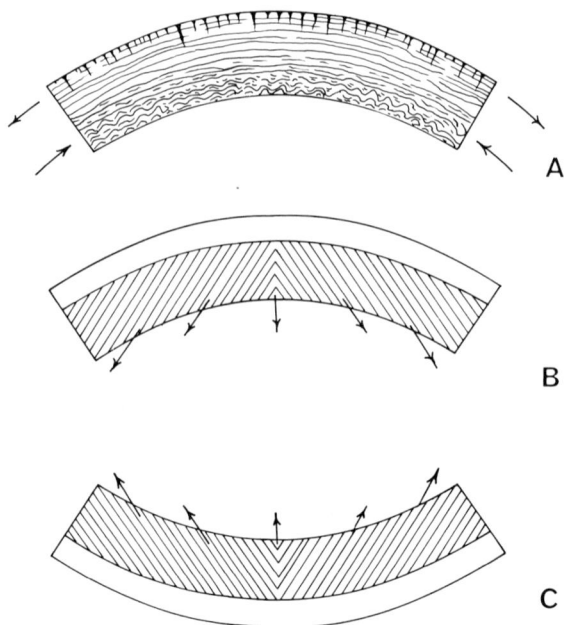

Figure 153. Diagram to show some effects that accompany flexure folding. A: A plastic body with small folds forming on the concave side and tension gashes on the convex side. B and C: A brittle body breaking along subparallel fractures that act as planes of movement (After Stočes and White, 1935, Figs. 227c, d, and e).

in this manner is probably unusual in natural rocks. A brittle substance, instead of being attenuated and lengthened, breaks into small fragments which move relative to one another. The shortening or lengthening can take place by very small movements along numerous subparallel planes (Figs. 153B, 153C).

SPATIAL RELATIONSHIPS OF SLIP AND FLEXURAL-SLIP FOLDS

Many geologists would agree that it is a matter of general observation that the Earth's nonmetamorphosed mantle of sedimentary rocks is not characterized by slip folds, but by an abundance of flexural-slip folds. These rocks contrast strongly with the schists and gneisses of the metamorphosed infrastructure, in which slip folds have an important role. Between these extremes sericite schists, phyllites, slates, and similar rocks from the lower grades of regional metamorphism commonly exhibit slip folds, while intercalated more-competent members (siltstones, sandstones, etc.) show flexural-slip features.

The style and intensity of folding can sometimes be related to specific granitic intrusions. For example, Hietanen (1961B) found that the intensity of deformation and metamorphism increases towards the northern contact of the Idaho Batholith, Idaho, U.S.A. The style of folds and the distribution of axial-plane foliation in lithologically-similar units are uniform in each metamorphic zone, but change from one zone to another. Hietanen considered that these changes are similar to those in more extensive areas of regional metamorphism, and, on a still larger scale, at successively deeper levels in the Earth's crust.

Fourmarier (1932, 1953A, 1953B, 1953C) suggested that the boundary between the upper layers of the Earth's crust characterized by flexural-slip folds and the zone of slip folds below occurs at roughly 4,000-6,000 meters under the surface; this conclusion was based on studies in many different parts of the world.

Thus, it might be suggested that there is a close relationship between the style of folding, orogeny, and metamorphism. This in turn would suggest a gross zonation of the structural and metamorphic phenomena in orogenic belts. In any particular area the successive zones may have a greater vertical or a greater lateral rate of change, depending on the rate and direction of advance of the migmatization front within the prism of sedimentary rocks undergoing deformation. Although superficial review might lead to these conclusions, some evidence suggests that slip folds have considerable importance in sedimentary rocks deformed at the most superficial levels.

De Sitter (1954, 1956A, 1957; in Fourmarier, 1956) and de Sitter and Zwart (1960) argued that the influence of temperature, introduction of water, and the "acceleration of the deformative stress" are equally important as confining pressure in defining this boundary. If Fourmarier's contention is correct, only deeply eroded mountain chains would show slip folds; however, de Sitter (1956A, p. 600) found that the Devonian rocks in the axial zone of the Pyrenees have slip folds exclusively, although they apparently never had a cover of more than 1,000 meters, and probably only half this thickness. In the field the boundary is commonly represented by a transition zone of variable width whose limits are difficult to define. Lithology is an important factor in controlling the width of the transitional zone; under a prescribed set of conditions shales, for example, are much more likely to develop foliation and slip folds than is a more competent sandstone. De Sitter (1954) and de Sitter and Zwart (1960) attempted to express these relationships graphically, and Figure 154 is based on their interpretation.

Maxwell (1962) studied the Martinsburg Slate in the Delaware Water Gap area, New Jersey and Pennsylvania, U.S.A., and concluded that the folds and axial-plane foliation (slaty cleavage) developed:

(a) under nonmetamorphic conditions because of the abundance of illite reported in the slates by Cuthbert (1946) and Bates (1947), and

(b) at shallow depths of less than 4,000 meters because of the stratigraphic evidence that the folding, foliation, uplift, and erosion involved, at most, late Maysville and early Juniata time;

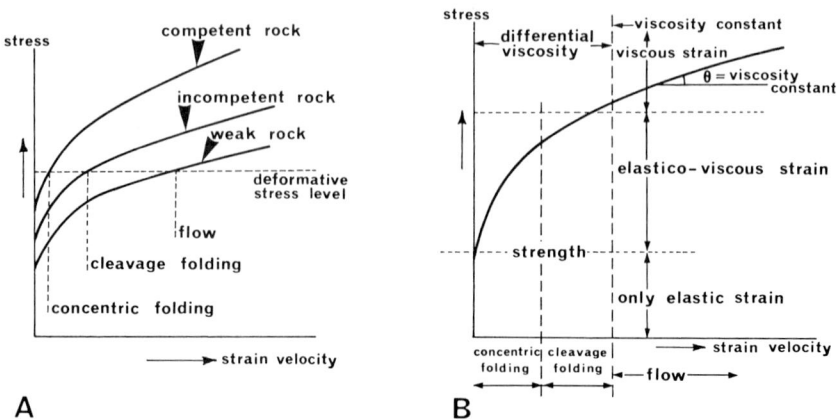

Figure 154. Schematic diagrams of the relationships between concentric folding, cleavage folding, rock flowage, stress intensity, and rock competence (After de Sitter and Zwart, 1960, Fig. 1).

sedimentation was renewed in latest Ordovician time (Juniata time).

Maxwell's hypotheses are considered in more detail in Chapter 9.

If research in other areas establishes that Maxwell's (1962) conclusions have general application, concepts about the depth control of slip folding may require considerable revision. Although, in many rock assemblages, either flexural-slip or slip folds occur alone, Figures 130, 131, and 133 draw attention to the intimate association of these two types of fold under many low-grade regional metamorphic conditions. The lack of quantitative information about the spatial distribution of these structures prevents any definite conclusions being drawn at this time.

OTHER VARIETIES OF FOLDS

Study of current structural geology texts shows that several additional varieties of folds are often given prominence. A few of these are referred to below, although, as will be shown, most are special cases of flexural-slip or slip folds described already.

FOLDS DUE TO VERTICAL MOVEMENTS

Billings (1954, p. 92) invoked differential vertical movements, unassociated with any fractures, as the cause of a wide variety of folds in the outer shell of the Earth. There are some grounds for placing these folds in a separate category, although, in fact, little attention has been given to their mode of formation and much of the deformation can be included with flexural-slip, slip, and flexure folds.

Billings (1954, p. 231) included folds associated with intrusion of magma in laccoliths and lopoliths, and with intrusion of salt domes (p. 232). Among very large structures he cited the domes (Cincinnati and Ozark uplifts) and basins (Michigan and Illinois basins) of the North

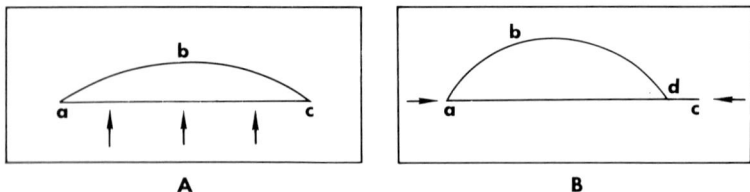

Figure 155. Folds resulting from vertical movements. A: By doming: the initial *ac* becomes *abc*. B: By flexural-slip folding: *ac* becomes *abd* (After Billings, 1954, Fig. 79).

American craton (p. 232); Billings suggested that in folds of this type points on the same stratum do not move closer together (Fig. 155); this is a point of similarity with slip folds, and of contrast with flexural-slip folds. In cratonic basins and domes the size factor is important in masking details of the deformation; if a small domain were considered, the deformation would probably be similar to that of other folded terranes.

Detailed studies of the fold styles and fabric of halite in salt domes has demonstrated that slip folds are dominant. Balk (1949) and Muehlberger (1959; 1960) described the slip folds displayed beautifully in the Grand Saline salt dome, Texas, U.S.A. Clabaugh and Muehlberger (1962, p. 149) made a petrofabric investigation of these rocks and demonstrated that:

> The internal structure of the dome consists of long, mostly isoclinal, nearly vertically plunging folds that were formed by flow of the salt. The axial planes of these folds macroscopically identify the planes in which crystal gliding has taken place and are not symmetrically related to the petrofabric patterns.

These folds appear to have formed by nonaffine viscous strain in a rheid (cf. Carey, 1954, pp. 78 ff.); the section through the salt dome near Heide, NW Germany, supports this contention (Fig. 124).

Until more detail is known about the other folds included in this group, it would seem to be unwise to erect a special category.

FLOW FOLDS

Both Billings (1954, pp. 90-91) and de Sitter (1956B) strongly stressed the importance of flow folds, but they did not give any specific examples. The term can lead to confusion because of the imprecise use of "flow" in this context. In standard usage mobility similar to that of a viscous fluid or a plastically-flowing solid produces **flow,** (i.e., continuous deformation in space without loss of cohesion; cf. Turner and Weiss, 1963, p. 37). In this sense flow can be applied to most flexural-slip and slip folds. However, Billings and de Sitter seem to have reserved flow fold for "wild" folds without regular orientation. The use of flow in this manner seems to be misleading and inappropriate. De Sitter (1956B, p. 182) considered that flow folding:

> ... is a kind of distortion in which orientation of shear-planes to a stress-direction is lost. The internal movement itself is no longer oriented but can take any direction, with the result that it can no longer be represented adequately in a section. It is typical of very weak rocks like salt, or of very high confining pressures or high temperature, and is the only true plastic deformation of rocks.

This does not provide a clear statement of the geometric and kinematic picture involved, although de Sitter and Zwart (1960, p. 253) suggested:

> A more irregular folding than in phyllites or mica-schists takes place in gneisses, which are either existing synkinematic augen-gneisses or newly formed by feldspathization of mica-schists or amphibolites. Frequently these rocks look like a stirred porridge, with fold axes pointing in various directions. This style of folding is typical for rock flowage . . . and is always accompanied by the disappearance of lineations and also of schistosity, due to continuing unoriented recrystallization. . . . Although this flow folding is not necessarily present in every late-kinematic gneiss, it seems to be very common and many migmatites show this feature very clearly. . . . The cause of this flow folding is possibly an increase in water content, . . .

Billings (1954, p. 91) admitted that in many respects, if sufficiently small volumes of metasedimentary rocks are considered, flow folds do not differ in appearance from flexural-slip folds, although he thought that associated minor folds are more abundant. Goguel (1962, p. 290) suggested that well-formed cylindroidal folds are not the norm of structures in the Earth's crust — to him they are a peculiarity of suprastructure deformations, while in general basement structures show ". . . much more irregular behavior." However, detailed investigation in most migmatized areas shows that the folds are essentially cylindroidal, although the fabric may be very complex because of repeated superposed folding. Small domains commonly possess homoaxial cylindroidal folds, whereas larger volumes (perhaps of the size of an outcrop) may simulate "stirred porridge."

With flexural-slip and slip folding of a sequence of competent and incompetent rocks, the incompetent members (e.g., calcareous beds, evaporites, shales, etc.) commonly flow more readily and develop more complex fold styles than the competent units, even at superficial levels in the crust. Hills and Thomas (1954, p. 125) described an example from the Ordovician laminated argillaceous and arenaceous rocks at Daylesford, Victoria, Australia.

Wynne-Edwards (1963) suggested that flow folds have equal status with flexural-slip and slip folds, and that these three types represent end members in a continuously variable series. Such folds were claimed to have identical geometry to slip folds, except for the absence of both axial-plane foliation and visible planes of movement. Turner and Weiss (1963, p. 481) also used flow fold in this sense, but added that slip folding is a plane deformation on specified slip surfaces while flow folding need not be plane and is not referred to any particular system of slip. While Wynne-Edwards considered that flexural-slip folds are most common in unmetamorphosed rocks, and slip (or shear) folds in low grade metamorphic rocks, he claimed that flow folds are characteristic of (a) metamorphic rocks of the upper amphibolite facies or higher, and (b) salt, gypsum, and glacier ice. In the so-called flow folds of the Grenville Province, Canada, Wynne-Edwards showed that refolding about the same *B*-axes is common, and that major folds persist for long distances and have parallel axial planes. Strain was said to be a product of nonuniform unsteady viscous flow. At the present time there seems to be little justification for recognizing this additional class of folds.

FOLDS IN CERTAIN UNCONSOLIDATED AND NEAR-SURFACE ROCKS

Small folds developed in unconsolidated lake and other surficial deposits, and certain slump phenomena, commonly result from rheid or plastic flow. Many examples of creep, or gravitational flow, in near-surface rocks could be cited. Ten Haaf (1959, pp. 49-50) described a good example from the Italian Apennines in which stratification can still be traced across the crumpled structures. Individual disturbed zones are a few decimeters or meters thick, and comprise mudstone sequences within which sandy seams have been dragged into folds as a result of concentrated differential movement beneath a thin overburden (Fig. 156). Figure 157 shows similar small folds developed in an allochthonous mass of Late Cretaceous-Eocene black shale near Passo della Cisa. Northern Apennines, Italy, described by Page (1963). Thin-bedded limestones and sandstone members were folded in response to the flow of the whole rock under the influence of gravity. It is typical of this kind of deformation that the folded blocks of the competent rocks are not in contact; in general, the gravitational sliding origin of such structures can be determined only by detailed regional field studies.

Contemporaneous and penecontemporaneous deformation of newly-deposited sedimentary units commonly results in small folds. Whittington and Kindle (1963) illustrated well-developed isoclinal folds produced by

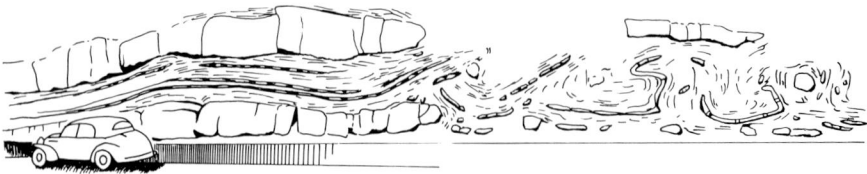

Figure 156. Folds resulting from slumping of Picene flysch units, F. Vomano, Italian Apennines (After ten Haaf, 1959, Fig. 35).

Figure 157. Folds developed in thin-bedded limestone and sandstone members within Late Cretaceous-Eocene black shale, near Passo della Cisa, northern Apennines, Italy. These folds appear to have formed by gravitational sliding within the near-surface rocks. The exposure illustrated is about 55 meters wide (After Page, 1963, Fig. 3).

0   feet   1

Figure 158. Isoclinal folds produced by contemporaneous deformation in thin limestone units of middle Table Head Formation (Ordovician), Table Point, Newfoundland, Canada (After Whittington and Kindle, 1963, Plate 2, Fig. 2).

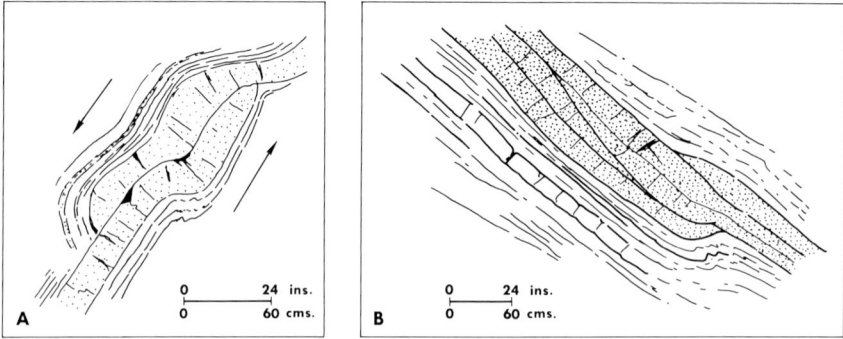

Figure 159. Wedges within sedimentary units of the central Appalachian fold belt, U. S. A. A: A thick limestone member in shales has been sheared, wedged, and "telescoped" in the west limb of an anticline in Wills Creek Formation, Maryland. B: Wedges of sandstone accumulated in a pile and enclosed in shale, E of Romney (After Cloos, 1964, Figs. 2 and 7).

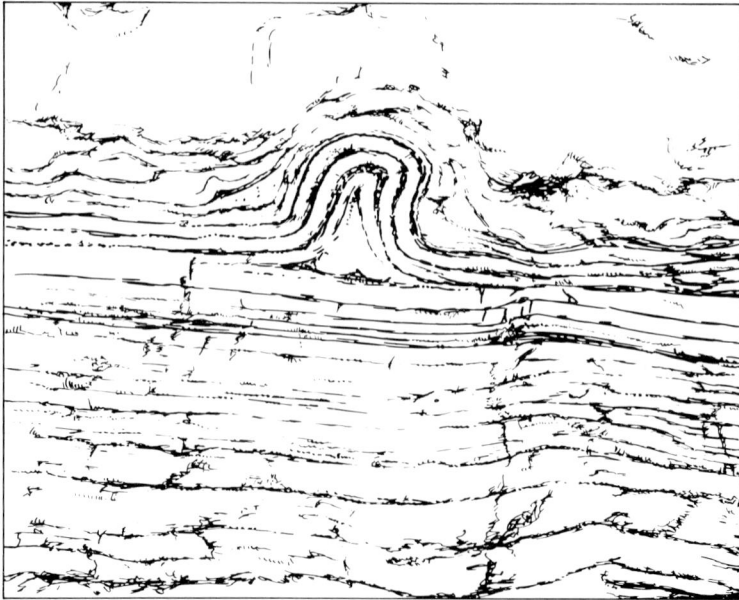

Figure 160. Small fold in lacustrine silts near Manitouwadge, Ontario, Canada (After Sutton, 1963, Plate 1).

submarine sliding within the middle Table Head Formation (Middle Ordovician), western Newfoundland, Canada (Fig. 158). Cloos (1961, 1964) drew attention to the common occurrence of bedding slips and **wedges** within the sedimentary rocks of the central Appalachian fold belt. These structures were attributed to gravitational slip along bedding planes at a time when the particular units were not deeply buried; this slip caused some arenaceous units to ride up over themselves to cause local thickening of the succession (Fig. 159). Similar wedges appear to be common within other lithologies (e.g., thinly bedded limestone) and other geosynclinal sequences. As Cloos suggested, such initial breaks in individual units, and the local thickening of the layers, are probably important in controlling loci of subsequent folding (cf. Cloos, in Cloos and Murphy, 1958, pp. 74 ff.).

Figure 160 shows a small fold (or *involution*; cf. Sharp, 1942) developed in horizontally-bedded lacustrine silts near Manitouwadge, Ontario, Canada. This is an example of one of many varieties of disturbance that can occur in soil zones, and Sutton (1963) thought this structure represents deformation consequent upon the periodic formation and decay of ice bodies in the soil zone during periglacial conditions.

Prelithification structures of these several types are not uncommon, and in folded sequences of metasedimentary rocks, there can be severe problems in differentiating them from other types of folds.

FOLD TERMINOLOGY

From the above discussion it will be clear that flexural-slip and slip folds form under a wide range of dissimilar conditions. The requisite plasticity is not restricted to deep-seated or high-grade metamorphic environments; folds of this type can be found in near-surface and surface rocks too. The multiplication of various special names for slightly different varieties of fold is unnecessary and confusing. The only safe method of description is to record the orientation of the fold axis, the axial surface, and the profile of the structure. Within a single hand specimen or outcrop these characteristics may vary widely; in consequence the size factor is important in describing the geometry of a fold. However, a very large number of descriptive and genetic terms occur both in structural geology texts and in the geological literature in general; Fleuty (1964) recently provided a useful review of the descriptive terminology. In order to understand this literature the meaning of these terms must be understood, and a considerable glossary will be found in Chapter 15.

Chapter 7

Disharmonic Folds and Transposition Structures

Disharmonic folds were introduced in Chapter 2, where it was shown that they occur in many folded banded rocks ranging from completely unmetamorphosed sedimentary rocks to high grade metamorphic rocks. Disharmonic folds are characterized by an abrupt change in profile in passing from one S-surface, or lithic unit, to another within a folded sequence. Parasitic folds associated with flexural-slip folds provide one example in which different lithologies have dissimilar strain under similar geological conditions. However, in many cases, much more dramatic disharmonic structures than parasitic folds develop. In this chapter the nature of disharmonic structures affected by more intense deformation is examined in detail.

The relative competence of different lithologies changes sharply in different geological environments; for example, limestones "flow" readily in the superficial rocks, and are readily attenuated and squeezed out, but in the deeper orogenic zones (although subject to extensive deformation by internal slip) they tend to remain structurally-discrete and cohesive units. However, in each particular environment, the dissimilar response to stress of the several rock types tends to result in disharmonic structures; the unmetamorphosed sedimentary rocks of SE England (West, 1964) and the migmatized marbles of southern Greenland (Wegmann, 1938) provide two examples (Figs. 161 and 162).

The widespread localities mentioned might suggest that descriptions of the structures abound in the literature. This is not the case, and most of the available descriptions are essentially qualitative. Virtually nothing is known about the quantitative aspects of the nature, variability, and spatial distribution of these structures. A huge field of enquiry awaits exploration.

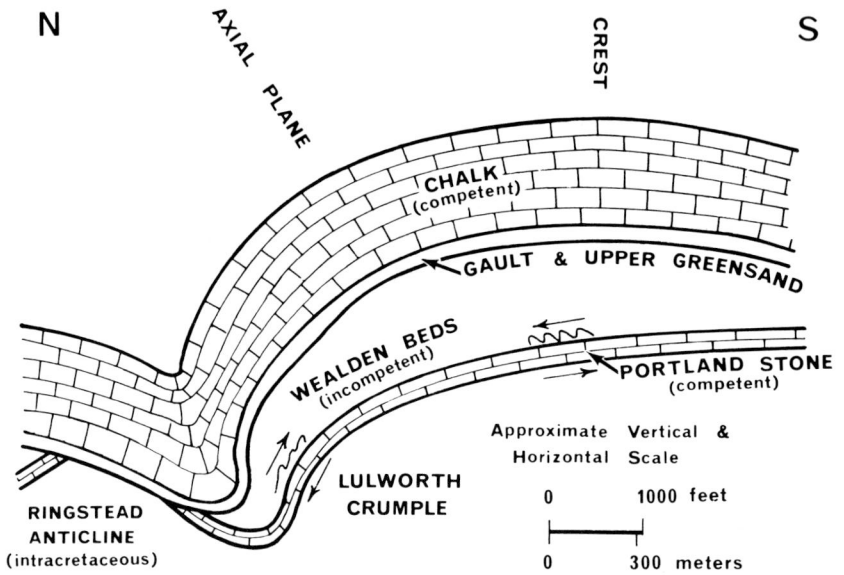

Figure 161. Diagrammatic cross section to show the deformation of the incompetent beds in the Purbeck anticline, SE England (After West, 1964, Fig. 1).

Figure 162. Disharmonic folds in migmatized limestone sequence, Ikerasak Sound, S Greenland (After Wegmann, 1938, Fig. 4).

TRANSPOSITION STRUCTURES

In a laminated sequence of rocks, successive layers tend to have slightly different competence. During progressive flexural-slip folding flexures of the more competent layers become progressively more "closed," appressed, and isoclinal, so that flanks of isoclinal folds become parallel to the major slip directions (which are parallel to the axial planes of the developing flexures). As this deformation continues the original S_0-surfaces (bedding) eventually become coincident with S_1 (foliation parallel to the slip directions), except at the closures of the folds (Fig. 163).

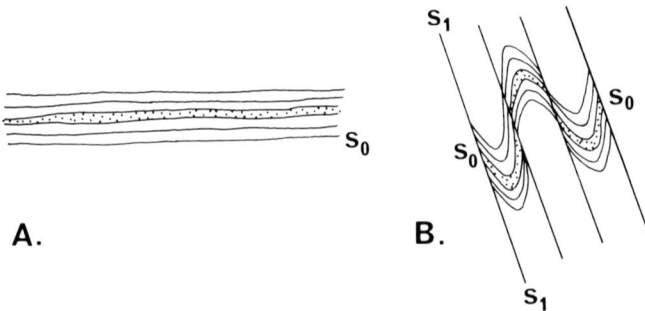

Figure 163. Development of transposition structures. A: Original bedded sequence (S_0). B: Transposition of S_0 as a result of slip along successive planes (S_1); eventually S_0 becomes completely transposed and parallel to S_1 (Adapted from Knopf, 1931, Fig. 6).

Continued translation along S_1 causes further attenuation of the fold flanks. The fold hinges become sharper as the folds become more closely appressed, and eventually strong differential slip may tear the fold closures apart along the stretched limbs. Eventually the only relicts of the original folds are crescent-shaped attenuated closures; these are "rootless," because the flanks of the isoclines are completely sheared out. Although the original S_0-surfaces may be completely obliterated, the new S_1-surfaces are merely transpositions of the old S_0, and not a structure crossing S_0. At this stage, the deformed rocks frequently have a phyllitic appearance and, because the folding (transposition) brought the lithological units into parallelism with S_1, the rocks appear to show schistosity parallel to bedding. However, the successive "beds" are separated by tectonic breaks; in a section normal to S_1, some of the competent laminae are inverted with respect to their neighbors, and the sequence therefore

Figure 164. *Ausweichungsclivage* resulting from transposition of original *S*-structure in the Tessin Gneiss, Urlaun Peak, Ticino Canton, Switzerland. Scale is 2 cm. (From new photograph supplied by Dr. E. B. Knopf from original published by Heim, 1900, Fig. 2, and Knopf, 1931, Fig. 8).

represents **pseudo-bedding,** rather than true bedding. Although flexural-slip folding can produce foliation parallel to bedding, the majority of examples of foliation parallel to bedding described in the literature probably represents transposition structures in which pseudo-bedding is parallel to schistosity.

Transposition structures can be considered an extreme case of disharmonic folding. The recognition of transposition structures largely depends on identification of fold closures, or even of an isolated closure, within the surrounding, less competent rocks. In some areas such closures are not preserved because lithologically-distinctive horizons were lacking, or because of repeated refoliation of the rocks.

This type of folding with associated transposition features is found in structures of all sizes ranging from hand specimens to major structures extending for several kilometers. The parallelism of the lithologic units may give a superficial resemblance to a normal sedimentary succession on a regional scale; transposition structures, and the accompanying tectonic breaks between successive units, make the true stratigraphy difficult to interpret.

Knopf (1931, p. 16) pointed out that the slip planes associated with transposition structures have often been called slip cleavage, strain-slip cleavage, or fracture cleavage by American geologists. Heim (1878) described this foliation that follows the direction of slip as **Ausweichungsclivage** (Figs. 164 and 165). Figure 166 shows two excellent examples of partial transposition described by Balk (1936) from New York, U.S.A. In the final stage when the S_0 are completely transposed the foliation was called **Umfaltungsclivage** by Sander. Although transposition structures have received little attention for many years, Van Hise (1896, pp. 649 ff.) wrote:

> *On opposite limbs of a fold the cleavage usually dips in opposite directions. Upon opposite sides of an anticline the cleavage usu-*

Figure 165. *Umfaltungsclivage (Ausweichungsclivage* with complete transposition) resulting from transposition of original *S*-structure in Tessin Gneiss, Urlaun Peak, Ticino Canton, Switzerland. Scale is $\frac{1}{2}$ cm. (From new photograph supplied by Dr. E. B. Knopf from original published by Heim, 1900, Fig. 4, and Knopf, 1931, Fig. 10).

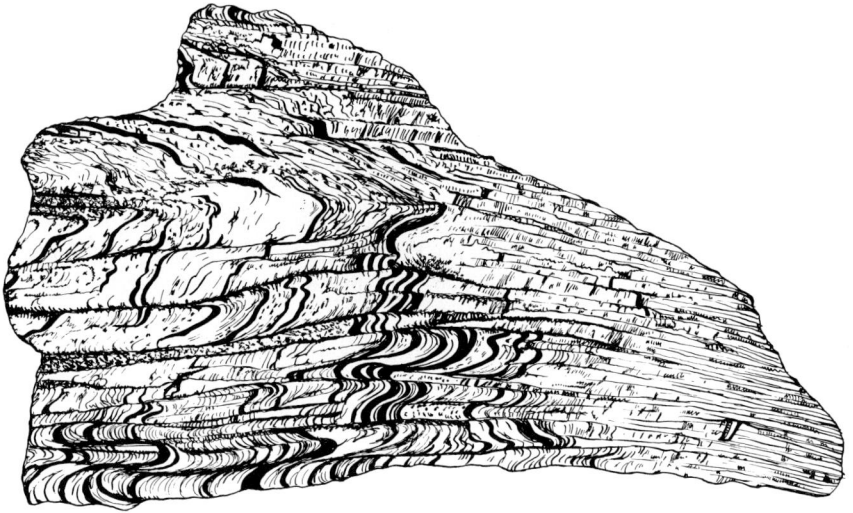

A

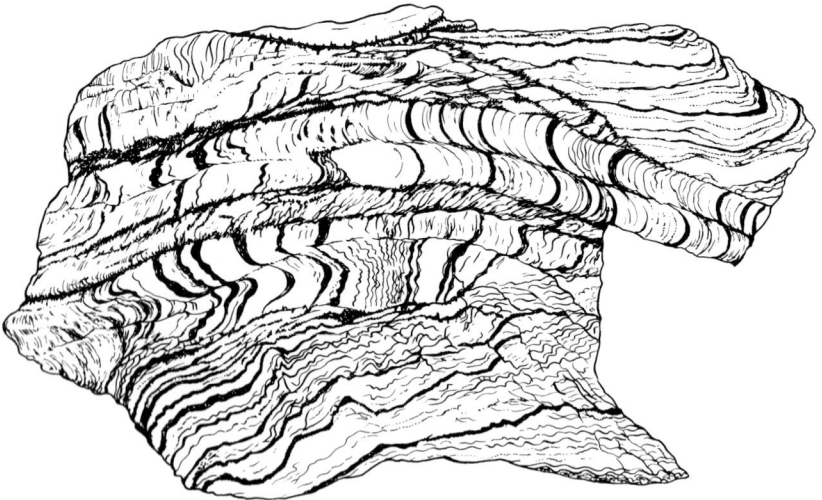

B

ally diverges downward, and on opposite sides of a syncline it
usually converges downward ... cleavage ... dip will be steeper
than the dip of the strata; hence, *on opposite sides of a fold the*
variation in the dip of cleavage is less than the variation in the
dip of bedding. ... If the compression goes so far as to form
isoclinal folds the bedding and cleavage may nearly correspond
upon the limbs of the folds, but the two will be at right angles to
each other upon the crests. ... In many districts where cleavage
has been described as everywhere according with bedding, and
the two do approximate but not exactly accord in direction
upon the limbs of the folds, a close examination shows that on
the crests of the anticlines and in the troughs of the synclines the
two structures intersect each other.

At the beginning of the process it may be noted that the
shortening is at right angles to the bedding. At the end of the
process the shortening is parallel to the bedding.

It is interesting to notice that in 1932 Stark and Barnes (1932, pp.
477-78) worked in the Sawatch Range, Colorado, U.S.A., and found that:

... schistosity in all the mapped areas is apparently parallel to
the bedding. As this was contrary to expectations, particular
attention was given to the relationship and in no place was the
schistosity found at an angle to the bedding. The parallelism
might be explained in part by the isoclinal folding which is the
dominant type throughout the range. ...

However, with justification, Knopf (Knopf and Ingerson, 1938, p.
190) observed that:

... prevalence of this process of transposition has been scarcely
recognized in English literature, and a failure to appreciate its
presence in a given area may lead to entirely erroneous interpretation of finely layered structures in schists or phyllites as
bedding.

For example, although transposition structures are now well known in the
area (e.g., Engel, 1949), in 1929 Buddington concluded that "... the
foliation of the Grenville formations is everywhere parallel to the bedding
in the northwest Adirondacks."

Foliation is discussed more fully in the next chapter, but it may be
noted here that the foliation (Ausweichungsclivage) which follows S_0 (as a

Figure 166 (facing page). Partial-transposition structures seen on polished samples
of phyllite, Clove area, New York, U. S. A. A: At lower right microlithons are so
closely spaced that the original S_0 is obliterated. Siliceous layers shown in solid
black: biotite porphyroblasts are developed along the shear planes. B: Garnet is
developed along the shear planes, but is absent from the phyllite microlithons
(After Balk, 1936, Figs. 15 and 16).

result of transposition) has long been differentiated (e.g., by Sander) from axial-plane foliation (slaty cleavage) which is characteristically developed oblique to S_0. Knopf (1931, p. 16) pointed out that these two types correspond to what Van Hise and Leith referred to as "fracture cleavage" and "flow cleavage," respectively. These terms, and especially "fracture cleavage," have been used in dissimilar ways and, as a result, they are liable to misinterpretation; Ausweichungsclivage results from dislocation (fracture) along planes of extreme slip induced by the folding, and, in the next chapter, it is defined as **crenulation foliation.**

TRANSPOSITION STRUCTURES: SMALL EXAMPLES

In a detailed structural study of Dutchess County, New York, U.S.A., Balk (1936) recognized disharmonic folds in a variety of lithologies. He described the manner in which pseudo-bedding was produced in pelitic rocks, and in describing the deformation of marbles he wrote (1936, p. 720):

> The limbs of the isoclinal folds in the marble have been further deformed by plastic flow along the planes of the limbs. The original thickness of each limb (if it had, perchance, retained its original thickness to that stage of the folding) has been reduced; apices of many isoclinal folds have been destroyed in favor of through-going planes of gliding; and, where the mineral composition of these planes varies, they appear now as subparallel "layers." In view of this complicated history, the term foliation planes seems more appropriate than the term stratification planes. If the "layers" are likened to layers of original deposition it should be borne in mind that they can be, at best, the emaciated remnants of original, isoclinally folded layers; and even this need not be true in every instance, for the foliation planes may be the reproduction of fracture cleavage planes that cross-cut the original beds of deposition; they may also be caused by silication by thermal solutions, . . . As these possibilities do not seem to have received enough attention, observations in support of this origin are given in some detail. There are three lines of evidence: (1) Straight, seemingly undeformed layers of marble enclose isoclinally folded and dismembered fragments of stronger rocks; (2) some apices of folds within the marble are intensely sliced and disturbed by slip planes (older than the recrystallization), parallel to the foliation planes nearby; (3) fracture cleavage, older than the recrystallization, crossing the apices of the deformed folds, also strikes and dips parallel to the foliation planes.

A fourth criterion, the orientation of linear parallelism on the foliation and cleavage planes, was discussed later in Balk's paper (1936, pp. 734 and 760) in relation to isoclinally folded quartzites.

The type of structure mapped by Balk (1936, Plate 10) north of Patterson, New York, U.S.A., is shown in Figure 167. The isoclinally folded fragments of more competent rocks enclosed within the marbles are analogous to the tectonic inclusions of the Grampian Highlands, Scotland, and the Turoka area, Kenya, described below (see Figs. 180 and 181).

Engel (1949) gave one of the most complete descriptions of small transposition structures in a detailed study of the Precambrian gneisses, schists, and marbles of the Grenville Series, northwest Adirondack Mountains, New York, U.S.A. (Fig. 168). The interlayered metasedimentary lithologies within the marbles show different degrees of deformation. In the pure carbonate rocks bedding surfaces, folds, and other structural features commonly became obscured early in the deformation as a result of continuing recrystallization. Where pure marbles were attenuated between the smooth surfaces of competent rocks both the slip surfaces in the marble and the marginal foliation of the competent rocks are remarkably uniform. The rocks commonly give a false impression of local unconformities because the marble deforms before the competent units and it also develops very uniform shear surfaces. For example, at a stage where a quartzite bed or a quartzose zone is strongly folded, but is still clearly recognizable, the bedding in adjoining marble may have been completely obliterated by the flow of the carbonate materials (Engel, 1949, p. 779). In Figure 169 the crumpled surfaces in the quartzite were interpreted as relict bedding, and the prominent surfaces in the marble as transposed *S*-structures (pseudo-bedding) that simulate bedding. The fold closures at A and B (Fig. 169) confirm the transposed nature of this lithic unit, but as Engel emphasized, the isoclines may be folds either of bedding or of younger shear surfaces produced by an earlier phase of deformation. Engel (1949, p. 775) also showed that:

> Where the folding and refoliation process has continued, the flank areas of the isoclines are subsequently folded, commonly along the same axes or axes only slightly askew from those of the initial folding. In these younger, essentially accordant folds, the cleavage which had its inception with the evolution of the initial isoclines clearly wraps around the nose of the younger fold. In some instances the closely spaced, prominent earlier cleavages near bedding facilitate interlayer slip and no strong axial plane cleavage evolves in the younger folds until they too approach isoclinal form.

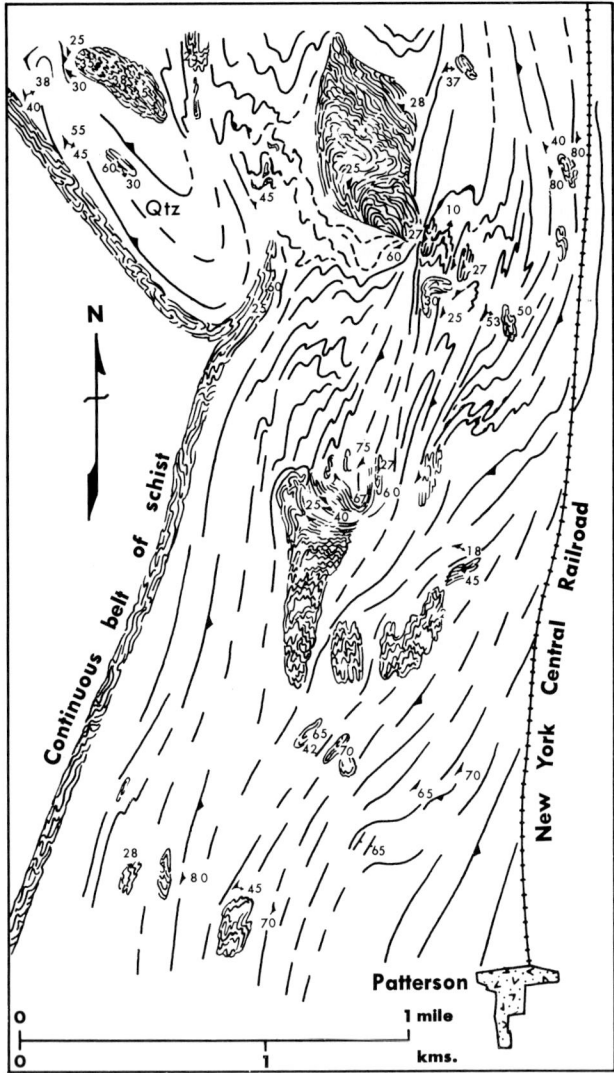

Figure 167. Map of schist tectonic inclusions in marble, Patterson, New York, U. S. A. (After Balk, 1936, Plate 10).

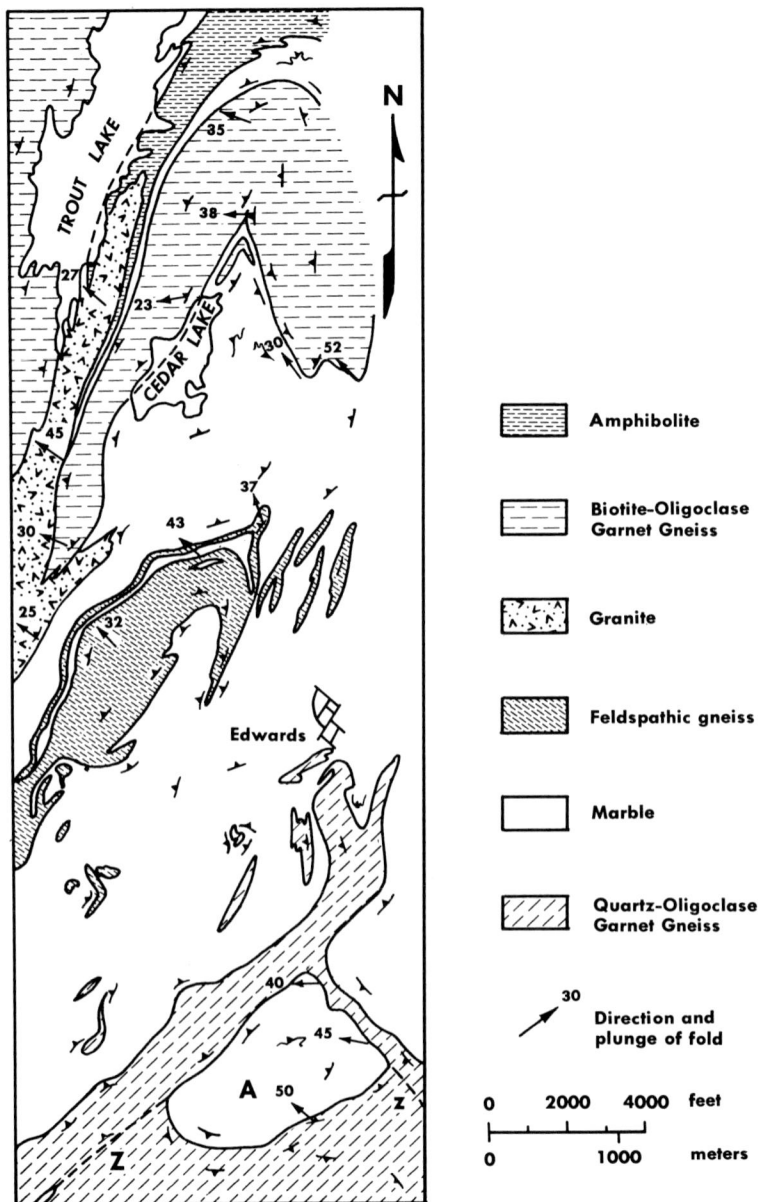

Figure 168. Metasedimentary rocks of the Grenville Series near Edwards, NW Adirondacks, New York, U. S. A. Folds of two sizes and with subparallel axes (plunging NW at about 35°) are represented; hook-shaped outlines of the feldspathic gneiss and biotite-oligoclase-garnet gneiss are believed to be parasitic folds on the nose of the large dome of area A (south margin of map). The east flanks of the parasitic folds are very attenuated. Marble at A is a remnant of interlayered unit in the quartz-oligoclase-garnet gneiss, whose position before attenuation is indicated by broken lines at Z; less-extreme attenuation is seen in the NW of the map (After Engel, 1949, Fig. 9).

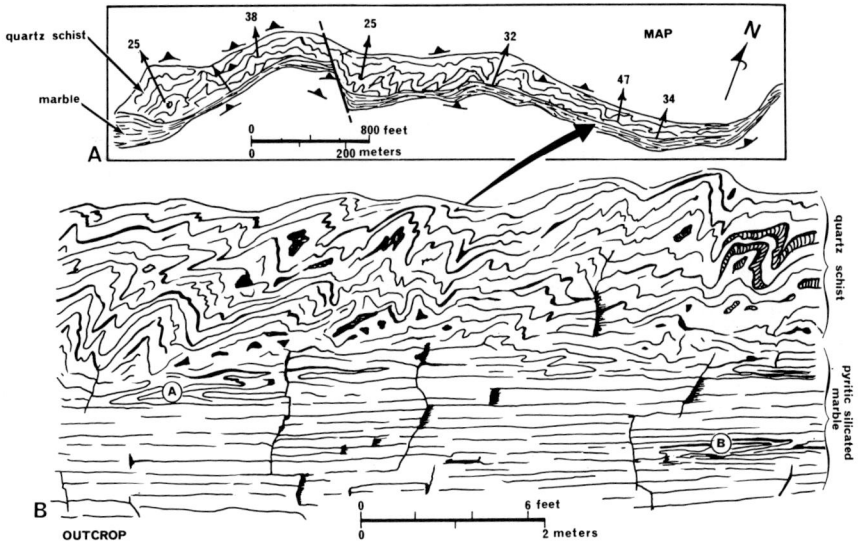

Figure 169. Relationships between quartz schist and pyritic silicated marble sketched by Engel from an outcrop of the Grenville Series, NW Adirondack Mountains, U. S. A. Plicated foliation in the schist is probably bedding, but surfaces in marble are defined by flow; relict folds are seen at A and B. The fold axes in the schist and the marble are approximately parallel (After Engel, 1949, Fig. 10).

Although pseudo-unconformities are sometimes produced by this disharmonic folding, Engel (1949, p. 779) emphasized that:

> ... the shearing out of bedding in more mobile layers, with the development of secondary surfaces essentially parallel to the trend of the adjoining harder rocks in which the foliation is not visibly folded, can completely mask stratigraphic unconformities. In such places, the tectonic contacts between contrasting lithologic units, zones of displacement probably measurable in tens or hundreds of feet, appear superficially to be as conformable and undisturbed as bedding in a conformable sequence of sedimentary rocks. In general, the shearing-out processes tend to produce many more pseudo-conformable than discordant relations.

Structural relationships of this type can be very perplexing until the transposition structures are recognized; Greenly (1930, p. 177) described such a situation (Fig. 170):

On the sea-cliff at the chasm southwest of Hir-fron . . . one meets with a phenomenon which is not a little disconcerting. The bedding of the South Stack Series is absolutely clear, the dips here being steep or even vertical, while the cross-foliation of the massive green grits . . . is at a lower angle than usual. Yet we find . . . between the planes of cross-foliation, 3-inch bands of a fine, hard, silicified grit, sharply distinct in character from the great green grit, running right across it. So the accepted criterion for distinguishing between divisional planes of deposition, from those of dynamic origin (namely, that the former are bounding-surfaces of differing materials, whereas the latter are not) has at last failed us. Nevertheless, we are able to hold by the fact that unequivocal stratification is general throughout the South Stack Series, while this phenomenon is but local, and is related to an indubitably dynamical structure.

Similar examples in pelitic and semipelitic rocks from widely-scattered parts of the world have been described. Knopf (Knopf and Ingerson, 1938, Plate 19) illustrated a quartz phyllonite from Patscherkofel, near Innsbruck, Austria, in which advanced transposition of original *S*-surfaces developed pseudo-bedding with thin laminae in which it is almost impossible to recognize folds (Fig. 171). Similar structures were illustrated by Weiss and McIntyre (1957, Plate 1A; see also p. 579) from the Loch Leven area, western Scotland (Fig. 172).

Transposition structures are much more prevalent than has commonly been supposed, and the recognition of rootless fold closures invariably implies that the kinematic picture is complex. Within the large roof pendant, or screen, of metasedimentary and metavolcanic rocks exposed in the granitic terrane near Mineral King, Sierra Nevada, California, U. S. A., many of the metavolcanic rocks appear to be even-bedded. However, in 1957 Christensen and Whitten observed occasional tightly-appressed fold closures which established that these rocks exhibit pseudo-bedding resulting from transposition; Christensen (1959; 1963) described these structures in detail.

King and Rast (1956A) described the structural styles produced during severe deformation of varied metasedimentary rocks in the Dalradian sequence of the Scottish Grampian Highlands. They compared and contrasted the behavior of common rock types such as quartzite, pelitic schist, and marble. It is instructive to examine some of the features exhibited.

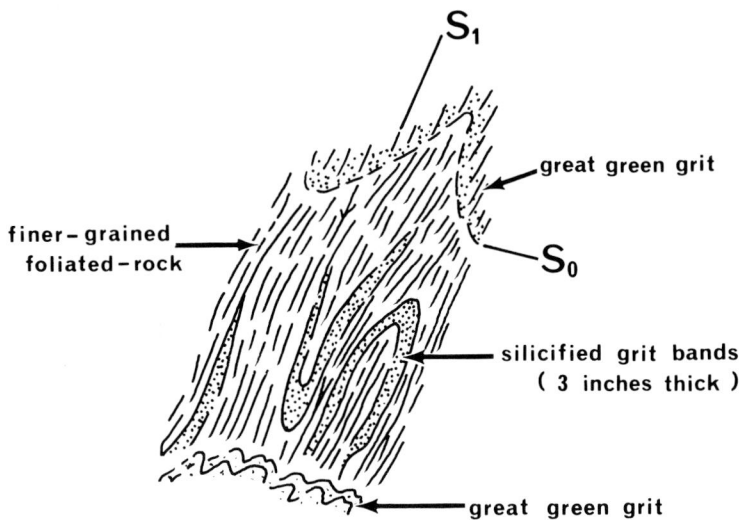

Figure 170. Plan view of transposition structure in Mona Complex, near Rhoscolyn, Anglesey, NW Wales (After Greenly, 1930, Fig. 10).

Figure 171. Quartz phyllonite from Patscherkofel, near Innsbruck, Austria, showing pseudobedding resulting from transposition of original S-surfaces. Arrow indicates a fold closure. Scale is 3 cm. (New photograph supplied by Dr. E. B. Knopf).

QUARTZITES

Original sedimentary features, such as bedding, pebbles, current bedding, etc., are more frequently and more faithfully preserved in quartzites than in other rock types. The complexity of the folds can be overlooked easily, because intense folding was commonly restricted to certain bands separated by more simply-deformed units. Weiss and

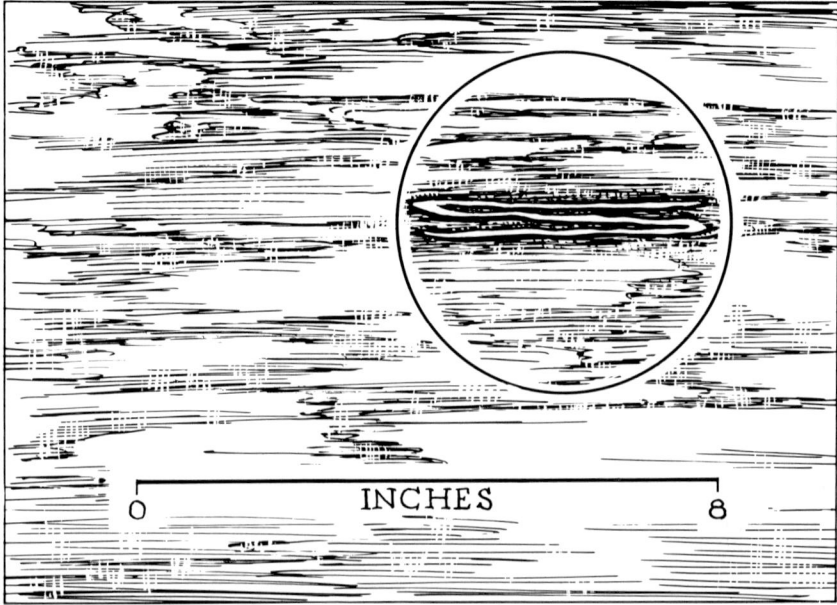

Figure 172. Pseudobedding resulting from advanced stage of transposition of bedding (S_0) in a pelite. Occasional more competent members preserve isoclinal folds (in circle). Based on photograph of Dalradian schist, NW of Kinlochleven, Scotland, by Weiss and McIntyre (1957, Plate 1A).

McIntyre (1957) drew attention to such phenomena in the massive quartzites south of Loch Leven (western Grampian Highlands, Scotland). Recrystallization of individual thick layers at fold closures commonly produces massive white quartzite in which all the original sedimentary structures have been eliminated. For example, Engel (1949, p. 775) noted that:

> ... In the described examples of more coherent and stronger beds, the initial and most thorough effacement of bedding, and of any foliation parallel with bedding, commonly occurs in the axial or short flank areas of folds. This very common destruction of folds in bedding in their axial areas is in marked contrast to the quoted [Lahee, 1941, p. 280] statement of Van Hise that axial regions of folds in highly deformed metasedimentary rocks are the best places to search for relict beds [Van Hise, 1896].

Inherited *S*-structures in quartzites are often difficult to discern after penetrative deformation accompanying folding, and later planar surfaces have sometimes been misidentified as bedding. Figure 173 illustrates the fold style at Tromie Bridge, Inverness-shire, Scotland, where quartz segregations developed parallel to an axial-plane parting (S_1). In most of the exposures this structure (S_1) can be, and has been, mistaken for the original sedimentary S_0-structure.

It may be debated whether fold closures in quartzites were thickened during folding, or whether they show the original bed thickness which was attenuated in the limbs; it is usually difficult to reach an unequivocal answer. The dissimilarity in bed thickness can be recognized in folds of all sizes, and allusion has been made already to small examples at Ord Ban, mid-Strathspey, Scotland, described by McIntyre (1951A). On a regional scale the Perthshire Quartzite at Schichallion, Scotland, shows immense flexures in which the limbs are extremely attenuated, but the closures appear to have been thickened (Fig. 330).

PELITIC SCHISTS

The approximate coincidence of schistosity with the general disposition of adjacent formations has sometimes led correctly, and sometimes wrongly, to the conclusion that bedding and schistosity are coincident. Numerous examples show that where pelitic formations contained psammitic or calcareous layers, the latter tend to preserve the best clues to the real structural geometry.

King and Rast (1956A) observed that in the Central Scottish Highlands the schists intervening between marbles and quartzites south of Braemar, and in the Killiekrankie Schists to the southeast of Ben a'Chuallaich (near Kinloch Rannoch), abound in structures like those of Figure 174. The commonness of such structures in many parts of the

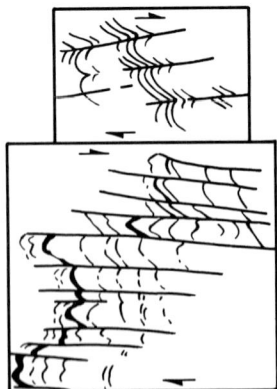

Figure 173. Profiles of small folds in Dalradian quartzite, Tromie Bridge, mid-Strathspey, Central Highlands, Scotland. The poorly-preserved S_0 tends to be near vertical and strongly-developed S_1 is nearly horizontal (After Whitten, 1959A, Fig. 2).

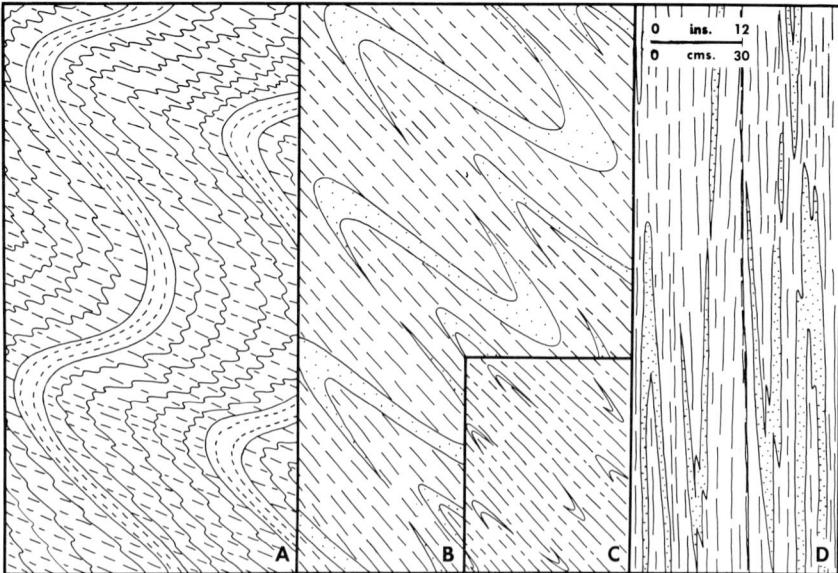

Figure 174. Diagrammatic profiles of fold structures in the pelitic schists of the Braemar-Kinloch Rannoch area, Central Highlands, Scotland. A: Initial folding with open flexures in quartzite layers and axial-plane schistosity in crenulated pelitic members. B: Limbs of quartzite folds attenuated so that only apices remain as indicators of the tectonic significance of the schistosity. C: More advanced stage of attenuation. D: Extreme elongation and flattening of quartzitic members so that schistosity and banding are almost, but not precisely, coincident: the oblique intersection reveals, even in the absence of fold closures, that the general layering is of tectonic origin (After King and Rast, 1956A, Fig. 2).

world means that the tectonic style must always be established before the sedimentary origin of observed lithologic transitions in highly deformed rocks can be accepted. King and Rast (1956A, p. 249) were so impressed by these structures that they questioned whether any transitions in the Central Highlands can be proved to correspond directly with original sedimentary boundaries.

Engel (1949, p. 783) had earlier been impressed by almost identical relationships. For example, he found that:

> ... in many areas of the northwest Adirondacks a map drawn with the strikes and dips of the diverse, major rock surfaces and layers plotted under a single symbol does little practical violence to the facts, or to gross structural and stratigraphic reconstructions. The thick units of gneiss and marble remain as distinctive masses which roughly outline most major structures. But the evolution of existing features, which include this tendency toward parallelism of many prominent primary and secondary rock

surfaces must be appreciated in unravelling the complex deformational history of these rocks, and in precluding marked over-
simplifications of their metamorphic history.

MARBLES

Carbonate rocks that contain closely-spaced siliceous members commonly reveal the structural geometry very clearly, because of the mark-
edly-different response of the two lithologies. Engel (1949, pp. 771 ff.)
found that such assemblages in the Adirondack Mountains yield the most
extreme examples of different rock competence during deformation. The
intercalated siliceous bands apparently had little "stiffening effect" on the
whole formation during deformation; they were broken into separated
fragments that commonly serve as excellent markers by which the kinematic picture for the enclosing marble can be elucidated. King and Rast
(1956A) described similar relationships from the Scottish Highlands,

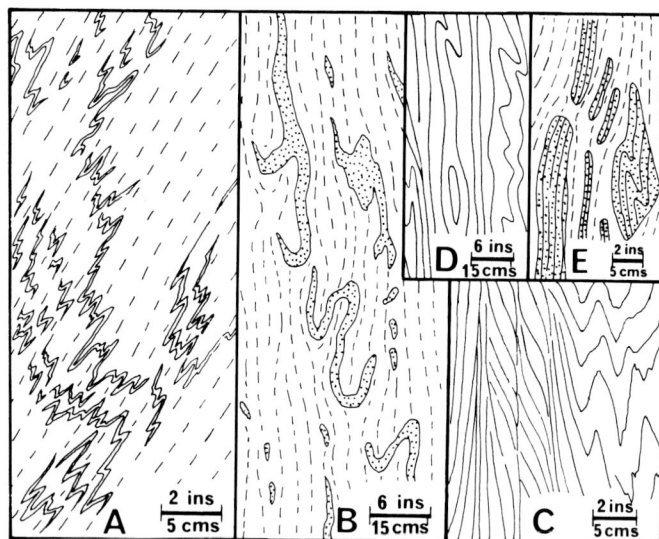

Figure 175. Tectonic styles in Dalradian marbles SW of Braemar, Central
Highlands, Scotland. A: General banding of marble is approximately parallel to
the axial planes of tiny folds in thin siliceous ribs. B: Original siliceous ribs
(stippled) preserved as discontinuous fold closures in marble. Banding of the
marble is locally deflected by large siliceous fold closures, but it is mainly parallel
to axial surfaces of the folds. C: Closely compressed folds in homogeneous
marble; fold closures are sheared out in left side of sketch. D: Folds in homogeneous marble; note eyed fold (above D) and that axial planes of the folds dip in both
directions (reflecting different rotation sense in some layers). E: Tectonic inclusions resulting from disruption of siliceous layers within marble (After King and
Rast, 1956A, Fig. 4).

where siliceous ribs also became isolated tectonic inclusions within marbles. The plasticity of the carbonate caused folds in the marbles to be sheared out rapidly and the tectonites to develop pseudo-bedding due to the pronounced schistosity. Excellent exposures of the Blair Atholl

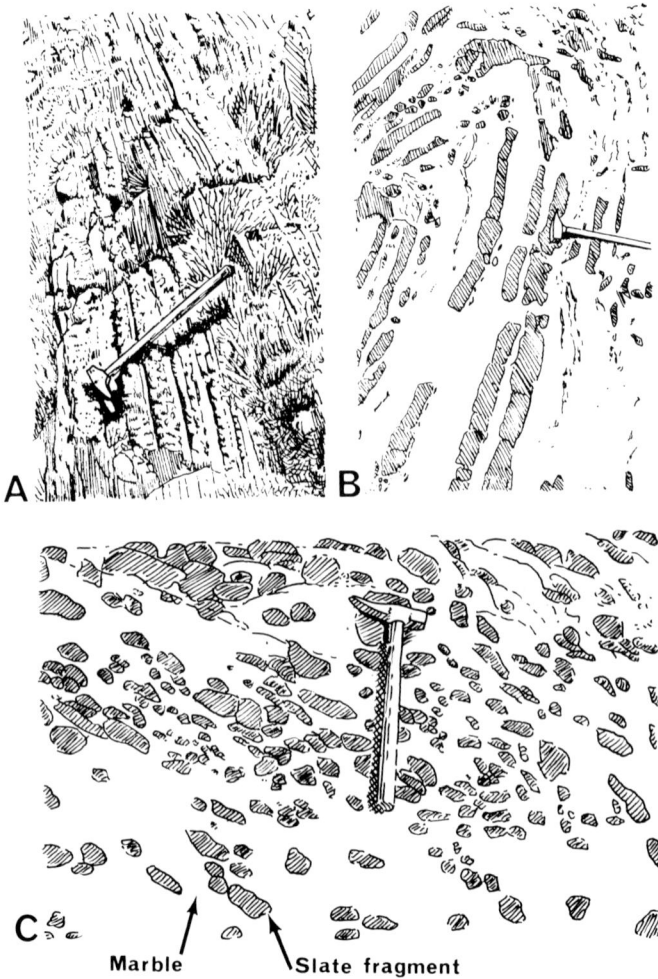

Marble   Slate fragment

Figure 176. Progressive stages in fragmentation of slate members and flow in recrystallized limestones during folding, Valle de Arán, Central Pyrenees. A: Undisturbed foliation is well developed in slate, but is vague in limestone. B: Initial fragmentation and flow. C: Chaotic arrangement of slate fragments in limestone (After drawings from photographs by Kleinsmiede, 1960, Fig. 32).

Limestone in Glen Tilt, south of Braemar, and north of Schichallion (Scottish Central Highlands) show structures of this type (Fig. 175). Such structures are extreme expressions of disharmonic folding; similar deformation in mid-Argyllshire, Scotland, was mentioned by Peach (Peach, *et al.*, 1909).

In the above examples transposition played an important part in isolating competent fragments. Competent lithic units can also be fragmented without transposition during folding by flow of a plastic interstratified unit such as marble. Figure 176 illustrates the fragmentation of slate

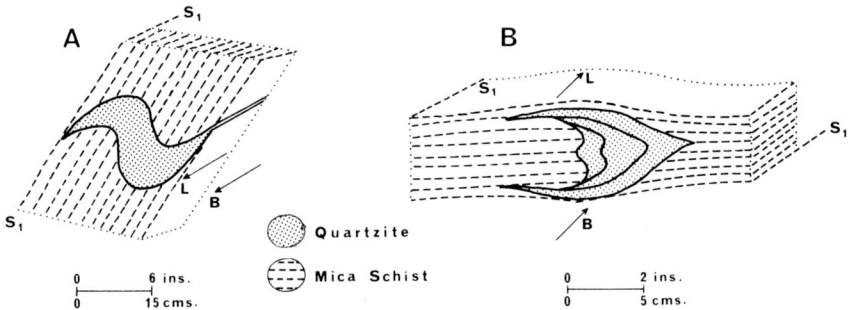

Figure 177. Diagrammatic sketches of tectonic inclusions, Perthshire, Scotland: relict fold closures isolated during formation of transposition structures. S_1 is axial-plane schistosity, B is parallel to local fold axis, and L is the direction of maximum elongation of tectonic inclusion (After Rast, 1956, Figs. 1a and b).

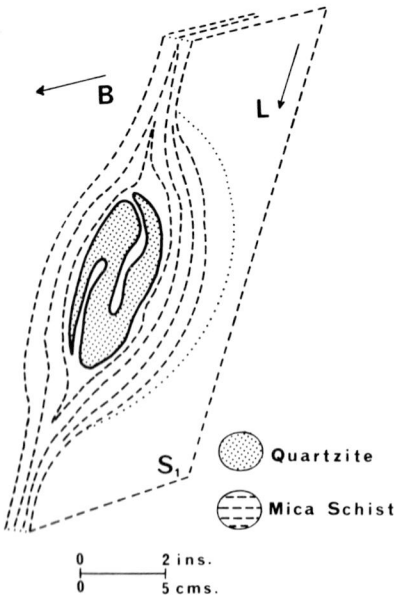

Figure 178. Diagrammatic sketch of tectonic inclusion, Perthshire, Scotland: contorted relict of fold closure affected by considerable rotation and deflection of S_1 (axial-plane schistosity). B is parallel to local fold axis and L is direction of maximum elongation of tectonic inclusion (After Rast, 1956, Fig. 1c).

members in limestone in the Valle de Arán, Central Pyrenees, described by Kleinsmiede (1960).

Rast (1956) suggested that the nomenclature of rootless folds and other small tectonic inclusions resulting from compressional forces should be more precise, so as to avoid confusion with boudinage phenomena (see Chapter 9). Following a detailed study of the area between Kinloch Rannoch and Dunalistair, Perthshire, Scotland, Rast classified tectonic inclusions as:

(a) *Relicts of folds or fold closures:* rootless fold closures isolated during the formation of transposed structures (Fig. 177). Such relicts have greatest dimensions parallel to the B-geometric axis.

(b) *Contorted relicts:* as shown in Figure 178, these have undergone considerable external rotation, which causes the schistosity to be deflected round the inclusions, and commonly the tectonic inclusions have little extension parallel to B.

Not all pseudo-bedding structures in metamorphic rocks are solely due to transposition. Metamorphic differentiation frequently develops alternating quartzo-feldspathic and micaceous laminae at the expense of compositionally-isotropic rocks (Turner, 1948, p. 151; Turner and Weiss, 1963, pp. 99-100). This lamination frequently develops parallel to schistosity (and thus parallel either to bedding or to pseudo-bedding resulting from transposition) in regionally-metamorphosed rocks. The low-grade Otago Schists, New Zealand, are tectonites that exemplify this purely metamorphic banding or pseudo-stratification (Turner, 1941); where extensive contacts between schists of markedly different composition have been mapped, the planes of contact are parallel to the schistosity and to the small laminated structures. However, there was considerable internal mobility within each lithologic unit, because commonly the pseudo-stratification (produced by metamorphic differentiation) has been obliterated locally by transposition structures during later stages of the deformation (Turner, 1941, p. 5).

EXAMPLES OF LARGE
TRANSPOSITION STRUCTURES

When large disharmonic structures occur, the regional tectonic style can be appreciated only after study of the folds in structures of all sizes. In the Scottish Grampian Highlands pioneer work by officers of the Geological Survey, and by Sir Edward Bailey (e.g., 1913; 1922, 1934; Bailey and McCallien, 1937) in particular, resulted in maps of the main stratigraphic units over large tracts of country. Others (e.g., Anderson, 1956) have now completed virtually all of the basic lithologic mapping of this region, so that a start towards visualizing the broad tectonic pattern has been possible (Fig. 326).

Within the Grampians, deformation of the metasedimentary rocks increases northwestwards across the Dalradian and into the Moine rocks, so that effects of compression and recumbent structures become more prominent northwestwards. Most of the rapid variations in formation

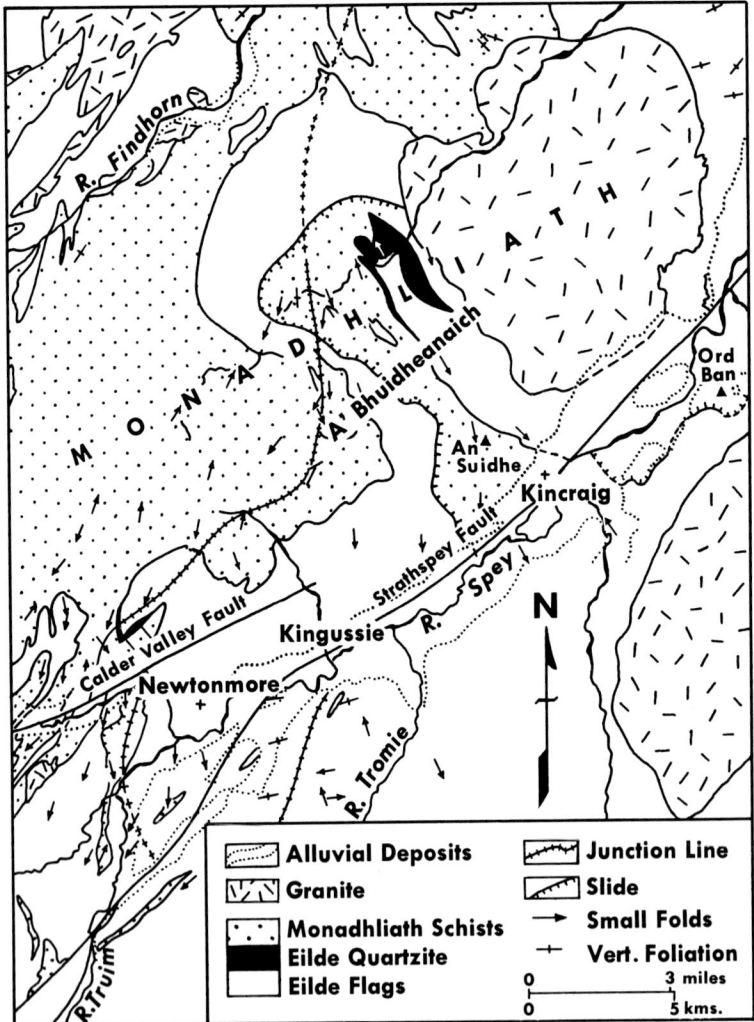

Figure 179. Geological map of the Dalradian metasedimentary rocks and Caledonian granites of the Monadhliath and mid-Strathspey, Central Highlands, Scotland (After Whitten, 1959A, Fig. 1).

thickness, and also the striking apparent stratigraphic discontinuities revealed by the mapping, seem to be primarily due to tectonic features. On a regional scale, **slide tectonics** resulted in large tectonic discontinuities between major stratigraphic units (see Chapter 14). Much of this structural and stratigraphic complexity developed because dissimilar structural styles arose in each lithology. The distinctive physical characters of each stratigraphic unit caused them to retain discrete identity during disharmonic folding. Detailed mapping has shown that quartzite units (e.g., the Eilde Quartzite, Fig. 179) which were originally a hundred meters or so thick, but which were bounded by pelitic rocks, may be wholly or partially attenuated in the limbs of major isoclinal folds, so that a significant thickness is preserved in the fold closures only. Regional transposition structures are developed, and in the process huge "rootless" fold closures of competent and cohesive rocks become isolated as tectonic "fish" within pelitic rocks. Although the size is changed, such structures are similar in all other respects to the structures of outcrop size described earlier.

The type of regional pattern that can develop under these conditions was demonstrated by McIntyre (1951B), who showed that both the Grantown and the Kincraig Series, Scottish Grampian Highlands, are giant isolated tectonic inclusions (Figs. 89 and 180). Weiss (1959A) proved an analogous tectonic style in the Turoka area, Kenya, where large tectonic inclusions of quartzite and marble became isolated within undifferentiated gneissic rocks (Fig. 181).

Development of tectonic inclusions (tectonic "fish") inevitably involves considerable mobility *within* individual lithologic units (cf. Engel, 1949; King and Rast, 1956A). However, during deformation leading to the complete isolation of tectonic inclusions (cf. Figs. 180 and 181), the main lithologic units appear to remain discrete. Thus, in the Monadhliath (Fig. 179), the rock-types and their sequential stratigraphic disposition accord with those of the less highly deformed areas of the Scottish Highlands to the west, described by Anderson (1956). In the broadest sense, the present apparent succession across the schistosity is stratigraphically valid, although a tectonic break occurs *between* each major stratigraphic unit, and marked internal mobility occurred *within* most of the lithologic units.

Penetrative deformation continued over a long period in the pelitic and semipelitic Dalradian and Moine rocks of the eastern Grampian Highlands, and during this time direct and indirect componental movements were important. The general nature of the fold styles, and the apparent lack of recumbent folds in the pelitic and semipelitic rocks, erroneously suggest that schistosity is sensibly coincident with original

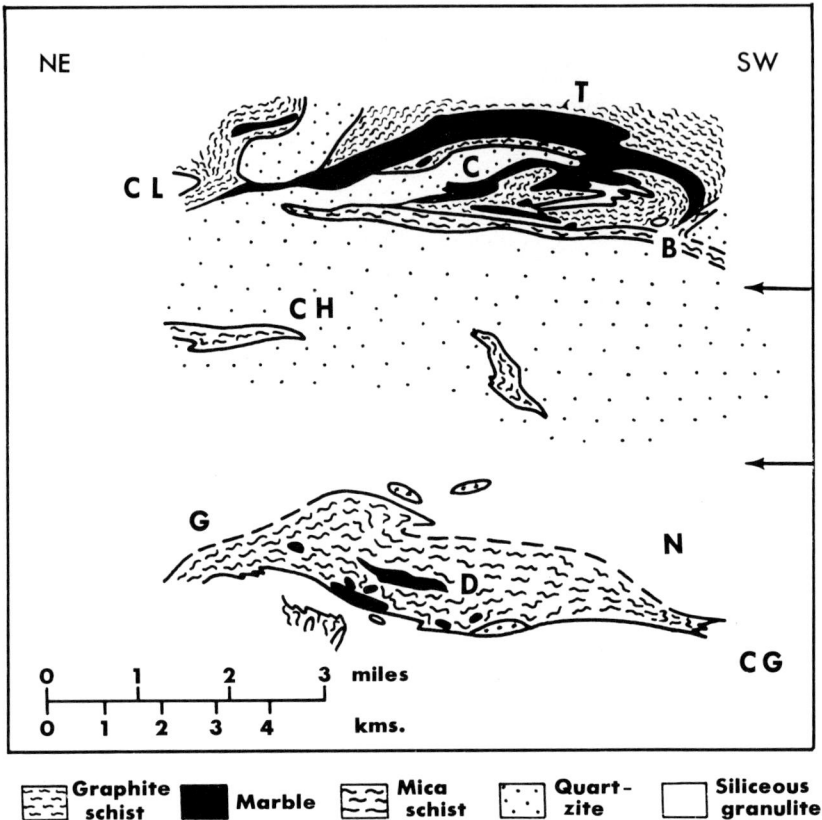

Figure 180. Profile of the homoaxial structure in the Grantown-Tomintoul area, Central Highlands, Scotland, shown in Figure 89. Projection plane strikes 030 and dips NW at 64 degrees. For convenience, the profile above the upper arrow is displaced 2.82 kilometers to SW relative to the part below; a similar displacement has been made at the level of the lower arrow. B indicates Bridge of Brown; C, Cnoc Lochy; CG, Craig Gartan; CH, Cromdale Hills; CL, Carn Liath; D, Dulnan Bridge; G, Grantown; N, Nethy Bridge; T, Tomintoul (After McIntyre, 1951B, Fig. 10).

bedding over most of the area. In the Monadhliath, bedding (S_0) has been completely obliterated in the pelites; even the S_0-fold closures have been eliminated. The micaceous schists developed successive new axial-plane schistosities, while the continuing deformation caused isoclinal folding of each earlier axial-plane S-structure in turn. Gentle folds of the youngest axial-plane schistosity surfaces preserve the impress of the last phase of deformation (Fig. 182). The existing folds in semipelitic rocks only rarely have axial-plane schistosity, and even when present, it is less well developed than the schistosity parallel to the pseudo-bedding (Fig. 183).

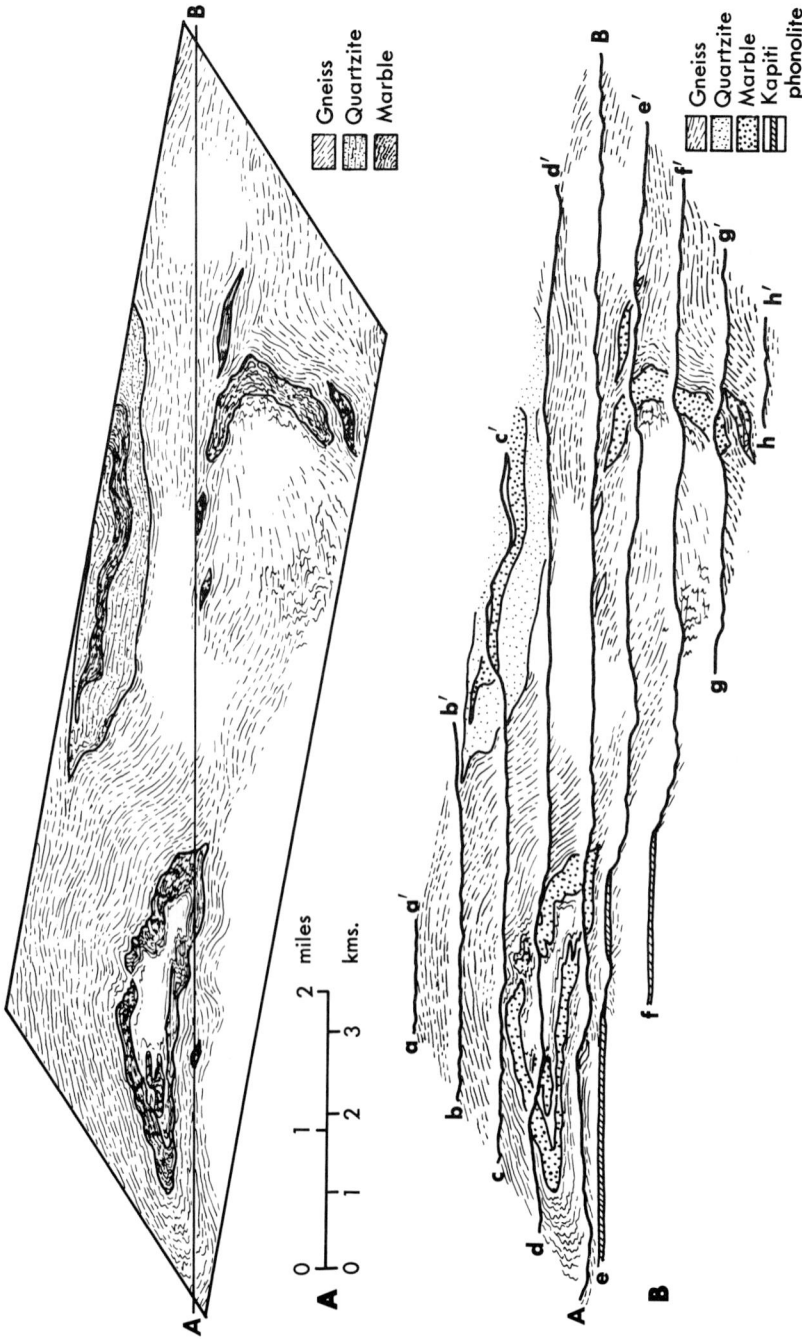

Figure 181. Tectonic inclusions of quartzite and marble in the Turoka area, Kenya. A: Transverse profile of area; plane of projection strikes 332 and dips 70° W. B: *Coulisse* profiles; topographic profile *A − B* is parallel to *A − B* on A (After Weiss, 1959A, Plate V).

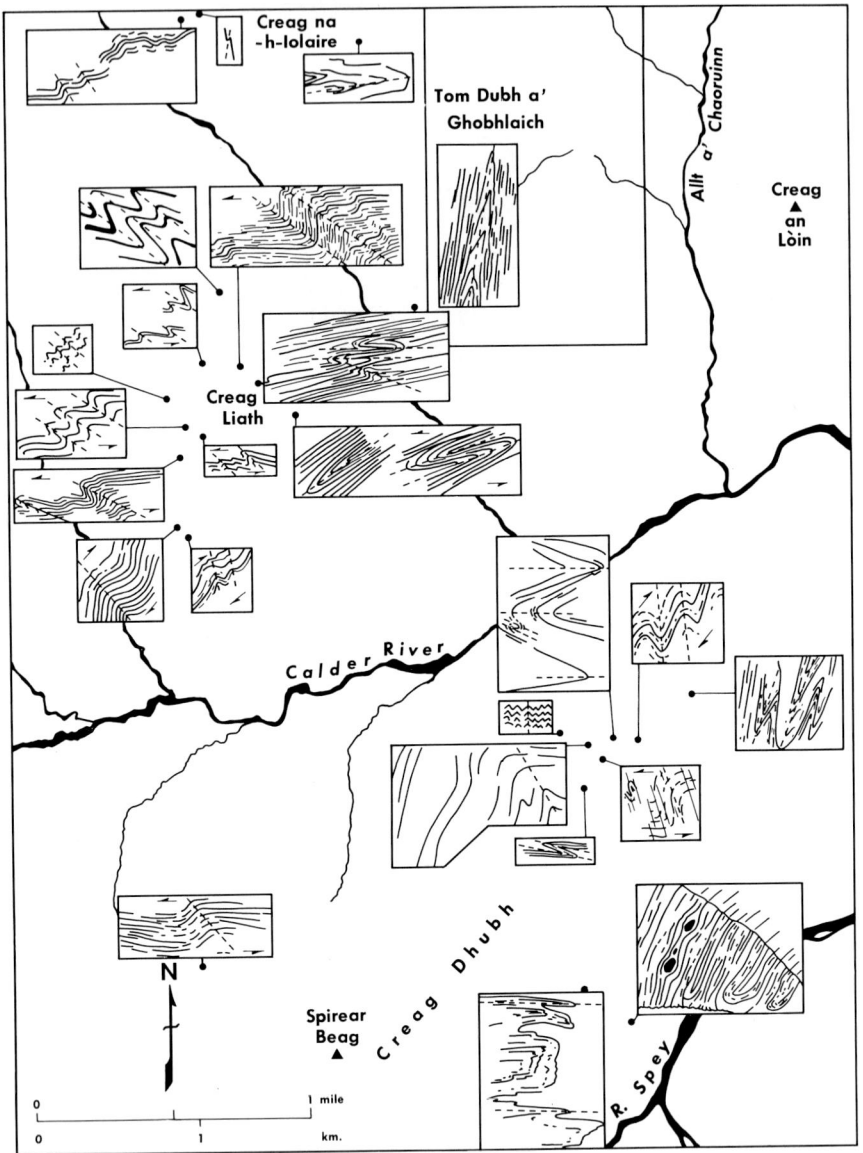

Figure 182. Fold styles, NW of Newtonmore, mid-Strathspey, Central Highlands, Scotland. All profiles drawn looking in a northerly direction along fold axes. Inset frame at north margin of map (about Tom Dubh a' Ghobhlaich) is location of Figure 183 (After Whitten, 1959A, Fig. 3).

Within the mid-Strathspey area the more internally-cohesive rocks (quartzites and marbles) retained traces of bedding that help to define recumbent isoclinal folds; schistosity is strongly developed but it is not

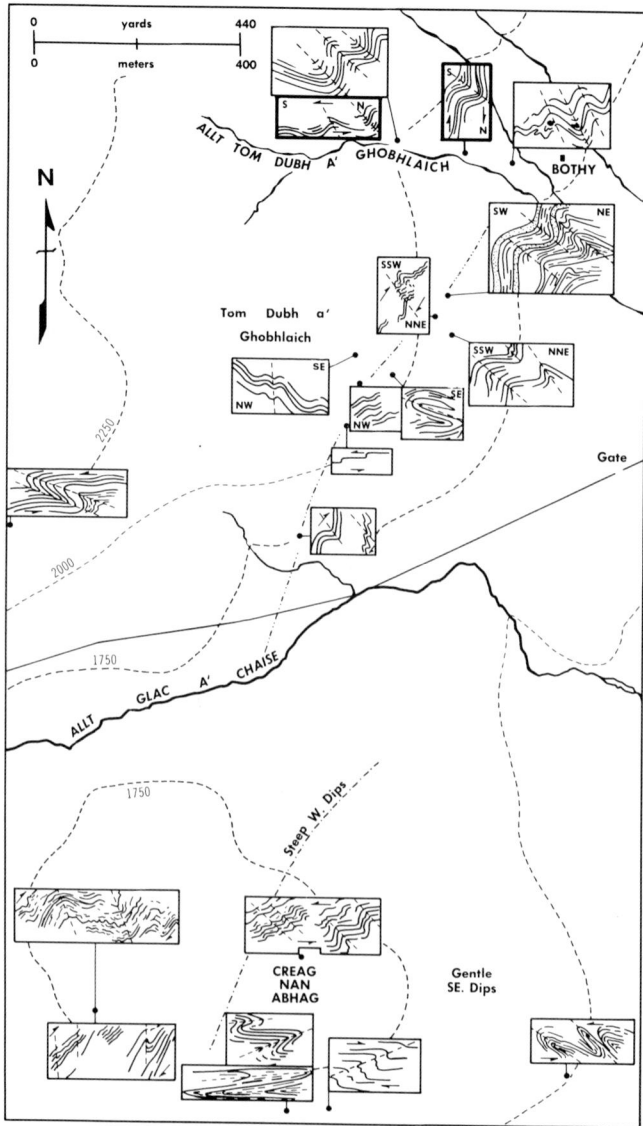

Figure 183. Fold styles, NW of Newtonmore, mid-Strathspey, Central High-lands, Scotland—inset area of Figure 182. All profiles drawn looking in a northerly direction along fold axes (except where the geographic orientation is labelled). Contours in feet (After Whitten, 1959A, Fig. 4).

very obvious macroscopically in these rock types. McIntyre (1951A), Weiss *et al.* (1955), and Whitten (1959A) described examples from An Suidhe (Figs. 179 and 184) and Ord Ban (Fig. 179).

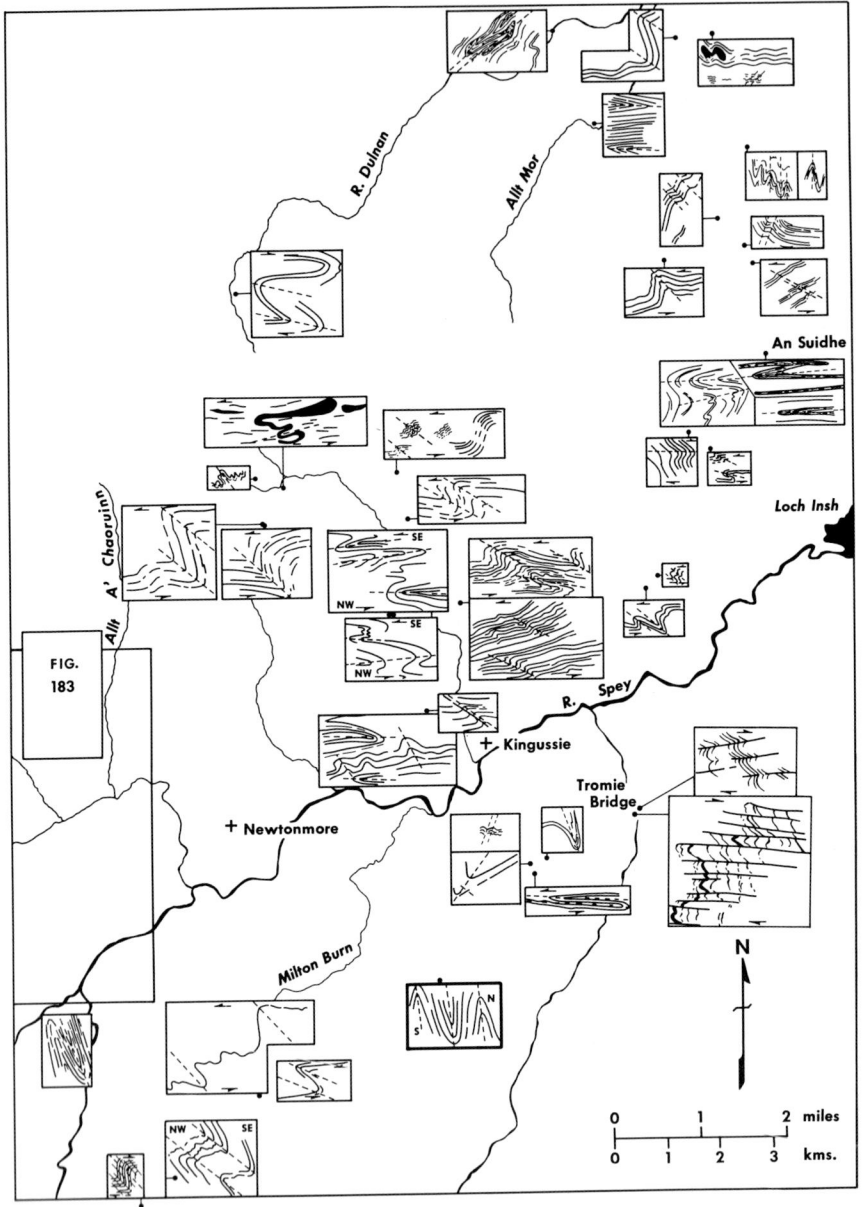

Figure 184. Styles of small folds in the Monadhliath and mid-Strathspey, Central Highlands, Scotland. The folds occur in several different lithologies (see Fig. 179); all profiles are drawn looking northwards along axes, except where geographic orientation is labelled. Small inset box shows location of Figure 183, and the large inset box shows location of Figure 182 (After Whitten, 1959A, Fig. 2).

McIntyre and Turner (1953) and Weiss *et al.* (1955) noted that a foliation is parallel to bedding in the marbles and quartzites at Ord Ban; this rather unusual feature is a result of the lithologies (quartzite and marble) involved, which behave very differently from pelitic and semipelitic rocks. Weiss (1954, p. 58) reported similar relationships in tectonites from a small area in the Mohave Desert, California, U.S.A.; interfolded marbles retained sedimentary bedding and a foliation parallel to the bedding apparently had no mechanical significance during folding. Again, in the Turoka area, Kenya (Weiss, 1956, p. 312), marbles and quartzites retained bedding relics, although in associated tectonites, and especially in the micaceous gneisses, the foliation was a structure in kinematic equilibrium during deformation; that is, it was "... a structure continually being formed, folded, transposed, and reformed during progressive deformation."

The recumbent folds in each of these areas imply very marked disharmonic folding between quartzites and adjacent members of the stratigraphic succession. The original bedding surfaces (S_0) between lithologic types became "unstuck" during development of transposition structures. The bounding surfaces of the isoclinally-folded lithologic units are broadly parallel to the schistosity of the pelitic rocks. Seemingly, the forces that promoted folding in the Monadhliath area retained a similar orientation over a very long period, because folds of both bedding and schistosity are essentially homoaxial (Whitten, 1959A). In this area folds developed in dissimilar lithologies possess geometric and kinematic *B*-axes that are essentially homoaxial. Despite this homoaxial deformation, there is evidence for another generation of folds in the Monadhliath region (Chapter 11).

As mentioned above, axial-plane schistosity and pseudo-bedding, developed at an early stage of deformation, can be folded during a later phase of the same orogeny to produce a new generation of cylindroidal folds. Under such conditions several sets of *S*-surfaces commonly occur within a single area, and because each set has a different function and age, they are not combined on structural maps. In his classic study of the tectonics of the Ketilides, southern Greenland, Wegmann (1938, pp. 52 ff.) described complex structures of this type and mapped the following sequence of *S*-surfaces:

(a) original sedimentary surfaces (or flow layers in lavas, etc.);
(b) surfaces of kinetic origin formed during the first or succeeding generations of folding;
(c) surfaces generated during migmatitic movement and migmatite development; several separate generations may be involved;
(d) postmigmatitic surfaces developed mainly in the mylonitic phase.

Wegmann found that in tracing one set of *S*-surfaces of tectonic origin, relicts of earlier *S*-surfaces are commonly found. Southward from Unartok Fjord well-defined folds are visible; these are not folds of S_0 but of younger tectonic *S*-surfaces, and he concluded that the earliest *S*-structures have often been refolded several times. Wegmann also described migmatized examples of similar structures from Angnalortok Island in which fragments of the original metasedimentary rocks are embedded in a fine-grained matrix; the rocks superficially resemble a breccia (Fig. 185). In less intensely deformed zones, the matrix comprises the most highly kinematically-transformed part of the original rock. Folds can be traced within the matrix where the deformation is less advanced.

A. G. Jones (1959) drew attention to disharmonic refolded structures within the Shuswap terrane, British Columbia, Canada. Folds in these rocks are not readily recognized, because strong axial-plane shearing and thrust faulting commonly obscure the fold crests and produce a misleading regularity of layering. However, where exposures are favorable, as in the Monashee Group in Hunters Range, refolded isoclines—a product of continued homoaxial deformation—are exposed clearly (Fig. 186). Jones recognized some major fold structures (Fig. 187) on the basis of the numerous minor structures that include many small isoclinal flexures.

Figure 185. Folded schist of the Sermilik Group, Angnalortok Island, S Greenland. Folded fragments of more competent units are enclosed in fine grained highly kinematically transformed matrix of quartz, biotite, and feldspars (After Wegmann, 1938, Fig. 19).

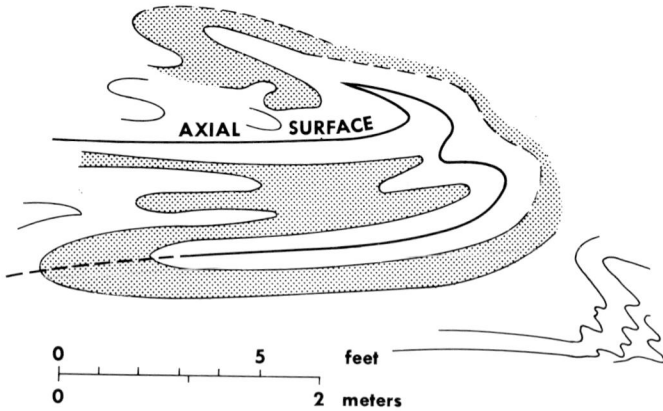

Figure 186. Homoaxially refolded isocline in Precambrian Monashee Group, Hunters Range, Shuswap terrane, British Columbia, Canada (After A. G. Jones, 1959, Fig. 16).

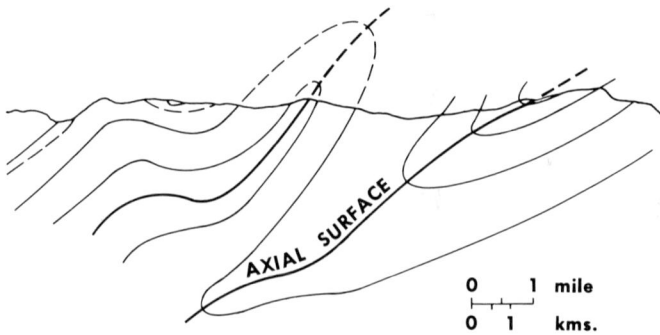

Figure 187. Simplified section through Fosthall Mountain, Shuswap terrane, British Columbia, Canada. The younger folding of the isoclinal structure was recognized and interpreted on the basis of numerous minor structures and small isoclines (After A. G. Jones, 1959, Fig. 15).

Disharmonic structures are particularly common in the Archean basement rocks. Many examples could be cited, but those at Pargas, SW Finland, are convenient examples that have been thoroughly studied (e.g., Metzger, 1945, 1947, 1954, etc.). Most of the geometry shown in Figures 188 through 192 has been proved by either quarrying or diamond drilling down to the 250 meter level.

Many of the major stratigraphical anomalies of orogenic belts can be interpreted in terms of the limited stratigraphical significance of the junctions between lithologic units. Vast successions are sometimes cut out along slide planes without any direct evidence of the precise surface

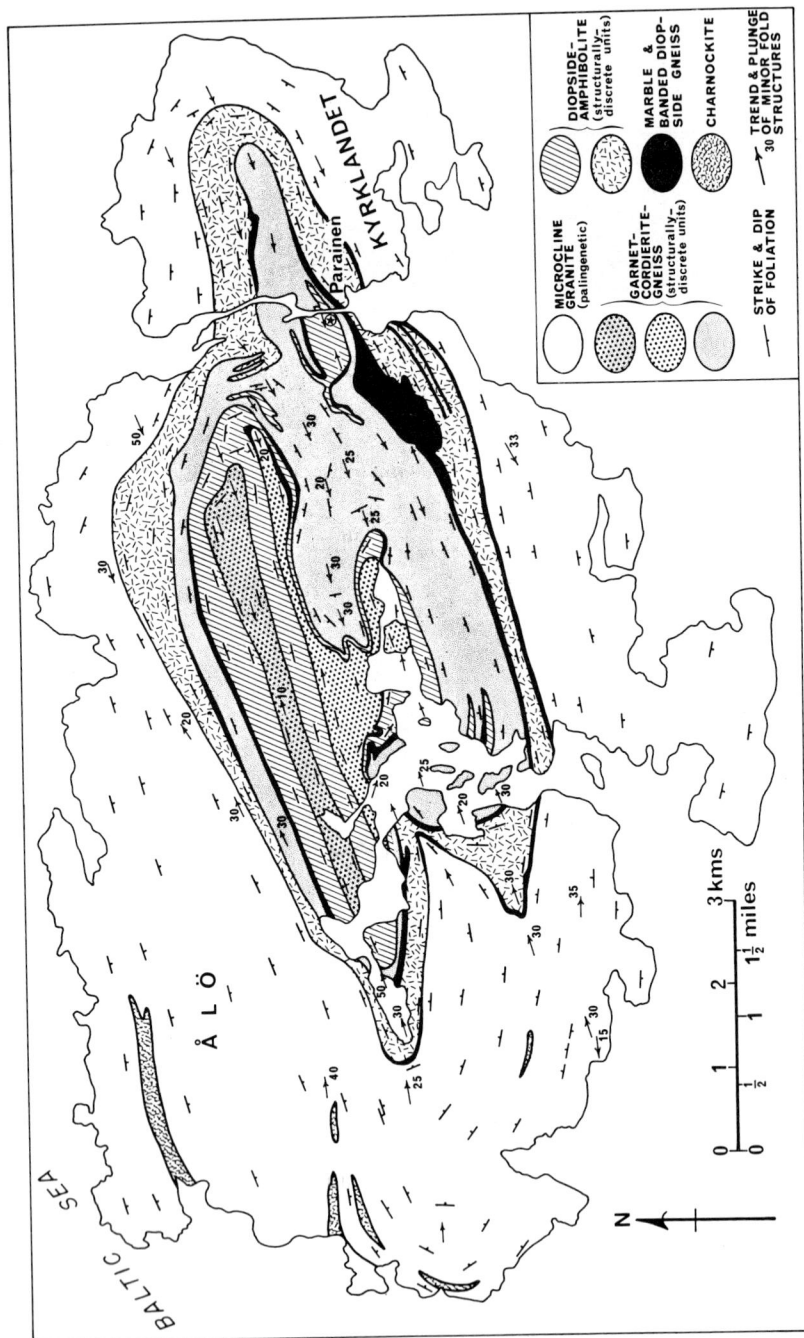

Figure 188. Structural geology of Ålö and Kyrklandet, Parainen (Pargas), SW Finland (simplified from manuscript map by Prof. Dr. A. A. T. Metzger).

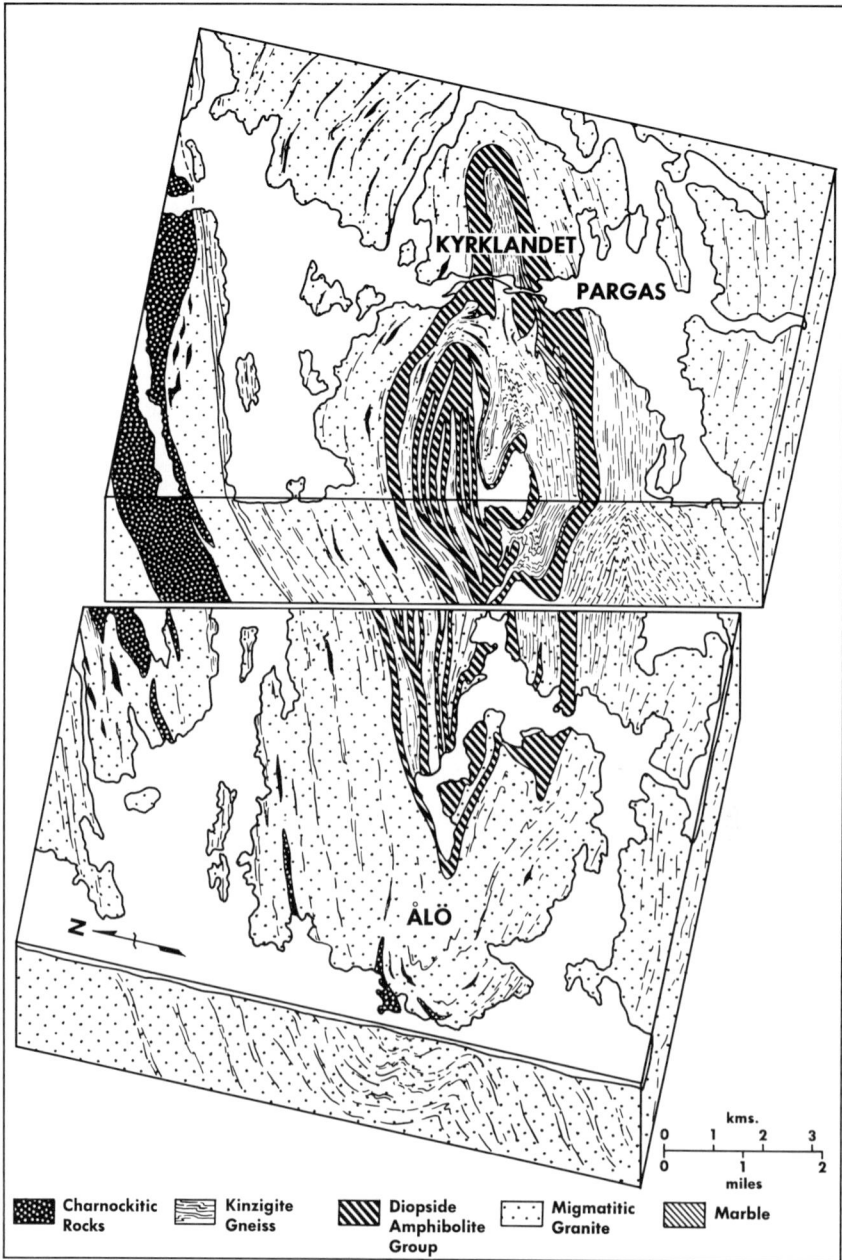

Figure 189. Stereographic block diagram of the Svecofennidic structures of Ålö and Kyrklandet Islands, Pargas-Parainen, SW Finland, based on work by V. Hackman and A. A. T. Metzger (After Metzger, 1945, Plate 18).

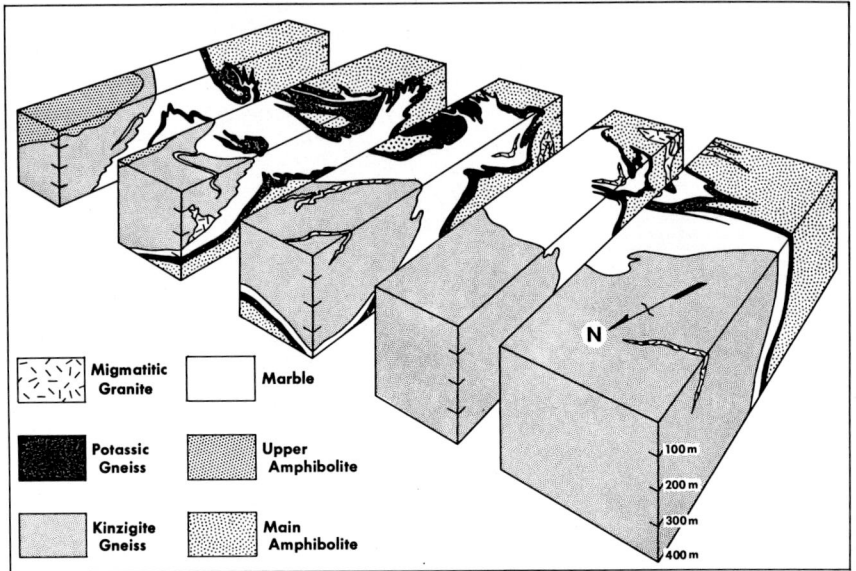

Figure 190. Stereographic block diagram of the Limberg Marble Quarry area, Pargas-Parainen, SW Finland, to show the geometry of Svecofennedic fold styles proved by quarrying and drilling (After Metzger, 1945, Plate 22).

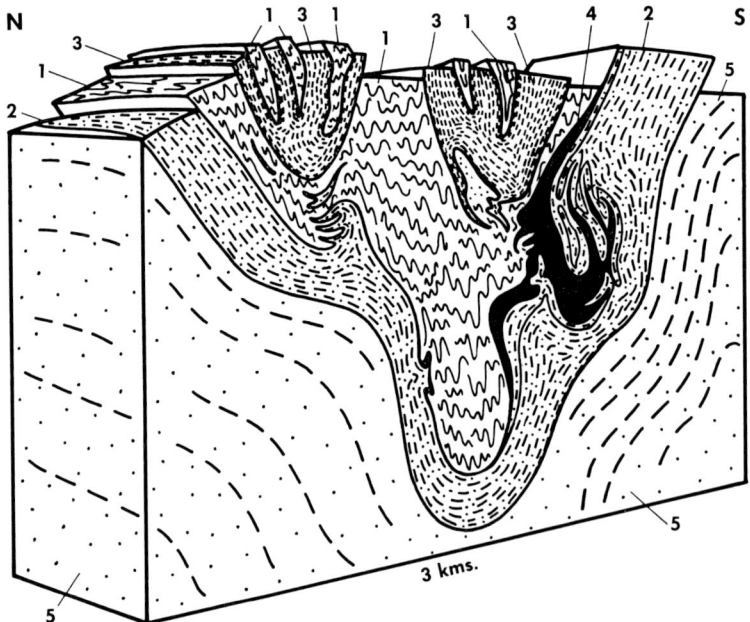

Figure 191. Section through the area quarried at Pargas, SW Finland; 1 = pelite: 2 = main amphibolite: 3 = upper amphibolite: 4 = marble: 5 = palingenetic granite (After Metzger, 1947, Fig. 1).

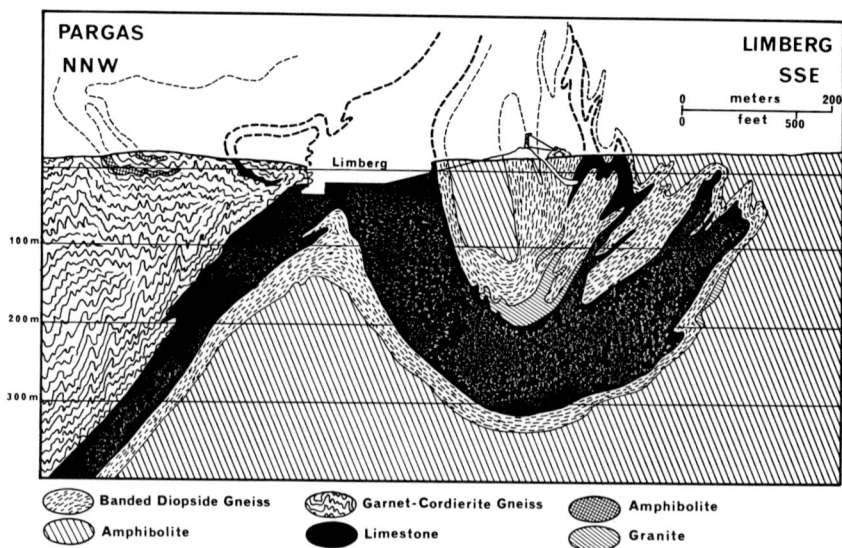

Figure 192. Profile of Precambrian fold structures seen in Limberg Quarry and adjacent area, Parainen (Pargas), SW Finland (After Metzger, 1954, Fig. 21).

along which the movement occurred; these are the enigmatic **tectonic slides** invoked by Bailey (1909, 1910, etc.). Ballachulish, Argyllshire, Scotland, is one classic area in which Bailey (1910) mapped a large slide. The trace of the slide crosses well-exposed ground, particularly on the foreshore of Loch Leven (Figs. 193 and 335), where the formations brought into juxtaposition have similar lithologies. The slide is not marked by brecciation or crushing so it is only possible to surmise its exact location on the ground.[1]

Mapping in the Monadhliath, mid-Strathspey, established (Whitten, 1959A) a slide which causes repetition of the Dalradian metasedimentary succession in the An Suidhe area (Figs. 179 and 195). Development of this slide was apparently penecontemporaneous with the main deformation, because the slide surface appears to be cylindroidally folded and coaxial with the minor structures that plunge SSE (Fig. 179). Such coaxial and penecontemporary relationships may be unusual. Sometimes development of slides may be unrelated to the local minor folds, and represent an entirely separate kinematic event. For example, detailed mapping in the Ballachulish area (Vogt, 1954) showed that the slide

---

[1]Recently, Voll (1965) alleged that these slides and nappes are nonexistent in the Scottish Dalradian rocks; his new evidence will require further careful investigation on the ground.

tectonics bear no obvious geometric relationship to the minor structures (folds, etc.) which seem to have been produced during different phases of the orogeny (Fig. 194).

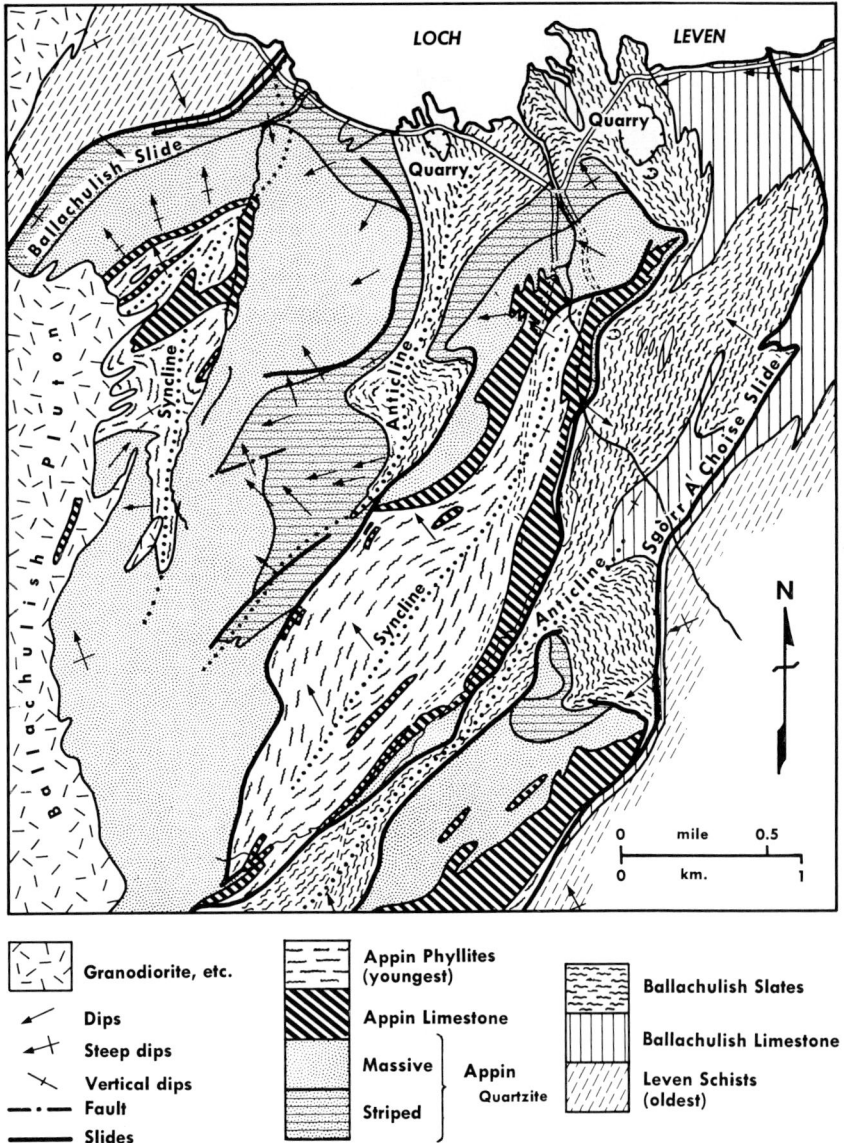

Figure 193. Geological map of the Ballachulish area, Argyllshire, Scotland, showing the Ballachulish and Sgorr a' Choise Slides (After Bailey and Lawrie, 1960, Plate III).

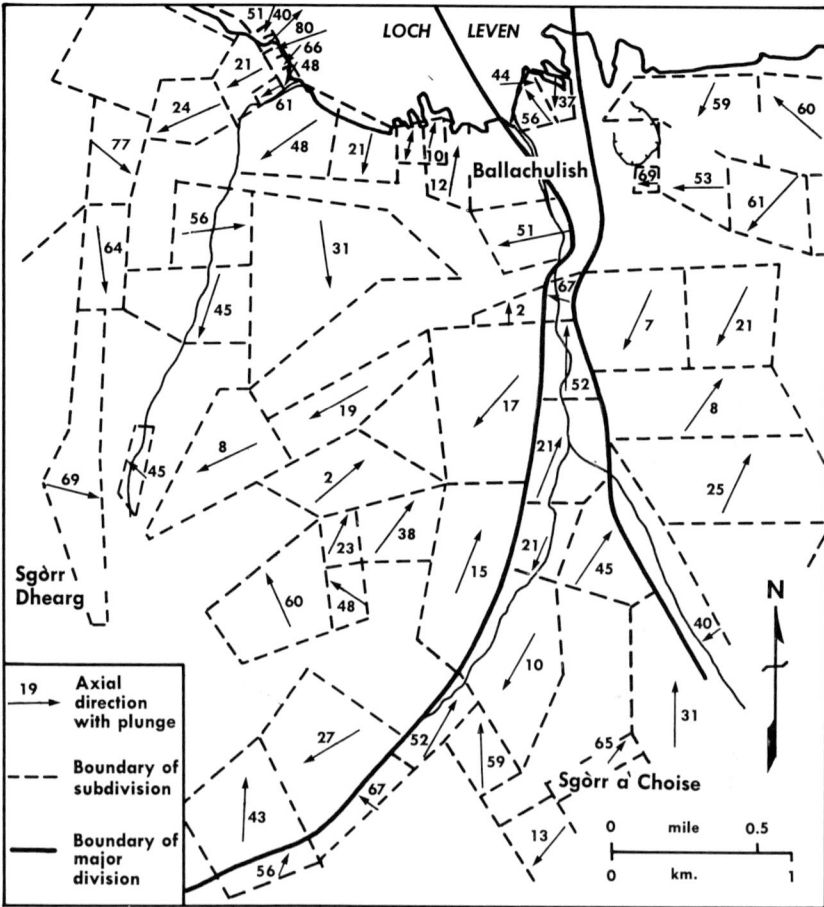

Figure 194. Subareas based on small-fold axial orientations within the Ballachu-lish area, Argyllshire, Scotland (see Fig. 193) (Based on part of unpublished map by Vogt, 1954, Map VI).

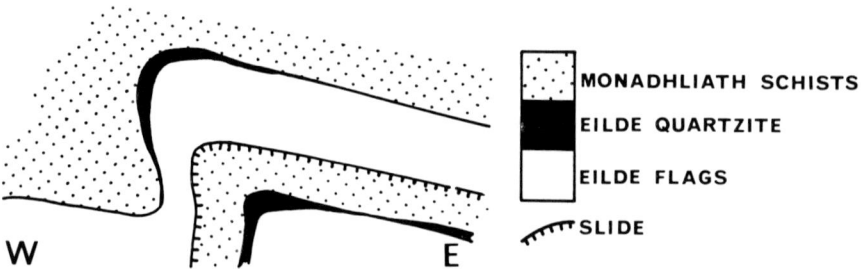

Figure 195. Diagrammatic section across the An Suidhe area, mid-Strathspey, Central Highlands, Scotland, to show the inferred slide tectonics (cf. Fig. 179) (After Whitten, 1959A, Fig. 6C).

Chapter 8

Foliation

Secondary planar and linear structural elements commonly develop during the folding of rocks. A complete description and evaluation of folded rocks includes detailed information about these elements. The planar and linear elements are significant in a geometric analysis and synthesis of an area of folded rocks. Hence, it is appropriate to devote this chapter to secondary planar structures and the next chapter to linear elements.

THE TERMINOLOGY OF FOLIATION AND ROCK CLEAVAGE

A pall of mystery seems to surround **foliation, cleavage,** and **schistosity**. The confusion has been intensified by the complex terminology employed in the literature.

Parallel planar elements give rise to **foliation** (Fairbairn, 1949, p. 5). Foliation may coincide with an original *S*-surface, such as bedding in a micaceous shale. However, secondary foliations only are considered in this chapter, and these may comprise either *S*-planes or *S*-surfaces. Commonly, a deformed rock has a single set of subparallel secondary *S*-structures, but, when long continued or repeated deformation has occurred, this structure may be S_1, S_2, . . . , or S_n. Two or more secondary foliations are sometimes preserved in the same domain.

The preferred orientation of the mineral grains which define secondary foliation structures results from componental movements. However, foliation has been used in a number of completely different senses. Commonly, most of the structures included as foliation in the above definition have been described as cleavage (rock cleavage) in recent texts. Rock cleavage, however, embraces numerous structures (e.g., fracture cleavage of many authors) which are not directly related to foliation.

Use of foliation, as defined above, would remove many confusing aspects of present-day terminology. Knopf (1941, p. 353) claimed that:

In America, "foliation" appears to have now largely supplanted "rock cleavage," and this use of the term foliation is a reversion to an earlier practice that goes back at least as far as Scrope (1825) and harmonizes with the etymologic derivation of the term.

Unfortunately, this statement is misleading because, although it applies to some workers (e.g., Fairbairn, 1935, 1949), "rock cleavage" was used by such authorities as Mead (1940), Swanson (1941), and Billings (1954). Many American authors probably derive their concepts ultimately from Van Hise (1896, p. 633), who defined cleavage in rocks as:

...a capacity present in some rocks to break in certain directions more easily than in others.... The term cleavage is from a property in minerals, and is here confined to a strictly parallel usage.

Van Hise further described **fissility** in rocks as the structure:

...by virtue of which they are already separated into parallel laminae in a state of nature. The term fissility thus complements cleavage, and the two are included under cleavage as ordinarily defined. Where a rock is finely fissile it may be called foliated.

Apparently many geologists follow Harker in employing foliation in a manner similar to the fissility of Van Hise, but in a sense entirely different from that defined at the beginning of this chapter. Harker (1932, p. 203) stated that:

To be carefully distinguished from schistosity ... is *foliation*. The two structures are very commonly found in association.... Darwin [1846, p. 141], who first used the term clearly distinguished this type of structure from schistosity (cleavage), while recognizing a close relation between the two. Owing, however, to the concordance shown by these two directional structures when associated, and to the fact that foliation conduces, together with schistosity, to the characteristic fissile property, a certain laxity of usage is often met with. *Foliation consists in a more or less pronounced aggregation of particular constituent minerals of the metamorphosed rock into lenticles or streaks or inconstant bands, often very rich in some one mineral[1]* and contrasting with contiguous lenticles or streaks rich in other minerals. All these show, at any one place, a common parallel orientation, which agrees with the direction of schistosity, if any, and is manifestly related to the same system of stress and strain in the rock.

The nomenclature controversy thus goes back to Scrope and Darwin in the first half of the nineteenth century. Scrope (1825, p. 233) intro-

[1]Italics supplied.

duced, and quite clearly defined, the term "foliated" in the following words:

> Humboldt asserts the *oldest granites* (that is the *lowest*) to contain most quartz and least mica; the increase of mica creating a foliated structure by the pallets beings placed in a parallel position, and then the rock passes into gneiss, or foliated granite. . . .

Scrope (1825, p. 234) also wrote that:

> . . . protrusion of the foliated rocks, gneiss, mica-schist, clay-slate, etc. was chiefly occasioned, as was observed above, by their peculiar structure; the parallel plane surfaces of their component crystals, particularly the plates of mica, sliding with facility over one another. . . .

Twenty-one years later Darwin (1846, pp. 141 ff.) used foliation in a completely different sense, and wrote:

> . . . that by the term *cleavage,* I imply those planes of division which render a rock, appearing to the eye quite or nearly homogeneous, fissile. By the term *foliation,* I refer to the layers or plates of different mineralogical nature of which most metamorphic schists are composed; there are, also, often included in such masses, alternating homogeneous, fissile layers or folia, and in this case the rock is both foliated and has a cleavage.

Darwin concluded (p. 167):

> . . . I must maintain that in most extensive metamorphic areas, the foliation is the extreme result of that process, of which cleavage is the first effect. That foliation may arise without any previous structural arrangement in the mass, we may infer from injected, and therefore once liquefied, rocks, both of volcanic and plutonic origin, sometimes having a "grain" (as expressed by Professor Sedgwick), and sometimes being composed of distinct folia or laminae of different compositions.

The italicized portion of Harker's definition reproduced above has been quoted very frequently, and the American Geological Institute's *Glossary* (Howell, 1957) attributed to Harker (1932, p. 203) its prime definition of foliation, namely: "The laminated structure resulting from segregation of different minerals into layers parallel to the schistosity. . . . British writers commonly follow this usage." As a second definition it was stated in the *Glossary* that "Foliation is considered synonymous with "flow cleavage," "slaty cleavage," and schistosity by many writers to describe parallel fabrics in metamorphic rocks and considerable ambiguity attends their current use."

These two dissimilar definitions in the *Glossary* reflect the early controversy, but neither is adequate. In this book the term *foliation* is used for planar structures defined by the preferred orientation of component grains or fragments in deformed rocks.[2]

In reviewing the literature extreme care must be used because there are several different varieties of foliation and the definitions used in descriptive accounts are often ambiguous, and the type of structure involved is not clearly and unequivocally stated. For example, Bonney (1919, p. 198) could have meant either of the dissimilar structures referred to in the *Glossary* definitions when he wrote: "... foliation denotes a structure due to a more or less parallel ordering of certain of the mineral constituents in a rock which is not a direct consequence of its stratification."

Holmes' (1928, p. 62) simple definition that cleavage in rocks is the "... property of rocks such as slates, which have been subjected to orogenic pressure, whereby they can be split into thin sheets. ..." is not acceptable; cleavage is a proper description of a common property of most minerals. It is often possible to cleave metamorphic rocks which possess a well-developed foliation. With increase of grain-size the foliations (slaty and slip or shear cleavages) have commonly been referred to as **schistosity,** but such structures are still foliations.

An additional problem is that, although the development of foliation is enigmatic, it is difficult to dissociate genetic and descriptive terminology. Sometimes a structure has been named and classified, often on the basis of uncertain genetic bases, before its geometric and kinematic nature has been adequately described. Although it is difficult to resist the natural desire to understand the reason why a phenomenon occurs, de Sitter (1956B, p. 104) was correct when he wrote:

> Unfortunately, we are not yet in a position to understand why in one case slaty-cleavage develops, in another fracture-cleavage, and again in others sub-vertical schistosity or sub-horizontal schistosity. No doubt variations in the combination of confining pressure, deformative stress, and temperature are the determining factors, but how the system works cannot yet be defined.

Again, in his foreword to a symposium on rock deformation, Hubbert (Griggs and Handin, 1960, p. xi) suggested that perhaps "... it will not be too optimistic to hope that before very long there may emerge an une-

---

[2]This definition agrees with the recent usage of Turner and Weiss (1963, p. 97) who wrote: "Following usage widely prevalent in the United States we use the term foliation to cover all types of mesoscopically recognizable *s*-surfaces of metamorphic origin. Lithologic layering, preferred dimensional orientation of mineral grains, and surfaces of physical discontinuity and fissility resulting from localized slip may all contribute to foliation of different kinds. ... British petrologists restrict 'foliation' to *s*-surfaces defined by lithologic layering and use 'schistosity' and 'cleavage' to cover other types of *s*-surface."

quivocal settlement of the century-old question of the origin of common slaty cleavage."

In discussing the usefulness of foliation in the interpretation of deformed rocks, Mead (1940, p. 1008) concluded:

> ... it is frequently a matter of belief or disbelief, faith or doubt, instead of judgment based on discrimination observation and understanding of the factors and principles involved. Much of the confusion (amongst geologists) is occasioned by failure to realize that there are several types of cleavage or foliation, developed in a variety of manners.

This conclusion echoes Van Hise (1896, p. 636) who wrote:

> ...under the term cleavage two entirely distinct structures of different origins have been confused. Theories which explain or partly explain one of these structures have been extended to cover both of them, because it was not understood that they are different.

The nature of these foliations is investigated below, but it must be recognized that during penetrative deformation all the minerals in a domain develop a fabric. Although foliation of platy minerals (e.g., micas) may be the most obvious, analysis with the aid of a universal microscope stage readily shows that quartz, feldspar, and all other anisotropic minerals also have preferred orientations. This fact has not been widely appreciated (e.g., Kvale, 1948, p. 16) and, because the foliation planes that are obvious in the field are commonly determined by the cleavage of the oriented micas alone, the term "cleavage" has unfortunately tended to enter into common use.

CLASSIFICATION OF FOLIATION STRUCTURES

As mentioned above, it is difficult to dissociate genetic from descriptive terminology in any discussion of foliation. Laudably, Kvale (1948, p. 16), Billings (1954, p. 338), and others advocated descriptive rather than genetic nomenclature for foliation. The majority of recent general descriptions of foliation used the term cleavage or rock cleavage, despite Knopf's (1941) and Turner and Weiss' (1963) claims that foliation is the accepted term in America. Many authors (e.g., Hills, 1963A) followed Leith (1905, 1923) in recognizing two generic types—flow cleavage and fracture cleavage. Many authors (e.g., Mead, 1940; Swanson, 1941; Wilson, 1946; and Billings, 1954) have recognized about four major categories of cleavage. Although modification of the current terminology and concepts is necessary, it is convenient to maintain this fourfold framework; the following categories are considered here:

1. Axial-plane foliation
2. Crenulation foliation
3. Foliation parallel to bedding or lithologic layering
4. "Fracture cleavage"

1. *AXIAL-PLANE FOLIATION*

The planar structure here called **axial-plane foliation** has commonly been referred to as **slaty cleavage.** This is the **flow cleavage** of many authors (e.g., Leith, 1905, 1923; Mead, 1940; Swanson, 1941), or **axial-plane cleavage.** Slaty cleavage is commonly, but not necessarily, parallel to the axial planes of associated folds; there would be some justification for recognizing slaty cleavage as the general type without specification of orientation and axial-plane foliation as a special case. However, slaty cleavage is very commonly approximately parallel to, or only slightly divergent from, the axial plane.

Axial-plane foliation is characterized by five features (cf. Fairbairn, 1949, p. 239):

(a) The foliation is subparallel to the axial surface of flexural-slip and slip folds. In flexural-slip folds the foliation commonly, though not invariably, tends to fan out away from the axial surface so that the foliation converges towards the concave side of the fold. In passing through lithological layers of dissimilar competence the foliation is commonly slightly refracted (Fig. 196); this phenomenon has been described by numerous authors and Furtak (1964) made a detailed geometric analysis of some examples from NW Europe.

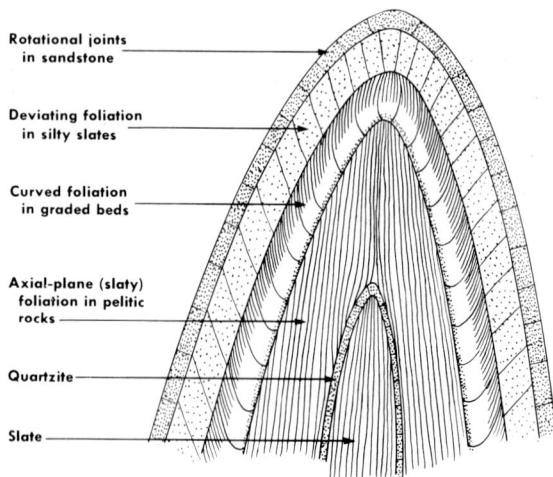

Figure 196. Diagrammatic sketch of foliation attitudes in beds of different lithologies (After Kleinsmiede, 1960, Fig. 44).

(b) The foliation is developed uniformly in the limbs and closures of folds; this applies whether open or isoclinal folds are involved.

(c) Within a given domain the foliation is developed uniformly throughout rocks of the same mineral composition and grain-size; the coarser the grain size of the rock the wider the spacing of successive foliation planes.

(d) The foliation results from neocrystallization; however, recent work by Maxwell (1962) described below suggests that neocrystallization is not always involved.

(e) Characteristically the foliation contains a linear structure. Commonly the linear structures are parallel to the fold axis of the flexures and thus to the intersection of the bedding (or pseudo-bedding) and foliation planes; sometimes the linear structures are perpendicular to the fold axis.

Sedgwick (1835) and Phillips (1844) recognized the general parallelism of the strike of both foliation and bedding in areas where the fold axes are subhorizontal, and that the regional orientation of axial-plane foliation is commonly constant despite local undulations and contortions of the folded bedding planes. Phillips (1844, pp. 60-61) described the mechanical deformation of fossils associated with foliation. From Wales, Devon, and Cornwall, Sharpe (1847) also described fossils flattened in the plane of foliation; he (1849) described flattened volcanic fragments in the green slates of Cumberland and Westmoreland (NW England). Haughton (1856) made the first numerical calculation of strain on the basis of fossils deformed in association with axial-plane foliation. Peach, Muff, and Wilson (in Peach, *et al.,* 1909, pp. 43-46) described deformed pillow lavas from mid-Argyllshire, Scotland, in which the "pillows" are flattened in the plane of the axial-plane foliation.

Early critical observations of this type stimulated hypotheses to explain the origin of axial-plane foliation (slaty cleavage). As long ago as 1847 Sharpe claimed that rocks affected by slaty cleavage have suffered compression perpendicular to the plane of foliation and expansion in the direction of "cleavage dip" (i.e., $\perp B$, in current terminology). Other views were advanced; for example, Fisher (1884A, 1884B) invoked gravitational forces resulting in internal movement parallel to the foliation planes. However, Harker (1885A, 1885B) attacked such views and supported Sorby and Sharpe in the claim that axial-plane foliation forms perpendicular to the largest principal stress during great lateral compression of the rocks. This opinion has found widespread acceptance (e.g., Born, 1929; Fourmarier, 1951, 1952; de Sitter, 1954; Goguel, 1962).

Fairbairn (1949) and de Sitter (1954) drew attention to the fact that axial-plane foliation has often been described as a product of simultane-

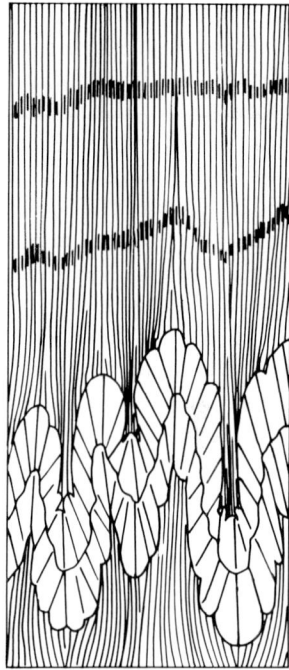

Figure 197. The effect of different lithologies on development of foliation near Ilfracombe, Devon, England. Closely-spaced foliation occurs in fine-grained, dark-colored, shaly slate in which bedding is shown by bands of coarser grain and lighter color. The highly contorted coarser-grained unit is light colored, sandy, slate with less perfect foliation (After Sorby, 1853, and Bailey, 1935, Fig. 5).

ous slip on intersecting shear planes inclined at roughly 45° to the largest stress (i.e., homogeneous irrotational strain of Becker, 1904, and pure slip of Eskola, in Barth, *et al.*, 1939). Sander (1934) described mica schists from the Tyrol in which contemporaneous movement on nonequivalent sets of plane slip surfaces were claimed to have produced the complex movement picture (cf. Fig. 119); contemporaneity in such cases is difficult to prove. De Sitter suggested that two dissimilar kinematic models for the formation of foliation have been advocated by different authors because foliation has often been produced experimentally under conditions differing from those of nature.

Despite the considerable interest in, and understanding of, axial-plane foliation in the nineteenth century, its formation is poorly understood and has been the subject of an extensive controversy in the literature. Despite debate for a century, Wilson (1946, p. 263) wrote, with reasonable justification: "Many British geologists tend to consider cleavage as an unnatural hazard put there 'to make it a bit more difficult' . . .we often hear the palaeo-stratigrapher's lament that the rocks were badly cleaved. . . ."

As a result of microscopic studies, Sorby (1853, 1856) suggested that bodily rotation of flat particles is more important than other mechanisms in the genesis of foliation (Fig. 197). This recognition of two possible processes for the formation of axial-plane foliation marks the first understanding of what, since the work of Sander (1930), have come to be known as indirect and direct componental movements. In many ways it was Heim's (1878) description of Ausweichungsclivage that marked the beginning of the modern understanding of the dissimilar foliation structures in rocks; Ausweichungsclivage is described later in connection with crenulation foliation.

For the Shuswap terrane, British Columbia, Canada, Daly (1915, 1917) claimed that the foliation resulted from recrystallization during static load metamorphism, but careful analysis by A. G. Jones (1959, p. 78) led to the conclusion that the axial-plane foliation can only be explained by a movement hypothesis. Both Swanson (1941) and Fairbairn (1949) reviewed the principal hypotheses about componental movements in foliation formation. Swanson attempted to arbitrate between recrystallization hypotheses (which he attributed to Leith, 1923, and Harker, 1932) and movement hypotheses (which he attributed to Sander, 1930, as interpreted by Knopf and Ingerson, 1938). Actually, Harker (1886) invoked mechanical rotation and flattening for the formation of axial-plane foliation, and then, at a later date, he (e.g., 1932) recognized the importance of neocrystallization. Although Swanson favored the recrystallization group of hypotheses, his argument is not very convincing.

Movement along axial-plane foliations is commonly not visible in thin sections (cf. Breddin, 1931, Mosebach, 1951, Hoeppener, 1956) because of the lack of visible S_0. In a limited number of cases slight compositional variability has permitted the amount of slip to be measured (e.g., Hills, 1963A). Voll (1960, p. 551) wrote that, in common with Breddin (1931), Mosebach (1951), Engels (1955), Hoeppener (1956), and Wunderlich (1959A), he thought that axial-plane foliation started forming as shear fractures. Voll suggested that:

> ... constant presence of more than one set of s_1 planes, the division into two different partial fabrics (mica-films and lenticles of matrix between them) and often the attitude of s_1 forbids the view that slaty cleavage forms purely by growth of platy minerals normal to the maximal compressive stress. Mosebach (1951) has shown that the s_1 planes form an open system in which muscovite is preferentially enriched, whereas the matrix lenticles between the mica films form a closed system with muscovite + chlorite. Mosebach attributes the orientation of muscovite in the films to selection of suitably oriented seeds. During deformation the distance between s_1 mica films is reduced by pressure-solution of clastic quartzes between them.

In most discussions of axial-plane foliation it has been assumed that the recrystallization and reconstitution of the rocks accompanied deformation at elevated metamorphic temperatures and high confining pressures. De Sitter (1954, p. 435) concluded, on the basis of a detailed and microscopic study of folded sericite schists from Rio Lladorre, Spanish Pyrenees, that development of schistosity is favored by uniform fine-grained sediments and that foliation develops earlier in the hinges than in the flanks of folds. De Sitter also found that foliation in pelitic bands between psammitic layers develops at an earlier stage than the general axial-plane foliation in the sequence.

However, on the basis of a detailed study of the Ordovician Martinsburg Formation in the Delaware Water Gap area, Pennsylvania and New Jersey, U.S.A., Maxwell (1962) suggested that axial-plane foliation in slates is not necessarily a metamorphic phenomenon related to folding at great depth; rather, he thought the formation of axial-plane foliation depends on factors essentially independent of the deep burial and elevated temperatures responsible for regional metamorphism.

Maxwell's hypothesis, if correct, has important implications. Therefore, those features of the Martinsburg slates that led Maxwell to his conclusions are worth careful consideration.

Within the Martinsburg slate scattered interbedded lenticular, thin-bedded, and poorly-sorted sandstones occur. The whole sequence has been affected by slip folding, and the penetrative axial-plane foliation (slaty cleavage) is essentially parallel over large areas and throughout 1,000 to 3,300 meters of slate. The lack of distortion of both shearing and foliation at the sandstone-shale contacts (cf. Fig. 196) was taken to indicate the incompetent behavior of the sandstones during folding. Because small sandstone dykes, branching from sandstone beds along

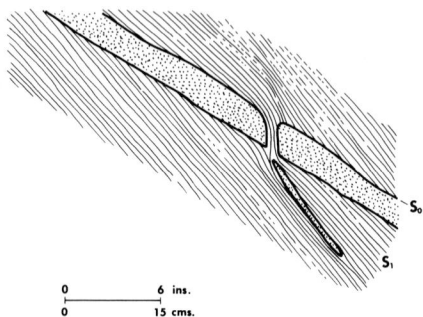

Figure 198. Sandstone dyke parallel to foliation of Martinsburg slate, Martinsburg formation, south of Columbia, New Jersey, U. S. A. (After Maxwell, 1962, Fig. 4A).

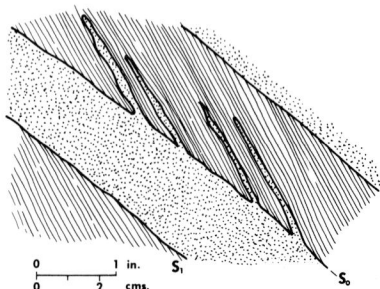

Figure 199. Sandstone dykes parallel to foliation of Martinsburg slate in Hudson River Group, west of Poughkeepsie, New York, U. S. A. (After Maxwell, 1962, Fig. 4B).

foliation planes, were found at two places (Figs. 198 and 199), Maxwell thought the sands were unconsolidated at the time foliation developed; he (1962, p. 288) suggested that sandstone-slate relationships illustrated by Dale *et al.* (1914, Plate 14) from Vermont, U.S.A., and by Morris and Fearnsides (1926) from the Dorothea Grit, North Wales, also indicate that interstratified sandstone members were unconsolidated at the time the foliation formed.

Axial-plane foliation in pelitic rocks is commonly marked by the preferred orientation of micas and chlorites which grew in response to metamorphic conditions. However, for the Lehigh-Northampton belt of the Martinsburg slate, Cuthbert (1946) and Bates (1947) reported an abundance of illite in the fine fraction of the slate, rather than sericite or chlorite.[3] Illite is very unstable under conditions of rising temperature and pressure, and, as Maxwell (1962, p. 299) pointed out, if the dominant mineral of a slate is sericite, then metamorphism probably is a necessary assumption. However, he believed that the large plates of illite in the Martinsburg slate developed from original sedimentary illites as a result of extremely low-grade, or even nonmetamorphic conditions. However, mica (illite?), quartz, carbonate, and rutile grains are elongate with their longest axes in the foliation plane and perpendicular to B. Behre (1933, p. 177) reported that some exceptional quartz grains in these rocks have axial ratios as great as 5:1 and he attributed this to secondary growth. However, Maxwell (1962, p. 298) thought that most of the grains are purely clastic (this seems to be questionable) and that their rather perfect orientation must be attributed to mechanical rotation during development of the foliation; thus, he thought that there is abundant evidence for transport of material parallel to the axial-plane foliation at both the microscopic and hand-specimen level.

Maxwell (1962, p. 295) believed that in the Delaware Water Gap area both folding and axial-plane foliation developed in the slates during the Late Ordovician (Maysville or younger), and that uplift and erosion occurred before sedimentation was renewed to the west in latest Ordovician (Juniata) time. He wrote:

> If the ages assigned are correct, the folding, uplift, and erosion involved, at most, late Maysville and early Juniata time. It seems improbable that any large thickness of rocks existed above the Martinsburg at the time of deformation. It is therefore quite reasonable to assume that the sands and shales of the Martinsburg were water bearing and poorly consolidated as deformation began.

[3] Even today the unequivocal identification of illite is hazardous, and too much reliance must not be placed on data collected in 1946-1947.

Deformation occurred at depths of burial:

> ...probably not exceeding 12,000 feet, hence at no greatly
> elevated temperature. Under these circumstances water (in clays
> and sands) was the plasticizing agent—indeed the only one
> capable of providing the degree of plasticity indicated by the
> perfect orientation of minerals and the bulk transport of material
> parallel to cleavage [axial-plane foliation] without formation of
> discrete planes of shearing or fracture. The shales and sands,
> dewatered as the movements progressed, became brittle and,
> in post-Devonian deformation at greater depth of burial, were
> refolded, fracture-cleaved, thrust-faulted, and brecciated.

Hence, Maxwell suggested that escape of pore water from the
impermeable shaly sediments was so slow that abnormally high pore
pressures developed as deformation commenced; this pressure may have
exceeded the lithostatic (confining) pressure and caused flow perpendicular to the maximum pressure. This flowage could have resulted in the
near-parallel orientation of the clay particles; Maxwell (1962, p. 301)
wrote:

> The essence of the process is the extreme plasticity, even
> fluidity, permitted by relatively large pore-water volumes and
> high pore pressures. During the late-stage escape of the water,
> large volumes must have flowed parallel to the cleavage [folia-
> tion], providing ideal conditions for such diagenetic changes as
> recrystallization and growth of illite and chlorite crystals, move-
> ment of limited amounts of calcite and quartz in solution, and
> deposition of some chalcedonic quartz and perhaps secondary
> carbonate.

Meade (1964) reviewed the meager and conflicting evidence about
the development of fabrics in clays during compaction; clearly there is
little evidence to support or deny Maxwell's hypothesis, and Meade
concluded that the influence of natural compaction on the fabric of
clayey sedimentary rocks needs to be evaluated in conjunction with the
many likely influential factors. Maxwell considered the process he proposed would permit slate to develop between unfoliated shales and
mudstones. He cited several examples described in the literature including (a) mudstones occurring below hard slates in Penrhyn quarry, North
Wales (Smith and George, 1948), and (b) Ordovician red slates associated
with, and underlain by, black graptolitic shales in the New York-Vermont
belt, U.S.A. (Dale, 1899).

Maxwell (1962, p. 308) concluded that appropriate conditions for the
formation of axial-plane foliation occur at two extreme conditions, namely
(a) when pelites have a relatively high connate water content and thus
behave plastically, or (b) after deep burial when high temperatures and

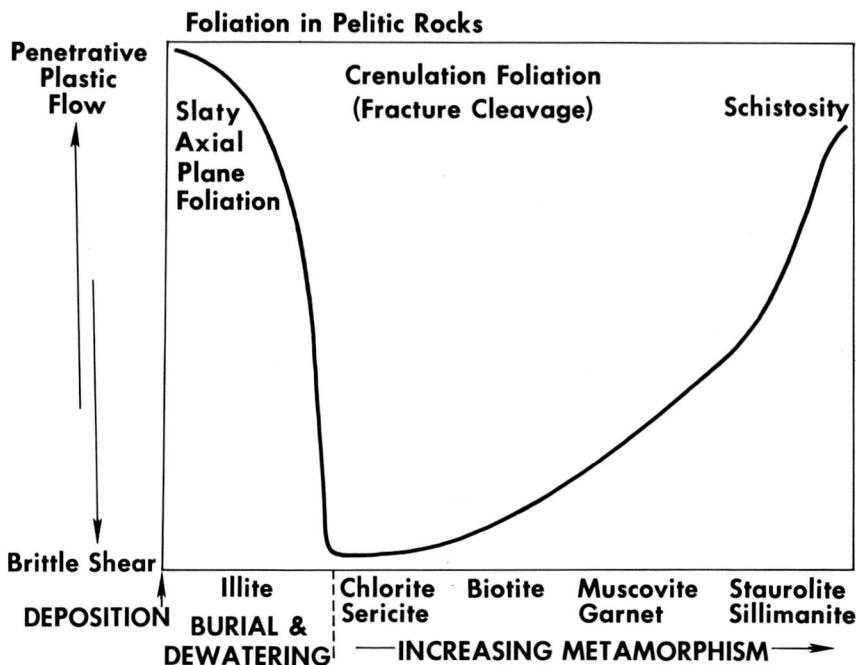

Foliation in Pelitic Rocks

Figure 200. Conditions suggested by Maxwell for development of foliation in pelitic rocks, Dutchess County, New York, U. S. A. (After Maxwell, 1962, Fig. 12).

confining pressures permit large-scale plastic flow. Figure 200 illustrates the relationships visualized by Maxwell for the pelitic rocks of Dutchess County, New York, U.S.A.

Probably both direct and indirect movements are important in the development of axial-plane foliation in different environments (or even possibly in the same domain). However, it is recognized generally that further research is required urgently before any definite pronouncements can be made (cf. Hubbert, in Griggs and Handin, 1960, p. xi; Meade, 1964).

Obvious examples of axial-plane foliation in the field tend to be limited to the more pelitic rocks. However, petrofabric analyses show that most anisotropic minerals are affected by neocrystallization during folding, so that nonpelitic rocks develop a preferred orientation too. Foliation is often difficult to detect in nonmicaceous rocks if the grain-size and composition are similar throughout a sample. However, original phenoclasts (pebbles, etc.) commonly bear a permanent record of the total strain that affected them. For example, pebbles deformed into lensoid shapes and flattened in the plane of foliation are illustrated in Figure 201; the longest axes of these deformed pebbles define a linear structure parallel to B. Similarly, Agron (1963A, Fig. B-6) found that boulders in the Pennsylvanian conglomerate east of Newport, Rhode Island, U.S.A., are flattened in the axial plane and that their longest axes have a preferred orientation parallel to *B*.

Cloos (1941, 1943, 1947A, 1953B) followed the earlier work of Heim (1878) and Loretz (1882) on strained ooliths with a detailed quantitative study of deformed Cambro-Ordovician oolitic limestones in the Appalachian fold belt between Hagerstown and South Mountain, Maryland, U.S.A. On the assumption that the ooliths were originally

Figure 201. Deformed Precambrian conglomerate, Coal Creek Canyon, Front Range, Colorado, U. S. A. Arrows point to bedding S_0 and two lines lie in the bedding (left margin of picture). Elongated cobbles define the axial-plane foliation.

spherical, and that their volume remained constant during deformation, Cloos demonstrated that the spheres were flattened in the plane of the foliation, and that the long axes of the strained ooliths have a preferred orientation normal to the fold axis, i.e., $\perp B$.

The dissimilar orientation of linear elements within the foliation — either parallel to, or normal to, B (as described, for example, by Agron and Cloos) — reflects a fundamental problem in structural geology that is discussed more fully in Chapter 9.

2. *CRENULATION FOLIATION*

The second major category of foliation structures has frequently been called **slip cleavage** (e.g., White, 1949; Brace, 1953; Billings, 1954), a term attributed to Dale (1896, pp. 560 ff.); Dale's illustrations of slip cleavage in phyllites near Rupert, Vermont, U.S.A., are reproduced in Figures 202 and 203. Mead (1940) and Wilson (1946) used **shear cleavage.** Heim (1878, p. 53) was probably first to describe this type of foliation in detail; he called it **Ausweichungsclivage,** which was translated as **strain-slip cleavage** by Bonney (1886, p. 95), as **cataclastic cleavage** by Knopf (1931, p. 17), and as **transposition cleavage** by Weiss (1949, p. 1696).

White (1949, p. 590) noted:

As used by Dale, the term, slip cleavage applies to the microfaults that develop on the limbs of tiny crinkles in foliate rocks

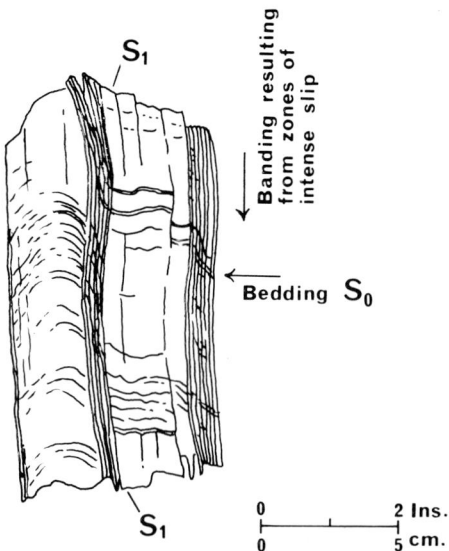

Figure 202. Crenulation foliation: "slip cleavage" in phyllite near Rupert, Vermont, U. S. A., as illustrated by Dale who described "cleavage banding" (S_1) developed along the shanks of the fold plications giving the appearance of stratification across the bedding (S_0) (After Dale, 1896, Fig. 90).

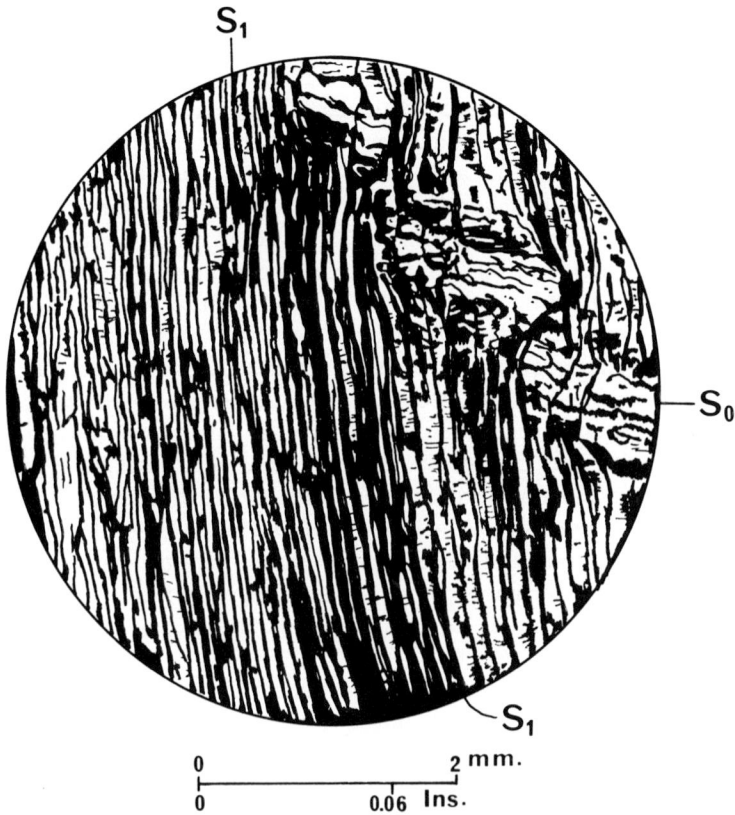

Figure 203. Part of a thin section from intensely foliated (360 slip planes per inch) portion of the muscovite-quartz phyllite shown in Figure 202; bedding (S_0) relicts run from left to right across the picture (After Dale, 1896, Fig. 91).

(Dale, 1896, pp. 566-567) ... the crinkles are essential to development of the microfaults, or slip cleavage, and determine the orientation of this cleavage. This is shown by the fact that the axial planes of the crinkles have the same attitude, whether or not their limbs are faulted. ... The crinkles are the independent part and the slip cleavage the dependent part of the phenomenon. In practice, in the rocks here described, the average attitude of the axial planes of the crinkles may be taken without error as the attitude of the slip cleavage.

White clearly implied that the structure is initiated by flexuring rather than microfaulting.

More recently Knill[4] (1960B, p. 318) and Rickard (1961, p. 325) utilized **crenulation cleavage** in descriptions of Precambrian rocks in Argyllshire, Scotland, and Donegal, Ireland, respectively; this term is a translation of **runnzelsclivage** introduced by Hoeppener (1956). The use of terms such as strain-slip and slip cleavage have unfortunate genetic implications, and, as Rickard pointed out, the structure can almost certainly originate in several different ways. At the same time, the structure is foliation in the sense employed in this chapter, so that **crenulation foliation** can be adopted as the most satisfactory term.

Knopf (1931) published a useful review of the European literature dealing with this type of foliation. She described the manner in which parallelism of the axial planes of small crinkle folds in micaceous rocks produces a parallel structure that crosses the original S-structure (e.g., bedding). As folding increases in intensity, slip along the fold flanks is accentuated and the fold crests become relatively thicker. While the original S-surfaces may be destroyed completely, the new S-structures constitute a transposition of the old structure. Heim (1900) described the progressive developmental stages of this type of foliation (Ausweichungs-clivage) in the Tessin gneiss, Switzerland (Figs. 164 and 165). Figure 204 shows diagrammatized versions of similar structures based on draw-ings by Hoeppener (1956).

Based on a detailed study of the Precambrian gneisses, schists, and marbles of the Grenville Series, northwest Adirondack Mountains, New York, U.S.A., Engel (1949) described the development of foliation. In the more siliceous bands, open folds commonly develop faint fracture and slip S_1-surfaces (Fig. 205, point A) parallel to the axial surface. Small dis-placements sometimes occur along S_1[5] in a manner similar to that in slip folds. Commonly S_1 is defined by faint corrugation, but at this early stage only a small proportion of the minerals is "... grown, rotated, or flattened distinctly parallel to ..." S_1, although such parallelism is slightly more marked in the axial portion of folds. This S_1-structure is a crenulation foliation according to the terminology advocated in this book. On the fold flanks, although little deformation is apparent in hand specimens, amphi-bole, mica, and quartz grains have their longest axes "... somewhat athwart the bedding, and roughly parallel to the axial plane of the fold" (Engel, 1949, p. 772).

[4]Knill's intention is difficult to interpret because he (1960B, p. 318) stated: "... it is suggested that a term such as 'crenulation cleavage' is preferable to 'strain-slip cleavage' in view of the generic implication of the latter," but subsequently, in a tabulated summary of his suggested classification, he (1960B, p. 322) used "strain-slip cleavage."

[5]The foliation is designated S_1 and the lithologic bands S_0; although S_0 suggests that the lithologic bands are defined by original sedimentary layers, this is unlikely in this area of intense deformation and metamorphism.

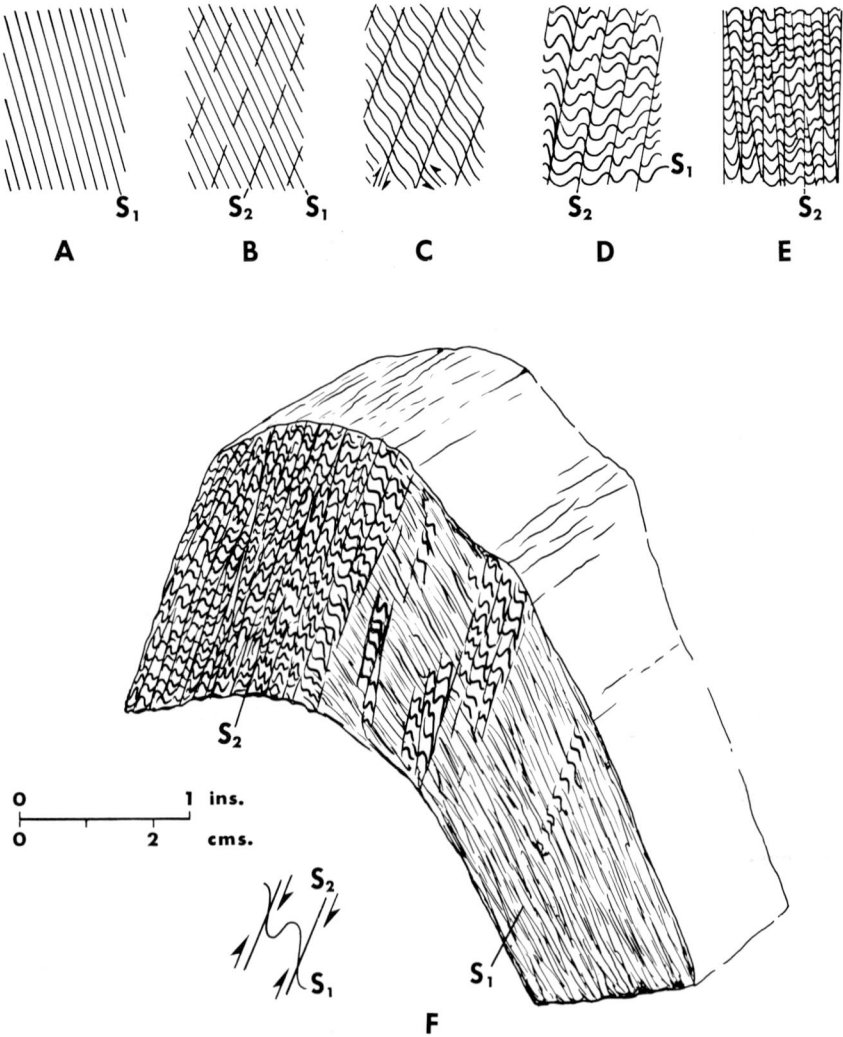

Figure 204. Development of crenulation foliation: A through E: Diagrammatic representation of progressive development of S_2-foliation by transposition and crenulation of pre-existing S_1-foliation (After Hoeppener, 1956, Fig. 24). F: Deformed slate, SW Sierra Nevada Foothills, California, U. S. A., showing the nature of crenulation foliation S_2 and its development by transposition of an older S_1 (After Best, 1963, Fig. 6).

As the folds become more closely compressed (Fig. 205, point B) S_1 becomes more prominent, and S_0 less distinct; while numerous mineral grains have a strong preferred orientation parallel to the axial plane. This type of foliation in the siliceous bands tends to grade directly into axial-plane foliation developed concomitantly in the interbedded pelitic horizons.

With more intense deformation the folds approach an isoclinal form (Fig. 205, point C), and the axial surfaces gradually and progressively rotate to be more nearly parallel to the major lithologic trends. The S_1-foliation takes the form of schistosity approximately parallel to the remaining S_0-fragments (Fig. 205, points C and D); these fragments are usually flanks of folds that are progressively obliterated by slicing and shearing out. In the Adirondack Mountains — as in many other areas — most exposures are of these sheared flank areas of folds, where foliation and bedding relicts — pseudo-bedding — are virtually parallel as a result of complete transposition. Continuing deformation caused folding of these pseudo-bedding structures and a consequent development of a new generation of foliation structures.

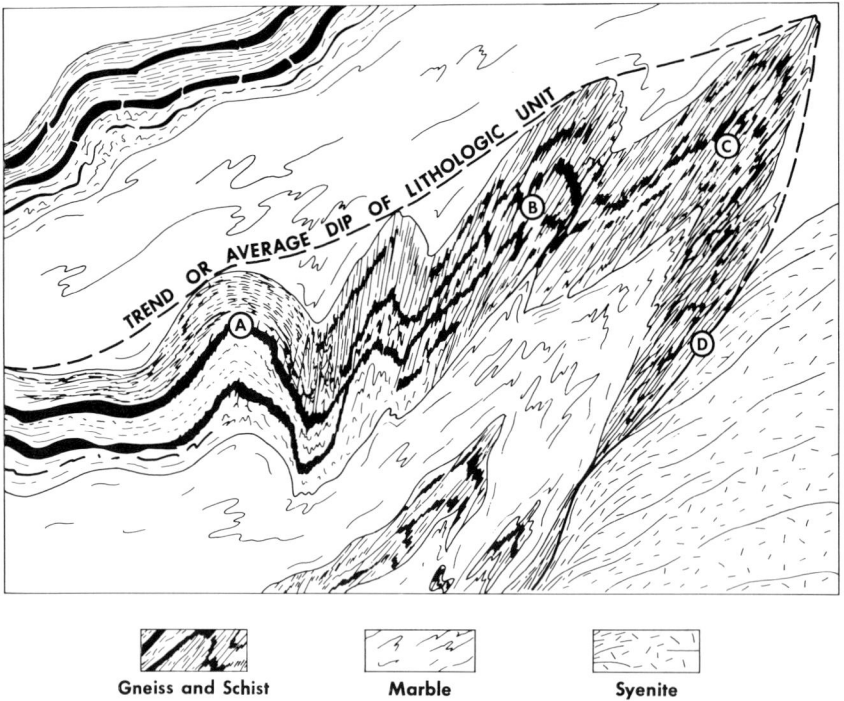

Gneiss and Schist   Marble   Syenite

Figure 205. Schematic examples of foliation development in metasedimentary rocks of the Grenville Series and associated syenite, Adirondack Mountains, New York, U. S. A. See text for significance of points A, B, C, and D (After Engel, 1949, Fig. 3).

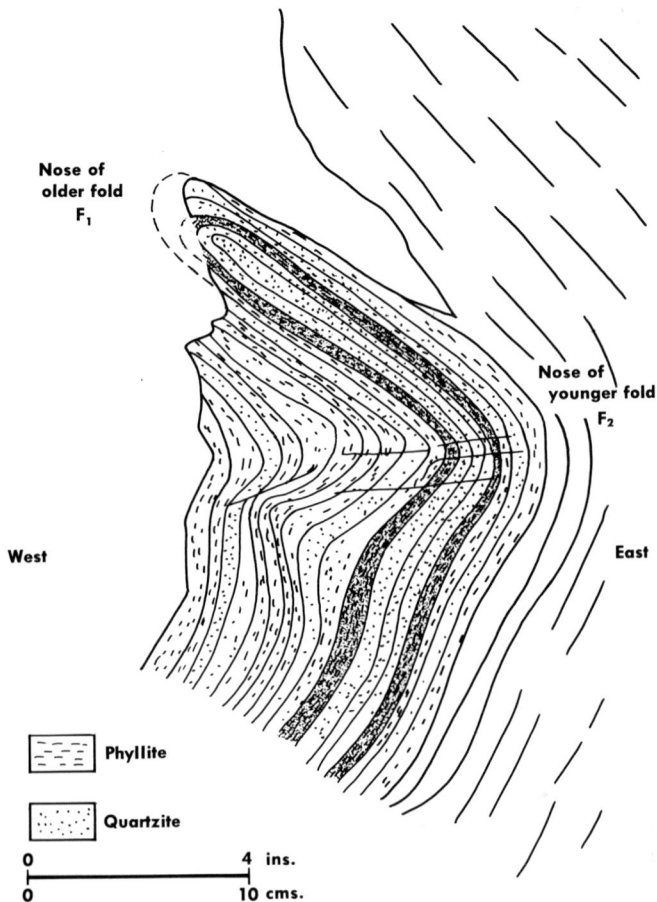

Figure 206. Field sketch of refolded isoclinal fold in laminated phyllite and quartzite series; incipient crenulation foliation is developed parallel to the axial plane of the second fold, Brattleboro, Vermont, U. S. A. (After Moore, 1949, Fig. 5).

Such continuing deformation is common, and numerous examples have been described in the literature. From the Keene-Brattleboro area, New Hampshire and Vermont, U.S.A., Moore (1949) illustrated a refolded isoclinal fold traversed by incipient to well-developed crenulation foliation parallel to the new axial plane (Fig. 206). Woodland (1965) also described bedding transposed during early folding that was accompanied by a strong axial-plane schistosity, S_1, in the Burke Quadrangle, Vermont, U.S.A.; the S_1 was folded during a later phase, and commonly crenulation foliation occurs parallel to the new axial planes. Woodland published an excellent series of photographs illustrating these crenulation structures.

Engel (1949), in common with White (1949), Rickard (1961), and Woodland (1965), believed that, in general, many of the crenulation foliation slip surfaces developed from tiny asymmetric crumples of the S_0 surfaces, and that discontinuous displacement is subordinate to flow and recrystallization. Although this is a common relationship, others can occur and each example must be examined carefully. Beavis (1964) clearly demonstrated that small folds in Ordovician rocks from Victoria, Australia, resulted from movement on the crenulation foliation surfaces.

The relationship between axial-plane foliation and crenulation foliation is an interesting problem which requires considerably more study. Long ago Van Hise (1896, p. 656) commented that, where:

> ... a rock series is composed of layers of different lithological character, and is in the zone of combined fracture and flowage, the deformation includes the development both of cleavage and normal plastic flow and of fissility in the planes of shearing, in both homogeneous and heterogeneous rocks, and perhaps of all gradations between the two.

Axial-plane foliation is essentially penetrative, it develops most readily in structurally and compositionally uniform rocks, and it tends to obliterate earlier structures. Crenulation foliation is less penetrative, it is commonly a feature of structurally nonuniform rocks, and it can frequently be seen crossing the axial-plane foliation developed at an earlier phase. Workers in the Adirondack and northern Appalachian Mountains have been able to trace a specific relationship between these two dissimilar types of foliation in the field; however, it is unknown how widespread such close relationships are, because of the complete lack of quantitative data about the spatial distribution of these structural elements. While sometimes intimately related, the two types of foliation frequently seem to reflect totally separate tectonic events.

An example of the close relationship between axial-plane and crenulation foliation can be illustrated from the Mount Ida group, Shuswap terrane, British Columbia, Canada. A. G. Jones (1959, pp. 68 and 78) reported[6]:

> The foliation in the argillaceous and quartzitic rocks corresponds with the "slip cleavage" of geologists in other fields. Most volcanic members and sericite- and chlorite-rich sedimentary rocks exhibit parallelism of the platy minerals solidly throughout the rock instead of along discrete shear planes. This type of foliation is identical to "flow cleavage" but as the two types are

---

[6]In this quotation slip cleavage corresponds to crenulation foliation as used in this book, and flow cleavage to axial-plane foliation.

parallel in adjacent beds the essential difference between them must lie in the lithology of the rocks affected and not in mode of origin. Apparently the soft chloritic beds sheared throughout, whereas movement in the argillites and quartzites was restricted to definitely spaced planes.

*   *   *

All foliation in rocks of the Mount Ida group lies parallel with bedding over wide areas and clearly originates through movement and shearing. Both flow cleavage and slip cleavage by Swanson's definitions are present and are unquestionably the result of a single process, for in adjacent rocks of different lithology has been found mutually parallel cleavage that in one rock is flow cleavage and in the other slip cleavage. The distinction between flow and slip cleavage in the Mount Ida group is strictly arbitrary as the two cleavages coalesce and grade in spacing, appearance, and origin.

It seems to be a general rule that crenulation foliation can develop only at the expense of an earlier S-surface. In a majority of cases the earlier S is of tectonic origin; in such cases the crenulation foliation is associated with later tectonic events, or episodes of superposed folding (cf. Rickard, 1961, p. 329). However, it is not uncommon to find crenulation foliation developed at the expense of original bedding S_0-surfaces defined by micaceous layers, etc.; hence it is incorrect to follow Kleinsmiede (1960, p. 194) and others in the use of **secondary cleavage** as a synonym for crenulation foliation. When the deformation or metamorphism is sufficiently intense during the development of crenulation foliation, it can grade into a definite new schistosity.

So little is known about the nature of crenulation foliation and its quantitative distribution that it is difficult to discuss in detail the variations which can occur. However, it appears to occur in four principal ways, although existing descriptions only seem to refer to the first three:

(a) parallel to previously- or penecontemporaneously-developed axial-plane foliation;

(b) obliquely crossing earlier foliation of tectonic origin;

(c) as conjugate sets at approximately 45° to an earlier foliation of tectonic origin; or

(d) obliquely crossing original sedimentary S_0-surfaces.

King (1956, p. 99) erected a classification which essentially corresponds to the first three of these members, and commented that he based his ideas upon perusal of actual field descriptions as opposed to textbook definitions. As will be gathered from the sequel, relatively few complete qualitative descriptions of these categories are available in the literature.

Crenulation foliations developed as conjugate sets (c) appear to comprise a distinctive group of structures, whereas (a) and (b) overlap with each other. Where crenulation foliation develops astride an older foliation, the new structure is commonly associated with flexural-slip folding and forms parallel to the axial surfaces of the new folds. In any one area it is common to find that several successive phases of deformation have each given rise to foliations, and under such conditions a particular structure may belong to both groups (a) and (b). The following examples, drawn from the literature, illustrate relationships that are widespread.

(a) CRENULATION FOLIATION DEVELOPED PARALLEL TO PREVIOUSLY- OR PENECONTEMPORANEOUSLY-DEVELOPED AXIAL-PLANE FOLIATION

Rickard (1961, p. 327) described a good example from the Dalradian schists of Donegal, Ireland, in which micas are aligned parallel to the axial surfaces in the limbs of both large and small folds and produce a schistose arrangement resembling axial-plane foliation. Rickard found that crenulation foliation is often preserved in the cores of the minor folds and that it grades into genuine axial-plane foliation; this relationship suggested the latter has been derived from, or developed at approximately the same time as, the crenulation foliation. Some of the banded pelitic rocks show a transitional stage in which a closely-spaced axial-plane crenulation in micaceous bands can be traced as an axial-plane foliation through semipelitic bands.

Comparable examples occur in the plicated Precambrian schists of Coal Creek Canyon, Front Range, Colorado, U.S.A. (Fig. 207), and from the Bosost Dome, Pyrenees, France (Fig. 208); it seems likely that such relationships are of widespread occurrence. Beavis (1964) gave a detailed description of crenulation foliation developed parallel to slightly older axial-plane slaty foliation in Ordovician pelites, central Victoria, Australia.

The caption of this section probably should be extended to include crenulation foliation parallel to the axial surface of folds in nonmetamorphic or very low grade rocks in which axial-plane foliation is not developed. Crenulation foliation showing this relationship is probably widespread, although descriptions are scanty in the literature. However, Hills (1945) described an excellent example in detail; his specimens came from the Silurian sandstones and slates along the Mitta Mitta River, Victoria, Australia (Figs. 130 and 131). Examination of the arenaceous components of the laminated rock showed that flexural-slip folding was accommodated by bedding-plane slip. Intercalated argillaceous layers yielded slightly by mass flowage, and by development of crenulation foliation which produced slip folds in these incompetent layers; the mean orientation of the foliation planes is parallel to the axial surfaces of folds in the competent beds. In these Australian rocks the crenulation foliation is nonpenetrative in the samples studied, but it appears to be a stage in the development of genuine axial-plane foliation, although the latter type of foliation does not occur within the rocks described. Both Balk (1936, p. 707) and Hills (1945) commented on the fact that, as is common in such lithologies, the foliation planes are more widely spaced in psammitic than in the more pelitic layers. Balk (1936, p. 707) found that in some pelitic rocks of Dutchess County, New York, U.S.A., crenulation foliation planes: ". . . coincide almost everywhere with the axial planes of the local folds. . . .Where folds belong to larger, fairly symmetrical synclinoria, the planes are arranged in a V-shape, dipping toward the axis of the large structure."

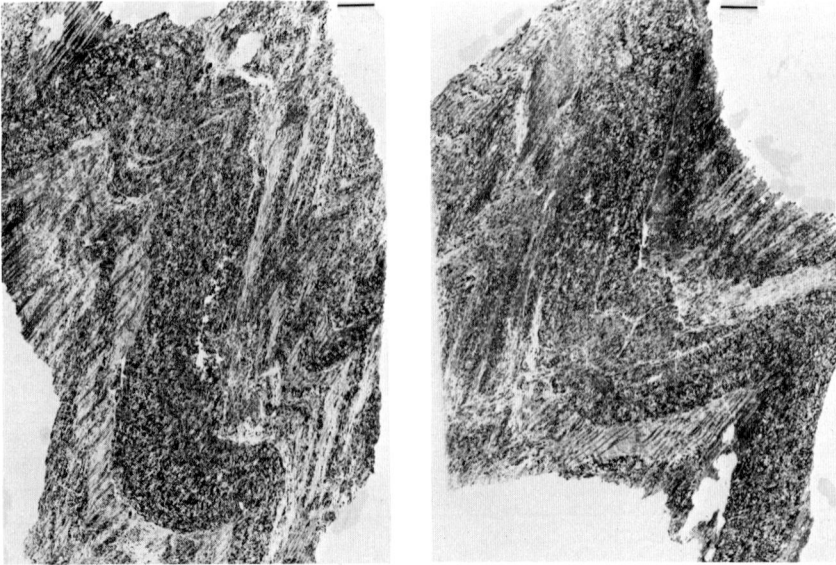

Figure 207. Precambrian mica schist, Coal Creek Canyon, Front Range, Colorado, U. S. A. Folds in original bedding defined by quartzose layer; crenulation foliation is developed in highly micaceous member and is parallel to axial-plane foliation of small folds. Photographs of thin sections (crossed polars) cut perpendicular to *B*; scale is 2 mm. (Photographs by Mr. S. B. Upchurch).

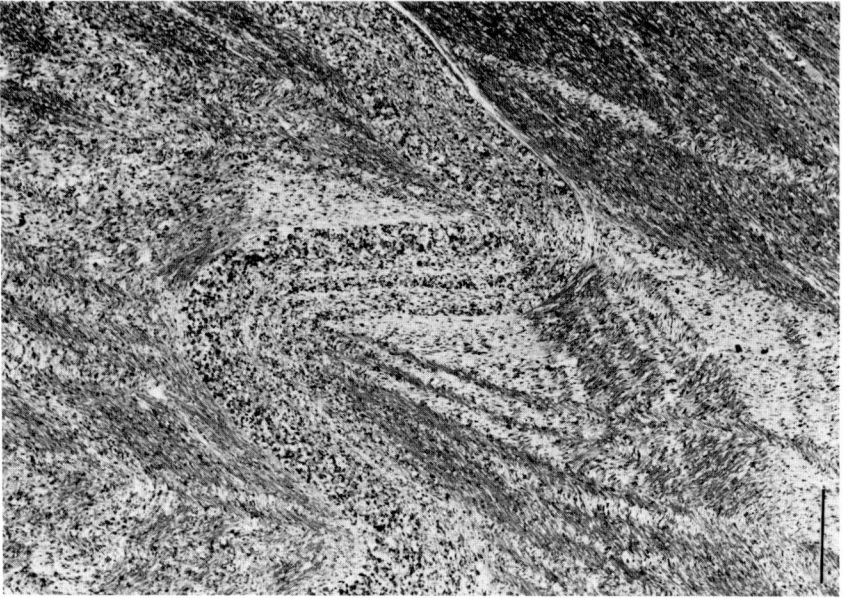

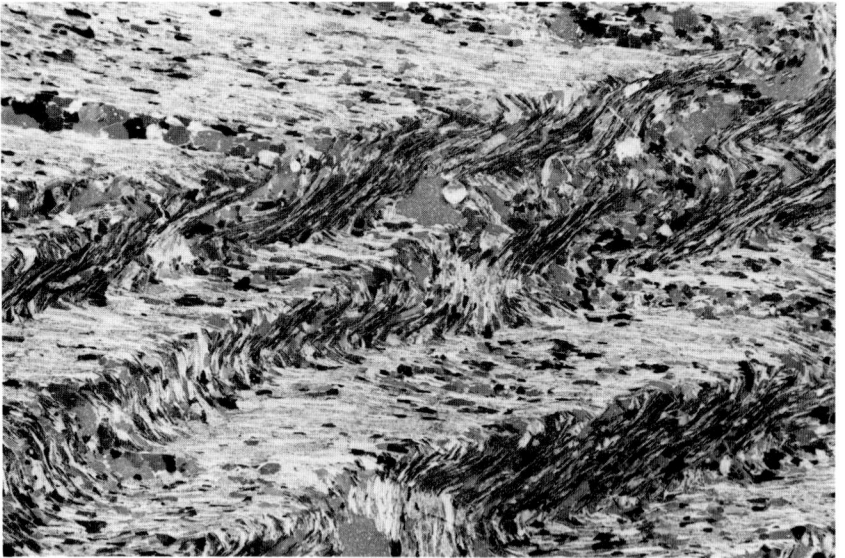

Figure 208. Folded mica schist with bedding defined by quartzose member, Bosost Dome, Pyrenees. Penecontemporaneous crenulation is developed parallel to fold axial plane, but near the folded quartzose band the foliation is parallel to bedding. Photographs of thin section (crossed polars) cut from hand sample supplied by Dr. H. J. Zwart. A (top): Note additional oblique crenulation foliation developed at lower right; scale is 4 mm. B (bottom): Enlargement of crenulation structure in A. (Photographs by Mr. S. B. Upchurch).

Crenulation foliation and "Gleitbrett" structures parallel to the axial planes of isoclinal folds in interlayered quartzites and graphitic phyllites of Keno Hill, Yukon Territory, Canada (Figs. 116 and 118), were described by McTaggart (1960, p. 23); Gillott (1956) and Simpson (1963) also drew attention to crenulation foliation parallel to the axial planes of second phase folds developed in the Manx Slate Series, Isle of Man, Great Britain.

Another example was described from the metasedimentary rocks of the northwest Adirondack Mountains, New York, U.S.A., by Engel (1949, p. 774), who found that:

> ... transitions in type and form, from the descriptive point of view, exist between the S_2 type of surfaces, which appear in some open folds in beds and the strongly developed, more schistose types of foliation surfaces associated with close folds. In some places where beds of contrasting composition and different competence are involved, the S_2 "fracture cleavage" types [crenulation foliation] cutting one bed merge abruptly into "flow cleavage" [axial-plane foliation] types in an adjoining bed of quite different composition and different susceptibility to shearing and recrystallization processes.

White's (1949) study of foliation in Vermont, U.S.A. yields a clear picture of similar relationships. Initially axial-plane foliation developed more or less parallel to the mapped formation boundaries, but for clarity this foliation is omitted from the map (Fig. 209). This early foliation is parallel to the axial-planes of minor folds, but small flexures of this type are rarely preserved in the area. Foliation formed in a later stage of deformation is present or dominant throughout most of the area, and transects or even obliterates the initial foliation. This younger structure strikes northwest-southeast, and comprises a crenulation foliation in the east and a penetrative schistosity (i.e., axial-plane foliation) in the west (Fig. 209). Since this crenulation foliation transects the first axial-plane foliation, it could be classed with group (b); however, during the second phase of deformation that produced the northwest-southeast foliation, the structure equated with group (a), because crenulation foliation developed parallel to the second axial-plane foliation.

The gradual transition zone between these two younger parallel foliations coincides with the eastern border of the staurolite zone of regional metamorphism (Fig. 209); this caused White to suggest that temperature was a significant control in determining which type of foliation developed. According to White (1949, p. 590):

> ... study of rocks from the zone of transition suggests that the latter schistosity has formed in part by mechanical rotation and smearing out of the earlier schistosity, and in part by the growth

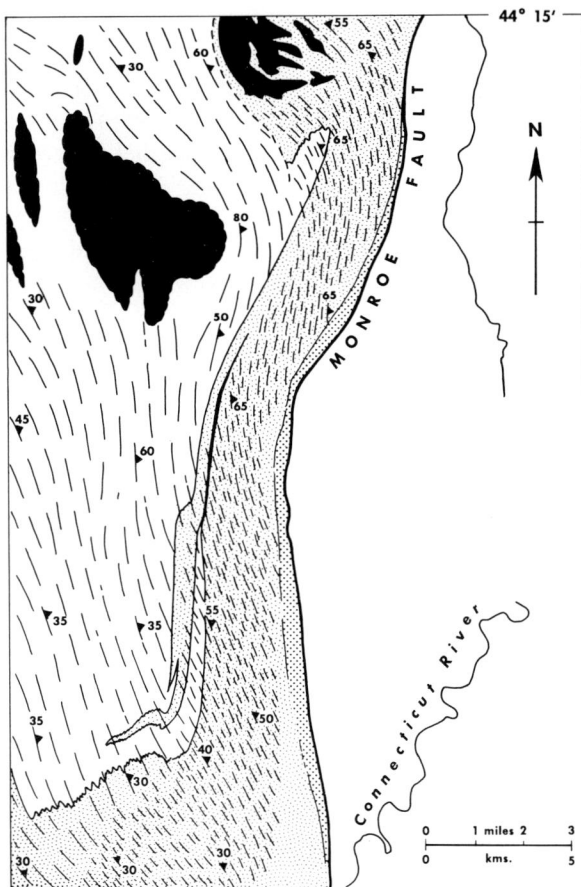

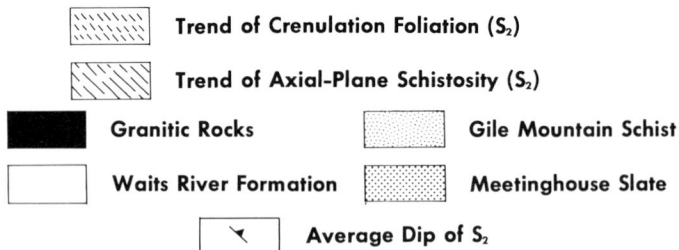

Figure 209. Map showing the younger foliation within the western part of Woodsville Quadrangle, Vermont, U. S. A. (After White, 1949, Fig. 4).

of new mica flakes oriented parallel to the later cleavage [foliation].

Since the orientation is not deflected in passing from the second axial-plane foliation (schistosity) to the crenulation foliation, White concluded that both structures formed with a similar, if not identical, angular relationship to the kinetic axes (of the younger deformation). This suggested that, with respect to the gross deformation of the rocks, both types of foliation are mechanically equivalent. Brace (1953, p. 87 ff.) discussed White's evidence, and some of the difficulties in the way of explaining the apparent parallelism and synchrony of crenulation and axial-plane foliations. A. G. Jones (1959) drew a similar conclusion to White with respect to rocks of the Shuswap area, British Columbia, Canada.

In a situation similar to that described by White, Rickard (1961, p. 327) found a late superimposed crenulation foliation in NW Ireland that was initiated as a series of flexures of the earlier axial-plane and crenulation foliations. As traced towards the contact of the "Older Granite" of Donegal, the superposed crenulation foliation passes into an axial-plane foliation (schistosity). A concentration of mica along the foliation planes apparently resulted from metamorphic differentiation processes, and locally developed a regular closely-spaced banding across the bedding.

Zen (1961, p. 323) also referred to finding all gradations between visible axial-plane and crenulation foliation in the northern Taconic Range, Vermont, U.S.A.

It would be very significant to know how widespread are the relationships described by White and Rickard. Before the interrelationships of axial-plane and crenulation foliation can be fairly appraised, it is essential that a detailed, objective, and quantitative picture of their areal distribution and variation be obtained.

(b) CRENULATION FOLIATION OBLIQUELY CROSSING EARLIER FOLIATION

Examples of this relationship are probably widely distributed in rocks which have undergone a protracted period of deformation or multiple phases of deformation.

In relation to the slates of Pennsylvania, U.S.A., Behre (1933, p. 60) suggested that a general condition favoring development of crenulation foliation is "A marked change in the direction of compression with respect to the previous direction, especially if such pressure is so abruptly applied as to preclude readjustment to the new direction." If this is a realistic kinetic picture, and the change occurs relatively late in the deformation sequence, an oblique set of crenulation foliation planes could develop. The structure could be an incipient axial-plane foliation, and is usually associated with small crumples (flexures).

Greenly (1930) described crenulation foliations developed parallel to the axial planes of folds formed by the folding of axial-plane foliation in pelitic rocks of the Mona complex, Anglesey, North Wales. Woodland (1965) illustrated the widespread occurrence of similar structural relationships in Burke Quadrangle, Vermont, U.S.A. In a detailed petrofabric study of a 10 mm. fold in quartz-sericite phyllite, Central Shikoku, Japan, Suzuki (1963) showed that quartz grains in quartzose members follow around the folded S_1, while micaceous members possess a crenulation foliation, S_2, parallel to the axial plane of the folded S_1 structure. As suggested by King (1956, p. 99), an oblique set of crenulation foliation planes appears to have the same kinematic significance as axial-plane foliation, and will develop parallel to the axial planes of an incipient generation of new folds.

From Dutchess County, New York, U.S.A., Balk (1936, p. 708) described small contortions which are commonly superposed on larger folds in a series of black phyllites; a crenulation foliation (Fig. 210) now coincides with the axial planes of the local crenulations (but not necessarily with those of earlier and larger isoclinal folds).

While emphasizing that axial-plane and crenulation foliations in the slates of the Martinsburg formation, New Jersey, U.S.A., must be differentiated, Broughton (1946, p. 7) gave evidence to substantiate his claims that "... genetically there need not be, and probably is not, any difference between the two...." He demonstrated that flexural-slip folding of the pelitic rocks occurred during the Taconic disturbance and that slip within the bedding planes (S_0) gave way to development of a penetrative foliation (S_1) parallel to the fold axial planes. During later deformation (Appalachian Revolution) flexural-slip folding of the axial-plane foliation (S_1) in the slates produced a foliation (S_2) parallel to the new axial planes.

This second deformation was weak in the north and progressively more intense southward towards Delaware; this regional pattern permitted study of the successive stages in the development of S_2. In the north S_2 comprises microscopic gashes (commonly some 3 mm. in length) developed perpendicular to S_1. These fissures are associated with a wavy crinkling of the axial-plane foliation (S_1). Southwards, S_1 has been thrown into flexures easily seen in hand samples and a crenulation foliation (S_2) is developed parallel to the axial planes; Broughton (1946, p. 9) considered this foliation "genetically related to folding." Locally, the appearance of S_2 in hand samples has become so similar to normal axial-plane foliation that it could be confused with undeformed S_1. S_1 remains as a relict structure in the railroad cut at Manunka Chunk, but the crenulation foliation (S_2) has lost its characteristic appearance and seems to be

NW   **SE**

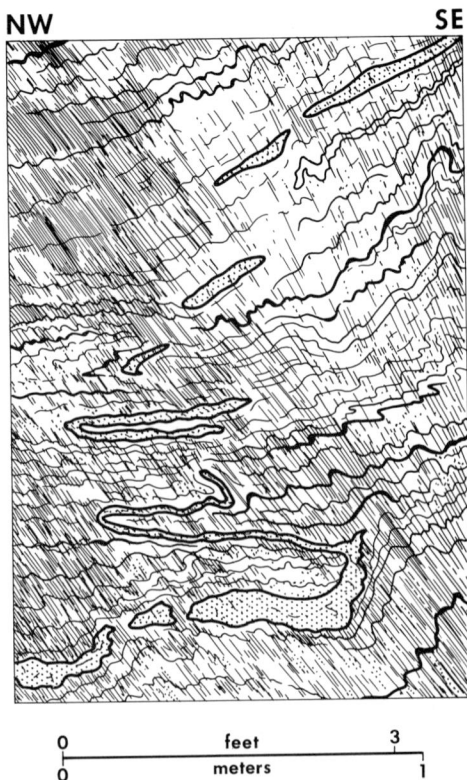

Figure 210. Small folds in black phyllite traversed by disrupted quartz veins (solid black). Younger minor folds and crenulation foliation cut the older isoclinal structures, WNW of Chestnut Ridge, Dutchess County, New York, U. S. A. (After Balk, 1936, Fig. 9).

transitional between crenulation and axial-plane foliation (Broughton, 1946, p. 13). The uniform elongation of quartz grains in S_1 observed farther north has disappeared as a result of continued and stronger deformation that developed a mineral orientation and a resultant younger axial-plane foliation along the S_2 crenulation foliation planes.

The complex nature of a sequence of superposed crenulation foliations on Rosguill, NW Ireland, was described by D. C. Knill and J. L. Knill (1961). By detailed geometrical analysis they determined that an early phase, during which regional tectonic slides developed, was followed by folding about north-south axes; bedding (S_0) and a foliation parallel to bedding (S_1) were mapped. During subsequent folding about axes plunging gently to the southwest, an axial-plane foliation (S_2) in pelitic rocks formed in association with overturned folds in southern Rosguill. D. C. Knill and J. L. Knill (1961, p. 280) suggested that this axial-plane structure (S_2) dies out northwards, but that it is presumably equivalent to a weakly developed, steeply dipping, crenulation foliation (S'_2) mapped in semipelitic rocks west of Derryhassan (Fig. 211).

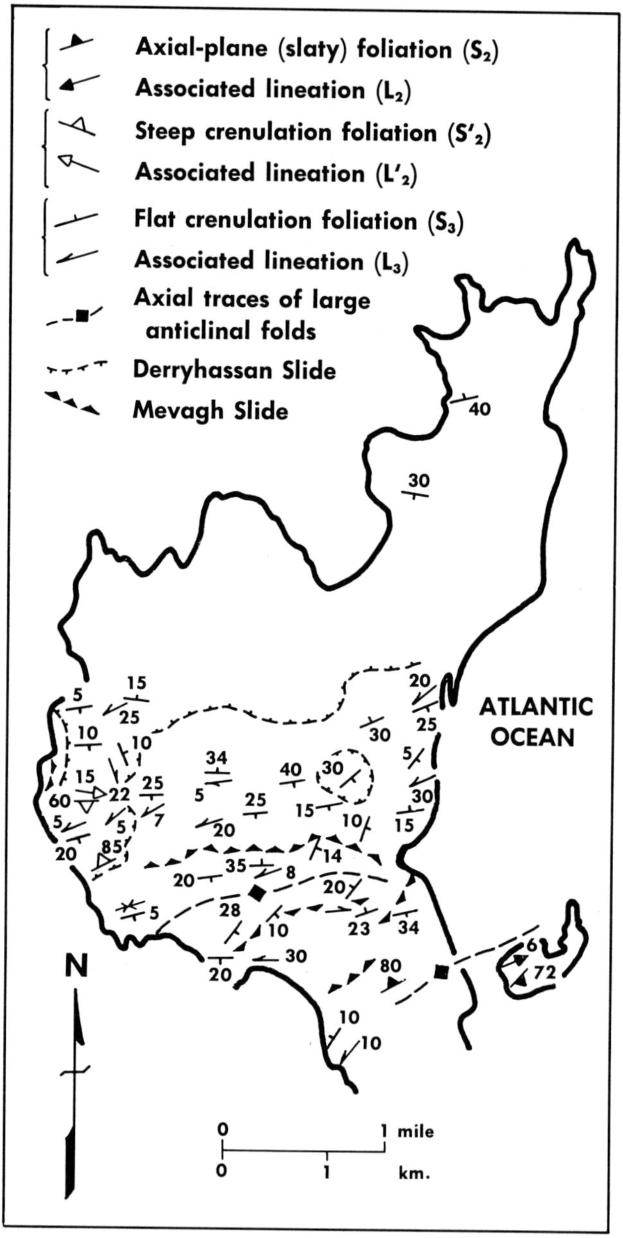

Figure 211. The older superposed foliations and lineations in Dalradian rocks at Rosguill, Donegal, NW Ireland. Values plotted are a representative selection from about 1,000 measurements made in the field (After D. C. Knill and J. L. Knill, 1961, Fig. 5).

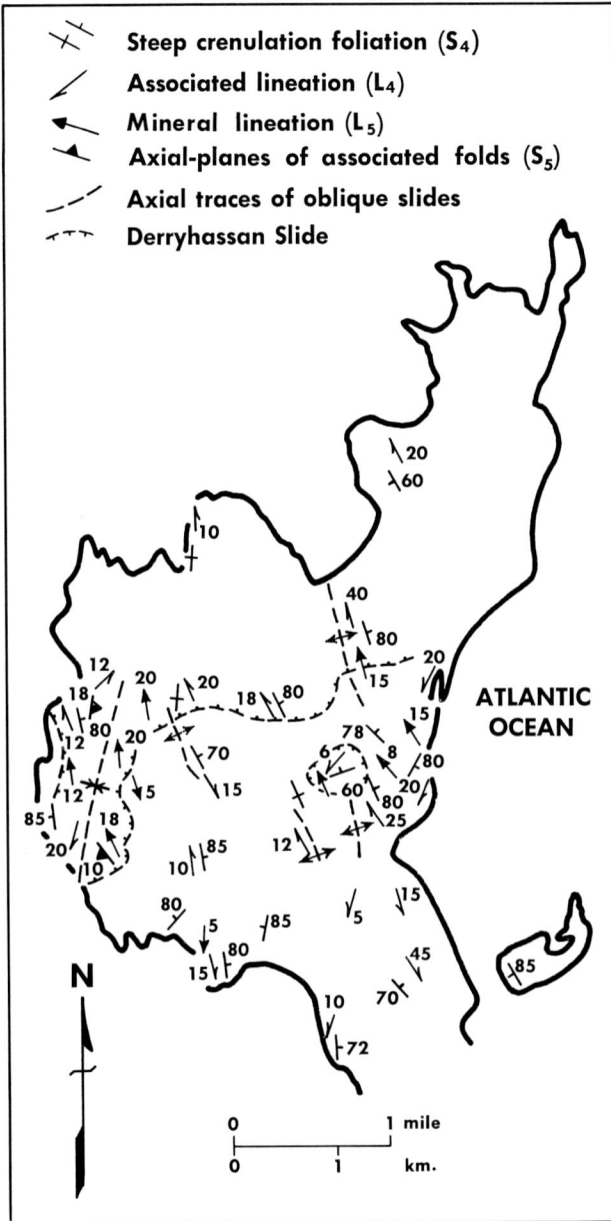

Figure 212. The younger superposed foliations and lineations in the Dalradian rocks of Rosguill, Donegal, NW
Ireland. Values plotted are a representative selection from
about 1,200 measurements made in the field (After D. C.
Knill and J. L. Knill, 1961, Fig. 7).

All of these structures in Rosguill were refolded later with the development of small folds with gently inclined axial planes. A flat-lying crenulation foliation (S_3) defines an axial-plane structure to the innumerable S- and Z-shaped folds of this generation. Commonly S_3 is associated with a prominent mineral banding, which locally grades into a new schistosity. This crenulation foliation is rather weakly developed in southern Rosguill, where it cuts the axial-plane foliation (S_2), but it is better developed to the north. Within this small promontory Knill and Knill mapped an additional crenulation foliation (S_4) which is generally parallel to the steeply inclined axial planes of both small and large folds that trend roughly north-south (Fig. 212). Following this complex sequence of events the tectonites were invaded by a group of granitic and migmatitic rocks.

It would be most surprising if complex deformational histories such as those at Rosguill were not of widespread occurrence. Similar relationships are probably common in most orogenic belts, and the study by D. C. Knill and J. L. Knill (1961) stands in bold relief when contrasted with the abysmal lack of quantitative information about crenulation foliation structures in other parts of the world.

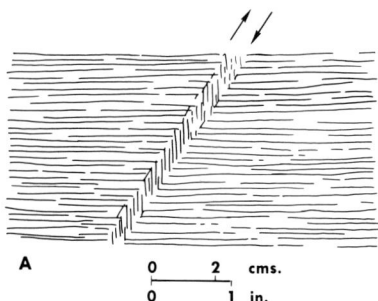

Figure 213. Kinked zone affecting foliation, SE Eifel district, Germany (After Hoeppener, 1955, Fig. 4).

Joint-drags appear to be a special variety of this group of crenulation structures; **kinked cleavage** has been widely used for this phenomenon (Fig. 213). Examples were described by Flinn (1952) from Unst, Shetland Islands, and by Knill (1961) from mid-Argyllshire, Scotland. In joint-drags widely-spaced pairs of planar dislocations cut and off-set an earlier foliation to produce small angular folds. Actual rupture of the earlier foliation is common; within the Valle de Arán area, central Pyrenees, Kleinsmiede (1960, p. 196) found that the structure is common in the fine-grained pelitic rocks which have strongly developed axial-plane foliation. Born (1929), Kienow (1934), Schenk (1956) and Teichmüller and Teichmüller (1955) all described structures similar to joint-drags from various German localities, while Simpson (1940) and Engels (1959) referred to intersecting sets of joint-drag features. Small-scale décollment sometimes gives rise to a very similar structure (Fig. 214) as shown by Hoeppener (1955). Knill (1961) noted that in Argyllshire microscopic examples commonly show that the micaceous minerals defining the earlier foliation are curved round into the drag planes without being broken; hence, they define an incipient foliation. Knill demonstrated that, within large areas of Argyllshire, joint-drags reflect a consistent sense of movement.

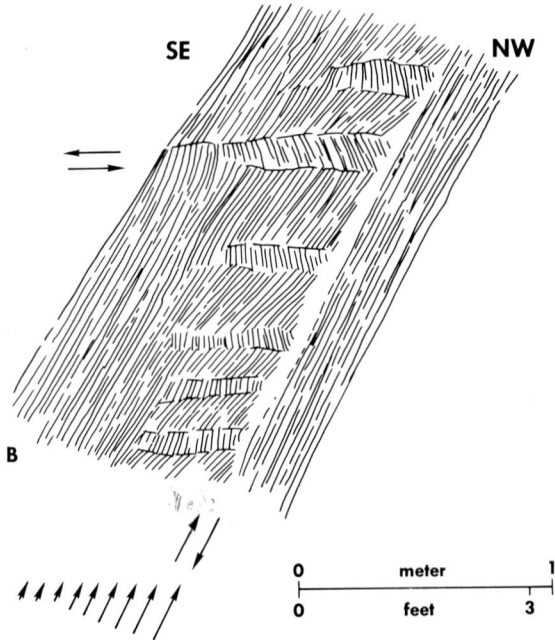

Figure 214. Kink zones developed by small décollment-like faulting in pelite (After Hoeppener, 1955, Fig. 5).

(c) CRENULATION FOLIATION DEVELOPED AS CONJUGATE SETS

One of the few descriptions of crenulation foliations in conjugate sets was given by Muff (in Peach, *et al.*, 1909) in an account of the Craignish phyllites, Argyllshire, Scotland. Muff described how the axial-plane foliation is intersected at approximately 45° by two crenulation foliations (one subvertical and the other subhorizontal or gently dipping), so that the three foliations all intersect in the fold axis (*B*). In interpreting these structures, Muff (p. 17) recorded that:

> ... the strain-slip-cleavage, the cleavage, and the folding are recognized as concomitant, though not simultaneous, effects of the same lateral pressure, and do not necessitate two or more

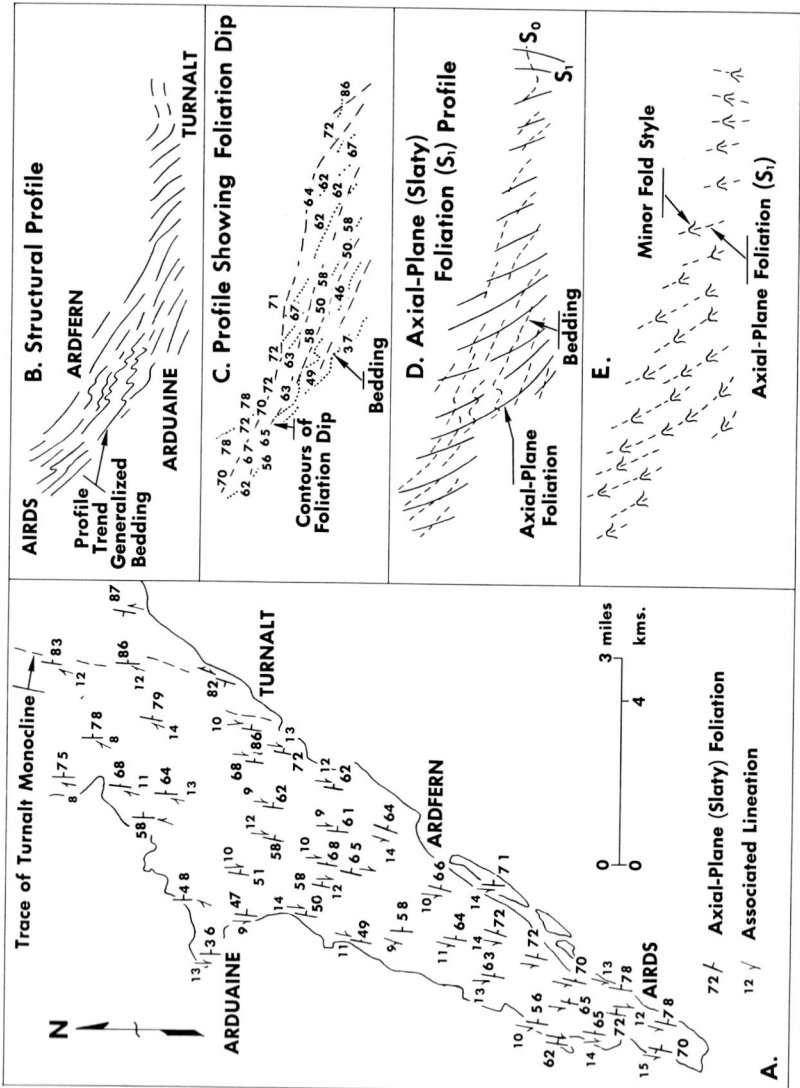

Figure 215. Main structural elements associated with flexural-slip folds in Dalradian rocks of the Craignish-Kilmelfort district, Argyllshire, Scotland. A: Map of axial plane foliation and linear structures for 43 subareas. B: Structural profile. C: Foliation dips projected on to a structural profile. D: Axial-plane foliation profile. E: Variation of small fold style across the foliation profile (After Knill, 1960A, Fig. 4).

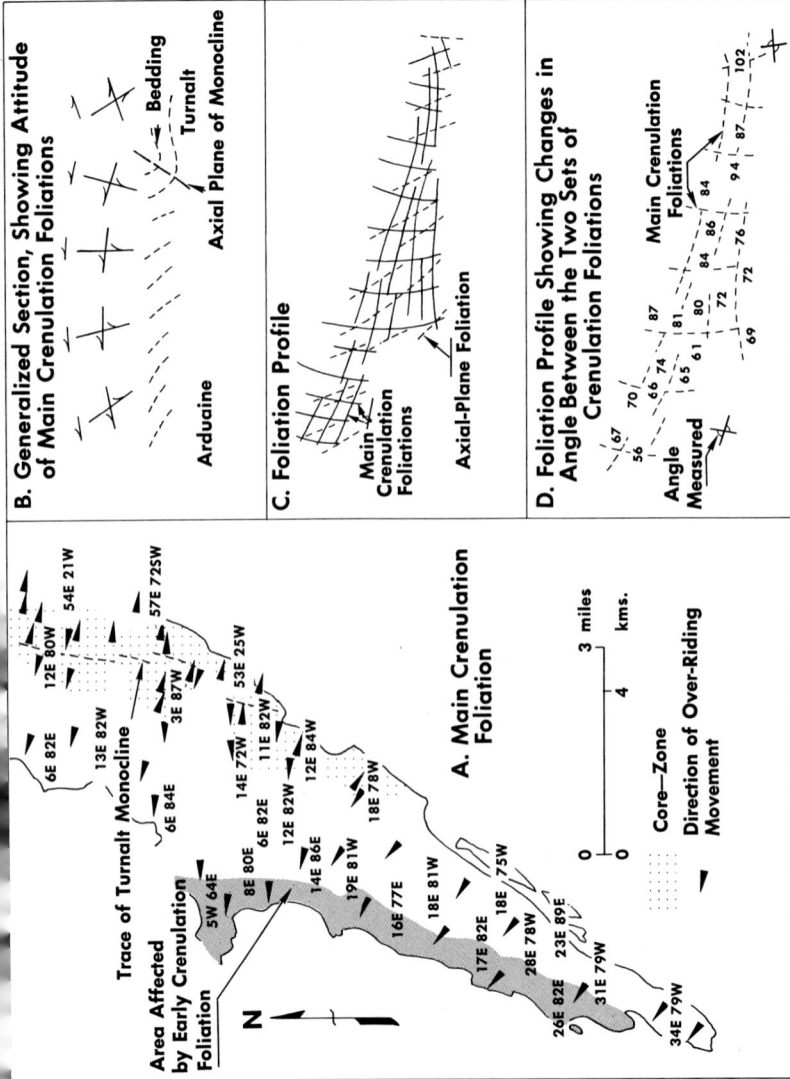

Figure 216. The early and second (main) crenulation foliations in the Craignish-Kilmelfort district, Argyllshire, Scotland. A: Dip of main conjugate crenulation foliations for 27 subareas, with westerly and easterly movement sense indicated by, for example 34E 79W (the foliation with westerly movement sense is given first); E and W imply the general direction of dip. Core zone is the area in which early crenulation occurs. B: Section through main crenulation foliations. C: Main crenulation foliation profile. D: Changes in angle between the two sets of the main crenulation foliation (After Knill, 1960A, Fig. 8).

distinct periods of earth-stress in the district. It should be pointed out, however, that a slight veering of the direction of maximum pressure towards the close of the period of earth-stress is required to account for the slight difference in strike between the strain-slip-cleavage and the other structures.

The complex structural history of the Dalradian rocks within the Craignish-Kilmelfort district was reexamined in detail by Knill (1960A). He showed that initial flexural-slip folding of the succession of phyllites, metamorphosed quartzites, limestones, and grits, and metadolerites produced the major structure (Fig. 215); slip along the bedding planes locally induced a bedding schistosity:

> At a certain stage, competent folding could no longer accommodate the applied stress and slip took place along curved cleavage surfaces which were superimposed upon the major structure. Folds on all scales, initiated during the competent movements, were accentuated and new folds were formed; during this phase there was strong extension within [the axial-plane foliation] ... (Knill, 1960A, p. 350).

The crenulation foliations were formed by distortion of the axial-plane foliation during continued folding, but four separate sets of crenulation foliation have been mapped. The first conjugate foliations comprise a gently dipping set (with northeast-southwest strike) and a steeply dipping set (with north-south strike) that intersect the axial-plane foliation about a horizontal or gently-plunging lineation; the two sets seem to have been formed approximately simultaneously. These structures are limited to the west of the area mapped, but the second or main conjugate set of foliations affects the entire area (Fig. 216).[7] The foliation planes usually occur in pairs and individual planes may be up to 2.5 cm. apart, or, at the other extreme, there may be 120 to 160 planes to the centimeter. Like Muff (in Peach, et al., 1909, p. 16), Knill demonstrated that the two foliation directions of the main phase formed at the same period. By microscopic examination Knill (1960A, p. 353) demonstrated that the crenulation foliation was initiated by flexure of the axial-plane foliation along parallel planes of slip which caused the preferred accumulation of phyllosilicates, particularly sericite, within the new foliation planes. Sometimes a strong mineral banding (up to 0.5 mm. wide) has been formed in this manner. It would appear that the attitudes of the main crenulation foliations were controlled by the orientation of the original axial-plane foliation (Fig. 216). During all of the tectonic events these

---

[7] In the field isolated exposures can be misleading because locally only one set of a conjugate pair of foliations may be developed, as in a case recorded by Knill (1960A, p. 352) with respect to the main set of crenulation foliations at Craignish.

metasedimentary rocks remained within the chlorite grade of regional metamorphism.

Although additional detailed descriptions of conjugate structures are difficult to find in the literature, they can commonly be observed in the field. From Great Britain, Harker (1932, Fig. 68) illustrated examples from Drws-y-Coed near Snowdon, North Wales, and Rosevanion, Cornwall.

Agron (1950) described conjugate foliations in the Peach Bottom Slate, Pennsylvania and Maryland, U.S.A. Here, one direction of crenulation foliation is more commonly developed than the other, but Agron recorded two intersecting sets from seven localities at which one horizontal or gently dipping set is associated with a steeply dipping set. Either set may be the more prominent, however, and there is no consistency about which accommodated the last movement. As Muff and Knill found in Argyllshire, the conjugate foliations are always symmetrically arranged with respect to the earlier axial-plane foliation. However, in the Peach Bottom Slate, Agron found that the subhorizontal set of crenulation foliation intersects the axial-plane foliation in B, and the steeply-dipping set intersects in B', where $B \wedge B' = 73°$. Thus, the slates were folded about both B and B', and the two axes are "largely contemporaneous, having formed during the same period of deformation, but rotation about B began slightly before rotation about B'."

Rickard (1961, pp. 331-32) suggested that crenulation foliations parallel to the axial planes of folds:

> ... develop under low to medium grade metamorphic conditions whereas deformation in conjugate planes of shear seems to occur only under very low grade conditions during the waning stages of tectonic activity. Since they imply different tectonic conditions the differentiation of single and conjugate sets is important, but the distinction may be difficult where only one of a conjugate set is developed.

Muff (in Peach, *et al.*, 1909, p. 16) thought that an important factor in promoting the change from axial-plane to conjugate crenulation foliation formation is a greater rigidity of the Craignish phyllites, or rather "... a loss of capacity to flow." However, as pointed out earlier, Sander (1934) described schists from the Tyrol, Austria, in which he found evidence for contemporaneous movement on nonequivalent pairs of slip surfaces, and he suggested that such deformation is more important in deep-seated metamorphic zones. At the present time there is little quantitative basis for assessing the distribution of conjugate crenulation foliations, although there seems to be growing evidence for Rickard's contention. However, the fact that conjugate crenulation foliations appear to form late may simply mean that, if developed earlier, they are readily overprinted and obliterated by more penetrative foliations.

It is generally agreed that crenulation foliations developed as conjugate sets appear to be a response to a distinctly different stress situation from that for crenulation foliations parallel to fold axial planes. However, Beavis (1964) believed that conjugate crenulation foliations in slates from Victoria, Australia, resulted from a stress system identical to that responsible for the axial-plane foliation; he thus concluded that the conjugate crenulation is a result of the main folding.

Although crenulation foliation is often initiated as minor flexures, large folds appear to be less commonly associated with crenulation than with axial-plane foliation. In rocks affected by crenulation foliation the neocrystallization has been nonpenetrative and thus restricted to the actual planes of slip. Knill (1960A) was at pains to point out that the direction of movement along each foliation of a conjugate set is consistently in the same direction, and this is probably a major point of difference from axial-plane foliations.

(d) CRENULATION FOLIATION OBLIQUELY CROSSING ORIGINAL SEDIMENTARY S_0-SURFACES

As pointed out previously, crenulation foliation appears to develop only at the expense of an earlier S-surface, but, contrary to the majority of reports, the earlier S-surfaces need not be of tectonic origin. Original bedding S_0-surfaces, defined in part by layers of phyllosilicates, can commonly be affected by crenulation structures (e.g., Beavis, 1964).

Literature references are virtually nonexistent, although Voll (1960, p. 553) found that, where sedimentary rocks were rich in clastic micas lying in bedding, the whole rock can sometimes be plicated and then the first foliation planes to develop have the character of crenulation foliations. They are more widely spaced than axial-plane foliation planes, and the whole picture simulates a secondary tectonic foliation.

Boswell (1949, pp. 97-98) described the microscopic changes visible in Middle Silurian rocks of North Wales in the zones of transition between concomitant and parallel crenulation and axial-plane foliations of neighboring areas; both foliations are essentially cylindroidal with respect to the B-axis. Where the crenulation foliation

> ... is ill-developed, no matter what the rock-type may be, the planes of parting behave as a system of closely-spaced joints, from 1 to 2 or 3 inches apart.... As the cleavage increases in intensity, a phacoidal structure appears, and this character is discernible even in the microscopic texture. Where it is most

strongly developed, the rock splits into "slates" about 1/12 inch in thickness, which ring under the hammer like roofing-slates. In these examples, the cleavage-planes can be detected under the microscope as sub-parallel traces down to .001 inch (about .025 mm) apart. But, except for a partial recrystallization of the originally muddy groundmass into micas, chlorite and quartz, oriented mineralogical reconstitution (the so-called flow-cleavage) is often difficult to detect. . . . Indeed, it is noteworthy that in many of these cleaved Salopian rocks the scattered grains of allogenic quartz, clastic mica and chlorite lying between the fracture planes have not even been turned so that their long axes lie in the cleavage direction. . . . I have experienced difficulty drawing a sharp distinction between "fracture-cleavage" and "flow-cleavage"[8] . . . the basis of the distinction adopted by earlier investigators appears to have been the presence in flow-cleavage of a mineralogical reconstitution, induced by pressure, which has resulted in a strongly marked orientation of the constituent minerals, . . . But . . . in Denbighshire . . . we find in the field a gradual passage from coarsely splitting to finely splitting slates . . . under the microscope, a similar gradation from fracture-cleavage to flow cleavage can be observed . . . as the cleavage becomes finer, the groundmass of clayey and micaceous material throughout which the quartz grains are scattered begins to be recrystallized and eventually reacts to the gypsum plate in such a way to show that most, if not all, of the fine-grained material has developed with similar optical orientation. The quartz-grains may still show no sign of orientation, however, although in a further stage of the process, these grains become oriented by re-crystallization.

Boswell (1949, p. 98) also noted that:

... some of the best roofing-slates from various well-known quarries elsewhere in North Wales and other localities are seen under the microscope to be but partly re-crystallized and re-oriented. They owe their splitting quality to the development of fracture-cleavage no less than to flow-cleavage.

In confirmation of Boswell's hypothesis, Shackleton (1954, p. 274), in reviewing the structural evolution of North Wales, drew attention to the way in which the intensity of foliation varies markedly from one area to another. Shackleton also mentioned that, although the nature of the foliation in North Wales is not well known, the axial-plane and crenulation foliations developed in different districts appears to fit into one regional pattern. Maxwell (1962, p. 300) suggested that the general picture of slip folding accompanied by extensive upward transport of

---

[8]Fracture cleavage and flow cleavage equate with crenulation and axial-plane foliation, respectively, according to the usage in this chapter.

plastic material proposed by Shackleton corresponds to the style of deformation within the Martinsburg slates of the Delaware Water Gap area, U.S.A. In any event, it seems likely that the crenulation foliation described by Boswell in the above quotation developed without the formation of an earlier axial-plane foliation.

Thin sections of numerous low-grade or unmetamorphosed sedimentary rocks show the development of incipient or weak crenulation foliation, which thus owe essential features to shearing and not to indirect componental movements only. Many quartzose sedimentary rocks of this type show a distinct sheen on foliation planes (due to the preferred orientation of phyllosilicates), although commonly few micaceous minerals can be seen developed along the nonpenetrative shear planes in thin sections. Because developed early, structures of this type are readily overprinted and obliterated by more penetrative foliations during any later phase of orogeny.

3. *FOLIATION PARALLEL TO BEDDING OR LITHOLOGIC LAYERING*

Bedding cleavage is almost always included as a major category of foliation in modern texts and papers; it seems to be included in deference to descriptions by Daly (1912, 1915, 1917) of static (latterly load) metamorphism in the Shuswap terrane, British Columbia, Canada. Daly (1915, 1917) quoted descriptions of numerous areas in eastern and western Canada, the western Alps, and many other parts of the world, to support his hypothesis that load metamorphism is a common phenomenon. Daly's (1917, p. 403) conviction stemmed in large measure from

> ... the exceedingly common parallelism between foliations or schistosity and the stratification. This fact is abundantly illustrated in the Canadian shield, in the Adirondacks of New York State, and in the Precambrian of the North American Cordillera, Scotland, Scandinavia, Finland, etcetera. Löwl (1906, page 50) has given a good statement of it in the following passage (translated): "The great majority of the crystalline schists are foliated, not across the bedding, but parallel to it. Their parallel texture must have been developed when the rocks lay undisturbed, and thus only because of the downward pressure of the overlying rocks, exactly as in the case of shale and most clay-slates, among which, indeed, transverse cleavage is not the rule, but the exception. It is not merely a case of the condensation of the buried rock by the dead weight of its cover. The load also causes foliation."

Read (1940, pp. 237-38 and 249) supported Daly's views and wrote:

... Daly (1917) has cited a score of regions where schistosity and bedding agree—in folded and horizontal rocks alike. These are the phenomena that Daly felt to be "truly inexplicable by pure dynamic metamorphism," an opinion with which I am in complete accord.... I believe that schistosity and stratification are coincident in a vast number of cases.... Whether the topic arises at all depends, of course, upon the validity of the argument that the banding of metamorphic rocks represents in most cases an original bedding, and I consider the validity unquestioned.

Deeply buried sediments remain unmetamorphosed unless igneous material gets access to them. In metasomatic metamorphism, original sedimentary textures can be reasonably preserved, mimetic crystallization can prevail, schistosity and bedding, even in violently folded strata can coincide. Finally, all those phenomena which Daly felt to be "truly inexplicable by pure dynamic metamorphism" are satisfactorily explained.

However, Leith (1923, p. 130) was correct when he observed that, within our zone of observation, load has not produced "static" foliation on any large scale; he considered it certain that foliation is commonly a result of movement. Buddington (1929, p. 82) also thought that:

... so far as the northwest Adirondacks is concerned, there is no justification for assuming that if the foliation and bedding is parallel it must have been formed by "load or static" metamorphism.... The writer would ascribe the foliation to recrystallization accompanying the deformation and igneous intrusions.

Admittedly a foliation is occasionally parallel to the real bedding (i.e., not pseudo-bedding), but, so far as is known, such structures do not result from the static or load metamorphism advocated by Daly (1917).

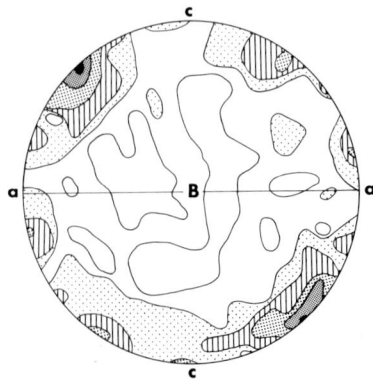

Figure 217. 330 c-crystallographic axes to quartz grains in hornblende-biotite schist, near Albert Canyon Station, Shuswap terrane, British Columbia, Canada. Contours at 0.3, 1, 2, 3, 5, and 8 per cent (After Gilluly, 1934, Fig. 3).

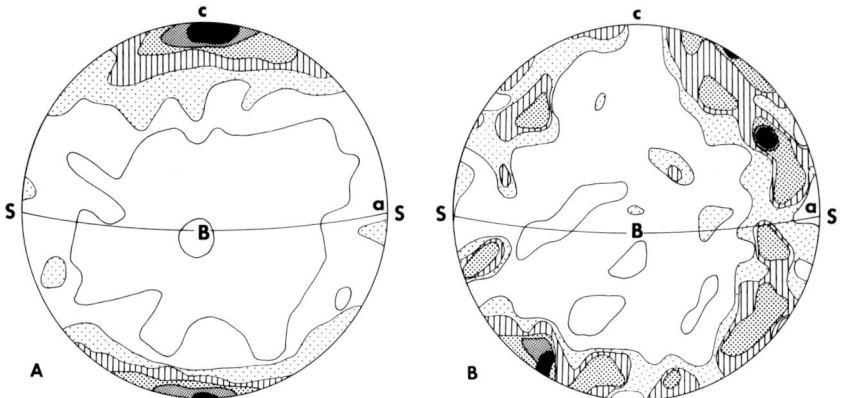

Figure 218. Fabric diagrams for biotite gneiss, Seymour Arm, Shuswap Lake, British Columbia, Canada. A: Poles to 473 biotite grains; contours at 0.2, 1, 3, 5, 8, and 10 per cent. B: 513 c-crystallographic axes of quartz; contours at 0.2, 1.0, 1.5, 2.0, 2.5, and 3.0 per cent (After Gilluly, 1934, Figs. 4 and 5).

Conclusive evidence concerning the Shuswap area described by Daly was advanced by Gilluly (1934) and A. G. Jones (1959). Gilluly showed that two representative specimens supplied by Daly have petrofabric structures with definite monoclinic symmetry (Figs. 217 and 218); this militates against the development of foliation parallel to bedding under the influence of static or load metamorphism, which would be expected to result in axial symmetry. Later A. G. Jones (1959) proved by detailed mapping that the foliation observed by Daly is actually parallel to the axial planes of recumbent isoclinal folds; pseudo-bedding, rather than true bedding, is parallel to the foliation in this area. Hence, this foliation certainly did not result from the gravitational load of the overlying rocks.

It would be desirable for bedding foliation (or cleavage) *due to load metamorphism* to be expunged as a category. However, under at least two different sets of conditions, foliations can develop parallel to the original bedding:

(a) Occasionally **belteroporic textures** develop in which minerals grow with a preferred orientation along some plane of easy access to solutions through the rock, such as selected bedding planes, shear planes, etc.

Neocrystallization accompanying diagenesis sometimes creates foliations of this type. To refer to but one example, Boswell (1949, p. 137) mentioned that many of the pale mica, chlorite, and hydromica grains in the Middle Silurian pelitic rocks of North Wales resulted from reconstitution and recrystallization of muddy sediments during diagenesis. In many cases these nondetrital grains have a strong preferred orientation parallel to bedding.

(b) Sometimes flexural-slip folding results in foliation parallel to the bedding of thin incompetent units interlaminated between competent ones. King (1956, p. 96) commented:

> That schistosity can develop on bedding is not a matter in dispute. It is typically displayed by thinly bedded competent formations involved in moderate folding (cf. Gair, 1950, p. 862), the intervening incompetent layers often showing axial plane schistosity.... Bedding schistosity of this kind is most marked on the limbs of folds, precisely where it is to be expected on the supposition that it is due to slip on bedding planes during flexure....

Sander (1930, p. 247) illustrated a rock from Passau, Germany, in which this type of foliation follows right round a fold in thin graphitic schist bands between aplitic layers (Fig. 140). Reference was made (p. 207) to bedding schistosities described in calcareous schists. Again, Knill (1960A, p. 346; 1960B, p. 323) described foliation parallel to bedding in the Dalradian low-grade metamorphic rocks of the Craignish-Kilmelfort district, Argyllshire, Scotland. Under the microscope, in rare cases, a schistosity can be traced around folds which now have, in part, axial-plane foliation. The relict-bedding schistosity was produced by the initial flexural-slip folding (not necessarily recumbent folding). The phyllosilicates forming this schistosity have been crinkled into the herringbone pattern of an axial-plane crenulation foliation, but deformation ceased before the schistosity was completely overprinted (Knill, 1960A, pp. 346 and 363).

D. C. Knill and J. L. Knill (1961, p. 280) also mentioned that a bedding schistosity is widespread throughout the Rosguill district, NW Ireland, where it was apparently developed during the formation of regional slide tectonics.

When flexural-slip and slip folding are closely associated in a series of rocks, successive lithological layers may show foliation parallel to the axial plane (e.g., S_1) and to the layering (e.g., S_0). For example, Figures 130 and 133 illustrate folds originally described by Hills and de Sitter that show two such foliation sets in different lithologic layers. Hietanen (1961B, p. 97) described similar relationships within metasedimentary rocks north of the Idaho Batholith, Idaho, U.S.A. Units rich in either mica or quartz alternate in the gneissic and schistose sequence. Quartzitic competent layers can be peeled off along micaceous laminae parallel to the lithologic banding; hence, this interlamination of nonuniform lithologic types has resulted in "bedding" foliation. The intervening thick micaceous bands tend to develop a predominant strong axial-plane foliation.

The various types of foliation parallel to bedding are very poorly documented, and, although numerous brief references may be found, much more attention needs to be paid to the structure.

4. "FRACTURE CLEAVAGE"—A VARIETY OF CLOSELY-SPACED JOINTING

Leith (1905, p. 119; 1923, p. 148), Billings (1954, p. 339), and de Sitter (1956B) pointed out that "**fracture cleavage**" is essentially a particular variety of closely-spaced jointing. Fracture cleavage has often been used in a very broad generic sense to include numerous dissimilar structures. This is misleading, and today fracture cleavage should not be used for foliations. Fracture cleavage differs from axial-plane and crenulation foliations in that the fracture surfaces are independent of, and *not* determined by, the preferred orientation of the mineral grains. In common with Leith (1923, p. 149), fracture cleavage is considered here to be a structure that does not pervade the entire rock mass or affect all particles of the rock; that is, it is a nonpenetrative structure.

Many joints are developed later than fold structures and transect the entire folded sequence. However, fracture cleavage is a distinctive form of jointing in that the individual fracture cleavage planes have a specific geometric relationship to the competent lithologic units in each fold and they tend to intersect in a line parallel to *B* (Fig. 138). The joint planes are sometimes very closely spaced so that they mask the bedding and even, in exceptional cases, enable fine-grained rocks to be used as crude roofing slates. Fracture cleavage of this type was described from the Upper Devonian and Lower Carboniferous rocks near Ardmore, County Waterford, Ireland, by Dawson-Grove (1955).

This structure is well developed in the thick weakly-deformed, moderately-indurated, poorly-fissile, mudstones of the New England Geosyncline, Australia. Crook (1964) made a geometric study of these quasi-planar fractures which are 0.6-1.3 cm. apart (nonpenetrative in hand samples but penetrative in outcrops) and intersect bedding in an anastomosing set of lines parallel to the *B*-geometric fold axes. The joints are approximately parallel to the fold axial planes, though tending to fan about *B*. No neocrystallization is visible by eye. This morphological variety of structure in weakly deformed mudstones was called **reticulate cleavage.**

Although fracture cleavages are essentially joints, one reason for discussing them in connection with foliations is that there is sometimes evidence that both have similar mechanical origins. For example, in the Devonian rocks near Plymouth, SW England, Fyson (1962, p. 216) found that:

> In the folded beds, particularly in the hinge areas, well developed slaty cleavage crosses the thick argillite beds and fracture cleavage crosses sandstones and limestones. Both types of cleavage are present and often indistinguishable in the silty and calcareous argillites. Individual fractures extend along parallel mica flakes and around irregularly shaped quartz or calcite grains; they are thus both dependent and independent of the mineral arrangement. . . .
>
> Widely spaced fracture cleavage across sandstones and limestones tends to form perpendicular, or at large angles, to the bedding. . . . With a change in rock type to argillaceous beds, there is a gradation from the fracture cleavage which radiates out from the cores of folds, to axial-plane slaty cleavage. This gradation suggests a similar mechanical origin for the two cleavages.

Similar examples could be cited, but review of the literature is hindered by the terminological problems; for example in Evans' (1963, p. 74) study of Charnwood Forest, England, a similar gradation appears to be described, but, in this case, "fracture cleavage" may possibly be intended to imply crenulation foliation—thus:

> The cleavage in the Charnian rocks varies from a perfect slaty cleavage in the Swithland Slates . . . to a fracture cleavage in the volcanic breccias. In the finer-grained pyroclastic rocks, which are frequently graded, it shows the characteristics of slaty cleavage in the finest bands and of fracture cleavage in the coarse bands.

Leith (1923, p. 149) originally suggested that fracture cleavage also embraces **close-joints cleavage, false cleavage, fault-slip cleavage, fissility, rift,** etc.; he recognized that all of these terms are not strictly synonyms, but that each designates some aspect of his fracture cleavage. Earlier, Leith (1905, p. 119) had also included strain-slip cleavage and "ausweichungscleavage" (varieties of crenulation foliation in this book); this precedent should be avoided because crenulation foliations are completely dissimilar structures from fracture cleavage. However, as recently as 1942, Billings (1942, p. 230) suggested that for practical purposes it is not essential to distinguish fracture cleavage from flow cleavage, although the suggestion was subsequently omitted (Billings, 1954, p. 350). Wilson (1946), probably because of his associations with Wisconsin, perpetuated Leith's (1905, 1923) usage when he suggested it is necessary to erect separate categories for shear or slip cleavage (i.e., crenulation foliation) and fracture cleavage; Wilson (1946), Kvale (1948, p. 19), and Boswell (1949) included all these structures under the heading of fracture cleavage. It seems inappropriate to include fracture cleavage as a category of foliation. However, this view is not held by all; for example, in a specific attempt to erect a classification of cleavages (i.e., foliations), Knill (1960B, p. 322) enumerated slaty cleavage, fracture cleavage, and strain-slip cleavage.

Such confused terminology (cf. O. T. Jones, 1961) means that it is often difficult to know what type of structure is involved when an author refers to "fracture cleavage"; although a specific new name is required for the narrowly-defined fracture cleavage described in this section, an appropriate term does not seem to have been proposed.

CONCLUSION

We may conclude that, with the exception of fracture cleavage, all of the structures commonly referred to as (rock) cleavage are varieties of foliation. Several distinctive varieties of foliation have been described above. It is important that precise operational definitions of the several varieties of foliation be adhered to rigidly, in order that quantitative estimates of their nature, abundance, and distribution may be obtained. At the present time it would appear that the conjugate sets of crenulation foliations are distinctive and separable from the other categories which all grade into each other through transitional types.

The distribution of these several types of foliation in space and time is very imperfectly known; before foliations can be understood and used effectively a quantitative appraisal of their distribution must be made for a number of different terranes. Fourmarier (1932, 1953A, 1953B, 1953C), de Sitter (1960A), de Sitter and Zwart (1960), and Knill (1960B) briefly and qualitatively discussed some of these distributional aspects. They suggested that axial-plane foliation is generally developed during an early phase of folding of sedimentary rocks. It has been claimed that during a late kinematic phase axial-plane and crenulation foliations are generated; both types are often intimately associated and this appears to indicate a close interrelationship of competent and incompetent conditions during folding. In relation to the "Martic Overthrust" area and the Glenarm Series of Pennsylvania and Maryland, U.S.A., Cloos and Hietanen (1941) suggested that axial-plane foliation always forms during an earlier event than crenulation foliations. Knill (1960B) and Rickard (1961, p. 330) suggested that conjugate sets of crenulation foliation always occur last in the sequence and under low-grade metamorphic conditions. This succession represents a progression from penetrative to less-penetrative structures. It would seem that these conclusions require some modification to agree with actual observations.

From the evidence discussed above, it is clear that axial-plane and crenulation foliations can develop under a very wide range of pressure, temperature, and compositional conditions. In some cases water-rich argillaceous sediments apparently develop axial-plane foliation during initial dewatering (e.g., Martinsburg slate, described by Maxwell, 1962); under analogous nonmetamorphic conditions crenulation foliations can develop at the expense of S_0-structures, as in the Silurian slates of North Wales (Boswell, 1949). At the other extreme, foliations readily develop under high-grade (e.g., sillimanite grade) regional metamorphic conditions. It seems reasonable to conclude that foliations can be generated under all conditions within these limits. During the development of an axial-plane foliation, or a penetrative crenulation foliation, all earlier minor structures tend to be obliterated, so that earlier foliations are destroyed to a greater or lesser extent.

Although quantitative data are lacking, it appears that the competence and degree of lithological uniformity of the rocks within a domain are critical in determining the type of foliation that develops under a given set of stress conditions. As metamorphic conditions change, the physical properties of each rock type vary and thus, at each stage, different original lithologies are amenable to development of foliation. The spatial location of a rock within a deforming orogenic sequence is important because the confining pressure also affects the physical properties of the rocks and the facility with which neocrystallization can be accomplished. Any model for the genesis of foliation will need to be quite complex, because the process factors — stress, metamorphic environment, lithology, time, etc. — vary within wide limits. Also, most areas in which foliations are developed have been subjected to prolonged deformation under changing conditions. Since stresses tend to be maintained over long periods the rocks behave as rheids (Carey, 1954), and it seems likely that in the core of an orogenic belt rheid flow (with the development of slip folds) will be significant in most lithologies.

Chapter 9

Linear Structures

Cloos (1946, 1953A) published an excellent review of linear structures. He (1945, p. 661) pointed out that from 1839 to 1944 linear structures and/or their orientation were referred to in 287 papers, but that generally the information given was "...meager, insufficient, and frequently inaccurate or incomplete."

Cloos (1945, 1946) and McIntyre (1950A) used **lineation** as a descriptive nongenetic term for any kind of linear structure within or on a rock. As Cloos emphasized, lineation is one of the most significant and important fabric elements, and it should be included in all complete structure maps.

Cloos recognized fifteen kinds of lineation, but, as A. G. Jones (1959, p. 95) pointed out, they were unfortunately distinguished principally on the basis of mode of origin. Because there are so many dissimilar types of linear structure that have quite varied genetic significance, it is important to use only clear, objective, and quantitative descriptions whenever possible. Recognizing that the origin of many structures is not known, Jones used a purely descriptive classification for the linear structures of Shuswap terrane, British Columbia, Canada.

Commonly fold axes are not included among lineations (e.g., Billings, 1954, p. 355) although such axes are linear elements and are considered lineations by some (e.g., Cloos, 1945). The question is essentially one of size. The axes of minor crenulations of an *S*-surface are lineations, whereas the axes of folds with amplitudes of several tens of meters are commonly excluded. Like A. G. Jones (1959), Turner and Weiss (1963, p. 101) restricted the term lineation to linear structures penetrative in hand specimens and small exposures; larger linear features such as fold axes, mullions, rods, and elongated pebbles were called linear structures.

It seems useful to use lineation in this narrowly-defined sense, and to consider lineations to be one category of linear elements and structures. On this basis the four main varieties of linear structure that are lineations comprise:

(a) intersections of *S*-surfaces

(b) aggregates of minerals, deformed ooliths, deformed pebbles, etc., that are elongate and subparallel

(c) prismatic or tabular mineral grains with linear preferred orientation; some lineations are not immediately visible and require microscopic petrofabric analyses to identify them (e.g., preferred linear orientation of the *c*-axes of quartz grains in a quartzite)

(d) axes of minor crenulations of *S*-surfaces

Both penetrative and nonpenetrative linear elements can occur in the same outcrop or specimen. The most obvious nonpenetrative linear structures are striations or slickensides — parallel striae (superficial furrows) parallel to the last direction of movement on fault or joint planes. Striae also commonly occur on *S*-surfaces normal to the fold axis of flexural-slip folds (Fig. 219); they resemble slickensides in being superficial grooves parallel to the last local slip direction on the *S*-surfaces. In sedimentary rocks joint and bedding planes intersect in lines that are nonpenetrative linear elements.

Within a single outcrop a number of different linear elements can commonly be observed; in recording and mapping their orientations it is important to keep the readings for each type separate. In most cases linear structures are systematically related to the local fold structures; this is not necessarily so if a particular linear element resulted from stresses independent of those produced by the obvious local folding (cf. Billings, 1954, p. 356).

It is important to keep all quantitative and qualitative measurements, descriptions, and map symbols for structural elements nongenetic. In this way the observed data retain objective and lasting qualities, rather than being obscured by subjective, and probably ephemeral, hypothetical

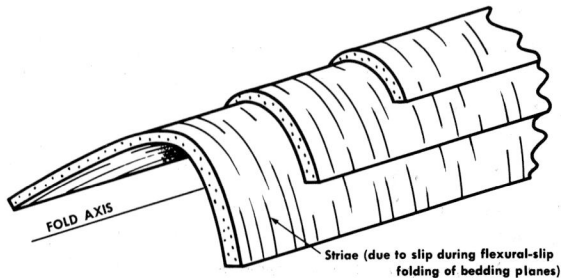

Figure 219. Diagrammatic flexural-slip fold with nonpenetrative linear structures (striae) developed perpendicular to *B*.

FOLD AXIS

Striae (due to slip during flexural-slip folding of bedding planes)

interpretations. Subsequent workers can interpret objective data in the light of new concepts, but this is difficult if records contain numerous subjective factors.

Linear elements have sometimes been used as the sole criteria for identifying the fabric axes prior to sythesizing the geometric and kinematic picture for the whole rock (e.g., Billings, 1942; Anderson, 1948). This unwise procedure is becoming less common.

It is also unwise to assume that elongate grains with preferred orientation are of necessity stretched; there has been a tendency to make this ascription automatically since the work of Naumann (1839). The concept is prominent in much recent writing. To cite but three examples, Billings (1937, p. 479) suggested that many pebbles are somewhat "stretched" in the Ordovician Ammonoosuc volcanic conglomerates, New Hampshire, U.S.A.; Sturt and Harris (1961, p. 697) concluded that elongated micas define a linear element in the "... direction of mineral stretching within the *ab* plane ..." near Loch Tummel, Perthshire, Scotland. In mid-Strathspey and Strathdearn, Central Scottish Highlands, Anderson (in Hinxman, *et al.,* 1915, p. 22; Anderson, 1923, p. 439) described the striations visible on most of the *S*-surfaces in the quartzitic metasedimentary rocks. Minute corrugations (with axes parallel to the striations) of micaceous laminae, without any mineral elongation, define the striations that "... probably coincide with the direction of shear" and that were "... probably caused by stretching."

Little work seems to have been done in comparing the nature and complexity of structual elements in rocks recrystallized in different metamorphic grades. However, due to the varying competence of rocks under different metamorphic conditions, dissimilar patterns are to be anticipated. Under some conditions structures are more penetrative and tend to obliterate elements developed during earlier tectonic phases; with less penetrative deformation earlier structures can be partially preserved. A. G. Jones (1959, p. 101), for example, found that in high-grade metamorphic rocks of the Monashee group, British Columbia, Canada, linear elements in directions other than the regional trend are rare, whereas in low-grade rocks linear structures in two or three directions within a single specimen are common. These relationships are characteristic of those in many other areas.

Linear structures are particularly important in unravelling complex superposed folding (see Chapter 10), although as indicated above, a particular linear element should not be referred to any specific geometric orientation (e.g., parallel to, or perpendicular to, *B*) without a complete geometric analysis of the rock. Penetrative linear structures parallel to the *b*-geometric axis have been described much more frequently than those perpendicular to *B* or parallel to the *a*-axis (cf. Phillips, 1950, 1951; Kranck, 1960; etc.).

Commonly one of the most difficult problems in areas of superposed folding is that of distinguishing homogeneous domains. A map showing the trend and plunge of linear elements is frequently one of the best initial guides in isolating such domains within a map-area. Hence, linear structures are particularly important in structural analysis (see Chapter 10). In areas in which actual fold closures are not seen, linear elements can sometimes be used to define the *b*-axis more accurately than is possible with β-diagrams, or even π-diagrams, provided that the total geometry is not too complex (cf. Ramsay, 1964).

IMPORTANT VARIETIES OF LINEAR STRUCTURE

LINEATIONS

No attempt will be made to describe all of the possible varieties that occur. For example, the nature of lineations defined by minor crenulations of *S*-surfaces is sufficiently obvious from the previous discussion of folds. Five varieties of particular interest are considered:

1. INTERSECTION OF *S*-SURFACES

Frequently the most obvious linear structures in tectonites are intersections of *S*-surfaces. When the bedding, or other *S*-surface, is folded with concomitant development of an axial-plane or crenulation foliation (S_1), the intersection of S_0 and S_1 defines a prominent linear structure (Fig. 220). In cylindroidal folds the intersections are approximately parallel to the *B*-geometric axis. Either the trace of S_1 on S_0, or that of S_0 on S_1, can be measured; when foliation is well developed (e.g., in slates) it is easier to measure the trend and plunge of the trace of S_0 on the foliation (S_1).

Several dissimilar *S*-surfaces may be present in rocks subjected to successive or continuing deformation events. The mutual intersection of each pair of surfaces defines a separate linear element, provided that at least one of each pair of surfaces is still planar.

In Figure 220 a set of *ac*-joints is illustrated. On a statistical basis the intersection of these joints with the S_1-planes defines a linear structure. Because S_0 is a flexed surface its intersection with the *ac*-joints is not a linear structure in a domain the size of the fold in Figure 220; however,

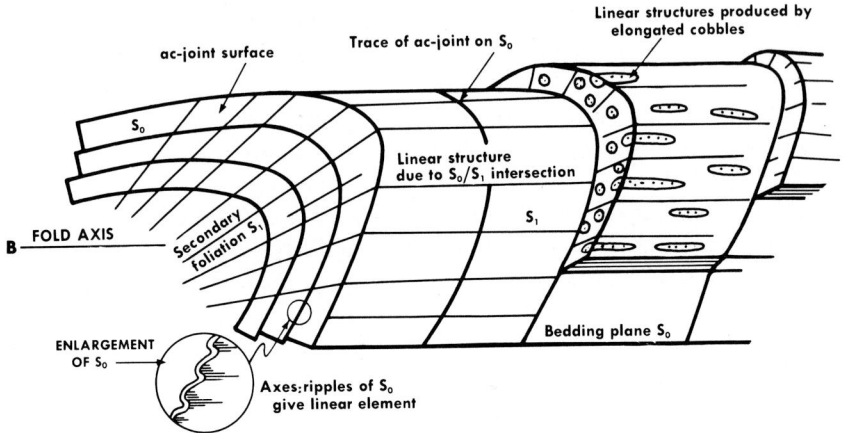

Figure 220. Various linear elements developed in a flexural-slip fold.

this intersection is a lineation in a smaller domain such as a single planar limb of the fold. Similarly, oblique or cross joints define linear structures where they intersect either S_1 or the *ac*-joints.

2. ELONGATE PARTICLES OR GRAINS

Penetrative deformation of rocks causes neocrystallization of the constituent grains and concomitant strain of both the whole rock and its included fragments, pebbles, ooliths, fossils, etc. In Figure 220 the individual deformed cobbles in the conglomerate horizon have a preferred orientation that defines a lineation. Occasionally, cobbles in an undeformed conglomerate (e.g., a glacial till) are inequant and possess an original preferred orientation within the sedimentary rock. However, a very strong linear pattern, as in Figure 220, can almost always be attributed to penetrative deformation.

The geometrical orientation of lineations defined by elongate components has provoked a very vigorous controversy. The disagreement seems to have arisen because most authors have assumed that all lineations of this general type always have the same orientation with respect to folds. However, strained fossils, ooliths, pebbles, etc., define lineations which, in different environments, are commonly either

(a) within the axial-surface and normal to B, or

(b) parallel to B.

Both of these orientations are common, but each is distinctive of a different environment. Voll (1960, p. 551) briefly alluded to the two possibilities. However, review of almost all recent literature would lead to the conclusion that each geologist believes there is a unique geometric orientation; some have supported one orientation, and some the other. Kvale (1945), Cloos (1947A), Anderson (1948), and Brace (1953), for example, championed orientation perpendicular to *B*. By contrast, Turner and Weiss (1963, p. 88) advocated the following rules, *inter alia*, for *all* tectonites:

> Fabrics dominated by a prominent planar structure *S*: *S* = *ab*; any regular lineation in *S*, especially if normal to a plane of symmetry of the fabric = *b*;
> Fabrics dominated by a strong lineation: Lineation = *b*; any direction normal to *b* = *a* (preferably lying in a planar structure).

Unfortunately, little is known about the quantitative distribution of these two classes of lineation. However, within the Appalachian fold belt, Cloos (1943, 1945, 1946, 1947A, 1953B, etc.) for the Cambro-Ordovician oolitic limestones and Maxwell (1962) for the Martinsburg slate advanced considerable evidence for slip folds in poorly consolidated sedimentary rocks and lineations perpendicular to *B*. It is possible that elongated components normal to *B* characteristically develop during the initial deformation of incompetent rocks. During flexural-slip folding of more competent units, and especially of units which are thoroughly lithified, analogous lineations commonly develop parallel to *B* (Fig. 220).

With both orientations there is a superficial appearance of stretching, but it is unwise to attribute the strain to any particular agency without thorough geometric and kinematic analysis of the whole rock (cf. Cloos, 1958). At present observations remain essentially qualitative and no wholly satisfactory process-response models for the genesis of elongate particles have been developed. In fact, it is difficult to erect a process model (i.e., a model for the causative factors) when the response models defined by the measured attributes are so poor. Some of the many different types of particles that define lineations are briefly reviewed below.

Elongated pebbles have been noted by so many authors that reference can be made to only a few; additional references were given by Cloos (1946, 1947A, 1953A). The first description of pebble elongation in North America appears to have been by Hitchcock, *et al.* (1861) in connection with Vermont and Rhode Island localities. Reusch (1887, 1888) described metamorphosed conglomerates from western Norway, in which the long axes of the deformed pebbles have a preferred orientation parallel to the fold axis defined by minor folds. Rüger (1933) for the Geröll Gneiss, Erzgebirge, Germany, and Billings (1937, p. 534) for the Ammonoosuc volcanic conglomerate, Littleton-Moosilauke area, New Hampshire, U.S.A., noted similar relationships. Sometimes quite extreme elongation occurs. Conglomerate portions of a quartzite in the

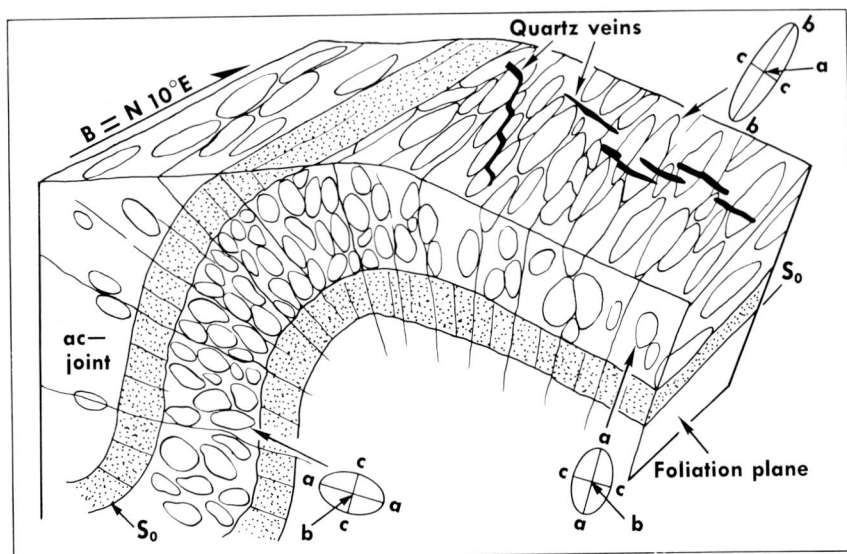

Figure 221. Diagrammatic sketch of a strained boulder bed, Rhode Island Formation, E of Newport, Rhode Island, U. S. A. Note that the flat *ab*-planes of the boulders lie parallel to the foliation, and the longest axes of the boulders lie parallel to the *B*-geometric fold axis (After Agron, 1963A, Fig. B-6).

Bergsdalen Quadrangle, western Norway, contain pebbles shaped, for the most part, like walking sticks and pencils (Kvale, 1945); the ratio of principal axes in these pebbles can reach 1:3:100.

Characteristically, the longest axes of such pebbles define a strong lineation parallel to the *B*-geometric fold axis. Agron (1963A, 1963B) made a detailed study of the quartzitic conglomerates deformed during the folding of the Pennsylvanian Rhode Island Formation at Purgatory, east of Newport, Rhode Island, U.S.A. (Fig. 221). The constituent boulders are 8-10 cm. to over 1.2 meters long (one is nearly 4 meters); most of the boulders are quartzitic, and some contain considerable mica and feldspar, while occasional granite and felsite boulders also occur. The average axial ratios of the boulders are 1:1.5:4.1, although in the zones of most intense shearing, ratios reach 1:2.4:5.7. The flat surfaces of the strained boulders are parallel to the foliation (S_1) and oblique to S_0, which is defined by occasional more sandy horizons (Fig. 222). Many of the boulders have elongated rhombic outlines with the straight sides parallel to shear planes cutting through both the boulders (Fig. 223) and the rock as a whole. Many boulders show indentations (like large crescentic scars formed by flaked off portions) where appressed against adjacent boulders.

Occasionally such lineations are perpendicular to *B*. In the Loch Tummel area, Perthshire, Scotland, Sturt (1961B, pp. 144, 156) considered the recumbent isoclinal F_1-folds to have formed by gravity sliding at a very low grade of regional metamorphism. Although most linear elements are parallel to B-geometric axes in this area, several examples of mineral lineation produced at the same time as the early folds are approximately perpendicular to B. Within a small conglomerate lens in the southwestern Sierra Nevada Foothills, California, U.S.A., pebbles (axial ratios 1:2:4) in a slaty matrix have axes plunging subvertically, whereas a weak lineation formed by intersection of S_0 and S_1 in adjacent slates plunges at $35°$ southeast (Best, 1963, p. 124). Dr. H. J. Zwart (personal communication) mapped elongate pebble lineations perpendicular to B in conglomerates from southern Sardinia, Italy.

A different style of deformation is shown by the metamorphosed pre-Middle Old Red Sandstone Funzie polymictic conglomerate of Fetlar, Shetland Islands (between Norway and northern Scotland), studied by Flinn (1956); 548 pebbles, collected at 36 sampling stations, showed the quartzitic pebbles to be flat, elongate, well-rounded, and to have an overall orthorhombic symmetry. Figure 224 shows that the quartzite pebbles are less deformed than pebbles of other rock types; these data are based on measurements of 3 to 33 pebbles at each station, although Flinn (1956, p. 486) thought that about 30 pebbles should have been measured at each site because "... variation in shape of the pebbles within a sample is greater than the variation of the averages of all the samples." Figure 224 suggests that pebble shape varies systematically across the outcrop area (Fig. 225). The approximate parallelism of the longest pebble axes defines a distinct lineation; increasing elongation of pebbles is correlated with increasing perfection of preferred orientation of the longest axes. Occasional pebble-free bands define S_0, but the schistosity/bedding intersection which might be expected to be parallel to a B-geometric axis, commonly deviates from the pebble lineation by $10°$, and by as much as $90°$ in the extreme north. Flinn concluded that the conglomerate was not deformed by the folding, that the pebbles were not rotated during the deformation, and that the pebble-free bands did not play an active role during deformation of the conglomerate. Flinn (1956, p. 499) thought the strain was due to orthorhombic compression (Einengung of Sander, 1948, p. 70) in the ac-kinematic plane, because two axes of the pebbles have been shortened; accompanying extension parallel to the kinematic b-axis is parallel to the lineation defined by the longest quartzite pebble dimensions.

Clear-cut geometric relationships like those described by Flinn and Agron do not always occur, and several different patterns will probably be revealed by further research. With conglomerates and agglomerates there is always considerable uncertainty about the original shape of the compo-

Figure 222. Deformed quartzitic conglomerate in folded Pennsylvanian Rhode Island Formation at Purgatory, near Newport, Rhode Island, U.S.A. A (above): A sandy member that defines S_0 within the conglomerate; the hammer handles are parallel to the axial-plane foliation (S_1). B: (opposite page) Conglomerate boulders flattened within the axial-plane foliation (S_1).

nents prior to deformation. Nevertheless, the dimensions of distorted fragments often give one of the few direct clues to the strain of the total rock. Where cobbles are scattered through a matrix of finer-grained material (as in a glacial till, for example), individual cobbles probably are not deformed to the same extent as either the total rock or the matrix materials; Mehnert (1939) provided evidence of this in a detailed study of a metaconglomerate from the Erzgebirge, Saxony, in which the longest diameters of strained quartzite blocks are up to ten times the least diameters, while in the associated micaceous graywacke the ratio of longest to shortest diameters of strained grains rises to 45:1.

Successive stages of deformation can be illustrated from Cambro-Ordovician oolites in the South Mountain fold of the Appalachian folded belt, Maryland, U.S.A., studied by Cloos (1943). Figure 226 shows an

essentially undeformed oolite with dolomitic ooliths set in a fine-grained calcitic matrix; Figure 227 shows the preferred orientation of ooliths that have undergone extension amounting to 24 per cent of the diameter of a sphere of volume equivalent to the deformed ooliths. Extension of 50 per cent is shown in Figure 228, in which small mud phenoclasts have also been oriented to some extent.

Even in carbonate rocks components of slightly dissimilar competence commonly show different strain. Cloos (1947A, pp. 877 ff.) showed that two kinds of elements can be distinguished in the oolites: (1)

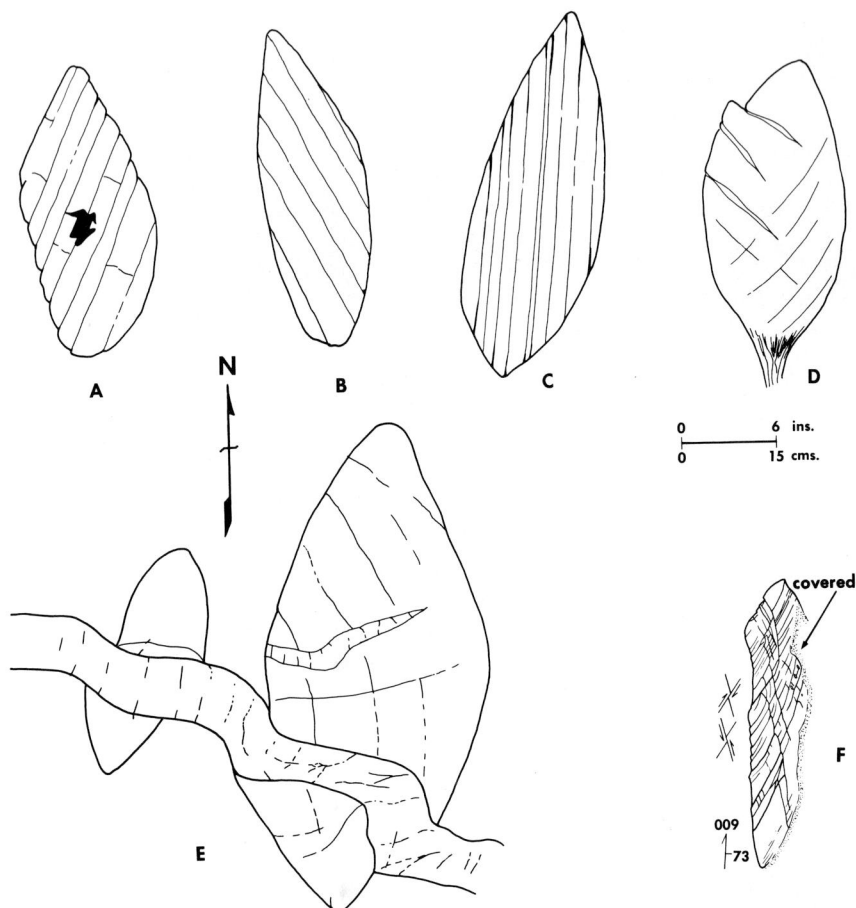

Figure 223. Elongated boulders from folded Rhode Island Formation, E of Newport, Rhode Island, U. S. A. The more prominent fractures are shown. Note offset by quartz vein in E, and the movement along conjugate fracture planes in F (After Agron, 1963A, Figs. B-7 and B-8).

competent constituents such as chert nodules, detrital sand grains, dolomitic spheres, and some fossil fragments, that remain relatively undeformed although surrounded by strongly deformed constituents (Figs. 229 and 230); (2) less competent constituents, such as calcitic ooliths, mud pellets, and calcitic matrix, that tend to be equally deformed within the limits of a single thin section. This emphasizes again that the strain of one component (e.g., an oolith or a pebble) can be of a wholly different magnitude from the strain of the total rock. Hence, it is dangerous to use triaxial ellipsoids defined by deformed ooliths to portray strain of the total

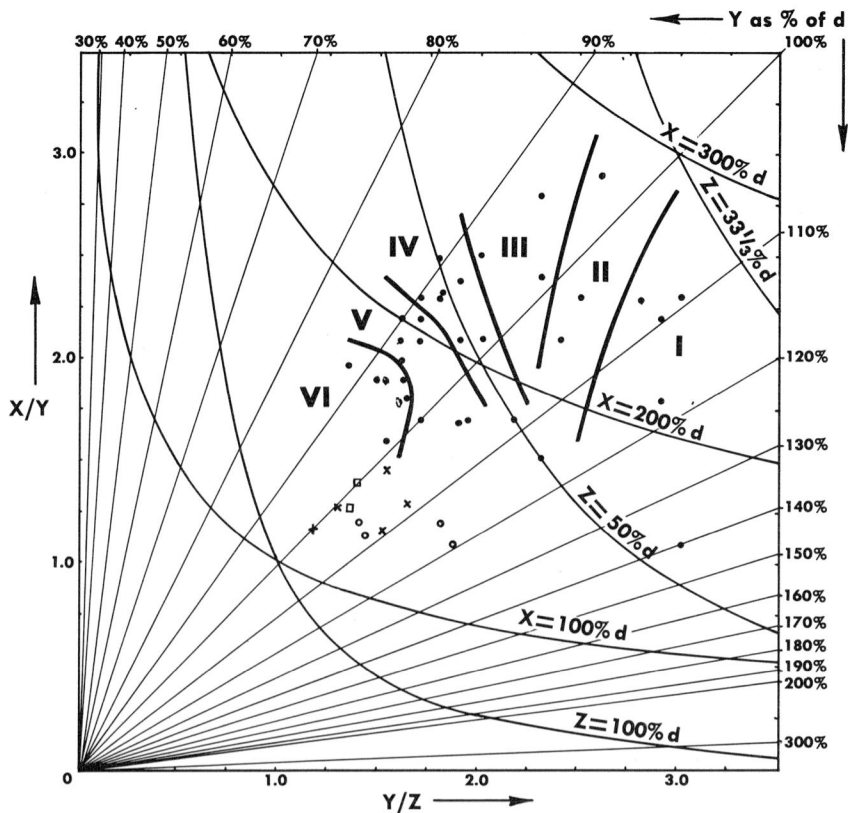

Figure 224. Variation of pebble shape with geographical location in the Funzie Conglomerate, Fetlar, Shetland Islands. • indicate pebbles; x clastic quartz grains of matrix; ∘ quartz grains of quartzitic pebbles; □ pebbles of modern beach and undeformed sedimentary rocks. X, Y, and Z are major axes of pebbles, where $X > Y > Z$. $d = \sqrt[3]{XYZ} =$ diameter of sphere with volume equal to that of pebble. Roman numerals refer to geographical subareas in Figure 225 (After Flinn, 1956, Fig. 4).

rock. Cloos (1943, p. 276) pointed out that oolites should not be used as indicators of the total regional deformation because the physical properties of limestones are markedly dissimilar to those of the associated sandstones and shales. Cloos (1947A, p. 906) also suggested that the strain of ooliths probably indicates a minimum for the total rock deformation, but it would seem that equally-strong arguments could be advanced for considering it a maximum value.

The South Mountain anticlinal fold system of Maryland, U.S.A. (Fig. 231) was interpreted as a large slip fold (Cloos, 1947A, p. 897; 1953B)

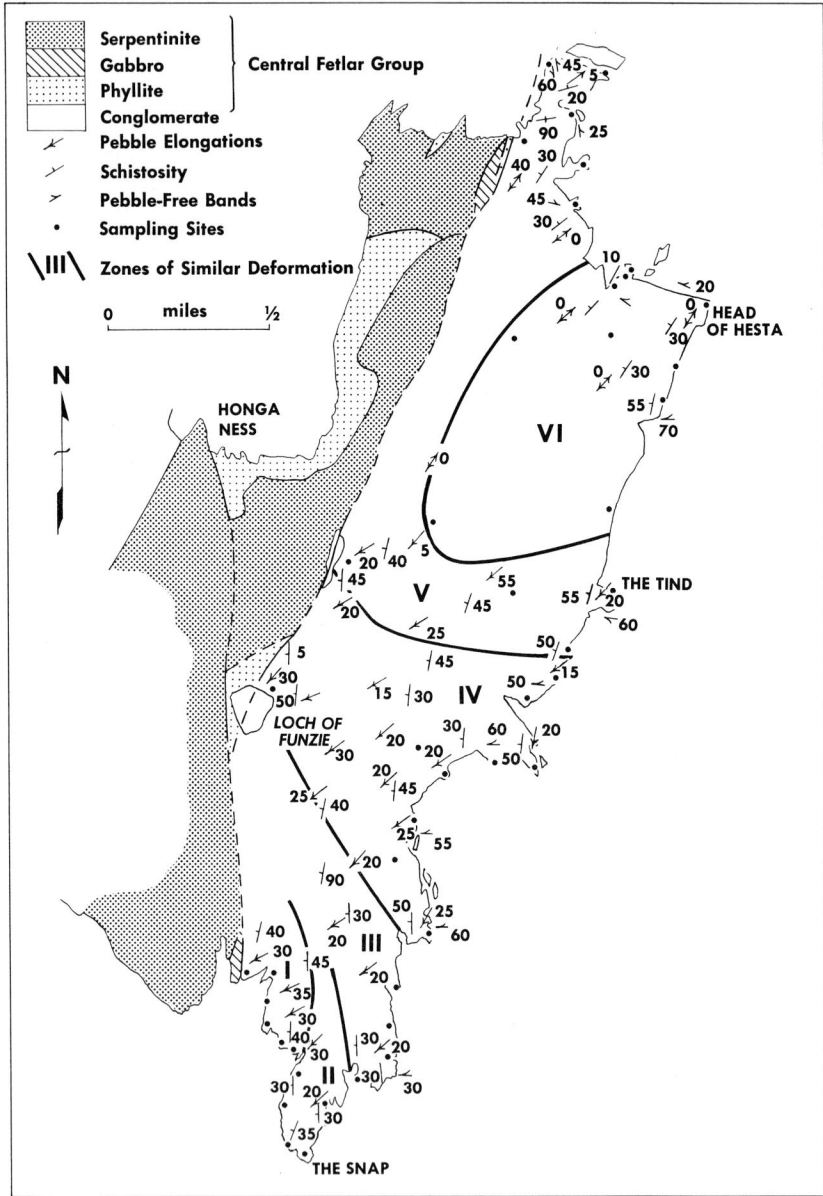

Figure 225. Map of linear structures defined by deformed pebbles in the Funzie Conglomerate, Fetlar, Shetland Islands (After Flinn, 1956, Fig. 2).

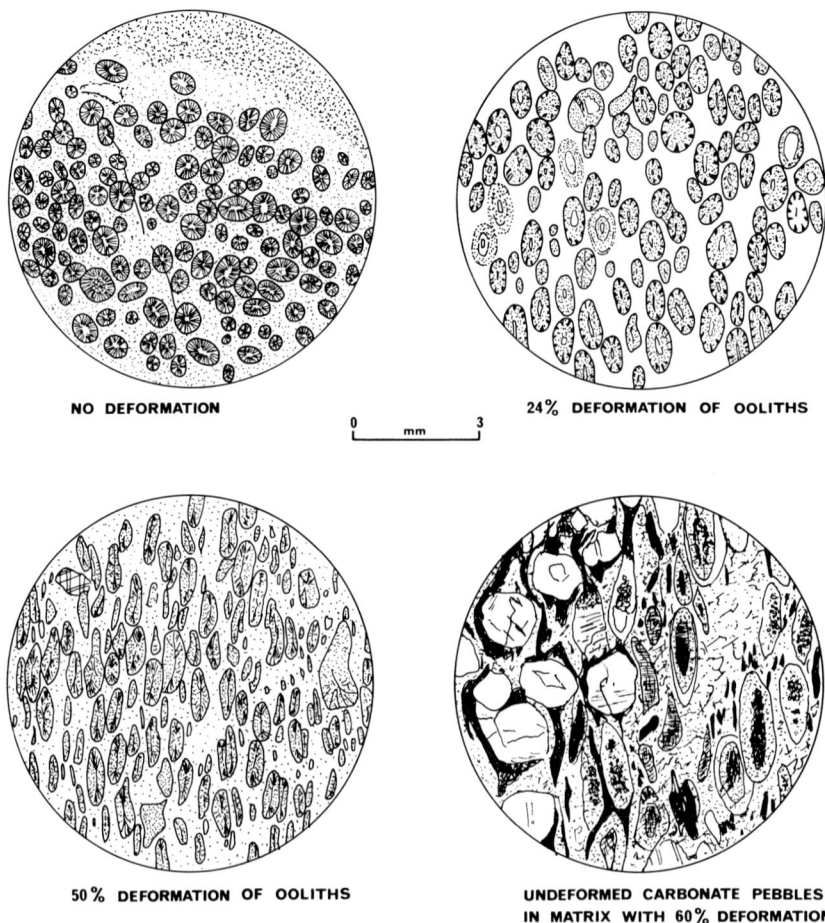

NO DEFORMATION

24% DEFORMATION OF OOLITHS

0 — mm — 3

50% DEFORMATION OF OOLITHS

UNDEFORMED CARBONATE PEBBLES
IN MATRIX WITH 60% DEFORMATION

Figure 226 (above, left). Undeformed oolite at contact of Rochester and McKenzie Formations, Tyrone Quadrangle, Pennsylvania, U. S. A.; note the random orientation of elongated ooliths (After Cloos, 1943, Fig. 1).

Figure 227 (above, right). Cambro-Ordovician oolite deformed during folding, Hagerstown Quadrangle, Maryland, U. S. A.; ooliths have dissimilar sizes but similar axial ratios (After Cloos, 1943, Fig. 2).

Figure 228 (below, left). Deformed Conococheague oolite, north of Sharpsburg, Maryland, U. S. A. (After Cloos, 1943, Fig. 3).

Figure 229 (below, right). Deformed limestone, south of Stoughstown, Pennsylvania, U. S. A. Undeformed carbonate pebbles and single calcite grains (white) occur in deformed matrix of crystalline carbonate, spherulitic ooliths, mud pellets, and shaly bands (After Cloos, 1947A, Plate 7, Fig. 1).

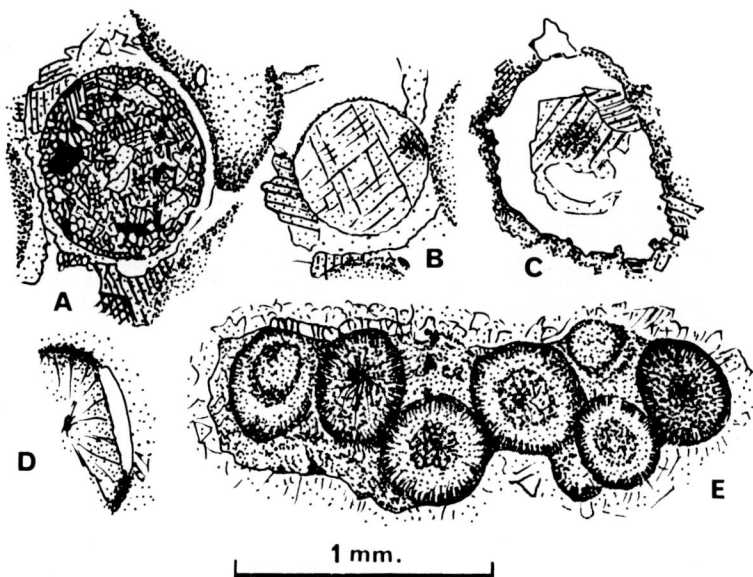

Figure 230. Details of ooliths in Conococheague Limestone, NW of Altenwald, South Mountain fold belt, Maryland, U. S. A. A: Well-rounded cluster of small carbonate crystals in matrix of mud pellets and newly-grown carbonate. B: Single well-rounded dolomite grain in matrix of secondary carbonate and spherical mud pellets. C: Chert replacing grain of carbonate. D: Chert replacing outer layer of an oolith. E: Large pebble of oolite containing several well-layered ooliths in a muddy matrix: calcite fringe grown on the pebble in undisturbed, clear grains (After Cloos, 1947A, Plate 8, Fig. 8).

and there is ample evidence that the main component of oolith extension was developed by laminar flow when the rocks were relatively unconsolidated and not significantly affected by flexural-slip folding. However, Nissen (1964B) showed that twinning of calcite in the ooliths reflects a minor late strain that was superposed on the main deformation of the rock. The fold axial planes dip gently southeast, and the B-geometric axes are subhorizontal and trend northeast-southwest. The foliation is essentially parallel to the axial plane, but it forms a fan which opens towards the anticlinal crest.

The maximum oolith axes (X) have a strong preferred orientation that plunges southeastwards within the foliation plane and normal to B (Fig. 232). Both the X-axis and the intermediate Y-axis of the ooliths invariably lie in the foliation; either X or Y may be parallel to B, but it is much more common for X to be perpendicular to B (Cloos, 1947A, p. 883). The oolith deformation increases from 20 per cent to over 100 per cent towards the core of the South Mountain fold, although the lower

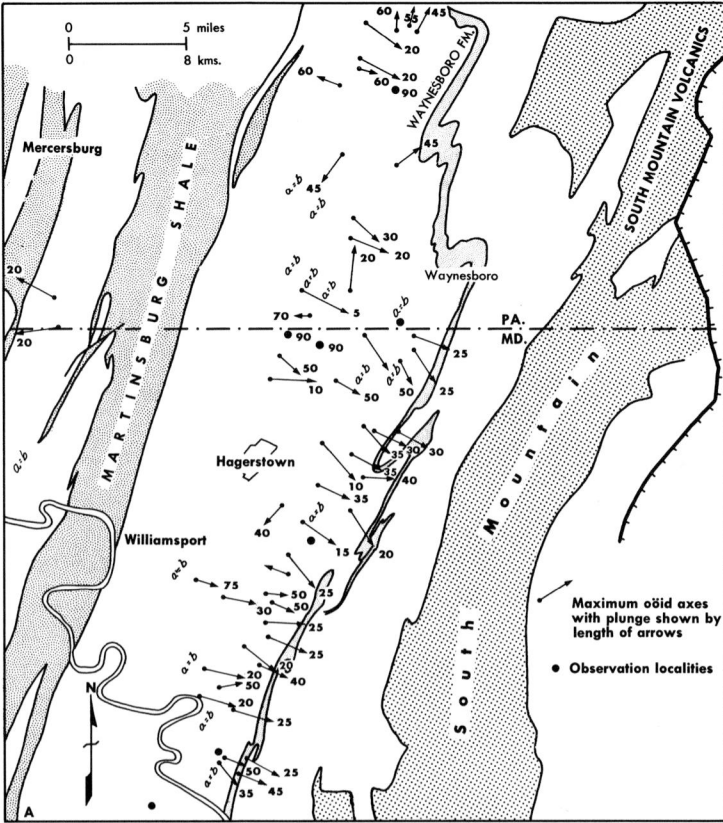

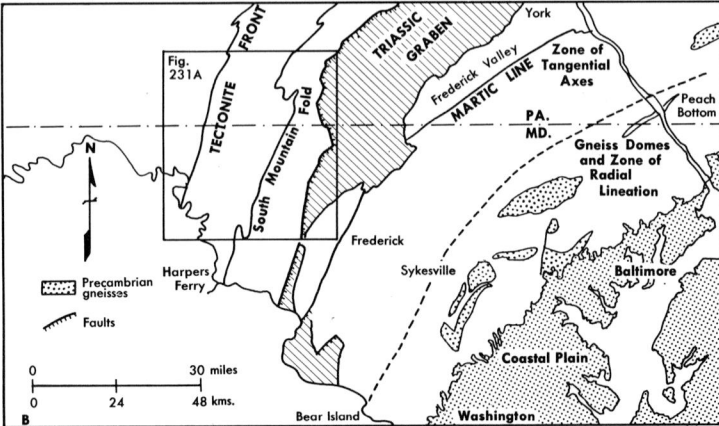

Figure 231. The South Mountain anticlinal fold system, Maryland and Pennsylvania, U. S. A. A: Map of oolith strain features. B: Regional tectonic setting of the South Mountain structures (After Cloos, 1947A, Plate 10 and 1953B, Fig. 7).

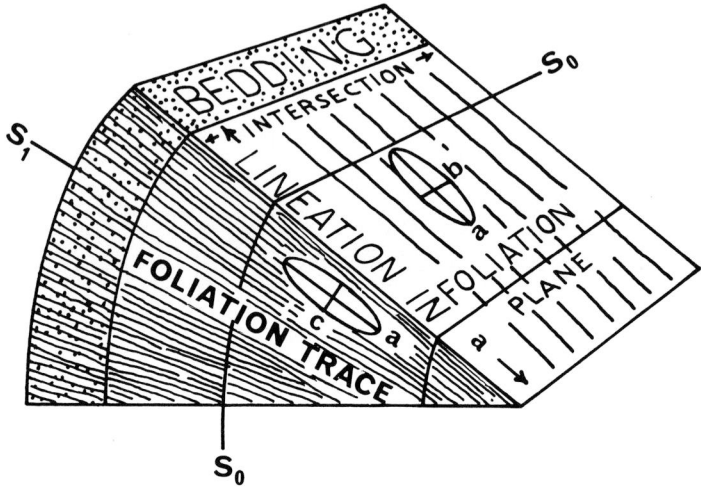

Figure 232. Strain pattern in the South Mountain fold system: see text for explanation (After Cloos, 1947A, Fig. 15).

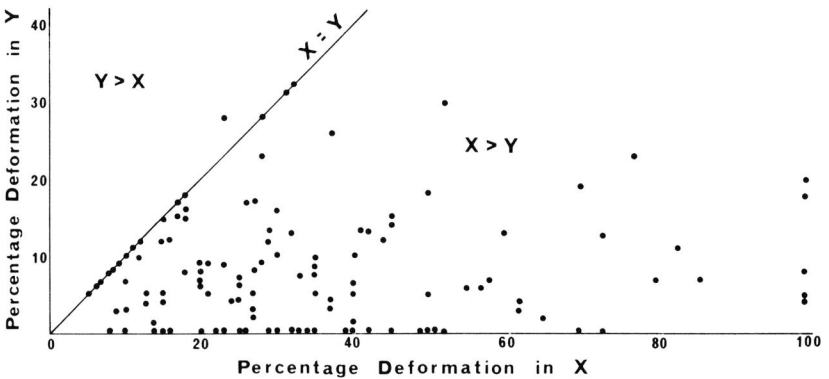

Figure 233. Deformation of ooliths in the South Mountain anticlinal fold system, Maryland, U. S. A. The maximum (X) and the intermediate (Y) axes of the strained ooliths lie in the foliation plane, with Y, in most cases, parallel to the B-geometric fold axis (After Cloos, 1947A, Fig. 12).

limbs of the major and minor folds have suffered the most intense deformation. Cloos established that the Y-axes of the ooliths have been elongated too. The elongations of X and Y are independent, and Y (which is, in most cases, parallel to the B-geometric axis of the fold) is elongated much less than X (Fig. 233).

In the core of the South Mountain structure volcanic rocks also show strong linear elements normal to B defined in part by elongated amygdules.

Cloos based his analyses on the measurement of 35 to 50 ooliths from each thin section, but in many cases he found it necessary to use two or three mutually perpendicular sections to determine all three axial ratios for a single sample. Unfortunately, Cloos (1947A, p. 877) suggested that "Irregularities can be overcome by measurement of many oöids and elimination of freaks. Only the best material should be used." This is a wholly incorrect method of sampling, because selection of "freaks" and "the best" relies upon subjective intuition; studies of this type should be grounded upon objective quantitative sampling techniques if sound results are to be obtained (see Chapter 3).

Cloos (1947A, p. 892) believed that the unequal axial extension of ooliths proves that stretching was not responsible for the strain, and that the intense lineation parallel to the *a*-kinematic axis is due to "movements

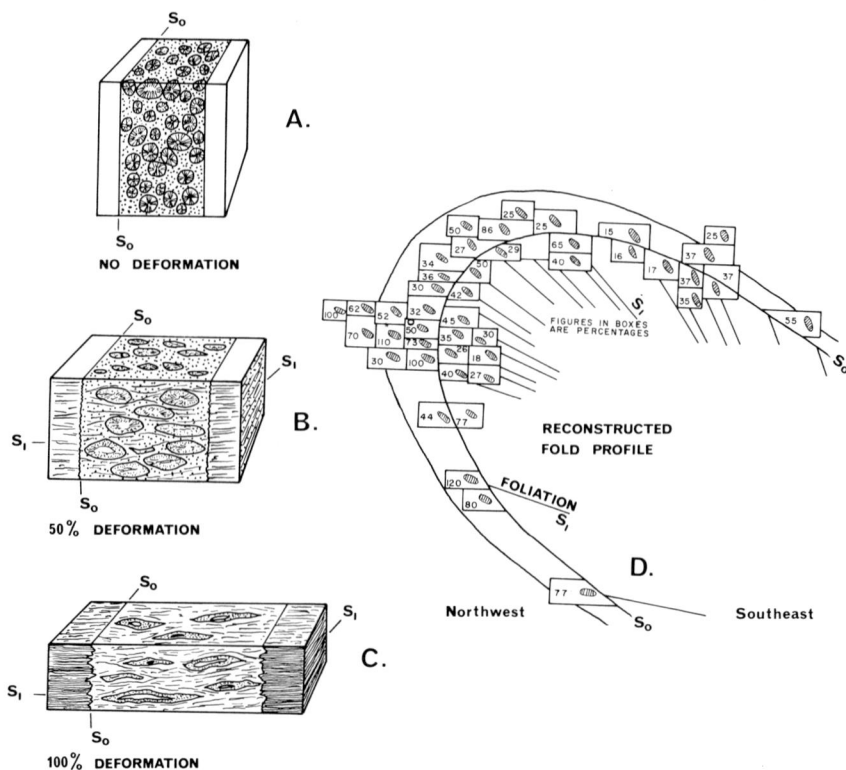

Figure 234. Varying intensity of strain within a fold: schematic diagram based on structures of the South Mountain fold belt, Maryland, U. S. A. A: Undeformed oolite; S_0 is bedding. B: Deformed oolite with development of foliation (S_1). C: 100 per cent deformation of oolite. D: Reconstructed fold profile showing foliation fan and varying degree of strain exhibited by ooliths (After Cloos, 1947A, Figs. 8 and 21).

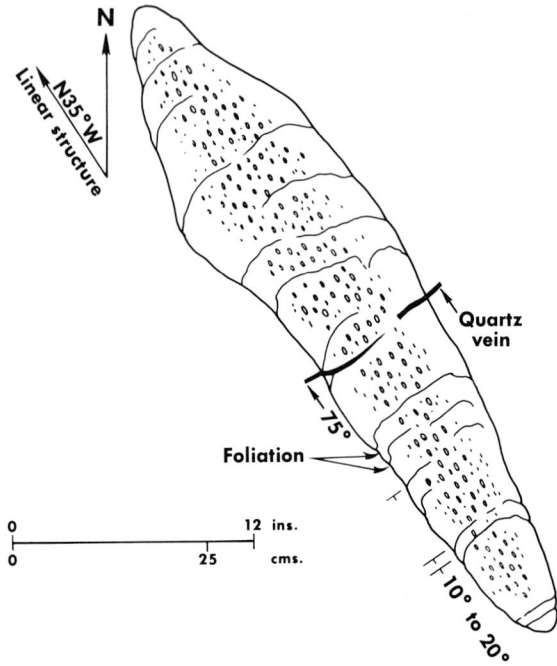

Figure 235. Plan of an elongated Upper Devonian pillow lava with deformed steam vesicles, Trebarwith Strand, Cornwall, England (After Wilson, 1951, Fig. 5).

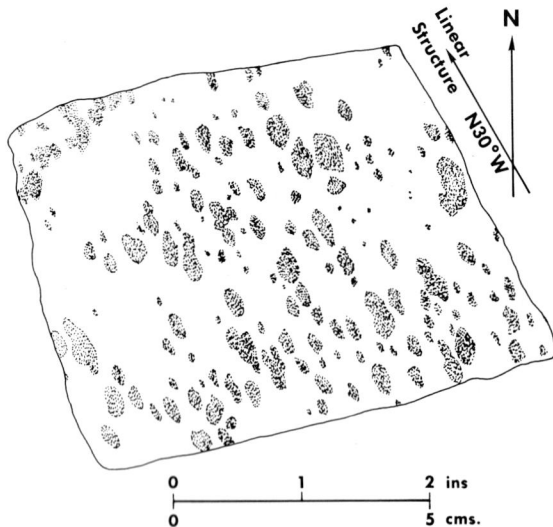

Figure 236. Spotted Woolgarden phyllites with deformed spots (pseudomorphs after cordierite), Tintagel, Cornwall, England (After Wilson, 1951, Fig. 8).

in that direction." However, it is probably premature to formulate genetic conclusions while the quantitative data are still scanty. Cloos (1943, 1947A) also assumed that the ooliths essentially maintained constant volume during the slip folding, and that the total rock was deformed equally with the contained ooliths. These assumptions may be somewhat in error, but, even after allowing for a possible large error factor, it is clear that Cloos was correct in claiming that the deformation was associated with very marked distortion of the thickness of the stratigraphic succession (cf. Fig. 234).

Linear features have commonly been described from deformed volcanic rocks. Elongated agglomerate fragments, lava bombs, gas vesicles, amygdules, etc., can define a lineation. Wilson (1951) described good examples in the Upper Devonian metavolcanic rocks near Tintagel, Cornwall, England (Fig. 235). From the same area he described dolerite intrusions (now coarse-grained chlorite schists) that induced cordierite spots in the adjacent pelites (Woolgarden Phyllites). The spots are now chlorite-quartz aggregates pseudomorphous after the cordierite. Subsequent to the thermal metamorphism the deformation caused elongation of the spots (Fig. 236), which now define a linear structure approximately parallel to the rough linear elements in the adjacent chlorite schists (metadolerites).

In favorable circumstances fossils can provide excellent quantitative estimates of the strain because they have known original shapes. Cloos (1947A) and Hellmers (1955) used crinoid stem ossicles at South Mountain, Maryland, U.S.A., and in the Rhine geosynclinal belt of Germany, respectively; Bryan and Jones (1955) used radiolaria. Breddin (e.g., 1956A, 1956B, 1957, 1958A, 1958C) accurately measured the distortion of brachiopods, lamellibranchs, goniatites, etc., in mildly folded and virtually unmetamorphosed sedimentary rocks in Germany and Switzerland (see Chapter 5). He found that the axis of maximum elongation lies in

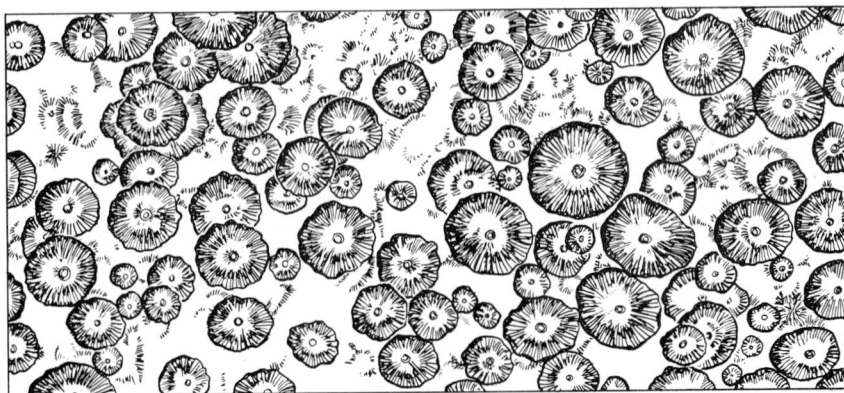

Figure 237. Crinoid ossicles whose circular cross sections are strained into elliptical shapes, Middle Devonian sandstone, near Cologne, Germany (After Kurtman, 1960, Fig. 1).

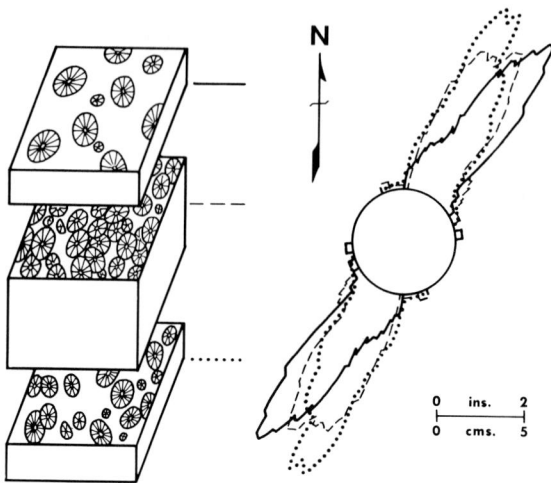

Figure 238. Orientation roses for the long axes of crinoid ossicles in three successive lithologic members that were strained during Variscan folding of Middle Devonian rocks, near Cologne, Germany (After Kurtman, 1960, Fig. 7).

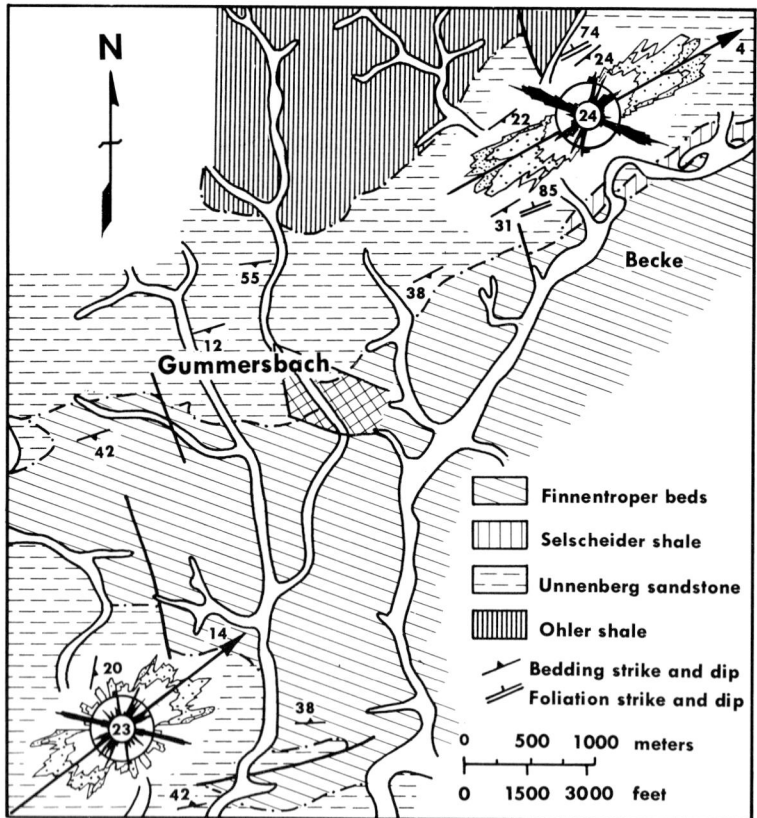

Figure 239. Map showing the preferred orientation of deformed crinoid ossicles in Middle Devonian rocks, Gummersbach, near Cologne, Germany; key to the orientation roses is included in Figure 240 (After Kurtman, 1960, Fig. 12).

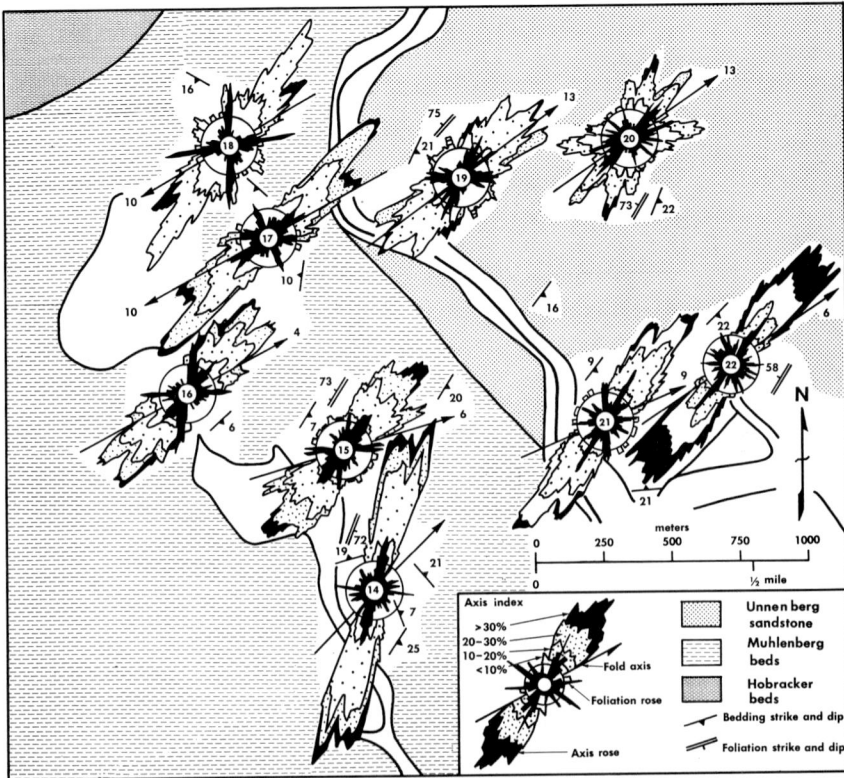

Figure 240. Map of the preferred orientation of deformed crinoid ossicles in the Middle Devonian rocks east of Lindlar, near Cologne, Germany (After Kurtman, 1960, Fig. 13).

the fold axial plane normal to the *B*-geometric axis (Figs. 101 and 102), while Hellmers established ossicle elongation parallel to *B*. Kurtman (1960) and Nissen (1962, 1964A) analyzed the deformation of single calcite crystals comprising crinoid ossicles in the Middle Devonian sandstones near Cologne, Germany. These ossicles apparently originally had oval sections that permitted alignment by current action during sedimentation. The ossicles were deformed by Variscan folding and the present dimensional parallelism of the oval sections (Figs. 237 and 238) reflects a compromise between the inherited primary orientation and an imposed strain due to twinning of the calcite under stress (Nissen, 1964A). The resultant major axes of these ovals have a distinct preferred orientation that can be mapped as a regional linear structure contained within the fold axial planes (Figs. 239 and 240). Furtak and Hellermann (1963) demonstrated the geometric methods by which the structures of fossil plants (Calamites, Lepidodendron, and Sigillaria) can be used to calculate accurately the strain in deformed coals.

In Chapter 8 the slip folds and associated axial-plane foliation described by Maxwell (1962) in the Martinsburg slates were discussed at some length. In these rocks, which were apparently deformed in an essentially unconsolidated state, a preferred habit orientation normal to B is well developed. Cloos (1943, p. 273) suggested that the Pottsville sandstone farther south within the Appalachian fold belt was only partially or poorly consolidated at the commencement of folding, and that consolidation was largely a result of folding. Cloos supported Fellows (1943A; 1943B) in thinking that under these conditions individual sand grains are free to move without severe deformation, and that they can develop a habit orientation within the axial-plane foliation normal to B. Fellows' reconnaissance study was based on some 800 square kilometers of the central Appalachian fold belt, ranging from little-disturbed sandstones of the Appalachian Plateau of West Virginia and Pennsylvania to the highly-deformed rocks of the Piedmont of Maryland and Pennsylvania, U.S.A.; the sedimentary rocks studied range from Precambrian through Mississippian in age. Fellows (1943B, p. 1430) recognized that elongated grain lineations are characteristically parallel to B in highly-deformed schists and gneisses, but he suggested that the quartz-rich "unmetamorphosed" rocks have developed lineations perpendicular to B. Fellows (1943A) contended that in unmetamorphosed rocks, habit orientation is achieved before lattice orientation (i.e., that lattice orientation represents a more intense degree of deformation than habit orientation alone).

Gangopadhyay and Johnson (1962) described two superposed sets of slip folds in mylonites close to the Moine Thrust Zone north of Glen Carron, NW Scotland. Careful geometric analysis established the orientation of the a-kinematic axis within the axial plane of the second generation of moderately appressed slip folds. Penetrative movement along the axial plane surfaces of these relatively simple second folds (Fig. 241) apparently caused recrystallization of quartz grains to give a preferred orientation of their longest dimensions and their c-axes parallel to the a-kinematic axis.

These varied examples suggest that a more detailed study of the fabric in little-metamorphosed folded sedimentary rocks and in slip folds in all environments is urgently required. At the present time there is little evidence concerning the extent of slip folding in either metamorphosed or unmetamorphosed rocks. Similarly, the data available in support of well-authenticated lineations that are not parallel to the B-geometric axis is meager—this presents a challenge, and in particular it suggests that

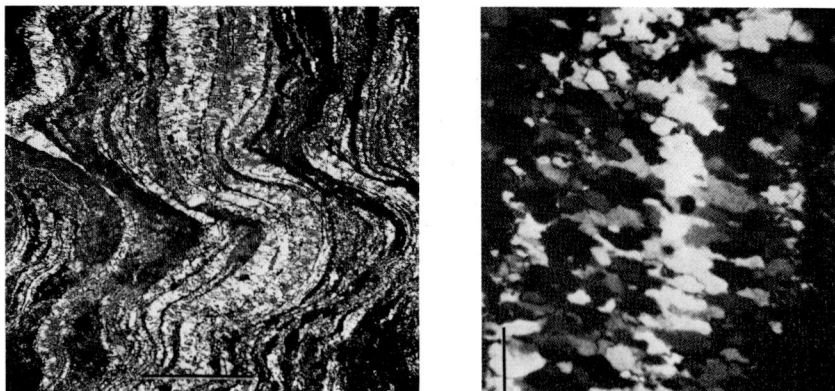

Figure 241. Superposed slip folds in Lewisian mylonite close to Moine Thrust Zone, north of Glen Carron, NW Scotland. A: F_2-slip fold; S_2 is horizontal in this figure and the scale is 2.5 mm. B: Enlargement of part of a quartz vein in A to show elongation of quartz grains within S_2 parallel to the kinematic a-axis; scale is 0.2 mm. (From photographs reproduced in Gangopadhyay and Johnson, 1962, Plate VII, Figs. 3 and 4).

considerably more work is needed on folded unmetamorphosed (or poorly metamorphosed) sedimentary rocks. Slip folds in general have received relatively little attention in the literature, despite the fact that recent observations indicate that slip folding has played a major role in orogenic processes. It would be interesting to compare the geometry of slip folds, and the lineations associated with them, developed in poorly consolidated sedimentary rocks and within deep zones of high stress.

3. ROTATED ELEMENTS

Axes of rotation are useful linear elements that can be observed in rotated elements of widely different sizes.

Large grains rotated during continuing crystallization and deformation commonly permit rotation axes to be measured, even when actual folds are not visible. When porphyroblasts grow during penetrative rotation, relicts of S-surfaces can be enclosed and continuously turned into spiral forms (Fig. 242). While feldspars, staurolite, and many other minerals commonly show this structure, garnets sometimes develop spectacular spiral habits. McLachlan (1953) discussed the bearing of rolled garnets on the concept of b-lineation in the Moine schists of Scotland. Peacey (1961) made a detailed study of rolled garnets in the Moine schists of Morar, Inverness-shire; the spiral inclusions prove that the garnets were rotated in different directions on opposite limbs of individual flexural-slip folds.

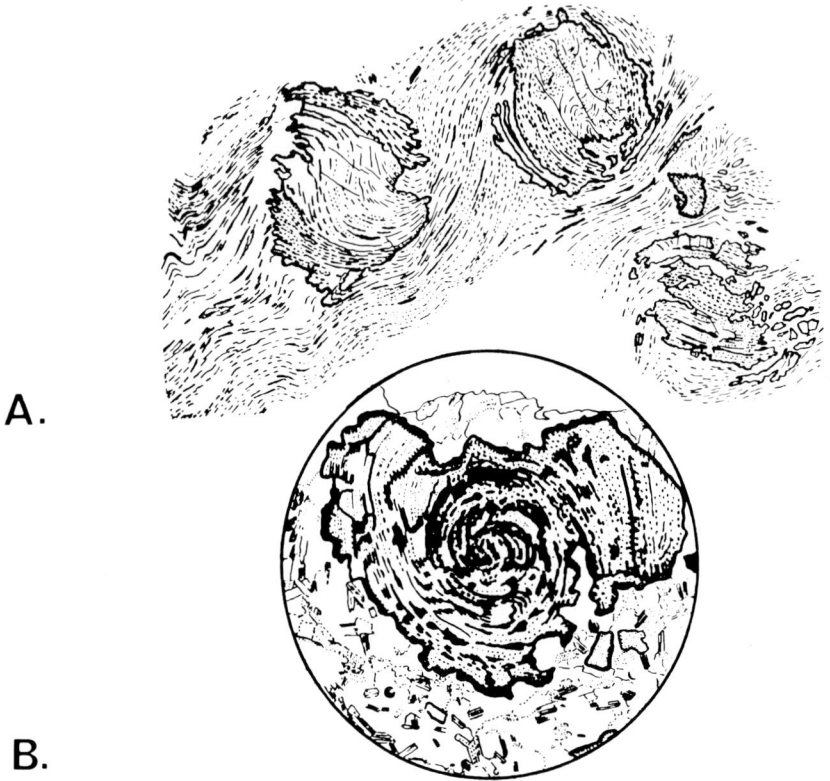

Figure 242. Rotated *S*-structures enclosed in garnet porphyroblasts grown during penetrative rotation and deformation of schists. A: Twisted garnets after Krige. B: Snowball garnet after Flett (After Read, 1949, Fig. 3).

Outcrop conditions are often sufficiently poor that actual fold closures are not seen over wide areas. Either large folds are not present, or the outcrop pattern does not expose the fold closures that do occur. Under such conditions rolled porphyroblasts, or any other evidence of the rotation axes, give useful evidence about the *B*-axis. Commonly foliation planes show minor wrinkles or ripples (actually tiny folds) whose axes are approximately parallel to *B*. In the northern Monadhliath (Central Highlands, Scotland) micaceous granulites show this structure well (Fig. 184), although small folds are rarely visible. Traced southward this ripple-like linear structure is shown to occur parallel to the *B*-axes on the flanks of the folded foliation surfaces that define the folds visible in individual outcrops. A. G. Jones (1959, pp. 100-101) described somewhat similar relationships from the schists of Shuswap, British Columbia, Canada.

Crenulation foliation (especially joint-drag structures) intersecting an earlier *S*-surface — either foliation or bedding — can be responsible for a well-marked linear structure. The crenulation foliation is associated with rotation, and the axes of rotation have a preferred orientation parallel to the linear structure defined by the intersection with the pre-existing *S*. Should more than one crenulation foliation be developed each results in a linear structure where it intersects the principal *S*-surface.

4. GROWTH OF MINERALS

Planes of easy access are available to circulating or permeating solutions, etc., when a directional fabric, or a joint system, is present in a rock. These solutions can cause growth of minerals with a preferred orientation that reflects, or imitates, the geometry of the access-planes. Introduced hornblende, for example, can develop a preferred linear orientation within the pre-existing foliation planes of a schist. Similarly, microcline microperthite megacrysts have grown with a linear alignment within the pre-existing foliation of the "Older Granite" of Donegal, NW Ireland (Whitten, 1957B). Such linear structures are **belteroporic fabrics** (Sander, 1950, p. 119).

Mimetic crystallization includes analogous structures, such as the growth of quartz veins in *ac*-joints. The joints provide planes of easy access and the quartz is free to develop a fabric in the veins that is totally unrelated to that of the host rocks. Many other minerals (e.g., calcite) can also produce mimetic structures.

5. LINEAR ELEMENTS DEFINED BY PETROFABRIC ANALYSES

Neocrystallization during deformation of rocks results in both habit and lattice orientations of the component grains; many mica schists provide good examples. Some habit orientations were discussed in connection with elongate particles or grains that define linear structures. Characteristically, all the grains of arenaceous and argillaceous sedimentary rocks become elongate during penetrative deformation accompanying folding; the resulting lineations can be either parallel to, or normal to, *B* in different tectonites. Lattice orientation of component grains is also important, although this is not always reflected by any obvious elongation of the mineral grains. Some quartzites and siliceous granulites are good examples; in such cases this linear structure cannot be mapped in the field.

It is not intended to deal extensively with petrofabric analysis, but the technique has traditionally involved measurement with a microscope universal stage of the three-dimensional orientation of such properties as the *c*-axes of quartz grains of the normals to biotite (001) planes. Commonly, a complete census of all the grains within a contiguous homogeneous area of a thin section is made. Some 100 to 300 grains must be measured and the directional properties are plotted on the lower hemisphere of a Schmidt equal-area net. Such a census for minerals like quartz can be made by more rapid methods, although these have not yet been used extensively. For example, the orientation of each individual grain can be estimated with a photoelectric cell (cf. Zimmerle and Bonham, 1962) and the results digitized and plotted automatically by computer. In this manner the preferred orientation of the measured lattice properties can be detected, and lineations and planar structures can often be identified from the contoured plots.

Turner and Weiss (1963) gave a full review of the petrofabric techniques in common use.

In most of the described examples in which a lineation is clearly visible in the field, the structure is parallel to the *B*-geometric fold axis.

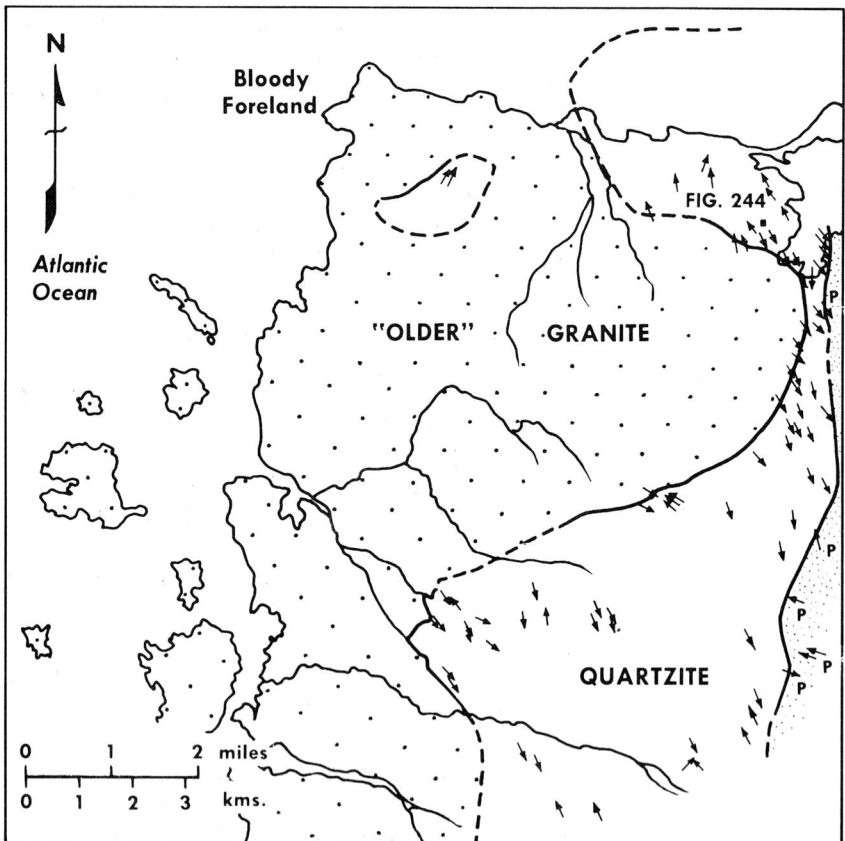

Figure 243. The Errigal Quartzite in extreme NW Donegal, Ireland. Trend and direction of plunge of linear structure defined by elongate grains are shown (arrows); P = Pelitic rocks. Point labelled Fig. 244 is location of sample used for fabric analysis in that figure.

For example, the Errigal Quartzite, NW Ireland, has a distinct grain elongation that can be mapped easily in the field (Fig. 243). The whole rock has been affected by neocrystallization, and petrofabric analyses show that quartz grains have a strong preferred orientation of their *c*-axes (Fig. 244).

Analogous linear structures parallel to *B,* defined by elongate grains within tectonites, were recorded on many 1:63,360 maps prepared for the Scottish Highlands by the Geological Survey of Great Britain. Some of these structures are shown in Figure 245. Linear structures were plotted on some sheets but omitted from others, so that the elements are developed more evenly than might be anticipated from Figure 245. Phillips' (1937) pioneer petrofabric study of a large area in the Northwest Highlands showed that the dominant linear structure (*b*-axes of the petrofabric girdles) defined by the lattice orientation of quartz (and often of micas too)

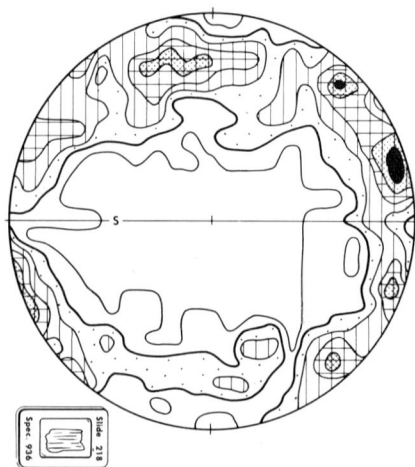

Figure 244. 301 *c*-crystallographic axes of quartz grains from Dalradian Errigal Quartzite, Derryconor, Donegal, NW Ireland; section was cut normal to lineation and rodding observed in the field. Contours at ¼, ¾, 1½, 2½, 3½, and 4½: maximum value 5.

plunges gently southeast, parallel to the linear elements mapped in the field. Some of Phillips' results are shown in Figure 246. Subsequent work (see Chapter 11) demonstrated that the Highlands do not have the simple homoaxial southeast-plunging *B*-structure originally supposed. Such linear structural patterns should not, therefore, be used in isolation for extrapolation to the regional structural geometry.

Since habit and lattice orientations are essentially ubiquitous in penetratively deformed rocks, many similar examples could be cited. Sturt (1961A), for example, recorded a lineation developed parallel to the fold axes of the deformed nepheline syenite gneisses from Sørøy, Norway (Fig. 92). Although there was probably a random orientation in the undeformed rock, the individual crystals now have a preferred habit

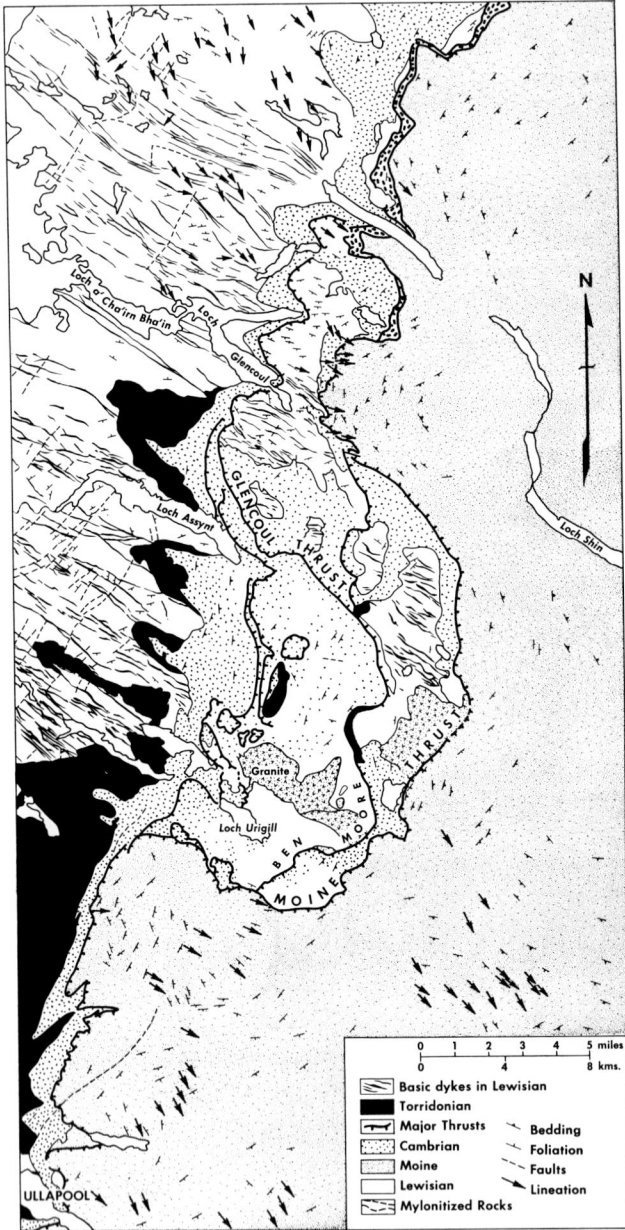

Figure 245. The Assynt area, Northern Highlands, Scotland, showing the complexities of the Moine Thrust Zone and some of the linear elements developed in the area (Based on 1 inch to mile Geological Sheets 101, 102, 107, and 108 of the Geological Survey of Great Britain and Cloos, 1946, Plate 9).

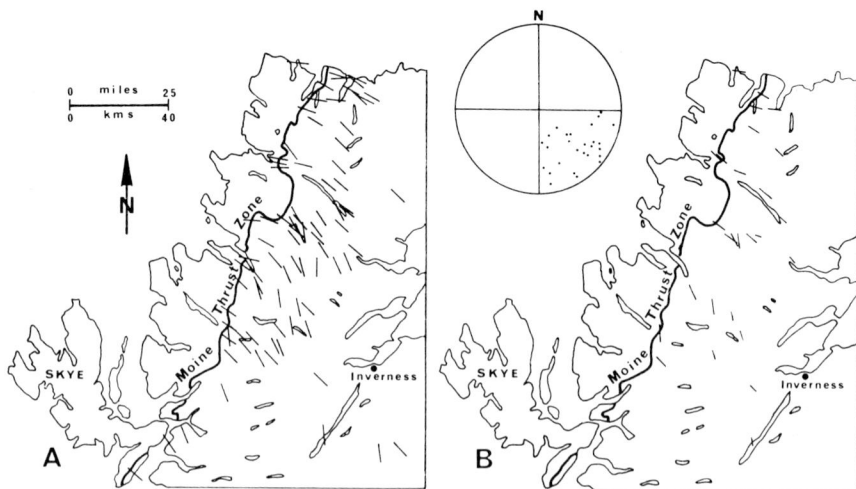

Figure 246. Linear structures within Moine rocks of the Northwest Highlands, Scotland. A: General map of linear structures observed in the field and believed to represent *B*-geometric fold axes. B: Trend of axes of petrofabric girdles for quartz (and often also of biotite and muscovite at the same locality) – the south-easterly plunge is inversely proportional to the length of the line. *Inset*: Schmidt projection of the petrofabric data shown in B (After Phillips, 1937, Figs. 1, 2, and 5).

orientation parallel to the *B*-geometric fold axes. Also the *c*-crystallographic axes of the nepheline grains show a strong maximum coincident with the fold axis.

OTHER LINEAR STRUCTURES

The preceding descriptions show that linear structures can result from a very wide range of dissimilar structural features. The intersection of *S*-surfaces, elongate particles or grains, rotated elements, and the growth of minerals have been discussed above, in addition to slickenside-type linear elements. Several additional more specialized structures such as boudinage, rodding, and mullions commonly define linear structures. Each is described individually below.

1. BOUDINAGE

Boudinage is a common feature in both tectonites and mildly-deformed sedimentary rocks. It is characterized by attenuation of a more competent layer or unit and concomitant flow of less competent adjacent units into the "neck" zones resulting from the attenuation. Continued attenuation commonly results in separation of the competent rock into isolated cylinder-like units, each of which was referred to as a **boudin** by Stainier[1] (1907 and in Lohest, *et al.*, 1908, 1909), in classic exposures of Lower Devonian quartzites and schists at Bastogne, Belgium; these structures (Fig. 247) were described in detail by Lohest (1910).

It is of historic interest that similar structures were referred to much earlier, as, for example, by MacCulloch (1816, Plate 15) in Glen Tilt, Central Scottish Highlands. From North Wales, Ramsay and Salter (1866, pp. 75-76, Fig. 18) and Harker (1889) described a quarry in which a competent rock unit bulges and thins within folded Silurian slates; meter-long bulged portions are separated by vein quartz. Harker (1889) referred to limestones in slates in Ilfracombe, SW England, and Gosselet (1888) to quartzites in slates in the Ardennes, Belgium, which show similar structures. Reyer (1892) described structures, which would have been called boudinage now, that are parallel to *B* on the flanks of anticlines.

Despite the detailed account by Lohest (1910), interest in boudins developed slowly and in 1923 Quirke (pp. 659-60) wrote that the boudinage of Bastogne is:

> ...a rare geological phenomenon. There is no mention of this type of deformation in any works in English, so far as the writer is aware, nor does it seem to have had consideration from any but the geologists of Belgium.... There seems to be nothing in the arts or in nature which can be compared in mechanical origin to boudinages, which makes them the more interesting and the more worthy of study.

Within the following two decades boudinage received little attention; but it was described by Balk (1927, p. 29), Holmquist (1931), Corin (1932), Wegmann (1932; 1935B), Read (1934B; 1936), and Walls (1937B). Although Quirke (1923) thought boudinage was formed by lateral compression under enormous loading, Corin (1932), Read (1934B), and Wegmann (1932) held that it results from extension of a competent bed during deformation. Wegmann (1932, 1935B) illustrated magnificent boudins aligned in one direction only from highly metamorphosed rocks in Norway, Finland, and east Greenland; however, he also drew attention to perpendicular boudinage structures ("tablettes de chocolat" structure) at Bastogne (Wegmann, 1932).

Soon after Cloos' (1947B) review, interest in boudinage multiplied rapidly, and it is now recognized as one of the more common structures of folded rocks. One is impressed by the commonness of boudinage in most

---

[1]Stainier (in Lohest, *et al.*, 1908, p. 472; 1909, p. 371) noted that "... sur l'initiative de M. Lohest on a fréquemment utilisé, pour la facilité du langage, des néologismes de boudiner et de boudinage."

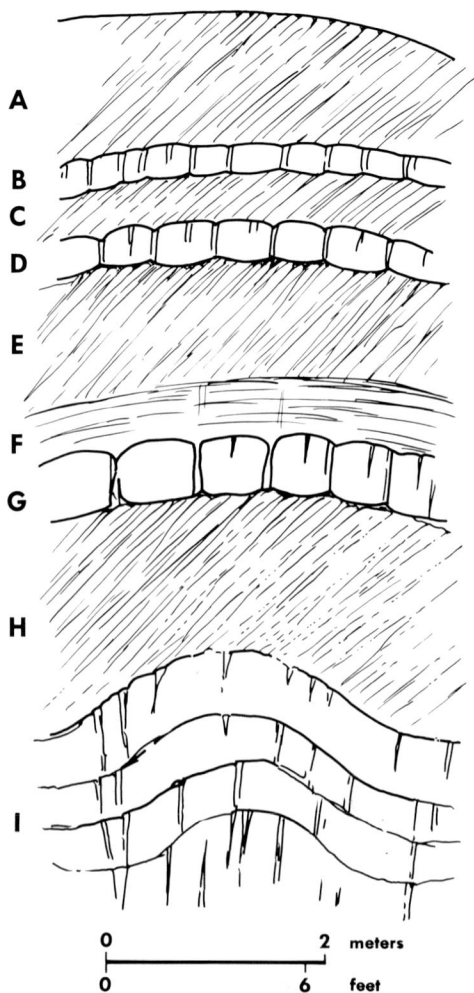

Figure 247. The original boudinage structures described from the Lower Devonian quartzites and schists, Bastogne, Belgium. A, C, E, and H: Shale with foliation clearly independent of stratification and without quartz veins. B: Slightly boudinaged grit (sandstone) with numerous veins. D and G: Boudinaged grit (sandstone) with numerous quartz veins. F: Stratified schistose sandstone without foliation but with thin quartz veins. I: Folded grit cut by numerous large veins. (After Lohest, 1910, Fig. 1).

regionally-metamorphosed terrains, ranging from low-grade schists (e.g., in Caledonian shale-conglomerate-sandstone-limestone complexes at Tröndelag, Norway) to high-grade gneisses such as the granulites of west Greenland (cf. Ramberg, 1955, p. 512).

Described examples abound. McIntyre (1950B) described boudins roughly parallel to the *B*-geometric axes of small recumbent folds in Moine Struan Flags at Dalnacardoch, Central Scottish Highlands. From the Upper Devonian and Carboniferous (Culm) rocks exposed in the sea cliffs of Tintagel, SW England, Wilson (1951) described many boudinage and incipient boudinage structures. Each lithology responded to deformation differently (Fig. 248), but detailed work permitted the geometric relationship of boudinage to the other minor structures to be determined (Fig. 249). A general shearing movement towards the NNW produced (a) slip along bedding planes with development of slickenside striations parallel to the local direction of movement, and (b) fine crenulations and boudins parallel to the axis of rotation.

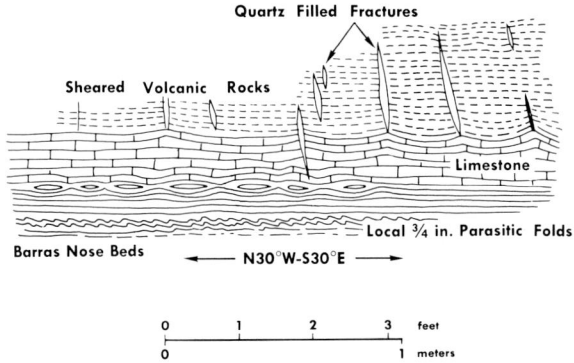

Figure 248. Tension fractures and incipient boudinage structures, near Tintagel, Cornwall, England (After Wilson, 1951, Fig. 3a).

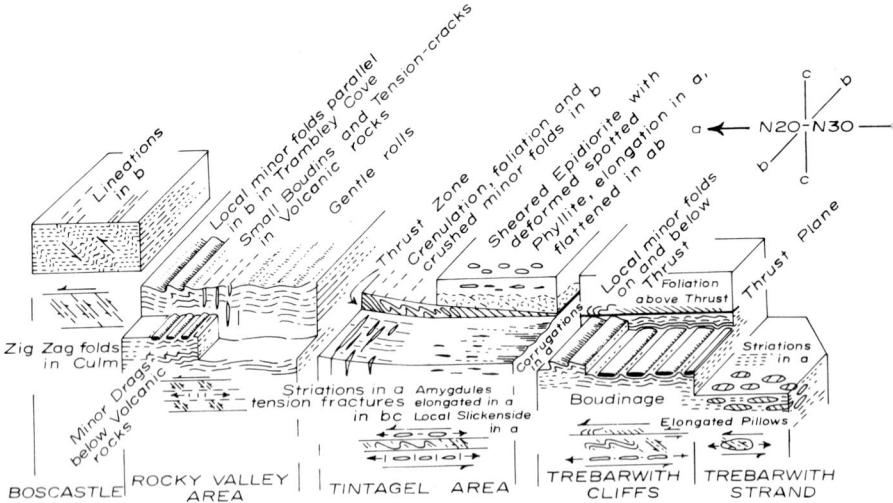

Figure 249. Synopsis of geometrical and kinematic relationships displayed in the Tintagel area, SW England; some late folding and normal faulting has been omitted (After Wilson, 1951, Fig. 10).

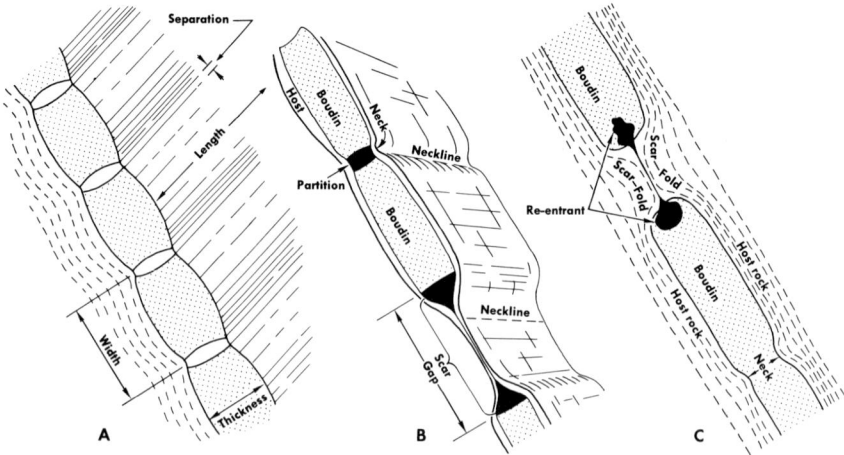

Figure 250. Descriptive terminology in current use for boudins. A after Wilson, 1961, Fig. 34A; B and C after A. G. Jones, 1959, Figs. 23 and 24).

Figure 250 shows the current descriptive terminology as formalized by A. G. Jones (1959) and Wilson (1961).

According to Ramberg (1955) the thinnest boudins are about 1 cm. thick (e.g., calcite veins in slates at Tröndelag, Norway), while the thickest are at least 20 m. thick (e.g., dolerite dykes in western Greenland); Wegmann (1935B, Fig. 7) photographed equally large boudins in Christian X's Land, northeast Greenland. However, Scheumann (1956) published a series of beautiful photographs and photomicrographs of microboudins within the amphibolites of Rosswein, Granulitgebirge eastern Germany; many boudins in these finely-banded rocks are much less than 1 cm. thick, and several have complex structures with minute boudinage developed within thicker boudinaged units.

During folding the relative competence of associated rocks is likely to have differed markedly from those of the same rocks at exposure. As a result, a large variety of rock types shows boudinage. In Greenland, Ramberg (1955) found boudins developed in amphibolites, hornblendites, diopside skarns, pyroxenites, and other caferromagnesian rocks enclosed in quartz-feldspar gneisses and mica schists. However, he also described (a) quartz-feldspar pegmatite boudins in micaceous quartzites of the Wind River Mountains, Wyoming, U.S.A., and (b) boudins formed in limestones, calcareous sandstones, and coarse-grained conglomerate layers in low-grade pelitic schists of the Caledonian belt of Tröndelag, Norway. Whitten (1951) illustrated boudinaged grossularite-idocrase members within Dalradian marbles at Bunbeg, NW Ireland (Fig. 251). By contrast,

Figure 251. Boudinage in competent grossularite-idocrase member of a Dalradian marble, Bunbeg, Donegal, NW Ireland. Less competent units (rich in calcite, diopside, and wollastonite) show compensational flowage. Scale is 2 cm. (After Whitten, 1951, Fig. 2).

Scheumann (1956) studied plagioclase amphibolites from Rosswein, E Germany, in which leucocratic plagioclase-rich layers developed boudins and interlaminated melanocratic material flowed in between them. Boudins in thick quartzite units occur within schistose thin quartzose layers containing argillaceous materials in Ruanda, Africa (Aderca, 1957). In the Tazin Group, Saskatchewan, Canada, Christie (1953) found that granitic rocks and quartzites were boudinaged during intense folding (cf. Fig. 252), while amphibolites and other mafic rocks, especially cordierite-amphibolites and biotite schists, were less competent, and dolomitic rocks were the least competent. In the Shuswap terrane, British Columbia, Canada, abundant boudinage structures occur in quartzitic units, veins of quartz, sills of pegmatite, and dykes, but they are commonest in amphibolites (A. G. Jones, 1959). Incompetent calcareous beds at Langø, west Greenland, were thrown into small folds (Berthelsen, 1957, 1960B); later these units gained in competence and were boudinaged (Fig. 253). The increased relative competence was either due to conversion of the original sediment into calc-silicate rock, or, since the change was relative, to change in the physical properties of the enclosing rocks. Hence, it is unwise to assume with Reitan (1959, p. 193) that under all environmental conditions basic rocks are less plastic than the more acidic rocks.

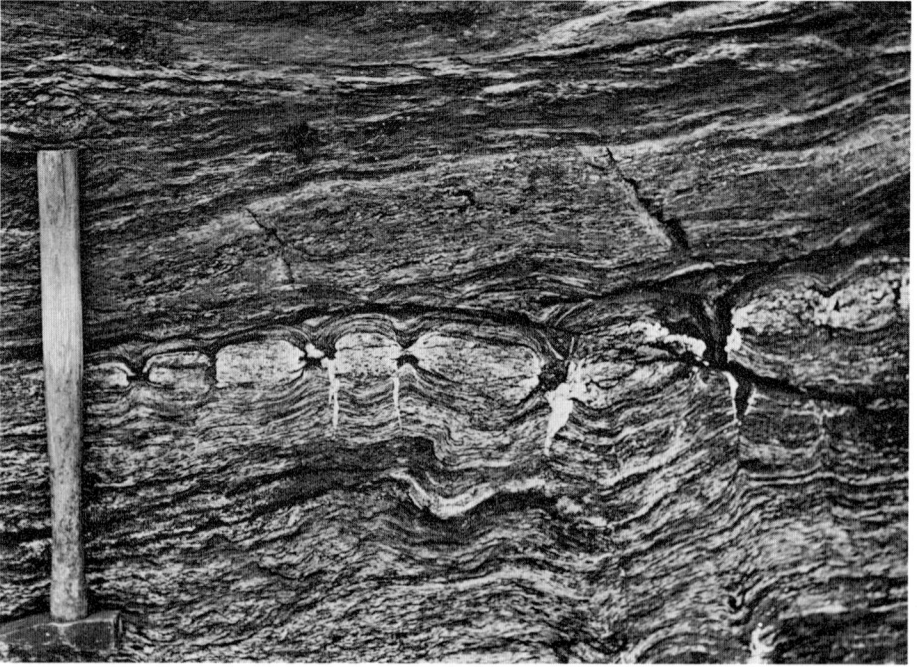

Figure 252. Boudinage of a narrow quartz-epidote-rich band in amphibolite of the Precambrian Tazin Group, Saskatchewan, Canada; secondary quartz occupies veins between some of the boudins (After Christie, 1953, Plate V A).

The cross-section shape of boudins varies widely. Barrel-shaped boudins appear to be the most common. The occasional rectangular boudins with sharp corners must have been sufficiently brittle to resist plastic necking-down (e.g., Fig. 254). Lens-like and rectangular cross sections appear to correlate with slight and considerable differences in competence, respectively (cf. Coe, 1959). Occasionally, the edges of boudins adjacent to the scar are reentrant (see Fig. 255 and Wegmann, 1932, Fig. 1). Ramberg (1955) and Coe (1959) generalized that long boudins are formed in thick units and short boudins in thin units.

The scar between adjacent boudins is commonly partially or completely occupied by a **partition,** or mineral-filled scar, of quartz, quartz and feldspar, or calcite. The partition minerals probably segregated from neighboring rocks and accumulated in the low stress scar zones. Some limited replacement of the boudins by introduced minerals can occur (cf. Ramberg, 1952, Fig. 122a). De Sitter (1956B, p. 87) suggested that the partition minerals are recrystallized portions of the competent units that produced quartz when the competent bed was sandstone, and calcite

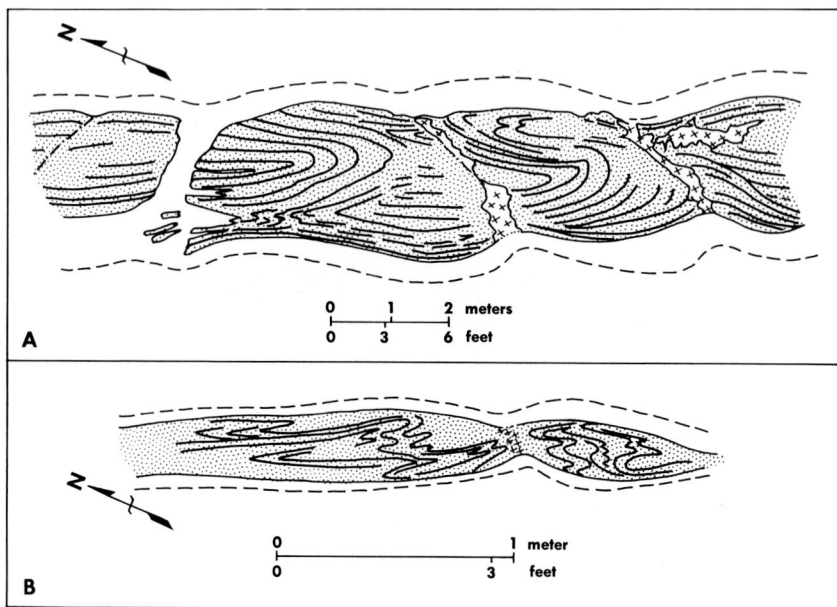

Figure 253. Boudinage developed in highly folded calcareous units (now calc-silicate rocks), N of Nordnor, Langø, W Greenland (After Berthelsen, 1957, Fig. 14).

when limestone. De Sitter described an unlocalized boudinaged limestone unit with chert layers in which recrystallized calcite occurs opposite the limestone, and recrystallized quartz opposite the chert bands (Fig. 256).

The ubiquitous **scar folds** represent plastic creep of the less competent rocks towards the scar or boudin neckline (Fig. 257). The scar folds sometimes have minor folds on their flanks. In sections perpendicular to the local *B*-geometric axis in Shuswap, A. G. Jones (1959, p. 112) found that the minor folds have the same shear sense on opposite sides of the boudin (Fig. 258), whereas, in sections parallel to the *B*-axis a symmetric pattern is common (Fig. 259; cf. Waters and Krauskopf, 1941, pp. 1381-2). The movement of the host rock past the boudin sometimes appears to develop a reentrant (cf. Fig. 255).

Rast (1956) and Wunderlich (1959B) described instances of boudin neck-lines perpendicular to, and oblique to, the inferred least principal stress. Rast suggested that the thickness/width ratio of quartzite layers in Perthshire, Scotland, controlled the orientation of the boudinage neck-lines with respect to the stress geometry; he drew comparisons with ductile steel sheets that develop necks perpendicular to the least principal stress if thickness/width < 1/7, whereas, if the ratio is > 1/7, oblique necking results (cf. Nadai, 1951). This suggestion requires further study because

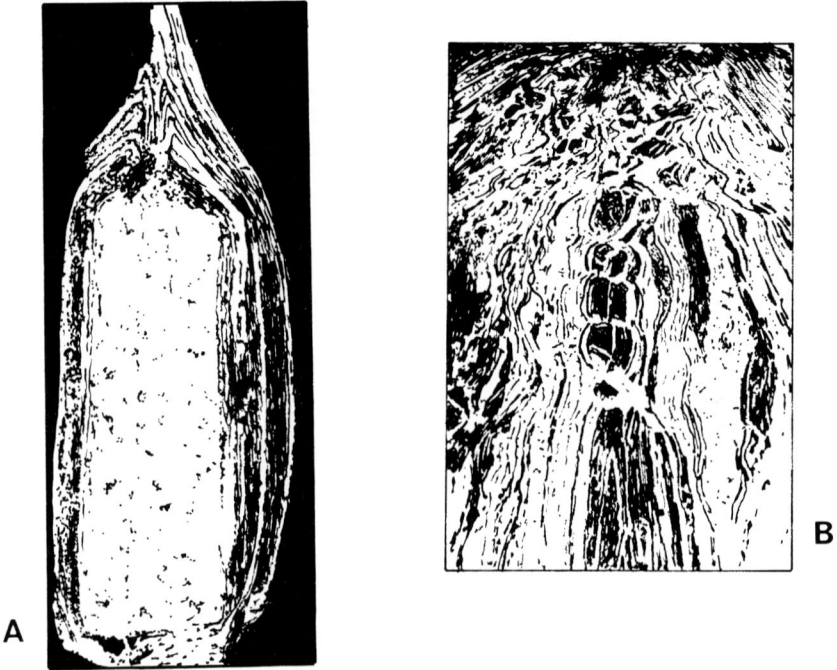

Figure 254. A: Boudin of pegmatite (17 cm. long) in micaceous quartzite, Wind River Mountain, Wyoming, U. S. A. B: Boudins of amphibolite in heterogeneous gneiss, W Greenland; the boudins are about 40 cm. thick (Drawn from photographs in Ramberg, 1955, Plates 1 B and 3E).

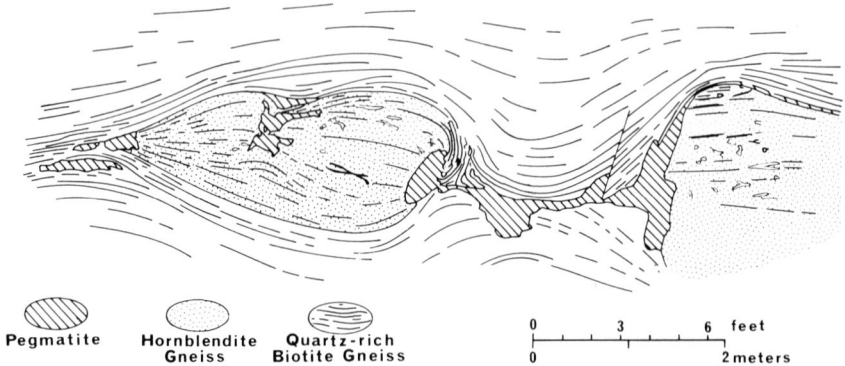

Figure 255. Boudinage of hornblendite member (originally a gabbro sill) in biotite gneiss, Three Valley Lake, Shuswap terrane, British Columbia, Canada (After A. G. Jones, 1959, Fig. 25).

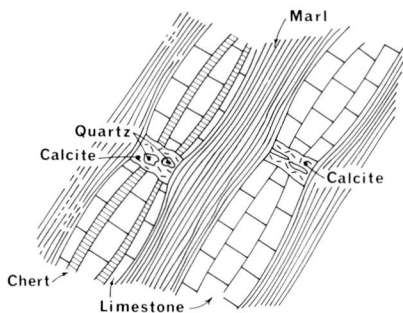

Figure 256. Boudinage in a limestone member with chert beds (After de Sitter, 1956B, Fig. 57).

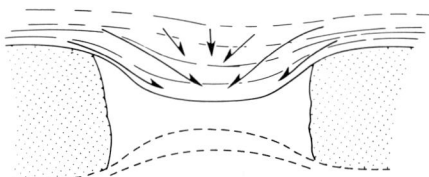

Figure 257. Symmetrical scar fold on flanks of boudins (After A. G. Jones, 1959, Fig. 27).

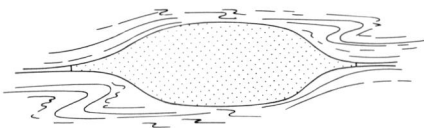

Figure 258. Asymmetrical scar fold on flanks of boudins (After A. G. Jones, 1959, Fig. 28).

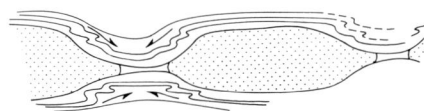

Figure 259. Drags on limb of scar fold showing direction of host rock flowage (After A. G. Jones, 1959, Fig. 29).

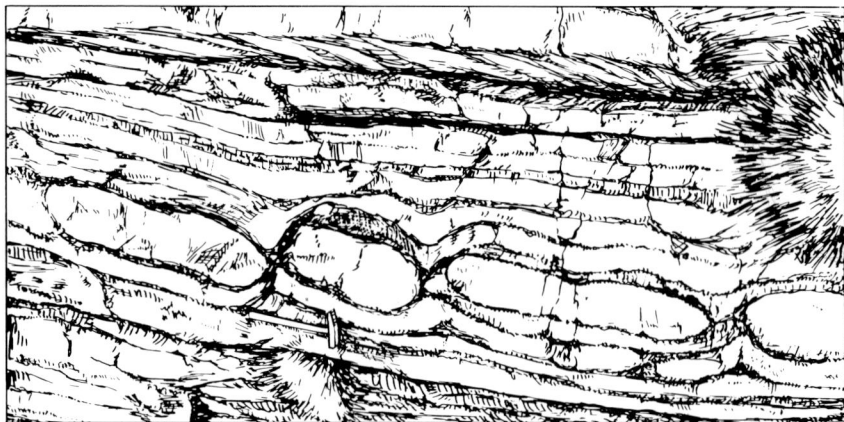

Figure 260. Boudins with slight rotation about their length, Moine Series, Ross of Mull, W Scotland; hammer shows scale (After Wilson, 1961, Plate IIIA).

Wilson (1961, p. 501-2) recorded that D. J. Shearman mapped boudins in Devon, England, whose lengths are always oblique to asymmetric elongated domes; Wilson suggested that because these boudins are parallel to local "tension gashes," the stresses that produced the boudins were oblique to the external forces that induced the folding.

Occasionally, boudins are rotated about their length with respect to the foliation (Fig. 260), but this probably results from a secondary rotation (due to continued or superposed stress) not comprising an essential part of the stress system that produced the initial boudinage. Reid (1957, 1963) described and illustrated boudins in the Pony and Cherry Creek Gneisses, Tobacco Root Mountains, Montana, U.S.A., that were rotated by later phases of deformation; foliation in the boudins is now oblique to that of the enclosing rocks.

Two perpendicular boudinage neck-lines were mentioned above in connection with the "tablettes de chocolat" structure of Wegmann (1932). Perpendicular sets of boudins in the same sample are by no means uncommon. The limitations of outcrop commonly make such relationships difficult to observe. A. G. Jones (1959) found that two sets of approximately perpendicular neck-lines are parallel to, and perpendicular to, the linear structures and fold axes in parts of the Monashee Group, Shuswap, British Columbia, Canada; few neck-lines are exposed, but the consistent arrangement of those seen makes the relationships reasonably certain. In detail the boudins are somewhat irregular in shape and their distribution within a layer is not highly regimented, because the neck-lines are not consistently perpendicular and each set is not rigidly parallel. Although essentially rectangular in plan, some are triangular, and most have irregular edges. Carbonate partitions only occur in those neck-lines approximately perpendicular to the regional linear structures. Figure 261 shows a sawn block in which a boudinaged hornblendite layer is enclosed in hornblende-biotite-garnet gneiss.

In the Bantry Bay area, SW Ireland, boudinage occurs with great regularity in the Upper Paleozoic Coomhola Series of sandstones, shales, and siltstones; boudins occur in sandstone beds, slumped sheets of silt, and dyke rocks. Two perpendicular sets of barrel-shaped boudins are commonly seen (Coe, 1959). Figure 262 shows a 45 cm.-thick lithified slumped siltstone sheet lying between flaggy siltstones that is boudinaged in two directions; lithification must have transformed the slumped silts into a competent unit. Somewhat surprisingly, the better-shaped and smaller boudins are elongated normal to the B-geometric axis; the less well-formed boudins are parallel to B and exposed on ac-sections. Fyson (1962, p. 214) described similar geometrical relationships in south Devon, England.

Top plan view of cut block

Left side view of block

Trace of cut

Right side view of block

Trace of cut

Trace of cut

Trace of cut

Front view of block

☐ Hornblendite

▨ Hornblende-biotite-
garnet gneiss

◼ Calcite

0 ___ 2 ___ 4 ins.

0 ___ 5 ___ 10 cms.

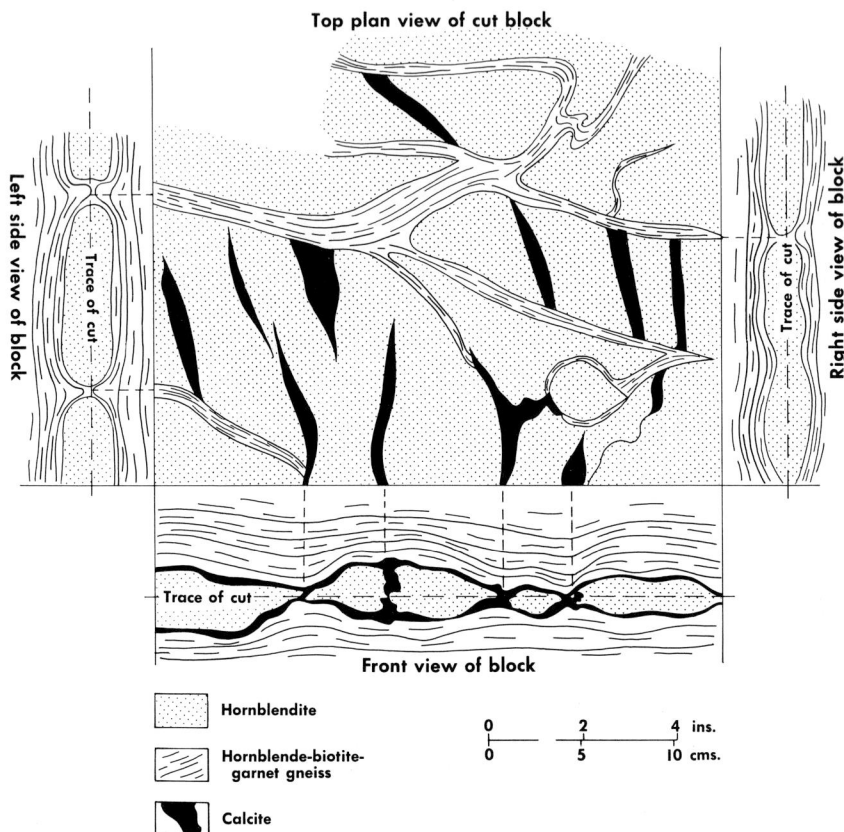

Figure 261. Boudinage of hornblendite seen in a sawn block; linear structure of the host rock is parallel to the front face of the block (After A. G. Jones, 1959, Fig. 26).

Jones' technique of sawing large specimens could yield valuable information from many areas. The three-dimensional geometry of boudins requires additional study before their genesis can be explained completely. Little is known about the relative abundance of the classical barrel- or cylinder-shaped boudins and those with intersecting neck-lines. It seems likely that each distinctive type of boudin is associated with a dissimilar type of folding, but the distribution of boudinage between flexural-slip and slip folds is not well known. Wilson (1961, p. 500) thought boudinage is commonly associated with slip folds in which the development of schistosity in the less competent units has been accompanied by plastic attenuation of the more competent strata on the fold flanks (Fig. 263). He considered that this "stretching" may even continue into the hinge zone of a fold, although not necessarily always.

Figure 262. Perpendicular barrel-shaped boudins in a lithified slumped siltstone of the Coomhola Series (Upper Paleozoic), Bantry Bay, SW Ireland (After Coe, 1959, Fig. 4).

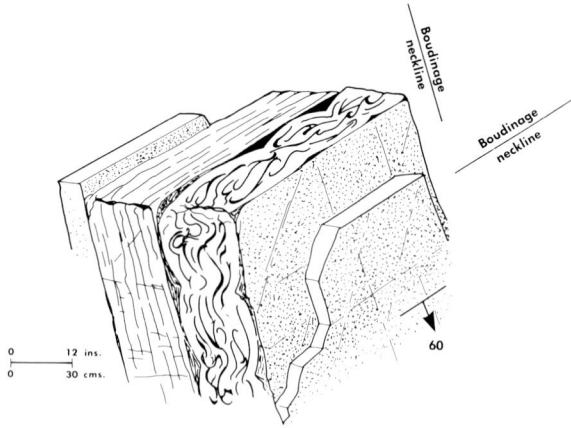

Figure 263. Diagrammatic representation of boudinage developed in competent unit on the flanks of a slip fold (After Wilson, 1961, Fig. 35).

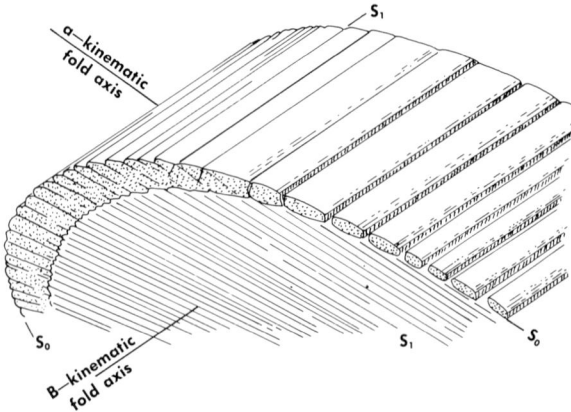

Figure 264. Fractured competent member in folded Dalradian Lettermackaward Limestone, Trawenagh Bay, Donegal, Ireland (After Gindy, 1953, Fig. 3B).

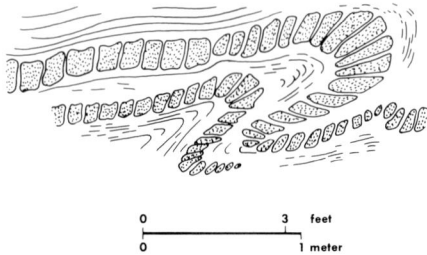

It is well known that boudins are commonly developed on the flanks of flexural-slip folds, but there is disagreement about the possibility of finding boudins around fold closures. Rast (1956) stated that there is ample evidence for uniform extension of competent layers in the limbs and closures alike; he cited as his sole evidence a structure (Fig. 264) in

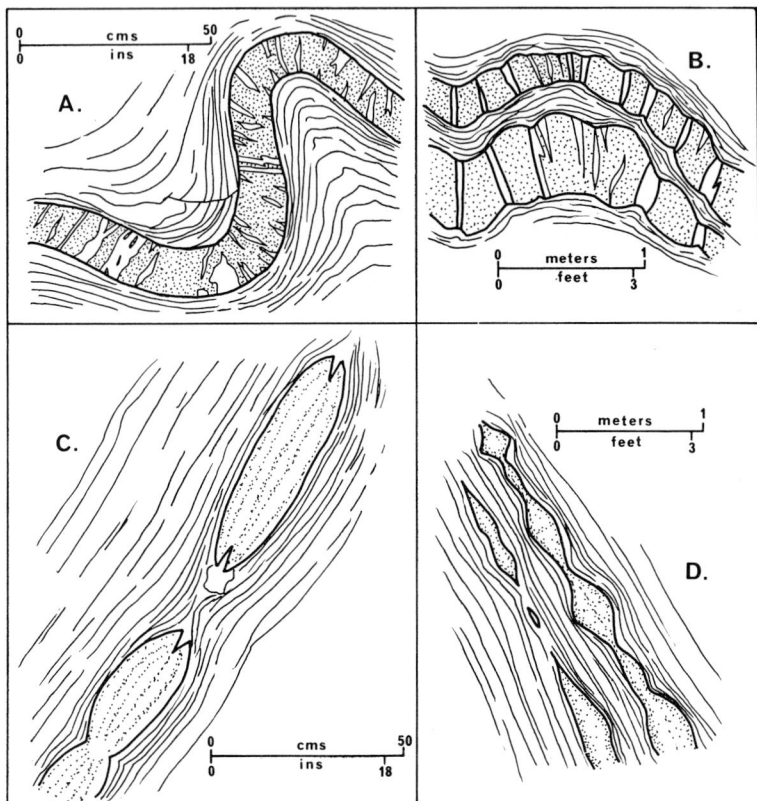

Figure 265. Stages in development of boudinage in competent units seen in *ac*-section. A: Incipient boudinage formation accompanying 50 per cent compression of the sequence, Rammelsberg, near Goslar, Harz Mountains, Germany. B: Development of fractured quartzite (between slate layers) into individual boudins associated with 25 per cent compression, Collignon, ENE of Bastogne, Belgium. C and D: Strongly developed boudins in Hettangien marble, Colonnata-Bedizzano road, Apuan Alps, Tuscany, Italy (After Wunderlich, 1962, Fig. 2).

the Lettermackaward Limestone of Trawenagh Bay, NW Ireland, described by Gindy (1953). Although Gindy (1953, p. 385) described boudinage, it is not clear that he considered the structure shown in Figure 264 to be boudinage; he (1953, p. 386) wrote, in descriptions of this structure, that:

> ... siliceous bands are first drawn out into crenulate dragfolds. ... the more rigid of the drag-folds are cleaved by closely spaced foliations parallel to each other. ... displacement along these cleavages results in the production of clearly cut fragments which may run loose in the matrix.

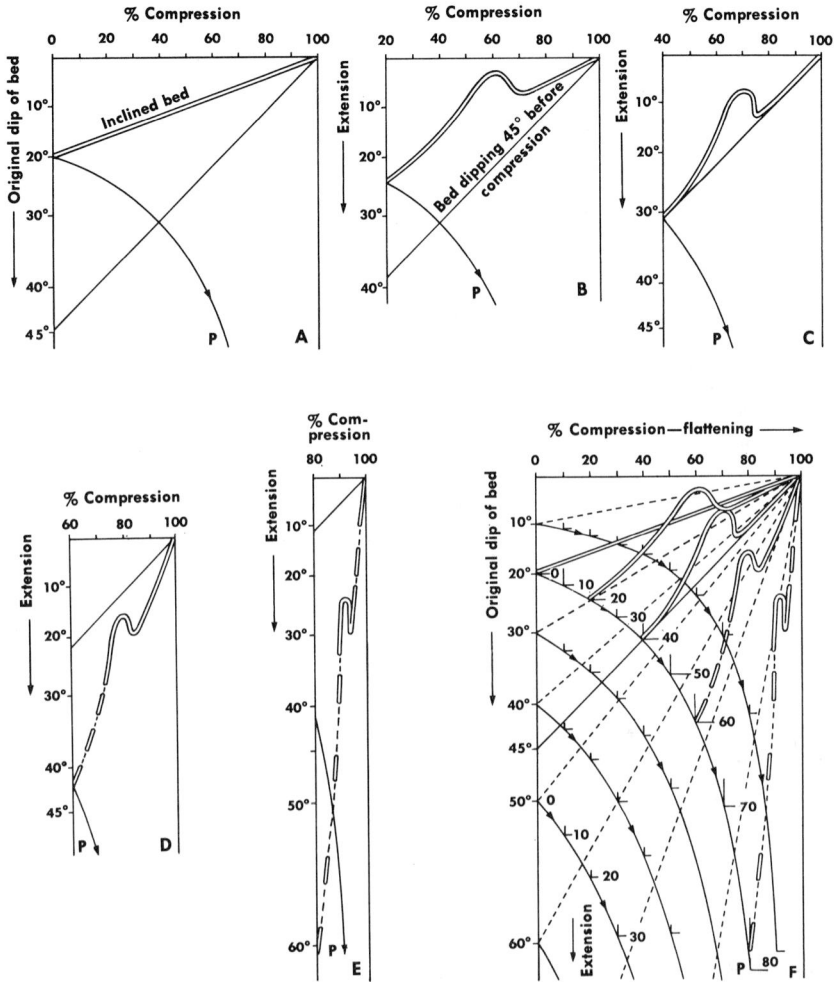

Figure 266. A model for the formation of boudinage. A: An undeformed bed with 20° dip. B: Initial horizontal compression with flexural-slip folding of the rock prism containing the bed; the lower end of the bed moves along the curve P and the orientation of a 45° dip prior to compression is shown for reference. C: Portions of the strained bed reach 45° dip. D: Continued horizontal compression and vertical extension is accompanied by boudinage formation. E: Extensive boudinage formation accompanying final strain. F: Synoptic diagram to show progressive strain, folding, and boudinage-formation for various initial bedding plane dips (Adapted from de Sitter, 1958, Fig. 10).

Wunderlich (1962) described a small example of boudins around the closure of concomitant folds in a quartzite within slates near Bastogne, Belgium (Fig. 265). However, superposed folding of a boudinaged layer could result in boudins around the closure of a second generation fold; the not unequivocal structures described by Ward (1959, p. 1451) in the Wilmington complex, Pennsylvania and Delaware, U.S.A., may be examples. Cloos (1947B) assumed that the relative movement of the enclosing, more plastic, beds during flexural-slip folding tends to induce boudinage in a competent layer; according to this model extension of the fold limbs should not be accompanied by extension of competent units around the fold closure.

Models incorporating elements of both flexural-slip and slip folding have also been proposed. For example, de Sitter (1958, p. 286) supposed that, in a "simple folding mechanism," boudinage is restricted to folds in which an initial stage of concentric or concertina folding caused an angle between the axial plane and the fold limbs of 45° or more. He proposed an ingenious model in which a horizontal rectangular prism of rock (comprising a layered sequence) is compressed laterally and transformed into a vertical prism. Under the constraints of this model, an original competent layer would initially develop vertical folds and subsequently be stretched (Fig. 266) because the diagonal of the deformed rectangle is a minimum when $\alpha = 45°$, since:

$$\text{length of diagonal} = \sqrt{\frac{\text{area of rectangle}}{1/2 \ \sin 2\alpha}},$$

provided there is no flow parallel to the fold axis, and α is the dip of the diagonal plane in the rectangular prism (cf. Wunderlich, 1959B). According to this model, boudinage would not form around the closures of major folds — unless the boudinaged layers are refolded subsequently. It is difficult to estimate whether rocks are deformed in the manner implied by Figure 266.

Wunderlich (1962) extended this hypothesis. After allowing for compaction effects, he showed that a considerable volume of material must be squeezed from a fold core during continued compression and flexural-slip folding (Fig. 267). Wunderlich thought this is one of three mechanisms that promote boudinage; he believed that expulsion of the less competent rocks from a fold core can stretch the competent units in both the fold flanks and the closures. Wunderlich drew support for his hypothesis from the fact that extension parallel to the B-geometric axis and boudinage in the aB-plane sometimes appears to have resulted from compression of a fold core (Fig. 268). On the basis of theoretical and experimental studies, Ramberg (1959, p. 108) concluded that boudins normal to the B-geometric fold axis can result from pure strain (Fig. 269).

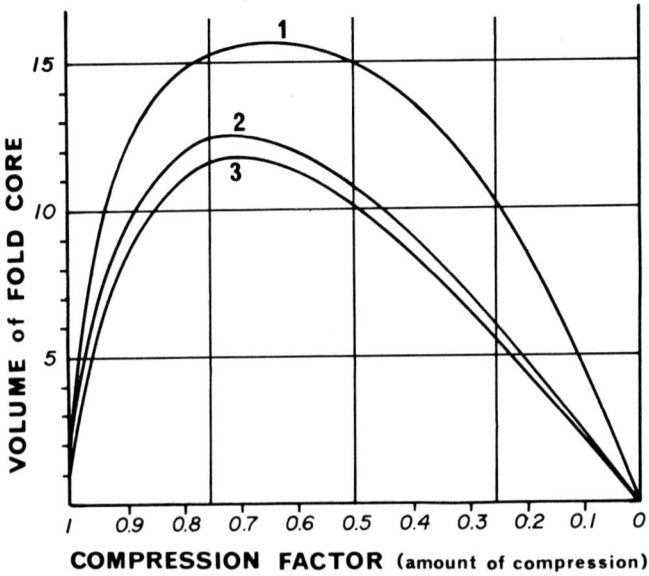

Figure 267. The changing volume of the core of a fold during continued flexural-slip folding. Curve 1: folds with half-elliptical profiles. Curve 2: concertina folds. Curve 3: folds with sine-wave profiles. The units of volume are proportional to the wave length and amplitude of the fold (After Wunderlich, 1962, Fig. 1).

It is commonly suggested that during the formation of boudins the least principal stress was perpendicular to their length, and that they either neck plastically or rupture along joints during folding. The precise mechanism by which adjacent boudins separate remains unclear. Ramberg (1955) thought the separation of individual boudins was caused by plastic elongation of the adjacent incompetent layers due to compression perpendicular to, or at an obtuse angle to, the competent layer. Coe (1959) suggested that two distinct types of boudinage can be defined: (1) boudins produced during folding with their length parallel to the fold axis, and (2) perpendicular sets of boudins (arranged perpendicular to, and parallel to, the fold axis) resulting from extension normal to the direction of maximum stress. While Walls (1937B) invoked the successive operation of stresses in two approximately perpendicular directions to account for perpendicular boudinage neck-lines in Aberdeenshire, Scotland, such sets appear to be penecontemporary in many cases. Ramberg (1955) thought that more or less equidimensional "tablettes de chocolat" boudins are to be expected with two-dimensional expansion in the plane of

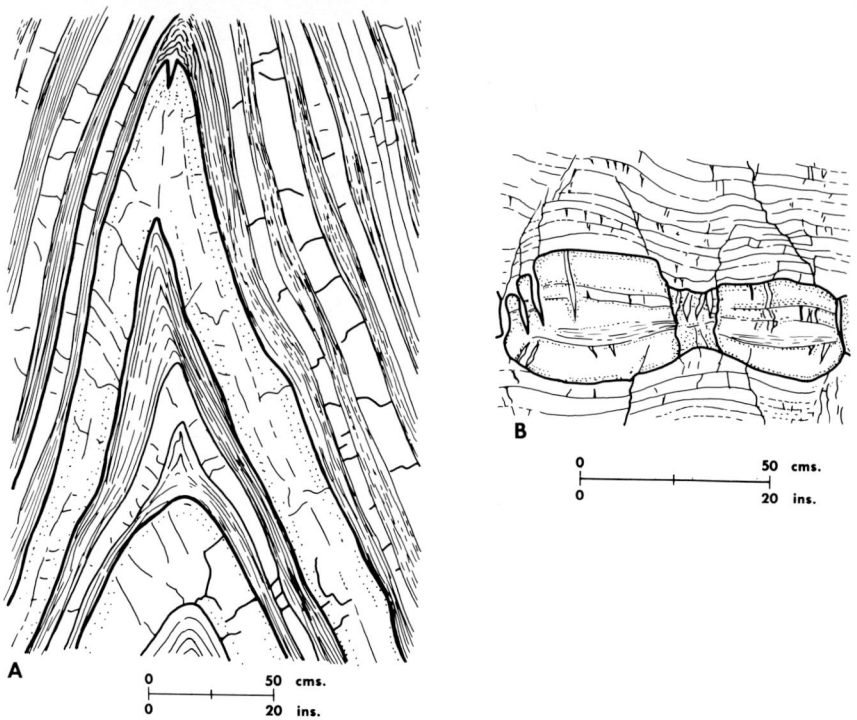

Figure 268. Initial stage in development of boudinage parallel to the *B*-fold axis; drawn from photographs of perpendicular sections through a fold in Flysch, Montalto, N of San Remo. A: Fold profile with negligible stretching. B: *aB*-section with incipient boudinage (After Wunderlich, 1962, Fig. 3).

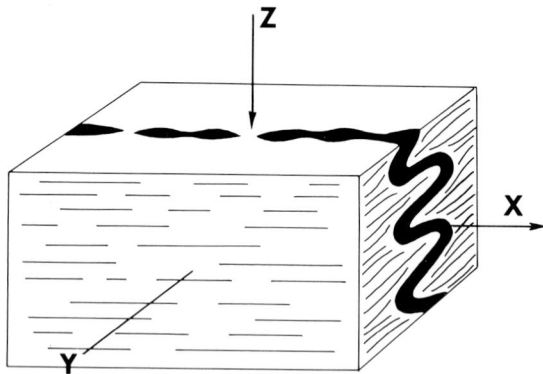

Figure 269. Minor folds and boudinage-like attenuations in a crosscutting vein produced simultaneously by pure shear under experimental conditions. X = maximum extension strain; Y = zero strain; Z = maximum compression. Axes of principal stress coincide with axes of principal strain (no rotation of coordinate axes during strain) (After Ramberg, 1959, Fig. 3).

schistosity, which, he suggested, is more common than is generally realized in gneissic areas.

This discussion suggests that there must be several different genetic types of boudinage, but it is impossible to arbitrate between the numerous genetic hypotheses until there is greater understanding of the geometry and kinematics of the structures that actually occur in nature. A more complex process-response model than any of those currently proposed is necessary to explain all types.

Boudins sometimes arise in rather special environments. For example, continued expansion of the Ardara granitic pluton, NW Ireland, resulted in tangential elongation of the aureole rocks and development of boudins in varied orientations within planes parallel to the granite contact (Akaad, 1956, pp. 283-84). Metadolerite and metalamprophyre dykes in marbles and sillimanite hornfels, quartz veins in andalusite hornfels, and siliceous and calc-silicate bands in sillimanite hornfels, all developed boudinage.

Rast (1956) noted that the term "boudinage" has occasionally been extended and used in a more general sense to include other types of tectonic inclusion. For example, relics of folds and rootless fold closures preserved during continuing isoclinal folding (e.g., Figs. 167, 174, and 180) are tectonic inclusions rather than boudins.

However, Cloos (1947B), Rast (1956) and Kalsbeek (1962) recognized the so-called lozenge-shaped variety as true boudins. Thick beds of edgewise conglomerate in the Conococheague Limestone of the Martic area, Pennsylvania, U.S.A., were broken into almost square prisms before rotation and their longest diagonal is now in the bedding plane (Cloos and Hietanen, 1941). Cloos (1947B) referred to similar relationships from Bear Island in the Potomac River gorge near Washington, D.C., U.S.A., where metamorphosed pelitic layers were more competent than the enclosing sandy units. Cloos (1947B) appeared to consider rotation an essential factor in the formation of boudins, and he found the barrel shape of the classical boudins "somewhat puzzling."

Rast (1956) traced the developmental stages of such lozenge-shaped structures in quartzitic members of the Dalradian Series, Perthshire, Scotland (Fig. 270); he suggested that rotation of individual segments took place by using the joints as surfaces of slip, and that the joints do not have such a role in the formation of barrel-shaped boudins.

Such differences appear to be sufficient grounds for not including these lozenge-shaped structures as boudins.[2]

[2]Gault (1945) used "boudin" to describe small schist xenoliths that were apparently pulled apart by flow after incorporation in the Pinckneyville quartz diorite magma, Alabama, U.S.A.; this usage seems to be inappropriate.

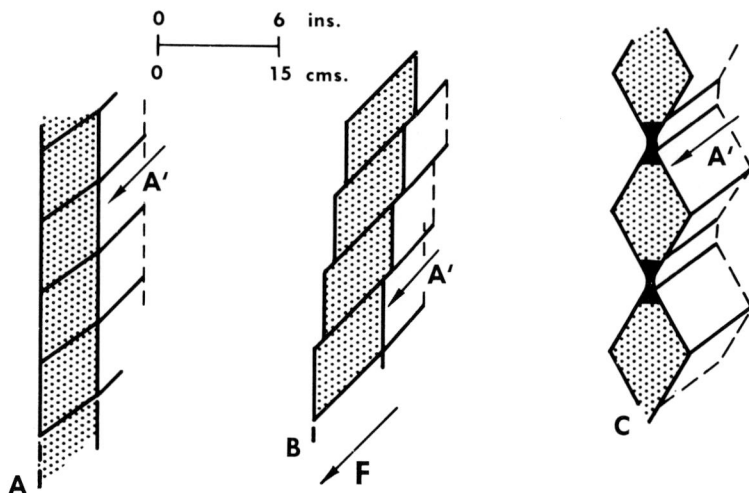

Figure 270. Stages in development of "lozenge-shaped boudins" in the Banded Group of the Dalradian sequence, Perthshire, Scotland. F = axis of local small folds: A = direction of maximum elongation of the boudins. A: Fractured competent unit. B: Slip along fractures. C: Resultant "lozenge-shaped boudins" (After Rast, 1956, Fig. 2C).

As with so many other structural features caution must be exercised to ensure that original sedimentation or compaction features are not interpreted as products of tectonism. Many pinch and swell features occur in unmetamorphosed sedimentary rocks. Some represent original features of sedimentation or concretion; others have been attributed to disruption of slightly more competent units during compaction. For example, Mc-Crossan (1958) described sedimentary "boudinage" structures from the calcareous shales and argillaceous limestones of the Upper Devonian Ireton Formation, Alberta, Canada. He thought the more competent limey strata were pulled apart by the relatively plastic shales during compaction. Greenwood (1960) arrived at similar conclusions for boudin-like structures in Cretaceous limestones of Zimapan, Mexico; in this area the rocks have been strongly folded, and, since the three-dimensional geometry of the structures was not determined, the exact mode of origin is uncertain. Numerous additional examples of sedimentary boudin-like structures, in which folding did not appear to have played any part, could be cited; it is unwise to refer to such structures as boudins.

2. MULLION STRUCTURE

Nolan and Kilroe (in Hull, *et al.*, 1891, pp. 53-54) described the compact quartzites of Gweedore and Errigal districts, NW Ireland, and noted that

> ... in some parts "mullion structure" was observed – a peculiar
> fluting due to the shearing of the rocks. ... South-west of Dun-
> lewy, ... "mullion structure" is also developed which, ... is
> attributable to intense movement ... which here ... affected the
> rocks during the second period of metamorphism.

This appears to be the earliest reference to mullions, although the term
was apparently well understood in 1891. Excellent mullions developed in
these quartzites, and in the granite close to the northwestern contact zone
of the Main Donegal Granite (Fig. 271), were described by Pitcher, *et al.*
(1959) and Pitcher and Read (1960).

Sander (1948, Figs. 44 and 45) illustrated good mullions, but the most
complete study was by Wilson (1953, 1961) who distinguished several
different types. Wilson used **mullion** for structures formed from the
country rocks, and **rod** for stick-shaped bodies, commonly composed of
quartz, enclosed in metamorphosed country rock; previously the terms
were used interchangeably for a general class of linear structures oriented
parallel to the local *b*-geometric axis, and both structures were included in
"linear foliation" by Peach and Horne (in Peach, *et al.*, 1907). Although
Bailey and McCallien (1937) preferred "corduroy structure" and Fermor

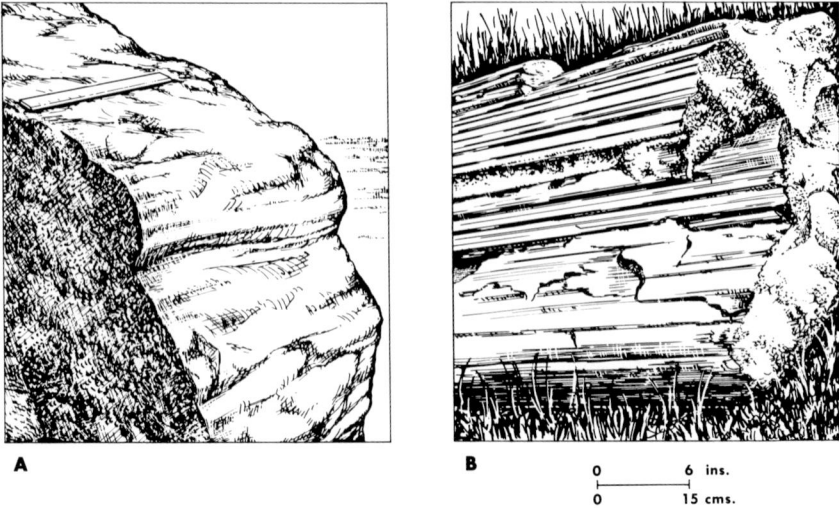

A   B
       0   6 ins.
       0   15 cms.

Figure 271. Mullion structures associated with the marginal shear structures of
the Main Donegal Granite, Donegal, NW Ireland. A: Mullions in the marginal
granite, Straboy; scale (one foot) is parallel to the linear structure. B: Mullion
in Dalradian quartzite, Lackagh Bridge (Based on photographs from Pitcher,
et al., 1959, Plate XI).

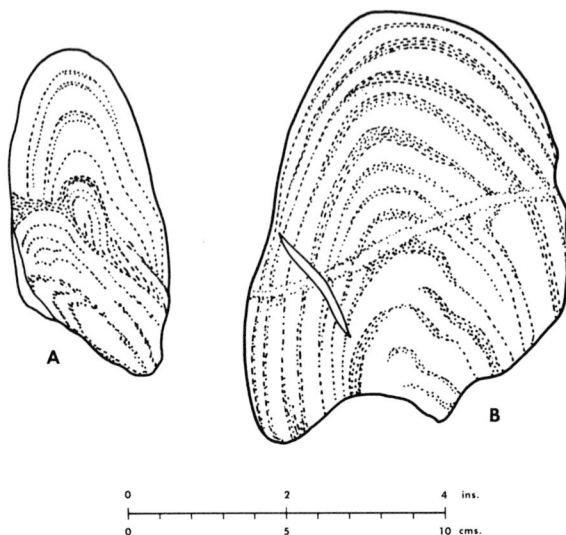

Figure 272. Profiles of fold mullions showing concordant and discordant structure between S_0 and the mullion periphery, Moine Series, Oykell Bridge, NW Scotland (After Wilson, 1953, Figs. 3b and 3c).

(1909, 1924) used "slickensides-grooving," mullion structure is to be preferred. Holmes (1928) wrote that mullion structure commonly has ". . . the appearance of the clustered columns which support the arches, or divide the lights of mullioned windows, in Gothic churches."

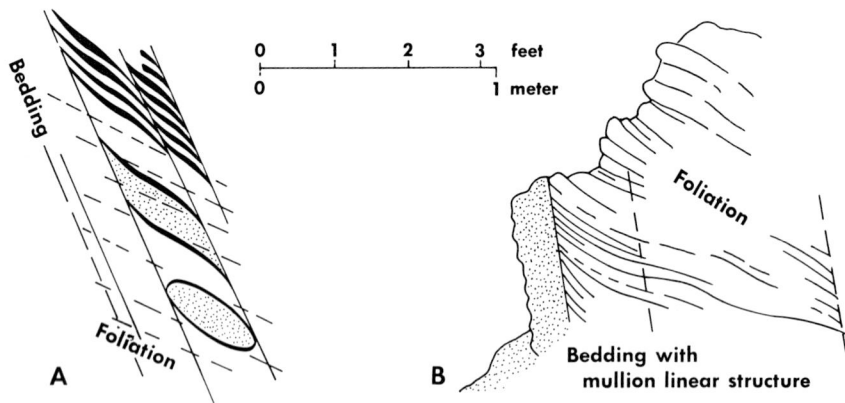

Figure 273. Foliation mullions (cleavage mullions of Wilson). A: Sketch illustrating mode of development. B: Sketch of actual foliation mullion in Moine rocks, Oykell Bridge, NW Scotland (After Wilson, 1953, Figs. 4a and 4b).

Mullions are completely different from the large slickensides, grooves, or furrows developed on fault planes that were referred to as mullions by Leith (1923), Gregory (1914), Billings (1942, p. 158), and Fairbairn (1949). Recent American textbooks (e.g., Billings, 1954; Turner and Weiss, 1963) followed Wilson's usage of mullion.

Since Wilson's (1953) detailed work, Oykell Bridge, NW Scotland, is the most convenient locality for defining the nature of mullions. Here, fine dark bands define original bedding in the Moine siliceous granulites and occasional semipelitic and pelitic layers occur within the succession. Petrofabric analyses show that the mullions are parallel to the *b*-geometric axis (Phillips, 1937). Four varieties of mullion were recognized by Wilson (1953, 1961):

(1) *Fold mullions:* Individual mullions are bounded in part by cylindroidally-folded *S*-surfaces (bedding or foliation planes). Commonly mullions comprise detached fold hinge lines (closures) with one side bounded by an *S*-surface and the other by a grooved surface cutting across the internal *S*-surfaces (Fig. 272). A. G. Jones (1959, Plate VIIIA) illustrated a good example in the Monashee Group, Fosthall Mountain, British Columbia, Canada, although he described it as "fold rodding."

(2) *Bedding mullions:* These are analogous to fold mullions, but occur where folding is less intense so that undulations of individual bedding planes (or other *S*-surfaces) are prominent. Such undulations range from pinch and swell structures to gentle flexures; they involve ribbing on the *S*-surface parallel to the local fold axis. Despite a superficial resemblance to fault-plane slickensides, there is commonly little or no movement along the length of adjacent mullions. Dalradian quartzites of southern Gweedore, NW Ireland, show good bedding mullions. Besides examples from Oykell Bridge and Portsoy, Banffshire, NE Scotland, Wilson illustrated examples (sometimes 100 meters long) from the Kandri manganese mine, Central Provinces, India. From the Tazin Group, Saskatchewan, Canada, Christie (1953, p. 23) noted mullions apparently of this type.

(3) *Foliation mullions (cleavage mullions of Wilson):* These long rock prisms are angular to rounded in section, and are bounded by intersecting *S*-surfaces (e.g., two foliations or bedding and a foliation). The bounding planes tend to be smooth or gently curving, but continued movement locally caused rounding of the prisms (Fig. 273). Pilger and Schmidt (1957A, 1957B) and Schmincke (1961) described mullions of this type in strongly folded Devonian sandstones and slates of the north Eifel region, Germany (Figs. 274, 275, and 276).

Figure 274. Foliation mullions in strongly folded Devonian sandstones, Dedenborn, north Eifel region, Germany; note hammer for scale (Based on photograph from Pilger and Schmidt, 1957A, Fig. 1).

Figure 275. Foliation mullions in strongly folded Devonian sandstones and slates, Dedenborn, north Eifel region, Germany; note hammer for scale (Based on photographs from Pilger and Schmidt, 1957B, Plate 4, Fig. 2).

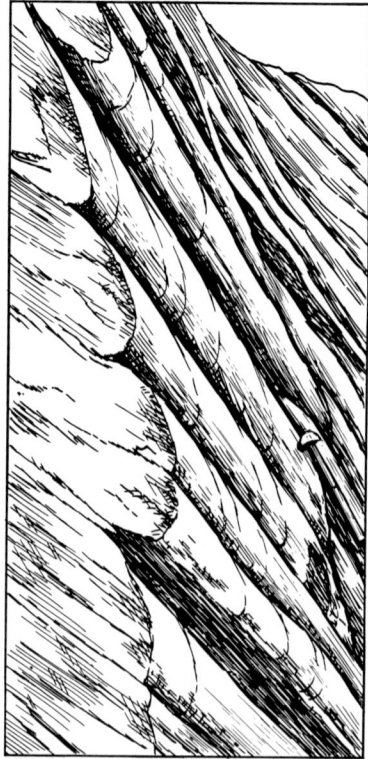

Figure 276. Foliation mullions in strongly folded Devonian sandstone, Dedenborn, north Eifel region, Germany; note hammer for scale (After Pilger and Schmidt, 1957B, Plate 3, Fig. 3).

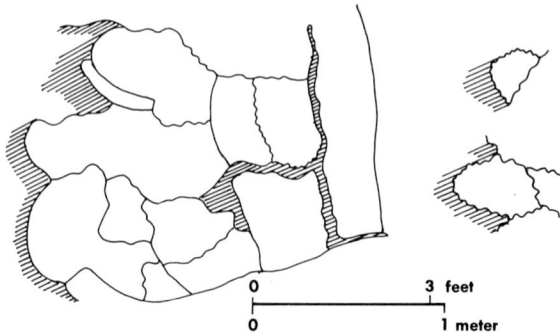

Figure 277. Diagrammatic sketch of profiles of irregular mullions in Moine rocks, Oykell Bridge, NW Scotland (After Wilson, 1953, Fig. 4c).

(4) *Irregular mullions:* At Oykell Bridge this type has very irregular cross sections due to the ribbing on the mullion sides which varies from fine striations to grooves comparable with corrugated-iron sheeting (Fig. 277). There is little relationship between external form and the *S*-surfaces (bedding and foliation), so the *S*-surfaces are truncated by the corrugated

surface. These rock prisms fit together like pieces of a jigsaw puzzle; excellent examples may be seen along the River Ling, southwest of Beinn Dronaig (McIntyre, in Wilson, 1953, p. 147), and in the Dalradian quartzites of Donegal, NW Ireland. Figure 278 shows an irregular mullion of chloritic gneiss from Villefranche de Rouergue, France, described by Collomb (1960, p. 6).

At Oykell Bridge, and in most other areas, a thin micaceous veneer ensheaths the length of mullions. Wilson (1961) reported manganese oxide sheaths at the Kandri mine, India. Such sheaths apparently developed after the folding and mullion formation, because they are continuous around the entire mullion.

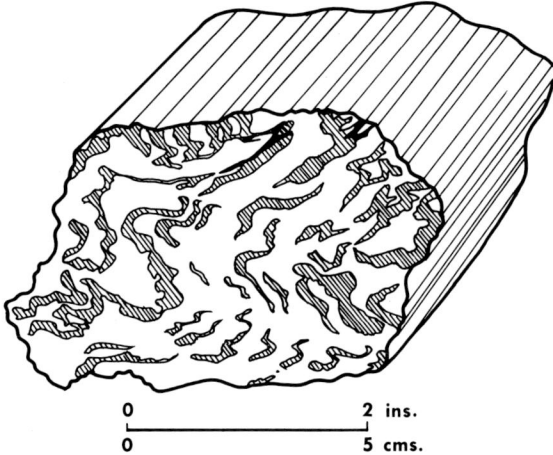

Figure 278. Irregular mullion in chlorite gneiss, Villefranche de Rouergue, France (After Collomb, 1960, Fig. 5).

Any or all types of mullion can be closely associated in the field. Wilson (1961, p. 514) believed that:

> ... in addition to a rotational or a compressional deformation, the mullioned rocks have also suffered a concomitant stretching parallel to their lengths. I thus disagree with Read's suggestion of two separated deformations; and consider that mullions are examples of *Einengung* (Sander, 1948; Weiss, 1954, p. 39 *et seq.*), in which the rocks have suffered a great squeezing normal to their lengths, but under conditions such that stretching parallel to their lengths was possible.

3. RODDING

As mentioned, there has been considerable confusion over the terms rodding and mullion structure. In many cases the two terms have been used interchangeably, but the terminology was systematized by Wilson

(1953, 1961).[3] **Rod** is a descriptive and nongenetic term for relatively thin cylindroidal bodies of quartz or some other mineral (e.g., calcite, pyrite, etc.). Rods are essentially monomineralic and differ from mullions in being composed of material different from that of the main mass of rock in which they occur (Wilson, 1961, pp. 514 and 517). Recently, Wilson (1961) included any elongated and rolled bodies (pebbles, segregations, etc.) as rods, although originally he (1953, p. 133) used rod in a genetic sense for products of segregation during folding, shearing, and metamorphism of country rocks.

At Ben Hutig (extreme northern Scotland) quartz segregations developed along bedding surfaces, axial-plane foliations, and oblique fissures

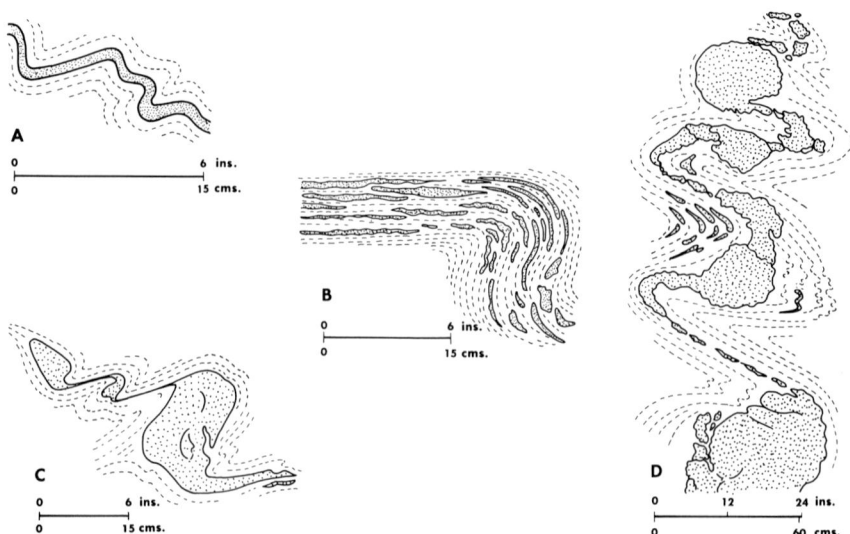

Figure 279. Profiles of quartz rods, Ben Hutig, Northern Highlands, Scotland. A: Quartz vein parallel to and folded with the lithological banding. B: Elongated and folded quartz lenses—initially conglomerate pebbles—parallel to stratification. C and D: Quartz rods formed from segregation quartz in hinge zones of parasitic folds (After Wilson, 1961, Fig. 40).

[3]Earlier Cloos (1946, p. 8) had defined rodding as follows: "... a term used mostly in the description of the Highland schists of Scotland, includes in it direction pencil structures, elongate pebbles, subparallel hornblende crystals, streaks, elongate particles, elongate calcite grains, magnetite blebs, pyrite clusters, and feldspar. Cross fractures perpendicular to rodding are frequently mentioned. (*See* Gunn, Clough, Hill, 1897; Peach and Horne, [*et al.,*] 1907)." He (1946, p. 20) also wrote that: "... Flowage, in cleavage planes and in *a,* grades almost imperceptibly into slippage and slickensiding in the same direction .... Rodding is partly slickensiding and partly flowage, and some of the lineation down the dip of cleavage planes may be striation."

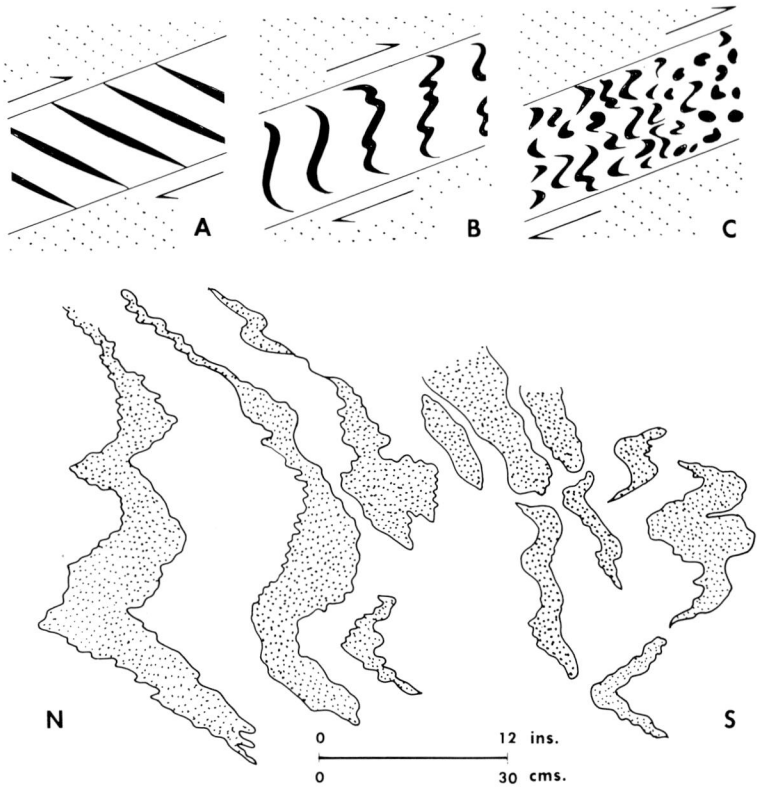

Figure 280. Genesis of quartz rods, Ben Hutig, Northern Highlands, Scotland. *Above:* A, B, and C show diagrammatically the progressive development of quartz rods by folding and deformation of quartz lenses. *Below*: Deformed quartz-filled tension fractures (After Wilson, 1953, Fig. 9).

of the Moine schists; some are undistorted and clean-cut, but others, together with some deformed conglomerate pebbles, were distorted or folded during continuing homoaxial deformation to form irregularly-cylindroidal rods ranging from 1.3 cm. to over 60 cm. in diameter with oval to nearly circular profiles. The long axes of the rods (0.6-1.2 m. long) have a preferred orientation parallel to the *b*-geometric axis. The conglomerate pebbles were of quartzite or vein quartz, but, because of the intense deformation, the pebbles and segregations are now extremely difficult to distinguish from each other. The undeformed segregations define linear structures, but the pebbles and those sheets segregated during deformation have been broken into rods by attenuation of the flanks of developing small folds (Fig. 279). Some segregations developed as minute saddle reefs at fold apices. "Tension gashes" have been mineralized also, and then deformed into rods (Fig. 280).

Waters and Krauskopf (1941, p. 1380) referred to feldspar augen rolled into spindle shapes within the gneissic and mylonitic marginal facies of the Colville Batholith, Washington, U.S.A., as rodding. Linear elements of this type abound in many gneissic terranes and there is every gradation into rotated linear elements described previously.

Although nongenetic terminology is to be commended, the propriety of extending the definition of rod to include features like deformed pebbles and rolled augen is questionable. Rods resulting from segregation concomitant with deformation are very common. Excellent examples may be seen along the coast of Start Point, SW England, where, as usual, the quartz rods define a strong linear structure parallel to the B-geometric axis (Phillips, 1950; 1951, p. 231; in Wilson, 1953, p. 147). Phillips (in Wilson, 1953, p. 147) also thought that the "... distinction of rodding from mullion-structure is valuable because the former appears to be a segregation phenomenon related to the regional metamorphism of the schists. ..."

Chapter 10

Superposed Folds

In zones of orogenic activity it is common to find that rocks have been subjected to more than one phase of folding. The stresses which cause folding apparently wax and wane spasmodically, and frequently they also change in orientation too. In a major orogenic belt, such as the Appalachians or the Caledonides, in which a pronounced structural grain extends for several hundred kilometers, the fold axial trend is homogeneous only when the attributes of very large geographic areas are considered. When smaller objects (geographic areas) are studied, the increased detail reveals a more complex kinematic picture. In many orogenic belts, stress operated about different axes, and with different intensities, at different points in the space-time continuum.

As deformation proceeds the physical-chemical nature of the rocks slowly changes. In consequence, at successive phases slip, flexural-slip, or flexure folding may dominate within a particular domain; different types of deformation do not necessarily succeed each other in any particular sequence. At a given "instant" of geologic time one domain may be dominated by one type of folding, another domain by a different type that may even be folded about a differently-oriented *B*-axis. The spatial distribution of type and style of folding can be similar or different in a succeeding "instant." Commonly, when successive tectonic events affect an area, superposed folding results; it is very common, and may develop in any tectonically-active region, and in innumerable dissimilar tectonic and metamorphic environments.

Superposed folds have been recognized for more than a century. Scrope (1862, p. 291) referred to the phenomenon and cited examples from many parts of the world. Clough (in Gunn, *et al.,* 1897) and Peach *et al.* (1907, p. 601) showed that more than one direction of linear structures (including fold axes) had been recognized in several parts of Scotland. Although records of similar structures were slow to appear

322

in print, superposed folds must have been noticed in the field. For example, Van Hise and Leith (1911, p. 123) described superposed folding in the Soudan Formation, Vermilion Iron district, Minnesota, U.S.A.:

> The cross folding of the district has been only less severe than the major folding. . . . Both the longitudinal and the cross folds are composite — that is, folds of the second order are superposed upon the major folds in each direction, and upon these folds are folds of the third order, and so on down to minute plications.

In 1913 Crampton (in Peach, Horne, Hinxman, *et al.,* 1913, p. 57) described a "double system of folding" (a term borrowed from Peach, *et al.,* 1907) in central Ross-shire, Scotland. At about the same time Martin (1916, pp. 96 ff.) described "tranverse folds" in the Adirondack Mountains, New York, U.S.A. Tectonists such as Argand (1912, 1915) were also describing crossing fold trends in the European Alps, and Kober (1923) and Staub (1924) showed that interest in superposed folds continued there.

During the past decade the number of references to multiple folding has increased rapidly and some very thorough studies have been published (see Chapter 11). Nevertheless, despite this growth of interest, the number of published quantitative geometric syntheses of actual areas is quite small. Many accounts are partially subjective because sufficiently-detailed quantitative, objective, and appropriate structural data are lacking; in such cases it is almost impossible to evaluate the geometry objectively, or to be sure whether the correct interpretation has been made.

In Chapter 6 it was demonstrated that, even with very simple folds, the kinematic picture can be deduced only after complete analysis of the geometry. It is even more true that the complex kinematic pictures associated with superposed folds can only be tentative until the geometry is wholly elucidated. This involves careful definition of the populations of objects and attributes involved, and the use of clear operational definitions in the identification and measurement of each attribute.

CLASSIFICATION

First it is desirable to identify some of the variables involved in the development of superposed folds.

Superposition implies tectonic events distributed in time, but the intervals between successive events may range from infinitesimally short (geologically) to several hundred millions of years. Within a single orogenic episode, folding may have been limited to one or two pulses, or have been effected by numerous phases. Careful geometric analysis can commonly enable each phase of folding to be recognized, and each phase to be located on an ordinal time scale. Geometric analysis alone cannot determine the magnitude of the time-interval between successive phases of folding. Sometimes radiometric, paleontologic, or stratigraphic techniques may add precision, but commonly the intervals between phases of superposed folding are shorter than the error factors inherent in these techniques.

Second, the kinematic nature of the folds can vary. Three categories were established for simple folds in Chapter 6: (a) flexure folds, (b) flexural-slip folds, and (c) slip folds, although the boundaries between these groups are not sharp. In superposed folds the initial flexures may have been any one, or a combination, of these three types; similarly, the second and any subsequent phases can result from deformation of any type. Since the rheidity and the metamorphic milieu (including the possibility of diagenetic neocrystallization) almost certainly change with time, different types of folding commonly occur in the same suite of rocks during the successive deformation phases.

Third, the orientation of the stress field, and thus of the resultant strain pattern, can change. Such a change may be slight or very considerable, and it is described in terms of the angle between the B-geometric axes and/or the axial surfaces of the fold systems. Commonly a single B-geometric axis is not sufficient to define a superposed fold set (as will be described below).

These complex variables demonstrate that the development of superposed folds is not governed by a single universal principle. Although this may appear obvious, it is contrary to the views of some geologists. For example, Rast and Platt (1957, p. 159) complained that "cross-fold" has not been adequately defined, and implied that the term should be reserved for a narrowly-defined and genetically-related group of structures. Again, Lindström (1957, pp. 7 ff.) implied that structural geologists subscribe to one or another of the several "schools of thought," each of which advocates dissimilar unique principles about superposed folding. Although the concept of "schools of thought" may have some reality in this field of endeavor, as in that of, say, granite petrogenesis, the implication is unfortunate. Several of the authors referred to by Lindström have recognized the diversity of superposed folds and thus they do not fit into his supposed "schools."

THE CLASSIFICATION OF SANDER (1948)

Sander (1948) erected two groups of superposed folds:

(a) $B_1 \perp B_2$: the penecontemporaneous development of folds with axes perpendicular to each other; and

(b) $B_1 \wedge B_2$: the development of unrelated superposed folds by rotation about axes (B_1 and B_2) that are at any inclination to each other (including perpendicular).

In both cases triclinic fabrics result, although, considered separately, the deformation with respect to both B_1 and B_2 is commonly monoclinic.

The differentiation of these categories of Sander is more easy when the time dimension is emphasized. Although "penecontemporaneous" and "synchronous" are employed commonly in discussing $B_1 \perp B_2$-structures, the terms can be somewhat misleading. Sander pointed out that, in $B_1 \perp B_2$-folding, the two deformations are not precisely simultaneous throughout; that is, when small units of time are considered, the strain with respect to B_1 and B_2 (considered separately) is not strictly contemporaneous or continuing, but is spasmodic. Occasionally these spasms with respect to B_1 and B_2 may overlap in time, but Sander did not visualize a triclinic movement picture developed in a single act during an "instant" of geologic time. Anderson (1948, pp. 120 ff.) correctly pointed out that simultaneous slip in two perpendicular directions can be resolved into a single monoclinic movement picture; such a kinematic (movement) picture is contrary to Sander's hypothesis, because it assumes that the total movement was homogeneous with respect to time.

A hypothetical example may be considered. Prior to $B_1 \perp B_2$-folding suppose that the S-surfaces or planes (bedding, for example) are planar. If flexural-slip folding occurs at the commencement of deformation, monoclinic folds about B_1 are likely to be produced. These initial flexures may be of any size, from a small ripple lineation to open folds several tens of meters in amplitude. At a slightly later time (although synchronous with respect to development of the total structure) compression, normal to that which caused the folding about B_1, may cause rotation about B_2, a geometric axis normal to B_1. Although, with this type of flexural-slip folding, B_2 will always be normal to B_1, the B_2-axes must have varying orientations within the plane normal to the compression responsible for B_2. The dip and strike of S vary from point to point after the initial flexure about B_1, and the orientation of B_2 within this plane depends on the local dip of S (Fig. 281).

The style, size, and amplitude of the flexures developed about B_1 and B_2 may be similar, but commonly they are dissimilar. With this model,

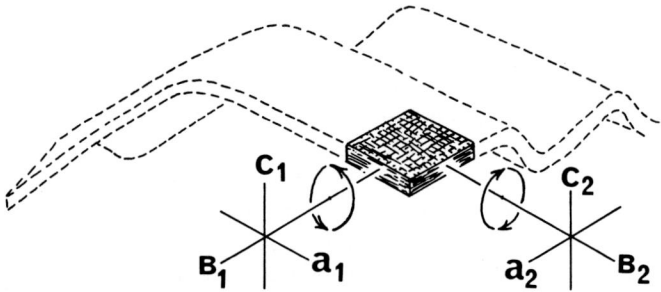

Figure 281. Resolution of triclinic strain into two mutually perpendicular monoclinic strains ($B_1 \perp B_2$) (After Weiss, 1959A, Fig. 13).

provided that a sufficiently small volume of rock is considered, both B_1 and B_2 are axes of monoclinic folding, although, in a larger volume of rock, the total deformation is manifestly triclinic. In a very small piece

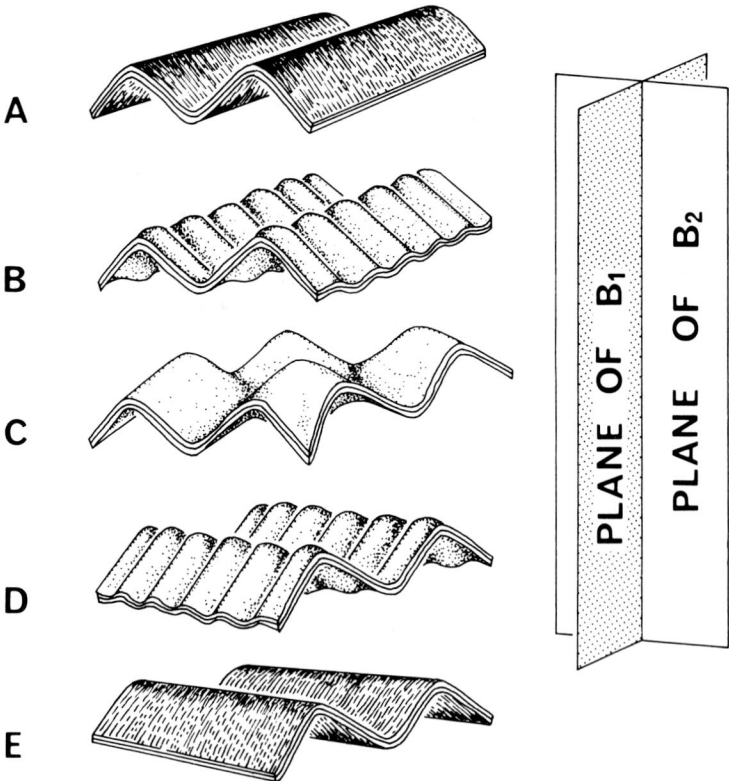

Figure 282. Geometric forms of $B_1 \perp B_2$-fold systems (After Weiss, 1959A, Fig. 12).

of rock within which S is still planar, B_1 is perpendicular to B_2 and both axes lie in S; hence, in a population of objects of this small size, $B_1 = a_2$, $a_1 = B_2$, and $c_1 = c_2$, as shown in Figure 281. When larger objects are considered, the plunges of both B_1 and B_2 vary; the variously-plunging B_1-geometric axes lie in a plane perpendicular to the plane containing the variably-plunging B_2-geometric axes. The mutual relationships of of B_1 and B_2 involved in $B_1 \perp B_2$-folding are illustrated in Figure 282. From the geometric point of view, the assigment of B_1 and B_2 to the two sets of axes becomes arbitrary, although, in the description of actual examples, it is common practice to designate the dominant set of folds as B_1.

To establish $B_1 \perp B_2$ it is necessary that folding about B_1 and B_2 was essentially synchronous on a regional-geographic and time scale. Weiss (1959A, p. 30) suggested that folds about B_1 and B_2 can continue to form synchronously once their general form has been established. In terms of large time units this may be possible, but with respect to small time units it is difficult to visualize how this could happen, or how it could be established with certainty.

With $B_1 \wedge B_2$ the unrelated superposed folds parallel to B_1 and B_2 are discretely separable in time, that is, folding about B_1 is completed before the B_2-folding is initiated; occasionally renewed, or posthumous, movement on B_1 can occur on a minor scale during subsequent deformation. Characteristically $B_1 \wedge B_2$-folding leads to very complex geometrical relationships.

Although reference has only been made to flexural-slip folds, neither $B_1 \perp B_2$ nor $B_1 \wedge B_2$ is restricted to a specific type of folding about B_1 or B_2.

THE PROPOSED CLASSIFICATION

Whitten (1959A) published a review of the described examples of superposed folds and developed a tentative, but unsatisfactory, classification. Sutton (1960A, pp. 153-54, 161) reviewed the superposed folds in the Scottish Highlands and drew up a complex and confusing pair of classifications.

It seems preferable to use a simpler scheme of classification, as follows:

A. $B_1 \perp B_2$-folds
B. $B_1 \wedge B_2$-folds
    1. B_1 and B_2 oblique but penecontemporaneously developed
    2. B_1 and B_2 superposed but developed during the same general orogenic phase

3. B_1 and B_2 superposed and developed during long-separated cycles of orogeny.

In each case the $B_1 \wedge B_2$-folds can be further subdivided according to whether:

(a) the preferred orientations of B_1 and B_2 are parallel, oblique, or perpendicular to each other, and

(b) folds about B_1 and B_2 are of flexural-slip, slip, or flexural type, or some combination of these types.

Many natural occurrences are, of course, transitional between these artificially-erected categories.

The interrelationship of these variables is shown in Figure 283. This three-dimensional classification system suffices when only two phases of folding were superposed, but additional dimensions are required to incorporate multiple superposition (i.e., three or more fold phases). For example, a third phase of folding requires a five-dimensional system; the fourth dimension would be divided into the same categories used for the second folding in Figure 283, and the fifth dimension would be divided

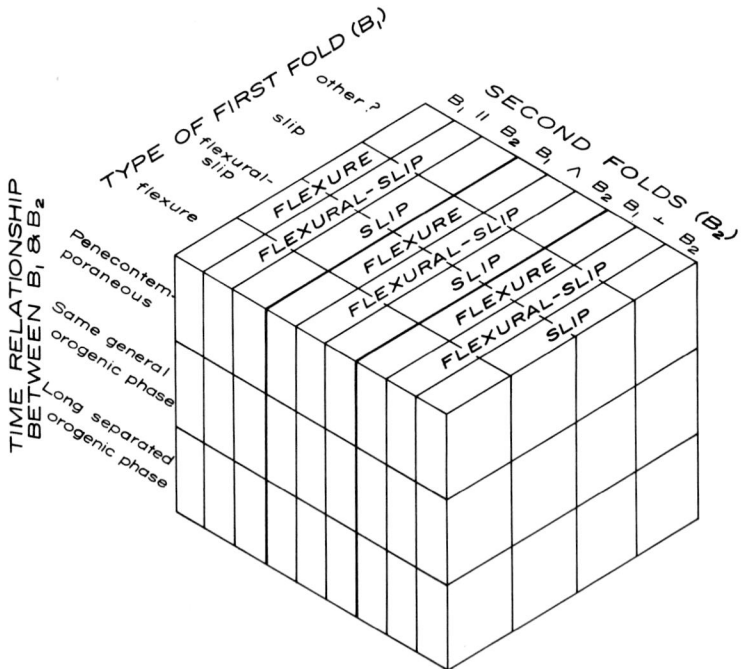

Figure 283. Possible interrelationships between two superposed phases of folding where each phase may be flexure, flexural-slip, or slip folding.

on the basis of the time relationships between the third and the earlier phases of folding.

Strong arguments for a genetic classification of superposed folds could be advanced. Apart from the general physical conditions during the folding (e.g., rheidity and the metamorphic milieu), the nature of the changing stress field would then be of paramount importance. Although a genetic classification might be an eventual aim, it is so difficult to determine the stress field for a folded sequence of rocks that, even when the strain picture is reasonably well known, it is impractical and dangerous to assay a genetic classification at the present time.

A variety of overlapping terms has been used recently for superposed folding. "Cross fold," "crossing fold," "transverse fold," "crossfold," "(cross-)fold," and "superimposed fold" have all been employed (see Whitten, 1959A). To avoid confusion with some of the earlier definitions (Nevin, 1949; Bhattacharji, 1958, 1959) "superposed fold" is used in this book (cf. Weiss, 1959B; Roach, 1960).

THE GEOMETRY OF SUPERPOSED FLEXURAL-SLIP FOLDS

The geometry involved when two sets of cylindroidal flexural-slip folds are superposed to produce $B_1 \wedge B_2$-folds is relatively simple. $B_1 \wedge B_2$-structures must be considered in three dimensions, and the most realistic current method is to use stereographic projection. The principles can be illustrated most easily with a hypothetical example.

In Figure 284 a single bed within an antiform is illustrated. It is assumed that the ground surface is horizontal. The fold axis plunges to 060 at 28° and the axial plane dips to 110 at 40°. In the east the beds dip to the east, while on the west flank the beds are overturned and dip southeast. These relationships are illustrated by stereographic projection in Figure 285. For simplicity the limbs of the fold are broken into short planar sections, each of which is plotted on the stereogram, which is a composite π- and β-diagram (Fig. 285). The great circle containing all of the S-poles is shown and $B_1 = \beta$.

Now assume that a small ripple-like linear structure parallel to B_1 is visible in most exposures of the map area (Fig. 284); alternatively, it may be assumed that numerous small parasitic folds are homoaxial with the main structure. It will be assumed that these minor structures can be identified after superposed folding.

It is reasonable to assume that the first deformation was caused by compression approximately normal to the axial plane. Suppose that a

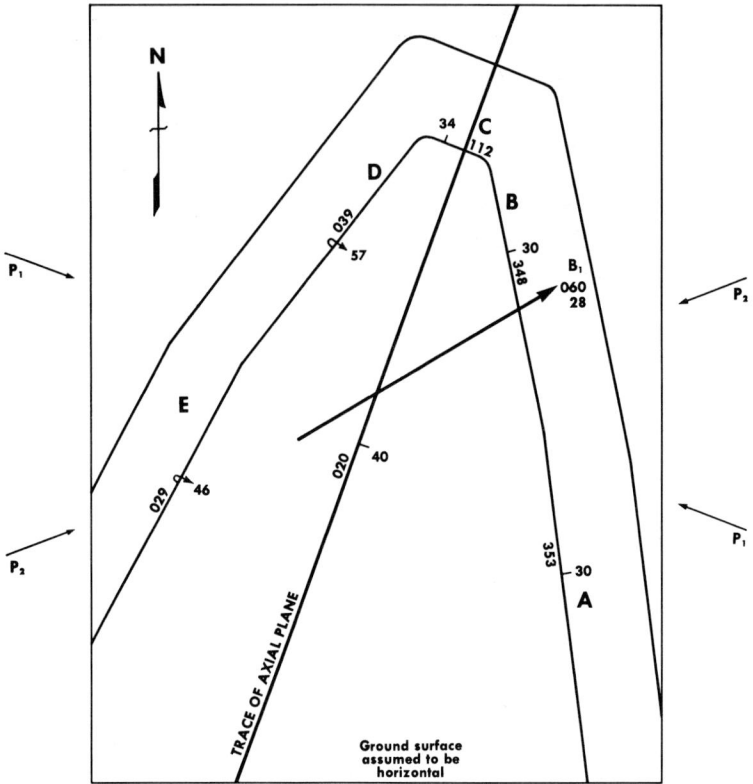

Figure 284. Schematic map of a plunging anticline with ground surface assumed to be horizontal.

second phase of compression acting horizontally and parallel to P_2 (Fig. 284) produces a second phase of flexural-slip folding so that $B_1 \wedge B_2$-folding results. Each bed of the earlier structure folds about a new axis (B_2) lying in the plane normal to the new compression axis, P_2. A separate second fold axis, B_2, develops for each bedding plane with a different dip and strike. Because the B_2-axes lie in the plane normal to the compression P_2, they plot on a great circle (cf. Braitsch, 1957).

A single S-surface from the original anticline (Fig. 284) can be used for illustration. In Figure 286 the bedding plane C (from the closure of the fold) is used; this bed had a strike of 112 and it dips north at 34°. The great circle normal to P_2 is a straight line in Figure 286, and the bedding plane C intersects it at B_2, the second fold axis with *respect to this particular S-surface (C)*. As the second phase of folding continues, the S-surface C rotates about B_2; this can be illustrated by a π-diagram or a β-diagram. The S-pole (π_{C_1}) for C rotates about B_2 during the second folding, and successive positions of π_{C_1} (designated π_{C_2},

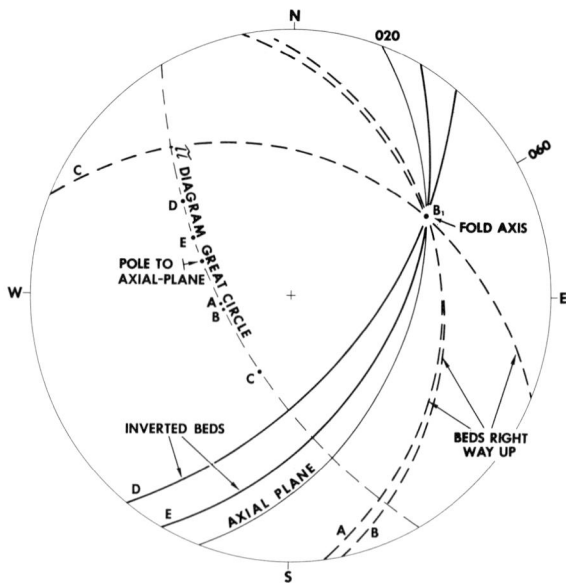

Figure 285. The geometry of the fold in Figure 284 shown in stereographic projection. The points A through D represent π-poles of the bedding planes indicated by the great circles (A through D) that intersect in B_1.

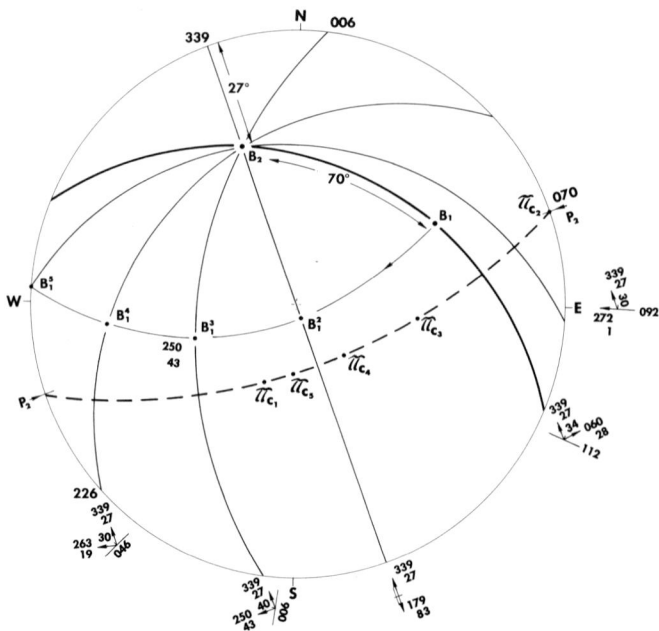

Figure 286. Superposed flexural-slip folding (B_2) of bedding plane C (from Fig. 284) as a result of compression about P_2. The great circle through B_1 and B_2 is the original bed C; the other great circles through B_2 are possible orientations of C after B_2-folding; poles to these rotated bedding planes (π_{C_1}, π_{C_2}, ...) lie on the great circle (broken line) through P_2. Early lineations parallel to B_1 (B_1^2, B_1^3, ...) lie on the small circle about B_2. Map symbols appropriate to the rotated C-planes are shown outside the margin of the projection.

$\pi_{C_3}, \ldots, \pi_{C_n}$ in Fig. 286) lie on a great circle because the angular distance between B_2 and π_{C_1} is 90°.

The linear structures parallel to B_1 (B_1 in Fig. 286) rotate with S during the second deformation. However, B_1 moves along a small circle because the angular distance between B_2 and B_1 is not 90°; the angle happens to be 70° in Figure 286. B_1 assumes positions $B_1^2, B_1^3, B_1^4, \ldots,$ B_1^n as π_{C_1} rotates to $\pi_{C_2}, \pi_{C_3}, \pi_{C_4}, \ldots, \pi_{C_n}$ (Fig. 286).

The field observations which would be associated with the successive positions of C are shown at the bottom of Figure 286. While the second linear structure (B_2) maintains a *constant* orientation, the first linear element has a very *variable* orientation ($B_1^2, B_1^3, \ldots, B_1^n$).

As another example, bed E, from the inverted limb of the anticline (Fig. 284) is shown in Figure 287; the strike is 029 and the dip 46° east. Figure 287 is primarily a β-diagram resulting from rotation of the S_0-surface (E) about the second fold axis B_{2_E}. The poles to the rotated S_0 are $\pi_{E_2}, \pi_{E_3}, \ldots,$ and the great circle containing them passes through P_2 (cf. Fig. 286). The second folding causes the linear elements B_1 (B_1 in Fig. 287) to move along a small circle to successive positions $B_1^2, B_1^3, \ldots B_1^n$; B_1 and B_1^2 are on inverted limbs, whereas $B_1^3, B_1^4, B_1^5,$ and B_1^6 are on S_0-surfaces that have returned from inverted positions to the correct way up. At the bottom of Fig. 287 field observations appropriate to several of the S_0-surface orientations are shown.

It is commonly helpful to compile synopses of the relationships exhibited by all the elements of the structure within an area. Figure 288 is a **synoptic diagram** based on operations like those used for Figures 286 and 287. Compression P_2 induced a second generation of folds with respect to each of the five planar sections (A, B, C, D, and E) of the original S_0-surface (Fig. 284). For each planar section a new fold axis has been developed that lies in the new axial plane normal to P_2. The five new fold axes are $B_{2_A}, B_{2_B}, B_{2_C}, B_{2_D},$ and B_{2_E} in Figure 288. It is a general property that each of the great circles defined by the poles to the refolded S_0-surfaces intersect in P_2. A further general property is that small circles drawn through the projections of rotated first linear elements (B_1) intersect at B_1 (Fig. 288); this is intuitively reasonable, because, for each position of S_0 the linear elements initially coincide with B_1, and during the second deformation they are progressively rotated from this direction. These relationships of the small and great circles on synoptic diagrams are useful in analyzing the geometry of an area.

For each generation of folds, the amplitude, style, and size of the flexures, and also the nature of the linear structures, are commonly

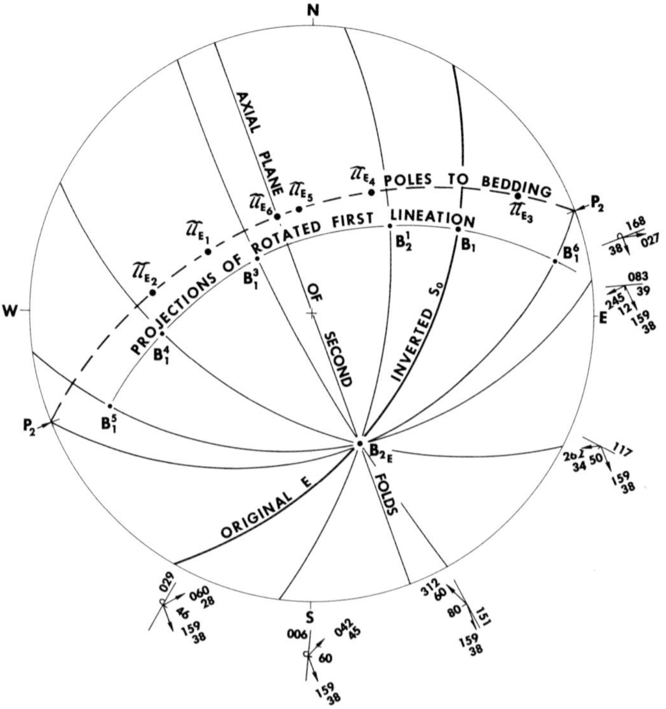

Figure 287. Superposed flexural-slip folding (B_2) of inverted bed E (from Fig. 284) as a result of compression about P_2. The explanation is the same as for Figure 286.

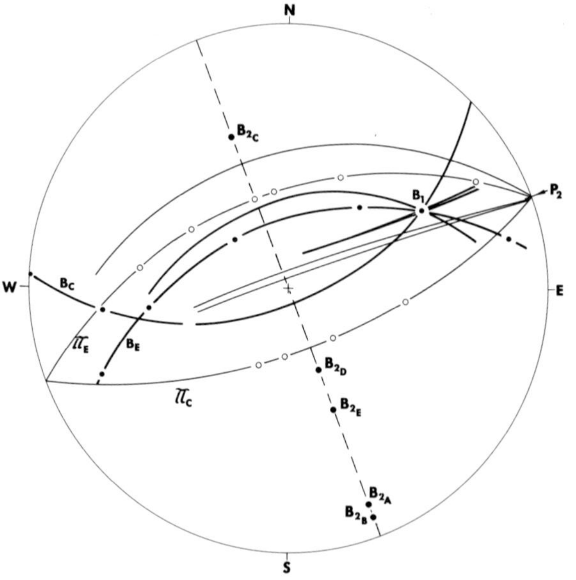

Figure 288. Synoptic diagram representing flexural-slip folds produced by compression P_2 affecting the fold B_1 of Figure 284. The great circles (light lines) intersecting in P_2 are defined by poles to the rotated S_0-planes. The small circles (heavy lines) intersecting in B_1 are defined by the rotated first fold linear elements that were parallel to B_1.

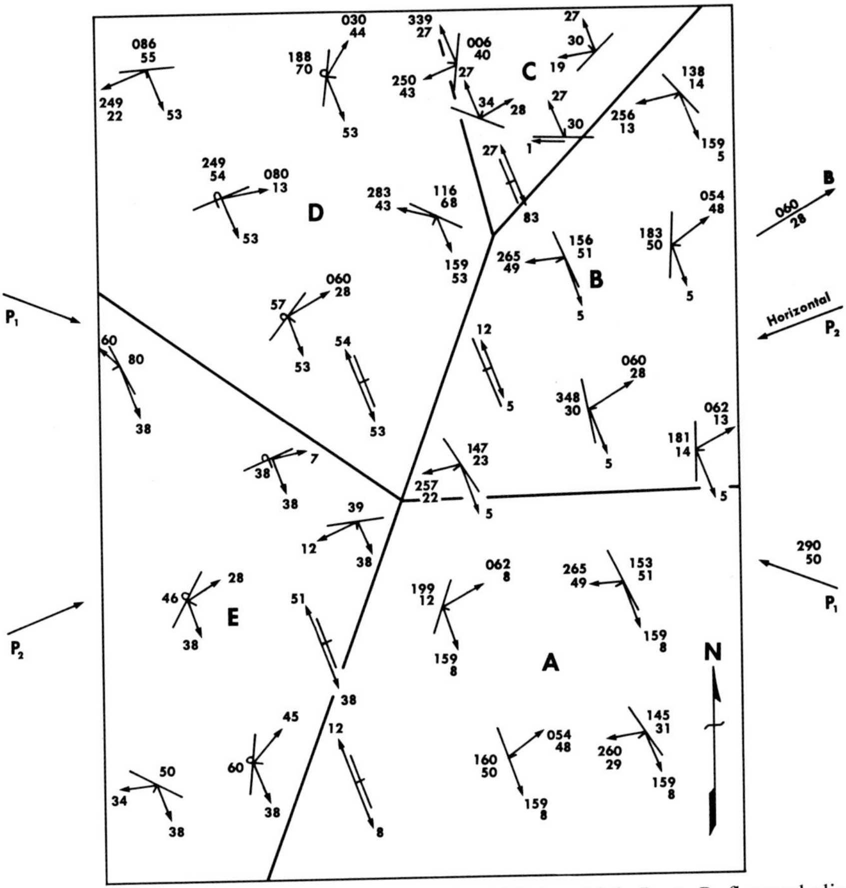

Figure 289. Schematic map of an area (Fig. 284) in which $B_1 \wedge B_2$-flexural-slip folding has occurred (the ground surface is assumed to be horizontal).

distinctive. Such differences can be very significant in analyzing superposed folds. Thus, the importance of observing and recording the distinctive characters of fold structures in the field cannot be overemphasized.

Even in those cases in which the fold and linear structures produced during both phases of $B_1 \wedge B_2$-folding are similar, the geometric relationships should enable the fold geometry and the sequence of events to be unravelled. The technique can be illustrated with the aid of the map shown in Figure 289.

The map of an area in which superposed flexural-slip folding occurred (Fig. 289) can be divided into subareas within which the S-surfaces were approximately planar at the close of the first folding. Theory dictates that the second deformation results in a parallel set of linear structures in each such subarea. For example, in the northwest part (area D) of Figure 289, one linear element plunges to 159 at 53° at each locality. Again, throughout the southern half of the map a linear structure trends at 159, but the plunge differs in each subarea (Fig. 289, areas A and E). Linear structures induced by the second deformation are constant in orientation in each domain. Linear structures of the first deformation now have variable orientations within each subarea.

In actual field studies, it is commonly quite difficult to delimit suitable geographic subareas. Current methods tend to be largely empirical, but one guide is to prepare a map of all the linear elements only. Geographic areas within which similar linear elements are approximately parallel (homogeneous) can then be isolated. Boundaries defined by this principle are marked on Figure 289.

If linear elements are not abundant, subareas can often be defined by locating domains within which S-poles lie close to a great circle on a π-diagram. This is strictly a trial-and-error process, and the boundaries of subareas are adjusted to include only S-poles that lie close to the great circle. Exposures at the domain boundaries can often be accommodated in either subarea.

The great circle suitable for an array of points is found by rotating the tracing paper over the Wulff graticule; as the great circles of the graticule pass beneath the plotted points, the one of best fit is chosen by eye and traced on the plot. Finding a small circle is more difficult, unless its center happens to lie on the primitive circle, when a small circle printed on the Wulff graticule can be traced onto the plot. A method appropriate for all other cases is illustrated in Figure 290 (which is based on the β-diagram of Fig. 287). Suppose the dots are projections of linear elements and that it is required to know whether they lie on a small circle about B_{2_E}. The angular distance from B_{2_E} to one of the clusters of points is measured along a great circle. This angle is then measured off from B_{2_E}, along several other great circles passing through B_{2_E}, to give the points a, b, c, d, ... The smooth curve passing through a, b, c, d, ..., is the required small circle. Since the circle passes through the second cluster of linear elements, both groups of projections lie on the same small circle. An additional small circle 20° from B_{2_E} has been plotted on Figure 290 by using the same method.

In the hypothetical example (Fig. 289), the structural elements from each subarea can be plotted on separate stereograms; for example, Figures 286 and 287 represent subareas C and E, respectively. When transferred to a synoptic diagram the small circles intersect at the location of the first linear element, B_1 (cf. Fig. 288); the intersection of the great circles defines the orientation of the second compression (P_2).

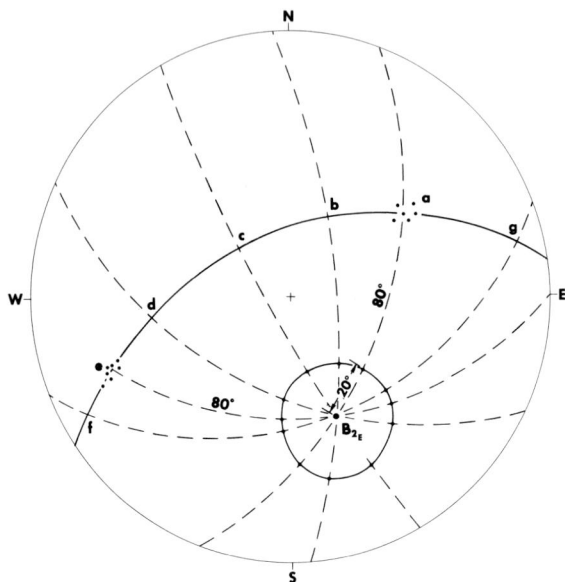

Figure 290. Construction of small circles about B_2 on a stereographic projection. For explanation see text.

Complication stems from the fact that commonly the second maximum compression axis (P_2) is not necessarily horizontal. When P_2 has a marked plunge, the trends of the second linear elements are no longer constant over the map area. However, the trend is constant within each subarea in which S_0 was planar at the close of the first deformation (e.g., within subareas A, B, C, D, and E of Fig. 284). The geometry can be illustrated by using the structure of Figure 284, but allowing P_2 to plunge at, say, 60° to the northeast. Figures 291 and 292 portray the resulting geometry of subareas C and E, respectively; these diagrams correspond to Figures 286 and 287. Figures 291 and 292 show that the trends of the second fold axes B_{2_C} and B_{2_E} are not parallel (whereas they are parallel in Figs. 286 and 287). However, as when P_2 was horizontal, the present B_{2_C} and B_{2_E} lie on a great circle normal to P_2. Some representative positions of S_0 are drawn as a β-diagram in Figures 291 and 292, and they are also plotted on a map (Fig. 293) for comparison with Figure 289.

For these geometric analyses it is not necessary to have a detailed map of the lithologic units. Naturally, however, this information is required if a complete picture of the style of the folding is to be obtained and if the original stratigraphic succession is to be interpreted. However, even where the lithologic units have been carefully mapped, and sufficient structural data are available for the geometric style to be interpreted, it is difficult to predict the behavior of a lithologic unit outside the area of the immediate survey. For example, Knowles, *et al.*, (1962) mapped the metamorphosed Precambrian iron formation in the Julienne Lake area, Labrador, Canada, and deduced the geometric style of the superposed

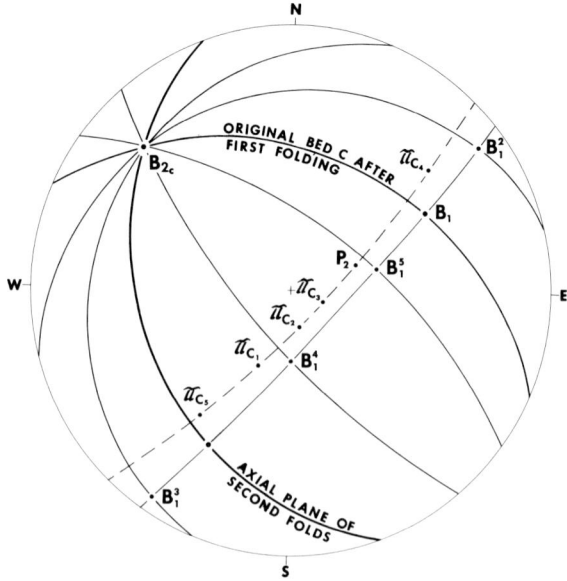

Figure 291. The geometry of subarea C (of Fig. 284) resulting from superposed flexural-slip folding due to compression P_2. The axial plane of the second folds is normal to P_2. The notation is as in Figure 286.

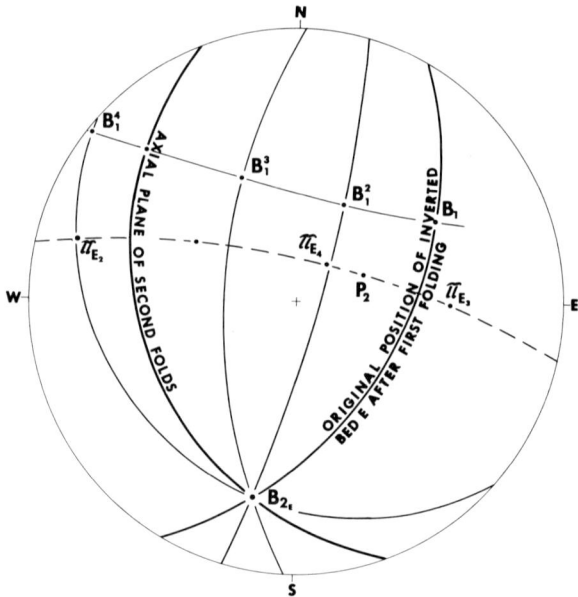

Figure 292. The geometry of subarea E (of Fig. 284) resulting from superposed flexural-slip folding due to compression P_2. The notation is as in Figure 286.

folding (Fig. 294). With the methods available at present, it is difficult to predict the precise location of such a doubly-folded ore body at depth or outside the limits of the immediate mapped area.

In the examples discussed above the original S_0 was assumed to be absolutely planar, and the two stress fields (responsible for B_1 and the

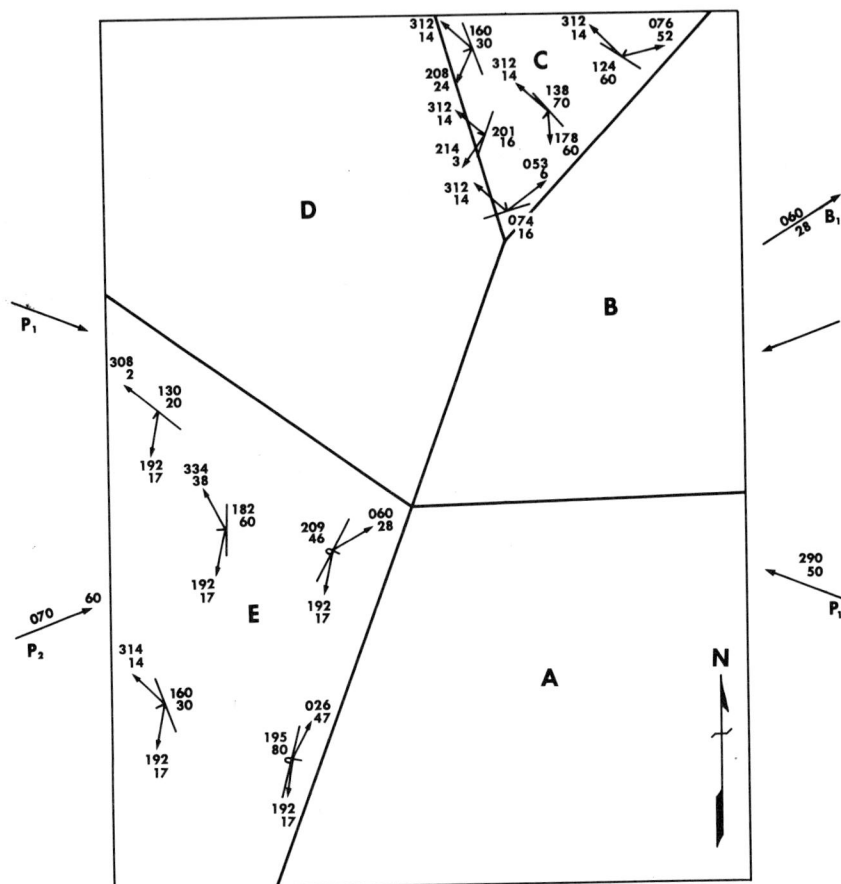

Figure 293. Schematic map of an area (Fig. 284) in which $B_1 \wedge B_2$-flexural-slip folding has occurred as a result of compression along an axis (P_2) plunging at 60°. See text for full explanation.

B_2-axes) were taken to be uniform throughout the map-areas. Thus, in Figure 289 the map has idealized geometry. In actual examples some variability occurs, and it cannot be anticipated that the field data will approximate too closely with great and small circles. However, by careful selection of the limits of subareas, domains that are approximately homogeneous with respect to specified attributes (such as a specified second lineation) can commonly be isolated. Even when small domains are carefully selected, a slight scattering of points is to be expected; that is, the areal distribution has subsystematic tendencies (see Chapter 3).

In all of these constructions it was assumed that the axial surfaces of the second generation folds are subparallel and planar. Undoubtedly this simplification does not always apply in natural examples, although

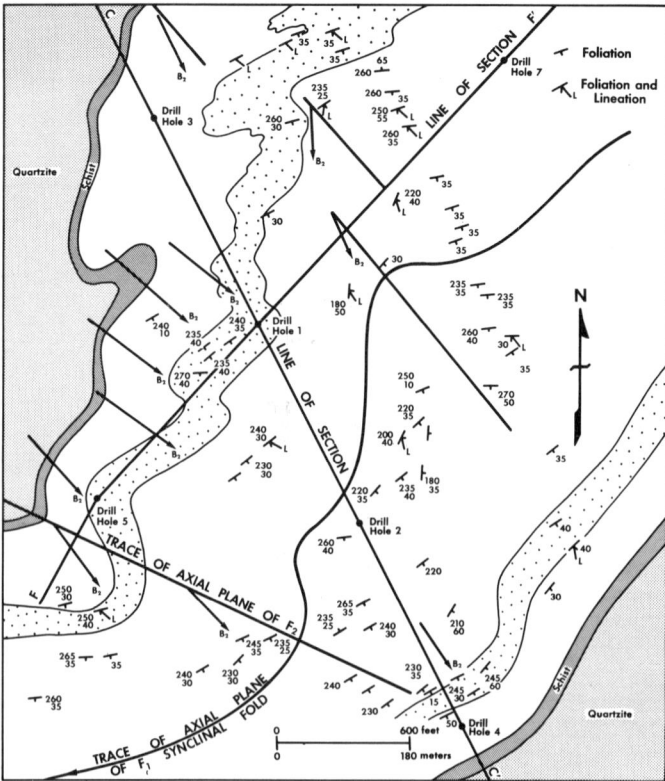

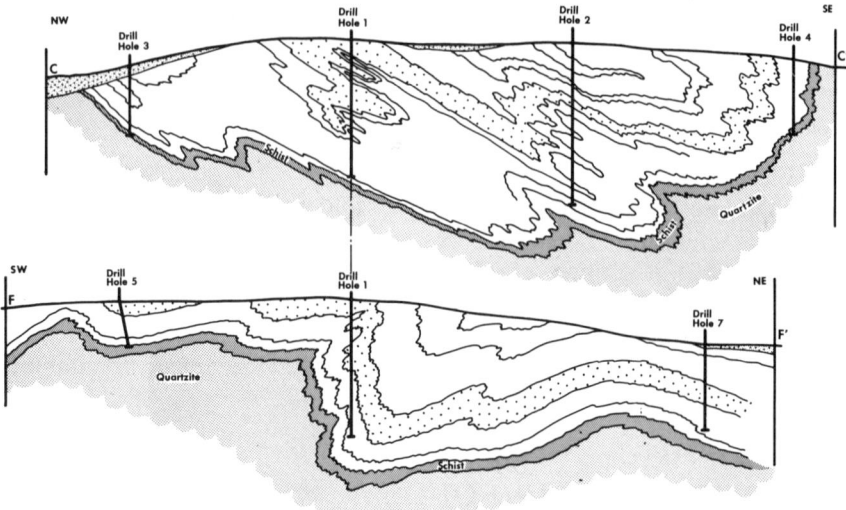

Figure 294. Superposed folds affecting the Precambrian iron formation in the Julienne Lake-Wabash Lake area, Labrador, Canada; the cross-sections show the fold styles as determined by mapping and drilling (Based on manuscript map dated 1960 and made available by Mr. D. M. Knowles).

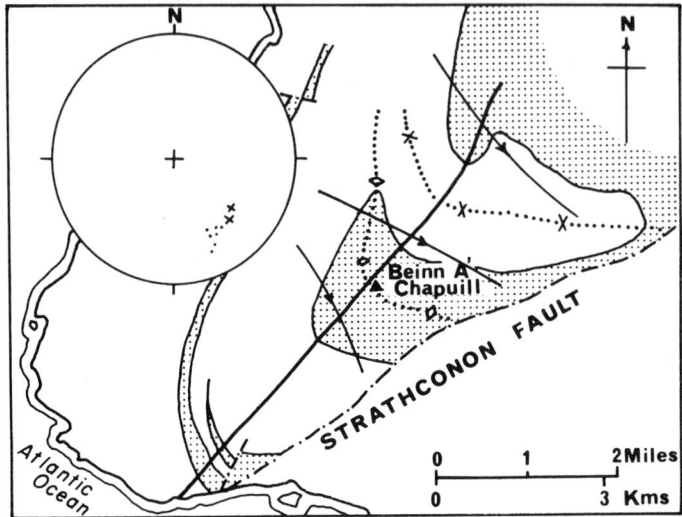

Moine schists

Lewisian gneiss

······◊······ Early antiform axial plane trace

······✗······ Early synform axial plane trace

─────── Axial plane trace of late fold

Figure 295. Map of folds in the Beinn a' Chapuill area, S of Glenelg, Northern Highlands, Scotland. Axes of both minor folds and of linear structures associated with the late folds plunge SE (represented by long arrows on map): note how they diverge from the antiform and converge on the synform. *Inset*: Late linear structures—dots are linear structures on NE limb of early synform and SW limb of early antiform: crosses are linear structures on the common limb of the two early folds (After Ramsay, in Clifford, *et al.*, 1957, Fig. 11).

as Weiss (1959B, p. 95) pointed out, such subparallelism has been described for several natural superposed fold systems. Sander (1948, pp. 177-78) referred to such features, and Weiss and McIntyre (1957, pp. 596-600) described examples from the Dalradian tectonites of Loch Leven, Scotland. Figure 295 shows the Beinn a' Chapuill region, south of Glenelg, W Scotland, in which Clifford, *et al.* (1957, p. 19) mapped a superposed set of folds with a constant axial plane orientation.

THE GEOMETRY OF SUPERPOSED SLIP FOLDS

In Chapter 6 the distinctive geometry of slip folds was described; a superposed second phase of deformation may be of this type. Slip folds may be superposed on either flexural-slip or slip folds. Figure 296 is a diagrammatic sketch-map that illustrates the changes in geometry as progressively more penetrative slip folding is superposed on flexural-slip folds.

The effects of superposed slip folding can be illustrated by reference to the fold in Figure 284. The S-surfaces S_A, S_B, ..., S_F resulting from the first folding are shown as a β-diagram in Figure 297. In this case $\beta_1 = B_1$. Superposed slip folding, resulting from slip on planes normal to P_2, will be considered; P_2 plunges with the same orientation as in Figure 293. In Figure 297, S' is the plane normal to P_2 and parallel to the new (second) axial-plane foliation. The kinematic axes during the second phase of folding can be designated a_2, B_2, and c_2; c_2 coincides with P_2, while a_2 and B_2 lie in S'. In Figure 297, a_2 and B_2 have been assigned arbitrary positions, but the location of a_2 controls the geometry of the second folding.

It follows from the discussion of slip folding in Chapter 6 that the second fold axis (β_2) lies at the intersection of S' and the original S, and that, in general, $\beta_2 \neq B_2$. Hence, the superposed folding produces a separate fold axis in each subarea. For example, in Figure 297, S_A yields the new fold axis β_{2_A}, S_B yields β_{2_B}, etc. As folding proceeds the S-surface in each domain is rotated about the new fold axis (β_2), and all first deformation linear elements parallel to B_1 rotate in the plane passing through B_1 and a_2. Thus, the projections of the rotated B_1-linear elements lie on the great circle containing B_1 and a_2; this is the same great circle for each domain. The geometry of this type of deformation was described by Weiss (1955, Fig. 1).

A few possible orientations of S_C following the second folding are shown in Figure 298; these were derived by rotation of rocks in subarea C (Fig. 284) about β_{2_C}. Possible rotated positions of B_1 are indicated by B_1, B_1^2, B_1^3, ... The corresponding geometry for subarea E (Fig. 284) is shown in Figure 299; each subarea yields a separate β-diagram, as shown synoptically in Figure 300. The results from Figures 298 and 299 are illustrated as a map in Figure 301.

In these diagrams none of the intersections of S and S' coincide with either of the kinematic axes a_2 or B_2. In a domain in which the intersection of S and S' is parallel to the a_2-kinematic axis, the S-surfaces are not folded because the slip is parallel to S (see Chapter 6). Should the intersection of S and S' be parallel to the B_2-kinematic axis, the second axis of folding (β_2) is parallel to B_2, and $B_2 = \beta_2$.

In the illustrations used above, the first folds were flexural-slip, so that $B_1 = \beta_1$. The geometry of the superposed folds would not be altered if

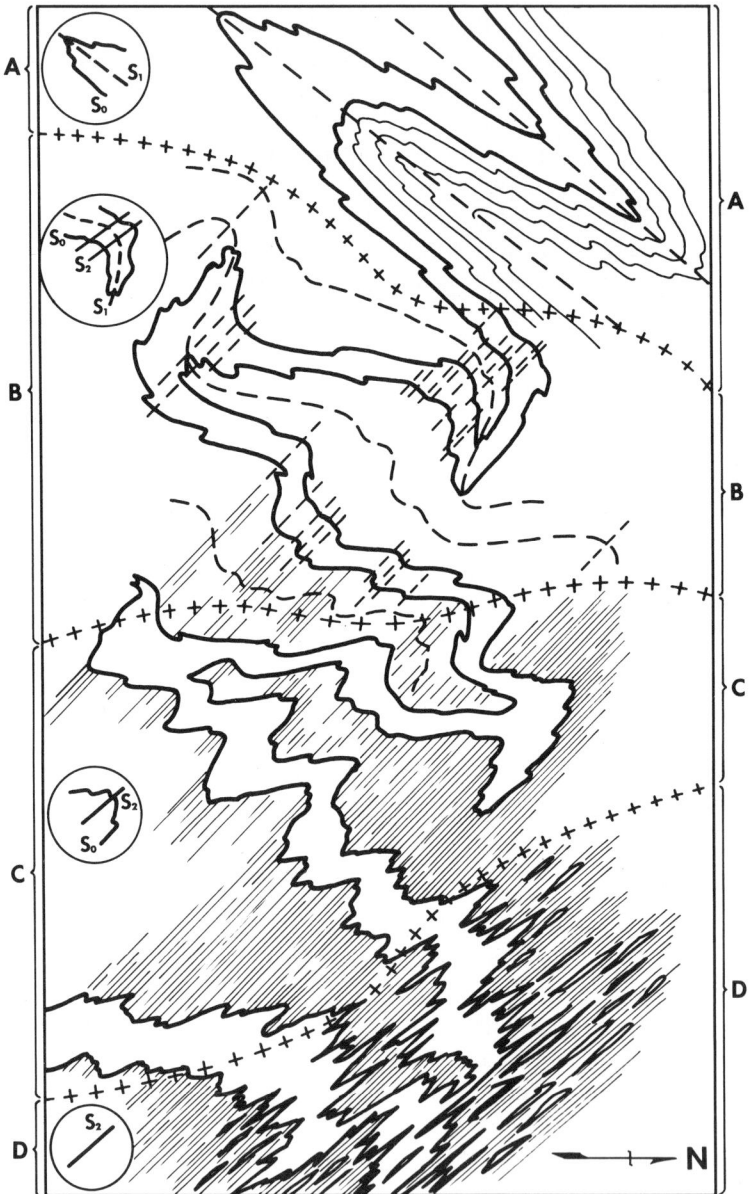

Figure 296. Hypothetical geological map showing transposition structures associated with superposed folds. Domains A through D show progressive stages of superposition of folds with axial planes striking NW; axial planes of the earlier folds strike NE. Transposition of the initial S-structure (S_0) into the secondary foliation S_2 is complete, on the scale of the map, in domain D (After Turner and Weiss, 1963, Fig. 5-26).

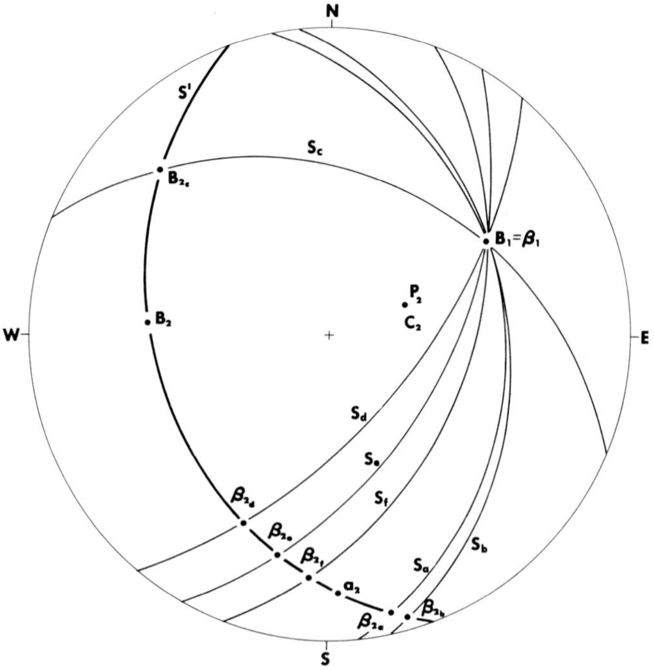

Figure 297. New fold axes resulting from slip folding superposed on the B_1-fold of Figure 284. The first fold is represented as a β-diagram; β_{2A}, β_{2B}, ... are the axes of second generation folds resulting from slip on S' normal to P_2. The orientation of the kinematic a_2- and B_2-axes within S' has been chosen arbitrarily.

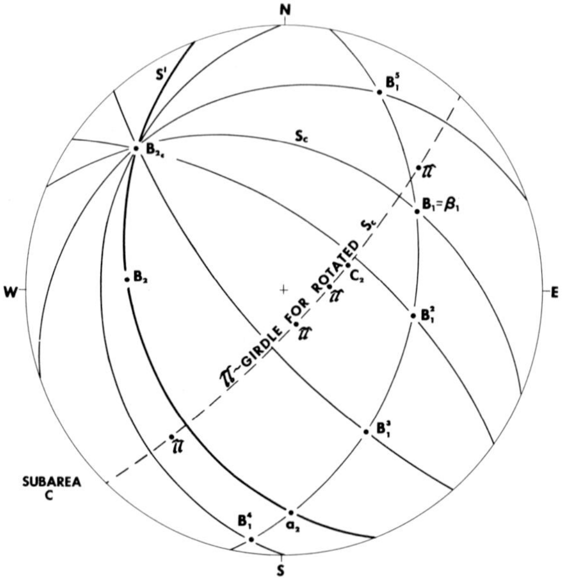

Figure 298. The geometry of the slip folds developed in subarea C (of Fig. 284) as a result of slip parallel to the a_2-kinematic axis.

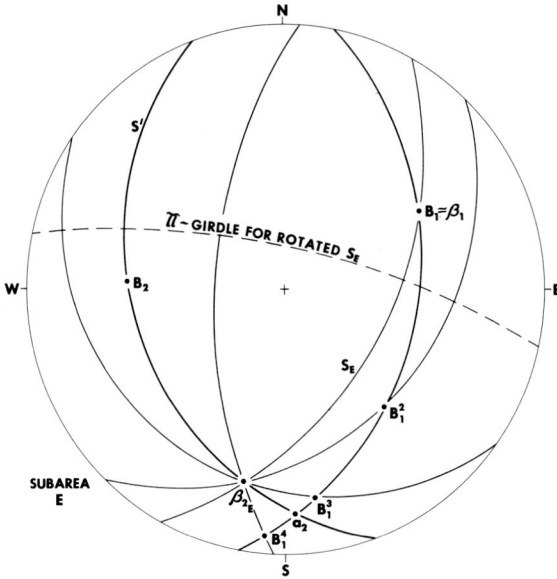

Figure 299. The geometry of slip folds in subarea E (of Fig. 284) produced by slip parallel to an a_2-kinematic axis normal to P_2.

Figure 300. Synoptic diagram showing the geometry of slip folds throughout the area of Figure 284 that result from slip parallel to an a_2-kinematic axis normal to P_2. The great circles (light lines) passing through β_{2A}, β_{2B}, ... represent the folded S-planes in each subarea.

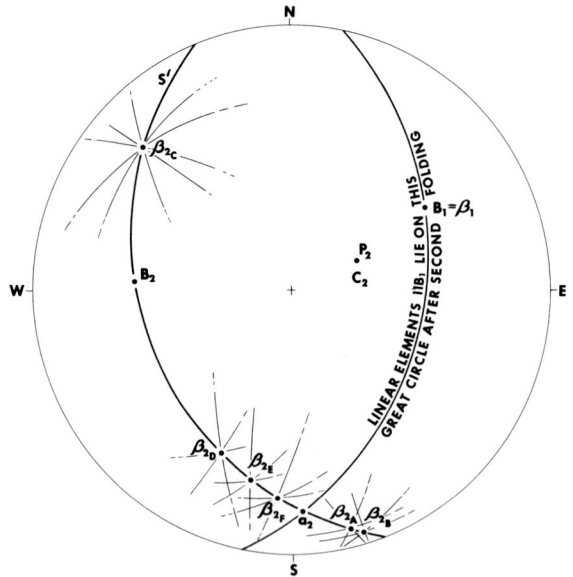

the initial folds were slip folds, although, of course, in the general case, the first fold axis β_1 would be oblique to the kinematic B_1-axis.

The solid geometry of some superposed slip folds is made more clear by Figures 302 and 303. In Figure 302 the plane of slip (the a_2B_2-plane) is perpendicular to the geometric B_1-axis of the first fold. Because the B_1-

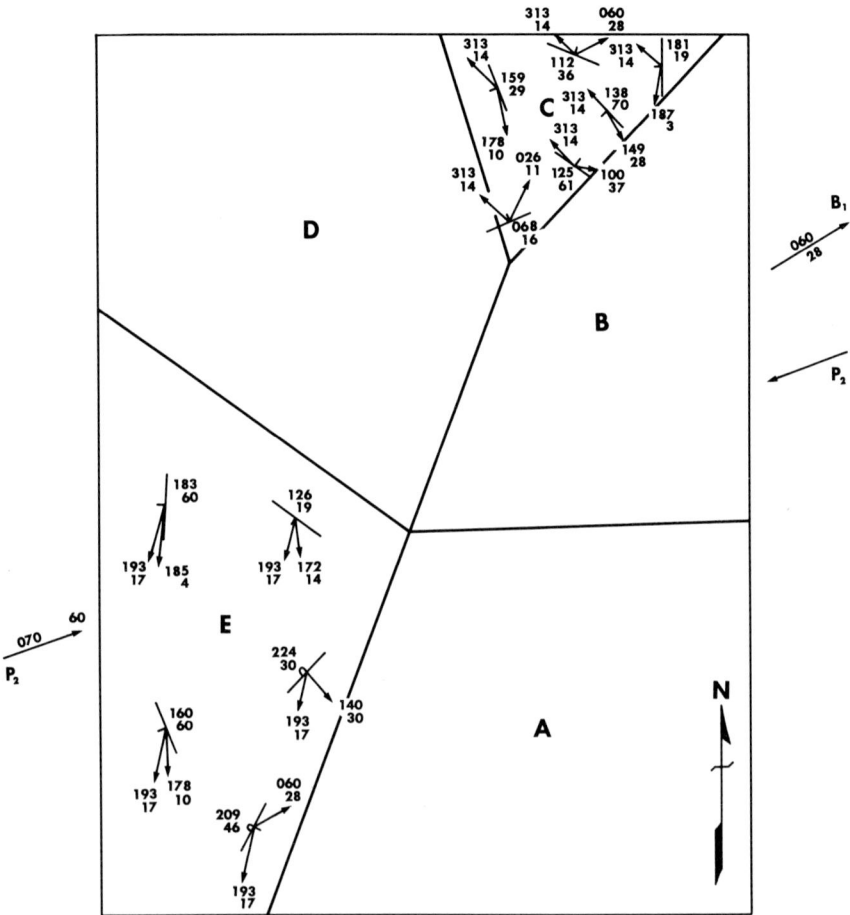

Figure 301. Schematic map of an area (Fig. 284) in which B_2-slip folding has been superposed on an initial B_1-flexural-slip fold system. See text for explanation.

axis of the synform (first fold) plunges towards the observer in Figure 303, B_1 is oblique to both c_2 and the a_2B_2-plane. With superposed slip folding the chances of the intersection of S and S' being parallel to either the a_2- or the B_2-kinematic axis are relatively small; such parallelism only tends to occur in very small volumes of rock.

As pointed out previously, the stereographic approach to folding has certain limitations because each fabric element is divorced from its actual U,V,W-location and moved to the center of the stereogram. In consequence, it is difficult to visualize the three-dimensional morphology of superposed slip folds from a stereogram. O'Driscoll (1962, 1964) illustrated experimentally-produced patterns for some geometries that result with superposed folding, while Ramsay (1962C) described the morphology to be expected with each possible orientation of the two sets of kinematic axes.

Ramsay (1962C) demonstrated that the morphology resulting from superposed slip folds (F_1 and F_2) depends on the:

    (a) morphology of the F_1-folds and the orientation of the F_1-axes,

    (b) orientation of the second kinematic axes with respect to the F_1-axes; three main geometric relationships occur (see Table 2).

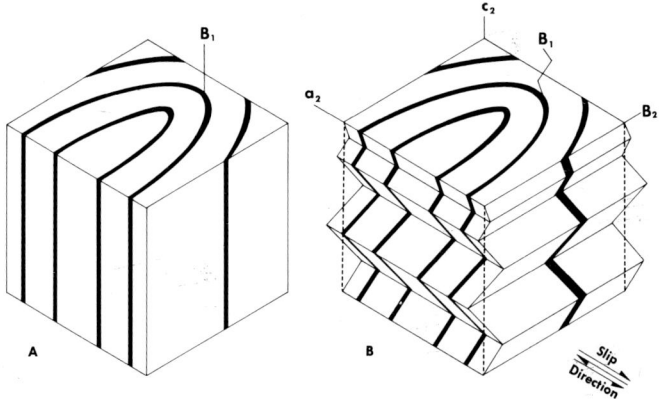

Figure 302. Diagrammatic block diagrams to show the geometry of a superposed slip fold with the plane of slip perpendicular to the first geometric fold axis. A: The first flexural-slip fold; B_1 is the geometric fold axis. B: Superposed slip fold; a_2B_2 is the plane of slip perpendicular to B_1.

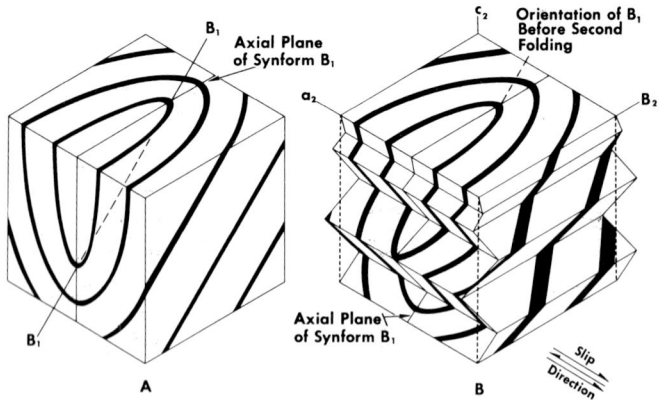

Figure 303. Block diagram to show effect of superposed slip folding, where the a_2B_2-kinematic plane is not normal to the B_1-geometric fold axis.

Figure 304. Main types of two-dimensional interference pattern resulting from superposed slip folding (see Table 2). Axial traces of F_1 were originally NE-SW and those of F_2 are N-S. A: Type 1 with movement direction of F_2 contained in F_1 axial plane. B: Type 1, general form. C: Type 1 with F_2 movement direction within a limb of a F_1 fold. D: Type 2. E: Type 3 (After Ramsay, 1962C, Fig. 14).

    (c) intensity of F_2-folding

    (d) amount of flattening accompanying F_2 (see Chapter 6), and,

    (e) orientation of outcrop surface with respect to F_1 and F_2.

Of these five factors, (a) and (b) are the most critical. Ramsay (1962C,

Table 2.—Geometric Relationships Resulting From Superposed Slip Folding

Type	Angle Between a_2-Axis and Axial Plane of F_1	Angle Between B_1-Geometric Axis and Axial Plane of F_2	Illustrated in Figure Number
1	a_2-kinematic axis lies close to axial planes of F_1	Any angle	304A, 304B, 304C
2	a_2-kinematic axis highly inclined to axial planes of F_1	Medium to high angle	304D
3	a_2-kinematic axis highly inclined to axial planes of F_1	Low angle	304E

A. FIRST FOLDS

B. SUPERPOSITION OF SECOND FOLDS —SHEAR COMPONENT

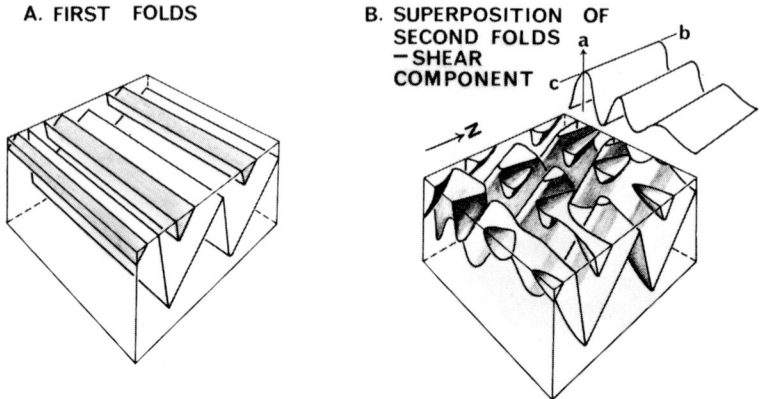

Figure 305. Interference structure resulting from superposed slip folds: Type 1 consisting of interlocking domes and basins (After Ramsay, 1962C, Figs. 1A and 1B).

p. 479) seems to have been correct in asserting that it is impossible for two sets of genuine slip folds to form simultaneously at exactly the same U,V,W-location.

The geometry of Type 1 (Table 2) is illustrated in Figure 305; domes result where antiformal axes cross and basins form where synformal axes cross. Figure 306 shows an excellent small-size example from the feldspathic sandstones (metamorphosed to the garnet grade) in the Moine Series, Loch Monar area, western Scotland; because (a) both fold sets have axial-plane foliation, and (b) the F_1-axial planes are slightly deformed and the F_2-axial planes are planar, it is reasonable to conclude that these fold sets formed separately, (i.e., not contemporaneously). Surprisingly steep domal and basinal folds result from superposed folds of this type; Ramsay reconstructed a scale model of an acute dome about 10 cm. across by cutting serial sections through a sample of banded hornblende-biotite-Lewisian gneiss from Glenelg (Fig. 307). With such geometry the change in plunge of individual folds is often greater than 90°.

Type 2 gives rise to crescentic and mushroom-shaped outcrop patterns (Fig. 308) that are commonly developed where tight recumbent folds of any size have been refolded. Ramsay illustrated an example from an outcrop 25 cm. across, in which isoclinal refolding of older isoclinal folds occurs within the Moine metasedimentary rocks at Ben Clachach, Loch Hourn area, western Scotland; folded metasedimentary rocks exposed in a 130 sq. km. map area near Marangudsi, Rhodesia, show an analogous pattern, although this example is many times the size of the Scottish example (Ramsay, 1962C).

The geometry of Type 3 (Fig. 309) is commonly developed in regions deformed during successive phases of the same orogeny, where the

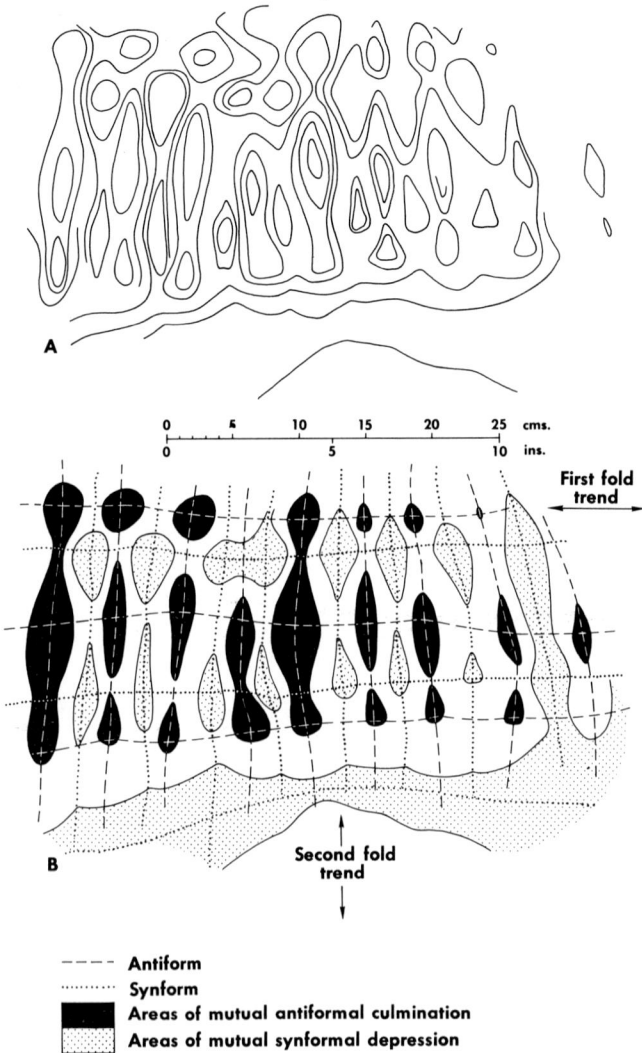

Figure 306. Closed fold forms resulting from superposed folds in feldspathic Moine sandstones, Loch Monar area, Northern Highlands, Scotland. A: Pattern of bedding planes on exposed surface. B: Analysis of domes and basins that are arranged in a regular manner along the trend lines of the two fold sets (After Ramsay, 1962C, Fig. 3).

B-geometric axes of each pulse may be parallel to subparallel (Ramsay, 1962C, p. 476). Closed outcrop patterns are not common with this geometry because B_1 lies close to the F_2-axial plane. Ramsay illustrated a small-sized example of this type of folding from the Lewisian horn-blendic gneiss near Arnisdale, Loch Hourn, western Scotland.

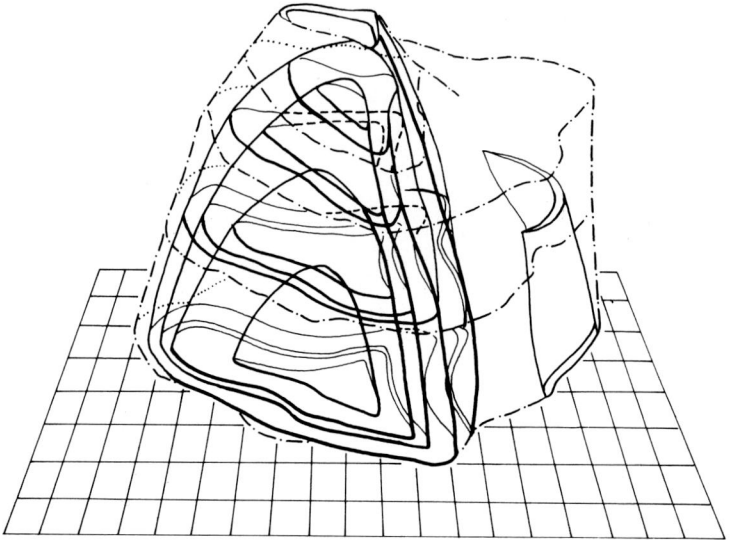

Figure 307. Diagrammatic reconstruction of an acute culmination dome in Lewisian banded hornblende-biotite gneiss, Glenelg, Northern Highlands, Scotland; the model is based on five serial sections (dot-dash lines) cut through a hand sample. Individual lithologic members are picked out by heavy solid lines; the model stands on a 1 sq. cm. grid (After Ramsay, 1962C, Fig. 6).

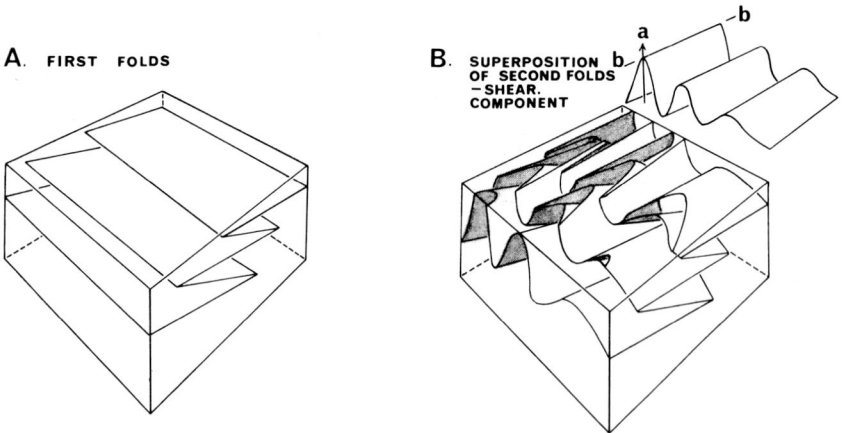

A. FIRST FOLDS

B. SUPERPOSITION OF SECOND FOLDS —SHEAR. COMPONENT

Figure 308. Block diagrams to show the crescentic outcrop patterns that result from superposed slip folding on recumbent first folds (Ramsay's Type 2). A: The first folds. B: Geometry resulting from superposed slip folds; the amount of second slip is indicated in the inset (After Ramsay, 1962C, Figs. 8A and B).

A FIRST FOLD

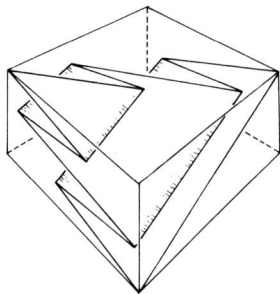

B SUPERPOSITION OF
SECOND FOLD -
SHEAR COMPONENT

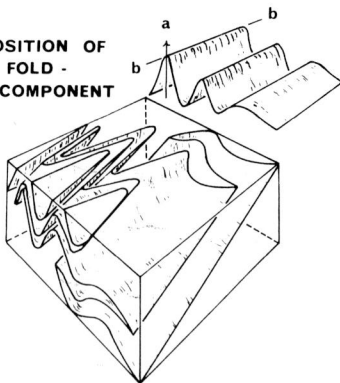

Figure 309. Block diagrams to show the outcrop patterns that result from super-posed slip folding of plunging recumbent first folds (Ramsay's Type 3). A: The first folds. B: Geometry resulting from the superposed slip folds; the amount of second slip is indicated in the inset (After Ramsay, 1962C, Figs. 11A and B).

It seems likely that additional factors may also be involved in the generation of some "eyed" or circular fold patterns; for example, Nichol-son (1963) found examples in the Sandvand Grey Marble, northern Norway, that are more irregular in form and areal distribution, and he invoked localized diapiric action (cf. comparable to miniature salt tectonics) superposed on a first generation of folds to explain their geometry.

When extensive continuous outcrops are exposed, an analysis of the symmetry surfaces drawn with respect to the folded oldest S-structure (e.g., bedding) can sometimes lead to a more direct understanding of the fold geometry (Carey, 1962A).

Relatively simple geometry results when the two superposed sets of slip folds have a common B-kinematic axis, but inclined aB-slip planes. In Figure 310A the original unfolded S_0-surfaces are shown. Figure 310B is the profile of first-phase slip folds (F_1), in which both the geometric and kinematic B-axes are normal to the page; the profile of an individual S_0-surface is shown in Figure 310C. Figure 310D shows the profile which would result from second-phase slip folds (F_2) affecting horizontal S_0-surfaces, such as those in Figure 310A. If slip with F_2-geometry is superposed on the F_1-folds (Fig. 310B), the complex pattern of Figure 310E results. Carey (1962A, p. 99) demonstrated that different geometric patterns develop if the kinematic patterns are reversed in time; thus, Figure 310F is the profile that results from F_2 (Fig. 310D) occurring before F_1 (Fig. 310C).

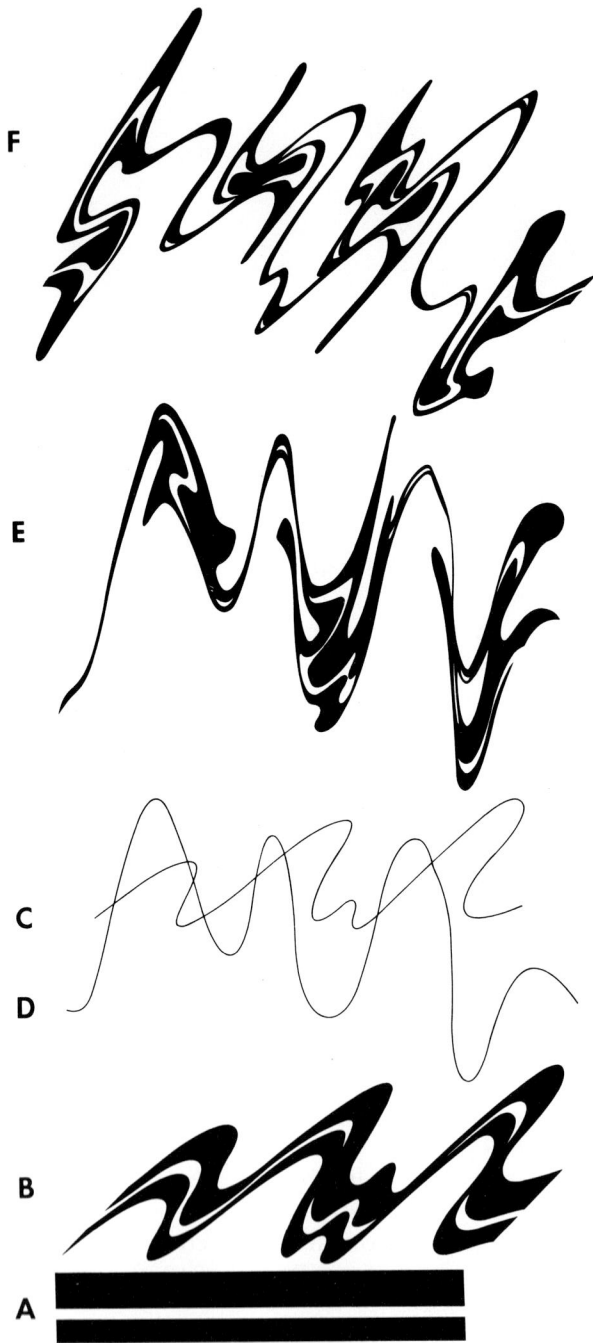

Figure 310. The geometry of superposed slip folds with common B-kinematic axes, but inclined aB-slip planes. A: Original unfolded S_0. B: Profile of F_1-slip folds. C: Profile of an individual F_1-fold. D: Profile that would result from F_2-folding of undeformed S_0. E: Profile resulting from F_2-slip folds superposed on F_1-folds. F: Profile that would result if F_1 geometry were superposed on an earlier F_2-fold system (After Carey, 1962A, Fig. 5).

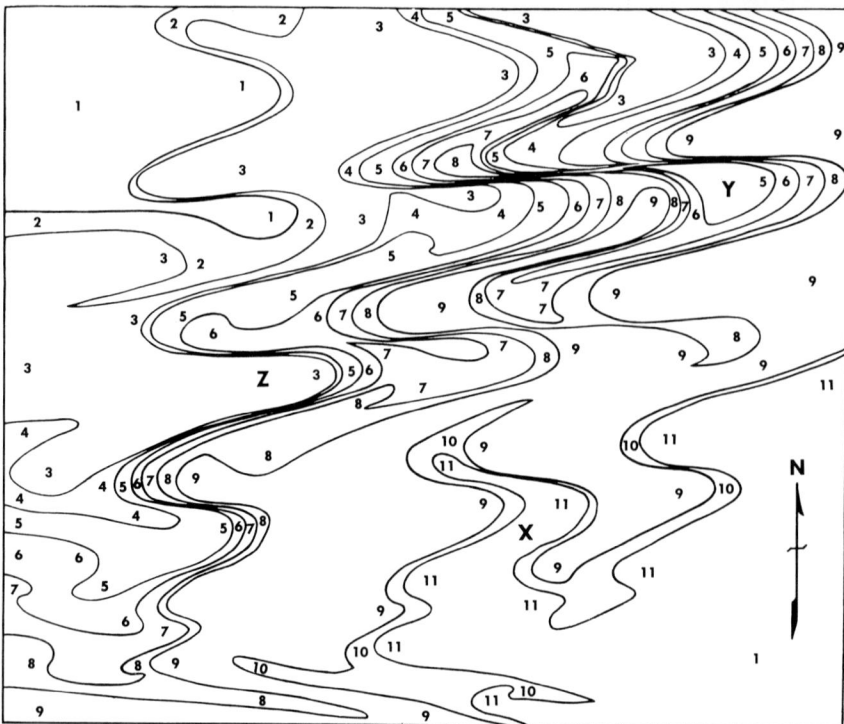

Figure 311 (top). Fold profile produced by two phases of slip folding: see text for explanation (After Carey, 1962A, Fig. 10). Figure 312 (bottom). Sequence of rocks derived from Fig. 311 (After Carey, 1962A, Fig. 11).

Carey (1962A) outlined a simple method for unravelling complex profiles like those of Figures 310E and 310F. Suppose Figure 311 is a profile with respect to two phases of folding (i.e., similar to those in Fig. 310). The successive lithologic units can now be numbered; the sequence of numbers in Figure 312 is consistent but implies relative age. Individual lithologic units may be sheared out, as at Z in Figure 312 where units 3 and 8 are almost in contact. With the aid of these numbered lithic units, symmetry axes can be traced across the map-area (Fig. 313); the symmetry referred to here is a crude numerical repetition which, for example, is shown at the point X which lies at the center of the sequence 9-10-11-10-9, and by Y that lies at the center of the sequence 9-8-7-6-5-6-7-8-9. By connecting up such symmetric points, two sets of traces can be drawn on the profile (Fig. 313), namely (a) a subparallel set of more or less straight lines and (b) a subparallel set of strongly flexed lines.

If it is assumed that both phases of deformation were dominated by slip folding, the more or less straight lines of symmetry must have been produced during the second or F_2-phase. Hence, the curved lines represent F_1-axial-plane traces refolded by F_2; if a composite tracing of these curved lines is made (Fig. 314C), the mean trace can be approximated (Fig. 314D). To remove the effects of F_2 the mean trace (Fig. 314D) is converted to a straight line by moving the structures parallel to the trace b of Figure 314. Figure 315B shows this operation completed for the original profile (Figs. 311 and 315A); unfortunately, this operation involves an arbitrary set of nonaffine slip movements because, although the line d (Fig. 314) must be made straight, it could be made perpendicular to b or any other arbitrarily-chosen angle to b. Without external evidence, which may be very difficult to find, the selection of a trace for the plane of symmetry of F_1 must remain arbitrary.

The F_1-folding can now be removed from the profile in Figure 315B. By making a composite tracing of the lithologic bands in Figure 315B, the mean fold profile is determined (Fig. 315C and 315D). The mean is transformed into a straight trace by slip movements parallel to the axial plane of F_1. As with removing the effects of F_2, the orientation of the straightened trace is purely arbitrary, unless there is some outside evidence for the original orientation. Hence, Figure 315F is only one possible interpretation of the pre-F_1 geometry.

In actual field examples the geometry is commonly more complicated to unravel, because it is unlikely that any random outcrop surface will contain the a-kinematic axis for both the F_1 and F_2 folds. In the examples described above, the a-kinematic axes are always in the plane of the paper. If a is oblique to the exposed outcrop surface for either or both phases of folding, an additional element of distortion is introduced.

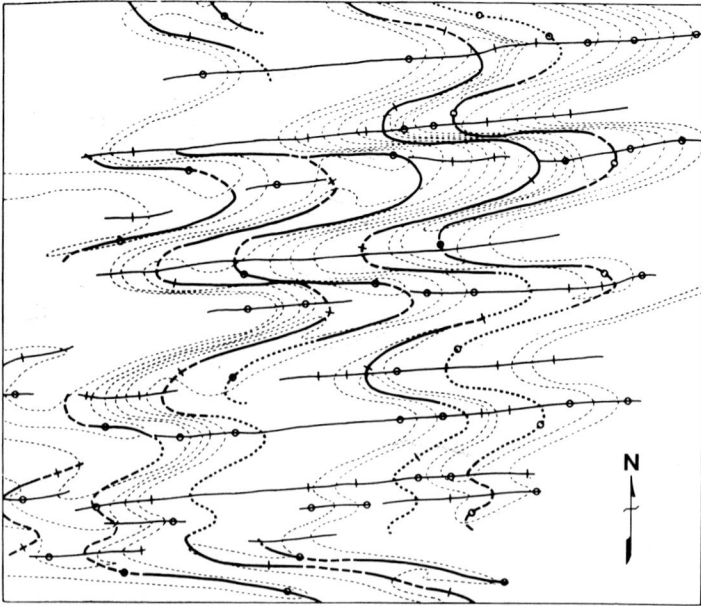

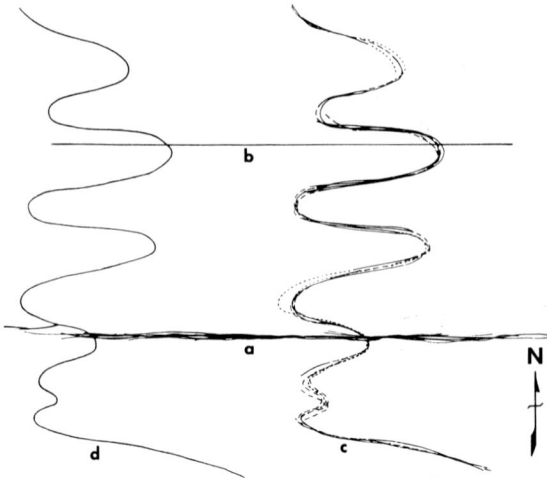

Figure 313 (top). Symmetry axes (fold axes) derived from Figure 312. The original bedding (light dotted lines) is receding and the involuted axial surfaces of the first folds, which act as *S*-surfaces for defining the second folds, become more prominent (After Carey, 1962A, Fig. 12).

Figure 314 (bottom). Determination of the mean directrix and mean fold profile of superposed folds recorded in Figure 313 (After Carey, 1962A, Fig. 13).

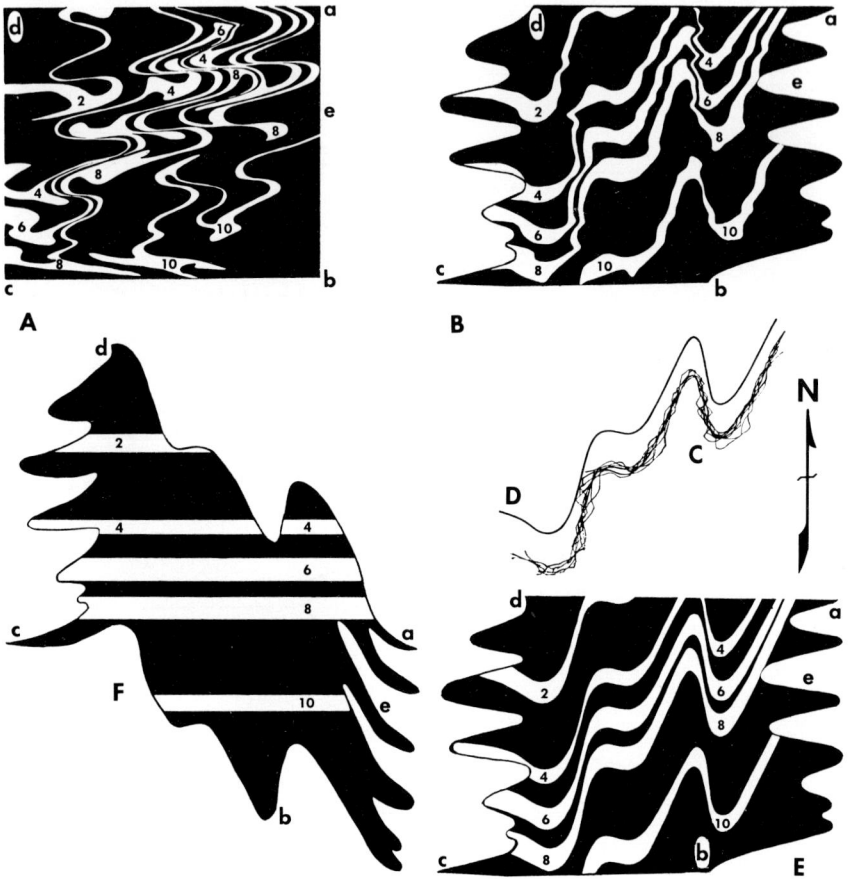

Figure 315. Resolution of superposed slip folds to the original beds. A: Final profile resulting from F_2 superposed on F_1-slip folds. B: Initial removal of F_2-folding. C: Successive bed profiles superposed from B. D: Mean fold profile from C. E: Profiles from B smoothed by use of the mean profile in D. F: Removal of D from profile in E to reproduce the original strata. a, b, c, and d mark the positions assumed by the original corners of A (After Carey, 1962A, Fig. 14).

These constructions suggested by Carey (1962A) provide considerable insight into the geometry of superposed slip folds, but their practical application is limited because of the required restraints on the geometrical relationships between F_1 and F_2. As shown above, many other geometric relationships commonly occur.

Ramsay (1962A, 1962C) suggested that "flattening" is an important factor in slip folding (see Chapter 6); if flattening effects are superposed on the F_2-slip folds, the geometry of both the F_1 and the F_2 structures will be distorted, and thus be more complex.

CONCLUDING REMARKS

As in the case of folding limited to a single phase, in superposed folds each period of deformation can incorporate elements of both flexural-slip and slip folding; interplay of the two types naturally complicates the geometric pattern developed.

After an area has been penetratively folded, it is possible for the same area to develop new folds at a subsequent period with an amplitude several times larger than the first folds. In terms of the regional gross stratigraphic units, such second folds can be cylindroidal, although, with respect to individual S-surfaces folded during the first phase, the superposed folds would tend to develop conical structures. Hence, when starting to investigate a new area, it is important to determine the size of the domains within which structures are cylindroidal. When structures associated with two periods of folding are both visible in exposures of the same size, it is a general rule that, where the largest folds are cylindroidal, the F_2-folds are always smaller than the F_1-folds, because they (F_2) can only form on the planar limbs of the F_1-folds (Weiss, 1959B, p. 98). This rule applies to both superposed flexural-slip and slip folds. In studies of progressive metamorphism and deformation, successive flexures have smaller amplitudes (see Chapter 11); however, very early structures are often small in size, but are obliterated by later, thoroughly penetrative deformations; folding episodes subsequent to such penetrative deformations commonly tend to be of smaller and smaller sizes.

Chapter 11

Examples of
Superposed Folding

In Chapter 10, idealized examples of superposed folds were described, and it was indicated that in nature the linear and planar elements tend to be somewhat dispersed; that is, natural rocks do not behave with geometric precision. Such dispersion means that a large volume of data must be accumulated from each area before the geometry can be determined accurately. Sir Edward Bailey (Bailey and Lawrie, 1960, p. 17) commented with justification that it is quite impossible for anyone to arrive at any true concept of the structure of a complex area on the basis of brief field visits. Unfortunately, descriptions of relatively few actual areas have been based on data adequate for unequivocal geometric analysis of the structure. However, the literature of superposed folding has grown to considerable proportions and in this chapter some representative examples are described.

$B_1 \wedge B_2$-FOLDS

In Chapter 10 $B_1 \wedge B_2$-folds were subdivided into three categories as follows:
1. B_1 and B_2 oblique but penecontemporaneously developed.
2. B_1 and B_2 superposed but developed during the same general orogenic phase.
3. B_1 and B_2 superposed and developed during long separated cycles of orogeny.

Each category was subdivided according to whether (a) B_1 and B_2 are parallel or oblique to each other, and (b) folds about B_1 and B_2 are dominantly of flexural-slip, slip, or flexural type, or a combination of these types.

Although apparently a rational and practicable scheme of classification, it becomes impossible to assign all of the described examples to one of these categories because the field relationships have not been described in sufficient detail. This is not surprising in view of the brief time during which the methods of geometric analysis have been available. Also, detailed geometric analyses require time-consuming and exacting study that has not been practicable in some investigations. However, several excellent geometric analyses have been completed recently, and these are referred to below.

In consequence, the classification is not followed in the sequel and all examples of $B_1 \wedge B_2$-folds are considered together. For convenience, the special case in which B_1 and B_2 are parallel is considered first, and then the more common situation in which B_1 and B_2 are oblique to each other.

B_1 AND B_2 APPROXIMATELY PARALLEL

When B_1 and B_2 are approximately parallel simple geometry tends to result. The simplest case involves long-continued cylindroidal flexural-slip folding about the same fold axis, so that the initial axial planes become folded about the second B_2-axis, and thus refolded about B_1. Commonly, under the appropriate metamorphic conditions, pelitic and semipelitic rocks undergo componental movements so readily that early-formed planar structures (S_1) are lost and overprinted by completely new, but homoaxial, S-structures. Examples abound in the metasedimentary rocks of mid-Strathspey, Inverness-shire, Scotland, and of County Antrim, NE Ireland (Whitten, 1959A). In the Dalradian Loch Tay Limestone just north of Torr Head, Antrim, the axial planes of isoclinal folds have been almost isoclinally refolded about WNW-trending axes that are essentially parallel to the axes of the initial isoclines. The Pitlochry Schists just south of Torr Head include coarse-grained sheared grits that are thrown into isoclinal folds with strongly-developed axial-plane schistosity. Some of the quartz segregation veins in these rocks are refolded. Such structures all suggest long continued homoaxial deformation.

Small examples of refolded isoclines are common; Figure 316 shows examples from the Spruce Pine district, North Carolina, U.S.A. Sutton and Watson (1959, pp. 238-39) referred to refolding about the same axis in Moine rocks of Glenelg, western Scottish Highlands, where the closures of many isoclinal folds are rolled over (Fig. 317). Figure 318 shows a similar common type of structure from the Central Scottish Highlands.

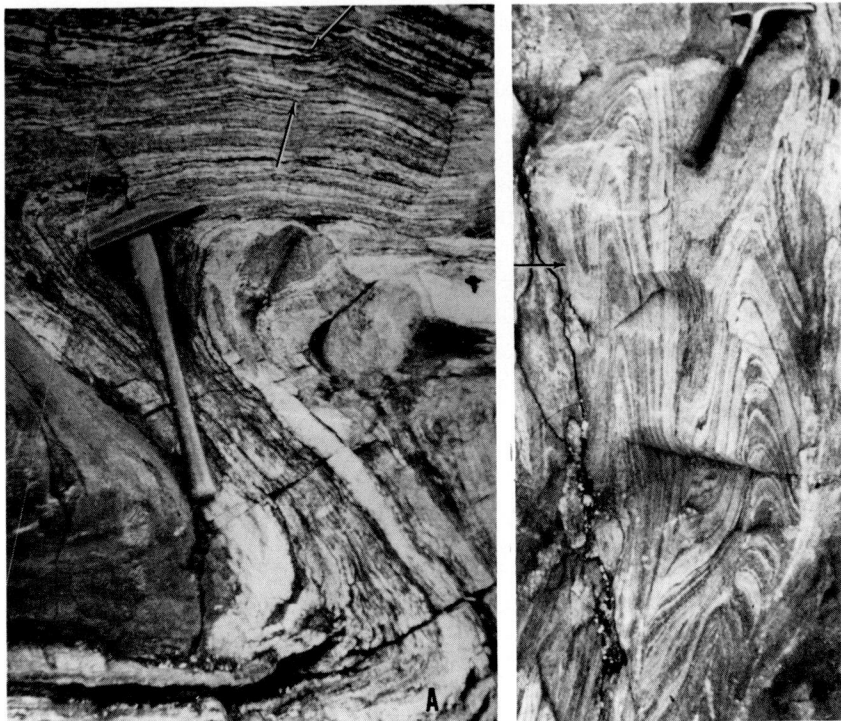

Figure 316. Refolded isoclinal folds in Precambrian metasedimentary rocks, roadcuts in Spruce Pine district, North Carolina, U. S. A. Arrows indicate closures of small folds.

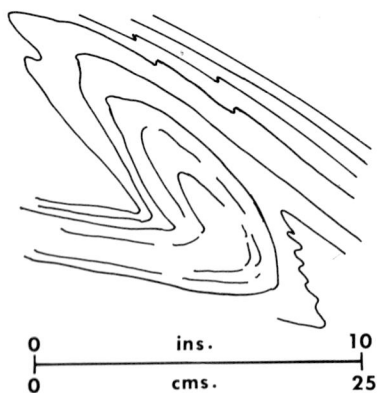

Figure 317. Isoclinal fold in Moine rocks refolded about original *B*-fold axis, Glenelg, western Northern Highlands, Scotland (After Sutton and Watson, 1959, Fig. 2E).

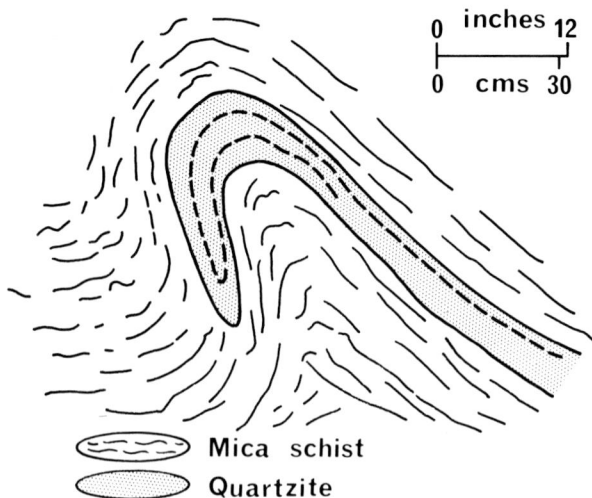

Figure 318. Refolded isoclinal fold, Creag Varr, Schichallion area, Central Highland, Scotland (After Rast, 1958A, Fig. 3).

Much larger examples have also been described. A good example is afforded by the Ben Lui district, SW Highlands of Scotland, which is well known from the mapping of Bailey (1922), Gilmour and McIntyre (1954), and Cummins and Shackleton (1955). According to Cummins and Shackleton the area was affected by at least two superposed phases of deformation; in many exposures both phases can be distinguished easily, although both are approximately coaxial. The earlier folds are isoclinal and some have an axial-plane foliation and others a schistosity that wraps round the fold closures; the latter relationship suggests that a still earlier generation of folds exists in this area (cf. Roberts and Treagus, 1964). The second folds are characteristically slip folds with crenulation foliation or closely-spaced parallel fractures parallel to their axial planes. The linear and planar structures mapped by Cummins and Shackleton are shown in Figures 319 and 320; Figure 321 is the profile based on this information and the sedimentary "way-up" data.

Numerous cases of renewed folding along old axes have been described in the literature. Craddock (1957) described a well-documented example from the Kinderhook Quadrangle, New York. Here folding commenced in pre-Middle Trentonian time, while stratigraphic evidence shows that the main phase was post-Trentonian and homoaxial with the earlier folding.

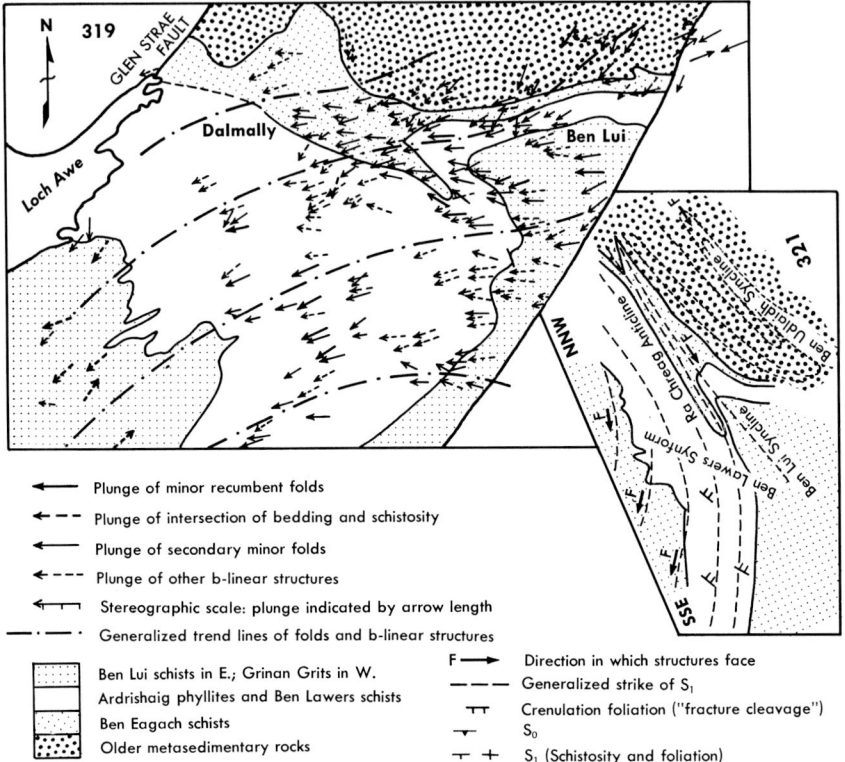

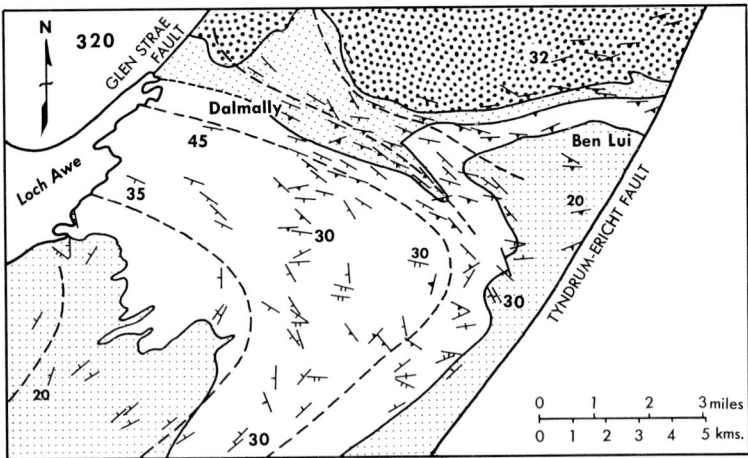

Figures 319-321. Structural geometry of the Ben Lui area, Central Highlands, Scotland. Figure 319: Map of minor fold axes and b-linear structures. Figure 320: Map of the planar structures. Figure 321 (*Inset*): Profile of the Ben Lui block based on a mean plunge of 15° along the elements shown in Figure 319; extension of the line forming the top of the figure across Figure 319 would mark the trace of the projection plane; vertical and horizontal scales are the same (After Cummins and Shackleton, 1955, Figs. 1, 2, and 4).

B_1 AND B_2 INCLINED

More complex geometry arises when B_1 and B_2 are not parallel. A spate of papers has recorded such structures from many different areas in the past ten years. Much of the modern pioneer work has been in the Caledonides, and this is discussed first; subsequently, examples from other parts of the world are considered.

SCOTLAND

Before describing the complex structures of the Scottish Precambrian and Lower Paleozoic rocks, it is necessary to review the broad interrelationships in space and time of these rocks (Fig. 322).

For many years the Moine Thrust dominated concepts concerning the kinematic development of the Moine Series. McIntyre (1954) reviewed the development of ideas about the thrust zone, and suggested that

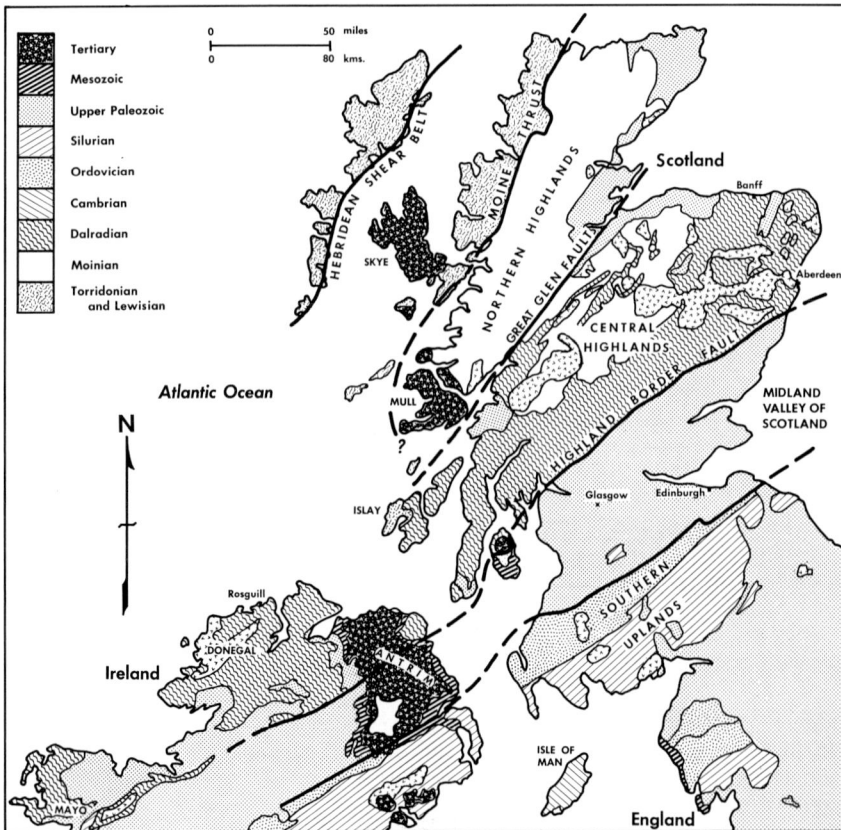

Figure 322. Structural sketch map of Scotland and northern Ireland; post-tectonic igneous rocks shown by v-pattern.

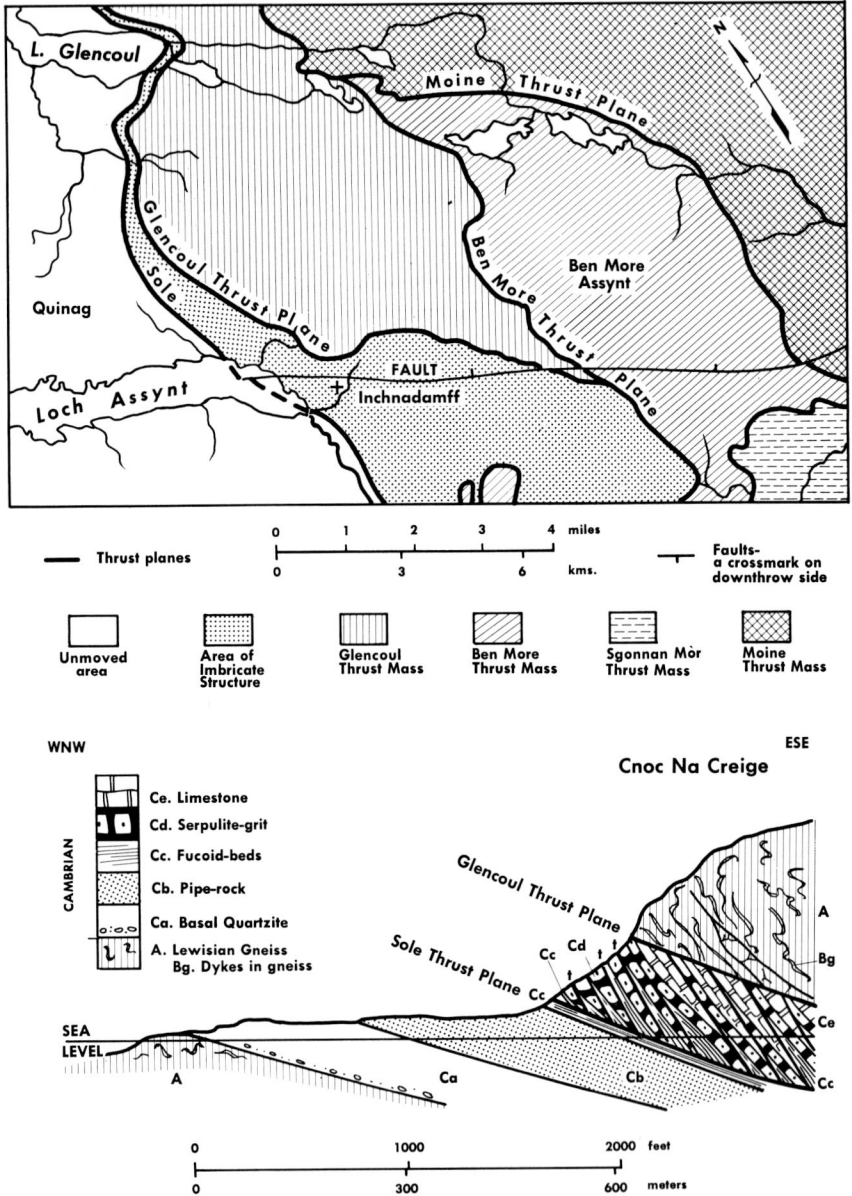

L. Glencoul

Moine Thrust Plane

N

Glencoul Thrust Plane

Sole

Quinag

Ben More Assynt

Ben More Thrust Plane

FAULT

Loch Assynt

Inchnadamff

| | 0 | 1 | 2 | 3 | 4 miles | |
Thrust planes

0   3   6 kms.

Faults-
a crossmark on
downthrow side

Unmoved area | Area of Imbricate Structure | Glencoul Thrust Mass | Ben More Thrust Mass | Sgonnan Mòr Thrust Mass | Moine Thrust Mass

WNW

ESE

Cnoc Na Creige

Ce. Limestone
Cd. Serpulite-grit
Cc. Fucoid-beds
Cb. Pipe-rock
Ca. Basal Quartzite
A. Lewisian Gneiss
Bg. Dykes in gneiss

CAMBRIAN

Glencoul Thrust Plane

Sole Thrust Plane

SEA LEVEL

A   Ca   Cb   Cc

0   1000   2000 feet

0   300   600 meters

Figure 323. Sketches of part of the Moine Thrust Zone, Assynt area, NW Highlands, Scotland. A: Map of the area around Inchnadamff. B: Sketch section across the map to show the imbricate thrust sheets (After Peach and Horne, 1914, Figs. 3 and 4).

its tectonic significance had been overemphasized. The thrust has a NNE-SSW outcrop trace and indicates thrusting from the southeast towards the northwest. This suggested that the major folding during the Caledonian orogeny in Scotland was produced by southeast-northwest compression and a northwesterly direction of movement. Recent mapping suggests that such an assumption is unwarranted because the thrusting reflects a younger event not directly related to formation of the tectonites — in fact, superposed on the main tectonites.

West of the Moine Thrust Zone lie Lewisian gneisses and schists. The Lewisian rocks were divided into the older Scourian and younger Laxfordian by Sutton and Watson (1951). Radiometric age dates suggest the Scourian metamorphism terminated some 2,460 million years ago, whereas Laxfordian metamorphism affected most the Lewisian rocks about 1,600 million years ago (Giletti, *et al.*, 1961). The Lewisian rocks characteristically comprise hornblendic and feldspathic gneisses, meta-sedimentary schists (including metamorphosed limestones), and meta-morphosed basic and ultrabasic igneous rocks.

The Precambrian Torridonian Sandstone rests with striking uncon-formity on the Lewisian rocks west of the Moine Thrust Zone. Cambrian and Ordovician sedimentary rocks (mainly orthoquartzites and dolomites) lie with equally marked unconformity on the Torridonian (Fig. 323).

Thrust over these Lewisian, Torridonian, Cambrian, and Ordovician rocks is the Moine Series. There has been considerable debate as to whether the Moine tectonites are metamorphosed time and/or lithological equivalents of the Torridonian rocks, or whether the Torridonian is younger than the Moine Series. The assembled evidence suggests that parts of the Torridonian and Moine sequences are time equivalents (cf. Sutton, in Johnson and Stewart, 1963, p. 254).

The Moine Series comprises a monotonous sequence of schists, granulites, and gneisses. Many attempts have been made to reconstruct the Moine succession, but over many hundred square kilometers, these tectonites pose an intractable problem. The problems stem from the original uniformity of the lithology; there are few reliable marker hori-zons, and the ubiquitous repeated folding has been accompanied by obliteration of most primary sedimentary structures. Sutton (1961) thought the Moine rocks represent an original succession of shales and sandstones that may have been as much as 6 to 7 kilometers thick; according to Long and Lambert (in Johnson and Stewart, 1963) they were deposited about 900 million years ago. Within such a succession it would be surprising if many of the original lithological sedimentation units were not lenticular and thus of rather local distribution.

Most of the recent progress in unravelling the Moine sequence has been made along (a) its lower margin in the west and northwest, where interfolded with the older Lewisian gneisses, and (b) its upper boundary in the south and southeast, where interfolded with the younger lithologically-varied Dalradian sequence (cf. Sutton, 1962). This work is described in detail below. It is unfortunate that along much of the Moine-Dalradian junction the Moine rocks are particularly uniform quartzitic schists.

In the Glenelg-Knoydart and Loch Morar districts of the western Highlands, however, a stratigraphic succession is apparent (Fig. 322). Work began there in 1897 (Clough, 1898) and the Moine stratigraphy has been debated intermittently but vigorously ever since. The history of this research was reviewed by Ramsay and Spring (1963); there have been many vicissitudes, and the magnitude of the stratigraphic correlation and interpretation problems in such a complexly-folded area makes it seem likely that their progress report will require some modification as detailed work proceeds in neighboring districts. However, Figures 324 and 325 show their postulated succession; in these maps the several Lewisian inliers are interpreted as anticlinal isoclines. More complete details are given in the sequel.

Isolated radiometric age determinations for complex polymetamorphic sequences must be interpreted with extreme caution (cf. McIntyre, 1963A), but Moinian pegmatites in Knoydart and Morar apparently formed more than 740 million years ago (Giletti, *et al.,* 1961), possibly during the first metamorphism of the Moine rocks. Many lines of evidence prove the Moine Series experienced at least two other major metamorphic events: the youngest occurred at least 420-430 million years ago. In the Central Highlands the Moine and Dalradian metasedimentary rocks are interfolded. A single major Lower or Middle Ordovician metamorphic event in the Dalradian schists had been suggested by the radiometric age dates (*ca.* 475 ± 15 million years ago; Giletti, *et al.,* 1961); however, additional major events have been deduced on the basis of geological mapping and the most recent potassium-argon dates (Fig. 435) suggest a major 420-430 million year event superposed on an older 490 million year metamorphism (e.g. Brown, *et al.,* 1965). Fossiliferous Lower Paleozoic sedimentary rocks appear to be interfolded with Dalradian rocks in the southeastern Central Highlands.

The sequence of metamorphic and tectonic events has produced some very complex structural patterns. The tectonites east of the Moine Thrust record several superposed deformations (described below). Although the composite thrust zone developed after most of these deformations, the thrusting produced localized new structures that were superposed on the rocks immediately above and below the thrust planes.

During the nineteenth century the "Riddle of the Highlands" concerned the essentially-unmetamorphosed Cambro-Ordovician rocks sandwiched between tectonites in the thrust zone area (Fig. 322). Work has continued steadily since that time, and structural geologists working in Scotland today start with the advantage of a wealth of accumulated knowledge and experience. Sutton (1960B, p. 371) pointed out that:

> A century of geological investigation by such men as Murchison, Geike, Peach, Horne, Clough, Greenly, Bailey, E. M. Anderson, Richey, Read, MacGregor and Kennedy, all at some time officers of the Geological Survey, has established a great body of fact recorded in many maps and memoirs. Large-scale topographic maps are available and the ground is often well exposed and contains a great variety of structure. It is possible to embark on investigations of a kind which might not be feasible in less well-known terrain.

Detailed quantitative techniques for determining structural geometry were introduced to Scottish geology by McIntyre, Ramsay, Sutton, Watson, and Weiss. Reference has already been made to the classic geometrical analysis of the Grantown-Tomintoul area in the Central Highlands by McIntyre (1951B). His work in this area suggested that the structures are homoaxial, although future research may demonstrate that the observed folds are superposed on, and have obliterated, most signs of earlier folding episodes. This study was a pioneer attempt at quantitative analysis of Scottish tectonites. The work of McIntyre in the Grantown-Tomintoul area, at Beinn Dronaig (1952), and in other areas established the reality of cylindroidal folds and the manner in which they can be studied. However, as more knowledge accumulated, it became clear that homoaxial conditions do not tend to extend over very extensive areas; this was clearly recognized by, among others, McIntyre (in Weiss, *et al.,* 1955; Weiss and McIntyre, 1959).

The evolution of concepts has similarities with the development of ideas about the Lewisian gneiss and schist inliers within the main outcrop of the Moine Schists of the Northern Highlands. Ideas about the supposed-inliers have evolved steadily with the growth of structural concepts. Sutton and Watson (1953) effectively established that the Lewisian inliers mapped by officers of the British Geological Survey (e.g., Flett, 1906; Peach, Horne, Hinxman, *et al.,* 1913; Peach and Horne, 1930) at Scardroy, Ross-shire, are actually unusual lithological varieties of Moine metasedimentary rocks. At that time, the evidence, which was largely based on fold geometry, appeared incontrovertible, but, as more detailed and extensive studies developed, the status of the supposed-Lewisian rocks had to be reappraised. In 1959, Sutton (in Clifford, 1960, p. 387) commented that he: "now agreed with Sir Edward Bailey and Dr.

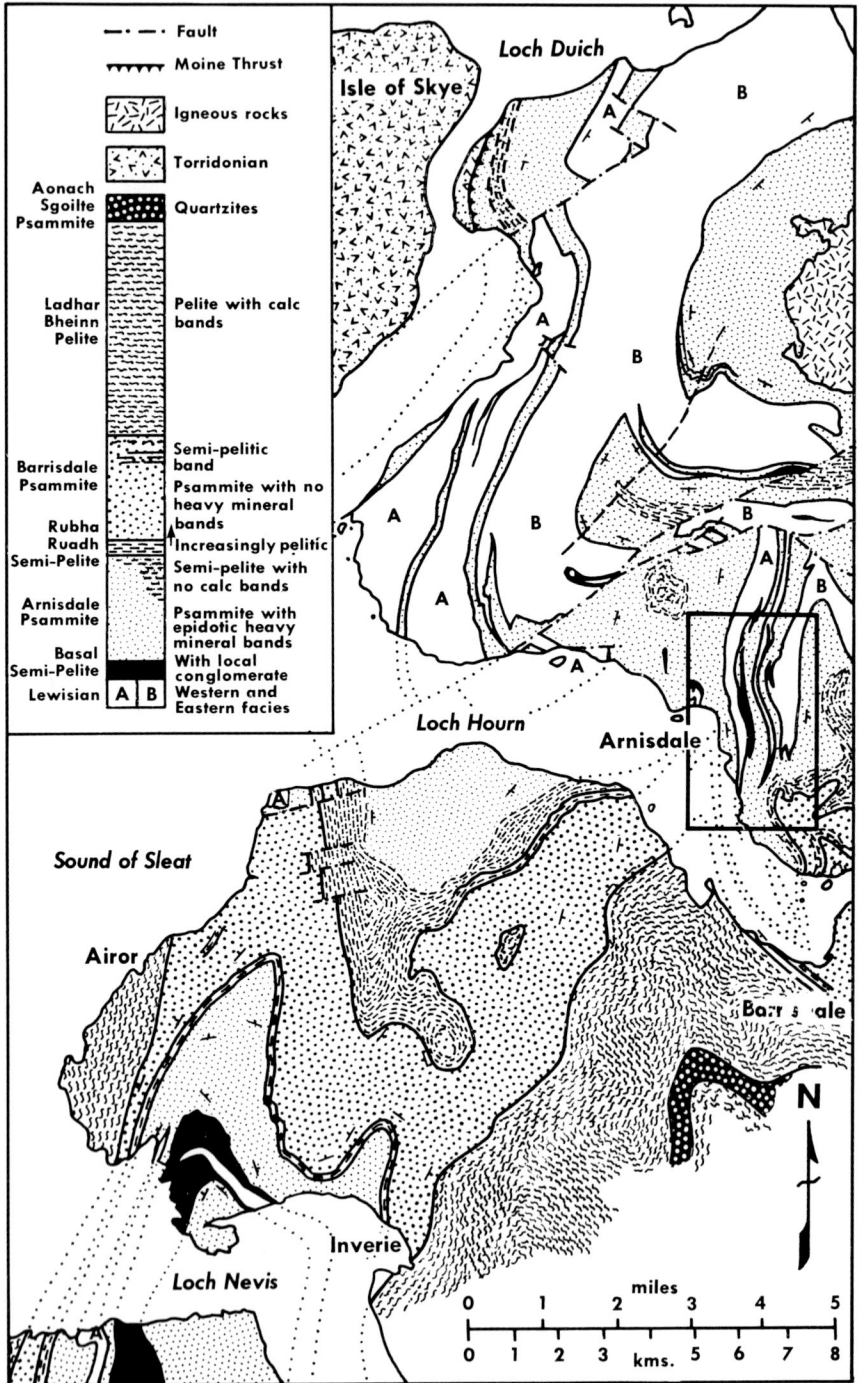

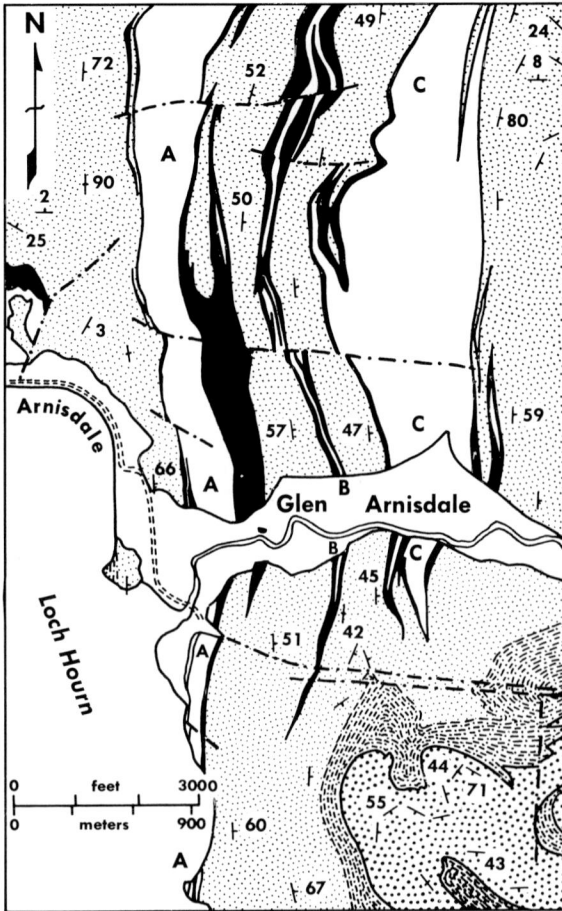

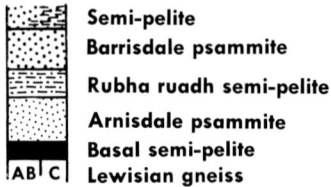

Figure 324 (facing page). Stratigraphy and structure of the Glenelg and Knoydart areas, western Northern Highlands, Scotland. Glenelg area is between Lochs Duich and Hourn, and Knoydart is between Lochs Hourn and Nevis. Inset at Arnisdale shown in Figure 325 (After Ramsay and Spring, 1963, Fig. 6).

Figure 325 (above). Stratigraphy and structure of the Arnisdale area, SE of Glenelg, western Northern Highlands, Scotland; enlargement of part of Figure 324 (After Ramsay and Spring, 1963, Fig. 7).

Lambert that the rocks at Scardroy were also Lewisian. It was clear that they formed a tectonic slice. . . ."

The status of the Lewisian rocks at Glenelg has changed even more radically as successive papers were published (see below). Because the techniques utilized in the study of tectonites have been developing rapidly, it would be surprising if new results did not entail reappraisal of earlier models and progress reports.

A basic problem in any large structurally-complex area like the Scottish Highlands concerns extrapolation from the population of objects actually studied (individual mapped areas) to a regional object of interest. This involves extrapolation from one size of objects to another size. Analysis of the Highland structures has required very detailed and careful evaluation of small-sized structures. This work emphasized the importance of cylindroidal folding, and as already mentioned, this led to the assumption that analogous attributes are possessed by objects (large volumes of rock or map areas) of regional extent. It was shown in Chapter 3 that many restraints apply to predictions about the attributes of objects of a different order of size. Thus, for example, although most of the areas actually mapped may have cylindroidal folds, objects larger than, say, 100 square kilometers might approximate conical, rather than cylindroidal, folding. Although it is valid to draw profiles across monoclinic fold systems, profile construction is fraught with difficulties when monoclinic symmetry is not present (cf. Ramsay, in Johnson and Stewart, 1963).

Mining experience has shown that caution is needed in applying the maxim that the smaller folds indicate the plunge of the large-sized features. Campbell (1958) cited several cases where the obvious uniform fold pattern mapped in small-sized areas (population of objects) is dissimilar to the regional fold pattern that emerged from study of objects of much larger size. That is, two generations of folds are present, but different sets of measurable attributes are observed in objects of different size. Such possibilities counsel considerable caution when regional profiles are based on small-sized structures observed in the field. It is dangerous to assume with Ramsay (1962C) that, because superposed slip folds seen in a 20 cm. Scottish exposure seem to have a superficial similarity with structures 7.7×10^4 times their size in Vermont, U.S.A., that, for all attributes, it is realistic to extrapolate freely between objects with such large size differences.

The Central Highlands of Scotland: The geological and regional setting of the Scottish Central Highlands is shown in Figure 326; Moine rocks have an extensive development in the northern belt and to the south Dalradian rocks occur exclusively. The Dalradian sediments accumulated during later Precambrian and Cambrian time. Knill and Rast (in

Johnson and Stewart, 1963, pp. 99 ff.) reviewed existing stratigraphic knowlege about these rocks, and Rast compiled a useful synthesis of the Dalradian structural evolution. A considerable volume of modern de- tailed work has been accomplished by a large number of geologists; this work was founded on a wealth of careful mapping by officers of the British Geological Survey dating from Clough's (in Gunn, *et al.,* 1897) recognition of superposed folding in Cowal (in the southwest) to the present day.

The **Iltay Nappe** occupies most of the Dalradian outcrop from the Banff coast in the east to Islay in the west. Bailey (1922, 1934, in Allison, 1941) considered the Dalradian rocks extending northeastwards from Ballachulish—the **Ballapel Foundation**—to be a separate structural and stratigraphic unit from the Iltay Nappe. Most of the modern structural work has been in the Iltay Nappe, and the precise relationships between this nappe and the Ballapel sequence still remain equivocal (cf. Voll, 1960, 1965).

An attempt is made below to draw upon recent studies to illustrate the techniques used in the study of superposed folding, and especially to

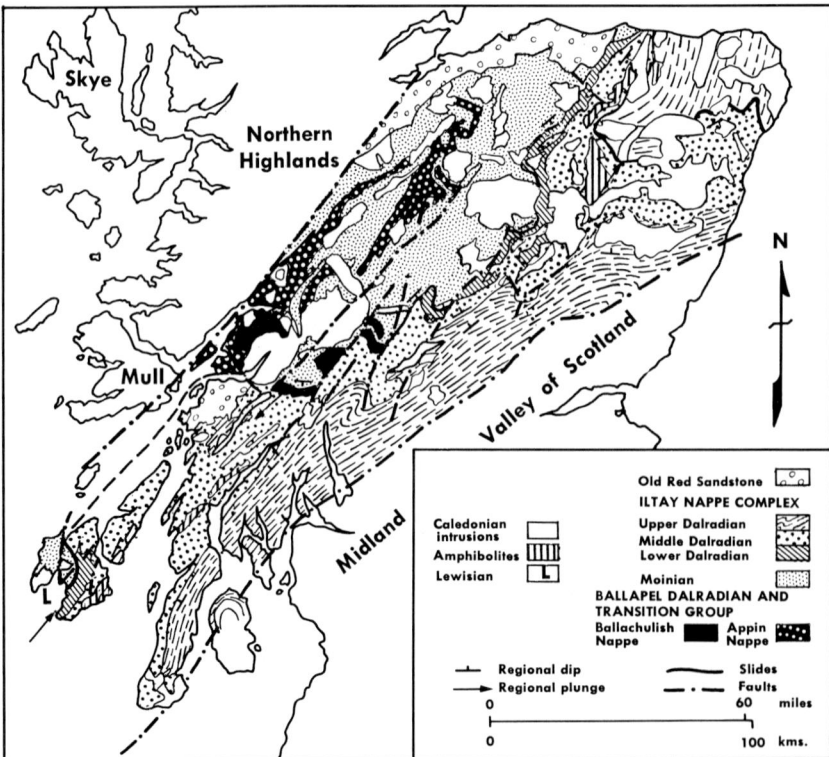

Figure 326. Geological sketch map of the Dalradian rocks of Scotland (After Rast in Johnson and Stewart, 1963, Fig. 1).

indicate the manner in which structural conceptual models can develop and have to be changed as research continues. When data become available it is legitimate to construct a conceptual model, but this requires testing in the light of new data. Commonly new data, and especially data made available by new techniques or the measurement of additional attributes, necessitate revision of earlier models.

First, attention is devoted to recent work on superposed folding within the Iltay Nappe; additional facets of the structural geology of this unit have been described in other sections of this book. No attempt is made to give a regional synthesis of the structural geology.

The area between Schichallion and Braemar has attracted considerable attention, and structural analysis has been aided by the varied lithology of the Dalradian metasedimentary rocks which were mapped in a masterly fashion by Bailey (1925) and Bailey and McCallien (1937). A

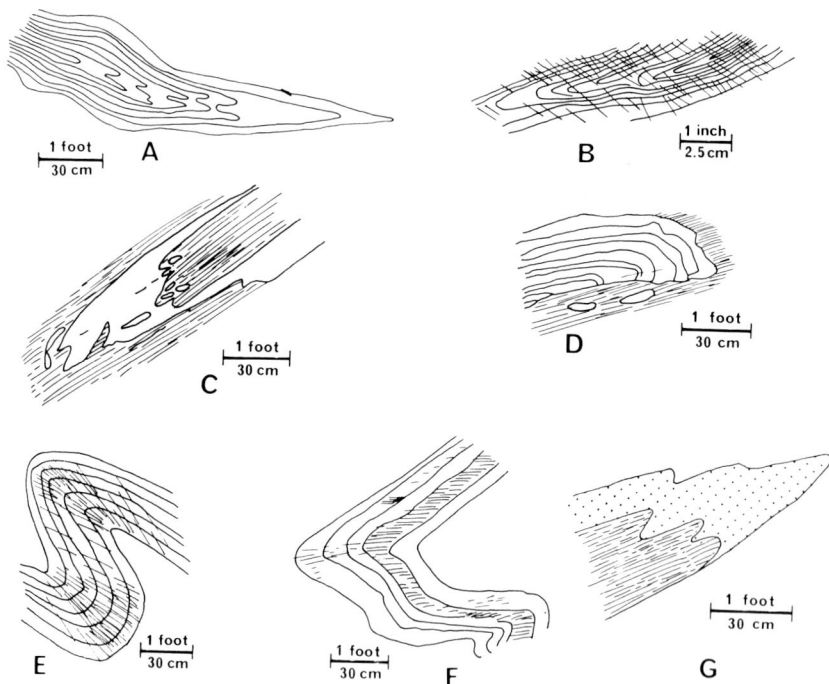

Figure 327. Examples of F_1-fold profiles ("Caledonoid" or main folds) in the Loch Tummel-Schichallion area, Central Highlands, Scotland. A: Recumbent fold in Ben Lawers Schist. B: Recumbent fold in schist horizon of Carn Mairg Quartzite refolded and foliated by F_3-movements. C: Ben Lui Schist. D: Ben Lui Schist: sheared-out lower limb. E: Ben Eagach Schist. F: Carn Mairg Quartzite. G: Recumbent isoclinal fold: psammite stippled and pelites ruled (After Sturt, 1961B, Figs. 2a-2f, and Rast, 1958A, Fig. 2a).

distinct tectonic slide-fault (the Boundary Slide) separates these rocks from the monotonous granulites of the Moine Series to the North (Fig. 326).

Within the Dalradian rocks eastwards from Schichallion, King and Rast (1956A) differentiated two discrete axial trends, namely:

(a) The "main" or "Caledonoid" NE-trending folds (F_1) that comprise recumbent folds with superposed coaxial folding and steeply inclined axial planes. More detailed work at Schichallion (Rast, 1958A), south of Loch Tummel (Sturt, 1961B), and in Glen Lyon (D. M. Ramsay, 1962) showed that tight isoclinal folds with gentle plunges characterize these folds (Fig. 327); strong foliation parallel to the axial planes is common, together with marked attenuation of the fold limbs.

(b) The so-called "cross folds" (F_2) that trend SE and which can be recognized in both the Dalradian and the Moine rocks on either side of the Boundary Slide. Styles associated with these folds are illustrated in Figure 328; particularly in pelitic members, crenulation foliation is developed parallel to the axial planes of the F_2-folds.

King and Rast (1956A) found that, although the trends of the axes vary locally, the two sets are always perpendicular in a given small area. In some areas the "cross folds" predominate, and in others the Caledonoid

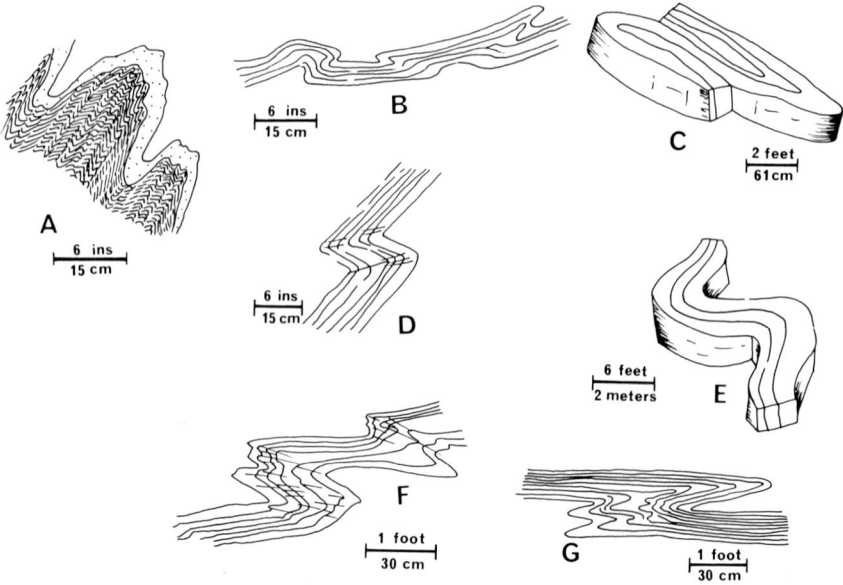

Figure 328. Examples of F_2-fold profiles (cross folds) in the Loch Tummel-Schichallion area, Central Highlands, Scotland. A: Schichallion area. B through G: Loch Tummel area: B: Conjugate folds in Farragon Group. C and E: Steeply-plunging folds in Carn Mairg Quartzite. D: Fold in schistose metadolerite. F and G: Pitlochry Schist (After Rast, 1958A, Fig. 2b, and Sturt, 1961B, Fig. 3).

folds occur exclusively. In yet other areas both sets are intimately associated as, for example, in the Ben Lawers Schists of Ben Vrackie and Glen Lyon; at Auchitbart and Creag Varr, Rast (1958A) found F_1-isoclines refolded by the "cross folds" (F_2) as shown in Figure 318. At a few localities both fold systems are represented in a single hand specimen.

These interrelationships have led to considerable discussion about the sequence of tectonic events in the Central Highlands. King and Rast (1956A, pp. 262 ff.) concluded that both fold systems developed during the "same general epoch of folding, despite the fact that their style shows that each has been produced by compressional stresses acting in directions transverse to the fold axes." The possibility of the "cross folds" (F_2) being younger than the Caledonoid folds was not entertained very seriously, although, apparently, in many localities a partial analysis of the structures was possible "by supposing the cross folding to be later, but this is only possible because of the limited recumbency shown by the major cross folds."

It is not unlikely that formation of the Boundary Slide (between the Dalradian and Moine rocks) was contemporary and coaxial with the isoclinal folding F_1, so that the trace of the slide was originally roughly NE-SW. However, the outcrop trace of the slide is now quite sinuous (Fig. 329); Rast and Platt (1957) interpreted the northerly projections as synformal cross folds, and the southerly projections as antiformal cross folds. This folding of the slide was used as strong evidence for the cross folds having attained their maximum development later than formation of the Caledonoid isoclinal folds. However, because minor Caledonoid and minor cross folds occasionally share the same schistosity, Rast and Platt believed that, at least over some interval of time, the main and cross folds evolved simultaneously, "... implying a causal relationship."

Rast (1958A) noted the wide variation in orientation of the F_2-folds; such variability would be anticipated if F_2-folds were superposed on S-structures of variable orientation following F_1-folding.[1] Probably sufficient data are still not available to yield an unequivocal answer, but Rast did not visualize the evolution of these structures in terms of F_2-structures superposed on F_1. Referring to the F_1- and F_2-folds at Schichallion (Fig. 330), he (1958A, pp. 41 and 45) concluded:

> It has been already pointed out (King & Rast, 1956A) that such a regular relationship indicates a genetic connexion. Indeed, unless it is assumed that generally contemporaneous causes gave rise to the F_1 and the F_2 folds it is difficult to understand why the

[1] F_2 superposed on S folded by F_1 could produce F_2-folds of varied trend and plunge (see Chapter 10), and in addition, if the F_1-folds were of small amplitude, adjacent F_2-folds (of dissimilar orientation) could share a common new foliation (Fig. 288).

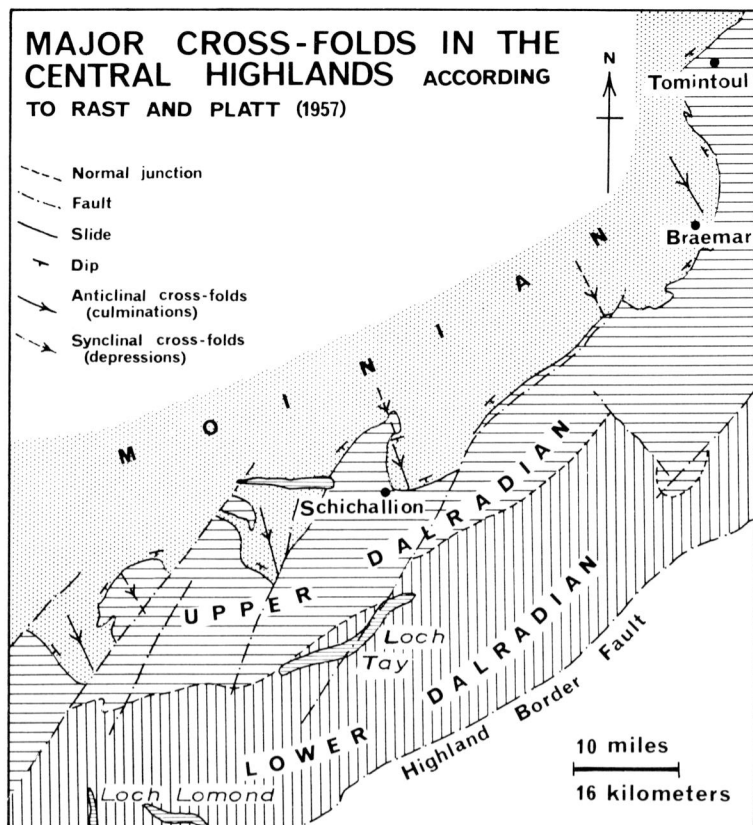

Figure 329. Caledonoid and cross folds in the Central Highlands, Scotland (Adapted from Rast and Platt, 1957, Fig. 1).

two sets of folds are practically always at right angles to each other. On a minor scale, in fact, there are folds ... which have a common megascopic schistosity with the F_1 folds, indicating the possibility of complete simultaneity. ... Where minor folds are concerned there seemed to be no doubt that the tight F_2 folds had refolded the recumbent F_1 folds. Nevertheless, in rare cases minor folds, with axial directions at right angles to each other, were found to share the same megascopic schistosity. In other words, very early "cross-folds" contemporaneous with the recumbent folds would have to be admitted.

In neighboring areas Sturt (1961 B) and Ramsay (1962 A) established that the F_2-folds are definitely younger than F_1. Sturt plotted all the linear structures on equal-area projections for each of seven subareas of the region south of Loch Tummel (Fig. 331). All the F_2-fold axes and

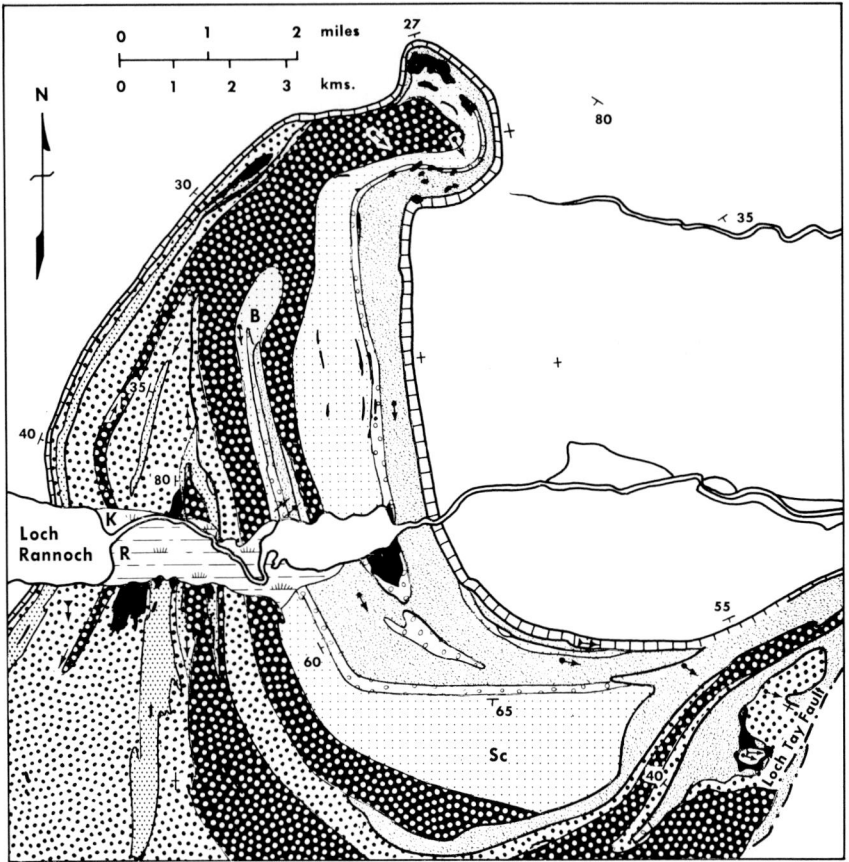

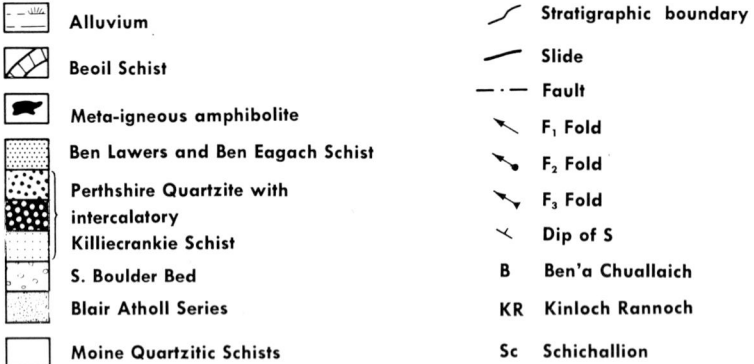

Figure 330. Map of the Schichallion complex, Central Highlands, Scotland. Belts of Blair Atholl Series in the eastern part of the area represent cores of recumbent anticlines refolded by the F_2-folds. Belts of the Ben Eagach and Ben Lawers Schists in the western area represent cores of F_1-recumbent synclines, of which the more easterly one is refolded by the F_2 and F_3 folds (After Rast, 1958B, Fig. 1).

lineations lie close to a great circle aligned NW-SE, while the F_1-linear elements are dominantly WSW-ENE. This relationship is particularly clear in subarea III (Fig. 331); the F_2-axes should lie on a great circle if the second folding is superposed on earlier F_1-folds (Chapter 10). In Glen Lyon, D. M. Ramsay (1962) also found that the plunge of F_2-folds is controlled by older F_1-structures, although the most prominent linear structures in this area (preferred orientation of minerals and pebbles, and intersecting S-structures) are associated with F_2.

Sturt (1961B, p. 153) concluded that:

> Although the second-phase structures apparently correspond to the cross-folds of King and Rast (1956A) there is no evidence that these folds were formed contemporaneously; on the contrary, all the evidence presented indicates that the second-phase structures were imposed subsequently.

Sturt (1961B, p. 156) considered that the recumbent isoclinal F_1-folds formed by gravity sliding towards the southeast when the rocks were at a very low grade of regional metamorphism. Farther to the west, Voll (1960) considered equivalent folding premetamorphic. Although most of the linear elements are parallel to the B-geometric axes, several instances were observed by Sturt (1961B, pp. 144, 156) in which a mineral lineation produced at the same time as the early folds is approximately perpendicular to B.

Rast (1958A), Sturt (1961B), and D. M. Ramsay (1962) all found ample evidence for a third phase (F_3) of large open folds with roughly east-west axial trends. Figure 332 shows some representative styles of these folds in the area south of Loch Tummel. The F_3-structures are associated with widespread development of crenulation foliations and broad warping of the older F_1- and F_2-structures. At Schichallion, Rast (1958B) thought that F_3 probably coincided with the highest temperatures of metamorphism, although Sturt (1961B, p. 153) found that F_3-movements are generally accompanied by retrogressive metamorphism at Loch Tummel. Later, Rast (in Johnson and Stewart, 1963, p. 129) revised his opinion to agree with Sturt, and wrote:

> Where they are well developed, the F_3 folds are associated with retrogression of the medium-grade metamorphic rocks. No such effects are noticed with respect to F_1 and F_2 folds, and, therefore, it is assumed that a fairly long time elapsed between the termination of the F_2 and the beginning of the F_3 movements.

The inset in Figure 331 is an interpretation of the Iltay Nappe (F_1) based on mapping by Bailey (1925) and Sturt (1961B); Shackleton (1958) established that all the minor folds face downwards in the closure zone of

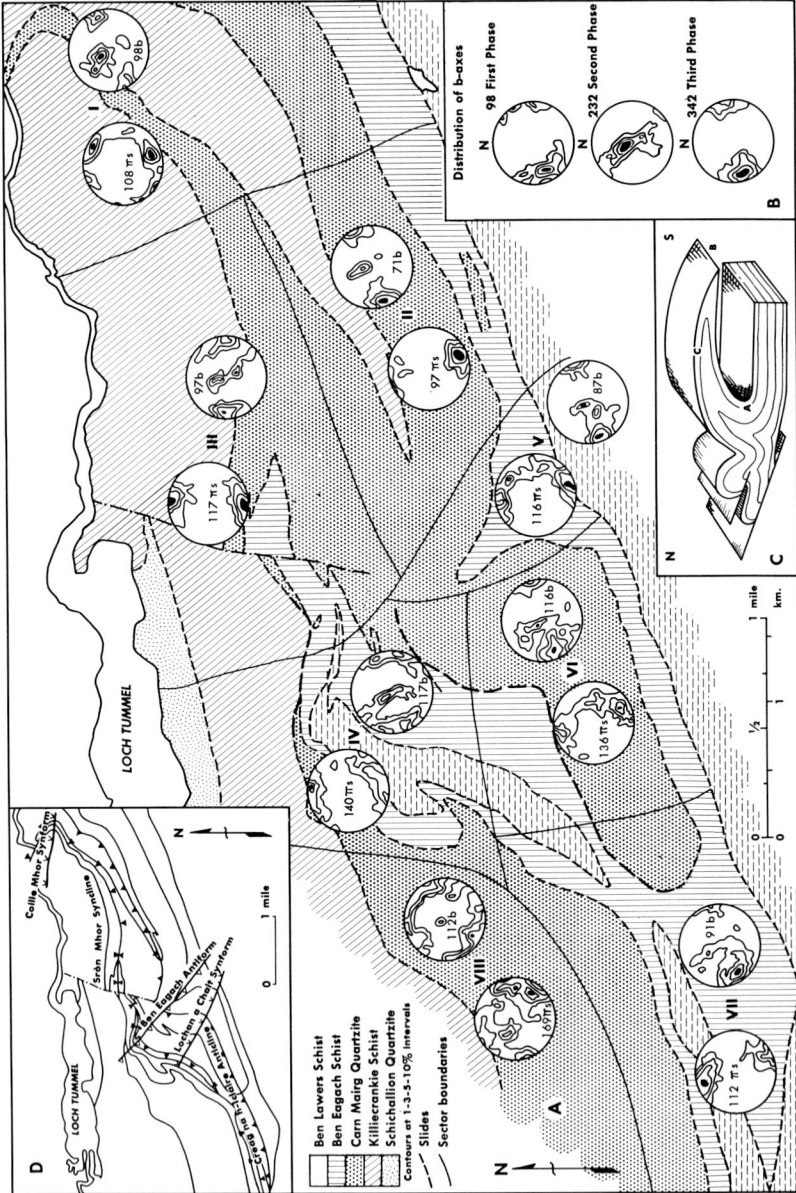

Figure 331. Structural geology of the area south of Loch Tummel, Central Highlands, Scotland. A: Geology and fabric elements in eight subareas. B: Diagrammatic sketch of the Iltay Nappe showing the Pitlochry-Kirkmichael recumbent syncline (A), the Aberfoyle anticline (B), and the flat belt (C). D: Axial traces of main fold structures in the map area (After Sturt, 1961, Plate VI and Figs. 1 and 12).

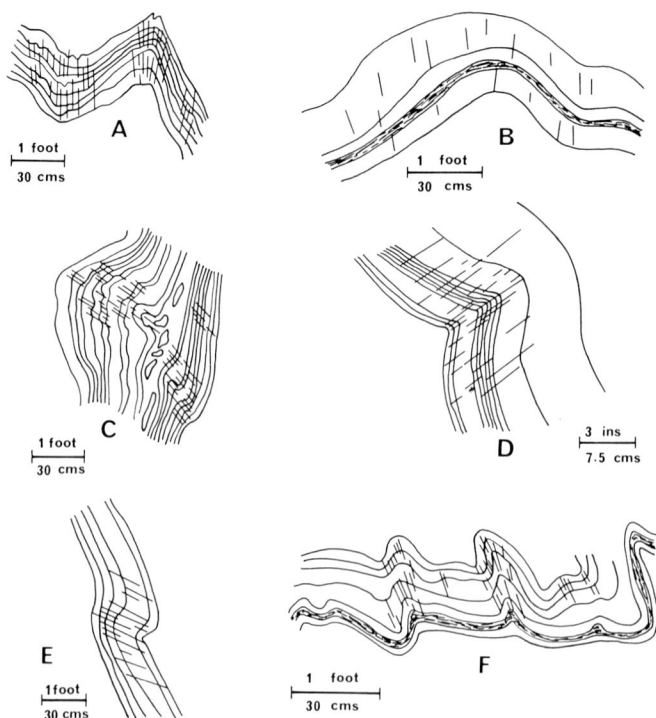

Figure 332. Examples of F_3-fold profiles from the Loch Tummel area, Central Highlands, Scotland. A: Ben Lawers Schist. B: Carn Mairg Quartzite. C: Farragon Group. D: Killiecrankie Schist. E: Schistose metadolerite. F: Ben Lawers Schist (After Sturt, 1961B, Fig. 4).

the Aberfoyle anticline exposed near the Highland Boundary Fault.

Recently, Rast (in Johnson and Stewart, 1963) suggested that the F_1-, F_2-, and F_3-phases of folding can be recognized throughout the Iltay Nappe, although major F_2-folds only occur in the lower tectonic levels of the Nappe (as exposed in Perthshire, for example). In making this synthesis, Rast recognized the discreteness of the three phases of folding and suggested that on a regional basis F_1 and F_3 are not uncommonly subparallel. He also pointed out that locally F_4- and F_5-phases occur too.

Thus, in the Central Highlands initiation of the modern phase of research led to progress reports that required revision with the progress of subsequent enquiry. In structural geology it has not always been appreciated that this is the natural development; controversy has often been furthered by attack—sometimes bitter—and subtle modification of a protagonist's own ideas without emphasis being given to the sequence of conceptual models that led to the current process-response model. The problems of the interrelationships of F_1 and F_2 are not unique to the Schichallion area; in fact, the development of ideas in this area has been described in detail because the problems are typical of those likely to be met in unravelling orogenic sequences.

To the north and east of Schichallion, the Moine and Dalradian metasedimentary rocks are intensely interfolded. The Dalradian rocks of mid-Strathspey are probably linked with the Ballapel sequence (which is more fully developed to the west near Ballachulish) rather than with the Iltay Nappe (Fig. 326). To the east, McIntyre (1951B) mapped homoaxial relationships throughout the Grantown-Tomintoul region, which lies astride the Moine-Dalradian boundary. He demonstrated that large tectonic inclusions of the younger Dalradian pelitic and calcareous schists (e.g., Grantown and Kincraig Series) occur within the older, more quartzitic, Moinian tectonites (Figs. 89 and 180). It is not unlikely that these SE-plunging structures should be equated with the dominant F_2-axes of the Schichallion-Loch Tummel area, and that obvious traces of the earlier F_1-folds in Strathspey have been obliterated by overprinting. This is not an unlikely situation; in fact, Saggerson, *et al.* (1960) suggested that the homoaxial folding described by Weiss (1959A) in the Turoka area, Kenya, represents superposed folds that locally obliterated all obvious evidence of the earlier regional main fold trend.

Although King (in Rast, 1958A, p. 44) believed that the structures implied in the areas adjacent to those studied by McIntyre (1951B) are "so wildly improbable," the homoaxial SE-plunge can be traced westwards into the Monadhliath (i.e., north of Schichallion). McIntyre clearly did not assume that homoaxial conditions extend indefinitely, because at Ord Ban (in mid-Strathspey, Fig. 179) he helped describe (in Weiss, *et al.,* 1955) two directions of folding. At Ord Ban intensely penetrative deformation resulted in recumbent folds in quartzite about north-south axes plunging gently northwards. Adjacent marbles, mica schists, and granulites show small monoclinal folds with east-west axes (Fig. 31); movement about these axes was not particularly penetrative and was not uniform throughout. Weiss *et al.* (1955) concluded that a distinct time interval separated the early north-south folds from the later east-west folds at Ord Ban, but suggested that Highland tectonics are too complicated to make generalizations from this local observation.

In the area west of Ord Ban, two homoaxial domains abut along a sinuous but sharply-defined line; in the west of this area folds trend NE-SW and to the east the trend is NNW-SSE (Figs. 179 and 333), while along the line of junction east-west folds are locally developed (Whitten 1959A). Although the style of folding is dissimilar in each lithology, the styles for each lithology are analogous in the eastern and western areas.

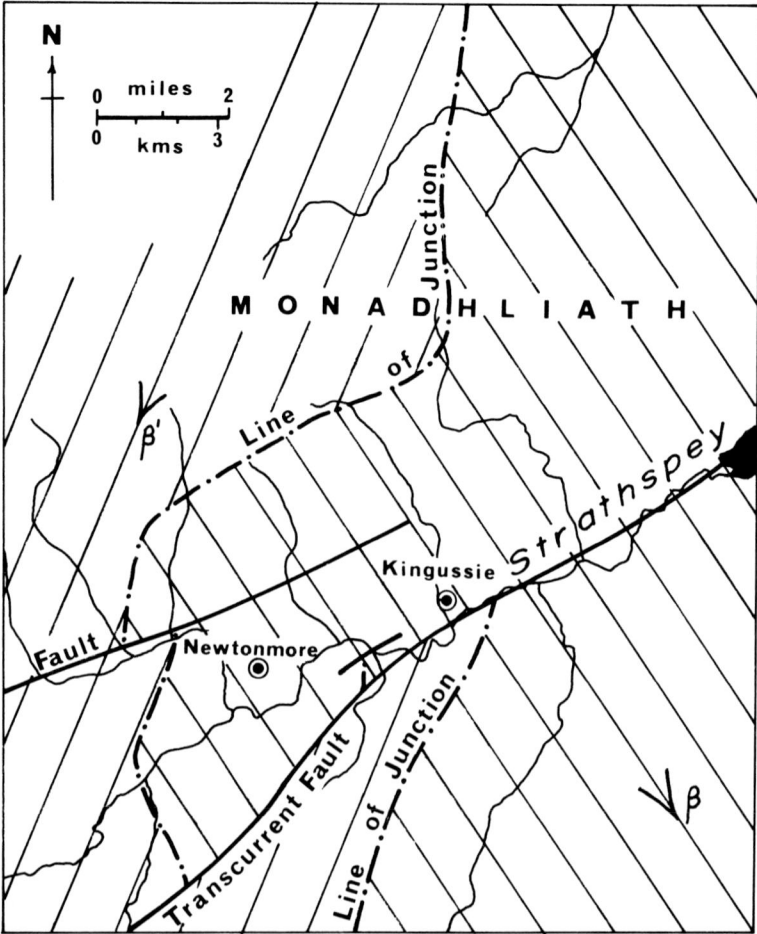

Figure 333. Map of the homoaxial domains (β and β') in the Monadhliath and mid-Strathspey, Central Highlands, Scotland (After Whitten, 1959A, Fig. 13).

The interrelationships of the folds in this area are still obscure. There is no suggestion of overprinting of the eastern or the western set of folds upon the other set. The map pattern of the major lithologic units might be expected to aid in interpretation of the critical relationships along the line of junction. Although structural data can be obtained from widely scattered parts of the Monadhliath, exposure is sufficiently poor over most of the area that realistic inferences about the locations of the major lithologic boundaries are impossible. Figure 179 is very similar to the original mapping by Barrow, *et al.* (1913), and Hinxman *et al.* (1915), which was only modified when positive evidence was available; new arbitrary boundaries were not drawn where there is no exposure. Inferences about the line of junction on the basis of the lithologic boundaries shown in Figure 179 are not justified (cf. Rast, in Johnson and Stewart, 1963, p. 137).

It is possible that the folding within this region was originally homoaxial, and that a late N-S compression caused rotation of the western area and development of the structural discontinuity of the junction line (Whitten, 1959A). The existing structural pattern (Fig. 333) is somewhat similar to that in the Loch Luichart area, Northern Highlands, where linear structures converge southwards (Fig. 334). Clifford (1960, p. 379) concluded that no simple system of interfering folds can produce the convergence of linear structures on an antiformal axis like that seen at Loch Luichart; of several hypotheses, Clifford preferred that which invokes rotation accompanied by distortion of the earlier structures.

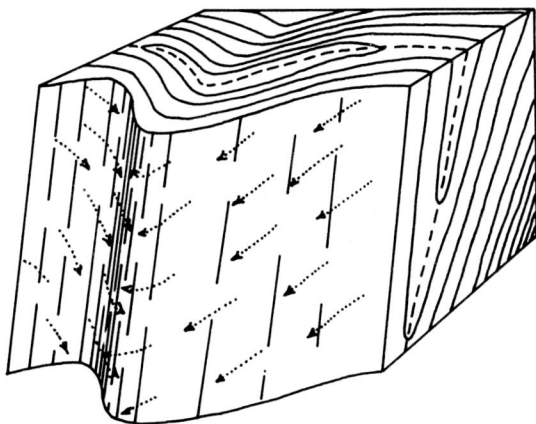

Figure 334. Diagrammatic representation of the deformation of early F_1-linear structures at Loch Luichart, Northern Highlands, Scotland, by F_3-folding (After Clifford, 1960, Fig. 11).

It was suggested above that the SE-plunging homoaxial structures of the Grantown-Tomintoul area may represent a superposed second fold system (F_2). An alternative hypothesis to that suggested above for the Monadhliath area is that the SW-plunging axes represent F_1, and that the SSE-trending structures are superposed F_2-folds which locally overprinted the F_1-structures. Evidence for or against this hypothesis is lacking at present.[2] As in the case of Ord Ban, it is unwise to develop an

[2]Since this section was written Rast's (in Johnson and Stewart, 1963, p. 137) critical review of McIntyre's (1951B) and Whitten's (1959A) papers was published, in which he suggested that the east-west folds in the Monadhliath-Grantown region may correspond to the F_3-folds of Perthshire.

elaborate tectogenetic hypothesis on the basis of mapping in this part of mid-Strathspey alone. The correct explanation will probably become apparent only when detailed stratigraphic and quantitative structural data are available for the surrounding areas. At present the areas to the north, west, and south are very poorly known; the first large-scale geological reconnaissance map of the area to the west (between Glen Roy and the Monadhliath) was published less than a decade ago (Anderson, 1956).

In the western Central Highlands superposed folds in the Ballapel sequence have been known for a considerable time. Bailey (Bailey and Lawrie, 1960) summarized the history of research for the Ben Nevis and Glen Coe area (Fig. 335) and recalled how, in 1908, he confirmed that the Ballachulish recumbent fold and slide are both refolded. Refolded structures are clearly shown in the Callert district, north of Loch Leven (Figs. 336 and 337). Similar structures were described from adjacent areas. The stratigraphic relationships carefully worked out by Bailey established that at the regional level recumbent folds and slides occur. The fold closures and the general map pattern indicate that the axial-plane traces of these structures trend northeast-southwest. Related minor structures have not been recognized and there are few clues from within the area about the axial trends of these folds. In consequence, at present, it is unwise to refer these major folds to any particular fold axial trend (cf. Weiss and McIntyre, 1959, p. 249), although, by extrapolation, King and Rast (1959) suggested that they have a "Caledonoid" (northeast-southwest) trend, and equated them with "Caledonoid" structures studied by them (a) to the south in Cowal, Argyllshire (1956B), and (b) to the southeast in the Schichallion region (F_1). Whatever the trend of these fold axes in the Ballachulish-Loch Leven area, they may be referred to as F_1; incomplete evidence suggests that they are products of premetamorphic gravity sliding (Vogt, 1954; Voll, 1960; Sturt, 1961B).

The metasedimentary rocks near Loch Leven (just east of Ballachulish) were subjected to detailed quantitative analysis on equal-area projections by Weiss (1956). The principal cylindroidal flexures preserved within the area involve large recumbent folds (F_2) whose geometric axes (B_2) plunge northwest (Fig. 338A). By dividing the area into twelve subareas, Weiss and McIntyre (1957) demonstrated that F_3-folds with symmetrical profiles (Fig. 338B) were superposed on the earlier NW-SE-trending F_2-folds. The youngest folds (F_3) have variable plunge, but fairly constant subvertical axial planes that strike dominantly northeast. This suggests that the F_3-folds resulted from subhorizontal compression normal (i.e., trends northwest) to the plane containing the B_3-axes of the F_3-folds.

Lack of uniformity of lithology and strains at Loch Leven caused F_2-folds to be dominant in some subareas and F_3-folds to be dominant in

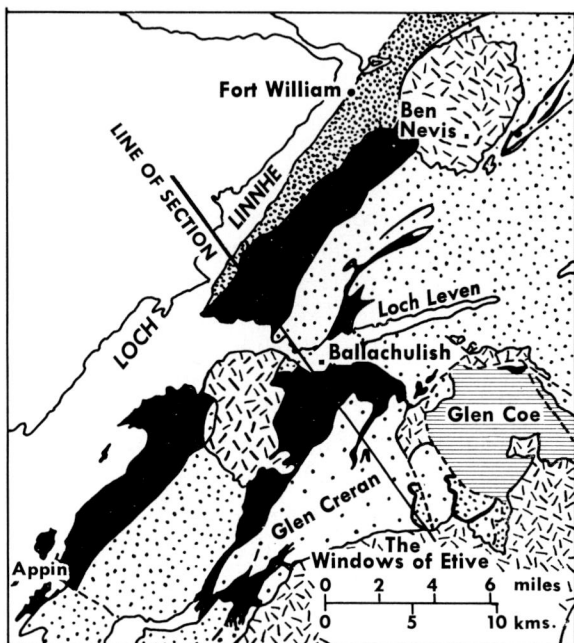

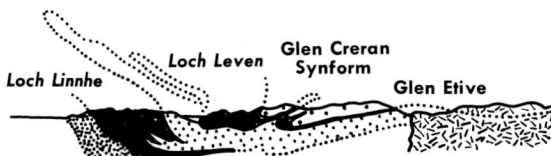

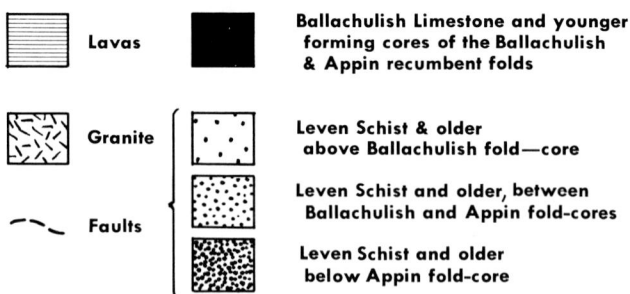

Figure 335. Map and section to show the nappe structure in the Ballachulish area, western Central Highlands, Scotland (After Bailey and Lawrie, 1960, Fig. 17).

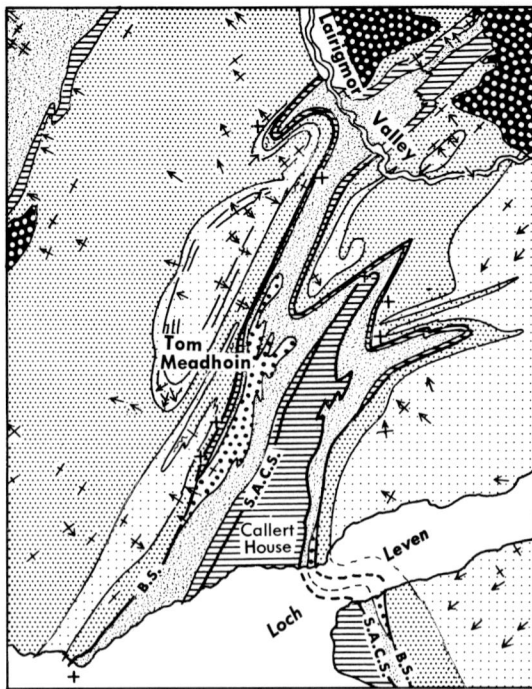

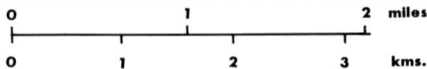

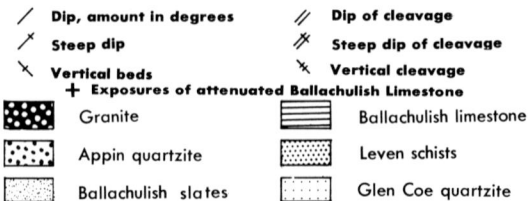

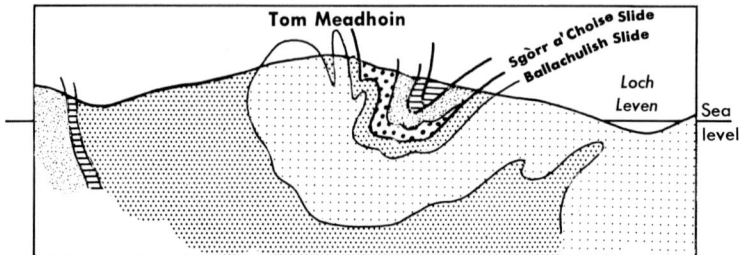

Figures 336 and 337. Slides and refolded folds of the Callert district, Loch Leven, western Central Highlands, Scotland; the metasedimentary rocks are listed in the legend in stratigraphic sequence, the Appin quartzite being youngest and the Glen Coe quartzite being oldest. Figure 336: Map of the area. Figure 337: NW-SE vertical section to show the relationships of the Ballachulish Slide to the Tom Meadhoin antiform (After Bailey and Lawrie, 1960, Figs. 8 and 9).

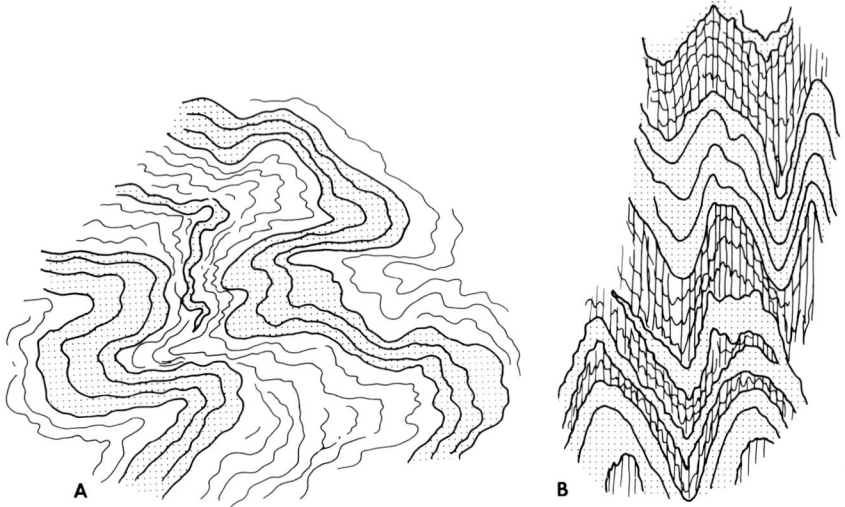

Figure 338. Fold styles in Precambrian Dalradian metasedimentary rocks near Loch Leven, western Central Highlands, Scotland. A: Profile of representative recumbent B_2-fold. B: Profile of superposed B_3-fold with symmetrical profile, showing refolded F$_2$-folds (After Weiss and McIntyre, 1957, Fig. 3).

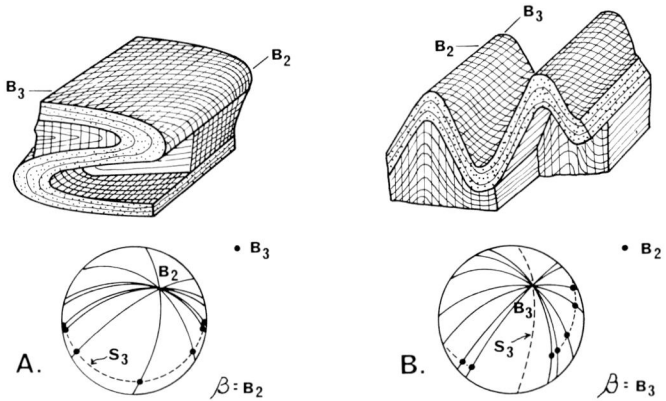

Figure 339. Relationships between B_2 and B_3 geometric fold axes, Loch Leven area, W Central Highlands, Scotland. A: Where B_2 is dominant and B_3 is subordinate. B: Where B_3 is dominant and B_2 is subordinate (After Weiss and McIntyre, 1957, Fig. 5).

others. Figures 339 and 340 illustrate these two cases, and Figure 341 shows the regional variability of the structures. By comparison with the subareas of the synthetic examples used in Chapter 10, those in Figure 341 are very irregular in shape. Domains commonly have to be defined after most of the field data have been collected, and their extent and shape

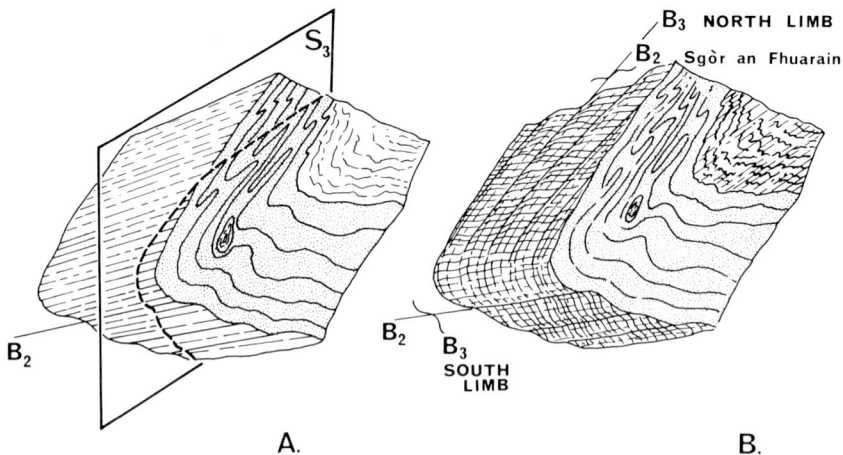

Figure 340. Geometry of the Sgòr an Fhuarain fold, Loch Leven area, Central Highlands, Scotland. A: Interpretation of the geometry after the F_2-folding. B: Geometry (diagrammatic) after superposed F_3-folding (After Weiss and McIntyre, 1957, Fig. 11).

is governed by the geometry of the earlier deformations, because these control the orientation of the S-structures prior to the youngest folding.

These detailed studies of the Central Scottish Highlands make it clear that any complete study of an area must involve (a) geometry based on extremely detailed quantitative areally-distributed data, (b) lithology, and (c) stratigraphic and structural relationships in adjacent areas.

To deduce the structure from a study of stratigraphy and lithology alone can lead to geometrical errors or, at least, to an incomplete geometric picture (cf. Bailey, 1959, and in Bailey and Lawrie, 1960). In southeastern Cowal, Argyllshire, King and Rast (1956B) distinguished three major folding episodes, but as Weiss and McIntyre (1957, p. 600) pointed out, the absence of published geometric data makes comparison of the geometries of Cowal and Loch Leven difficult or impossible. Similarly, the paucity of geometric data adversely affects the appraisal of many studies from various parts of the world.

Again, the results of geometrical analysis applied to cylindroidal structures within one area need not persist outside the field actually examined (cf. Weiss and McIntyre, 1957, 1959).

Again, geometrical analysis applied within a limited domain may not yield evidence of the earliest tectonic events in the region, because one metamorphic or tectonic event commonly obliterates, or largely obliterates, fabric elements of earlier phases. For example, a domain under study may be only several square kilometers in extent, and all of the metasedi-

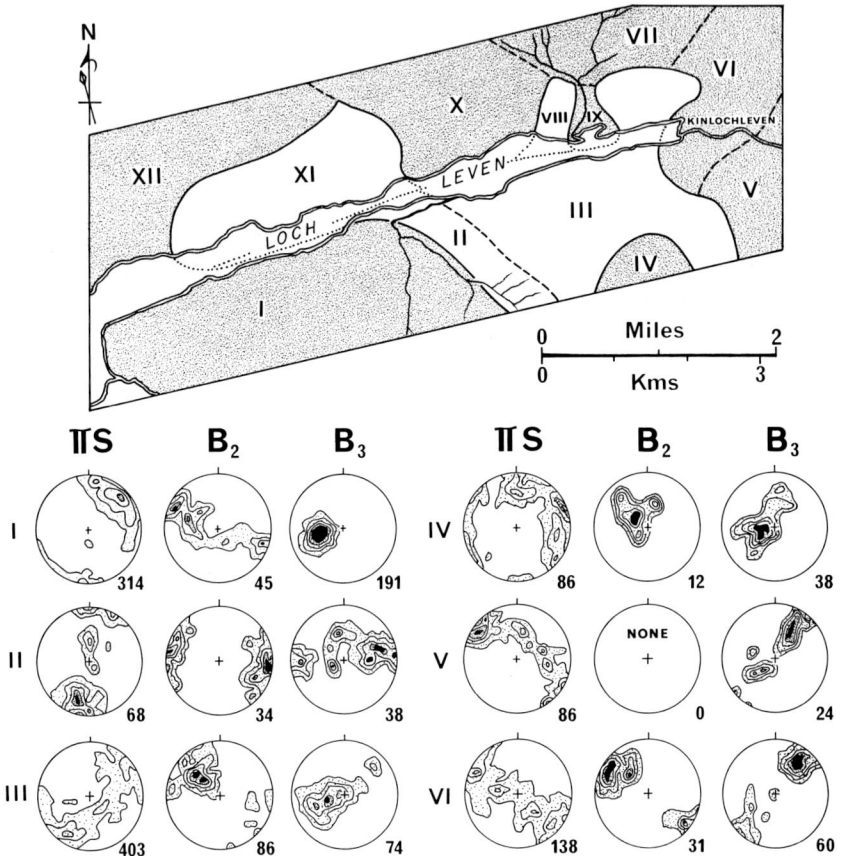

Figure 341 (above and facing page). Geometry of the B_2- and B_3-geometric fold axes in the Loch Leven area, western Scottish Central Highlands; the number of elements measured is shown beside each stereogram (Adapted from Weiss and McIntyre, 1957, Fig. 8).

mentary rocks involved may comprise one limb of a recumbent isocline (F_1) that was deformed by later folding (F_2, F_3, . . .). Thus, if S resulting from F_1 is approximately planar, evidence for F_1 is only likely to stem from regional stratigraphic relationships, or from chance relict fabric elements not obliterated by later events.

In areas of varied Dalradian metasedimentary rocks, stratigraphy aids materially in any complete study, whereas when only Moine rocks occur, stratigraphic control is virtually nonexistent.

The Northern Highlands of Scotland: Useful reviews of the Caledonide structures of the Scottish Northern Highlands were provided by Phemister (1948, 1958, 1960) and in Johnson and Stewart (1963); reviews

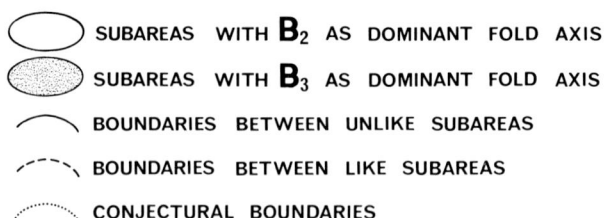

SUBAREAS WITH B_2 AS DOMINANT FOLD AXIS

SUBAREAS WITH B_3 AS DOMINANT FOLD AXIS

BOUNDARIES BETWEEN UNLIKE SUBAREAS

BOUNDARIES BETWEEN LIKE SUBAREAS

CONJECTURAL BOUNDARIES

CONTOURS: $1-3-6-9-12\%$ per 1% area

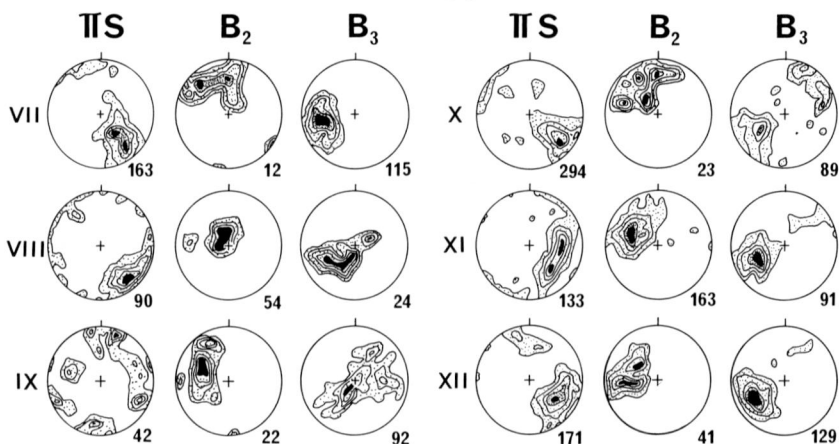

were also given by Bailey (1950), Kürsten (1959), Pavlovskiy (1958), Read (1956), Schreyer (1959B), and Sutton (1960B, 1962). In 1948 Phemister thought that the WNW-ESE-system of folds in the Moine metasedimentary rocks is younger than the regional NNE-SSW-system. This conclusion was drawn because the WNW-ESE-folds were thought to be associated with movements of the Moine Thrust Zone, and, near these thrusts, effects of the NNE-folding are apparently obliterated. In reviewing more recent work, which shows that the structures are considerably more complicated than was originally supposed, Phemister (1960, p. 23 ff.) noted that the WNW-folding might be older than the NNE-folding; he wrote: "Views on the deductions to be drawn from details of structural petrology, including the time incidence of major and minor folding and cross-folding, are, in fact, still in a state of flux."

A major controversy has centered around rocks at numerous locali-ties east of the Moine Thrust that the British Geological Survey (e.g.,

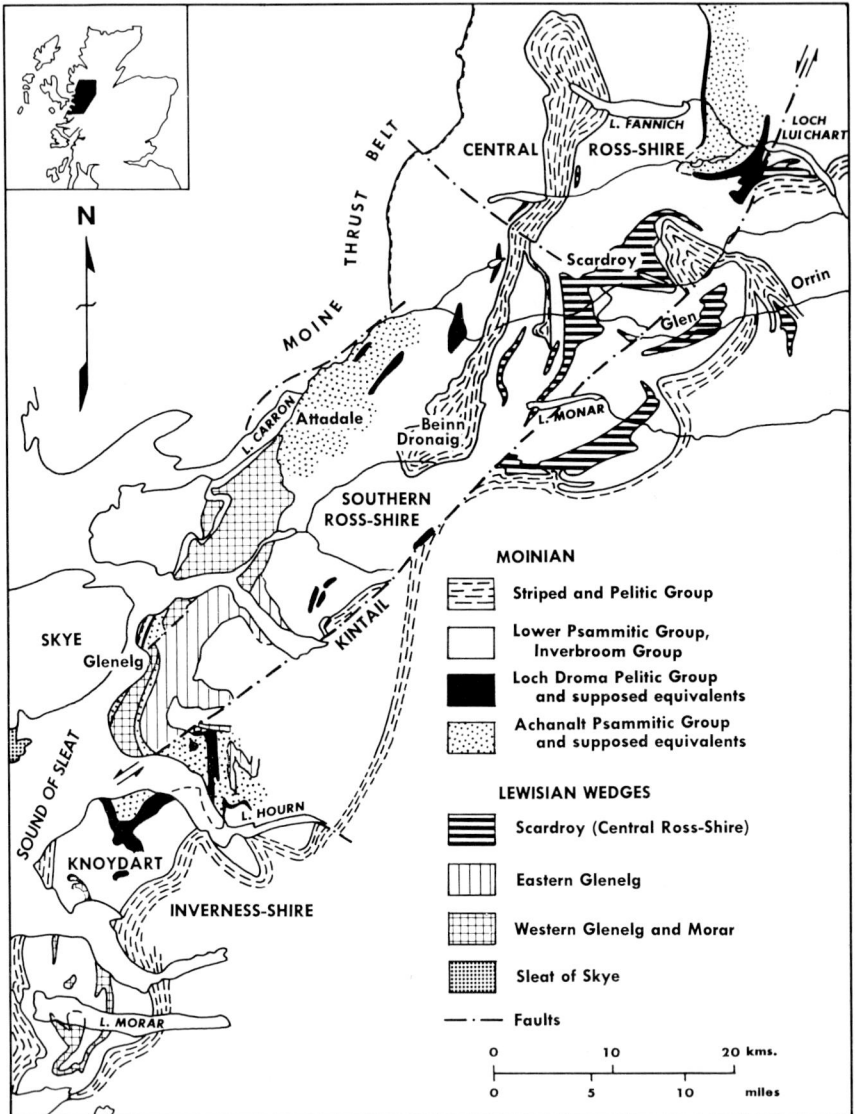

Figure 342. Distribution of Moine and Lewisian rocks in the western Northern Highlands, Scotland (After Sutton and Watson, 1962A, Fig. 1).

Flett, 1906) correlated on petrographic grounds with the Lewisian gneisses west of the thrusts (Fig. 342). The combined Survey work suggested that the Moine metasedimentary rocks rest unconformably on Lewisian basement in these inliers. Referring to the original memoirs

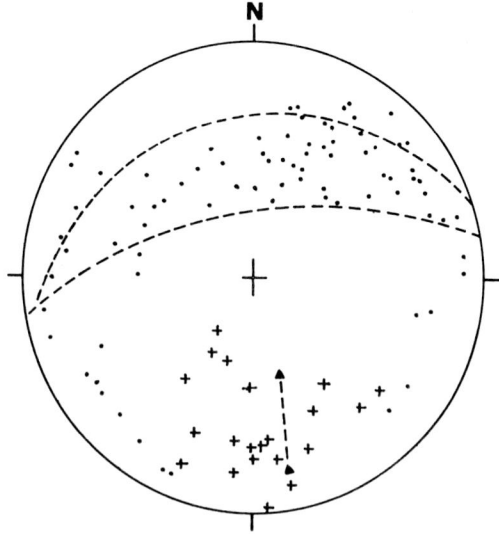

Figure 343. Composite stereographic projection for the Scardroy area, Northern Highlands, Scotland; dots: poles to foliation planes; crosses: projections of lineations (After Sutton and Watson, 1953, Fig. 3).

(Peach, *et al.,* 1910; Peach, Horne, Gunn, *et al.,* 1913), Peach and Horne (1930, p. 147) considered that this:

> ...view was strengthened by the detection of conglomerate rocks at the base of the Moine Series in contact with Lewisian Gneiss, by B. N. Peach in the area between Loch Carron and Loch Alsh, Ross-shire; by C. T. Clough, in the district of Glenelg...and by E. M. Anderson in Glen Strath Farrar, Central Ross-shire.

It was claimed that mapping indicated a progressive Moinian overstep across the Lewisian basement, and that some tectonic elements within the Lewisian inliers do not continue into the overlying Moine rocks. On the basis of such evidence all the Lewisianoid rocks east of the Moine Thrust Zone were interpreted as basement inliers exposed in anticlinal structures.

Read (1934A) challenged the Lewisian age of the so-called inliers. Hornblendic members of the Moine Series were being mapped in Sutherland (extreme north of Scotland), and these more basic rocks were assumed to be an integral part of the Moine succession. On this basis, Read thought it reasonable to assume that similar Lewisianoid lithologies could form part of the Moine sequence farther south.

Phillips (1937, 1945) made an extensive reconnaissance study of the microscopic petrofabric structure of the Moine schists in the Northern Highlands; he assembled impressive evidence to show that the tectonites are essentially homoaxial, and that, regionally, the B_1-axis plunges southeast (Fig. 246).

Since 1953 a series of geometric studies began by effectively demolishing the evidence for the major basement inliers, although more recent work has resulted in a completely new kinematic model that involves Lewisian gneisses within the Moine rocks. The development of these changing concepts is examined as a case history of unravelling complex superposed folds.

Sutton and Watson (1953, p. 122) claimed that the tectonites which include the supposed-Lewisian inlier at Scardroy, Ross-shire (Fig. 342) "... belong to a single, strongly folded series of dominantly sedimentary rocks of Moine age, modified by metamorphism and migmatization during a single orogenic period." The conclusion that the biotite-, hornblende-, and pyroxene-gneisses cannot belong to the Lewisian basement was based on what was, at that time, compelling structural evidence. Thus, π-diagrams of foliation (believed to define bedding) showed that both the "Lewisian" and the Moine tectonites have a common fold axis plunging southeast. Actually, the composite π-diagram (Fig. 343) does not define a simple great circle, but a triclinic girdle. Petrofabric analyses of Moine and "Lewisian" rocks also gave essentially similar subfabrics. Sutton and Watson (1953, p. 110) noted that, in principle, the petrofabric data confirm Phillips' (1937, 1945) findings, and imply that if any earlier Lewisian fabric existed it has now been entirely destroyed.

Sutton and Watson (1955) used similar compelling structural evidence to demonstrate that the supposed Lewisian basement of the Fannich Forest region, Ross-shire, is an integral part of the Moine succession. Several major structural studies have been completed in neighboring areas (Fig. 342). Ramsay (1958A) in the Loch Monar area, Clifford (1960) at Loch Luichart, and Fleuty (1961) in Glen Orrin all deduced very complex structural patterns, but demonstrated that rocks of the supposed-Lewisian inliers recorded by Peach, et al. (1912), Peach, Horne, Hinxman, et al. (1913), and Horne, et al. (1914) are really parts of the Moine succession, with which they have been intimately folded. For example, Ramsay (1958A, p. 273) showed that:

> ... the supposed Lewisian rocks appear at four separate stratigraphical levels in the Moine succession. No evidence is found of structural or metamorphic discordance between the supposed Lewisian rocks and the Moine Series.

In their several areas, Ramsay, Clifford, and Fleuty included the "Lewisian" horizons as part of the Moine Series, but, with justification, claimed that the validity of their *geometric* interpretations is not prejudiced by the tectonic or stratigraphic origins of the Lewisianoid rocks.

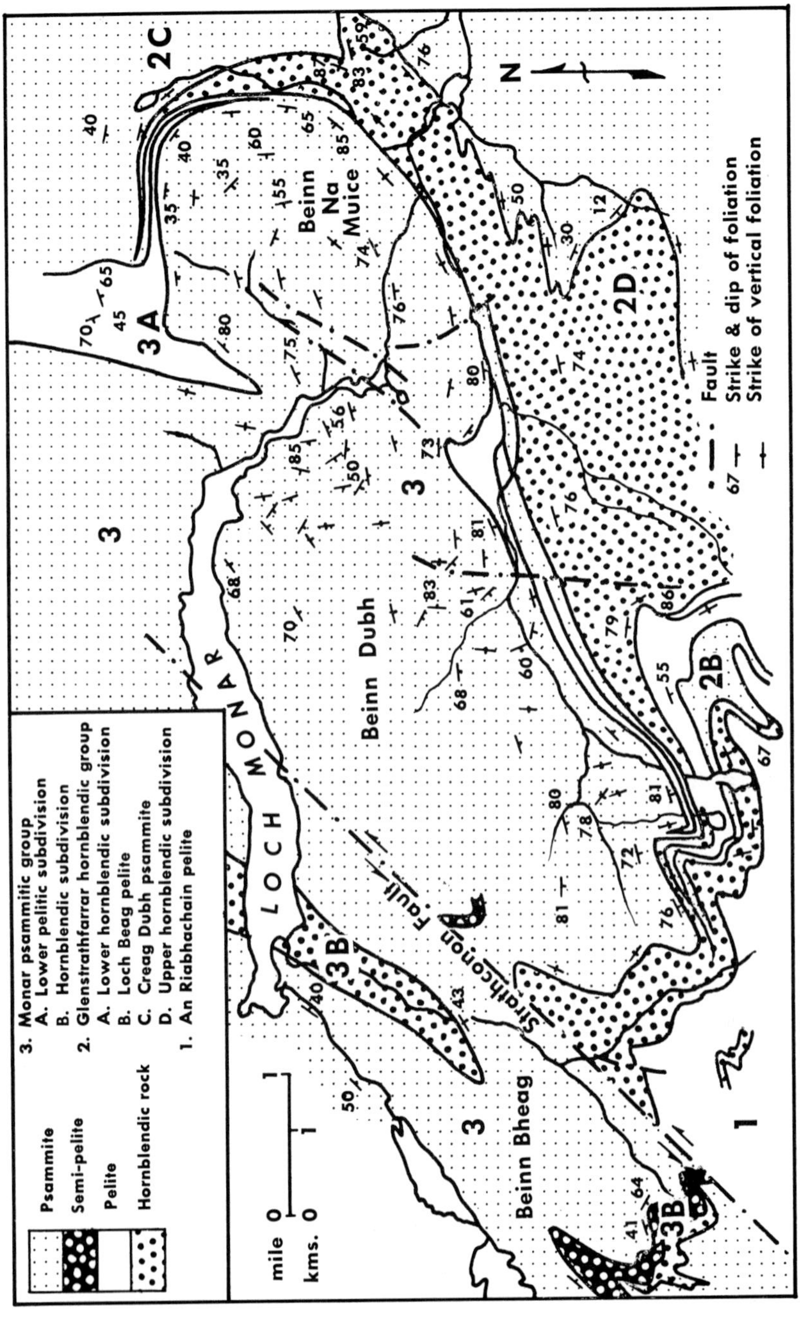

Figure 344. Geological map of the Loch Monar area, NW Scotland (After Ramsay, 1958A, Fig. 2).

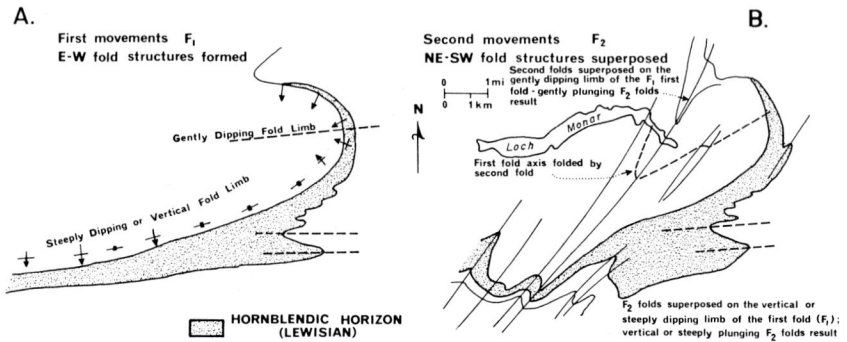

Figure 345. Simplified diagrammatic maps of the general effects of superposed folds in the Loch Monar area, Northern Highlands, Scotland. A: The first folds. B: The effects of the second folding (After Ramsay, 1958A, Fig. 5).

The structural geometry of these three areas was quantitatively studied with considerable care. Although the limitations of outcrop, etc., occasionally leave room for ambiguity, these studies represent unusually thorough and complete analyses of multiple fold superposition.

Within the limits of the Loch Monar area, Ross-shire (Figs. 344 and 345), Ramsay (1958A) found evidence for two major phases of folding. The first fold axes (F_1) originally plunged gently due west, and the folds were overturned to the north. The superposed second folds (F_2) have axial planes with a fairly constant NE-SW-strike (Fig. 346). Naturally, being dependent on the orientation of S following the F_1-folding, the F_2-geometric fold axes have variable trend and plunge. By dividing the 90 sq. km. into 22 subareas, Ramsay demonstrated that the major and minor

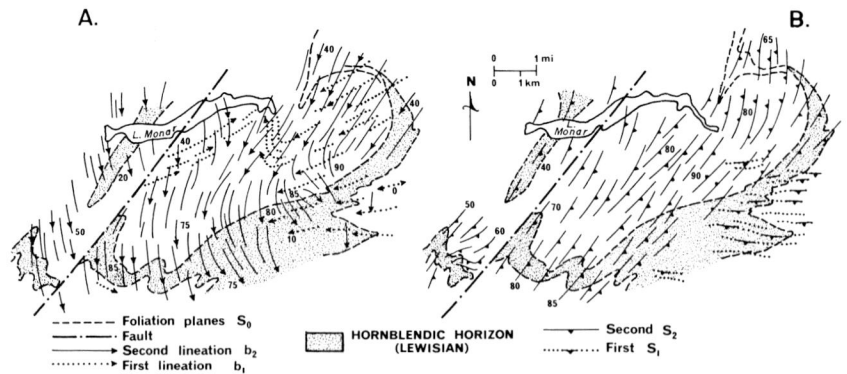

Figure 346. Trends of the structural elements in the Loch Monar area, Northern Highlands, Scotland. A: Linear elements. B: Axial-plane foliation planes with respect to first and second sets of folds (After Ramsay, 1958A, Fig. 6).

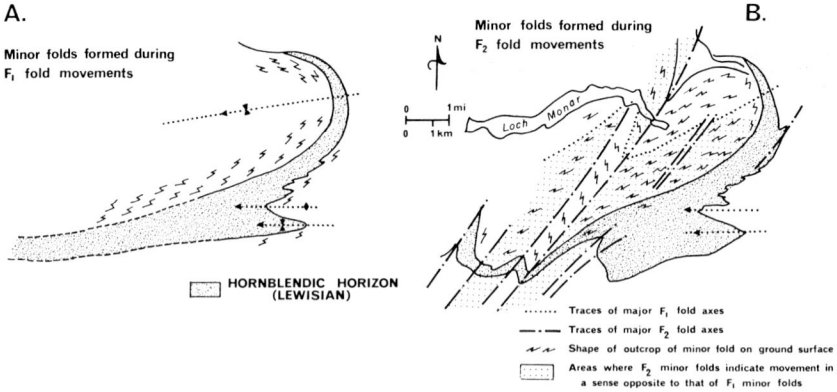

Figure 347. Geometry of minor folds formed during first and second phases of folding in the Loch Monar area, Northern Highlands, Scotland (After Ramsay, 1958A, Fig. 16).

structures behave in a consistent manner; two sets of minor structures were found over most of the area, and all the folds could be related to either F_1 or F_2. The F_1-folds have large amplitudes and small wavelengths, and the F_2-folds are generally sharp corrugations of small amplitude. Over part of the area both sets of minor folds have the same sense of movement, so that the first structures are accentuated by the second; in the remaining area there is an opposing sense of movement (see Fig. 347). Ramsay (1962C) showed that the rules relating parasitic folds to their major flexures can be extended to superposed folds; the characteristic relationships are shown in Figure 348. In addition to the foliation parallel to the lithological layers, Ramsay mapped foliations parallel to the axial planes of both F_1- and F_2-folds. It is striking, as Sutton (in Ramsay, 1958A, p. 307) pointed out, that at Loch Monar the attitude and form of almost every small structure can be related to some such larger fold.

Fleuty (1961) described the Glen Orrin synform and the Sgùrr na Fearstaig fold immediately northeast of Loch Monar (Fig. 349). Fleuty was unable to determine which is the youngest member of the apparent stratigraphic succession, and considered that the Lewisian gneisses mapped previously (Peach, Horne, Hinxman, *et al.,* 1913; Horne, *et al.,* 1914) are really part of the Moine Series. Careful analysis of the complex structures of all sizes suggested that the oldest recognized folds (F_1) had a north-south trend, and that the second (F_2) folds (Glen Orrin and Sgùrr na Fearstaig structures) were developed on one limb of the Strathfarrar antiform (F_1). A major slide (the Sgùrr na Cairbe slide) has a curved trace

Major F₁ Fold Structures

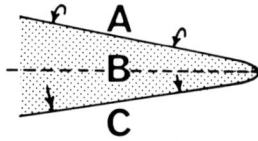

Superposed F₂ Structures

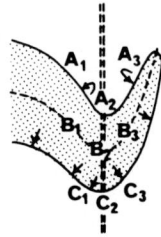

- - - - - Axial Plane Trace of F₁
===== Axial Plane Trace of F₂
↓   Dip of Beds
⌒   Overturned Beds

Minor Folds on F₁ Structures

Compound Minor Folds Produced by Two Phases of Folding

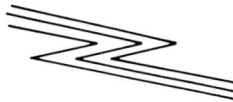

ZONE A   Z − FOLDS

A₁ S on Z   A₂ M on Z   A₃ Z on Z

ZONE B   Ƨ − FOLDS

B₁ S on Ƨ   B₂ M on Ƨ   B₃ Z on Ƨ

ZONE C   S − FOLDS

C₁ S on S   C₂ M on S   C₃ Z on S

Shapes of minor folds developed on limbs & hinges of a major fold structure.

Nine possible combinations of pattern which result from the interference of early & late minor folds.

Figure 348. Patterns assumed by superposed minor folds (After Ramsay, 1962C, Figs. 12 and 13).

through the middle of the Glen Orrin area (Fig. 349) and appears to have been active between the first (F_1) and second (F_2) phases of folding. Foliations and linear structures associated with F_3-folds, which have NE-SW vertical axial planes, pass across the Sgùrr na Cairbe slide without disturbance. A fourth (F_4) series of folds plunges to the ENE in the north of the Glen Orrin area. Individual folds of each successive phase of folding are smaller than those of the older sets.

Both Ramsay and Fleuty found that congruous structures near the hinges of early folds are commonly preserved during later folding, while similar structures in the fold limbs tend to be obliterated. When the later superposed folds are intense, the older structures may be almost completely destroyed. Thus, although Fleuty (1961, p. 475) found that older isoclinal major folds (F_0) are suggested by: "certain scanty evidence, it is nevertheless impossible to detect such folds with certainty on local structural evidence."

According to Ramsay (in Fleuty, 1961, p. 476) older, nearly isoclinal, major and minor folds (F_0) were also found by Mr. Merh between Glen Strathfarrar and Glen Cannich on the flanks of the oldest antiform (F_1) positively identified by Fleuty. What appear to be first folds (F_1) in Glen Orrin should probably be equated with F_1-folds mapped at Loch Monar.

The geometry of the Sgùrr na Fearstaig fold is distinctive (Figs. 350 and 351), and, since it closes sideways, it is neither an antiform nor a synform. Following the usage of Bailey and McCallien (1937) in Perth-shire, Central Highlands, such structures are **neutral folds.**

Near Loch Luichart, Ross-shire (Fig. 342), Clifford (1960) mapped evidence for three phases of folding. The oldest fold mapped is the almost-isoclinal Tarvie syncline that has a shallow plunge to the southwest (Fig. 352); the axial planes of the associated minor folds (Fig. 354) are parallel to those of the main syncline. The second phase of folding caused broad warping about an axis plunging at 30° SSE (Fig. 353); few minor folds are associated with this second folding, which has a vertical axial plane striking SSE. Clifford thought that formation of the Tarvie syncline and warping of its axial plane by the second folds was a continuous process produced by one fold movement; this interpretation corresponds to that of Sutton and Watson (1954) for the structures of Fannich Forest, referred to above.

A third phase of folding is well developed; rounded folds plunge steeply southwards and their axial planes dip at 80+° to the east. The postulated geometry is shown in Figure 334.

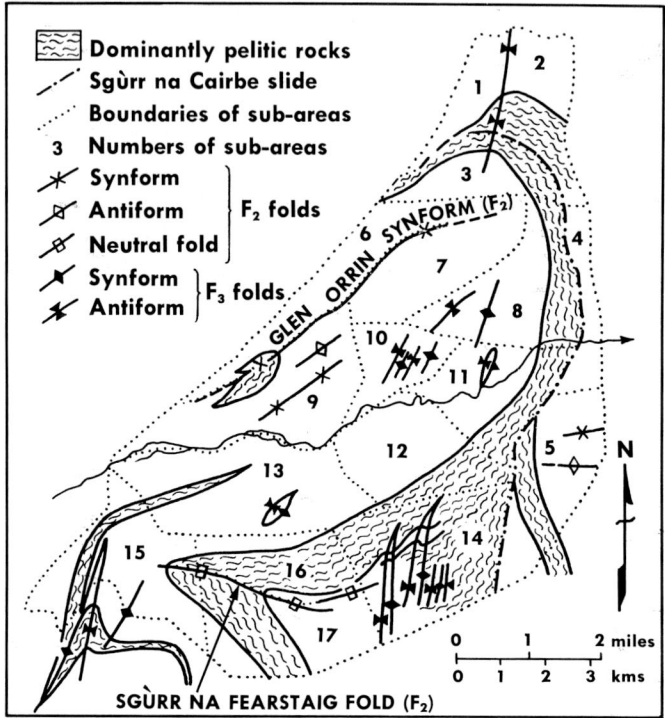

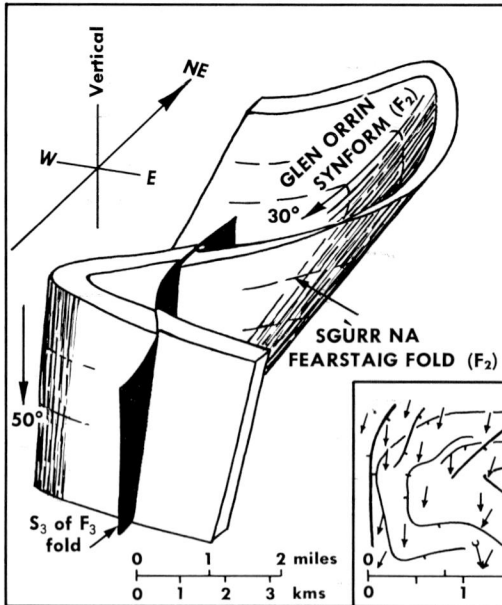

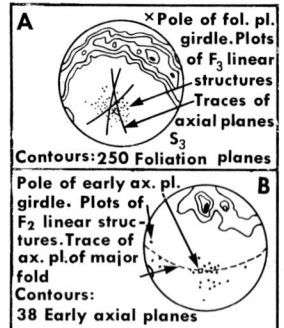

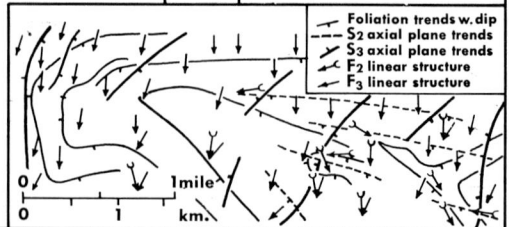

Figure 349 (top). Map of the major structural features and the subareas recognized for Upper Glen Orrin, Northern Highlands, Scotland; the Glen Orrin synform plunges SW and the axial surface dips SE; the Sgùrr na Fearstaig fold plunges south at about 50° (After Fleuty, 1961, Fig. 1).

Figure 350 (lower left). Diagrammatic block diagram of the Glen Orrin-Sgùrr na Fearstaig F_2-fold system shown in Figure 349; note the attenuation of the common limb between the folds. One of the small later F_3-folds is indicated by its axial surface (S_3) (After Fleuty in Clifford, *et al.*, 1957, Fig. 9).

Figure 351 (lower right). Structural map of the Sgùrr na Fearstaig fold, Upper Glen Orrin, Northern Highlands, Scotland. Synoptic equal-area projections of the structural elements are shown in A and B (After Fleuty in Clifford, *et al.*, 1957, Fig. 10).

Clifford (1960, p. 388) mapped the oldest schistosity as parallel to the lithologic units throughout the Loch Luichart area. He considered this schistosity to be essentially mimetic after the original sedimentary bedding, and correlated its genesis with development of the Tarvie syncline. Notwithstanding, it is possible, as in many other instances of supposed parallelism of schistosity and bedding, that this foliation is an axial-plane structure related to an older penetrative deformation. Direct published evidence is lacking on this point, but in reply to the discussion following the oral presentation of his paper, Clifford (1960, p. 388) suggested that:

... the earliest period of deformation, which is held responsible for the emplacement of the Lewisian, will become evident, not

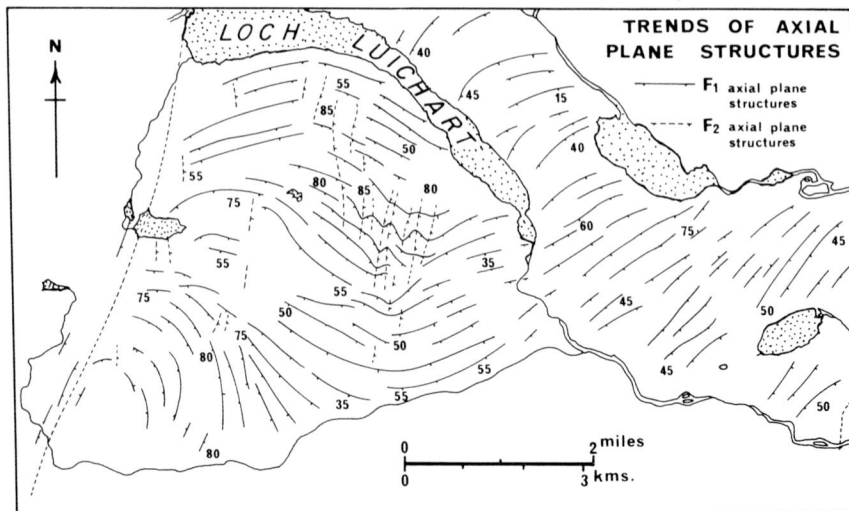

Figure 352. Planar structural elements of the Loch Luichart area, Northern Highlands, Scotland: the main F_1-structure is the almost-isoclinal Tarvie syncline (After Clifford, 1960, Fig. 5).

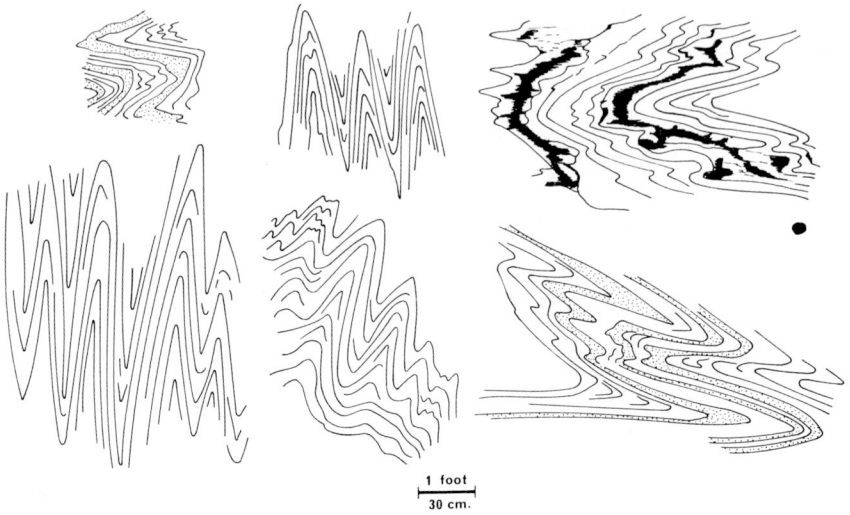

Figure 353. Representative profiles of minor F_1-folds associated with the Tarvie syncline, Loch Luichart area, Northern Highlands, Scotland (After Clifford, 1960, Fig. 6).

on the scale of the linear structure, but by the demonstration of a huge regional structure or structures.

The significance of Clifford's suggestion may not be immediately obvious. However, it implies that the "supposed-Lewisian" described in

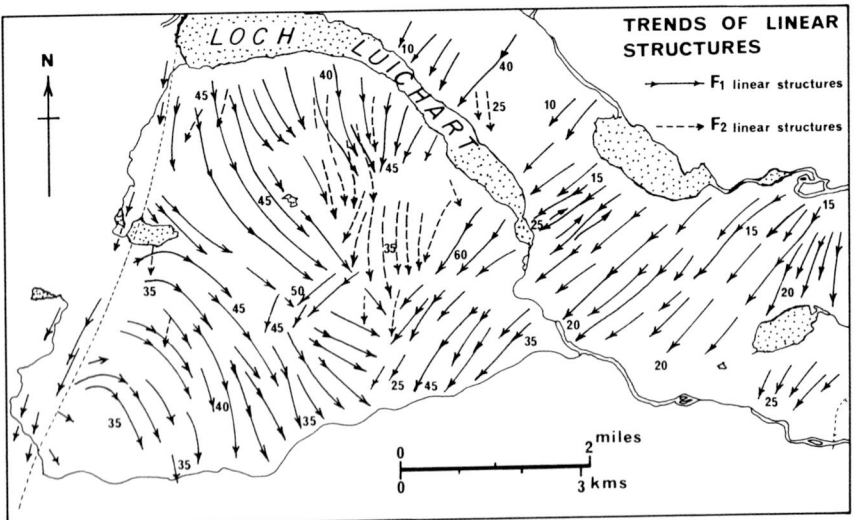

Figure 354. Linear structural elements of the Loch Luichart area, Northern Highlands, Scotland (After Clifford, 1960, Fig. 4).

his paper is now thought to be genuine Lewisian gneiss tectonically "emplaced" within the Moine rocks, rather than being an integral part of the Moine stratigraphic succession (as reported in the papers of Ramsay, 1958A, Clifford, 1960, and Fleuty, 1961). This gave a completely different status to the Lewisian-type rocks, and a return to a modified form of Horne's (1898) old hypothesis.[3] Although in 1955 widespread credence was not given to a thrust hypothesis, Sutton and Watson (1962A, p. 532) pointed out that the concept was:

> ... revived by Sir Edward Bailey and J. S. Turner (in discussion of Sutton and Watson, 1955) and by T. N. Clifford (1957). Kennedy was the first to invoke slides within the Moine Series itself when he suggested that in western Inverness-shire a Morar nappe of Moinian psammites and pelites had been thrust over a lower unit made up of much broken Lewisian basement with an autochthonous cover (1955, p. 371). Subsequently, T. N. Clifford (1957) advanced the interpretation of the Kintail region ... which involved the transport of Lewisian and Moinian rocks on an extensive dislocation.

On the same occasion that Clifford noted his revised concept about Loch Luichart, Sutton (in Clifford, 1960, p. 387), in discussing Clifford's paper, interpolated his revised opinion about the Lewisian-type rocks at Scardroy. As mentioned above, the Scardroy rocks were mapped as ordinary inliers by the British Geological Survey, and were later reinterpreted as an integral part of the Moine succession by Sutton and Watson (1953); Sutton (1960) now considers that these rocks comprise a tectonic slice of genuine Lewisian basement.

The nature of the thrusting was elaborated by Sutton (1962), Sutton and Watson (1962A), and Ramsay and Spring (1963). Sutton and Watson (1962A) suggested that, when allowance is made for the F_1- and F_2-folds that affected the area, it seems likely that the Lewisian bodies of central Ross-shire lie at approximately the same structural level and comprise portions of a single slice. At Scardroy, the slice of Lewisian basement is at least several hundred meters thick, although it tapers off rapidly to the NE and NW. They supposed that the Lewisian basement splinter wedged apart the Moine sequence, so that, beyond the limit reached by the wedge, the succession remains continuous. Thus, the Lewisian rocks must be bounded above and below by planes of movement (Fig. 355). The kinematics of this model were not fully explained by Sutton and Watson (1962A), although they reported a few folds within the Moine rocks that may have been associated with the thrusting. Old structures within the

---

[3]Horne (1898) explored the possibility that the Lewisian rocks at Fannich rest upon a thrust, but he (in Peach, Horne, Gunn, *et al.,* 1913) later rejected the suggestion.

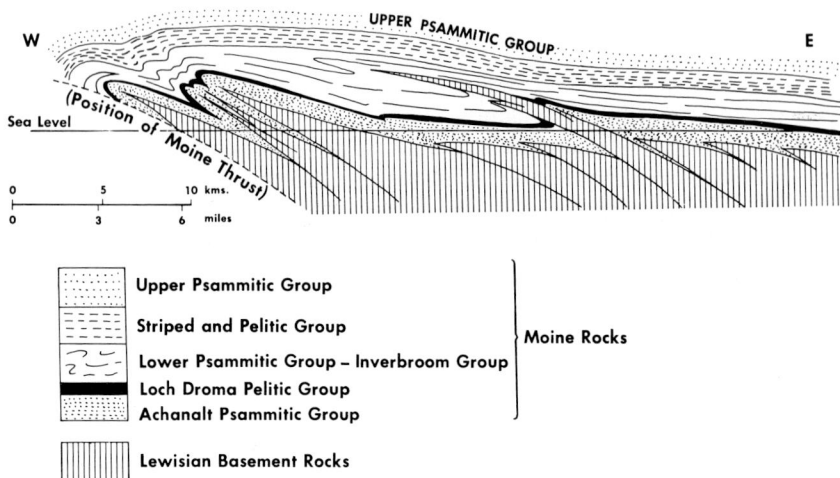

Figure 355. Conceptual model for the arrangement of the basement rocks at the close of the early period of deformation in the Northern Highlands of Scotland. This schematic composite section shows wedges with autochthonous cover (as at Glenelg) in the west, and a displaced wedge (*cf.* Scardroy) in the center. Effects of later phases of deformation are omitted (After Sutton and Watson, 1962A, Fig. 2).

Lewisian slice are scarce, and Sutton and Watson (1962A, pp. 532-4) wrote that:

> In most parts of the sheet, the banding and foliation, whether Lewisian or Caledonian in origin, are arranged more or less parallel to those of the adjacent Moines; the occasional structures dating from before the emplacement of the sheet are confined to the larger masses, those recorded by Hinxman west of Scardroy, for example (Peach, [Horne, Hinxman] and others, 1913), lying in one of the thickest parts of the Scardroy mass.

The revised hypotheses about these Lewisian rocks do not reflect adversely upon the quality of the earlier research; revised concepts are to be expected, and they emphasize that current interpretations should be regarded as progress reports only. The present model (Fig. 355) may well require modification when the stratigraphy of the Moine sequence in Ross-shire becomes better known; already there is debate. Thus, Ramsay (in Johnson and Stewart, 1963, pp. 143 and 166) concluded:

> ... it is not certain if the basement here is located in isoclinal fold cores or if it has been transported as great thrust masses. ... They might be parautochthonous folds, but some evidence seems to indicate that a few might represent the noses of nappes.

One cannot but agree with MacGregor (in Kennedy, 1955, p. 383) in his belief that ". . . the problems of the Scottish Highlands were so complex, and the causes of many recorded facts so imperfectly understood, that *all* published syntheses should be regarded as inherently speculative."

It was suggested in the previous section that the more obvious folds seen in the field at Loch Leven, Central Highlands, are relatively young and that they obscure the larger older structures; identification of these major regional structures depends on correlation and mapping of stratigraphic units over very large areas. Because of the lithologic monotony of the Moine rocks, regional mapping of this type has proved to be extraordinarily difficult. The possibility that older and larger structures exist was foreshadowed by Sutton (in Ramsay, 1958A, p. 307) in relation to the Loch Monar area when he suggested that:

> There seems to be no purely *structural* evidence for folds or thrusts earlier than the first of the three sets Dr. Ramsay has established. There may, however, turn out to be valid *stratigraphical* evidence, but that question must remain until much larger areas have been re-investigated.

Rutland (in Clifford, 1960, p. 387) also suggested that the most important period of deformation at Loch Luichart may have been earlier than any of those for which structural evidence has been advanced.[4]

Sutton (1962, p. 83) thought it possible that the earliest structures of the Northern Highlands antedate the regional metamorphism which produced the Barrovian zones, and that a Precambrian metamorphism (in addition to the more-apparent Silurian one) may have affected the Moine rocks (cf. Fig. 435).

Temporarily evading debate about a possible Precambrian metamorphism of the Moine rocks, it is likely that the earliest structures (F_0) over much of the Northern Highlands were major nappes. These folds (F_0) would have been premetamorphic or concomitant with the earliest and lowest-grade neocrystallization, as suggested (Voll, 1960; Sturt, 1961B) for the Dalradian rocks to the south (e.g., Ballachulish region). Thus, three major phases of folding are now recognized over some 5,000 square kilometers in western Inverness-shire and in Ross-shire, although the earliest phase (F_0) is partially obliterated over wide areas by the F_1- and F_2-phases (Ramsay and Sutton, in Johnson and Stewart, 1963). Ramsay suggested that a fourth phase (F_3) is localized in the neighborhood of the

---

[4]It must be emphasized that in the preceding pages the designation of phases of folding as F_1, F_2, F_3, etc., must be entirely provisional. Such nomenclature is based on the oldest folds observed within the area actually mapped, and subsequent research may reveal phases of folding not recognized at present.

Moine Thrust Zone and includes the conjugate folds described initially by Johnson (1956A), although these folds are not necessarily concomitant with the fourth (F_3) and fifth (F_4) folds recognized farther east; Sutton also alluded to the abundant but largely unpublished, evidence of later phases of movement east of a line from Loch Ailort to Loch Luichart. On the bases of these syntheses the correlations shown in Table 3 can be suggested.

Table 3.—Correlation Between the Phases of Folding Recognized in the Western Northern Highlands, Scotland

Regional Structure	Loch Monar Ramsay (1958A)	Glen Orrin Fleuty (1961)	Loch Luichart Clifford (1960)	Glenelg District
Initial isoclines and slides F_0		F_0		F_0
F_1	F_1	F_1	F_1	F_1
F_2	F_2	F_2	F_2	F_2
		F_3	F_3	
		F_4		

Farther west in the Northern Highlands Lewisian and Moine rocks are intimately associated in an area that extends over 50 km. from Loch Carron in the north to Loch Morar in the south (Fig. 342). The structural geometry of this region has been debated vigorously, and there has been continuing doubt about whether the Lewisian and Moine rocks are really all Moine, or are ever in unconformable relationship, or whether all of the contacts between Lewisian-type and Moine-type rocks are tectonic. Superposed folds are obvious throughout the area, but correct interpretations have been hindered by uncertainty about the stratigraphy. The exposures at Glenelg and Morar have been critical in the debate. The elucidation of these structures provides another excellent case history to illustrate the difficulties involved in establishing the correct structural history in a complex area.

The Glenelg area was originally mapped by officers of the British Geological Survey; their progress was reported by Clough (1898, 1900, 1901), Peach, *et al.* (1910), and Bailey (1955). Early in this work Clough (1900) suggested that two phases of folding are recorded at Glenelg. The careful and accurate lithological mapping by the Survey built up a picture of alternating bands of Lewisian and Moine rocks; isoclinal folds were thought to account for this pattern. Long ago Clough (in Bailey, 1910, p.

620) suggested that slides might occur in the Moine rocks east of the Moine Thrust in the western Northern Highlands, similar to those of the Ballachulish area of the Central Highlands. However, it was Kennedy (1955) who first suggested specific tectonic discontinuities on the basis of mapping east of the Moine Thrust Zone (cf. Sutton, 1962).

During the past decade the region has been reexamined in great detail and has passed through several conceptual revolutions. Concepts developed during work in the Scardroy-Loch Luichart area strongly influenced recent work in the western Northern Highlands; although intuitively reasonable, such influences in structural studies can lead to incorrect interpretations. Initial detailed geometric analyses suggested that the Glenelg-Knoydart region comprised a series of tectonic slices of basement and cover which has been folded twice (cf. Ramsay, 1958B). More recent work confirmed the original Survey concept that the Moine-Lewisian "stripes" at Glenelg mainly result from very early isoclinal folds with Lewisian cores (Ramsay and Spring, 1963; Lambert and Poole, 1964).

Kennedy (1955) believed the Morar Nappe moved westwards on the Morar Slide over a lower tectonic unit comprising normal Moine rocks interfolded with tectonic slices of basement Lewisian gneiss (Fig. 356). The upper tectonic unit (the Morar Nappe) was mapped as a simple mantle of Moine metasedimentary rocks with abundant cross stratification; Richey and Kennedy (1939) and MacGregor (1948) had originally mapped these rocks as an unconformable sequence on the highly folded "core." However, Kennedy (1955) called the lower tectonic unit a "window"—the window of Morar—opened by erosion of the Morar Nappe at a plunge culmination (the Morar antiform). Detailed stratigraphical and structural study within the window suggested tectonic contacts (Kennedy, 1955, pp. 361-2), although postdeformational neocrystallization was blamed for obliteration of the more obvious evidence of mechanical disturbance. The folds within the window are consistently overturned towards the west on both flanks of the Morar anticline; these folds include a broken recumbent anticline with an amplitude of at least 3.2 km. According to Kennedy (1955) folding of the upper tectonic unit—the Morar Nappe—was accompanied by folding of the basal slide and the rocks exposed in the window of Morar (Fig. 357).

Kennedy's model is illustrated in Figures 356 and 357, but it requires modification for at least two reasons:

1. R. St. J. Lambert (1958, 1959) showed that the supposed slide along the core-envelope boundary (separating Kennedy's Moine and Morar Nappes) is not a tectonic break. Wilson (in Kennedy, 1955, p. 386) initially suggested, and Lambert subsequently established, that this boundary is the limit of advance of a front of retrograde metamorphism

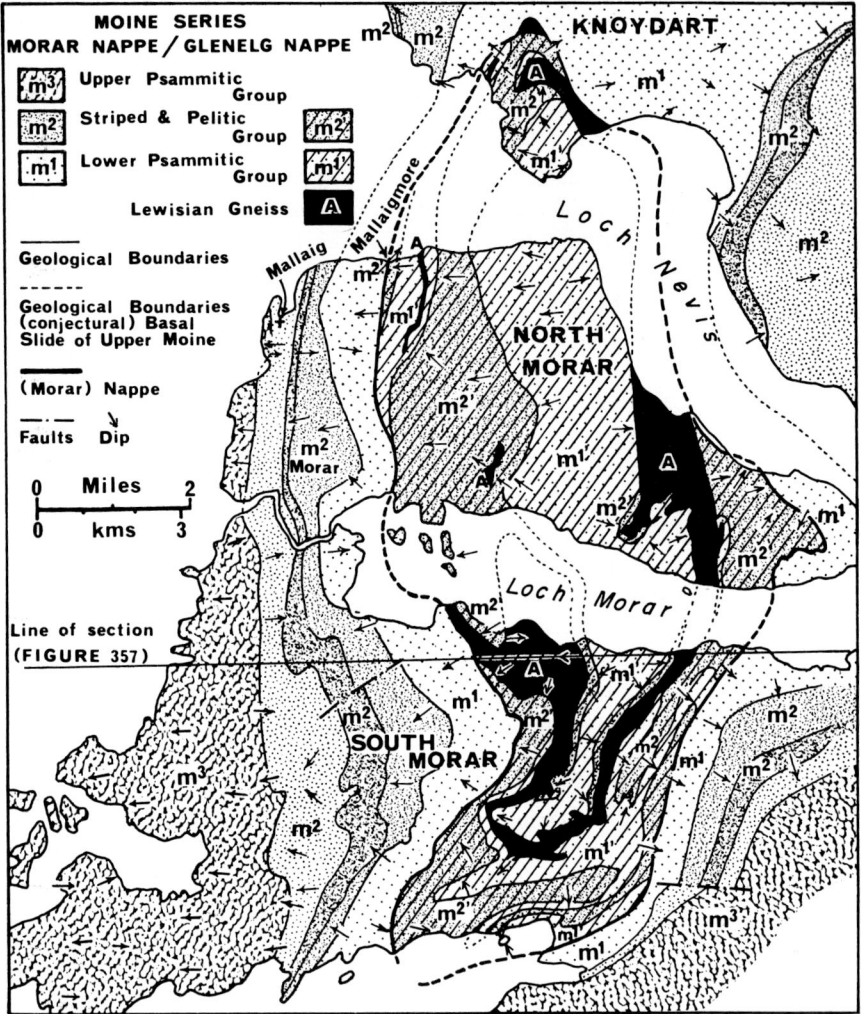

Figure 356. Geological map of the Morar area, western Northern Highlands, Scotland (After Kennedy, 1955, Plate 17).

that only affected the "core" rocks (i.e., the Moine Nappe of Kennedy). Ramsay and Spring (1963, p. 319) also advanced stratigraphic evidence to show that no tectonic break occurs within the Moine rocks at the horizon required by the supposed Morar Slide. In itself this revision does not affect the status of the Lewisian slice thrust *within* the Moine sequence of the "core" mapped by Kennedy (1955). Wilson (in Kennedy, 1955, p. 385) apparently erroneously, even thought that the Lewisian mapped by Kennedy is an integral part of the Moine succession, but Lambert believed that the Lewisian mass transgresses the Moine succession (Fig.

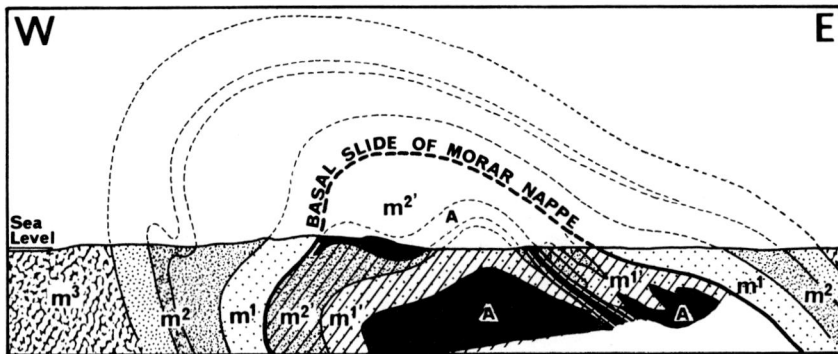

Figure 357. Vertical section across the Morar area (Fig. 356), western Northern Highlands, Scotland, to illustrate the tectonic model proposed by Kennedy (After Kennedy, 1955, Plate 17).

358). According to Lambert's interpretation, the Loch Morar structure begins to resemble the geometry suggested more recently at Scardroy and Loch Luichart (see above), where relatively late folds were superposed on Lewisian and Moine slices separated by early slides (cf. Fig. 355); Sutton and Watson (1962A, p. 540) thought that the relatively short displacement of basement slices at Morar may illustrate less advanced break-up of the basement than is seen at Scardroy in the inner part of the fold belt of Ross-shire.

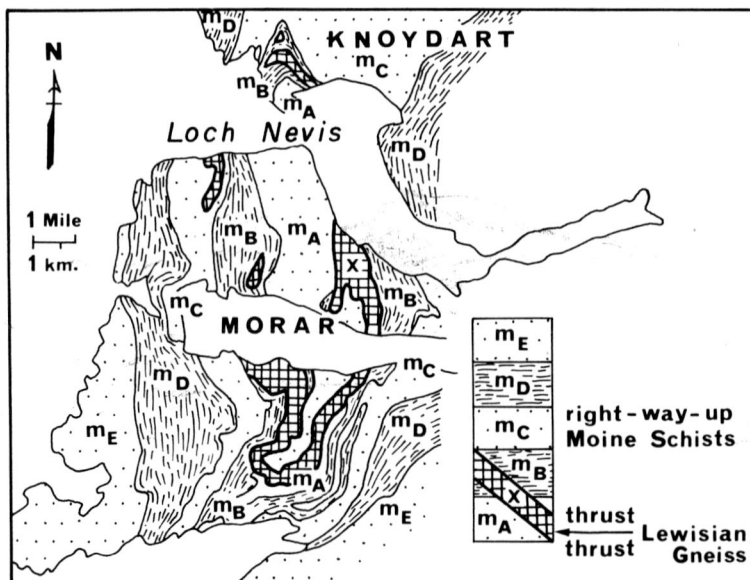

Figure 358. Interpretation of the stratigraphy of the Morar and S Knoydart areas, Northern Highlands, Scotland, largely based on mapping by Lambert (1958, 1959) (After Ramsay and Spring, 1963, Fig. 5).

2. Recent stratigraphic correlations in the Moine succession imply that the Lewisian rocks at Morar form anticlinal folds, and that the rocks lying beneath the Morar Lewisian are on the inverted limbs of the folds (Ramsay and Spring, 1963). The more recent mapping by Lambert and Poole (1964) also suggests analogous geometry in the envelope rocks, although the rocks at the gradation from Lewisian to Moine studied by them at Mallaigmore (Fig. 356) are dissimilar to the basal Moine rocks mapped by Ramsay and Spring in adjacent areas. Lambert and Poole gave evidence for refolded recumbent folds (with Lewisian cores and Moine mantles) that extend over at least 20 square kilometers.

Hence, recent work has supported two radically dissimilar models, and it is clear that the Morar structures will remain equivocal until the regional stratigraphic relationships are confirmed positively.

The Glenelg area (Fig. 342) has been critical in establishing the Lewisian-Moine relationships and the Moine stratigraphy. Ramsay (1958B) and Sutton and Watson (1959) confirmed that the extensively-developed Lewisian rocks comprise hornblende-, biotite-, and feldspar-gneisses, with bands of hornblende schist, marble, and other mafic rock types (Fig. 359). The parallelism between the banding of the Lewisian and that of the Moine rocks is most marked. The area is particularly interesting because at a dozen or more localities the Moine rocks preserve current bedding that helps to establish the Moine stratigraphy and to show that the Moine rocks young (or face) away from the Lewisian gneisses (Ramsay and Spring, 1963). In Glenelg, Clough (in Peach, et al., 1910), Bailey and Tilley (1952), and Ramsay (1958B) described very local basal conglomerates in the Moine rocks that lie unconformably on the Lewisian basement gneisses. In Gleann Udalain, north of Loch Alsh, Sutton and Watson (1959, p. 251) also confirmed the identity of a local basal Moine conglomerate. Ramsay and Spring (1963) suggested that the lowest Moine unit is the Basal Semi-Pelite, which, although never of great thickness, is a remarkably persistent group that forms an almost continuous envelope to the Lewisian rocks[5]; the conglomerates were interpreted as local developments in the Basal Semi-Pelite.

The revised stratigraphic succession erected by Ramsay and Spring (1963) for the Moine rocks between Glenelg and Morar is shown in Figure 324; working farther south, Powell (1964) was able to extend their stratigraphic succession. The proposed succession leaves several anomalies, but this stratigraphic model is a useful summary of current concepts.

---

[5] A detailed chemical, petrographic, and structural study (Lambert and Poole, 1964) revealed problematical rocks of a somewhat different character at the Lewisian-Moine transition zone at Mallaigmore, NW Morar (Fig. 356).

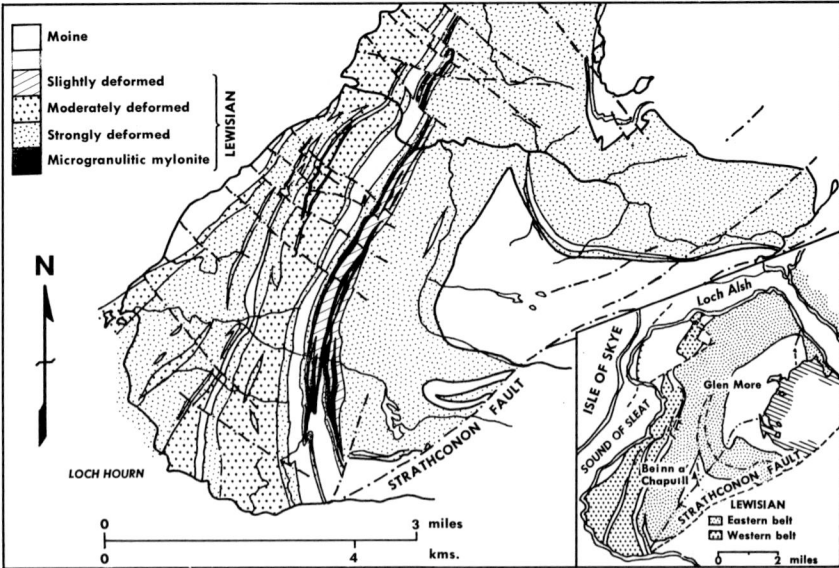

Figure 359. Detailed structural geology of the Beinn a' Chapuill area, Glenelg, western Northern Highlands, Scotland; *Inset* shows continuation of the structures northwards to Loch Alsh (oblique ruling is the Glenelg-Ratagan igneous complex) (After Ramsay, 1958B, Fig. 3, and Sutton and Watson, 1959, Fig. 1).

The structural interpretation has undergone considerable revision. Ramsay (1958B) and Sutton and Watson (1959, 1962A) concluded that the Lewisian rocks comprise tectonic slices with bounding slides roughly parallel to the lithological banding; they suggested that most remnants of the basal Moine conglomerate were destroyed by these dislocations. Mapping at Arnisdale, SE of Glenelg (Figs. 324 and 325), led Ramsay (1960) to revise his opinion about the Moine-Lewisian "stripes" mapped by him previously at Glenelg (Fig. 359). Although Sutton and Watson (1962A) and Sutton (1962) still thought the Lewisian rocks at Glenelg are relics of two thrust wedges of the basement, Ramsay and Spring (1963, pp. 310, 320) appear to be correct in considering that large-scale thrusting has not occurred there; they summarized their current interpretation thus:

> ...the Moine-Lewisian "stripes" were produced mostly by isoclinal folding.... The Lewisian basement is situated in the cores of these very early isoclines. The Moinian strata with distinct lithologies are symmetrically arranged around the Lewisian "stripes" and reading outwards from the anticlinal fold cores of this first generation of structures a consistent structural succession has been established.

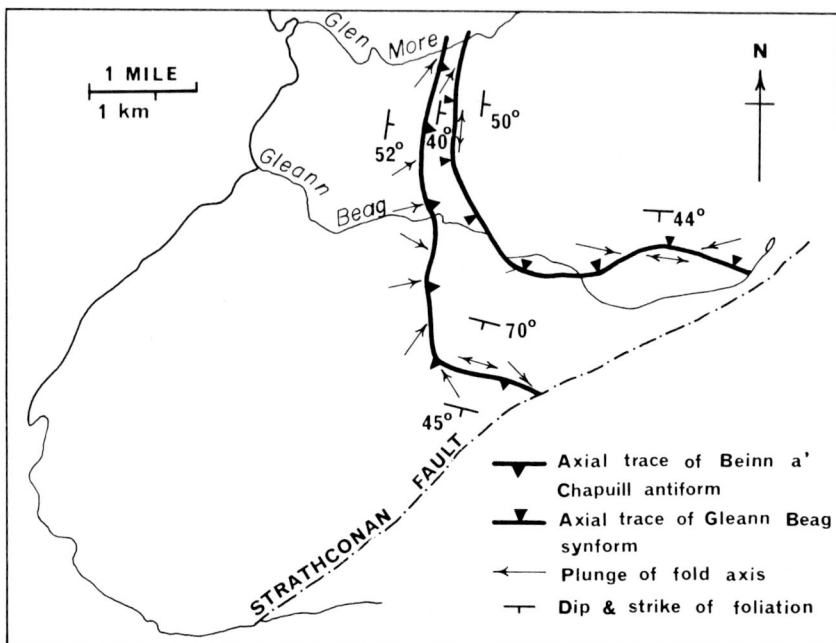

Figure 360. Axial traces and direction of plunge of the major F_1-folds, Glenelg area, Northern Highlands, Scotland (After Ramsay, 1958B, Fig. 6).

Hence, in the Glenelg region the first phase of folding (F_0) was apparently plastic and isoclinal, and regional slides or thrusts may have had little importance. It will be recalled that Ramsay and Spring (1963) suggested that a similar model applies to the Morar structures too.

Ramsay (1958B) and Sutton and Watson (1959) attributed two major structures — the Beinn a' Chapuill-Letterfearn antiform and the Gleann Beag-Loch Duich synform (Fig. 360) — to the second phase of folding F_1. Moine rocks occupy the core and Lewisian rocks the flanks of the Beinn a' Chapuill fold, although stereographic analysis of the foliation planes established that it is an antiform and the younger Moine rocks lie structurally below an envelope of older Lewisian gneisses (Fig. 361). Strongly-developed axial-plane foliation is commonly associated with the long-limbed almost-isoclinal folds of this generation. Structures in the Lewisian gneisses, whose original orientation was greatly at variance with the new axial direction (F_1), were probably so greatly deformed during F_1-folding as to become unrecognizable (Sutton and Watson, 1959, p. 236).

Because of the complications arising from the third phase of folding (F_2), the original regional disposition of the second fold structures is not yet clear (Sutton and Watson, 1959, p. 257). The axial planes of the major

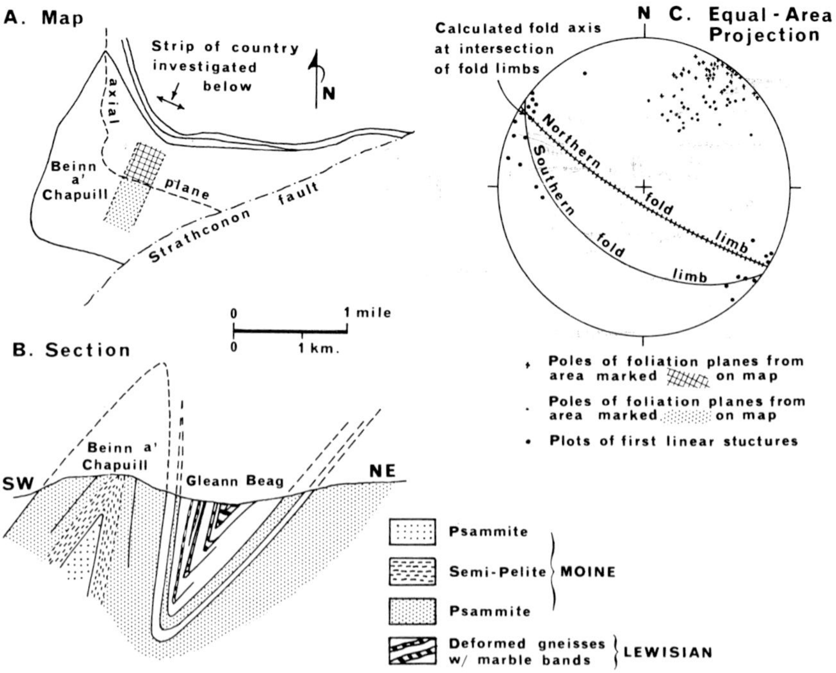

Figure 361. Structures of the Beinn a' Chapuill-Glen Beag fold structure, Glenelg area, Northern Highlands, Scotland (After Ramsay, 1958B, Fig. 7, and Ramsay and Spring, 1963, Fig. 8).

F_1-folds were bent by the F_2-folds and they now dip either east or SSW (Fig. 360). Ramsay (1958B, p. 508) showed that in the southern Glenelg area, poles to the axial planes of minor F_1-folds scatter along a great circle whose pole is the axis of F_2-folding (Fig. 362). The axial planes of the F_2-folds dip southeast, and the associated fold axes and linear structures also plunge southeast in the south, and more northeastwards in the north. These F_2-structures show a less plastic style and are commonly unaccompanied by new planar and linear structures. Minor folds of small amplitude are the dominant minor structures. Thus, the Lewisian and Moine rocks appear to have become progressively less plastic with the succeeding three movement phases.

Clifford (1957) thought two major tectonic elements — both containing typical Moine and Lewisian rocks — are separated by the Kintail Thrust (Fig. 363) in the Kintail area northeast of Glenelg and midway between Morar and Loch Fannich (Fig. 342). Stratigraphic mapping showed that the lower unit and the overlying nappe were isoclinally folded together about NE-SW F_1-axes (Fig. 364). Younger F_2-folds that plunge

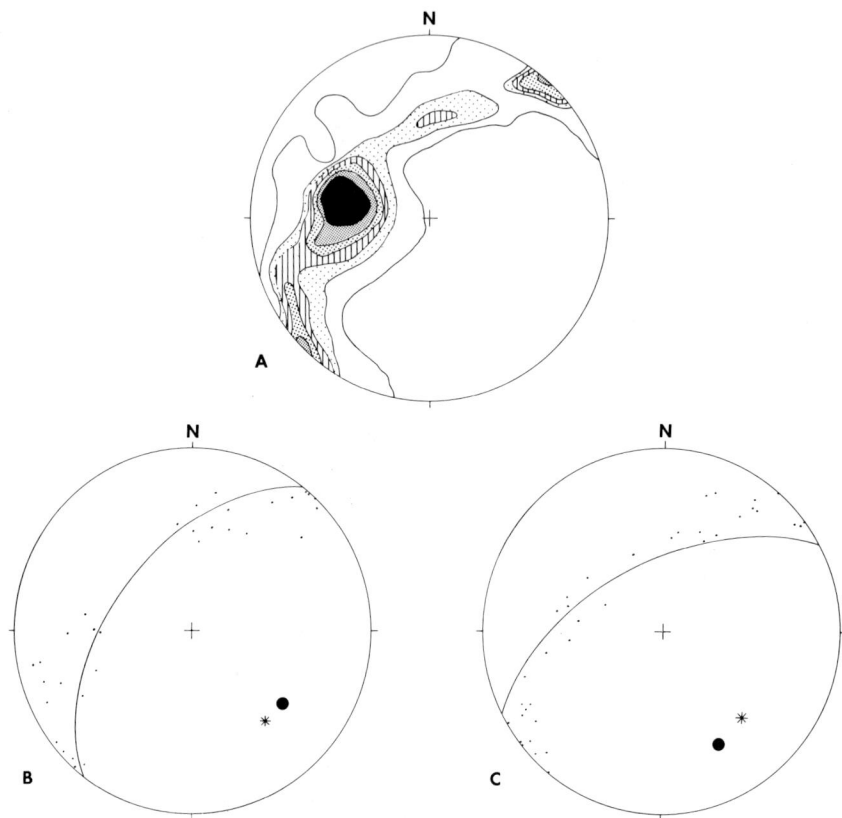

Figure 362. Structural relationships in the Glenelg area, western Northern Highlands, Scotland: poles to axial planes of minor F_2-folds. A: 249 poles for whole area, contoured at 1, 2, 3, 4, 5, and 10 per cent. B: NE limb of Beinn a' Chapuill fold. C: SW limb of Beinn a' Chapuill fold. • Axis of F_3-folding of foliation planes; * Axis of F_3-folding of F_2-fold axial planes (After Ramsay, 1958B, Fig. 11).

to the southeast can be recognized in the field, although folds mapped by McIntyre (1952) and Rutledge (1952) at Beinn Dronaig, north of Kintail, appear to be a better development of these folds. There is considerable doubt about the correlation of the Lewisian rocks at Kintail with those of neighboring areas; Sutton and Watson (1962A) thought their lithology indicates that they may be part of the Scardroy slice, but this is still an open question.

Lack of sufficiently-detailed mapping makes the correlation of phases of folding at Glenelg and adjacent areas difficult. However, the structures

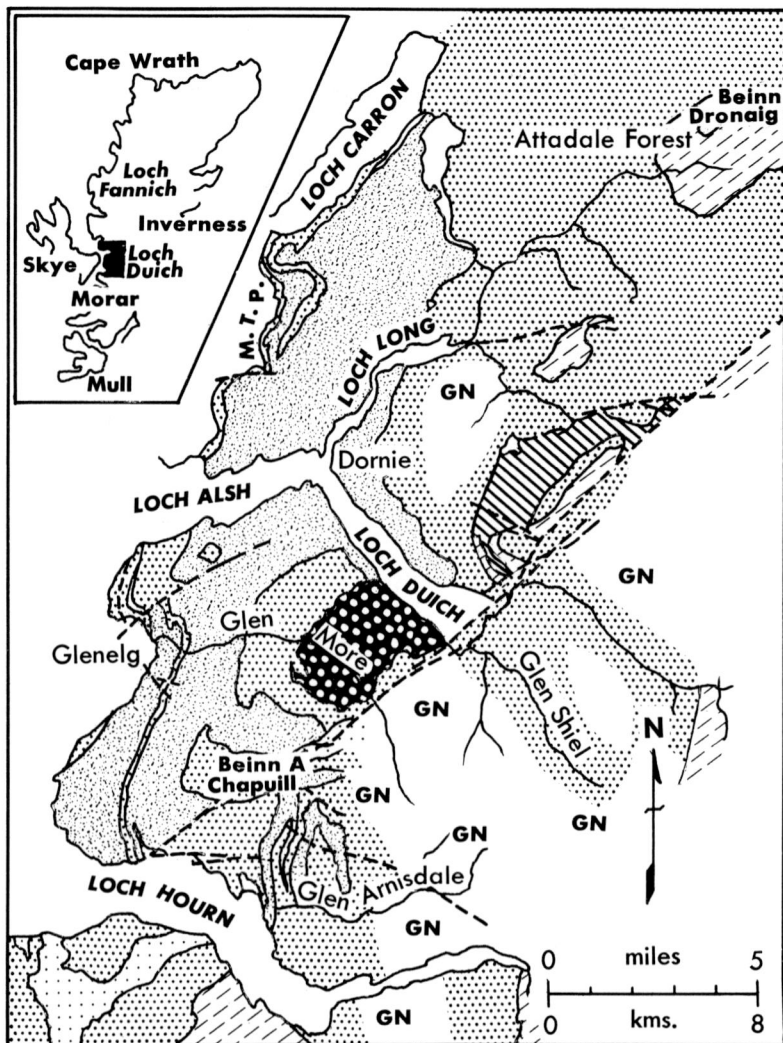

Figure 363. Generalized geological map to show the regional structural setting of the Kintail klippe, western Northern Highlands, Scotland; M. T. P. indicates Moine Thrust Plane or Zone (After Clifford, 1957, Fig. 1).

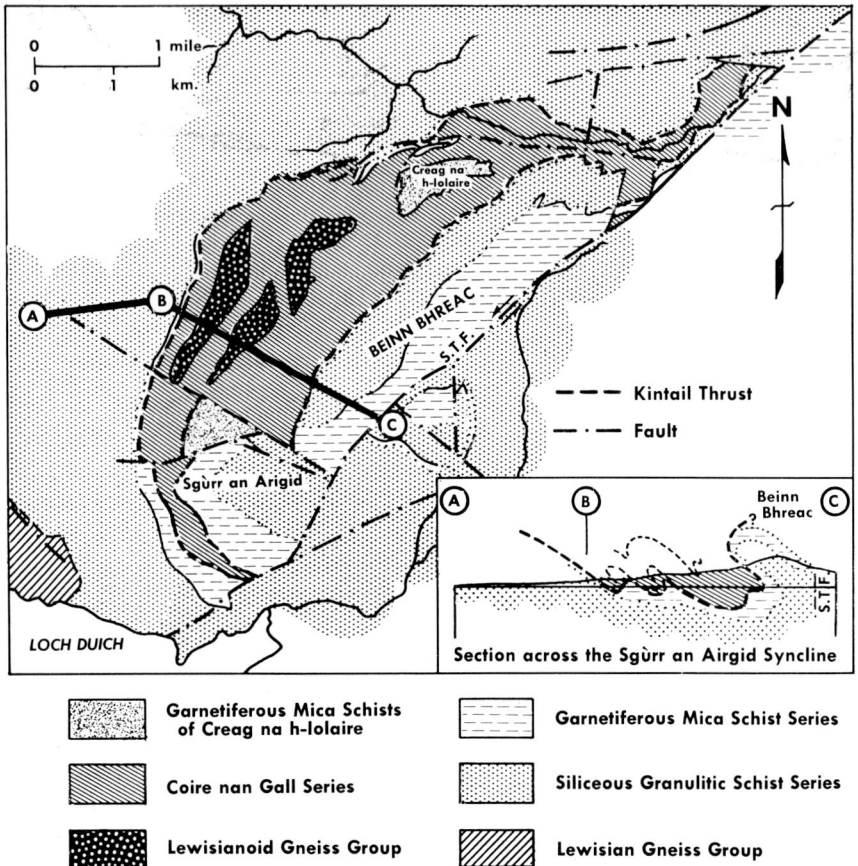

Figure 364. Geological map of part of western Kintail, western Northern Highlands, Scotland; S. T. F. indicates Strathconon transcurrent fault. *Inset*: Section along line A-B-C marked on map (After Clifford, 1957, Plate IX).

now recognized as second folds (F_1) at Glenelg may equate with the north-south folds (F_1) of Glen Orrin and Loch Monar; the third Glenelg folds (F_2) are probably equivalent to the F_2-structures at Glen Orrin and Loch Monar. The very poorly preserved F_0-folds briefly mentioned by Fleuty (1961) at Glen Orrin possibly equate with the isoclinal F_0-folds at Glenelg.

It is interesting to recall that the Laxfordian metamorphism of the Lewisian basement rocks occurred about 1,600 million years ago, and that the Moine sediments were apparently deposited about 900 million years ago. Throughout the Northern Highlands the Lewisian basement rocks were apparently in a highly plastic condition through the F_0-, F_1-, and F_2-folding that affected the Moine rocks. Ramsay (in Johnson and Stewart, 1963, p. 172) suggested:

> The cross-section shapes of these folds indicates that the main deformation[s] were produced for the most part by mechanisms of shear. This folding led to the partial or complete obliteration of the unconformities and all angular discordances.

Finally, complex superposed folds have been found associated with the Moine Thrust Zone, which comprises a series of subparallel thrust planes. In a detailed study of the superposed tectonic events, Christie (1956, 1963) showed that, in the north, phyllonites developed along the thrust zone contain north-south folds induced by the thrust movements. Johnson (1956B, 1957, 1960, 1961) made a very detailed geometric study of the thrust zone tectonics farther south in the Coulin Forest and Lochcarron areas of Wester Ross. These papers provide excellent examples of geometric analyses and of the complications that arise where the folding accompanying thrusting caused local folding of the regional Moinian and Lewisian fabrics. For each superposed set of folds produced by the thrusting, Johnson (1957, Fig. 8) mapped the irregular geographic pattern of intense and of poorly developed folding.

OTHER AREAS OF THE WORLD

The tectonics of the Central and Northern Highlands of Scotland are extraordinarily complex. It might be assumed that such relationships are anomalous, and that most fold belts are much more simple. However, rapidly-emerging evidence suggests that a long and complex sequence of overprinted tectonic events is the rule, rather than the exception, even in essentially unmetamorphosed parts of orogenic belts. In fact, any model that involves one or two simple homoaxial sets of folds for such regions is unlikely to stand the test of time, although, where the lithology is monotonous, or the degree or number of phases of metamorphism is considerable, the complexity may not be easy to prove.

In the Precambrian Shields of the world (e.g., Canadian, Scandinavian, Indian, African, etc.) intersecting major orogenic cycles have frequently been postulated. The Indian Shield is illustrated in Figure 365. Such maps are commonly based on broad regional syntheses of areas in which there is little detailed mapping. Although many North American geologists claim that orogenic belts develop around the margins of cratonic massifs, it seems likely that major tectonic belts have frequently crossed each other. However, more details of the consequent tectonic complexities would be desirable.

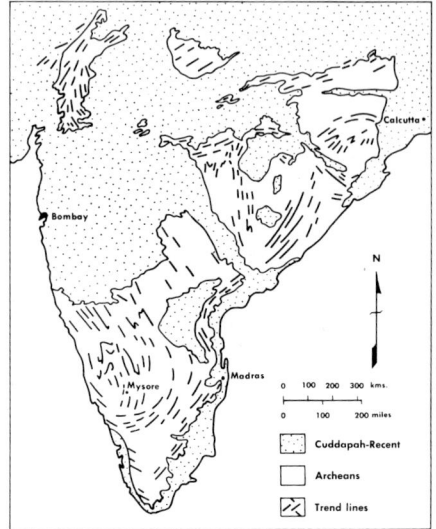

Figure 365. Tectonic map of Peninsula India showing main trend lines of the Archean formations (After Pichamuthu, 1962, Fig. 1).

Considerable attention has been focused on relationships between the Svecofennidic and Karelidic orogenies which had been assumed to cross one another in southeastern Finland (Fig. 366). Earlier accounts assumed that the Karelidic orogeny is distinctly younger. However, recent structural and geophysical evidence has revolutionized concepts

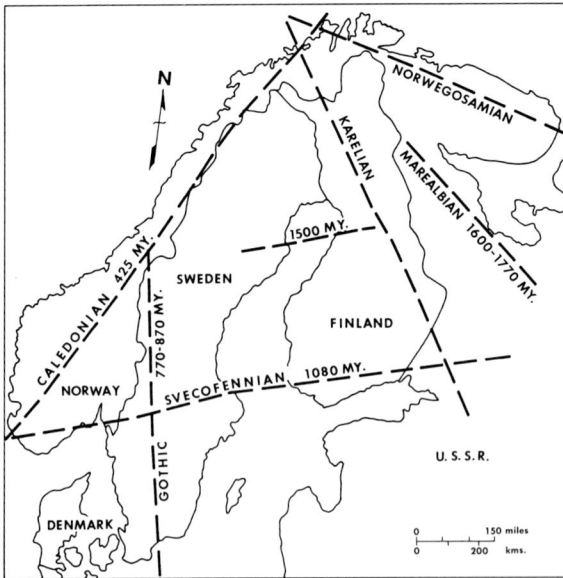

Figure 366. Diagrammatic sketch of early concepts about the supposed discrete orogenic fold belts of the Scandinavian Shield.

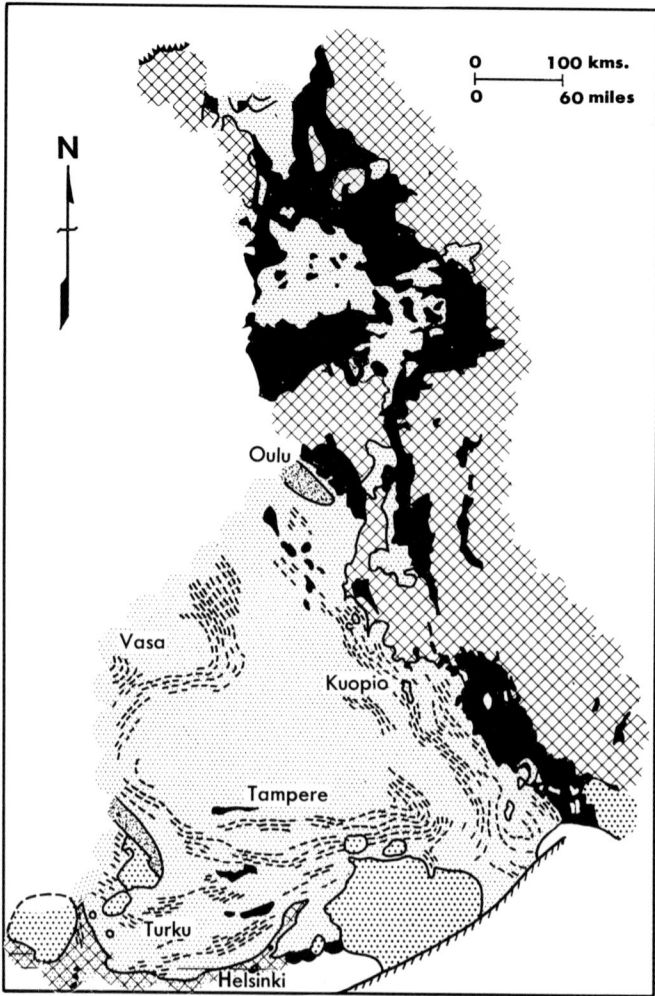

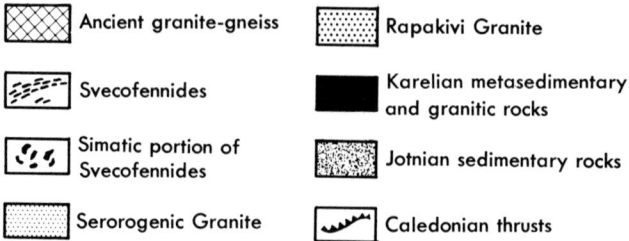

Figure 367. Structural trends of the Svecofennidic orogenic belt; the map approximately corresponds to the territorial boundaries of Finland (After Metzger, 1959, Fig. 1).

about the Svecofennides; in addition radiometric age determinations suggest that both orogenic systems formed about $1,800 \times 10^6$ years ago (Wetherill, *et al.,* 1962). In consequence, Metzger (1959) and Marmo (1962) suggested that it is no longer necessary to think of two separate orogenic systems (Fig. 367). More work is required on this and other pairs of supposed intersecting orogenic belts, but superposed orogenies definitely do exist in several regions; in such cases profound refolding has occurred. Examples are numerous. The Laxfordian orogeny was superposed on, and caused refolding of, Scourian orogenic structures formed over 1,000 million years previously in the Lewisian foreland of northwest Scotland (Sutton and Watson, 1951, 1962B, Dearnley, 1963, Park, 1964). In Peninsula India intersecting orogenic belts have been mapped for some time, and Pichamuthu (1962) incorporated the latest information (Fig. 365); some of the most detailed data relate to the Singhbhum area west of Calcutta where the Singhbhum orogeny (920 million years ago) transects, and to some extent refolds, the Iron Ore orogenic belt (2,000 million years old) (Sarkar and Saha, 1962, 1963). Equally-impressive examples of Alpine folds superposed on Hercynian orogenic structures have been described many times from southern Europe (e.g., by Oulianoff, 1953, for the Mont Blanc and Aiguilles Rouges basement rocks of the Swiss and French Alps, and by Richter, 1963, for the western Spanish Pyrenees). Caledonide structures superposed on the Precambrian orogenies in Greenland provide an additional example.

The Scottish Highlands were considered in considerable detail to emphasize the difficulties of evaluating the structural geometry even in areas studied intensively over a considerable period by a number of geologists. Below an attempt is made to review briefly some of the many other regions of the world in which superposed folds have been described. Most of these illustrations are based on limited mapping and geometric analysis. Because complex structural patterns can only be unravelled adequately after exhaustive geometric analysis of numerous fabric elements, many published results must be considered provisional progress reports; continuing research may well cause modification of the models proposed at the present time.

Ireland: In northwest Ireland Precambrian Dalradian metasedimentary rocks have extensive development. The individual rock units and the regional NE-SW-folds can be traced southwestwards from Scotland into northwest Ireland, but near the west coast the structures swing abruptly to a more NW-SE direction. The reason for this sudden change is enigmatic. Since 1947 a considerable amount of detailed evidence has been compiled for multiple phases of folding in northwest Donegal. Preliminary qualitative descriptions were given by Anderson (1954,

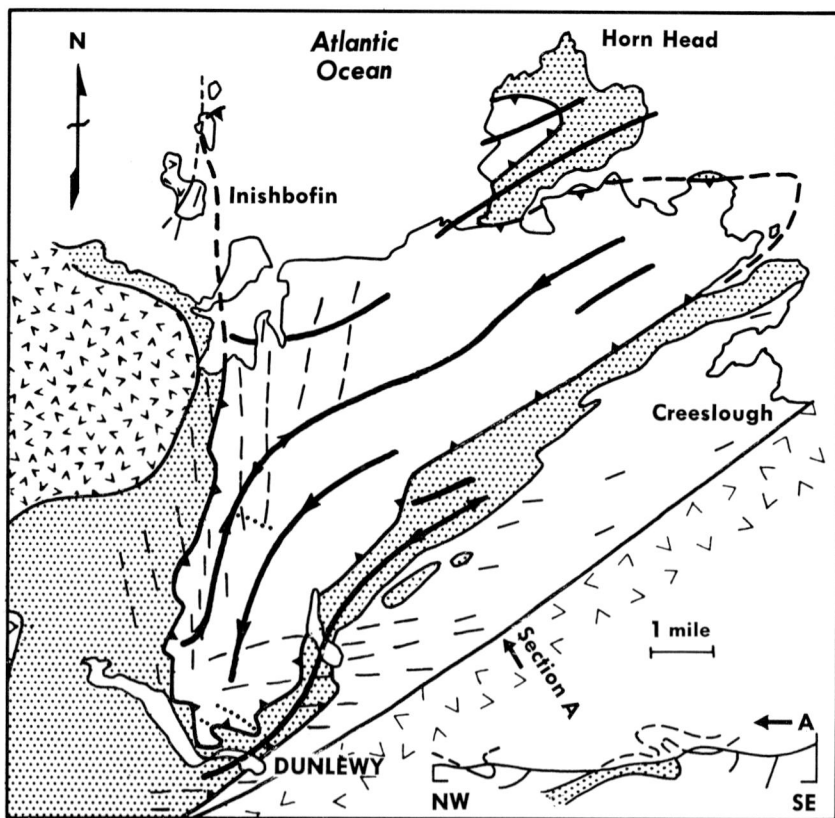

Figure 368. Superposed folds in the Creeslough-Dunlewy area, Co. Donegal, Ireland (Based on Rickard, 1962, Fig. 3, and 1963, Fig. 1).

1955). Reynolds and Holmes (1954) described Caledonoid folds superposed on older folds in the Dalradian rocks of Malin Head, while Knill and Knill (1958) and D. C. Knill and J. L. Knill (1961) made a detailed study of the neighboring Rosguill area on the north Donegal coast.

The stratigraphic succession farther south in the Errigal area was laid out carefully by McCall (1954) and Rickard (1962), and the structural framework is shown in Figure 368. The oldest folds are major SW-NE recumbent synclines and anticlines (Fig. 369), and a foliation parallel to

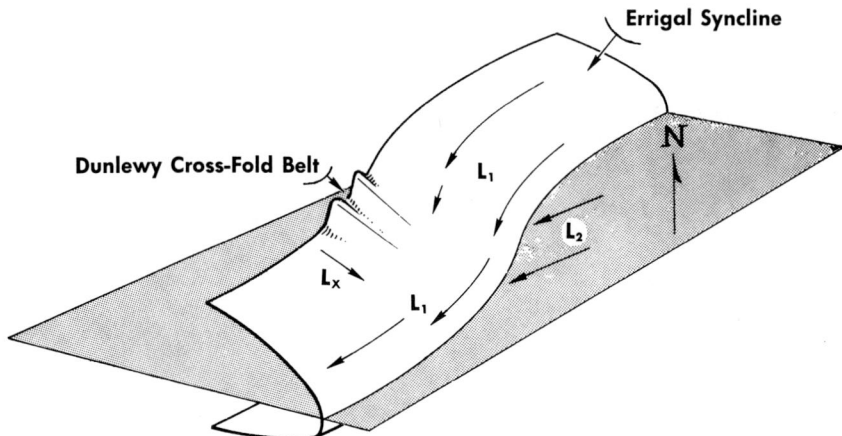

Figure 369. Diagrammatic representation of the main structural elements in the Errigal area of Donegal, NW Ireland; L_1 = early linear structures, L_2 = late linear structures, L_x = cross fold axes (After Rickard, 1962, Fig. 6A).

the axial-planes is strongly developed in the pelitic members. Although the foliation is commonly parallel to bedding, it is markedly discordant in the cores of the major and minor folds; it is commonly of crenulation type in the major fold overturns (Rickard, 1961). An important feature of the Errigal area is the association of recumbent folds and slides. The slides mainly occur at the margins of the massive Errigal Quartzite, and along the margin of the quartzite, on the normal limb of the Errigal syncline, more than 1,800 meters of the succession are cut out tectonically (Rickard, 1962) in what is an excellent example of a lag (see p. 564 for definition; cf. Bailey, 1934).

To the south, near Dunlewy, both limbs of the major Errigal syncline are folded about minor Z-shaped WNW-ESE-folds (Fig. 368). Farther southwest, in the Crockator area, this trend is more prominently developed (Fig. 370). Rickard (1962; 1963, p. 417) thought these folds are "cross folds" developed contemporaneously with the NE-trending main (Caledonoid) folds; he drew analogies with the small cross and main folds at Schichallion, Central Scottish Highlands, that were originally thought to be contemporaneous by King and Rast (1956A).

The Donegal Granites were emplaced later in the Caledonian orogenic cycle than the regional metamorphism and the major folding; Rickard demonstrated that emplacement of the granite produced minor folds and foliations in the envelope rocks near the contacts. Analogous effects were induced on two separate occasions because cross-cutting granite contacts show that the "Older Granite" to the west is older than

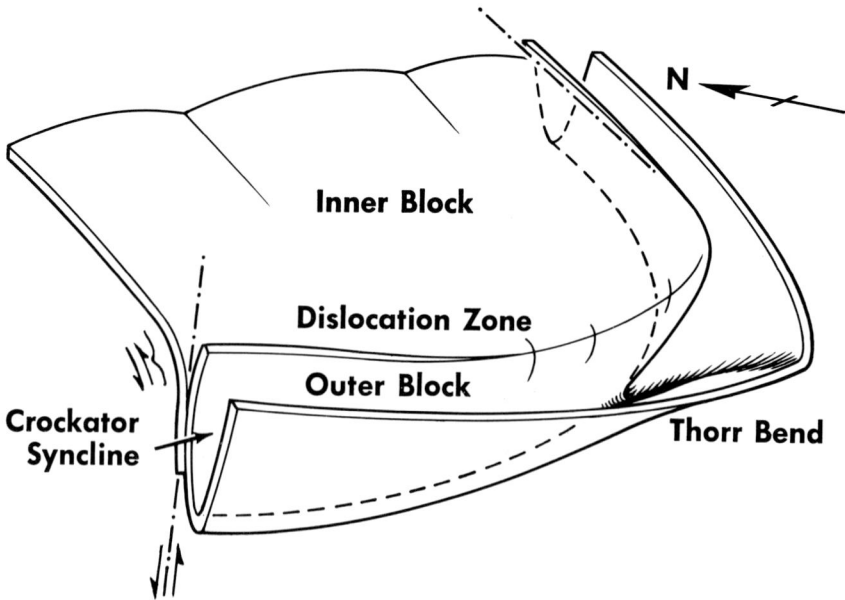

Figure 370. Geometry of the folds within the Errigal Quartzite, Crockator area, Donegal, NW Ireland; note scissor-like faulting along the dislocation zone (After Rickard, 1963, Fig. 4).

the Main Younger Granite in the east. Rickard found that along both contacts the new folds trend roughly parallel to the contacts and that they appear to be accommodation structures related to granite emplacement. In the east these folds are chevron-shaped and range from crenulations to flexures a few meters across; steeply-inclined crenulation foliation parallel to the axial planes is commonly developed. Mullion structure is associated with these folds (Pitcher and Read, 1960; Rickard, 1962, 1963). Locally, in the west (on Inishbofin), the earlier foliation is almost entirely obliterated by the younger crenulation foliation formed during granite emplacement.

It is difficult to be certain of the date of the successive fold episodes in Donegal. However, Dewey and Phillips (1963) made an interesting study of the multiple folding of the Dalradian metasedimentary rocks of Mayo on the west coast of Ireland, and demonstrated the manner in which the overlying Ordovician and Silurian sedimentary rocks of Mayo and Galway have been folded by some of the later movements. Evidence for four phases of pre-Arenig (? Cambrian) folding in the Dalradian rocks was found, and Figure 371 gives some impression of the complexity that developed in this province of continuing mobility. Dewey and McManus (1964) published a detailed geometric study of the Silurian metasedi-

A

?
↑

ENE Dextral and WNW Sinistral tear faults (Erriff Phase)
NE Sinistral and NW Dextral tear faults (Maam Phase)   ↑
F9 NNW Sinistral strain slip cleavage
F8 Conjugate strain slip cleavages   **Downthrow to North in H.B.F. Zone**
F7 Conjugate strain slip cleavages

    F6 N. Dipping slaty cleavage ⎱ **High angle reverse faulting**
    F5 Steep E-W schistosity ⎰ **and thrusting in H.B.F. Zone**
Concentric E-W folding   **Corvock Granite and Aplite Sills**   ↑

		Faulting	Stage
	Lamprophyres		**Post Wenlock**
SILURIAN	**Acid vulcanicity**		**Wenlock**
SILURIAN — Broad regional warping	**Acid & intermediate vulcanicity**		**U. Llandovery**
Concentric folding, thrusting & uplift			
ORDOVICIAN — Rise of Connemara Antiform	**Peleean vulcanicity**	**Maam Faults ?**	**Caradoc**
	Detrital Staurolite	**Subsidence of South Mayo Trough**	**Llandeilo** / **Llanvirn**
	Spilitic Suite	**Letterock Fault defines north margin of South Mayo Trough**	
CAMBRIAN — **F4 E-W Open recumbent folds**	**Peridotites**	**Removal of U. Dalradian south of H.B.F.**	**Tremadoc** / **U. Cambrian**
F3 E-W Isoclinal folds (-475m.y.)		**Large downthrow to North on H.B.F.**	
F2 N-S Recumbent folds	**Gabbros & Dolerites**		

B

Structural level of Clare Island

15 miles
Sea Level

Structural level of the mainland

N

F_4

F_2 Lineation

F_3

S_3

S_1

Axial plane of major F_1 recumbent fold

F_2

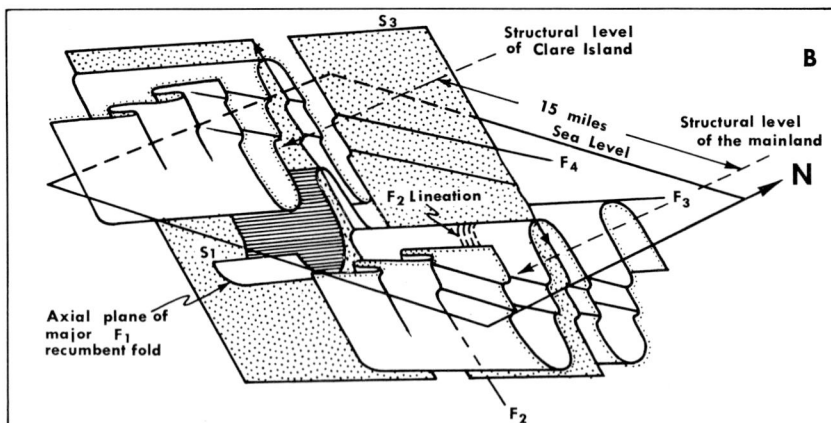

Figure 371. Structural history of the Dalradian, Ordovician, and Silurian rocks of south Mayo, W. Ireland. A: Tabular summary of tectonic events; H. B. F. indicates Highland Boundary Fault. B: Diagrammatic representation of F_3 isoclinal folds superposed on a major F_1 recumbent fold; black triangles indicate direction of facing of the F_3-folds (After Dewey and Phillips, 1963, Figs. 4 and 6).

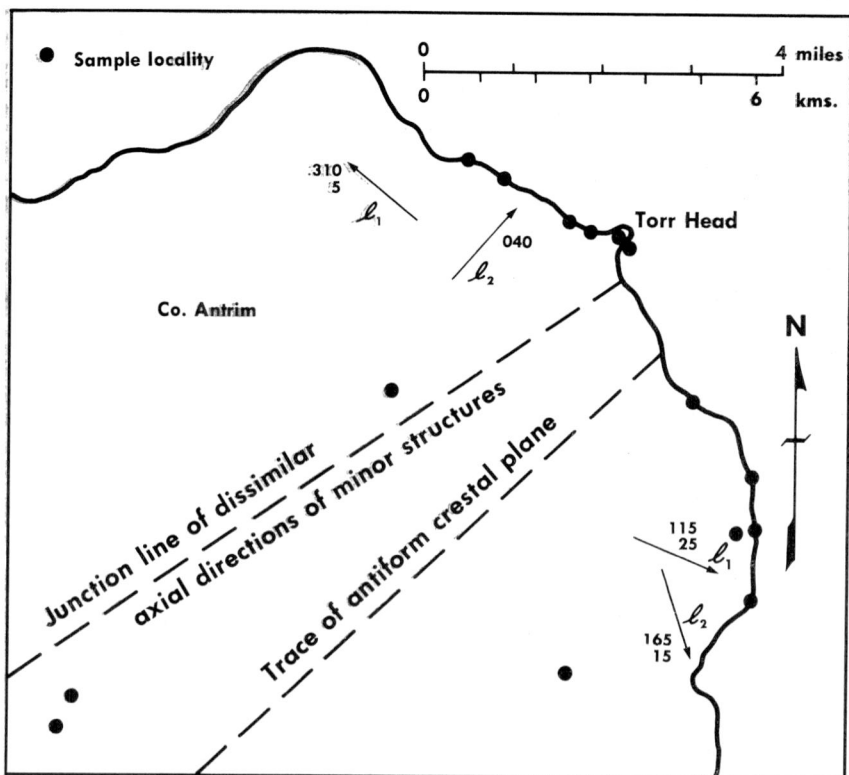

Figure 372. Structural elements of the Precambrian Dalradian rocks of NE Ireland (Based on Goldring, 1961).

mentary rocks of southwestern County Mayo. Biotite grade metamorphism accompanied the major F_1-folding; minor F_2-folds superposed on the Croagh Patrick synclinorium vary in plunge and facing direction according to the dip and facing direction of S_0 in the initial F_1-folds.

Preliminary details (Bailey and McCallien, 1934; Goldring, 1956) suggested that simple divergent fold trend relationships similar to those mapped in the Monadhliath, Central Scottish Highlands, occur in the Dalradian schists of northeast Ireland. However, the structural picture appears to be more complex (Goldring, 1961; Whitten, 1959A), and an earlier set of NW-SE linear structures (l_1) found throughout the area had a later set (l_2) superposed on them. The l_2-structures have different trends to the northwest and southeast of a line passing through Torr Head (Fig. 372), a geographical relationship analogous to that described for Strathspey, Scottish Highlands (Whitten, 1959A). The geometry is not completely clear, but if the deformation giving rise to l_2 had been more

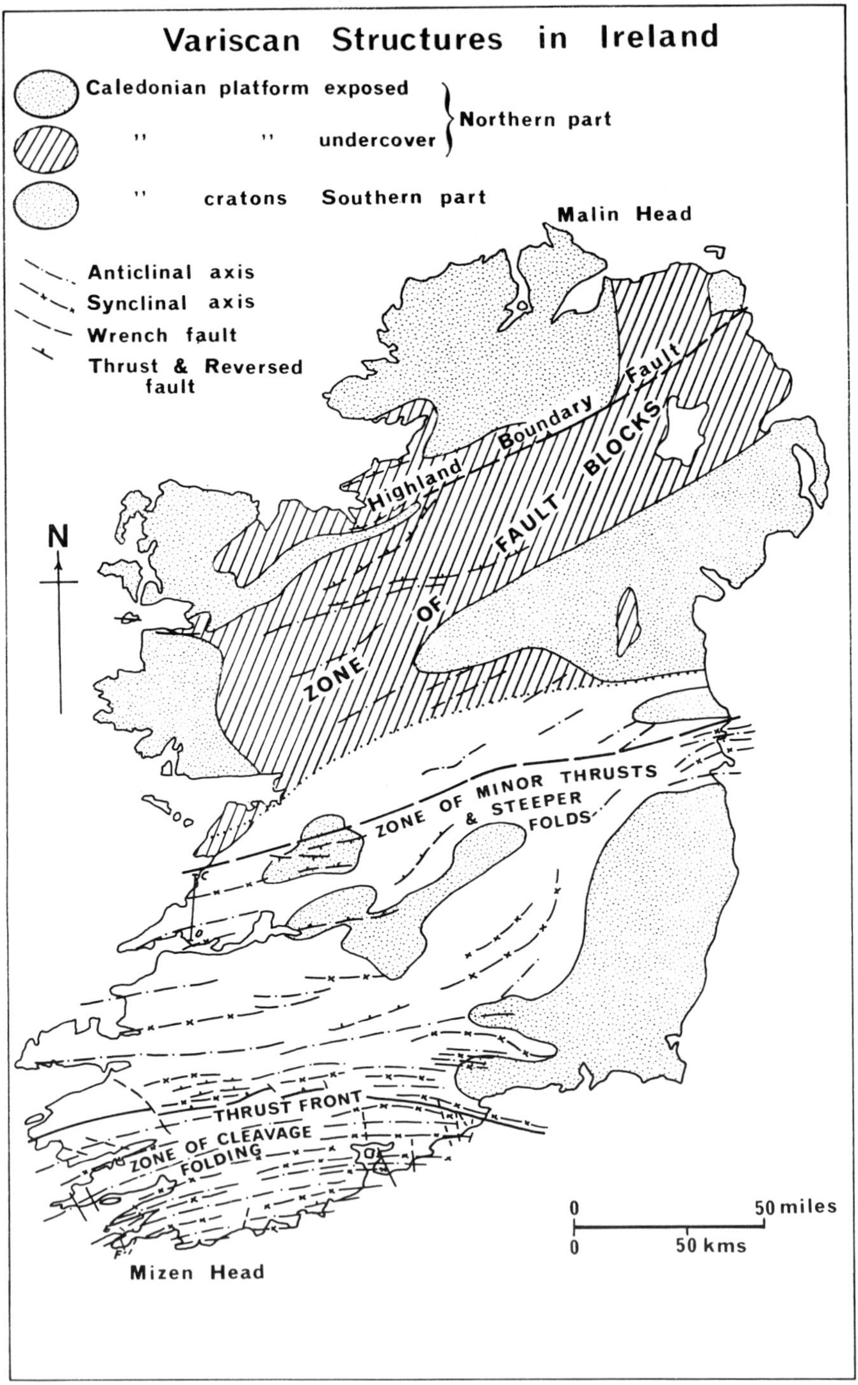

Figure 373. Variscan structures in Ireland (After Gill, in Coe, 1962, Fig. 3/I).

intense, the l_1-structures might have been completely overprinted to yield a closer apparent analogy with the Strathspey pattern.

Although most examples of regional superposed folds have been recognized in metamorphosed rocks, Gill's (in Coe, 1962) review of recent work in southern Ireland is a reminder that similar structures are common in unmetamorphosed noncratonic sedimentary sequences. Effects of the Variscan orogeny in Ireland become more· intense from north to south (Fig. 373). The belt through the center of Ireland is characterized by concentric folds. In the Slieve Ardagh coalfield, for example, plastic deformation of coal seams is associated with folding about northeast-southwest axes, which was followed by minor folding about east-west axes. This post-Westphalian folding produced intense crushing of the coal seams.

Farther south flexural-slip F_1-folds are prominent in the arenaceous members of the thick Old Red Sandstone and Carboniferous clastic sequence; axial-plane foliation developed in argillaceous members. Belts of minor F_2-folds with slightly divergent axes are common. The major F_1-structures plunge southwest in the Mizen Head area (Fig. 374); the trend of minor F_2-folds is 5 to 10 degrees from the major F_1-folds axes and they show plunge reversals over a distance of a few kilometers. The F_2-folding caused marked bending of the first foliation and deflection of the lithological boundaries defining the F_1-folds (Fig. 375). Locally, minor igneous activity occurred between the F_1- and F_2-folding; dykes transecting the F_1-folds were boudinaged during F_2-folding (Coe, 1959). Gill (in Coe, 1962) concluded that in the south of Ireland F_1-folding began in Nassauian times, and that the deformation continued into post-Westphalian time.

Great Britain (excluding Scottish Highlands and Ireland): Numerous incidental references to superposed folds occur in the literature, and only some of the more complete geometric studies are mentioned here. Greenly (1930) recorded axial-plane foliation refolded into isoclines (with a new foliation related to these second folds) in the Mona Complex of Anglesey, Wales. More recently, Weiss (1955) reinvestigated the geometry of superposed structures in some complex tectonites from Anglesey.

Despite the lengthy tectonic chronology established for the Scottish Highlands which involves gigantic folds with cores of plastically-deformed Lewisian basement, most of the extensive literature about the Caledonides to the south suggests that only a single phase of folding with an associated axial-plane foliation developed. However, it is possible that a single pattern of plastic flow extended from the gneissose and schistose roots in the northwest to the almost unmetamorphosed rocks of Wales in the south. Persistence of such a flow pattern would not imply tectonic

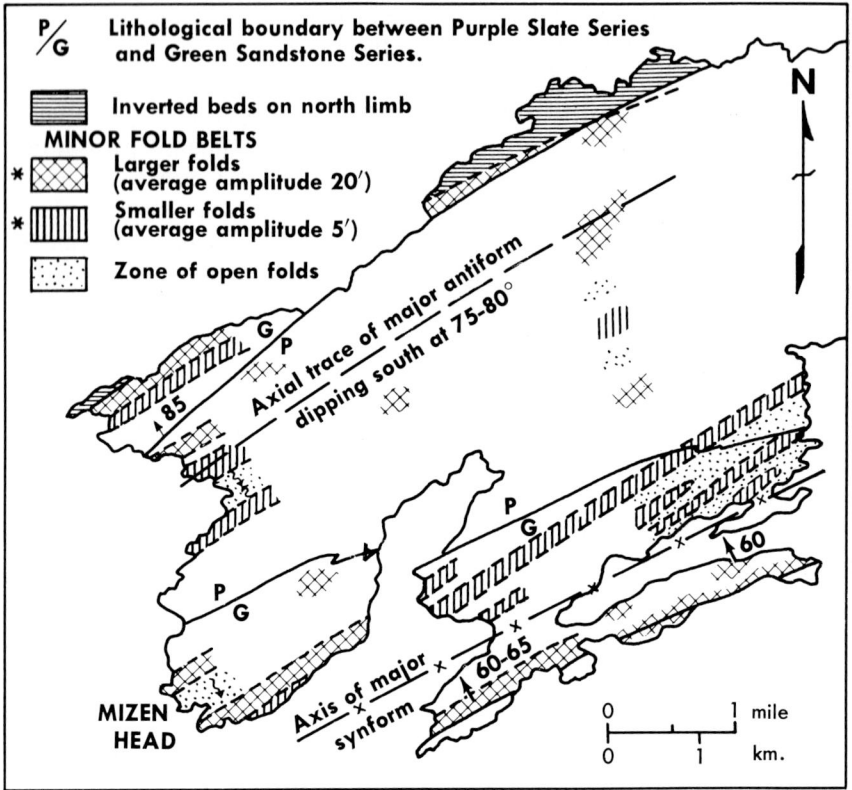

Figure 374. Fold structures in the Upper Paleozoic sedimentary rocks of the Mizen Head area, County Cork, SW Ireland (Based on work by T. A. Reilly; map after Gill in Coe, 1962, Fig. 3/V).

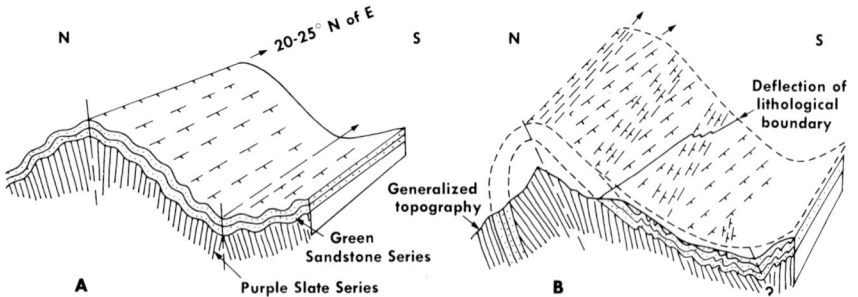

Figure 375. Structural interpretation of superposed fold structures in the Mizen Head area, County Cork, SW Ireland (Based on work by T. A. Reilly; diagrams after Gill in Coe, 1962, Fig. 3/VI).

transport over this great distance; Freedman, *et al.* (1964) postulated an analogous gradation across the Appalachian fold belt in Pennsylvania, U.S.A. Recent work suggests that polyphase Caledonide folding and concomitant superposed foliations extend a long distance south of the metamorphic terrain in the Scottish Highlands. At present precise spatial and temporal correlation between fold episodes in the Highlands and the area to the south cannot be assayed. Although the style of deformation may be broadly similar, the structures tend to be smaller southward; the structures also vary because of the differing lithology and the less severe stress and temperature conditions to the south.

Complex superposed Caledonide fold structures in the Silurian gray-wackes, siltstones, shales, and calcareous siltstones of the Southern Uplands are exposed along the Berwickshire coast of eastern Scotland (Dearman, *et al.*, 1962; Shiells and Dearman, 1963). The most complex structures involve folds of varied size and orientation in the Coldingham Beds; included are some good examples of eyed folds (see Chapter 10). Three quite distinct Caledonide folding episodes affected these rocks (Fig. 376). It is common to see a continuous passage from one fold to another at right angles, in which both folds share the same axial plane.

Considerably farther south, Simpson (1963) made a detailed geometric study of the Manx Slate Belt in the Isle of Man (Irish Sea). He demonstrated that these Cambrian sedimentary rocks experienced three Caledonide phases of superposed folding which attained a size and complexity comparable with those in parts of the Scottish Highlands; each phase (F_1, F_2, F_3) is clearly represented by both large and small folds. More recently Helm, *et al.* (1963) found that these three phases of folding can be recognized easily in the Cambrian through Ludlovian sedimentary rocks of North Wales, and in the Arenigian rocks of the English Lake District. They considered it inevitable that comparable Caledonide folds and foliations will be found in other parts of the Irish Sea region. The large F_1-folds and the associated axial-plane foliation (S_1) formed during the most severe phase of deformation in North Wales, although the deformation is less intense eastward; the F_1-folds had a Caledonoid trend and a southeasterly vergence.[6] F_2-folds occur sporadically in North Wales, but F_3-folding caused regional flexure of F_1-structures and development of congruous minor folds with steep axial planes (S_3). According to Helm, *et al.* (1963) these structures (and those in the Arenigian Skiddaw Slates of the Lake District) correspond exactly in geometry and style with those mapped in the Isle of Man by Simpson (1963).

[6]In recent British literature the term **vergence** has become popular; it is attributed to Cloos (1936) and simply refers to the direction up the dip of the axial surface of a fold.

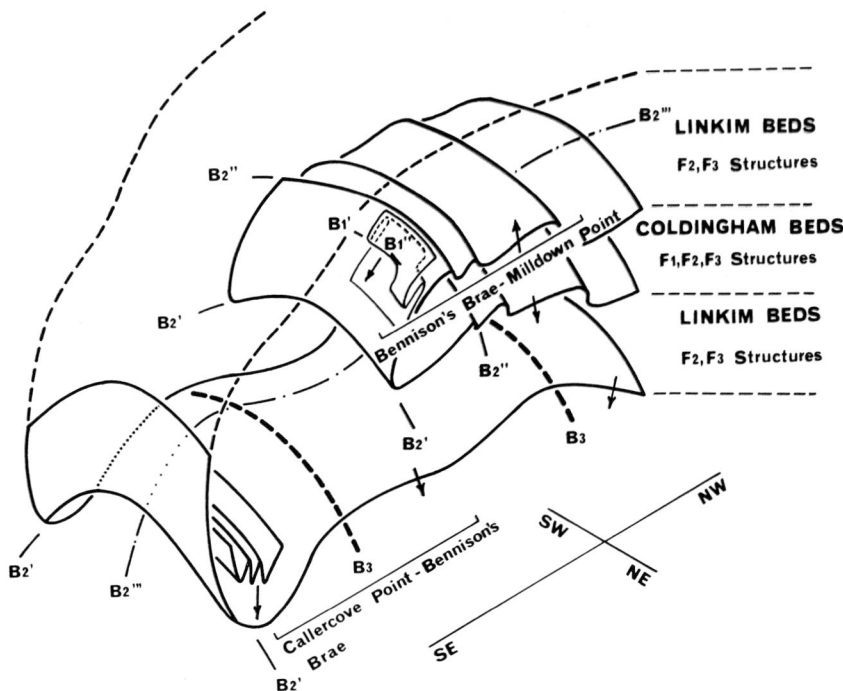

Figure 376. Superposed Caledonide folds in the Coldingham Bay area, Berwickshire, SE Scotland; the three superposed deformation phases are diagrammatically represented by:

F_1-folding phase that produced B_1'-isoclinal folds and B_1''-cross folds that refold B_1'.

F_2-deformation represented by a major B_2'-recumbent anticline, open B_2''-folds, and B_2'''-cross folding that produced culminations and depressions in the B_2'- and B_2''-structures.

F_3-deformation that gave regional flexures of Middle Old Red Sandstone age. (After Shiells and Dearman, 1963, Fig. 13).

Scandinavia: Crossing fold structures, formed during different phases of the same epoch of deformation, or in different periods of deformation, were described by Kvale (1948) in the Bergsdalen Quadrangle in the Caledonides of western Norway. Strand (1951) claimed that in the Sel and Vågå areas (southern Norway) NE-trending folds (parallel to the main Caledonide trend) formed first and that the rocks were subsequently refolded by folds with NW- or WNW-trends.

Bryhni (1962) also made a detailed geometric analysis of the superposed folds of the Gröneheia area, Eikefjord, western Norway (Fig. 377). F_1-folds are recumbent slip flexures with deformation penetrative at each level of sampling, but the fold axes form an arc closing to the west. F_2- and F_3-folding was less penetrative and the linear elements have variable

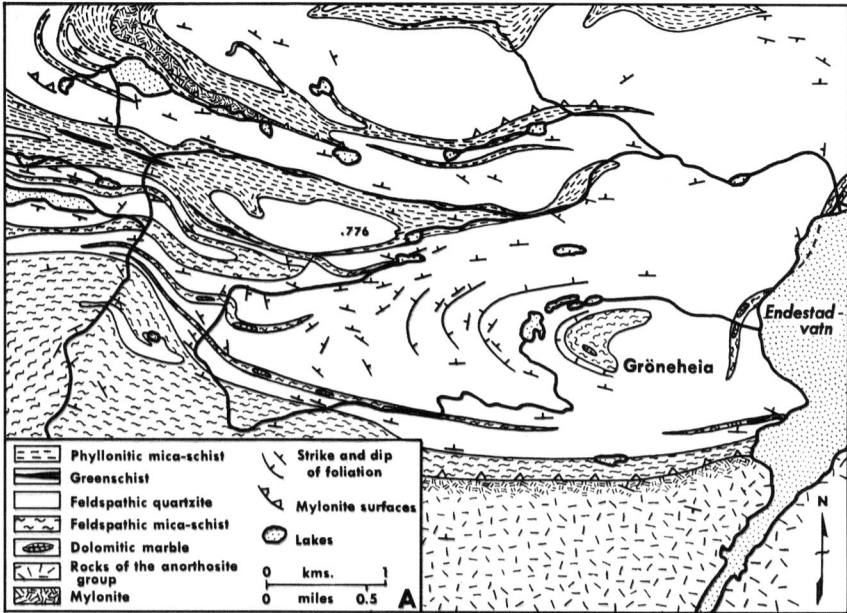

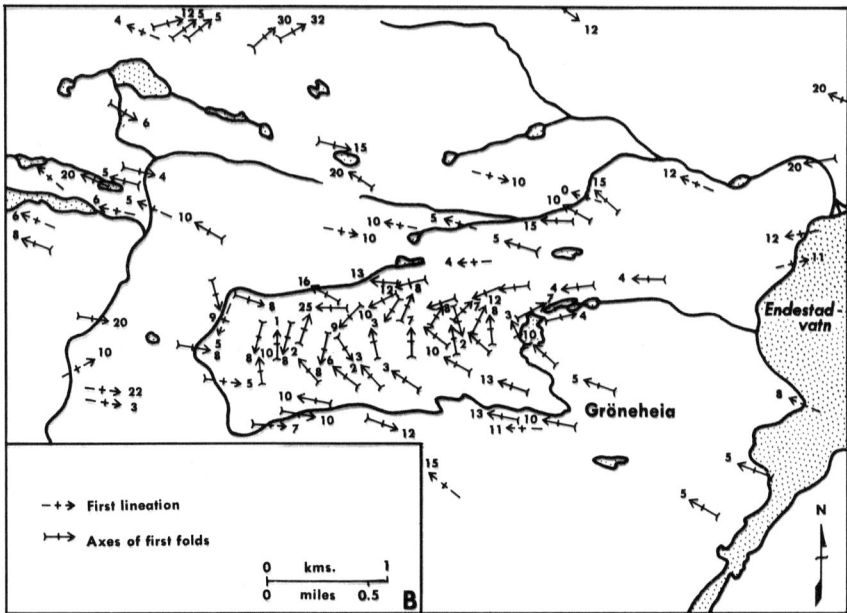

Figure 377. The Gröneheia area, Eikefjord, W Norway. A: Geological map. B. Map of first linear structures (After Bryhni, 1962, Figs. 2 and 7).

attitudes dependent on the orientation of the S-structures following F_1. By geometrical analysis of the relationships of F_3 to the earlier linear elements, Bryhni showed that F_3 are slip folds. However, the arcuate pattern of F_1-fold axes is somewhat enigmatic; it may suggest that F_0-structures were present, but have been largely obliterated. Bryhni (1962, p. 367) correctly claimed that it is not reasonable to draw general conclusions for the Caledonides from an isolated detailed geometric study of one small area; however, it is only with the aid of such detailed analyses that the significant structural elements can be identified.

Dons (in Barth and Dons, 1960) summarized details of work on superposed folding in southern Norway that includes numerous papers published between 1939 and 1957 by Michot (e.g., 1957) on the area between Sandnes and Vikeså, south of Stavanger and southern Rogaland. Michot concluded that the metasedimentary rocks of southern Rogaland were successively thrown into (a) recumbent folds and nappes with north-south axes, (b) recumbent folds and nappes with east-west axes, and finally (c) open folds with steep axial planes and east-west axes. Dons also mentioned the easily-recognized superposed folding in the Iveland-Evje area of southern Norway.

Recently detailed mapping has been carried out in some 900 sq. km. near Sokumvatn, in the heart of the Caledonide orogenic belt of northern Norway. Rutland (1959) showed that superposed folds can be demonstrated there very clearly because (a) several contrasting lithologic types occur, and (b) a phase of granite veining and granitization enables the fold episodes to be separated in time. An initial phase of intense isoclinal folding of the metasedimentary and metavolcanic rocks has been recognized (Rutland, 1959; Ackermann, *et al.,* 1960; Hollingworth, *et al.,* 1960; Rutland, *et al.,* 1960). Internal slip and major synmetamorphic slides were associated with formation of the recumbent isoclines and most of the formation contacts are now essentially tectonic. At Sokumvatn each isocline is confined to a single major lithologic unit, so that slides now separate the units. This deformation occurred during the main regional metamorphism (almandine amphibolite facies). There appears to have been a distinct time interval between the first and second deformations. The second folding took place during a period of granitization at a lower temperature and pressure than accompanied the main deformation. Minor folds associated with the two fold episodes differ in style, and the second set folded the early regional schistosity. The later folds are relatively open and generally indicate less plastic deformation than was associated with the first folds.

The present outcrop pattern is largely due to the second folds (Fig. 378). As would be expected, the second fold axes are variable in attitude

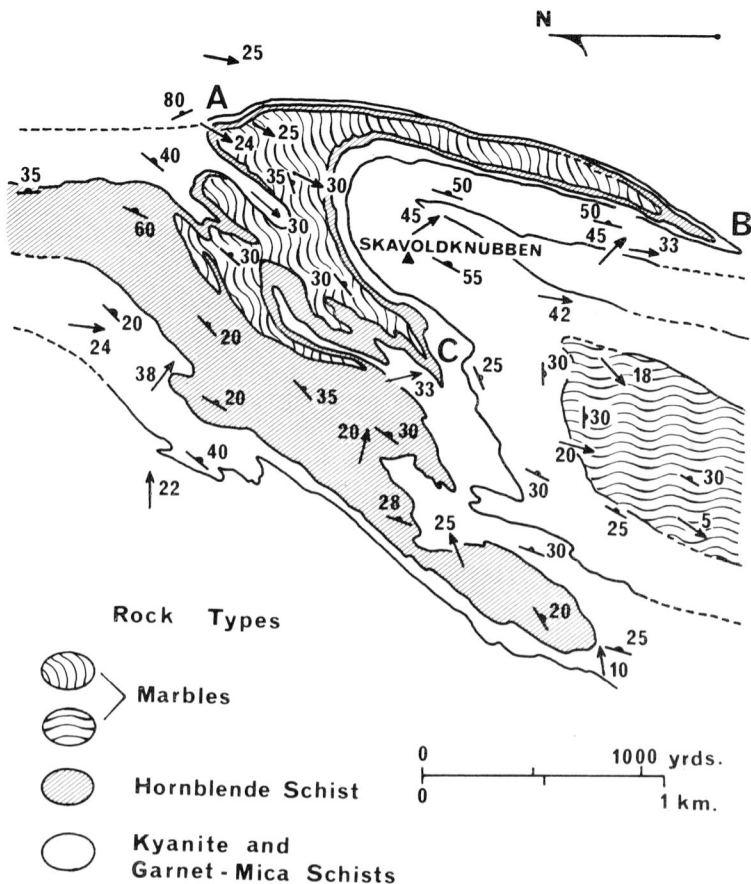

Figure 378. Geology of the Skavoldknubben area, near Glomfjord, Norway, showing the effects of superposed folds. Below is a simplified sketch showing interpretation of the structure and the nature of the crescent-shape outcrops between points A, B, and C (After Hollingworth, *et al.*, 1960, Fig. 2).

because their orientation depends on the attitude of the *S*-structures upon which they were superposed. In detail each lithology reacted to the stress differently; thus, Rutland (1959, p. 328) wrote:

The marbles have readily developed second minor folds and pelitic schists lineations with gentle plunges. But minor folds in more competent calc-silicate bands within the marbles often have steeper plunges than those of the marbles; and the thicker quartzite bands only develop warp structures with steep plunges.

Crenulation foliation was developed during the second folding at the expense of the first-phase regional schistosity. Some of the second-phase folding may have been partly controlled by uprise of the granite masses that abound in the area; Hollingworth, *et al.* (1960, p. 38) claimed: "... the trends of the folds near the granite margins are parallel to the latter but may change direction rapidly as the influence of the granites dies out away from the contacts."

Some of the banded marbles in the Sokumfjell Marble Group display small closed ("eyed") folds (Nicholson, 1963), as shown in Figures 379 and 380. Interlaminated pelites within the Marble Group show that porphyroblastic staurolite, kyanite, and garnet grew in the interval between the two main fold episodes. Many of the eyed folds belong to Types 1 and 2 of Ramsay (1962C) (see Chapter 10), but those in the Sandvand

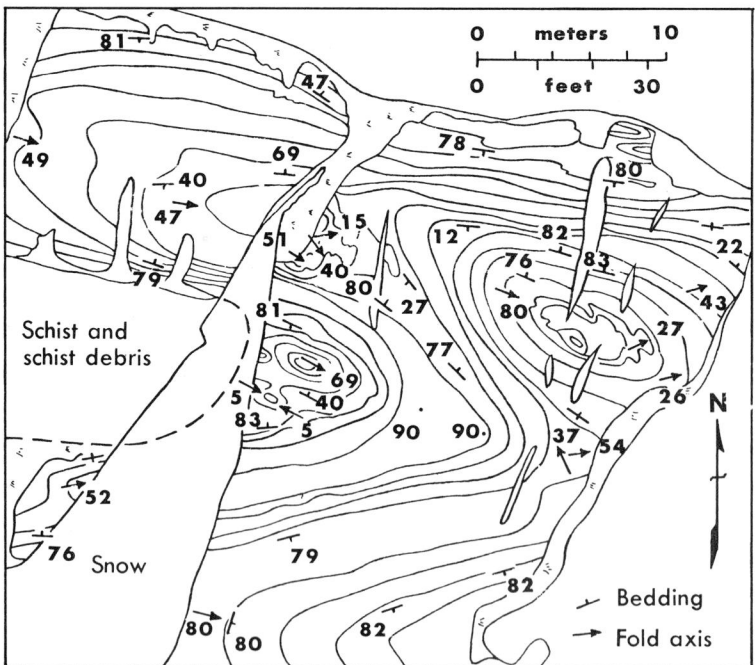

Figure 379. Closed fold structures in the Sandvand Grey Marble, N Norway (After Nicholson, 1963, Fig. 5).

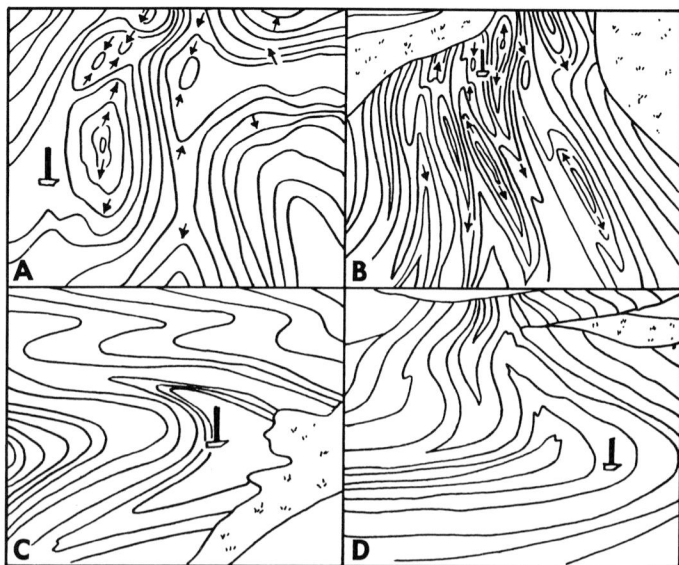

Figure 380. Superposed fold patterns from Sokumfjell Marble Group, N Norway (After Nicholson, 1963, Fig. 4).

Grey Marble are more irregular in form and areal distrbution. Nicholson invoked localized diapiric action (cf. salt tectonics) superposed on the first phase of folding to explain the latter.

Fifty kilometers northwest of the Caledonian Front in Sørøy, northern Norway, Ramsay and Sturt (1963) described the geometry of the repeated folding and metamorphism of the Eo-Cambrian sequence. In Sweden, Koark (1951) used microscopic petrofabric methods to investigate the refolding of initial isoclinal folds in the marble-leptite-amphibolite series exposed in quarries near Burträsk, Västerbotten. Lindström (1955A, 1955B, 1957) worked on the Caledonide structures of Swedish Lapland, but evaluation of the geometry of these complex superposed folds requires more detailed quantitative work.

Weiss (1953) presented well-analyzed data for an area in Vestspitzbergen; here the long time interval between the superposed deformations can be measured stratigraphically. The earlier folding affected only the Hecla Hook formation (main "Caledonian" axis of folding), and later folding affected the overlying Carboniferous rocks, probably during a Lower Tertiary deformation.

Greenland: The Caledonides of eastern Greenland provide ample opportunity for the study of superposed folding, as shown by the mapping of Fränkl (1953), Haller (1956A, 1956B, 1956C), Cowie and Adams

(1957), and Sommer (1957). This followed the important studies by Wegmann (1935A, 1935B, 1938) which included broad studies and interpretations of the remobilized and refolded infrastructure. In eastern Greenland the Upper Precambrian and Lower Paleozoic rocks possess a dominant northeasterly (Caledonide) trend, which is interrupted by east-west folds that produce axial culminations and troughs along the Caledonide trend. It is generally accepted that the east-west structures are older (pre- to early-Caledonide) and resulted from the rise of a migmatite front through the basement (Haller, 1955), although, in the Kempes Fjord area, at least, Haller (1957) believed that synchronous movement about both axes occurred.

These structures in the superstructural rocks are spectacular, but detailed work revealed much more complex superposed Precambrian structures in the infrastructure that are well exposed on the west and southwest coasts of Greenland. This is part of the Ketilide fold belt. In the Ivigtut region, for example, Berthelsen (1960A) recognized three distinct phases of Ketilide folding within the migmatitic and gneissic infrastructure, and less complex superposed folding within rocks of the superstructure too. The earliest deformation was premigmatite recumbent isoclinal folding; in the next phase the largest disharmonic folds developed concomitant with migmatization and gneissification; the third phase of Ketilide deformation comprised "semi-plastic refolding, twisting and bending."

According to Berthelsen the original stratigraphic sequence in the Tovqussap nunâ area was about 1,000 meters thick. The earliest recognized deformation (the Midterhøj phase) resulted in the Midterhøj and Kronehøj) isoclines (Fig. 381A). During the second (Smalledal) phase several recumbent or overturned folds developed at the expense of the Midterhøj structures (Fig. 381B). Next parasitic folds formed in the central portion of the Tovqussap dome during regional doming (Fig. 381C); in the ensuing Pâkitsoq phase of refolding, extremely complex structures resulted (Fig. 381D). These four phases represented the first main cycle of deformation, during which metamorphism reached the granulite facies. Hypabyssal diorites are post kinematic with respect to this main cycle, and these intrusions permit the younger second cycle to be recognized easily. Slip folding associated with almandine amphibolite facies metamorphism predominated during the relatively weak second cycle.

The sequence of superposed folds visualized in the Tovqussap nunâ area (Berthelsen, 1957, 1960B) is very complicated. Berthelsen admitted that his reconstructions are not wholly objective, but with the aid of his

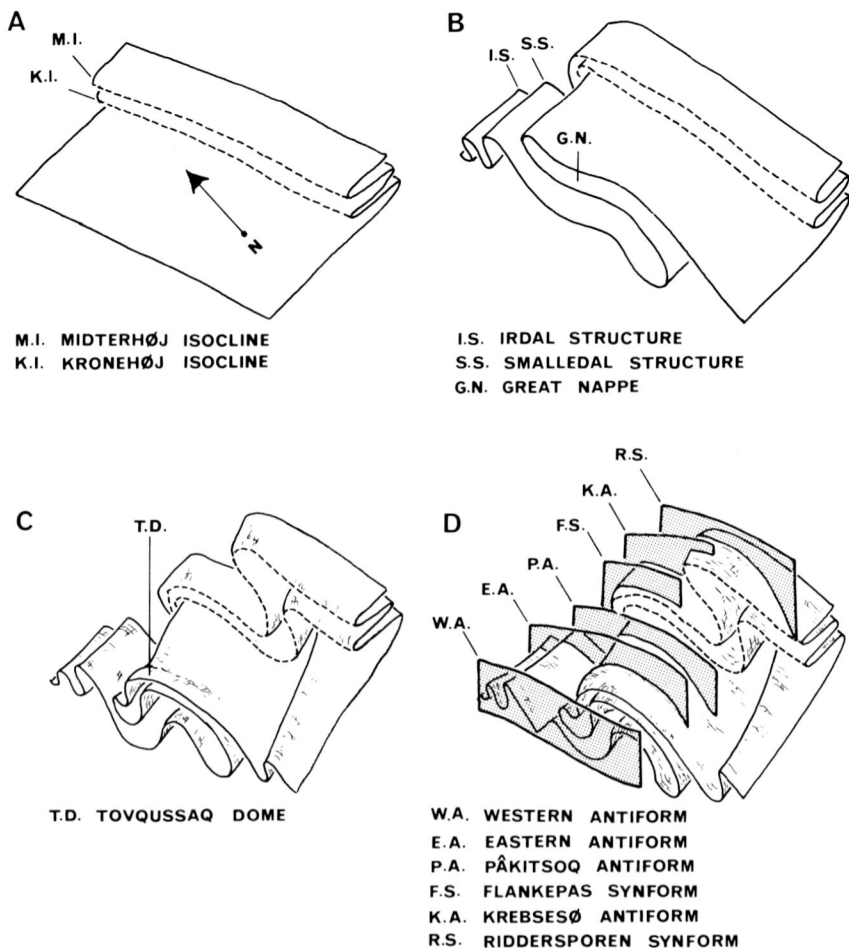

A

M.I.
K.I.

M.I. MIDTERHØJ ISOCLINE
K.I. KRONEHØJ ISOCLINE

B

I.S. S.S.
G.N.

I.S. IRDAL STRUCTURE
S.S. SMALLEDAL STRUCTURE
G.N. GREAT NAPPE

C

T.D.

T.D. TOVQUSSAQ DOME

D

R.S.
K.A.
F.S.
P.A.
E.A.
W.A.

W.A. WESTERN ANTIFORM
E.A. EASTERN ANTIFORM
P.A. PÅKITSOQ ANTIFORM
F.S. FLANKEPAS SYNFORM
K.A. KREBSESØ ANTIFORM
R.S. RIDDERSPOREN SYNFORM

Figure 381. Kinematic evolution (schematic) of the Tovqussap structures, Tovqussap nunâ area, W Greenland (After Berthelsen, 1960B, Fig. 78).

structure-contour-map method (Berthelsen, 1960C), he demonstrated the complexity of structure that can develop in orogenic belts. Figure 382 shows the Tovqussap nunâ area, and Figures 383 and 384 are structural stereograms of two of the complex structures. Equally impressive structures undoubtedly occur in many parts of the world, but without extremely good exposure, well-defined marker horizons, and adequate relief, very involved patterns cannot be unravelled as easily as along the Greenland coasts.

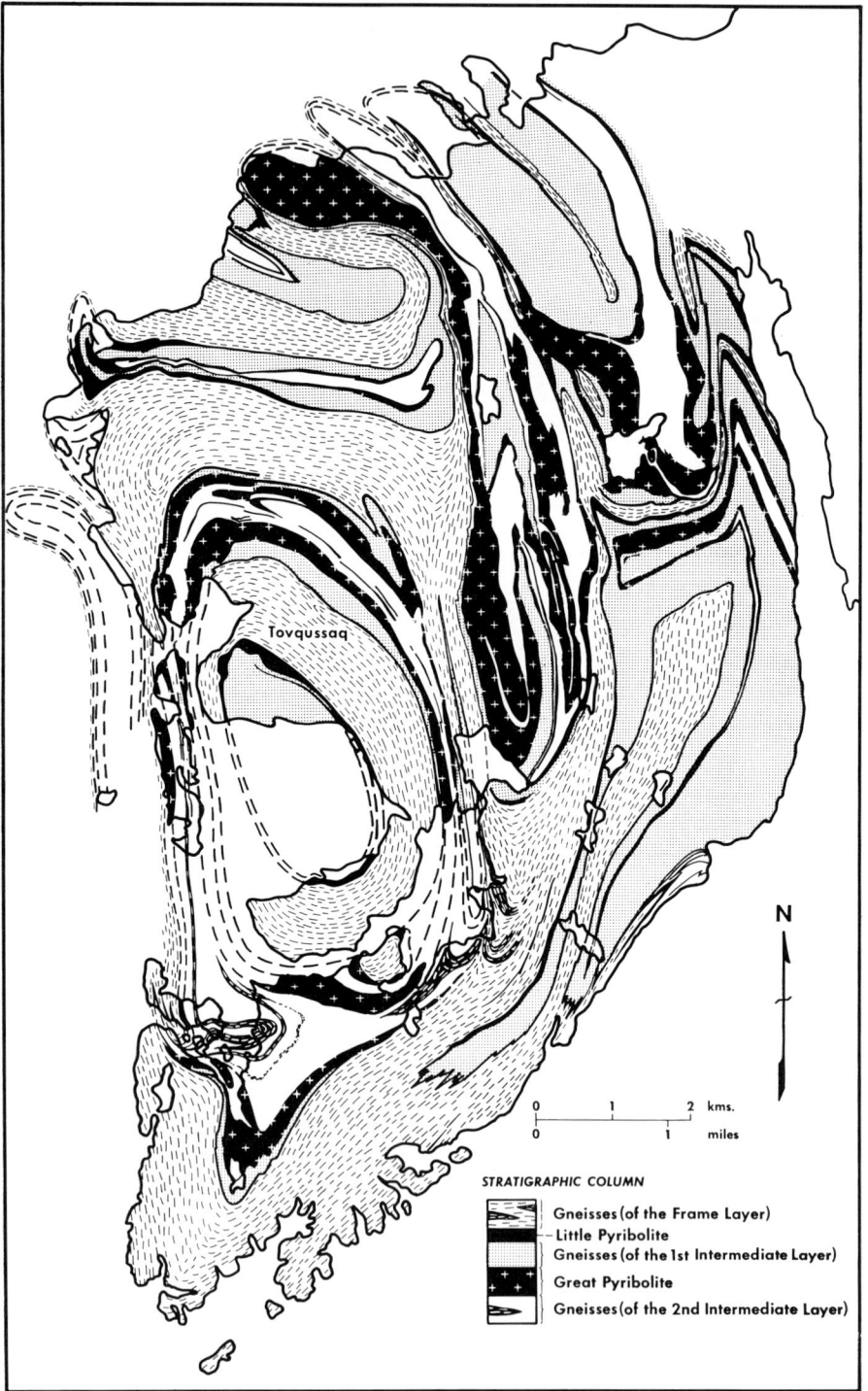

Tovqussaq

N

| 0 | | 1 | | 2 | kms. |
| 0 | | | 1 | | miles |

STRATIGRAPHIC COLUMN

Gneisses (of the Frame Layer)

Little Pyribolite

Gneisses (of the 1st Intermediate Layer)

Great Pyribolite

Gneisses (of the 2nd Intermediate Layer)

Figure 382 (facing page). Geological map of Tovqussap nunâ, Greenland (Based on Berthelsen, 1960B, Plates 1 and 4).

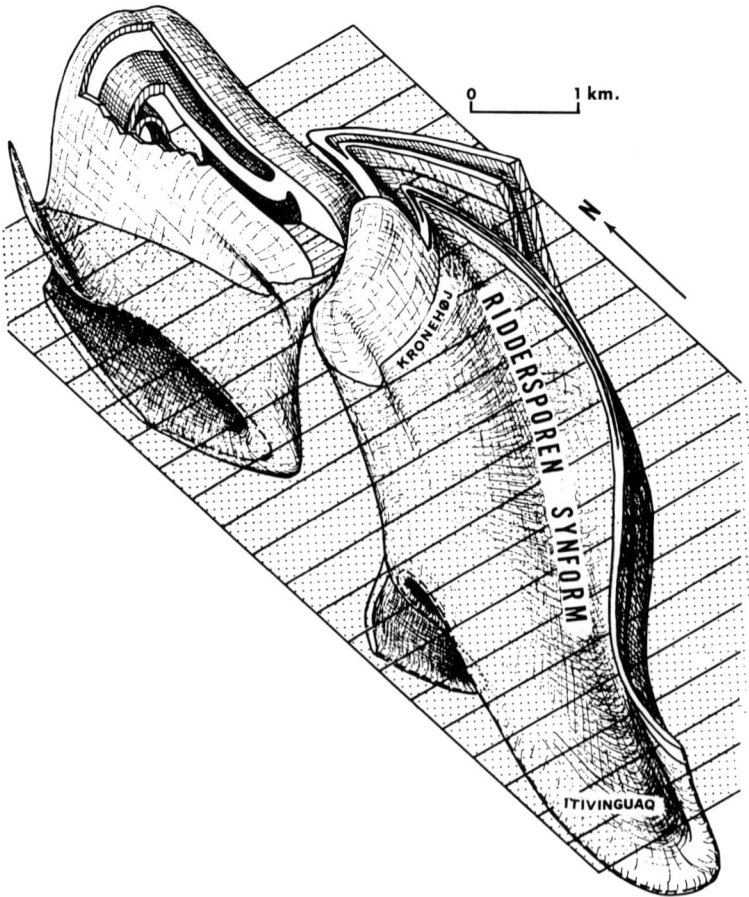

Figure 383. Structural stereogram of eastern Tovqussap nunâ, W Greenland; the stippled plane is the 250 meter level (After Berthelsen, 1960B, Fig. 64).

Continental Europe (excluding Scandinavia): Many examples of superposed folding might be quoted from this region, but the Alpine refolding of Hercynian structures is some of the best documented. Although deformed principally in the Hercynian orogeny, the Pyrenees experienced tectonic rejuvenation during the late Cretaceous and in the early Tertiary (de Sitter, 1956B, p. 322), and provide a good example of an older orogenic belt deformed during a later orogeny.

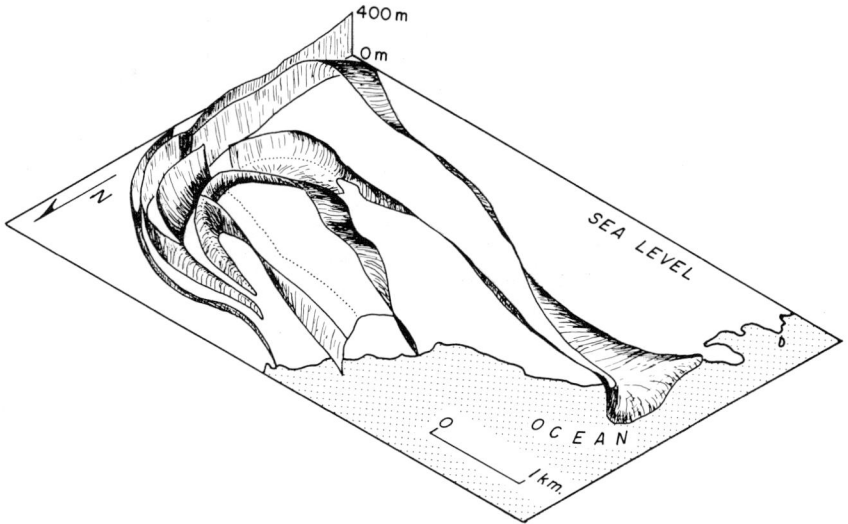

Figure 384. Structural stereogram of the Irdal and Smalledal structures, W Greenland (After Berthelsen, 1960B, Fig. 66).

In the Cantabrian fold belt, northern Spain (Fig. 385), there is sufficient stratigraphic control to allow quite accurate dating of the successive pulses of folding. De Sitter (1959; 1960B; 1962; in Coe, 1962) described and mapped the stratigraphic sequence and tectonics. A more or less continuous stratigraphic sequence occurs from the Lower Cambrian to the Upper Carboniferous, but during the Hercynian orogeny at least three major periods of folding occurred (see Table 4).

Wealden (Cretaceous) sands and pebble beds near Colle (Fig. 385) lie unconformably on both the Stephanian and the Devonian rocks of the F_1-nappe structures. The thrust planes were originally approximately flat-lying, but they have been refolded twice (Figs. 386 and 387). Figure 388 shows part of Figure 385 on a larger scale; the trace of the axial plane of

Table 4.—Sequence of Folding Episodes Recognized in the Cantabrian Fold Belt, Northern Spain

F_1	Nappes and folds of varying trends—WNW, W and WSW	post-Namurian	Sudetic	
F_2	WNW and NNE-folds		Asturian	Hercynian
F_3	WNW- to W-trending folds of Stephanian sedimentary rocks that lie unconformably on F_2-structures	between Permian and Triassic	Saalic	
F_4	E-W boundary flexures			Alpine

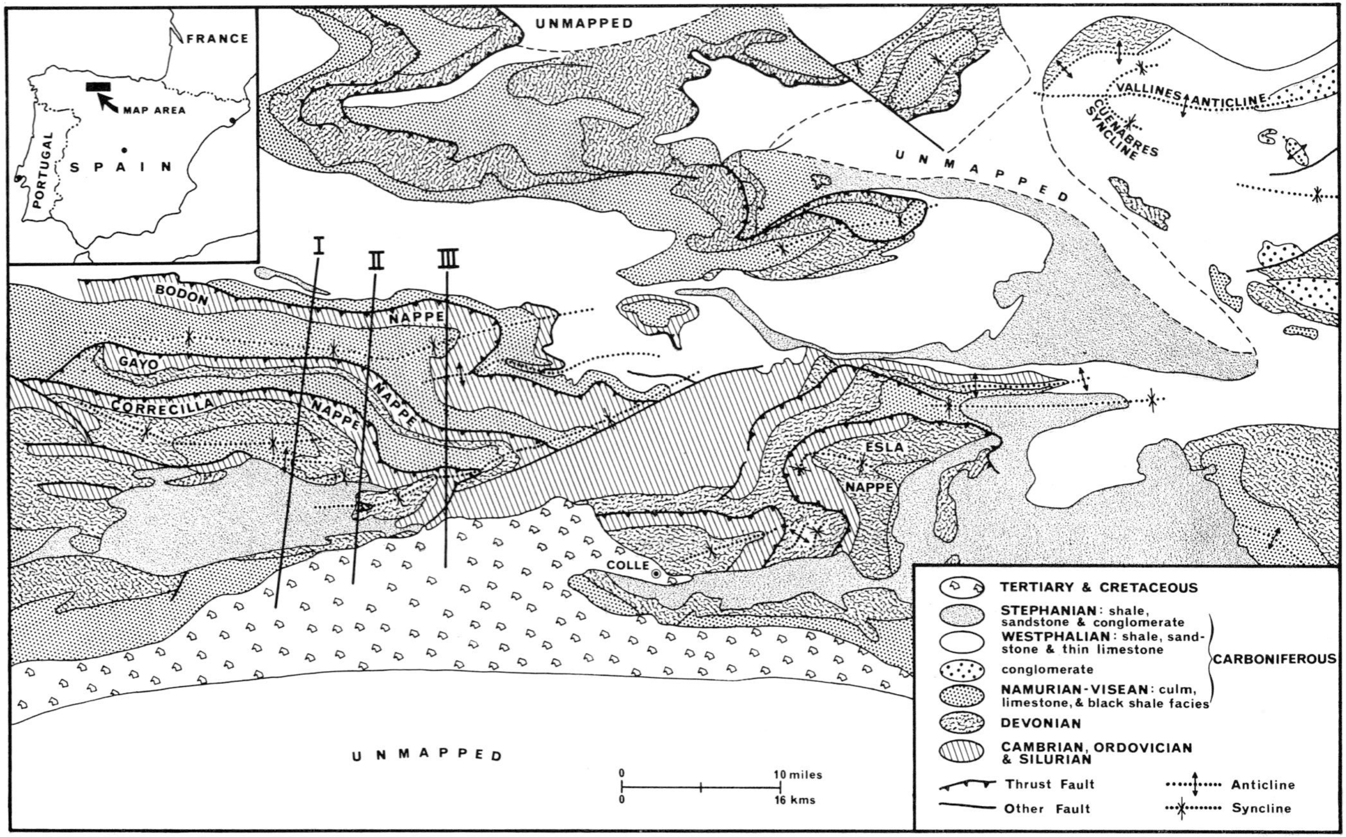

Figure 385. Structural geology of part of the Cantabrian fold belt, northern Spain; I, II, and III are lines of section in Figure 386 (After de Sitter, 1962).

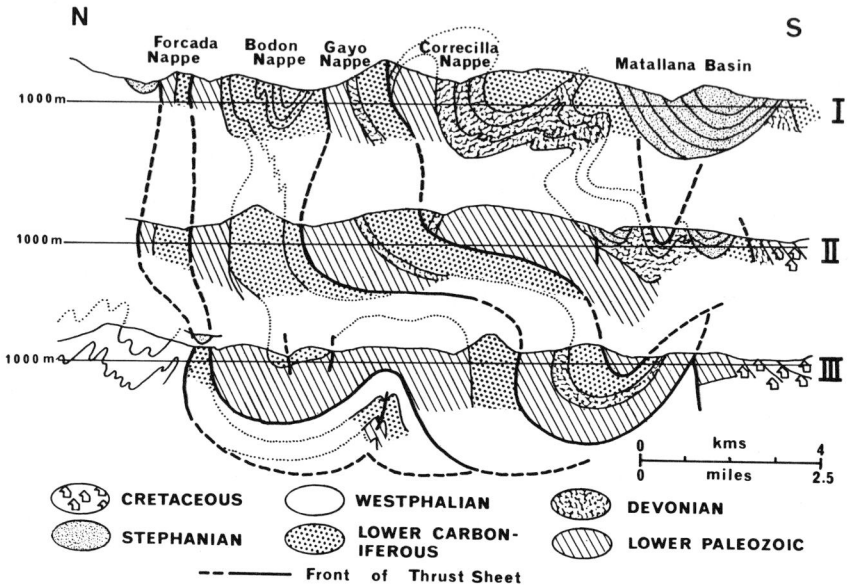

Figure 386. Vertical sections across part of the Cantabrian fold belt, northern Spain, to show the folded thrust sheets; lines of section are shown on Figure 385 (After de Sitter, in Coe, 1962, Fig. 1/IV).

the Cuenabres syncline (F_2) was originally N-S, but it was refolded during development of the Vallines anticline (F_3); intersecting foliations relating to F_2 and F_3 occur, but that associated with F_3-folds predominates.

The Tertiary F_4-folding (Alpine) was relatively unimportant in the Cantabrian area (Fig. 385), but, on the southern border of the Pyrenees, Tertiary refolding of Hercynian structures is more strongly developed (Fig. 389), and was described by Zwart (1953), Schulman (1959), Richter (1963), and others. The Pyrenean structures are discussed in more detail in Chapter 12.

The axes of approximately synchronous folds have variable trends within the Cantabrian area (Fig. 385); this must be anticipated with superposed folds, and a particular phase of folding cannot necessarily be recognized by a consistent mapped regional trend. It is a striking feature of the Cantabrian area that many blocks of country that were folded early, escaped the direct influence of later phases of orogeny (de Sitter, 1960B). In the same way that sedimentary units vary in three dimensions, so phases of folding occur at different times in different subareas; hence, strict synchrony between the different phases of Hercynian folding cannot be anticiapted throughout the Cantabrian region, the Pyrenees, and the Alps. In a carefully-documented and detailed account of 300 sq. km. in the Garonne Dome and Valle de Arán areas, Central Pyrenees,

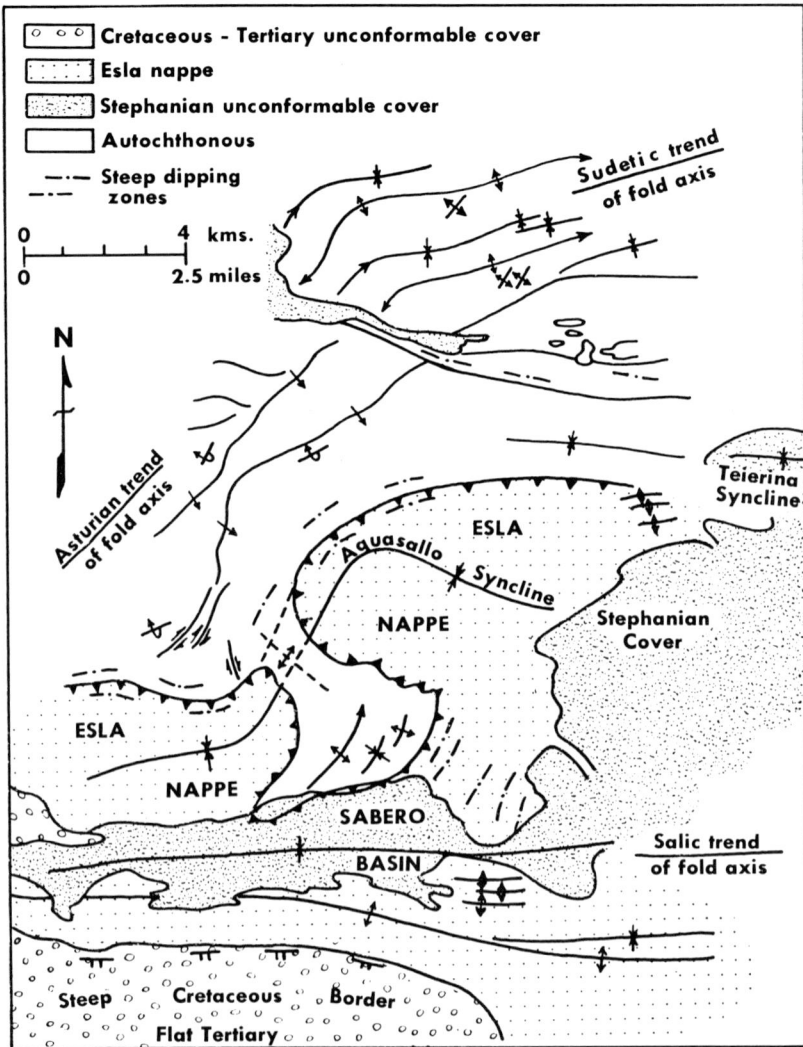

Figure 387. Structural geological map of the Rio Esla region, Cantabrian fold belt, northern Spain, showing traces of the fold axial surfaces folded by superposed structural events (After de Sitter, 1959, Fig. 3).

Boschma (1963) mapped the discrete areas in which the several Hercynian episodes of folding were recognized (Fig. 390); sometimes two phases are seen in a hand sample (Fig. 391).

Many additional examples might be cited. For example, Kvale (1957) used petrofabric methods to study Hercynian and superposed Alpine folds in the Gotthard massif and adjoining areas of Switzerland. In the

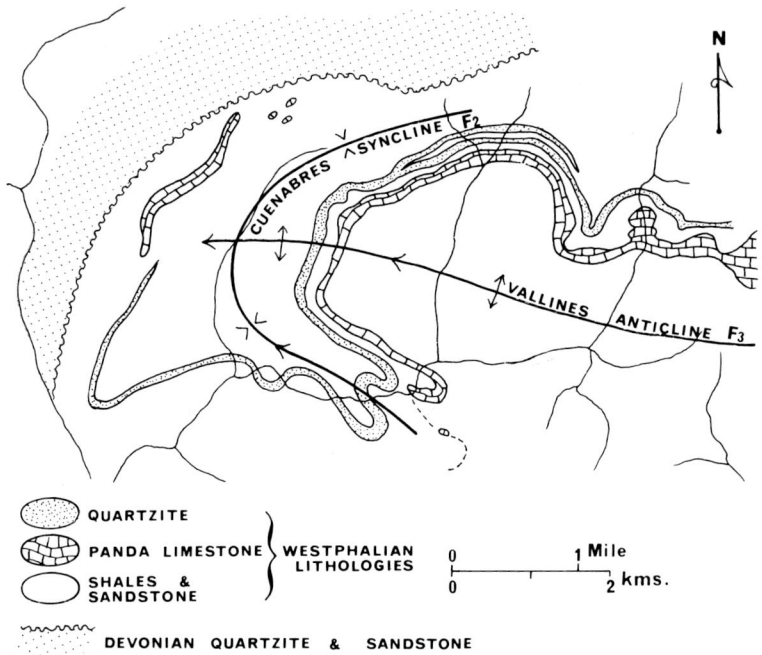

Figure 388. Part of the Cantabrian fold belt, northern Spain, showing the Cuenabres syncline (F_2) refolded by the Vallines anticline (F_3) (After de Sitter, 1960B, Fig. 3).

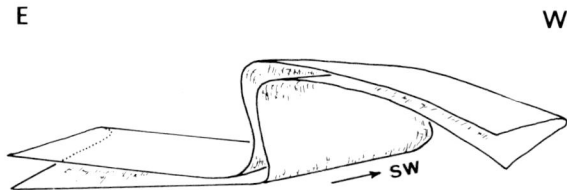

Figure 389. Diagrammatic sketch of a Tertiary F_4-fold affecting a Hercynian recumbent fold studied by Schulman (1959) in the Nogueras zone, southern Pyrenees, Spain (After de Sitter, 1960B, Fig. 5).

Hercynian basement of the Belledonne massif, French Alps, parallel schistosity and lithological banding are thought to have resulted from isoclinal folding; in a later phase of folding small folds formed that include eyed folds (Kalsbeek, 1962, Fig. VI-2) similar to those described by Ramsay (1962C) (see Chapter 10). Mapping in the Dolomites of northern Italy shows that Hercynian fold structures were refolded by the Alpine deformation; these structures were the subject of quantitative studies

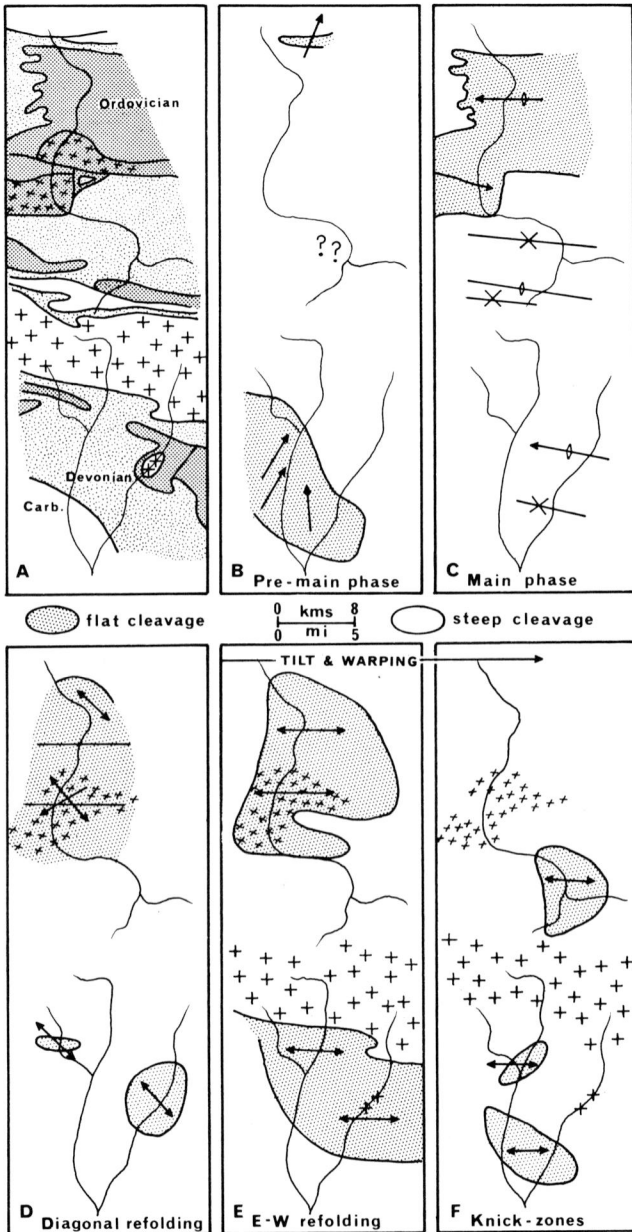

Figure 390. Distribution of successive phases of Hercynian folding in the Garonne Dome and Valle de Arán areas, Central Pyrenees. A: Schematic geological map. B-F: Successive fold phases with fold axes (arrows) and igneous rocks (crosses) shown (After Boschma, 1963, Fig. 101).

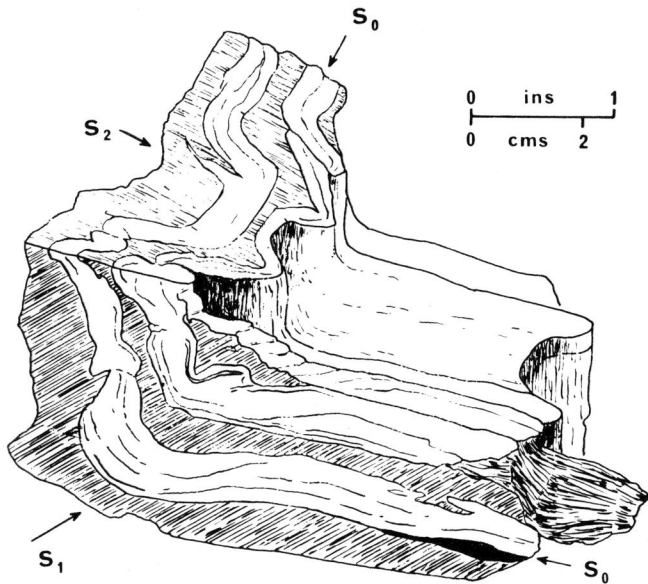

Figure 391. Hand sample showing effects of superposed Hercynian folding, north of Bohi, Central Pyrenees, Spain (After Boschma, 1963, Fig. 70).

(e.g., Agterberg, 1961, 1964B; D'Amico, 1964) that are referred to in Chapter 14 (Figs. 475 and 476).

In addition to refolding of the basement, the overlying sedimentary rocks were also subjected to severe folding during the Alpine orogeny and the structural geometry is often complex. Commonly folds have varied or perpendicular trends within small areas, and such relationships can extend over large regions—for example, in NW Italy from east of Genoa to Switzerland (e.g., Wunderlich, 1963A, Fig. 4). The complex Alpine folds characteristically represent a long sequence of successive folding events as, for example, in the Merano region, Bolzano, north Italy (Dietzel, 1960). Effects of superposed fold phases produced during the Alpine orogeny are displayed excellently in the Isle of Elba, Italy (Fig. 392), and were described by Wunderlich (e.g., 1963B).

In southern Sardinia, Italy, large E-W-folds formed during Upper Cambrian time, and associated N-S-folds owe their origin to deformation in the Hercynian orogeny. These folds produced an intricate interference pattern (Fig. 393) recognized by Teichmüller (1931), Vardabasso (1941, 1956), and Poll and Zwart (1964); the latter found that the complicated minor structures and microstructures seen in the field and under the microscope indicate that the Hercynian folding should be separated into three discrete phases.

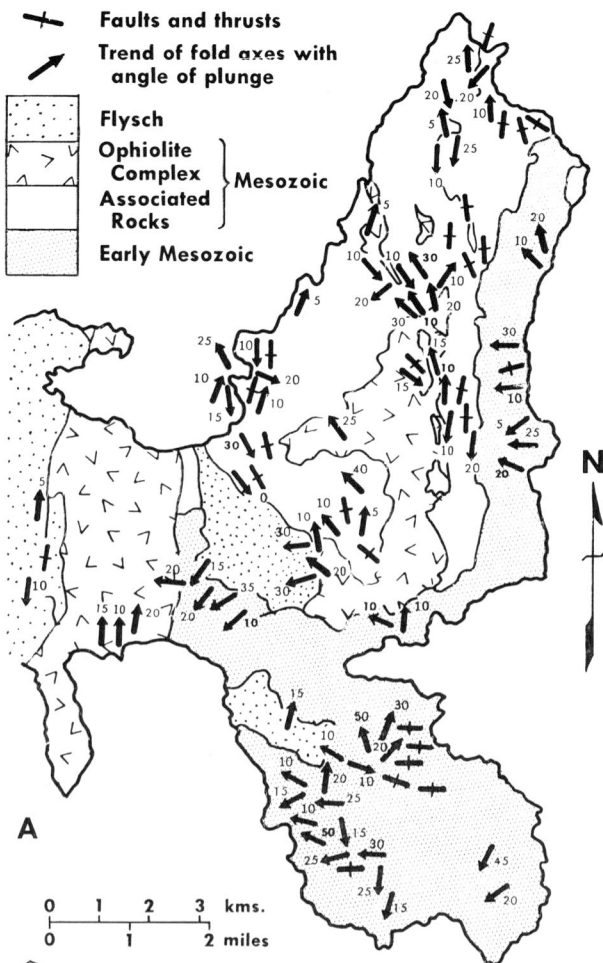

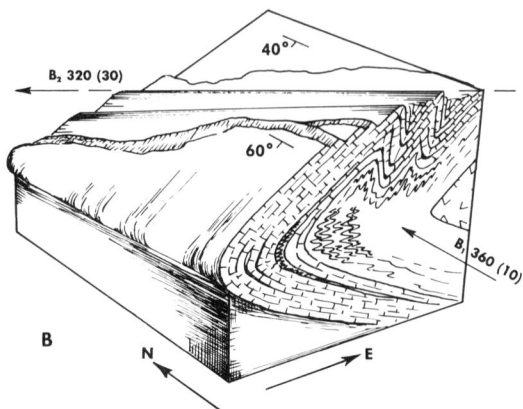

Figure 392. Superposed folds at east end of the Isle of Elba, Italy. A: Map of the fold axes. B: Schematic block diagram of two fold systems affecting the Mesozoic rocks of Elba (After Wunderlich, 1963B, Figs. 3 and 4).

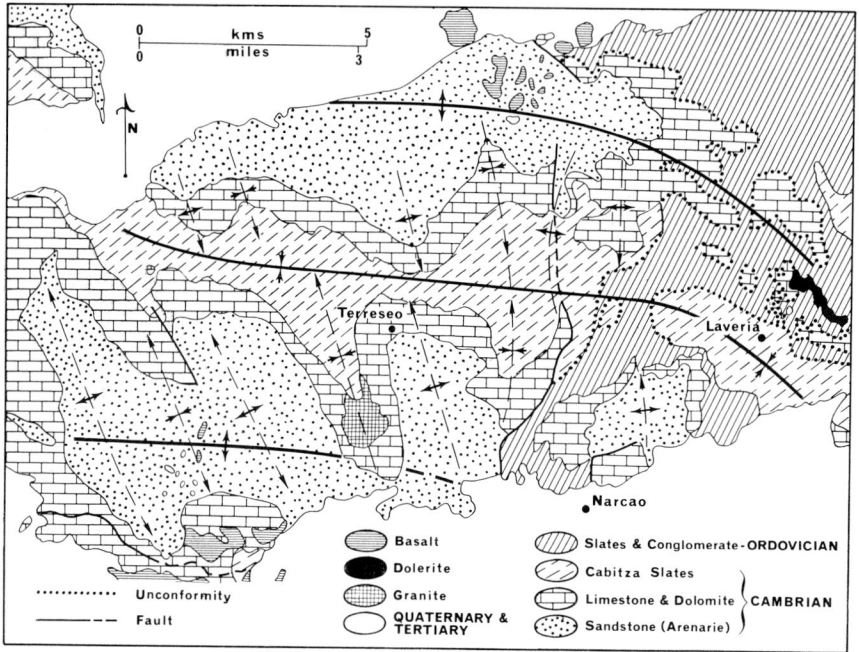

Figure 393. Geological map of the Rosas-Terreseo area, Sardinia, Italy, showing two sets of major folds (After Poll and Zwart, 1964, Fig. 1).

Belov, *et al.* (1960) worked in the Caucasus, Schreyer (1959A) in the southern Bavarian forest, Hoffmann (1961) in the Hohen Venns, and Hoeppener (1960) and Breddin (1962) made detailed studies in the Schiefergebirge; all these authors quote evidence for superposed folding. As detailed work continues, complex fold systems will be described more commonly.

North America: In the North American continent superposed folds are widespread, but the number of specific references to this type of deformation is surprisingly small. Few references describe detailed geometric analyses, although sometimes the gross stratigraphic mapping clearly demonstrates oblique fold trends superposed on an earlier generation. A few examples drawn from Canada and the conterminous United States are mentioned to illustrate the diversity of settings in which superposed folds have been noticed.

United States: Broughton (1946) studied superposed folds in the Martinsburg Formation of the Appalachian fold belt, New York (see

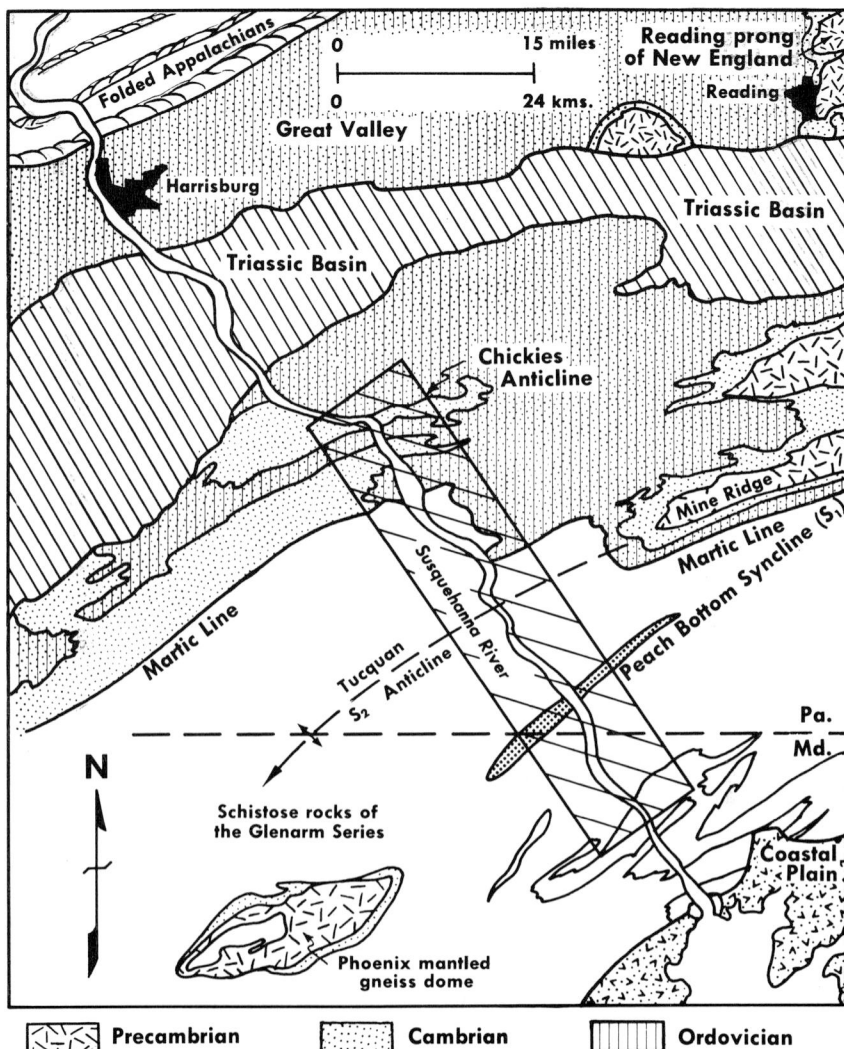

Figure 394. Regional setting of the Pennsylvania Piedmont area, U. S. A. Rectangular area (ruled) along the Susquehanna River is region of detailed structural analysis (After Freedman, *et al.*, 1964, Fig. 1).

Chapter 8), and Agron (1950) referred briefly to minor superposed folds in the Peach Bottom Slates along the Susquehanna River.

Freedman, *et al.* (1962, 1964) made a detailed geometric study along a 50 km. section through the Pennsylvania Piedmont and across the Martic Line (Fig. 394). They gave a lucid description of the clear evidence for three fold episodes in the area. During the F_1-folding the entire rock mass (including quartzites) apparently behaved as a single viscous unit in which S_1 corresponds to the planes of laminar flow (Fig. 395). The F_1-folds are isoclinal and S_0 is transposed and parallel to S_1, which represents a major schistosity produced under greenschist facies. Nappes and recumbent folds mapped by Geyer, et al. (1958) in the Great Valley (Fig. 461) can be traced (e.g., Wise, 1958) into the present area, which must represent the root zone (Fig. 395). Freedman, et al. (1964) thought uniform flow movements penetrated the whole rock mass, but this does not imply tectonic transport of rock for 50 kilometers. The second major fold event (F_2) resulted mainly from slip up or down along Gleit-brett to produce slip folds (Fig. 396); S_2 is commonly a crenulation foliation although complete recrystallization of mica occurred in some lithologies (e.g., Peach Bottom Slates). A third minor deformation was not equally developed throughout the area (Figs. 395, 396); in the north, crenulation foliation and rare folds contrast with abundant slip folds (F_3) in the south. The fourth phase is represented by planar zones (S_4; Fig. 396) spaced a few, or even a hundred or two, meters apart that bend earlier S-structures into S-shaped flexures.

Scotford (1955, 1956) introduced the term "axial-plane fold" for large secondary folds of older fold axial planes in the Poundrige area, New York. The term was used later by several writers (e.g., Ward, 1959, for the Wilmington complex).

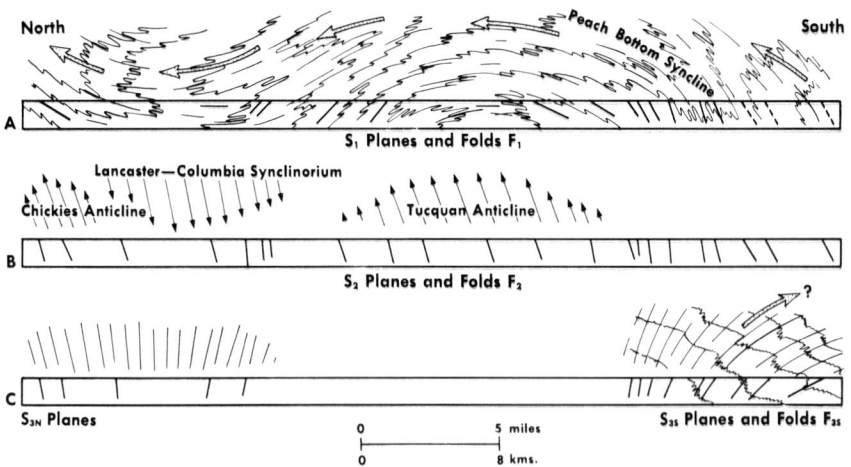

Figure 395. Diagrammatic section along the Susquehanna River, Pennsylvania, U. S. A., to show the orientation of the various S-structures (After Freedman, et al., 1964, Fig. 6).

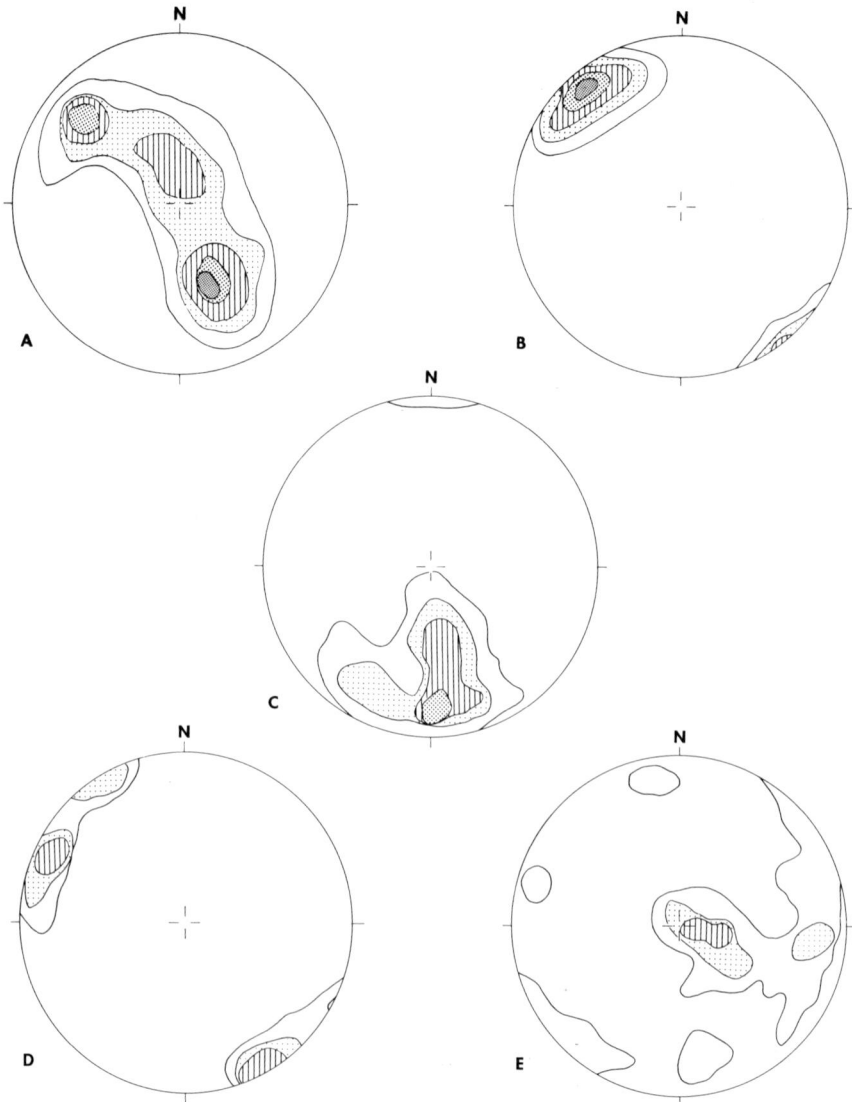

Figure 396. Synoptic diagrams of *S*-structures exposed along the Susquehanna River, Pennsylvania Piedmont, U. S. A. A: 443 poles to S_1-planes (contours at 1, 2, 4, 6, and 10 per cent). B: 357 poles to S_2-planes (contours at 2, 5, 10, 15, and 20 per cent). C: 253 poles to S_3-planes in the southern part of the region (contours at 2, 4, 6, and 8 per cent). D: 56 poles to S_3-planes in the northern part of region (contours at 2, 4, and 6 per cent). E: 86 poles to S_4-planes in the whole area (contours at 2, 4, and 6 per cent). (After Freedman, *et al.*, 1964, Fig. 4).

Superposed folds must be much more common in the southern and central Appalachians than would be supposed from current descriptions. To take one example, the outcrop pattern shown in the 1:24,000 geologic map of the Spruce Pine district in the Blue Ridge Province, western North Carolina (Brobst, 1962), strongly suggests significant superposed folds. Spectacular small examples of tightly folded axial planes of the earlier folds are abundant in the roadside outcrops of this area (Fig. 316).

In the Adirondack region, New York, superposed folds were mapped by Martin (1916) in the Canton Quadrangle. Later, Buddington (1929, 1934, 1939) mapped superposed folds in several Adirondack localities. The type of structure involved is shown by Brown and Engel's (1956) maps (e.g., Fig. 397) of the inferred refolding in the Balmat-Edwards synform. The gently-plunging northeast-trending folds are approximately parallel to major elements of the regional system. Although the two fold sets are ascribed to discrete periods of deformation, northeast structures merge into, and are blurred by, northwest-plunging superposed folds. Such relationships are of widespread importance throughout the Adirondacks, although detailed geometric studies are scarce.

Dale (1896, pp. 553-4) referred to transverse folds in his paper on the Green Mountain region of eastern New York; he suggested that two fold directions developed simultaneously near Williamstown, Massachusetts. In Vermont and New Hampshire superposed folds are common. Representative structures in Vermont were described by White and Jahns (1950). For the Keene-Brattleboro area, New Hampshire and Vermont, Moore (1949) attempted to outline homogeneous domains in order to elucidate the complex structures. In the Woodsville Quadrangle, Vermont and New Hampshire, White and Billings (1951) described schistosity formed parallel to the axial planes of minor folds (F_1) during the Taconic orogeny (end of the Ordovician) that was folded (F_2) during the Acadian (middle or late Devonian) orogeny; crenulation foliation developed parallel to the axial planes of the minor F_2-folds. Goldsmith (1961) published a preliminary account of a large refolded flexure in a sequence of metasedimentary and metavolcanic gneisses, southeastern Connecticut (Fig. 398); the folding was associated with sillimanite grade regional metamorphism that induced a granitoid appearance in some of the lithic units. Modern structural analysis is continuing in this region; Woodland (1965) published a good well-illustrated example from the Burke Quadrangle, Vermont, in which 12 subareas were isolated in a detailed study of superposed Acadian fold episodes.

Within the Precambrian Idaho Springs Formation of the Front Range, near Denver, Colorado, mapping by a number of workers demon-

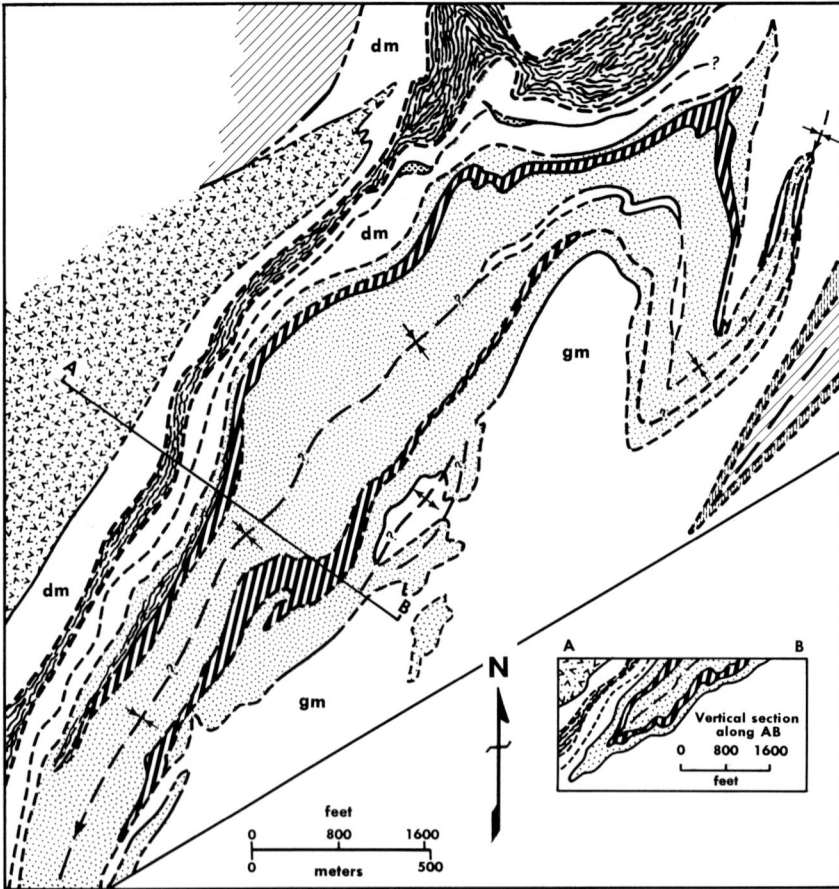

Figure 397. Geological map of inferred refolded Balmat-Edwards syncline of Precambrian rocks, NW of Edwards, New York, U. S. A. The trace of the syncline axial surface is shown (After Brown and Engel, 1956, Fig. 2).

strated superposed folding — for example, in the Idaho Springs-Central City area (Moench, *et al.,* 1962), and in the Chicago Creek area (Harrison and Wells, 1959). They suggested that the earliest recognizable deforma-

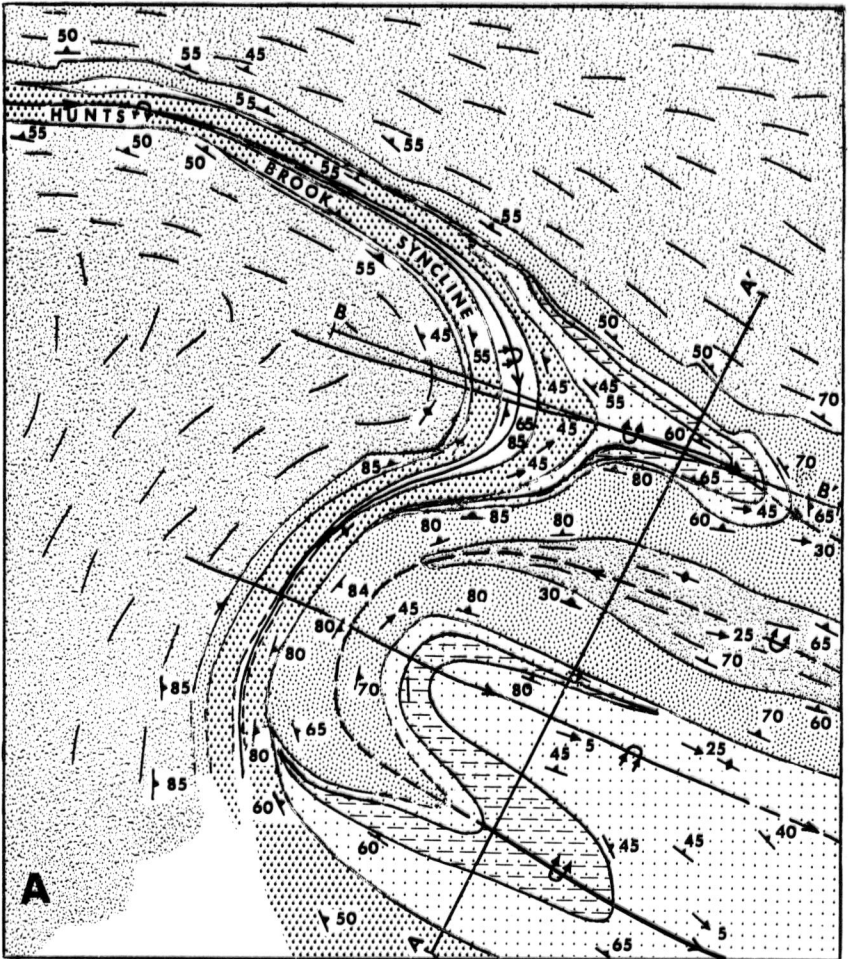

Figure 398. Geological map (A) and vertical sections (B, facing page) across an area of refolded metasedimentary rocks lying at the intersection of Uncasville, Montville, Niantic, and New London Quadrangles, southern Connecticut, U. S. A. (After Goldsmith, 1961, Fig. 169.1).

tion involved plastic folding that developed major disharmonic folds with sinuous subhorizontal NNE-axes. Younger superposed folds in relatively incompetent rocks appear to be contemporaneous with intense granulation in the more competent units. To the northeast, Wells, *et al.* (1961) also recorded a complex sequence of superposed deformations in the Coal Creek area.

Houston and Parker (1963) made a detailed petrofabric study of a Precambrian quartzite in the Medicine Bow Mountains, southern Wyo-

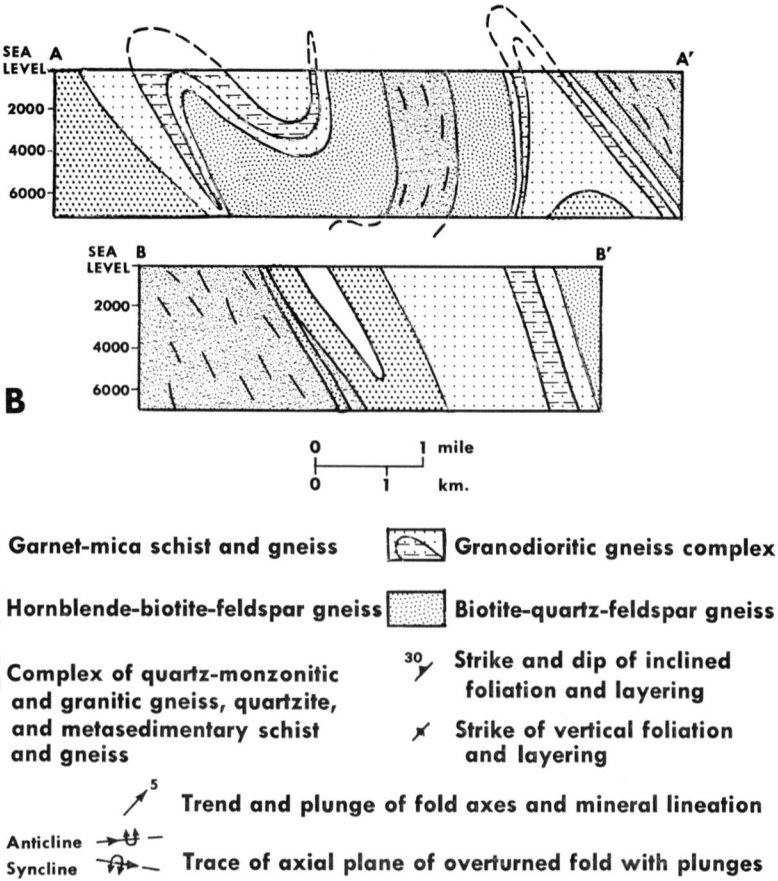

Garnet-mica schist and gneiss

Granodioritic gneiss complex

Hornblende-biotite-feldspar gneiss

Biotite-quartz-feldspar gneiss

Complex of quartz-monzonitic and granitic gneiss, quartzite, and metasedimentary schist and gneiss

^{30} Strike and dip of inclined foliation and layering

Strike of vertical foliation and layering

^{5} Trend and plunge of fold axes and mineral lineation

Anticline

Syncline

Trace of axial plane of overturned fold with plunges

ming. A β-diagram for S_0-surfaces in the French Creek fold defined a strong point maximum, whereas linear structures (e.g., crenulations in associated sericite schist layers, phyllites, etc.) have a more diffuse pattern interpreted as a small-circle distribution about β. This geometry (see Chapter 10) could result from a second phase of flexural-slip folding of a lineated quartzite.

Clear evidence for a definite time interval between two oblique phases of folding in the northern Black Hills, South Dakota, was presented by Noble, *et al.* (1949).

Throughout western U.S.A. evidence for multiple folding in widely separated terrains has been noted by many authors. Along the northwestern border of the Idaho batholith, Reid (1959A, 1959B) and Hietanen

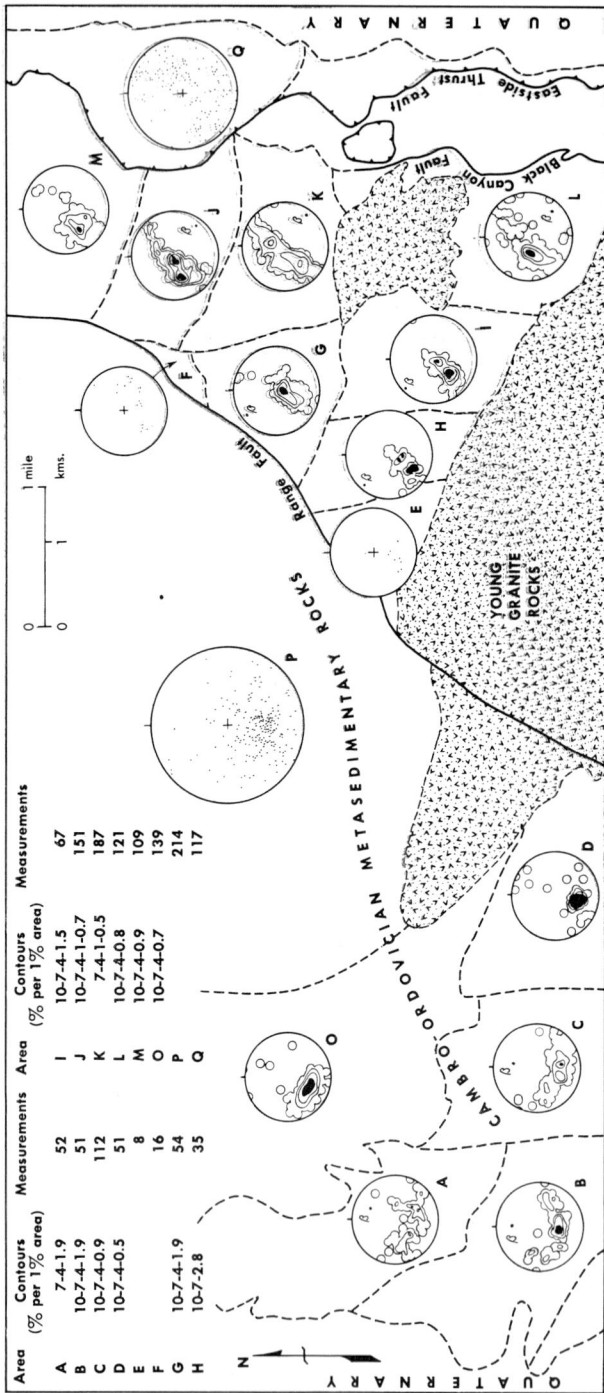

Figure 399. Geometry of π_{s_0} in Cambro-Ordovician rocks of the central Toiyabe Range, S of Austin, Nevada, U. S. A. (After Means, 1962, Fig. 5).

(1961C, 1962) gave detailed and liberally-illustrated descriptions of superposed folds in the high-grade Precambrian Belt medasedimentary rocks. Both Reid and Hietanen described three major phases of pre-Idaho batholith folding. F_1 and F_2 are differently oriented but both are commonly isoclinal; Hietanen (1961C) found that in some outcrops F_1-structures were deformed by F_2, whereas in others, F_2 are deformed by F_1. The actual style of folding associated with F_1, F_2, or F_3 depends on the lithology; mica schists are closely wrinkled and form folds parasitic to larger structures, whereas the competent units develop rounded folds. The F_3-folds are more open structures. While Reid (1959B) was uncertain of the orogenic cycle to which each set of folds should be assigned, Hietanen (1961C) ascribed them all to the Nevadan orogeny (? Jurassic).

Roberts, *et al.* (1958) showed that the structure of north-central Nevada is very complicated. From within this region, Means (1962) studied the Cambro-Ordovician rocks in a small section (about 100 sq. km.) of the Toiyabe Range south of Austin; with the aid of geometric analyses of subareas A through Q he showed that the obvious folds visible in the field are F_2-structures (Fig. 399). He found that within areas A through L (the Clear Creek area) π_{S_0}-diagrams define β_2-axes that are coplanar but that cluster in three maxima; the axial-plane foliation (S_2) of these F_2-folds is approximately parallel within this group of subareas (Fig. 400). The planar parallelism of S_2 shows that the β_2-axes were not folded from some common orientation by a late superposed folding — rather, the S_0-planes must have had a different orientation in subareas A-D, E-I, and J-L as a result of a F_1-fold system. Because there is no evidence for more than two phases of folding, Means assumed that F_1 formed from subhorizontal strata by folding about a subhorizontal axis, and because the attitude of S_0 after F_1 varied more from east to west than from north to south, it was not unreasonable to *assume,* as a first approximation, that F_1 had a north-south trend (Fig. 401). This reasoning led to the reconstruction of F_1 shown in Figure 402. Means found that π_{S_0} and π_{S_2} for subarea M define two pairs of maxima on a projection, and that the structures contributing to these maxima are irregularly distributed within the subarea (Fig. 403A). This geographic pattern suggested that S_0 was plicated about β_1 by the F_1-folding, rather than having been planar over broad areas (as in subareas A through L); each small fold limb with respect to β_1 gave rise to a β_2-axis corresponding in orientation to one of the two projection-maxima. Thus, area M is much too large for adequate geometric analysis; subareas, measured in three or four meter units would be necessary to define areas homogeneous with respect to S_0 (Fig. 403B).

Very well-developed small eyed folds can be collected from the Paleozoic (? Cambrian) marbles that were refolded by a major southerly-

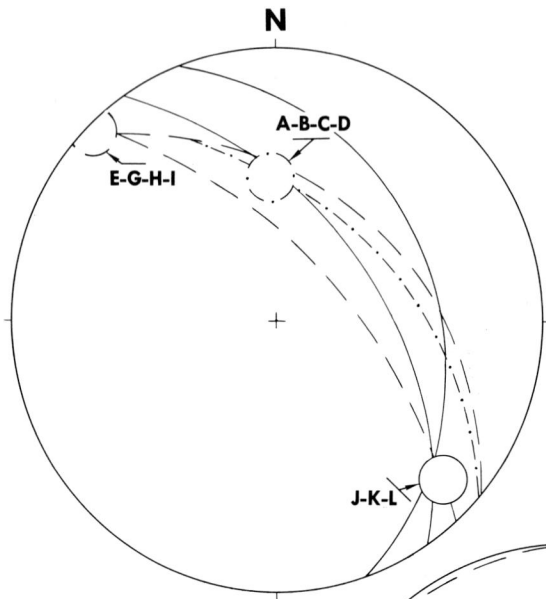

Figure 400. Synoptic geometry of the Clear Creek area, Toiyabe Range, Nevada, U.S.A.; the three little circles represent β_2-axes defined by π_{S_0}-diagrams, while the great circles represent foliation planes defined by π_{S_2} for each of the three sub-areas (From Means, 1962, Fig. 8).

Figure 401. Synoptic geometric interpretation for the Clear Creek area, Toiyabe Range, Nevada, U. S. A. The great circles represent attitudes on three parts of the F_1-folds. Geometric fold axes of the first and second generation folds are represented by β_1 and β_2, respectively (From Means, 1962, Fig. 11).

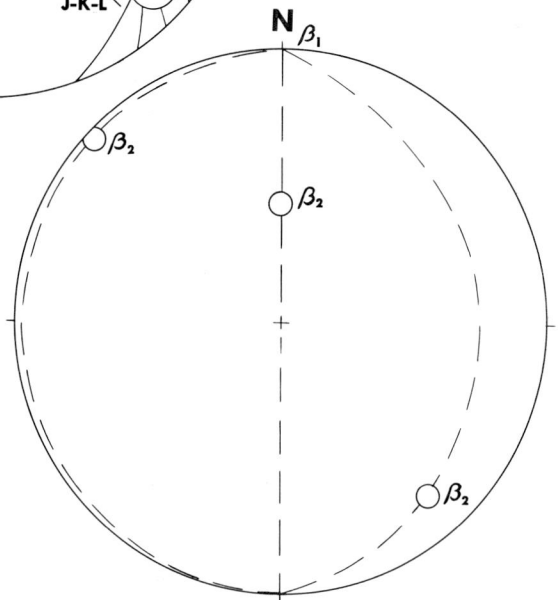

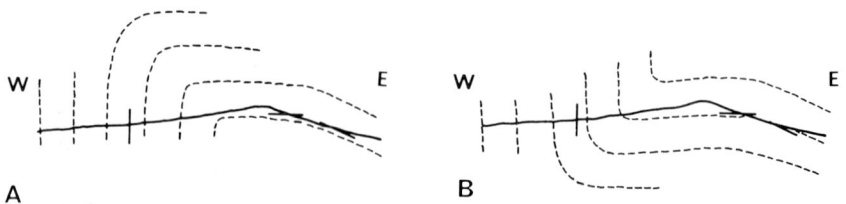

Figure 402. Two possible profiles for F_1-folds as reconstructed by Means. Short solid lines represent established attitudes of S_0; long solid line is the generalized ground surface (After Means, 1962, Fig. 12).

moving thrust, near the old Stardust Mine, northern Snake Range, eastern Nevada.

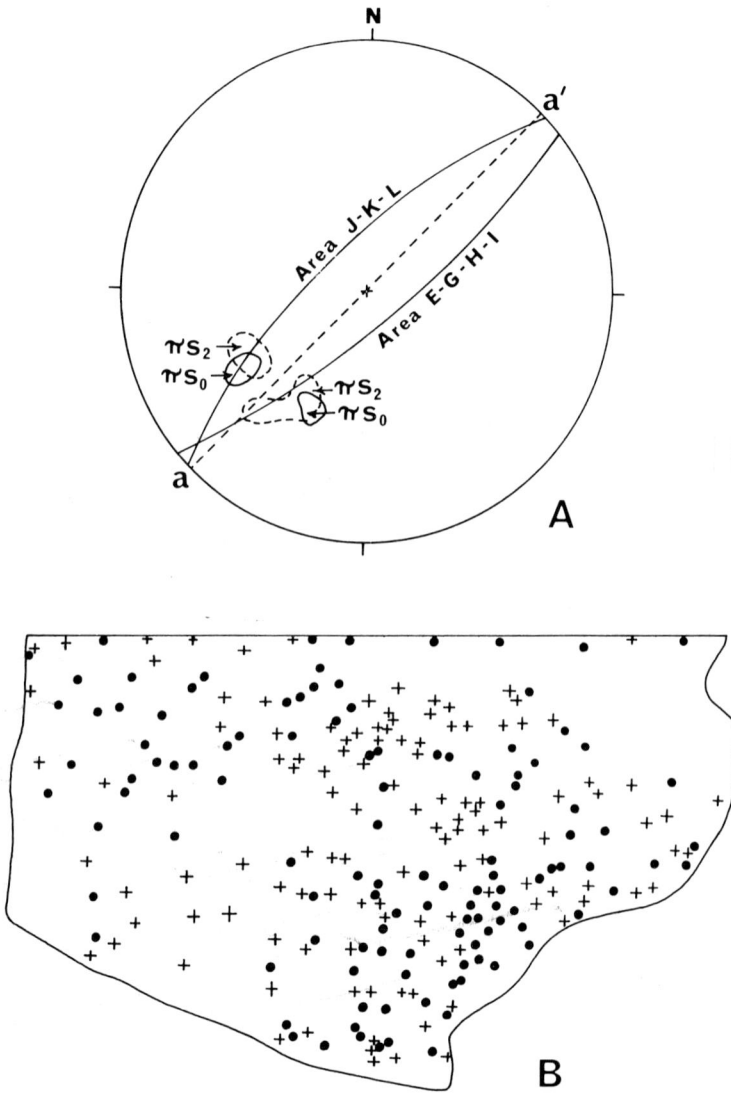

Figure 403. Geometrical features of subarea M, Toiyabe Range, Nevada, U. S. A. A: π_{S_0} and π_{S_2} fields from subarea M compared with π_{S_0} and π_{S_2} girdles from other subareas of the Clear Creek area. B: Map of subarea M showing locations of S_0 and S_2 attitudes that fall to left (dots) and right (crosses) of line $a-a'$ in A (After Means, 1962, Figs. 13 and 14).

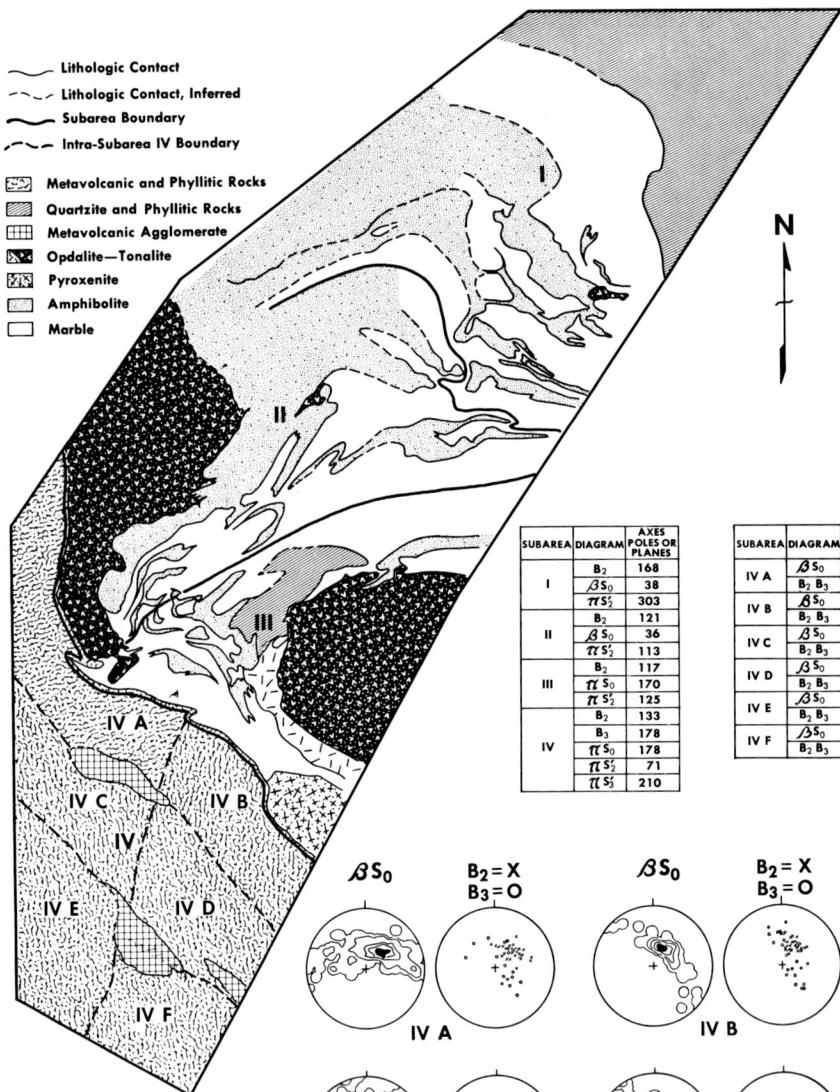

Lithologic Contact
Lithologic Contact, Inferred
Subarea Boundary
Intra-Subarea IV Boundary

Metavolcanic and Phyllitic Rocks
Quartzite and Phyllitic Rocks
Metavolcanic Agglomerate
Opdalite—Tonalite
Pyroxenite
Amphibolite
Marble

N

SUBAREA	DIAGRAM	AXES POLES OR PLANES
I	B_2	168
	βS_0	38
	$\pi S_2'$	303
II	B_2	121
	βS_0	36
	$\pi S_2'$	113
III	B_2	117
	πS_0	170
	$\pi S_2'$	125
IV	B_2	133
	B_3	178
	πS_0	178
	$\pi S_2'$	71
	$\pi S_3'$	210

SUBAREA	DIAGRAM
IV A	βS_0 / $B_2 B_3$
IV B	βS_0 / $B_2 B_3$
IV C	βS_0 / $B_2 B_3$
IV D	βS_0 / $B_2 B_3$
IV E	βS_0 / $B_2 B_3$
IV F	βS_0 / $B_2 B_3$

0 ____ 1 mile
0 ____ 1 kms.

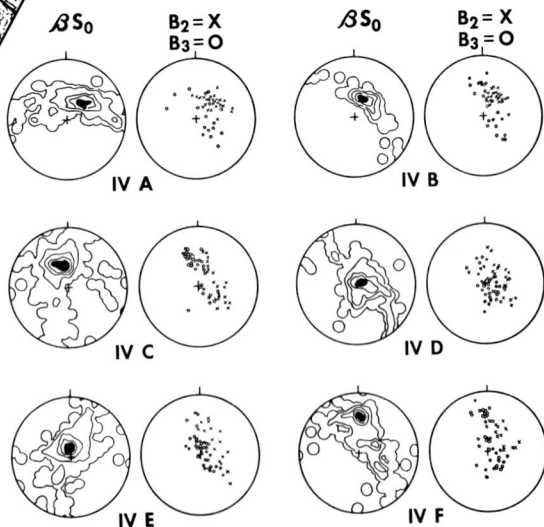

βS_0 $B_2 = X$ $B_3 = O$ βS_0 $B_2 = X$ $B_3 = O$

IV A   IV B

IV C   IV D

IV E   IV F

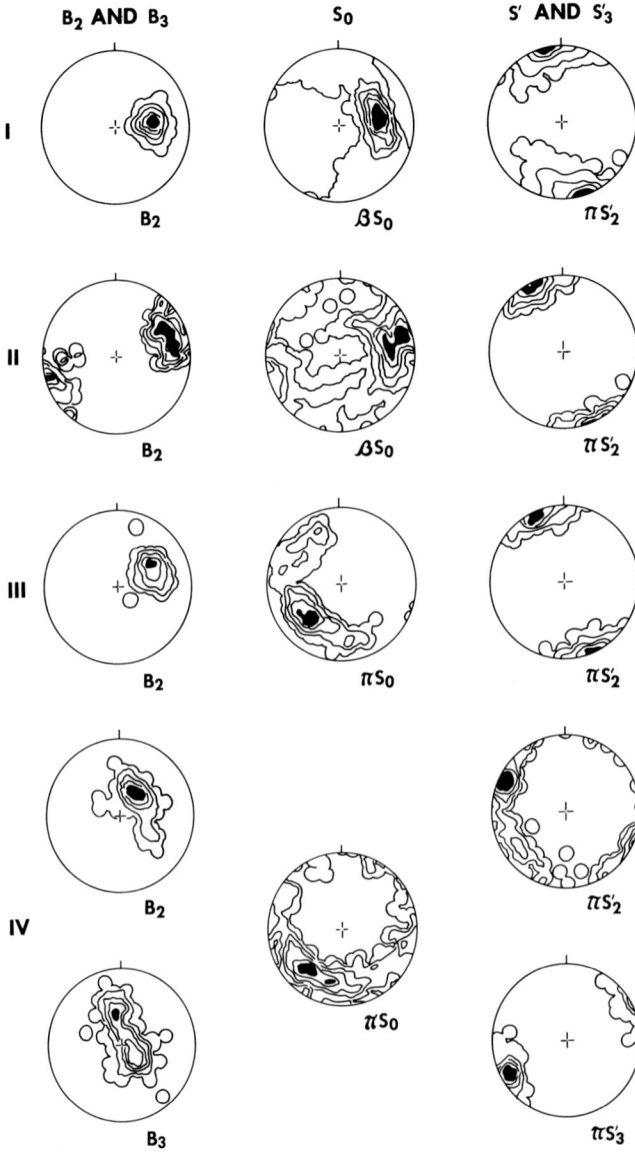

Figure 404. Map and structural elements of the subareas of the region in the central Sierra Nevada Foothills, east of the Mother Lode, California, U. S. A. (After Baird, 1962, Fig. 3).

In a reconnaissance study, Mayo (1941) mapped four different fold directions in the Sierra Nevada of California. As yet the time and geometric relationships of these structures are obscure. Parker (1961) made a detailed geometric analysis of the folds in some Triassic metasedimentary and metavolcanic rocks preserved as roof pendants in the granites of the Sierra Nevada. He concluded that initial flexural-slip folding (F_1) of bedding (S_0) was followed by slip folding (F_2) produced by penetrative slip on surfaces parallel to the vertical axial planes (S_1) of the F_1-folds.

Two of the most detailed analyses of the geometry of superposed folding in the U.S.A. were published by Baird (1962) and Best (1963) who studied small areas of the Sierra Nevada Foothills, California.

East of Merced, in the southwestern foothills, Best (1963) mapped two subparallel phases of NW-SE-folding. The major F_1-folds are rather open slip folds — up to 1.6 km. from hinge to hinge — in most lithologies, although flexural-slip folds characterize the siliceous argillites. In pelitic members axial-plane foliation (S_1) developed. The geometry of the sub-areas shows that second folding developed spasmodically within the area and caused refolding of bedding (S_0) and flexure of S_1; F_1-structures were almost completely obliterated where F_2-folds occur. A crenulation foliation (S_2) is developed at the expense of, and oblique to, S_1. Metasiltstones were folded (F_2) by flexural-slip, but F_2 was dominated by slip folding, as shown by the fact that (a) lineations originally parallel to B_1 now lie on a great circle (rather than a small circle), and (b) most folds of S_1 have a similar style. Best's (1963) clear account of the relatively simple geometry associated with this area makes it a useful case history for detailed classroom study.

Basing his work on the isolation of homogeneous subareas, Baird (1962) developed excellent well-documented evidence for three phases of superposed folding in the central Sierra Nevada Foothills east of the Mother Lode (Fig. 404). The earliest folds (F_1) recognized are only apparent on the regional scale (Fig. 405), and the regional form of the marble-amphibolite body is not controlled by the earliest generation of folds (F_2) visible within the area actually mapped. The strike of the axial surfaces of the F_1-folds is parallel to the Sierra Nevada range, and Baird believed this strike to be associated with major isoclinal folds about subhorizontal axes. Any minor structures originally associated with F_1 seem to have been destroyed by the later deformation (F_2 and F_3).

The first folds (F_2) visible in the map area yield indirect evidence of the earlier F_1-folds. Thus, (a) the axial surfaces (S_2')[7] of the F_2-folds strike

---

[7]There is no standard convention for labelling different S-structures; many systems appear in current literature. In this book successively-formed S-structures are called S_0, S_1, $S_2, \ldots S_n$, although it is sometimes convenient to use a different convention. For Baird's work, S' is used for the axial surface of a fold, and the subscripts refer to the fold generation, e.g., S_2' is the axial surface of a F_2-fold.

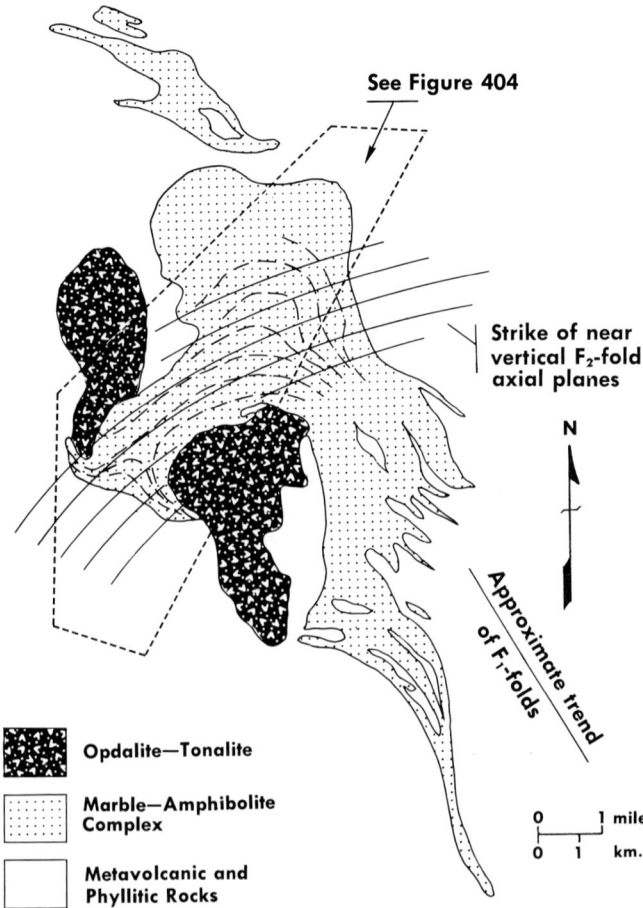

Figure 405. Relationship between the regional folding (F_1) of the marble-amphibolite body and the axial planes of F_2-folds, central Sierra Nevada Foothills, east of the Mother Lode, California, U. S. A. (After Baird, 1962, Fig. 11).

at right angles to the axial surfaces inferred from the regional distribution of the marble-amphibolite unit (Fig. 404), and (b) the F_2-fold axes (B_2) have a steep but variable plunge in folds ranging in size from hand specimens to large folds (cf. Weiss and McIntyre, 1957).

The F_2-folds are mainly upright flexural-slip folds of the lithologic layers. Movement on surfaces subparallel to the axial planes of these F_2-folds produced elongate lenticles of contrasting lithology (Fig. 406) and, in consequence, a persistent streaking parallel to the fold hinges. The axes (B_2) of the F_2-folds plunge steeply and trend perpendicular to the Sierra

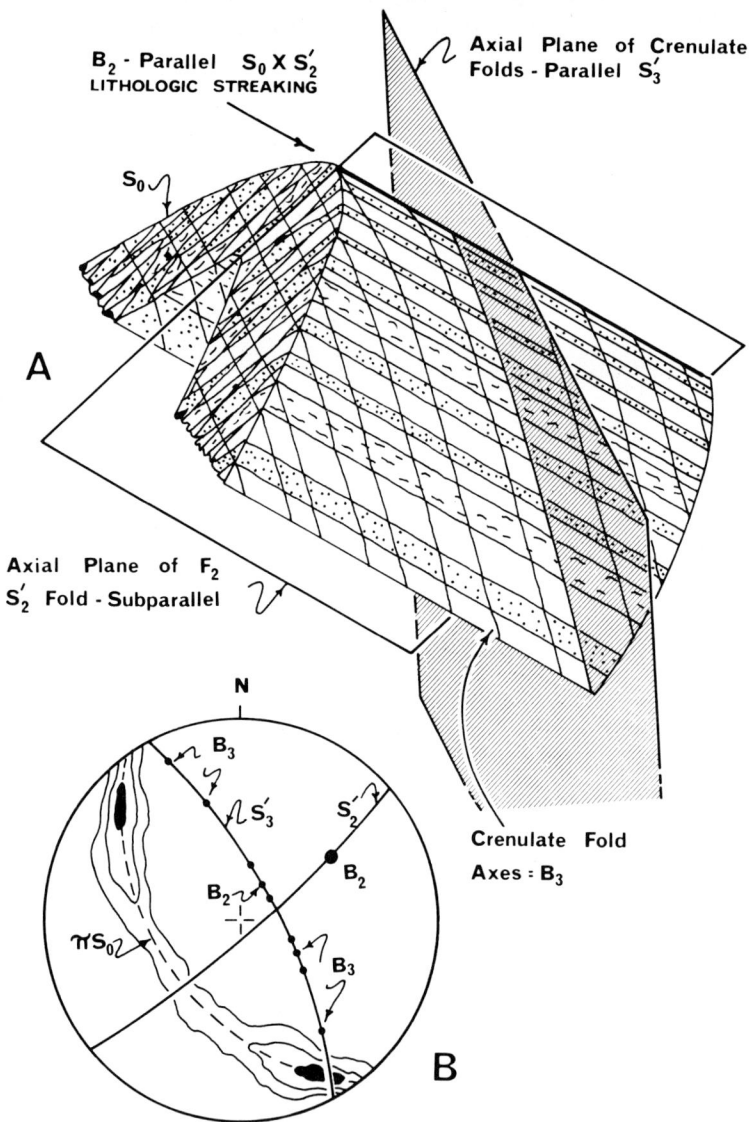

Figure 406. Sketch of the geometrical relationships between F_2- and F_3-folds, with F_2 dominant. A: Block diagram. B: Projection of relationships shown in A. See text for discussion (Based on Baird, 1962, Fig. 4).

Nevada; however, although the B_2-axes have variable plunge, they lie within S_2' (axial-plane surface of F_2). Continuing stress produced more folding about the B_2-axes and caused shear on the S_2'-surfaces; Figure 407 shows the type of deformation associated with disrupted quartzite layers in strongly foliated marble.

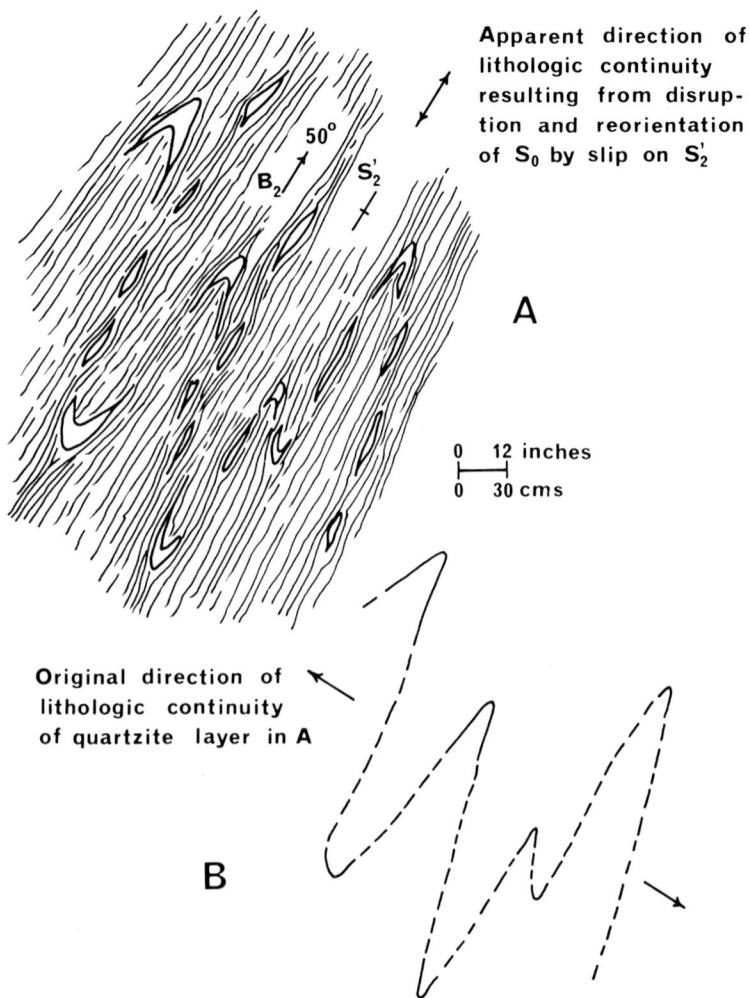

Apparent direction of lithologic continuity resulting from disruption and reorientation of S_0 by slip on S_2'

$50°$

B_2 S_2'

A

0   12 inches
0   30 cms

Original direction of lithologic continuity of quartzite layer in A

B

Figure 407. Type of deformation described by Baird as associated with disruption of quartzite layers in strongly foliated marble. A: Mapped relationships. B: Reconstruction of quartzite profile (After Baird, 1962, Fig. 6).

The third generation of folds (F_3) has strongly-developed axial-plane surfaces (S_3') that intersect those of the second generation (S_2') at high angles. S_3' produces linear structures parallel to the third fold axes (B_3) (Fig. 408). Superficially the hinges of F_3-folds resemble those of F_2-folds, but Baird showed that (a) the lenticles produced by disruption of the layering during the F_2-folding are folded round the F_3-hinges, (b) crenula-

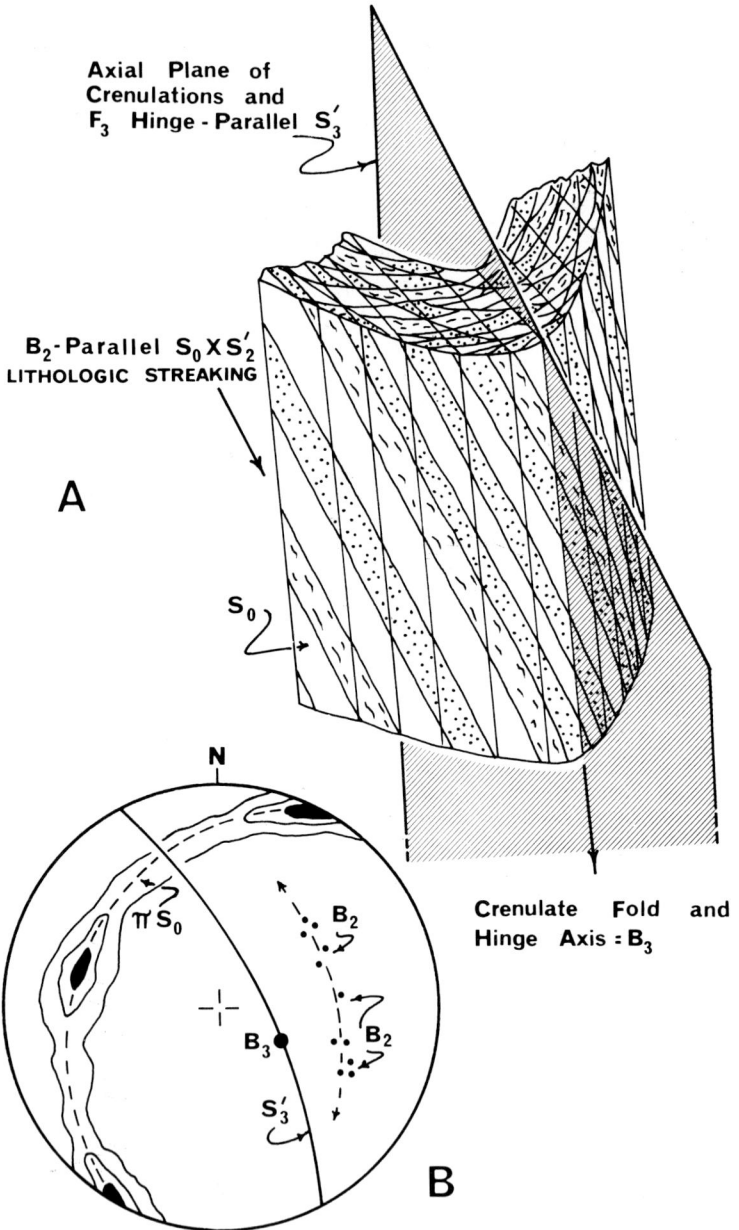

Figure 408. Sketch of geometrical relationships between F_2- and F_3-folds, with F_3 dominant. A: Block diagram. B: Projection of relationships shown in A. See text for discussion (Based on Baird, 1962, Fig. 5).

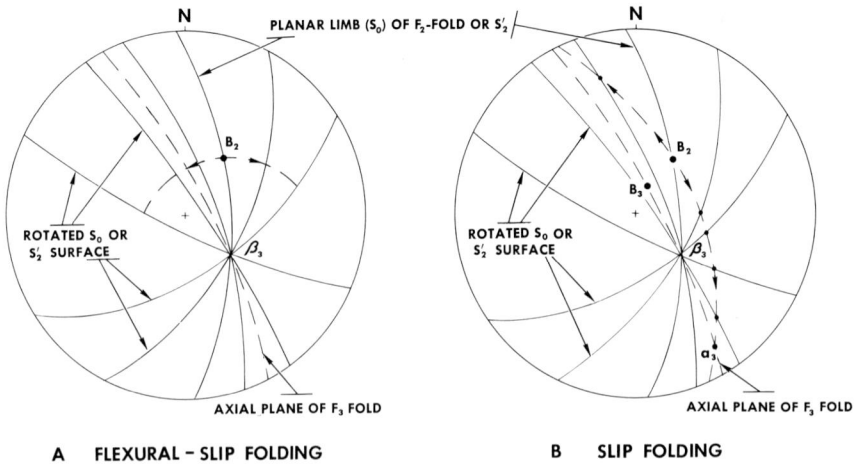

A   FLEXURAL – SLIP FOLDING   B   SLIP FOLDING

Figure 409. Contrasting geometry that would result from superposition of F_3-folds on the flexural-slip F_2-folds in the area shown in Figure 404. A: Superposition by flexural-slip F_3-folds. B: Superposition by slip F_3-folds. See text for full explanation (After Baird, 1962, Fig. 15).

tions associated with development of the F_3-folds appear to curve round the F_2-structures (Fig. 406), although they are parallel to B_3 of the F_3-folds (Fig. 408), and (c) these crenulations have axial planes parallel to the axial planes of F_3, but oblique to S_2'.

The F_3-folds are only developed locally and are commonly of small size, whereas the F_2-folds are of all sizes and occur in all rock types. Baird used the methods of Weiss (1959B) and Ramsay (1960) to determine whether flexural-slip or slip F_3-folds were superposed on the flexural-slip F_2-folds. Figure 409 illustrates a hypothetical structure in which a plane (S_0 or S_2') containing a B_2-lineation is refolded about a new axis (B_3). Both flexural-slip and slip folding would have the same effect on the planar surface (Fig. 409), but they would cause dissimilar rotations of B_2. With flexural-slip F_3-folding, B_2 rotates on a small circle (Fig. 409A) and keeps a constant angle with β_3. B_2 would rotate through the fabric along a great circle towards the direction of slip (a_3) if slip F_3-folding occurs.

In Figure 411 field data from Baird's subarea IV are plotted for comparison with Figures 409A and B. Three β_3-axes (F_3), corresponding to the two maxima on the B_3-diagram and the pole (β_3) to the great circle on the $\pi_{S_2'}$-diagram for subarea IV (see Fig. 404), are plotted together with S_3' (the F_3 slip surface) and a_3 (the "arbitrary" direction of slip in S_3' chosen by Baird). Baird took the orientation of the F_2-axis, B_2, prior to superposition of F_3 from domains IVA and IVB, where B_2-structures are

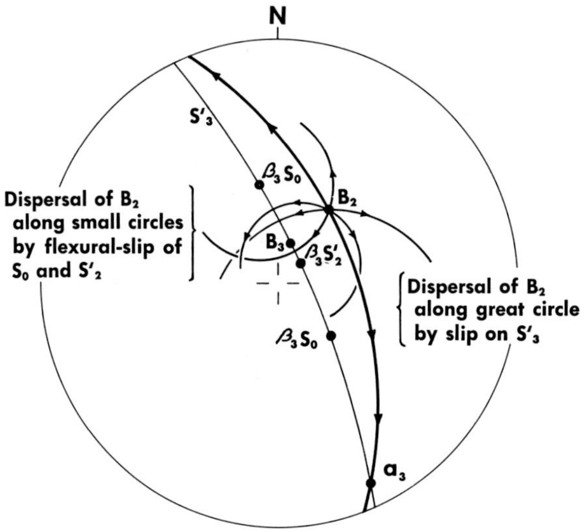

Figure 410. Geometry of limiting cases for subarea IV studied by Baird, central Sierra Nevada Foothills, California, U. S. A. Possible dispersal routes of B_2 by flexural-slip of S_0 and S_2' are shown by small circles. Possible dispersal routes of B_2 by slip on S_3' are shown by the great circle through $B_2 a_3$. See text for explanation (After Baird, 1962, Fig. 16a).

dominant (Fig. 404). Now, if flexural-slip F_3-folding rotated the B_2-axes, they would move along one or more of the small circles about the three β_3-axes plotted in Figure 410. With slip folding, B_2 would rotate along the great circle passing through B_2 and a_3. Rotated B_2-elements measured in subareas IVC to IVF are plotted in Figure 411. Despite some dispersion through B_3, there is a marked tendency for the B_2-elements to lie along the great circle corresponding to the trace of S_3'. Hence, Baird suggested that the dominant mechanism of superposition (F_3) was folding of S_0 and S_2' by slip on S_3' in the subhorizontal direction a_3, with internal rotation of B_2 in the plane defined by a_3 and B_2. The dispersion through B_3 (Fig. 411) might suggest that flexural-slip played some part in the F_3-folding. This is likely, because, most folds in metamorphosed rocks involve features of both flexural-slip and slip folding, although one mechanism is commonly dominant.

The Coast Ranges of California also contain complex structures which can be illustrated from the Healdsburg Quadrangle, north of San Francisco (Gealey, 1951). Here the folds are in general open and symmetrical; however, overturning occurs locally, and overfolding and thrusting are developed on a small scale. Gealey thought that the first period of folding occurred between the close of the Jurassic and the Upper Cretaceous, that the second folding occurred during the post-Upper Cretaceous, and the third in the Lower Pleistocene as a phase of the major Coast Ranges orogeny.

Gilluly (1963, p. 150) pointed out that the Pacific geosyncline is not particularly well known because of the abundant granitic massifs, the

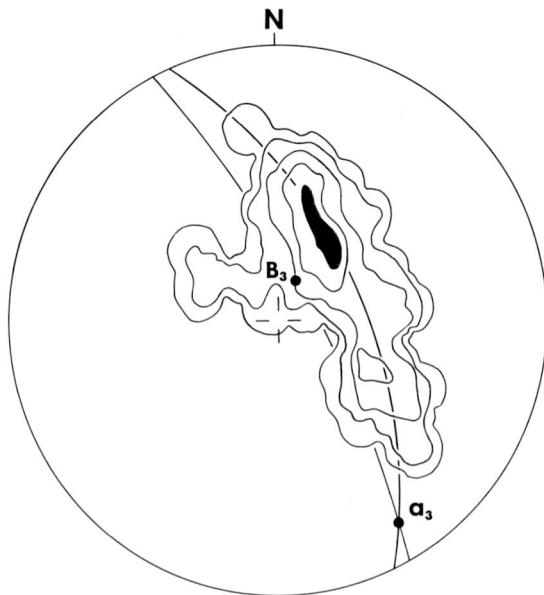

Figure 411. Contoured diagram of the rotated B elements in subareas IVC through IVF of the area studied by Baird, central Sierra Nevada Foothills, California, U. S. A. See text for explanation; contours at 0, 2, 6, 10, and 14 per cent (After Baird, 1962, Fig. 16b).

metamorphism, and the widespread cover of younger rocks. He wrote:

> Although there is abundant indirect evidence in the sedimentary record of crustal deformation through the western Cordillera, direct evidence such as is furnished by closely bracketed folds, thrusts, and unconformities is of only sporadic occurrence. Well-dated orogenic pulses took place in early, in middle, and (twice) in late Cretaceous time in California (Taliaferro 1943; [D. L.] Jones 1959), twice during the late Cretaceous in Nevada (Willden 1958; Hewett & others 1936; Nolan, Merriam & Williams 1956), and in the mid-Cretaceous in Oregon. In California, angular unconformities have been mapped at the base of the Palaeocene, between lower and middle Eocene, between middle and upper Eocene, at the top of the Oligocene, and at several other horizons in the early Cainozoic (Hoots, Bear & Kleinpell 1954). More than forty separate pulses are recorded during Miocene and later time in California, where the marine record is more complete than anywhere else in the region. ... Doubtless many other of the Pacific areas were similarly active, but much of the record has been lost through erosion or burial.

Of course, although concomitant superposed folding *may* have been occurring at depth, it is *not necessary* to assume that each of the deformations referred to by Gilluly is reflected by a recognizable fold event in the neocrystallized rocks beneath.

Canada: Precambrian metasedimentary and metavolcanic rocks in the Malarctic area, Quebec, occur in an isoclinal syncline with subvertical

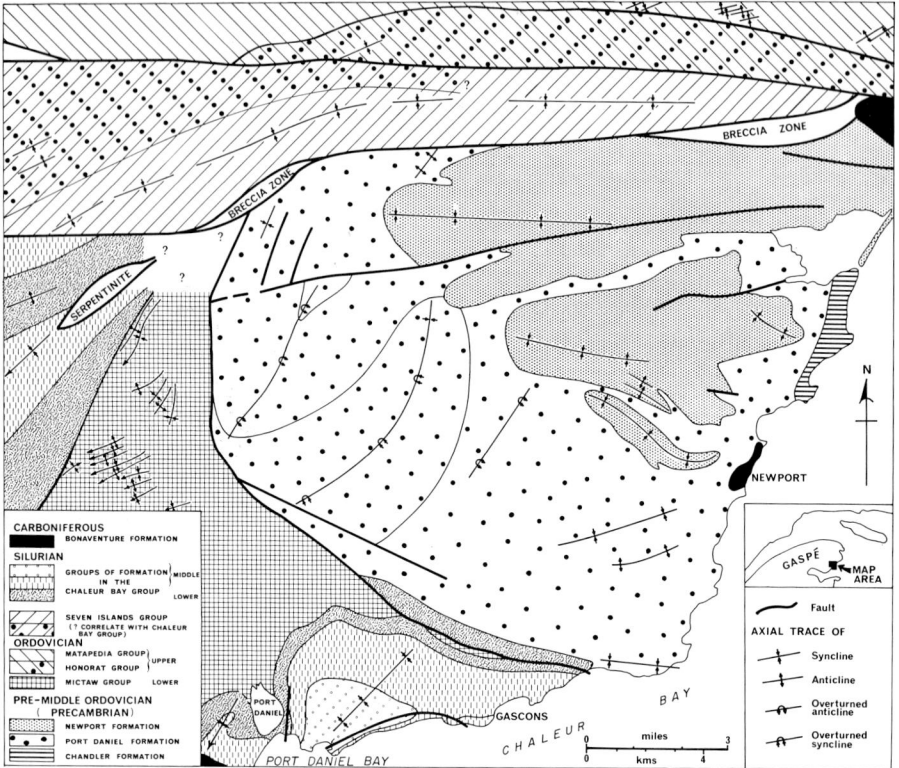

Figure 412. The Chandler-Port Daniel area, Gaspé Peninsula, Quebec, Canada (Simplified from Ayrton, 1964, Map 2).

limbs (Gunning and Ambrose, 1940). Mining established the existence of superposed Z-shaped minor folds, with subvertical axes almost perpendicular to the main fold axis.

The Wabush Lake and Julienne Lake areas of southwestern Labrador and eastern Quebec (Fig. 294) were studied extensively with modern quantitative methods because of the economic importance of the Precambrian iron ore formations (Gastil and Knowles, 1960; Knowles, *et al.,* 1962). At Wabush Lake, Gastil and Knowles (1960, p. 1252) found that: "... the structures of the Grenville orogenic province appear to truncate, deflect, and be superimposed upon the north-south structures [Labrador geosyncline]."

Ayrton (1964) made a detailed structural study of Precambrian, Ordovician, and Silurian sedimentary rocks of the Chandler-Port Daniel area, Gaspé Peninsula, Quebec (Fig. 412). Geometric analysis showed that the Precambrian Maquereau Group (originally graywackes, quartzites, and volcanic rocks) was strongly folded during the Gaspésian orogeny (? Precambrian) prior to deposition of the unconformable Middle

Ordovician sedimentary rocks. Pulses of deformation correlated with the Taconic and Acadian orogenies had little effect on the Precambrian rocks but they produced folds in the Middle Ordovician rocks. The Upper Ordovician and Silurian sedimentary rocks in this area were folded by the Acadian movements only. It seems clear that spasms of folding activity associated with the Taconic and Acadian orogenies occurred at different times in different geographic regions within the Appalachian fold belt.

Essentially unmetamorphosed thinly-bedded shales, siltstones, and limestones of the Horton Group (Lower Mississippian) in the Cheverie-Walton area, Nova Scotia, developed disharmonic flexural-slip folds towards the end of the Paleozoic (Fyson, 1964). The main folds with low axial plunges and the superposed folds with steep axial plunges both have very similar profiles and include many conjugate pairs of folds; this suggested that all the folds formed under similar physical conditions.

In Ontario, Quirke and Lacy (1941) mapped complex structures on Lookout Island, northeast shore of Georgian Bay, which they interpreted as products of simultaneous folding about two axes. Wynne-Edwards (1957) described successive but discrete deformations in the Grenville sub-province near Kingston. Saha (1959B) briefly mentioned two directions of folding within marbles and gneisses of the envelope rocks of the Wollaston, Deloro, and Chandos Lake plutons, southeastern Ontario.

Robertson (1953) and Kalliokoski (1953) referred to superposed folds in neighboring areas of Manitoba. Robertson mapped three northwest-trending synforms exposing large areas of Sherridon Gneiss in the Batty Lake map-area. Younger refolding is indicated by overturning of these broader folds to the southwest, and by later intersecting northeasterly-trending fold axes (e.g., west of Batty Lake and southeast of Nokomis Lake). Folds in the Amisk Series trend north or north-northeast between Weldon Bay and Flin Flon (Kalliokoski, 1953), but, during a later orogeny, they were bent to the west by superposed folding.

A discrete time interval separating major tectonic events responsible for superposed folding has been suggested by several other workers in the Canadian Shield—for example, Derry (1939) in Ontario and northern Quebec, Cooke (1948) in the Lake Huron-Sudbury area, Dunbar (1948) at the Porcupine ore deposit, northern Ontario, and Henderson (1948) between Great Slave and Great Bear Lakes, Northwest Territories. Derry, *et al.* (1948) discussed superposed folding that warps an overturned syncline at the Matachewan Consolidated Mine. A realistic understanding of such structures awaits more detailed mapping and geometric analyses. Work in this direction is being continued by the Geological Survey of Canada in many areas. For example, in his preliminary map accompanying the detailed lithological map of Mesa Lake area, District of Mackenzie (Northwest Territories), Ross (1959) cited widespread evidence for folding about three trends. Ross believed that many of these flexures are complementary and show overlapping time relationships.

In the Shuswap terrane, British Columbia, A. G. Jones (1959) found clear evidence for two phases of folding differing in nature, geometry, and tectonic orientation. The older group is typified by extensive isoclinal recumbent folds and intense shearing. The second phase was less severe and produced normal upright folds with much less intense warping and folding. In the extreme north of British Columbia, the regional northwest Cordilleran trend is interrupted by a belt of northeast-trending structures (Norman, 1962). These structures are comparable to those described by Buddington and Chapin (1929) from neighboring southeastern Alaska, where the main axis of pre-Tertiary folding affected the Mesozoic rocks, and superposed Tertiary folding also produced simple folds in Eocene strata with axes almost perpendicular to the main fold trend.

Other continents: Considerable progress has been made with mapping the Precambrian Shield in East Africa. In Kenya, for example, the major folds appear to be mainly parallel to the Mozambiquian orogenic belt (north-south), while synchronous cross folds and refolding episodes are said to be responsible for transcurrent trends (e.g., Schoeman, 1949, and J. M. Miller, 1956). Saggerson, *et al.* (1960) reviewed recent work on the complex fold systems being mapped in Kenya; widely differing fold styles are now recognized in different parts of the country, but it is uncertain whether one or more orogenies are represented by the crossing fold structures, each of which could reflect an individual phase in a single orogeny. In some districts the first folds dominate, and in other areas the second folds dominate and have apparently obliterated all earlier structures. McCall (in Saggerson, *et al.,* 1960) even suggested that the homoaxial structures described by Weiss (1959A) in his oft-cited geometric analysis of the Turoka area (Fig. 181) should possibly be referred to the superposed phase of folding.

Ackermann and Forster (1960) described how, in their opinion, the Muva System between Katanga Pedicle and Lake Nyasa was first folded by the Tumbide orogeny and subsequently by the Irumide orogeny. Again, Reece (1960) demonstrated that the Igara Group of southwest Uganda was isoclinally folded before being unconformably overlain by the Buhwezu Group. Subsequent deformation produced strong folding in the Buhwezu Group and concomitant refolding of the infrastructure (Igara Group). Axial-plane foliation developed in the Igara Group during the first orogeny was overprinted and obliterated in some areas by crenulation foliation developed during the second deformation. Forster (1963) used microscopic petrofabric and petrographic data to differentiate effects of intersecting fold systems in the "Irumide" Mountains of

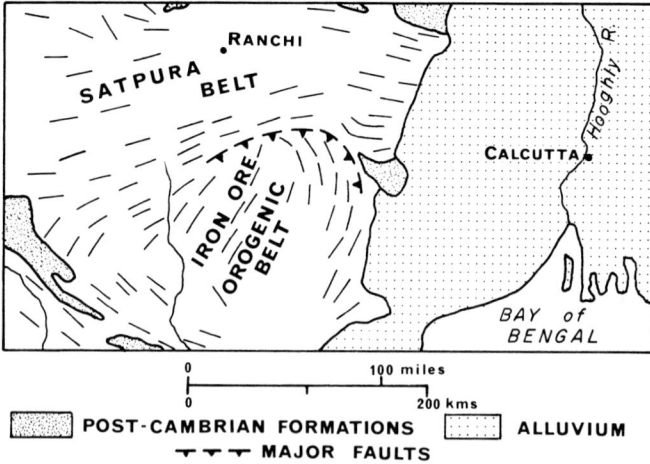

Figure 413. Generalized structural trends in the Singhbhum region and adjacent areas, India (After Sarkar and Saha, 1963, Fig. 1).

Zambia. In Chapter 10 reference was made to the superposed fold pattern described by Ramsay (1962C) from the pre-Karoo basement of the Marangudsi area, Rhodesia.

In the Moroccan High Atlas, de Sitter (1952; 1956B, pp. 315-7) mapped early folds in sedimentary rocks about ENE-axes of Pyrenic (or post-Eocene) age and a younger set of superposed east-west folds of post-Miocene age.

Work on superposed deformations and folds is being published in India. For example, Naha (1956) worked in Dhalbhum, Bihar; and Sarkar (1957) outlined the nature of some superposed folds in the Precambrian of the Bhandara-Drug-Balaghat area of Bombay and Madhya Pradesh. Sarkar and Mukherjee (1958) and Saha (1959A) also made brief reference to superposed folding in eastern Singhbhum, Bihar, while Sarkar and Saha (1962, 1963) synthesized a model of the tectonic events of this area on the basis of recent published and unpublished work (Fig. 413). Sarkar and Saha documented evidence for NNE-trending folds and an oblique sequence of superposed folds both generated by the Iron Ore orogeny that culminated with the formation of the Singhbhum granitic complex about 2,000 million years ago. To the north the Singhbhum orogeny is characterized by E-W folds (and locally-developed superposed folds); this orogeny (about 920 million years ago) caused some additional superposed flexures in the Iron Ore orogenic belt to the south.

Preliminary geometric analysis within very large areas of the Otago Schists, South Island, New Zealand (e.g., Means 1963) suggests that at least three phases of folding can be mapped in this unit. The work of Spry (1963B) and Spry and Gee (1964) on superposed fold episodes in the

Precambrian metasedimentary rocks of Tasmania, Australia, is discussed in Chapter 12.

$B_1 \perp B_2$-FOLDING

It is difficult to find many well-documented examples of $B_1 \perp B_2$-folding. Most of those referred to below require more detailed documentation before the precise geometry can be established; more complete analysis will probably show that several of these structures are not strictly $B_1 \perp B_2$-folds (as originally defined by Sander). This unsatisfactory situation should be viewed as a challenge to field geologists; the $B_1 \perp B_2$ model requires more exhaustive testing against field observations.

In the Canton Quadrangle, Adirondack Mountains, New York, U.S.A., Martin (1916) mapped the Pierrepoint fold that plunges northwest roughly perpendicular to the axes of the regional Precambrian isoclinally-folded schists. Engel (1949) re-examined this area but could find no evidence to suggest superposition of the cross folds on a regional set of earlier folds, lineations, or foliations; he concluded that the two fold directions developed synchronously as a result of deformation "of pronounced triaxial type."

$B_1 \perp B_2$-folding appears to have been inferred in several studies of the Scandinavian Caledonides. For example, Kvale (1948) considered that intersecting fold axes in the Norheimsund area of Norway commonly form in different phases of the same deformation, but that perpendicular fold axes occasionally developed simultaneously by triaxial deformation; intersecting linear structures occur in some areas, but other areas are dominated by one or other of the linear trends. Additional examples include the Tysfjord area (Foslie, 1941) and the Bygdin conglomerate (Strand, 1945) in southern Norway. The Bygdin structures, including the famous highly-elongate deformed cobbles, were stated to be related to transverse folds (cross folds) with geometric fold axes parallel to the regional a-kinematic axis (i.e., perpendicular to the assumed B-kinematic axis of the Caledonide mountain chain). These conclusions are equivocal, and the exact geometry of these rocks appears to be open to question; in particular, it is by no means certain that locally the B-kinematic axis for the Bygdin conglomerates is parallel to that of the Caledonide chain.

Koark (1952) emphasized the importance of $B_1 \perp B_2$-deformation in the development of superposed folds in some Scandinavian Archean and Caledonide ore bodies. In three areas of Swedish Lapland, Lindström (1955A, 1955B, 1957) described three or four separate superposed deformations, and a perpendicular pair of fold trends was attributed to each.

Lindström described the principal fold axes (B_2) as parallel to the *a*-kinematic axis, and the cross fold axis (B_1) as coincident with the *b*-kinematic axis. However, the orientation of the *a*- and *b*-kinematic axes was determined by extrapolation to the regional structure, rather than by analysis of the movement picture within the domain actually studied (cf. Strand, 1945). It seems unlikely that the postulated complex geometry will withstand detailed geometric analysis.

Several papers dealing with Scottish tectonics involve perpendicular fold axes, though not necessarily $B_1 \perp B_2$-folding in the strict sense. For example, after a preliminary study of the complex structures of Beinn Dronaig, Northern Highlands, Crampton (in Peach, Horne, Hinxman, *et al.,* 1913, p. 58) thought that the two superposed fold systems developed synchronously during the same tectonic movements. In the Fannich Forest district, the main structure comprises roughly north-south step-like folds consistently overturned towards the west (Sutton and Watson, 1954). An associated series of smaller NW-SE-folds shows no consistent direction of transport (as deduced by inclination of axial planes); apparently these folds mainly occur in the gently dipping limbs of the major folds (Fig. 414). Sutton and Watson found no superposition or rotation of one fold set by the other, and claimed that the smaller folds merely represent culminations and depressions in the plunge of the main fold axes. They concluded that the two fold sets are perpendicular, but the axes of the subsidiary folds trend NW-SE, because the major structures have a predominantly southerly plunge. This suggests $B_1 \perp B_2$-structures, although neither the perpendicularity nor the overall geometry seems to be definitely proved. The major folds only have a preferred north-south axial trend and Figure 415 also emphasizes the predominance of SE-trending linear elements. Emphasizing this striking anomaly, Phillips (in Sutton and Watson, 1954, p. 50) thought that the regionally-characteristic NW-SE linear structures must have more significance than the trivial one of being parallel to minor oblique wrinkles on the flanks of major structures.

In the Loch Luichart area, southeast of Fannich Forest, Clifford (1960) described the Tarvie isoclinal syncline which is deformed by a second phase of folding characterized by broad warping. Clifford thought development of the Tarvie syncline, inflexion of its axial plane, and the development of the second phase folds comprised a continuous process. Thus, the folds were considered products of one fold movement, and comparable to the structures developed at Fannich Forest.

King (1955) described a complex series of fold structures within the Scourian (Lewisian) schists of Clashnessie Bay, Sutherland (Northwest

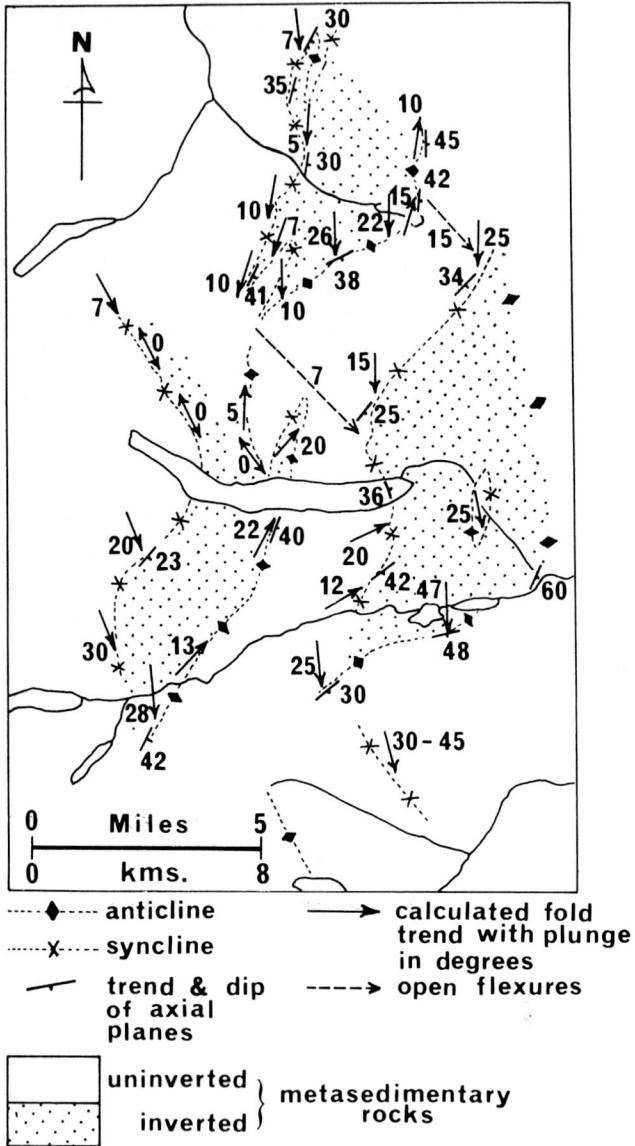

Figure 414. Major structures of the Fannich Forest district, Northern Highlands, Scotland (After Sutton and Watson, 1954, Fig. 4).

Highlands of Scotland). He found that these hornblende gneisses show simultaneous close folding about axes trending at 055 (F_1) and 155 (F_2), although the scanty published geometric data make this hypothesis difficult to evaluate. Since the regional disposition of the formations is largely related to folding on NE-SW-axes, King (1955, p. 80) contended that this

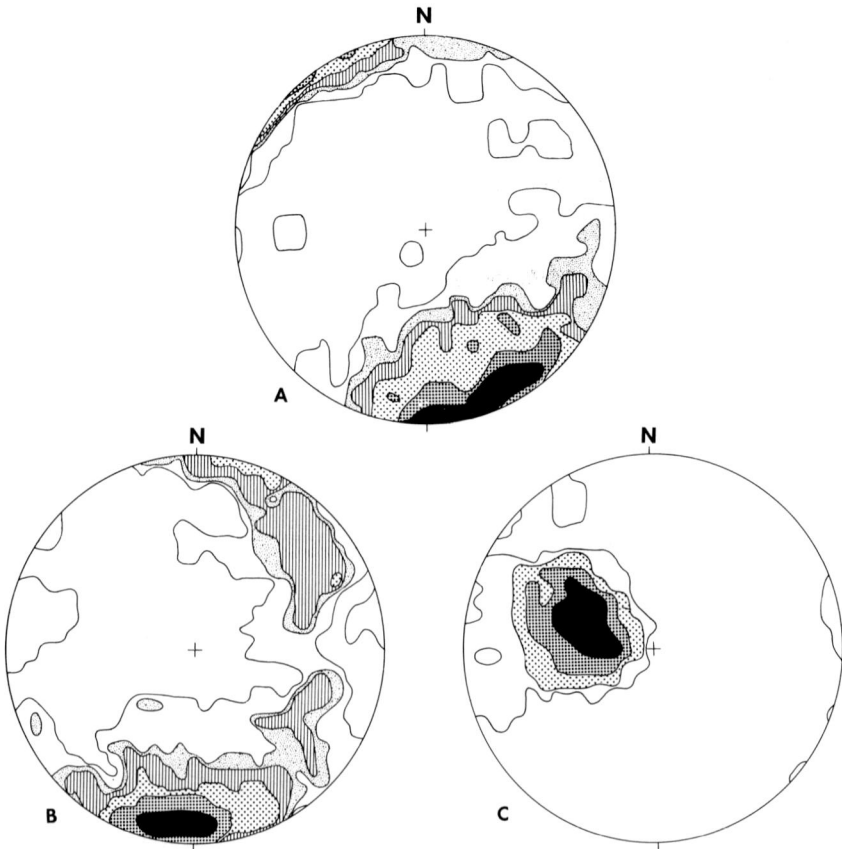

Figure 415. Structures in Moine rocks at Fannich Forest, Northern Highlands, Scotland. A: 498 linear structures (contours at 0.2, 0.6, 1, 2, 4, and 8 per cent). B: 381 axes of small folds (contours at 0.2, 0.6, 1, 2, 4, and 8 per cent). C: Poles to foliation planes (contours at 0.6, 2, 4, 8, and 16 per cent) (After Sutton and Watson, 1954, Fig. 9).

direction is probably the main tectonic b-axis, while the southeasterly-trending axes resulted from accommodation to: "... internal stresses consequent upon confinement of the orogenic belt in the direction of the tectonic b-axis." This is a relationship similar to that originally described for the Schichallion area, Central Highlands of Scotland, by King and Rast (1956A); however, as mentioned earlier in this chapter, more recent analyses showed that these fold sets were not contemporaneous, and thus that they are not $B_1 \perp B_2$-structures.

For the Gramscatho Beds at Helford River, Cornwall, England, J. L. M. Lambert (1959) described two sets of approximately perpendicular

(WSW-ENE- and NW-SE-trending) folds. Both sets are similar in form and size and are related to a common axial-plane crenulation foliation. The two fold sets are essentially synchronous and there is no evidence of refolding of the foliation or the fold axes. Each set of folds appears to occur in alternate zones which trend parallel to the strike of the foliation.

In discussing the Precambrian structures of the Kimberley area and the Median Belt of Western Australia, Hills (1946, p. 79) mentioned conjugate folds found in the Nullagine rocks. Because each fold is a simple structure, he thought the "cross-folding" could not be attributed to later diastrophism, but that both flexures, although trending at right angles, must be regarded as synchronous. Hills believed that while such structures could not result from regional compression, differential vertical movement (stemming from faulting in the Basement) might be responsible. More complex "cross-folds" in the Snowy Mountains of New South Wales were described by den Tex (1956, 1959). Here isoclinal folds in metasedimentary rocks show sharp plunge reversals that were attributed to perpendicular flexures resulting from penecontemporaneous crossed strain.

This brief review shows that it is not common to find unequivocal descriptions of $B_1 \perp B_2$-folds in the recent literature.

Chapter 12

Microtextures in Fabric Analyses

Traditional methods of structural analysis involving the study of structures and stratigraphic relationships mapped in the field do not always yield the information required to elucidate the geometry and structural chronology of an area. This is particularly the case when distinctive lithological units are not available, when the styles of the successive generations of folds are not distinctive, or when the area is not well exposed. However, additional evidence can be derived from microscopic analyses, and this work falls into two categories:

1. *Microscopic Petrofabric Analysis:* This involves study of the preferred lattice orientation of component mineral grains within homogeneous portions of a rock. Schmidt (1925) apparently published the first petrofabric diagrams, but Sander's (1930, 1950) books gave particular stimulus to the subject. Commonly populations of quartz, biotite, calcite, or dolomite grains are analyzed, although most common anisotropic rock-forming minerals have been used in particular cases. The technique involves measurement of the three-dimensional orientation of lattice directions (e.g., c-axis of quartz grains) of individual grains and plotting these on the lower hemisphere of a Schmidt equal-area projection. It is usual for a complete census of the grains of a particular mineral species within a specified area to be made; 200 to 300 grains are commonly necessary to define the preferred orientation accurately.

2. *Microtextural Analysis:* This involves analysis of palimpsests (or relicts) of older structural elements preserved within fabrics determined by younger orogenic events. Within an active orogenic zone the regional metamorphic conditions tend to change continuously so that one assemblage of minerals is commonly partially replaced by another; a clearly recognizable sequence of events can often be detected because of preserved palimpsests. Concomitant with the changes in the metamorphic environment are changes in the stress conditions, so that a sequence

of superposed phases of folding can be elucidated. In consequence, palimpsests of minerals grown at the earlier orogenic phases commonly preserve relicts of habit orientations and foliate structures that differ from those generated during more recent events. Relationships of this type have been studied for many years, but within the past decade particular use has been made of palimpsestic structures in analyzing fold systems.

In the preceding chapters the complex geometry that results from several epochs of superposed folding was described. Considerable success has been achieved in establishing the kinematic history by correlating metamorphic events with microtextural features. This technique promises to yield valuable results in future work, and, to illustrate the method more fully, the remainder of this chapter is mainly devoted to examples of the approach based on recent work in the Pyrenees (Spain and France) and the Central Highlands (Scotland).

MICROTEXTURAL ANALYSIS

Within a metasedimentary rock characterized by well-marked S-surfaces, porphyroblasts may develop under the stimulus of increasing grade of regional metamorphism. For example, small diopside and wollastonite crystals define a pronounced S-structure within a Dalradian calc-silicate marble at Bunbeg, NW Ireland; at a late stage in the regional metamorphism rhombicdodecahedral porphyroblasts of grossular garnet grew and poikiloblastically[1] enclosed many older silicate grains. These garnets transect the foliation (without deflecting the S-structures) and trains of diopside and wollastonite grains sometimes pass right through the porphyroblasts (Fig. 416). Later, changed metamorphic conditions caused idocrase to pseudomorph the grossularite partially (Fig. 417), but palimpsestic diopside and wollastonite grains (and the S-structures defined by them) are still preserved within the pseudomorphs. Massive bands dominated by grossularite and idocrase were more competent than the enclosing carbonate-rich material, and developed boudinage (Fig. 251) during a late phase of deformation (Whitten, 1951).

Simple observations of this type enable the detailed chronology of metamorphic and structural events to be analyzed. The diopside-wollastonite bands in the marbles at Bunbeg are parallel to boundaries of the lithologic units and to the foliation defined by micas in adjacent pelitic and

---

[1] An excellent review of the textural terms applied to metamorphic rocks is available in Harker (1932).

Figure 416. Thin section (crossed polars, negative photographic print) of calc-silicate rock, Bunbeg, Donegal, Ireland, showing subhedral grossular garnet porphyroblast (white) set in calcite, diopside, and wollastonite matrix (gray) with schistosity (left to right) picked out by diopside grains (black). Small diopside grains (black and gray) have been incorporated in the garnet without deflection during porphyroblast growth; large enclosures (black/gray) in garnet are pseudomorphous idocrase. Scale is 2 mm. (Photograph by Mr. S. B. Upchurch).

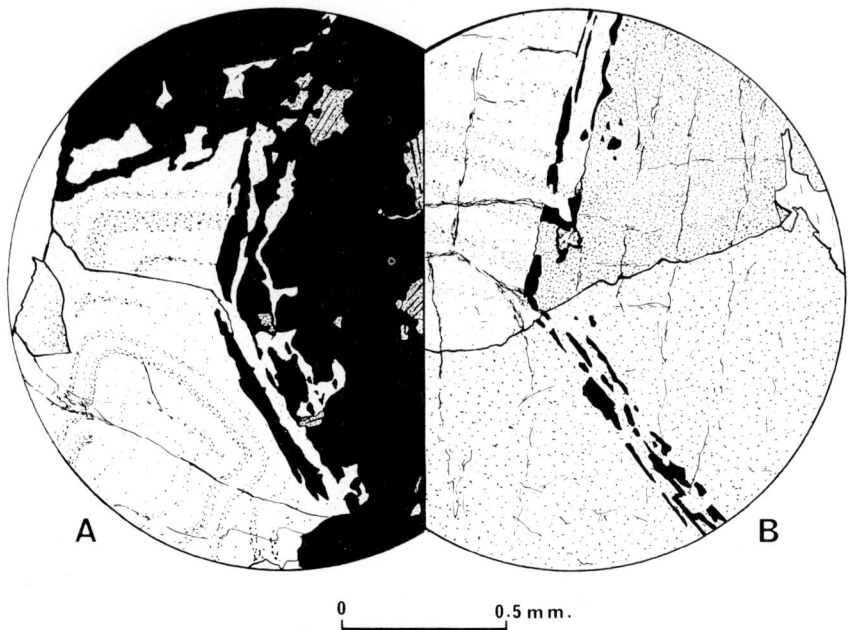

Figure 417. Idocrase partially replacing grossular garnet in a Dalradian calc-silicate hornfels, Bunbeg, NW Ireland. Garnet black; idocrase stippled. A: Early stage of differential replacement of garnet zones by zoned idocrase; diopside and quartz are poikiloblastically enclosed in idioblastic garnet. B: Palimpsestic idioblastic garnet within replacing idocrase (After Whitten, 1951, Fig. 6).

semipelitic metasedimentary rocks. These relationships suggest that intense deformation gave rise to isoclinal folding, and that this folding was concomitant with high-grade regional metamorphism. The grade of metamorphism is indicated by staurolite and pseudomorphs after sillimanite in the pelites. Subsequent to this folding, grossular garnet apparently grew under static conditions and developed euhedral porphyroblasts without distorting the foliation of the enclosing rock. Diaphthoresis (retrograde metamorphism) followed during which idocrase replaced grossularite. Finally, there is evidence of neocrystallization during a period of mild deformation that produced the boudinage. All of these events are recorded in a small raft of metasedimentary rocks enclosed within the Caledonian "Older Granite" of Donegal, and thus they represent pregranite phases of deformation and metamorphism. Correlation of textural characteristics from neighboring areas may well show that these tentative suggestions for Bunbeg need revision, but this example shows the manner in which a chronology can be built up.

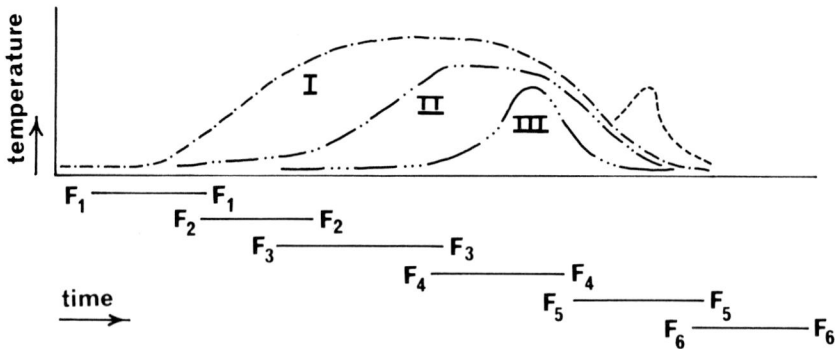

Figure 418. Schematic correlation chart for temperature and movement phases during metamorphism in southern Greenland. I: a highly migmatized area. II: a less migmatized area. III: an area with a single major intrusive event. IV: an area of late intrusions (granites, pegmatites and aplites) with offshoots from depth intruded after cooling of the migmatites. F_1: pre-granite kinematic phase. F_2: for I, the beginning of migmatization; for II and III pre-granite movements. F_3 and F_4: for I and II the migmatitic phase; for III, F_3 represents pre-granite movements and F_4 the granitic movements. F_5: transition from migmatitic to mylonitic kinematic activity. F_6: entirely mylonitic activity (After Wegmann, 1938, Fig. 15).

Wegmann (1938) made a detailed correlation of metamorphic temperatures and movement phases in southern Greenland, and his diagram (Fig. 418) constitutes one of the earlier examples of the type of synthesis described in this chapter.

The term **helicitic texture** refers to relics of precrystalline structures enclosed within porphyroblasts. Following Sander's terminology, the foliation of the host rock is S_e, and that inside a porphyroblast is S_i (Interngefüge). It is not uncommon for a post-tectonic porphyroblastic mineral phase to enclose minutely-folded S-surface relics. After enclosure of such relics (S_i), all traces of the early structures within the host rock may be obliterated during continuing neocrystallization; a new S might develop in the host rock. For example, large albite porphyroblasts in the Otago Schists, New Zealand, enclose strings of epidote and iron-ore granules that mark the trace of older S-surfaces (S_i) that were commonly strongly plicated prior to enclosure within the albite (Turner and Hutton, 1941, p. 233; Turner, 1948, p. 240). K. A. Jones (1961) published photographs of microfolds helicitically-preserved within albite porphyroblasts from western Perthshire, Central Highlands, Scotland.

Genuine helicitic textures must be differentiated clearly from spiral structures developed by para-tectonic crystallization of porphyroblasts. The so-called snowball garnets, for example, appear to have grown during

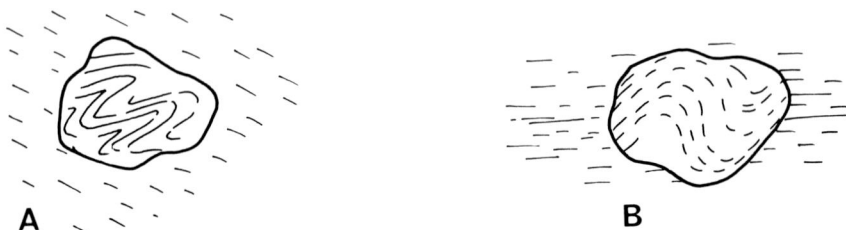

Figure 419. Garnet porphyroblasts. A: Helicitic texture. B: Para-tectonic crystallization of garnet.

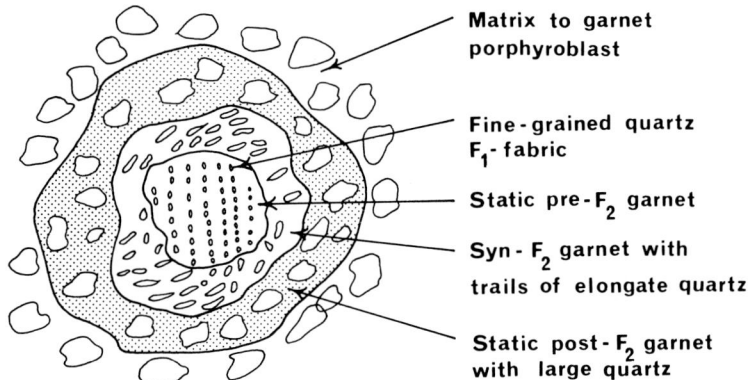

Matrix to garnet porphyroblast

Fine-grained quartz F_1-fabric

Static pre-F_2 garnet

Syn-F_2 garnet with trails of elongate quartz

Static post-F_2 garnet with large quartz

Figure 420. The three growth stages (diagrammatic) of garnet porphyroblasts in the Schichallion area, Central Highlands, Scotland; note that quartz grains in the outer zones have similar size and shape to those in the matrix (After Johnson, 1963, Fig. 3).

continued shear, so that during growth there was concomitant progressive rotation of the porphyroblast and its incorporated relict ground-mass materials (e.g., Schmidt, 1918; Spry, 1963A). The different geometry associated with helicitic and para-tectonic spiral structures is indicated in Figure 419. In some cases surprisingly-large rotations have been claimed for garnets; for particular examples, rotations of 60° (Zwart, 1960A), 173° (Schmidt, 1918), and 540° (Flett, in Peach, *et al.,* 1912) have been claimed. Although the rotation of garnet has been emphasized, most porphyroblastic minerals can show snowball and helicitic textures in appropriate environments.

In polymetamorphic terranes definite zonation can often be traced in garnets. Drawing on recent published work on the Scottish Central Highlands (see below), Johnson (1963) diagrammatically illustrated an example (Fig. 420) in which three phases of growth occur, namely: (a) the core of static pre-F_2 growth that enclosed palimpsests of fine-grained F_1-

fabric elements, (b) the middle zone representing syn-F_2 growth that enclosed curved trails of elongate quartz grains, and (c) the outer zone reflecting static conditions in which equidimensional quartz grains, similar in size and shape to those of the matrix, were enclosed within the growing garnet.

Ramsay (1962A, p. 323) interpolated a note of caution and showed that "flattening" (see Chapter 6) can result in a change in orientation of S_e without influencing the S_i in para- and pre-tectonic porhyroblasts (Fig. 421). Such relationships give a false impression of porphyroblast rotation. It was the frequent occurrence of parallel S_e-structures within neighboring porphyroblasts that led Ramsay to his appreciation of the significance of the flattening process in rocks.

The interpretation of the inclusions within garnets and other porphyroblastic minerals is beset with numerous problems; this has been illustrated by the recent debate about the evolution of quartz inclusions in garnet (see Harris and Rast, 1960A, 1960B; Galwey and Jones, 1962; Rast, *et al.,* 1962). Considerable care is required in arriving at any positive conclusions.

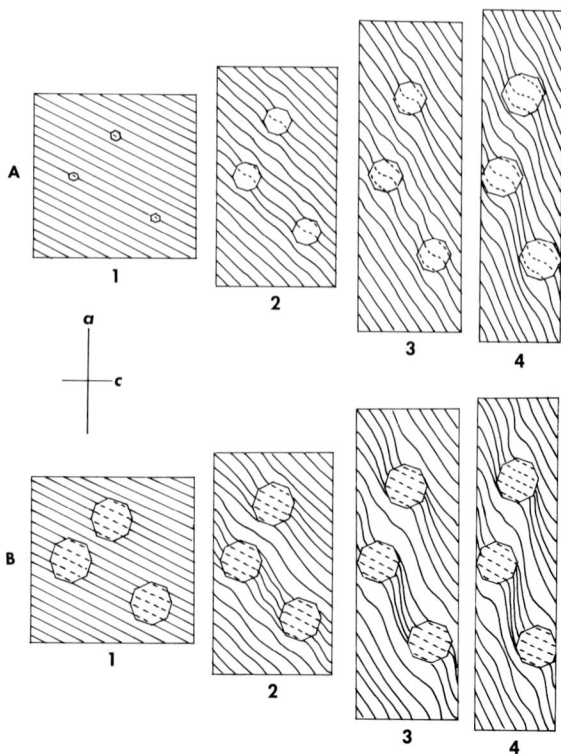

Figure 421. The possible effects of "flattening" on the relationships of S_e and S_i. A: Development of curved S_e within syntectonic porphyroblasts of rocks deformed by flattening (1 through 4). B: The effects of flattening on a rock containing large pretectonic porphyroblasts (After Ramsay, 1962A, Figs. 15 and 16).

CENTRAL HIGHLANDS OF SCOTLAND

Diablastic (sieve) textures of garnets have been recognized for many years; similarly, palimpsestic relicts have formed the basis of much purely metamorphic interpretation of the genesis of metasedimentary rocks. Many of these concepts were developed in the Scottish Highlands, and excellent accounts of the textures associated with different types of metamorphic rock were summarized by Harker (1932).

One of the earliest applications of textural analytical techniques was in the Upper Dalradian metasedimentary rocks of Banffshire. The rocks are highly folded, but the metamorphic boundaries do not reflect this folding, and the metamorphic zones do not follow the bedding in detail (Walls, 1937A; Sutton and Watson, 1955; Johnson, 1963). Thus, although the work and techniques outlined below are not entirely new, the systematic evaluation of the geometric significance of textural features has introduced an important new tool into structural geology. The new phase of work within the Scottish Highlands seems to have been initiated by Rast and Sturt (1957), who analyzed the crystallographic and geologic factors of garnet growth in central Perthshire. Complementary detailed accounts for the Schichallion (Rast, 1958B) and the Loch Tummel areas (Sturt and Harris, 1961; Sturt, 1961B) soon followed.

In Chapter 11 the four principal discrete structural events recognized in the Schichallion and Loch Tummel areas were described (see Figs. 326 and 330; also Table 5). A penetrative foliation, S_1, resulted from F_1. In response to F_2 and F_3, crenulation foliations (S_2 and S_3) developed that are defined by micas concentrated along shear planes parallel to axial planes of the microcorrugations of earlier foliations. The recognition of texturally-distinct syntectonic and post-tectonic (or static) garnets materially aided in erecting the geochronology of the Schichallion and Loch Tummel areas:

(a) *Post-tectonic garnets enclosing helicitic relicts:* At Schichallion an early generation of garnets grew under static conditions (post-F_1 tectonic phase) of metamorphism and enclosed relicts of syntectonic F_1-structures. In many areas these garnets were rotated during F_2, and in such cases a new foliation tends to wrap around the porphyroblasts, so that the S_i-fabric is not continuous with S_e; it is relatively easy to recognize that S_i and S_e are different generations. When later F_2-rotation did not occur, the helicitic relicts (S_i) are commonly continuous with the matrix structures (S_e) (Fig. 422). Sturt and Harris (1961) noted that S-shaped helicitic S_i-fabrics occur in those garnets that enclosed F_2 microfabrics during a post-F_2 static phase of growth; when rotated by F_3-movements these garnets simulate syntectonic porphyroblasts, except that the S_e is deflected around the porphyroblasts.

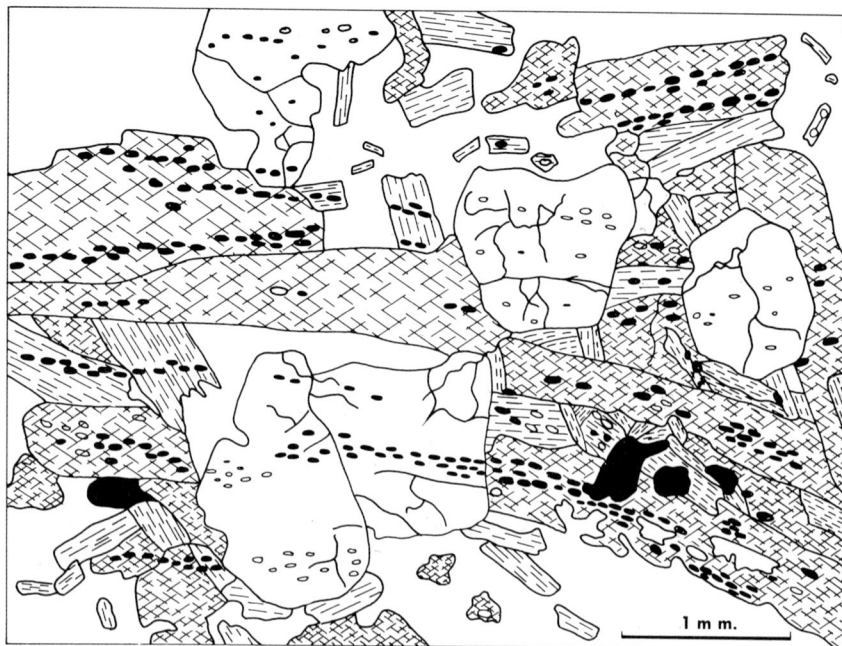

Figure 422. Post-tectonic crystallization with preservation of undeflected in-clusion trails in garnet, hornblende, and biotite grains, Central Highlands, Scotland (After Sturt and Harris, 1961, Fig. 1).

(b) *Syntectonic garnets:* These garnets have S-shaped or spiral S_i-fabrics that represent S planar fabrics incorporated and rotated during porphyroblast growth. At Schichallion such garnets are particularly characteristic of F_2 growth.

In Perthshire the growth of many garnet porphyroblasts apparently overlapped both static and syntectonic phases (Fig. 420). Rast and Sturt (1957), for example, described garnets with post-F_1 cores that enclose helicitic rectilinear S_1-fabric elements and with outer zones that contain S-shaped patterns reflecting para-F_2 crystallization. Again, Rast noted syntectonic (F$_2$) cores and post-tectonic margins in garnets at Schichallion. Many different zonation combinations occur; sometimes continuous, sometimes discontinuous, growth is indicated by the refractive index variations between successive zones of large garnets (Sturt and Harris, 1961).

Table 5 summarizes the kinematic and metamorphic episodes recognized in the Schichallion and Loch Tummel areas by Rast and Sturt; details of the geometry of the superposed folding were given in Chapter 11 (Fig. 330). Throughout much of the Iltay Nappe, Central Highlands,

Table 5. – Kinematic and Metamorphic Episodes at Schichallion and Loch Tummel, Central Highlands, Scotland

Structural Events		Metamorphic Events		
Phase	Event	Phase	Schichallion	Loch Tummel
F$_1$	Early recumbent folding with nappes; ENE or NE axial trends; boundary slide between Dalradian and Moine rocks initiated	Syntectonic	Pelitic sedimentary rocks converted to fine-grained phyllites	
		M$_1$ Post-tectonic	Occasional garnets grew in parts of area	
F$_2$ Folding on NNW-trending axes		M$_2$ Syntectonic	Garnets grew in parts of area	
		Post-tectonic	Most important period of regional metamorphism; garnet, tourmaline, staurolite, kyanite, and feldspar growth	Garnet and tourmaline grew and pegmatites formed
F$_3$ Gentle open folds about E- or ENE-trending axes		M$_3$ Syntectonic		Retrograde decay of porphyroblasts; chlorite grew
		Post-tectonic		Local muscovite and biotite developed
F$_4$ Loch Tay fault movements post-dating regional metamorphism		Syntectonic	Local retrograde metamorphism and development of chlorite	

the F_1-folding occurred when the grade of metamorphism was probably no higher than chlorite or biotite (i.e., greenschist facies). At Schichallion and Loch Tummel the first major phase of metamorphism (M_1) was pre-F_2, so that micas oriented during F_1 were folded in the F_2-phase, and after the F_2-event the micas recrystallized mimetically. Metamorphic intensity

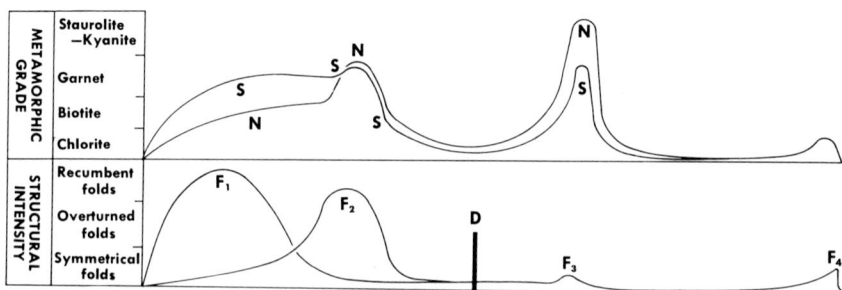

Figure 423. Structural and metamorphic events in the Schichallion district, Central Highlands, Scotland. F_1 through F_4 are the periods of intense deformation, D is a dolerite dyke intrusion phase, and the N and S curves refer to events in the north and south of the region respectively (After Rast, 1958B, Fig. 4).

increased contemporaneously with development of the structural F_1 and F_2 events (Fig. 423). Metamorphic grade may have declined temporarily after the F_2-folding, because, during this relatively quiescent phase, dolerite dykes were intruded (Fig. 423); Johnson (1963, p. 127) wrote:

> Rast (1962) has suggested that the P-T conditions after M_2 must have changed to allow the tension joints, followed by the dolerites, to be developed. It cannot be assumed however that this dolerite event necessarily implies a "pause" in the regional metamorphism, for a dolerite would chill against schists undergoing high-grade metamorphism.

Johnson believed that the clearest evidence for a nonorogenic interval is provided by the great gravity-layered gabbros of NE Scotland (Belhelvie, Huntly, Insch masses, etc.), which he considered to have been intruded after M_3 and before F_3 (Fig. 424). However, the age of these gabbros is not unequivocal (cf. Stewart and Johnson, 1960), and the correlation of structural and metamorphic events between NE Scotland and the main Iltay Nappe is open to debate (see below). What does seem clear, however, is that both in deeper metamorphosed, and in shallow unmetamorphosed, portions of orogenic zones definite evidence is emerging for spasmodic folding; during periods of less intense stress, subparallel sheets of igneous material are commonly intruded (cf. Tremlett, 1963).

Growth of the higher-grade metamorphic minerals (M_3) at Schichallion occurred during the static conditions between F_2 and F_3, or during F_3

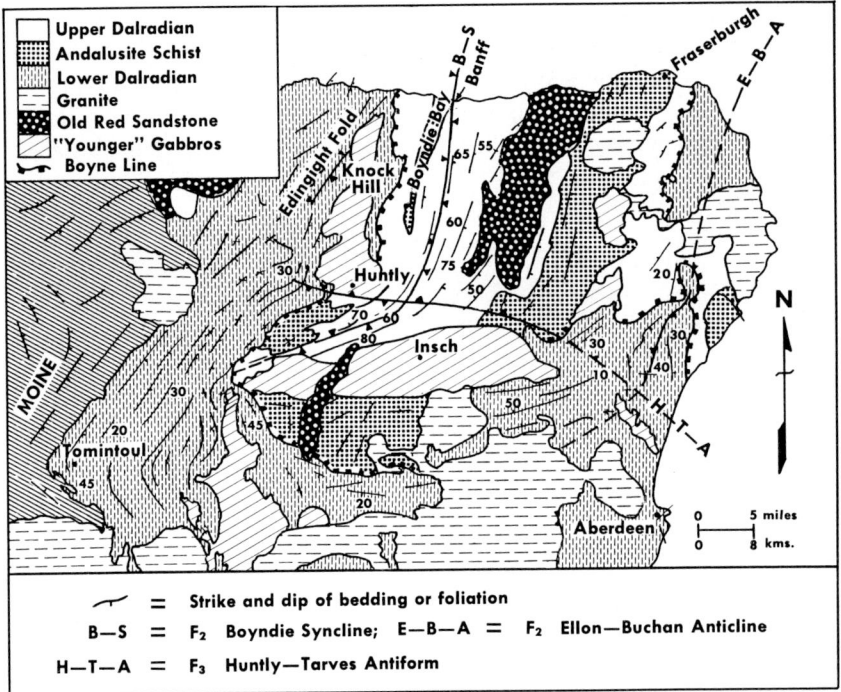

Figure 424. Structural geology of NE Scotland (After Johnson and Stewart, 1960, Fig. 1).

(Fig. 423). This static period coincided with renewed metamorphic activity that led to profound neocrystallization and to the appearance of staurolite, kyanite, and metasomatic feldspar in some areas; biotite grew independently of fabric. In some localities kyanite, plagioclase, garnet, and tourmaline grew in random orientations and helicitically enclosed F_2-structures.

The present-day garnet zone of regional metamorphism in the Schichallion area, as summarized on Kennedy's (1948) map, for example, is the result of M_2 (F_2) garnet-grade metamorphism in the central and southern area overlapped by the F_3 garnet zone northwards (Rast, 1958B). Rocks of the Loch Tummel area tend to conform to the staurolite-quartz subfacies of the almandine-amphibolite facies (Francis, 1956), but this grade was not attained until after the F_2-stage (except locally at Dunfallandy Hill). Thus, in this area, there is a southward overlap of an earlier M_2 garnet zone by the M_3 (post-F_2) garnet zone. Hence, in Perthshire the M_3 metamorphism appears to have affected a larger area than has M_2, so that M_3 tends to define the resultant garnet zones on an isograde map (cf. Johnson, 1963, pp. 127-8).

In the Loch Tummel area kyanite and staurolite are limited to a narrow zone of Pitlochry Schists; the relation of the kyanite and staurolite to small micaceous microfolds shows that these minerals were post-F_2 (i.e., post-M_2), although they grew before the principal period of garnet development during M_3. Successive crystallization events of the micas and amphiboles also permitted details of the deformational sequence in these areas to be elucidated by Rast and Sturt.

The correlation of kinematic and metamorphic events of the Iltay Nappe with those of NE Scotland has been a subject of controversy. Sutton and Watson (1956, pp. 106 and 119) stated that in the Upper Dalradian metasedimentary rocks of Banffshire (Fig. 424) porphyroblasts of andalusite, cordierite, and staurolite appear to have formed during a phase of regional metamorphism after folding. Specifically, at Knock Head a foliation developed later than the folding, and evidence for polymetamorphism was based on andalusite crystals that grew across this foliation and enclosed micas that lie parallel to the foliation. However, Johnson and Stewart (1960, p. 100) recently postulated that the structural history is more complex than supposed by Sutton and Watson (1956). Johnson and Stewart recognized a sequence — somewhat akin to that originally advocated by Read (1955) and Read and Farquhar (1956) — of three kinematic events which they considered to be in approximate accord with those mapped in Perthshire; Johnson (1963) described the associated metamorphic events on the basis of textural characteristics. The resulting chronology, which conflicts with that proposed by Sutton (1960B) and Read (1961, p. 666), involves (Fig. 424):

(a) Isoclinal E-W or SE-NW trending fold axes (F_1) with possible weak Barrovian-type metamorphism and development of a mica-defined foliation — Read (1955) and Watson (1963) pointed out that if this E- or ESE-facing fold structure is continuous with the Iltay Nappe, as appears likely, it extends for 350 kilometers in Scotland alone and comprises part of the largest Caledonide fold identified in Scotland;

(b) Minor and major (Boyndie syncline) NNE trending folds (F_2) followed by an important static post-F_2 growth of andalusite, staurolite, and cordierite oriented randomly across the F_2-foliations; and

(c) Flexure (F_3) on a regional scale about an ESE antiformal axis that caused distortion of the post-F_2 metamorphic minerals.

In many areas additional work will be required before the detailed local chronology can be elucidated. For example, in the Glen Clova area, Angus, Chinner (1961) studied the paragenesis of the aluminum-silicates in an area of complex multiple deformation. Here, kyanite often shows a

Figure 425. Field sketch of sillimanite veining, The Lunkard, Glen Doll, Angus, Scotland. *Stipple*: semi-pelite (quartz-feldspar-biotite-muscovite-garnet) gneiss. *Dashes*: quartz-feldspar-biotite-muscovite-garnet-kyanite pelitic gneiss. *Unshaded*: sillimanite-rich veins (After Chinner, 1961, Fig. 1B).

high degree of preferred orientation, and it may enclose grains of garnet, mica, quartz, or tourmaline, or be included in any of these minerals. Although sillimanite is post-tectonic with respect to the kyanite-bearing fabric (Fig. 425), in other cases Chinner recorded the sequence kyanite → sillimanite → kyanite. Johnson's (1963, pp. 133 ff.) discussion of Chinner's results shows that the chronology must remain uncertain until more details of the structural geometry in this area are known.

Ramsay (in Johnson and Stewart, 1963) and Johnson (1963, pp. 131 ff.) discussed the correlation of kinematic and metamorphic events of the Dalradian Series and the Moine rocks of the Northern Highlands. Johnson referred to Howkins' (1961) description of three stages of garnet growth in Moidart, SW Northern Highlands, which are reminiscent of the stages described from Perthshire. Figure 426 shows that Johnson implied a close contemporaneity between analogous events throughout the Central and Northern Highlands, although Sutton (in Johnson and Stewart, 1963) and Watson (1963) were, with justification, somewhat more cautious. Watson (1963, p. 254) thought that it could be "tentatively suggested" that the earliest phases of folding in the Dalradian and Moine rocks were "broadly contemporaneous," and that

> ...both are characterized by remarkably flat axial planes and both appear to be earlier than the climax of the metamorphism. On the other hand their controlling mechanisms may be different. The structures in the Dalradian die out downwards and appear to be controlled by gravity whereas the structures in the

Moines tend to die out upwards and appear to be related to compression in the underlying basement.

Such tentative correlations present an exciting and challenging picture of the development of an orogeny. Despite the possible contemporaneity of the metamorphic zones in the Dalradian and Moine areas, the zones cut across the stratigraphic boundaries and the limbs and axial planes of the primary recumbent folds, nappes, and slides (cf. Kennedy, 1948).

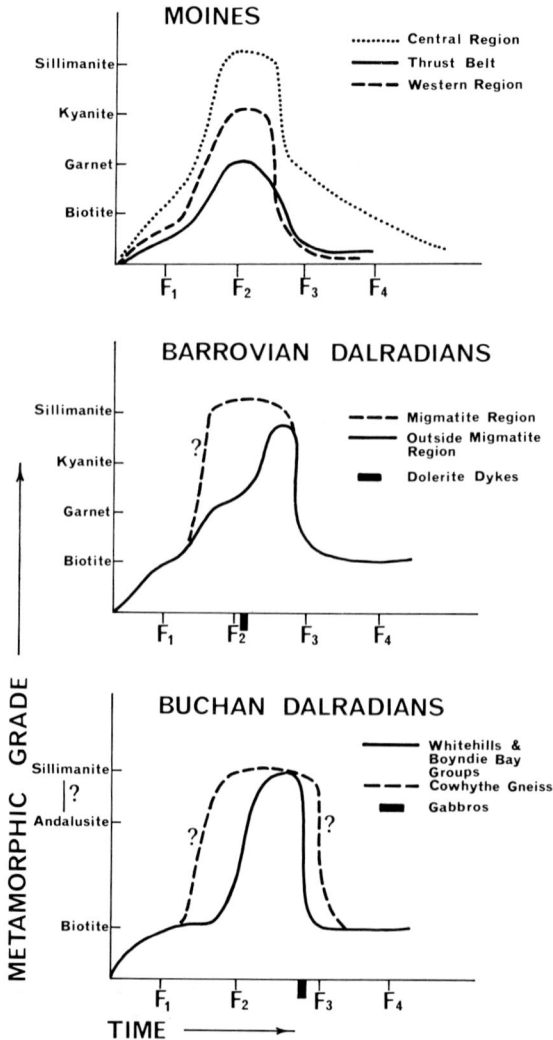

Figure 426. Variation of metamorphic grade with time in different parts of the Moine and Dalradian rocks of the Central and Northern Highlands, Scotland (After Johnson, 1963, Fig. 8).

CENTRAL PYRENEES, SPAIN AND FRANCE

It is a well-established principle that the sequence of metamorphic phases can commonly be unravelled by study of the texture of the rocks and, in particular, by study of the palimpsestic relicts. For example, van der Plas (1959) showed that the succession in time of glaucophane → blue-green amphibole → actinolite is common in the northern Adula area, Switzerland. This succession was observed in zoned crystals and also in armored relicts (e.g., glaucophane enclosed in garnet, with blue-green amphibole abundant outside the garnet crystals). Once such a mineralogical sequence has been established it can be most useful in disentangling the succession of structural events in an orogenic belt.

This approach was used in the central Pyrenees by Zwart and other geologists of Leiden University, Holland. On the basis of studies in the Bosost area, which is part of the Garonne dome in the Pyrenees (Fig. 427), Zwart (1963A) suggested that it is probably safe to assume that the start and termination of a folding phase was contemporaneous over an area of a few tens of square kilometers, but not over a major sector of the mountain chain. The total regional metamorphism was broadly contemporaneous with the penetrative Hercynian kinematic events, but Zwart (1963A, 1963B) demonstrated that, in detail, the metamorphic processes had little direct relationship to the various individual folding phases. In the Pyrenees the whole metamorphic sequence can be divided into zones

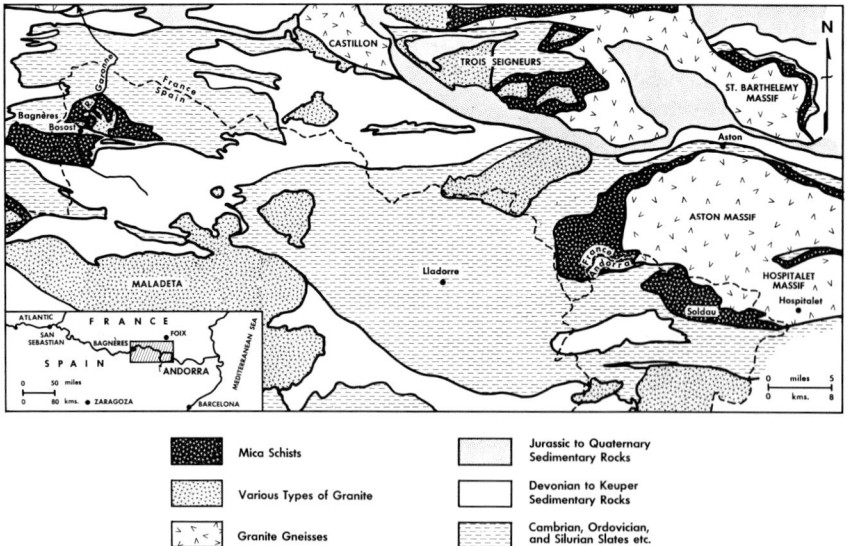

Figure 427. General geological map of the central Pyrenees Mountains in Andorra, France, and Spain (After de Sitter and Zwart, 1961, Plate 1).

that are apparently independent of the original sedimentary stratification (de Sitter and Zwart, 1959, p. 399), because the metamorphic zones commonly transect the structural and stratigraphic boundaries. The closest conformance in the Bosost area is between the F_2-kinematic phase and formation of muscovite-biotite-schists, but even in this case the biotite isograd is discordant with the stratigraphy (Fig. 428A).

The style of folding is dissimilar in different members of the Garonne dome (Fig. 428B). Flat-lying foliation is commonly parallel to the lithologic banding of the Cambro-Ordovician phyllites, while the Devonian rocks around the dome occur in tight folds with vertical axial-plane foliation. Thus, the Devonian rocks are folded disharmonically with respect to the Cambro-Ordovician rocks, and the incompetent Silurian slates acted as a transitional layer, with a subhorizontal contact below and steep pinched anticlines in the Devonian rocks above (Kleinsmiede, 1960; Zwart, 1963A).

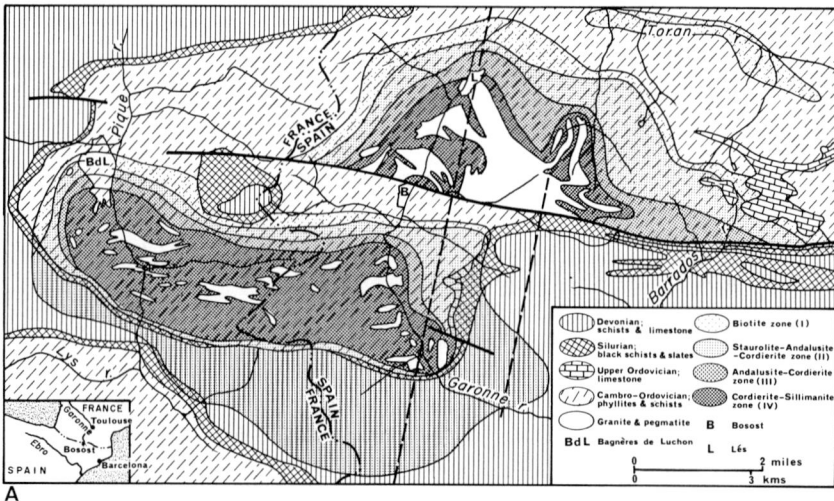

Figure 428. Structural geology and metamorphic zones of the Bosost area, central Pyrenees. A: Simplified geological map. B: North-south vertical section through the area (After Zwart, 1963A, Figs. 2 and 3).

Fold episode	F_1	F_2		F_3		F_4 - F_5	
Central part Hospitalet massif							
Trois Seigneurs massif					?		
Bosost dome (Valle de Arán)							
Western part Aston massif							
Western part Hospitalet massif (El Serrat)							
	pre-kinematic phase	first stage (mainfolding) early synkinematic phase		second stage		late synkinematic phase	post kinematic phase

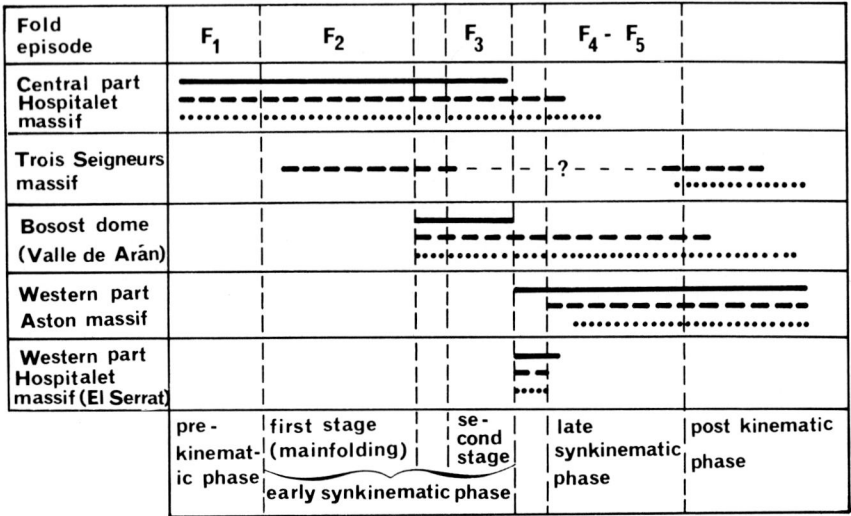

———— Staurolite   ▬ ▬ ▬▪ Andalusite   •••••• Cordierite

Figure 429. Chronological relations between crystallization and deformation in various mica-schist areas of the Pyrenees (After Zwart, 1960A, Fig. 11).

The first phase of Hercynian metamorphism apparently did not start long before the beginning of folding, and the ensuing metamorphic events continued until long after all orogenic movements ceased (Fig. 429). The highest grades of metamorphism occurred in the axial zone of the orogenic belt and successive higher grades moved outwards overprinting the lower-grade assemblages developed during earlier phases; palimpsests confirm that many rocks recrystallized in response to several of the lower grades of metamorphism before attaining the highest grade. Although a general chronology can be established for the whole region, Figure 429 shows that not all of the metamorphic and kinematic events can be recognized at each locality; in some cases overprinting may have obliterated evidence of earlier events, but it is also clear that commonly the effects of an orogenic event are not developed uniformly throughout a large area. An individual rock sample may have been subjected to metamorphism before, during, and/or after the local deformational phases (Zwart, 1963A, 1963B). Hence, a great variety of local circumstances prevailed during the Hercynian metamorphism; in extreme cases rocks transformed into mica schists were later affected by either syn-kinematic or post-kinematic metasomatism that converted them into gneisses or granitoid rocks in the infrastructure.

Detailed mapping (Guitard, 1955; Zwart, 1956, 1959A; de Sitter and Zwart, 1961) showed that early flexural-slip folding (F_1) was followed by four major kinematic phases (F_2-F_5) that can be recognized in both the supra- and the infrastructure. The four phases F_2 through F_5 can apparently be correlated over considerable areas (Fig. 429). Additional phases without regional distribution were recognized locally on the basis of detailed study of superposed folds in the Cambro-Ordovician through Carboniferous metasedimentary rocks. The principal metamorphic phases in the Pyrenees recorded by Zwart are described below; he (Zwart, 1960A) found that parts of the complete sequence of events can be traced with thin sections of the slates and gneisses, but that the complete story can only be followed in thin sections of the andalusite-staurolite-cordierite-mica schists.

1. PRE-KINEMATIC PHASE

Apparently only simple flexural-slip folds (F_1) developed in this so-called pre-kinematic phase, and they are difficult to identify on account of the younger superposed foliation. In the Hospitalet massif andalusite, staurolite and the cordierite porphyroblasts have been recognized as relics of this period of metamorphism (Figs. 427 and 429). These megacrysts grew before any foliation developed and they enclose irregularly-oriented quartz and biotite grains that do not define an S_i (Fig. 430A and 430B); porphyroblasts developed during ensuing phases contain well-defined S_i-structures.

2. EARLY SYN-KINEMATIC PHASE

Strike of axial planes (S_2 and S_3) and bedding (S_0) are parallel.

(a) *First or main kinematic stage* (F_2): Zwart called this the Main Phase of folding because it is the most widespread in the Pyrenees. It is characterized by E-W trending fold axes and linear structures, and by the development of pronounced foliation. Slip folds with subvertical foliation are developed strongly in slates and phyllites; the size of the folds is commonly difficult to determine because of the lack of distinctive marker horizons, although interstratified competent units define parasitic folds. When traced into zones of higher-grade metamorphism the phyllites were called mica schists at the first appearance of biotite; in these schists of the infrastructure, bedding (S_0) is isoclinally folded and the foliation is sub-horizontal. Progressive mineralogical changes up to the kyanite grade can be recognized in, for example, the Hospitalet massif (Fig. 429). Habit orientation of biotite and lattice orientation of the c-axes of andalusite and cordierite define lineations parallel to the B_2-fold axis. Syntectonic crys-

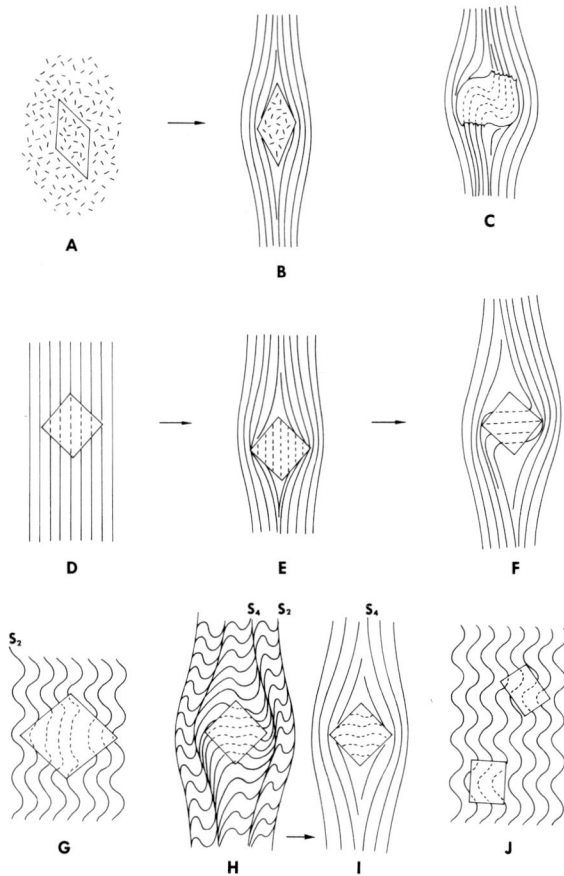

Figure 430. Relationships between S_e and S_i. A and B: Prekinematic meta-morphism with no S_i. C: Early synkinematic metamorphism. D, E, F: Develop-ment of S_i in porphyroblast and bowing out of S_e during later flattening. G: Late kinematic metamorphism, with S_i showing stronger folding towards the porphy-roblast rim. H and I: Porphyroblast with weakly folded S_i (S_2); S_e more strongly folded with development of S_4; in I S_4 is a completely new S_e-schistosity and S_2 is represented by S_i. J: Helicitic folds in porphyroblasts resulting from post-tectonic crystal growth (After Zwart, 1960B, Figs. 4 and 5).

tallization of staurolite and andalusite was established by enclosed spiral S_0-structures, with rotation axes parallel to the E-W B_2-fold axes (Fig. 430C). The foliation (S_e) is always deflected round these megacrysts because of continuing or later shearing. In other examples the schistosity formed, the megacrysts grew and enveloped planar S_i, and then the porphyroblasts were rotated by shearing, so that S_e was deflected and $S_i \neq$

S_e. Zwart (1960B, p. 209) suggested that in many cases the megacrysts grew so rapidly that the continuing deformation had no time to affect the crystal during growth.

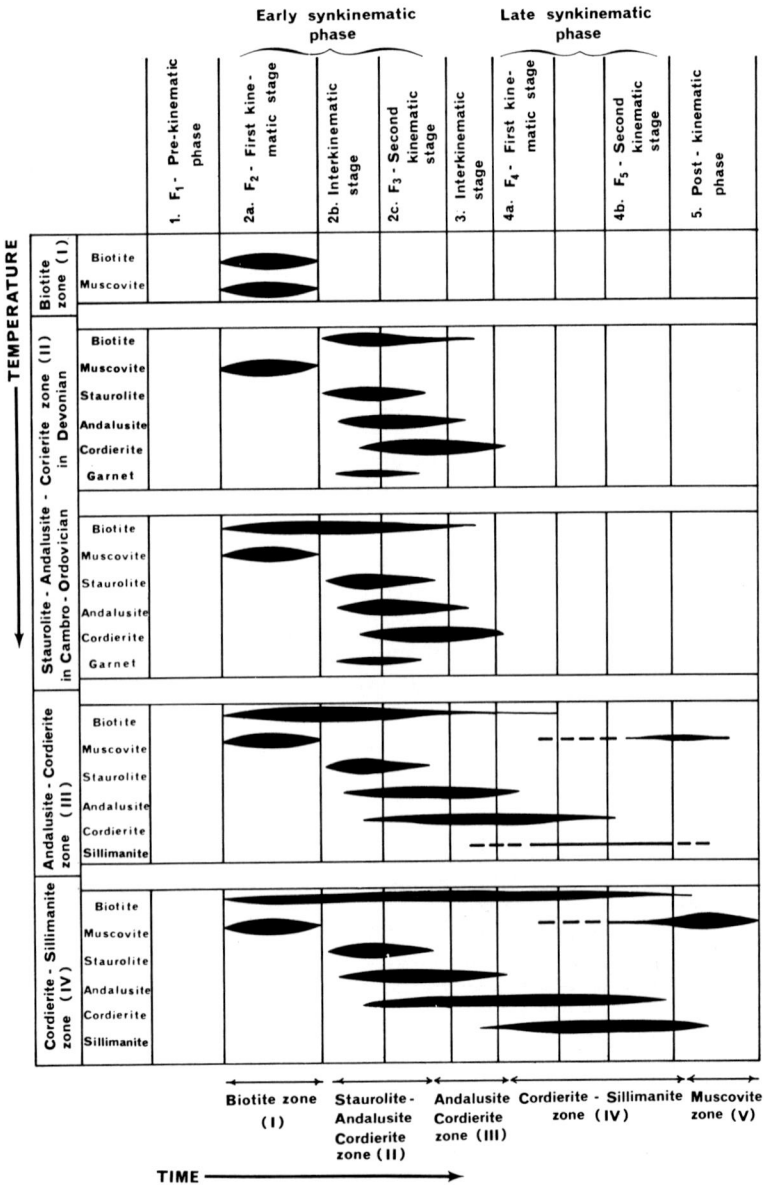

Figure 431. Time, metamorphic zones, and deformation phases in the Bosost area, Central Pyrenees (Adapted from Zwart, 1963A, Fig. 4).

In the Hospitalet massif the kyanite-grade mica schists pass down into massive gneisses with parallel foliation and linear structures, although fold closures have not been recognized in the gneisses. Zwart (1960A, p. 169) concluded that, in the low- as well as in the high-grade rocks of the Hercynian belt, folding and metamorphism proceeded simultaneously during F_2-time.

(b) *Inter-kinematic stage (between F_2 and F_3):* In the Bosost area the absense of crystals rotated about the B_2-axis proved that aluminum silicates were not formed during the F_2-stage. However, rocks of this area entered the andalusite zone before the second kinematic stage (F_3) because abundant andalusite, staurolite, and cordierite grains contain planar S_i and were rotated about N-S F_3-axes (Fig. 430, D, E, F). Some staurolite grains began crystallization during this inter-kinematic stage and then continued growth during the ensuing F_3-stage. Andalusite and cordierite had similar and concomitant paragenetic histories, although they both began crystallization a little later than staurolite. Such relationships are demonstrated by idioblastic staurolite enclosed in andalusite, and by staurolite or andalusite enclosed in cordierite; the reverse relationships are not observed (Zwart, 1963A, pp. 50 ff.). Zwart (1963A, 1963B) illustrated the details of these regional metamorphic sequences with the aid of the diagram shown in Figure 431.

(c) *Second kinematic stage (F_3):* This stage is recognized by relatively minor folding about N-S B_3-axes in the mica schists of the infrastructure only; the sense of movement is always from W to E (Zwart, 1963B, p. 144). Here B_2 and B_3 are perpendicular, but these fold episodes were clearly separated in time. Occasional compositional banding in the schists defines small isoclines and many thin sections show minerals rotated about N-S F_3-axes (Fig. 432B). Although Zwart (1960B, p. 206)

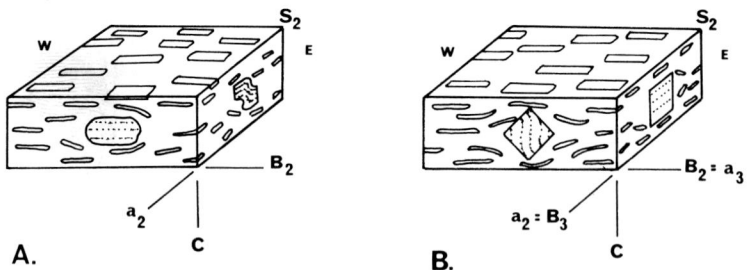

Figure 432. Schematic diagrams of fabric in a biotite schist from the Central Pyrenees. A: Porphyroblasts rotated about E-W B_2-fold axis of early synkinematic F_2-fold phase. B: Porphyroblasts rotated about N-S B_3-fold axis of early synkinematic F_3-fold phase (After Zwart, 1960B, Fig. 1).

considered the proof quite definite, the relative age of F_2 and F_3 can only be established on the basis of microscopic evidence. Zwart (1960B, p. 212) concluded that F_3 was dominated by isoclinal flexural-slip folding with limited slip in the fold limbs because rotation of porphyroblasts about B_3 is characteristically less than 90°; however, more recently he (Zwart, 1963B, pp. 144-5) concluded that the constant sense of slip indicated by the rotated porphyroblasts establishes that F_3 constitutes Gleitbrett or slip folds.

3. INTER-KINEMATIC STAGE

In the western part of the Hospitalet massif, and in other areas (see Fig. 429), andalusite, staurolite, and cordierite developed with planar S_i that was not rotated about either the B_2 or the B_3-fold axes; subsequent folding caused S_e to be deflected around porphyroblasts grown in this inter-kinematic stage.

4. LATE SYN-KINEMATIC PHASES

(a) *First stage (F_4' and F_4''):* This stage is represented by a conjugate fold set resulting from E-W compression (Zwart, 1963B, pp. 146-7); the dominant northwest-trending F_4'-folds and the less-abundant northeast-trending F_4''-folds both have vertical axial planes, and comprise penecontemporaneous small folds of the S_2-foliation. Crenulation foliation is commonly developed parallel to the new axial planes (S_4' and S_4''). Because of the variable attitude of S_2, the intersections of S_2 with the two S_4, and

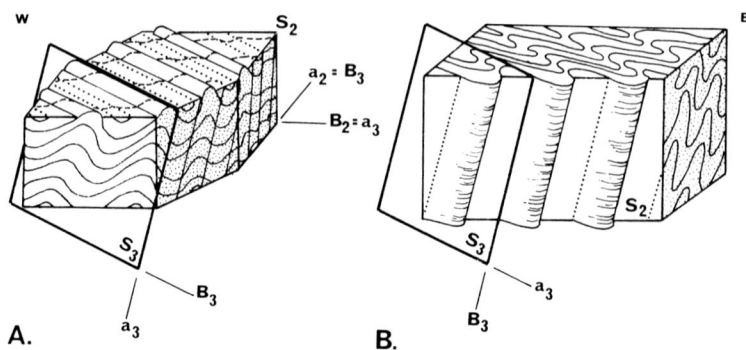

Figure 433. Schematic diagrams of superposed folds in biotite schists, Central Pyrenees; axial planes in both diagrams are parallel, but the fold axes vary in trend and plunge. A: Superposed NW-SE folds in schist with horizontal schistosity; broken lines on schistosity surface are traces of E-W B_2-lineation. B: Superposed NW-SE folds in rock with vertical schistosity (After Zwart, 1960B, Fig. 2).

the two new sets of minor fold axes (B_4' and B_4'') have variable plunges (Fig. 433). The F_4-folds are commonly post-crystalline, but they are occasionally para-crystalline. The latter relationship is illustrated by the mica schists of Soldeu, Andorra, which have a strong E-W lineation defined by the habit orientation of mica, andalusite, and cordierite; minor F_4-folds are superposed on this structure and a new generation of andalusite and cordierite developed a habit orientation parallel to both B_4' and B_4'' (Fig. 434).

Figures 430G, H, and I illustrate the characteristic S_i patterns developed by syn-kinematic F_4-crystallization. In the center of the megacrysts S_i is weakly folded or perfectly planar, but outwards the S_i-folds become steeper and then marginally they merge into S_e (which may be

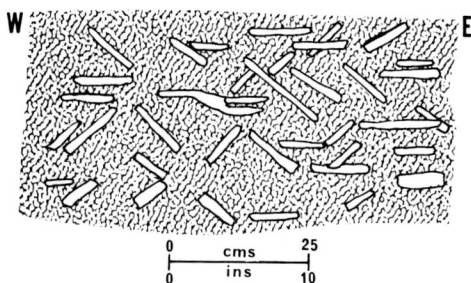

Figure 434. Oriented porphyroblasts in andalusite-cordierite-mica schist, Soldeu, Hospitalet massif, Andorra. Andalusite and cordierite porphyroblasts aligned E-W parallel to elongated micas define a first lineation. Andalusite and cordierite porphyroblasts aligned NW and NE are parallel to axes of superposed small folds (After Zwart, 1960A, Fig. 8).

even more strongly folded). These relationships suggest that the deformation steadily increased in magnitude during crystallization.

Chloritoid grew at several different periods during F_4. Some chloritoids have preferred orientation within the crenulation foliations (S_4); these planes have been deflected around some of the chloritoid porphyroblasts, and some of the chloritoids are slightly bent or broken which suggests postcrystalline folding. In other cases small folds of the S_2 foliation are helicitically enclosed within chloritoids, and this implies post-F_4 chloritoid growth (Zwart, 1959B).

Where strong F_4-folding affected mica schists, the associated neocrystallization commonly obliterated all earlier structures, although the resulting assemblages tend to show little or no preferred orientation. Competent units (e.g., quartzites) seldom show any response to this late-kinematic phase, although such rocks were strongly folded during the earlier phases.

(b) *Second stage (F_5):* Folds about E-W axes have been recognized within both the infra- and the suprastructure, but mainly outside the areas referred to above.

5. POST-KINEMATIC PHASE

The metamorphism of this phase is characterized by staurolite and biotite porphyroblasts that contain true helicitic folds congruent with the folds (S_e) outside the crystals (Fig. 430J). The S_e-planes are not deflected by the porphyroblasts.

GENERAL DISCUSSION

The type of structural analysis described above obviously has wide application in deciphering the sequence of tectonic and metamorphic events in other regions. However, there is a real danger of trying to simplify relationships in order to fit a varied sequence into a simple model. In particular, there is no compelling reason to assume that particular metamorphic climaxes were strictly contemporary over thousands of square kilometers; similarly, it is intuitively reasonable to assume that kinematic events may have followed the same sequence in neighboring areas without necessarily having been strictly contemporary at each stage.

Gilluly (1963, pp. 151 ff.) referred to the hazards of extrapolating the age of one kinematic event to a near neighbor. He pointed out that in the Los Angeles basin, California, U.S.A., the early Pleistocene beds of the Signal Hill anticline were tilted to dips of 60-80° in mid-Pleistocene time, while less than 5 kilometers away the parallel Wilmington anticline, whose structural relief of some 300 meters was acquired in mid-Pliocene time, remained completely dormant. Because graben subsidence spans the time of several compressional phases in southern California, Gilluly suggested that

> ... the assumption of world-wide episodes of compression alternating with others of tension is quite untenable ... [and is] not supported by present-day events nor by the geological record of Cainozoic time—the part of geological history we can read most clearly.

It is difficult to determine to what extent such concepts should be extrapolated to the deeper zones of an orogenic belt such, for example, as the Central Highlands of Scotland. It is highly unlikely that each phase of folding in the superficial sedimentary rocks reflects a phase of folding at depth. It is possible that phases of folding that affected the Dalradian rocks of Scotland were as ephemeral as those described from California by Gilluly, although, without the precision afforded by detailed stratigraphic marker horizons, exact time relationships are difficult to prove.

Reference to Table 5 and Figure 423 shows that Rast (1958B) interpreted the M_3 climax as contemporary with F_3 at Schichallion, Scotland, whereas Sturt (1961B) found that the climax of M_3 was pre-F_3 at Loch Tummel. Such a time difference is not unlikely, although, in his synthesis for the whole Central Highlands, Johnson (1963) defined M_3 as a pre-F_3 episode. Of course, further research may show that M_3 was strictly contemporary in Schichallion and Loch Tummel, and maybe the F_3 was contemporary in both areas too.

Although it is difficult to prove that kinematic and metamorphic episodes in neighboring areas are homogeneous with respect to time, there is still a tendency to assume contemporaneity over large geographic areas. Johnson (1963, p. 139) suggested that although "... caution is needed in setting up long range correlations of events in widely separated parts of the Highlands it is obvious that many local time scales are strikingly similar to one another...." The evidence cited by Johnson (1963) and Watson (1963) for the age of the Caledonide folding and metamorphism in Scotland and Ireland, together with a few additional data (including some radiometric dates from Kulp, *et al.,* 1960, and Giletti, *et al.,* 1961) are summarized in Figure 435; the columns in this figure are arranged in a NW-SE section across the orogenic belt. Watson (1963, p. 253) referred to the three main items of evidence for the Dalradian and Moine rocks, namely:

(a) It is "virtually certain" that Cambro-Ordovician rocks are involved in the Lochalsh fold in the Kishorn Nappe (beneath the Moine Thrust; see Fig. 342), while superposed folds are linked by several workers with later-generation folds affecting adjacent Moine rocks. If this reasoning is correct, folding of the Moine rocks commenced here after Lower Ordovician time.

(b) Folding of the Dalradian rocks in the SE Central Highlands was initiated after the Middle Cambrian if the Leny Limestone is accepted as an integral interfolded portion of the Dalradian succession.

(c) Ordovician Arenig rocks in western Ireland lie unconformably on Dalradian (Connemara) schists. Radiometric dates for the schists include one of 495 million years, and the low metamorphic grade Ordovician rocks were mildly folded and foliated during a late Caledonide event in Devonian time.

Johnson (1963, p. 140) alluded to similar dates and noted that, although some workers have suggested a Middle Ordovician age for the regional metamorphism in both the Moine and the Dalradian rocks, it "... must be concluded that there are real difficulties in the way of correlating the early events in the Moines and Dalradians despite the obvious similarities."

LOCATION / TIME	S C O T L A N D					WESTERN IRELAND	SE IRELAND AND NW ENGLAND	WALES
	MOINE WESTERN 1	MOINE CENTRAL 2	DALRADIAN 3	MIDLAND VALLEY 4	SOUTHERN UPLANDS 5	6	7	8
MY — DEVONIAN — 400		395	390 • NEWER			GALWAY 387	360 • WEARDALE LEINSTER 385 • SHAP	PRE-L.O.R.S.
SILURIAN — 450	ROGART 420 B 420- W3 430	W3 RS 425 KA 420 B 420 430	405 • 420 B 430 2 PHASE	UNDEFORMED • T LESMAHAGOW	400 • CRIFFEL	RS 404 L 420 KA	FOLDED SILURIAN	POST-LUDLOW T
ORDOVICIAN — 500	PHASE 446 3 L.ORDV. KA		475 W3		PRE-BALA FOLDS 460	W4 RS 475 ARENIG	L 444 METAMORPHISM 465 OF GRANITE	
CAMBRIAN 550	M RANGE IN MORAR KNOYDART AND S.ROSS 560 KA PHASE 2 B	562 560 KA B	490 B KA PHASE 1 LENY LMST.			495 W3	510 L RS OUGHTERARD POST-FOLDING	
— 600 — 650 — 700 PRECAMBRIAN 740 750	KNOYDART PEGMATITE W3 PHASE 1	B PHASE 1 METAMORPH- ISM				MIGMATITE L 725 POST-F2 PRE-F3 RS DALRADIAN FOLDING		

KEY
• POST-KINEMATIC INTRUSIONS
▨ STRATIGRAPHICALLY DEFINED LIMITS OF FOLDING AND METAMORPHISM
⊤ RADIOMETRIC DATING

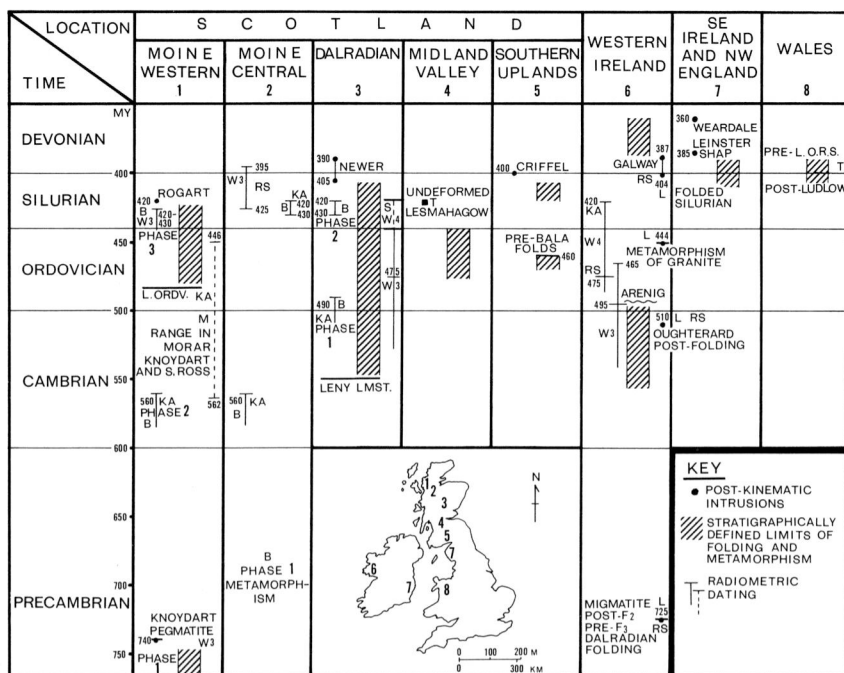

Figure 435. Some of the evidence for the age of folding and metamorphism in the British and Irish Caledonides. The figures are radiometric age dates in millions of years (KA = Potassium-argon; RS = Rubidium-strontium); phase 1, etc., refers to the phase of metamorphic neocrystallization. Some of the sources of the information are shown by the following abbreviations: B – Brown, *et al.*, 1965; L – Leggo, *et al.*, 1965; M – Miller and Brown, 1965; S – Soper, 1964; T – Tremlett, 1963; W3 – Watson, 1963; W4 – Watson, 1964.

Similarly, Watson (1963, p. 253) concluded that: "the discrepancies between these various possible dates may mean that some of the assumptions are wrong, or they may reflect real variations of timing in different parts of the fold belt."[2]

In an interesting review, den Tex (1963) suggested that the Alps, Pyrenees, and Caledonides show three outstanding common features:

[2] The most recent radiometric data suggest that a younger tectonic and metamorphic episode (420-430 million years minimum) was superposed on previously-metamorphosed Dalradian (> 490 million years) and Moine (< 560 million years) rocks in Scotland (Brown, *et al.*, 1965). Miller and Brown (1965) also published a map showing clearly-defined systematic variation in potassium-argon age dates in Morar, Knoydart, and southern Ross-shire, which they interpret as a product of migration of the locus of deformation and metamorphism with time. The general conclusion of this work is that thermal highs and the rise of migmatite fronts occurred at different times in different areas of the orogenic belt (cf. Johnson and Harris, 1965).

(1) Regional metamorphism and the growth of index minerals never commenced until after the first or main phase of orogenic deformation (nappes, recumbent folds, etc.) or later than the second set of fold movements. Niggli (1960), Wenk (1963), and Schuiling (1963) provided evidence to support this generalization.

(2) The climax of regional metamorphism coincides roughly with the interval between the penetrative kinematic phases (or with the last penetrative kinematic phase) immediately preceding the final ". . . brittle folding, faulting and retrograde metamorphism."

(3) The successive isograde mineral parageneses overlap in both space and time. In the high-grade zones (located in the deeper or more internal parts of the orogene) the critical mineral assemblages continued to grow longer than did those of the low grade zones.

Den Tex (1963) also thought that the successive isogradic mineral parageneses attained their climaxes almost simultaneously in the Caledonides, but at progressively later dates relative to deformation in the Pyrenees and the Alps. Figure 435 suggests that this may have been an incorrect conclusion so far as the Caledonides are concerned.

Figure 436 is a pictorial representation of the orogenic model proposed by den Tex (1963), in which he visualized a simple succession of waxing and waning metamorphic and kinematic episodes "wandering through space in the course of time." He thought that, while it is not to be expected that every single phase should be recognizable in all levels of an orogene, the metamorphic thermal fronts radiate outwards symmetrically from a roughly permanent focus. In consequence, rocks near the focus have higher-grade assemblages and were exposed to thermal recrystallization over longer periods than rocks farther from the focus. During waning orogenic activity the metamorphic front would retreat towards the focus. Den Tex (1963, p. 174) reported that:

> . . . in the Alps . . . it has become the concensus of opinion that every phase of deformation starts in the interior and progresses outwards, like a front of waves to become extinguished at the external border (Wunderlich, 1963 [C]) . . . because metamorphism dies out *towards* and deformation *away from* the central area . . . a mineral starting to crystallize before a specific phase of deformation in a high-grade zone, nearer the focus, could be older according to absolute standards than a similarly prekinematic mineral in a lower-grade zone, further away from the central area.

This model will undoubtedly require refinement as more detailed chronologies are worked out for the Alpine, Pyrenean, Caledonide, and other orogenic belts of the world. If with further work the data in Figure 435 prove to be more nearly correct than incorrect, there would seem to

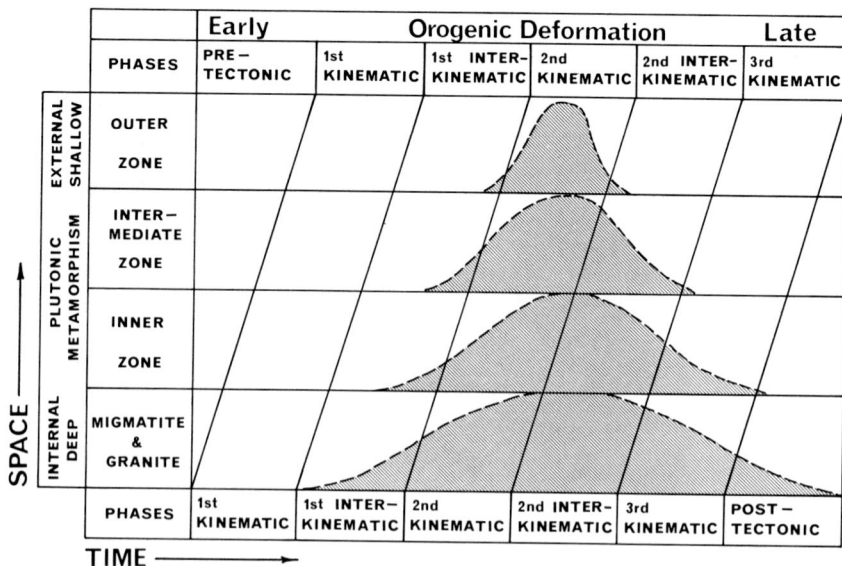

Figure 436. Model correlating orogenic deformation and metamorphism with time; maximum development of a metamorphic zone is indicated by the summit of each shaded area (After den Tex, 1963, Fig. 3).

be a real case within the Caledonides for the radiation outwards through time of kinematic and metamorphic events. However, in the meantime, it will be clear that detailed regional correlations between kinematic and metamorphic orogenic episodes must remain tentative. This is largely because of the difficulty of relating such events to the absolute time scale.

MICROTEXTURAL EVIDENCE CONCERNING PORPHYROBLAST GROWTH

The type of interpretation proposed in Perthshire, Scotland, and the Pyrenees appears to have considerable promise for unravelling the detailed structure of complex metamorphosed areas. It will be noted, however, that the concepts involve a reinterpretation of the forces involved during the growth of porphyroblasts. For many years Harker (e.g., 1932) recognized three different types of "force of crystallization": (a) the physicochemical "force" causing crystallization of a particular phase; (b) the crystallographic form energy; and (c) the force exerted by a growing crystal to dilate the host rock in order to create space for itself. The curvature of mica flakes around garnets in metamorphic rocks has commonly been used as evidence of dilation caused by porphyroblast growth.

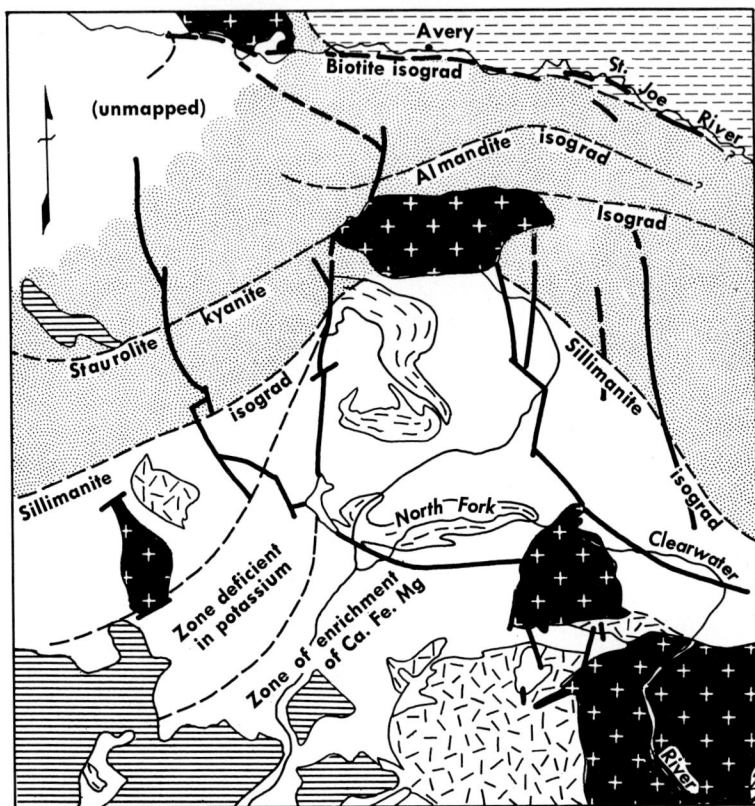

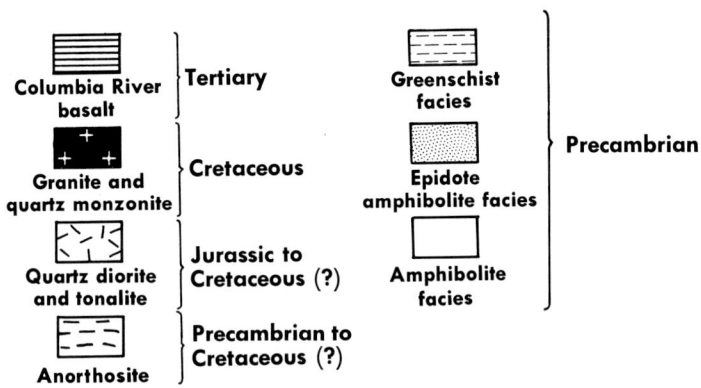

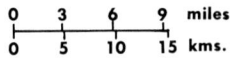

Figure 437. Map of the Precambrian Belt metasedimentary rocks that have been folded and metamorphosed during emplacement of the Idaho Batholith, area northwest of the batholith, Idaho, U. S. A. (After Hietanen, 1961A, Fig. 345.1).

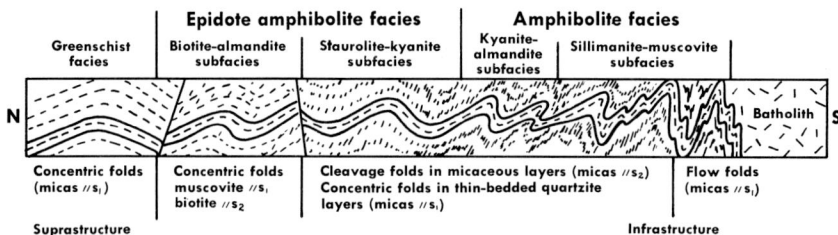

Greenschist facies	Epidote amphibolite facies		Amphibolite facies	
	Biotite-almandite subfacies	Staurolite-kyanite subfacies	Kyanite-almandite subfacies	Sillimanite-muscovite subfacies

N [diagram] Batholith S

Concentric folds (micas //s₁)	Concentric folds muscovite //s₁ biotite //s₂	Cleavage folds in micaceous layers (micas //s₂) Concentric folds in thin-bedded quartzite layers (micas //s₁)	Flow folds (micas //s₁)

Suprastructure   Infrastructure

Figure 438. Schematic N-S section across Figure 437 to show the changes in style of folding and in metamorphic facies, northwest of the Idaho Batholith, Idaho, U. S. A. (After Hietanen, 1961B, Fig. 6).

Ramberg (1952) discussed the formation of porphyroblasts by concretionary growth and by chemical replacement; the former mechanism must be involved to explain forceful dilation of the host rocks.

Study of several thousand porphyroblasts (mainly from the Pyrenees) convinced Zwart (1963A) that chemical replacement is the only process by which neocrystallization proceeds in metamorphic rocks, and that apparent dilation results from subsequent deformation and plastic flow of schistose material around grown porphyroblasts. Zwart found that the host rock is never distorted where it can be proved that the porphyroblasts resulted from post-kinematic crystallization. Additional work is required before these conclusions should be accepted as a valid generalization, although the work of Harris, Rast, Sturt, and Zwart all lends strong support.

MICROTEXTURAL ANALYSES IN OTHER REGIONS

Preliminary results from the Precambrian rocks of Idaho and Montana, U.S.A., indicate that the kinematic and metamorphic sequences are analogous to those described from the Scottish Central Highlands and from the Pyrenees.

Hietanen (1961A, 1961B) demonstrated that the grade of metamorphism of the Precambrian Belt Series increases southward towards the Idaho Batholith. Within 65 kilometers the metasedimentary rocks can be traced from the greenschist facies through to the sillimanite-almandine subfacies of the almandine amphibolite facies (Fig. 437). The gradual changes in style of folding and deformation in the successive zones of metamorphism are diagrammatically shown in Figure 438; a parallel sequence of mineralogical changes was established on the basis of poikiloblastic enclosures, and such textural features form the basis of Figure 439.

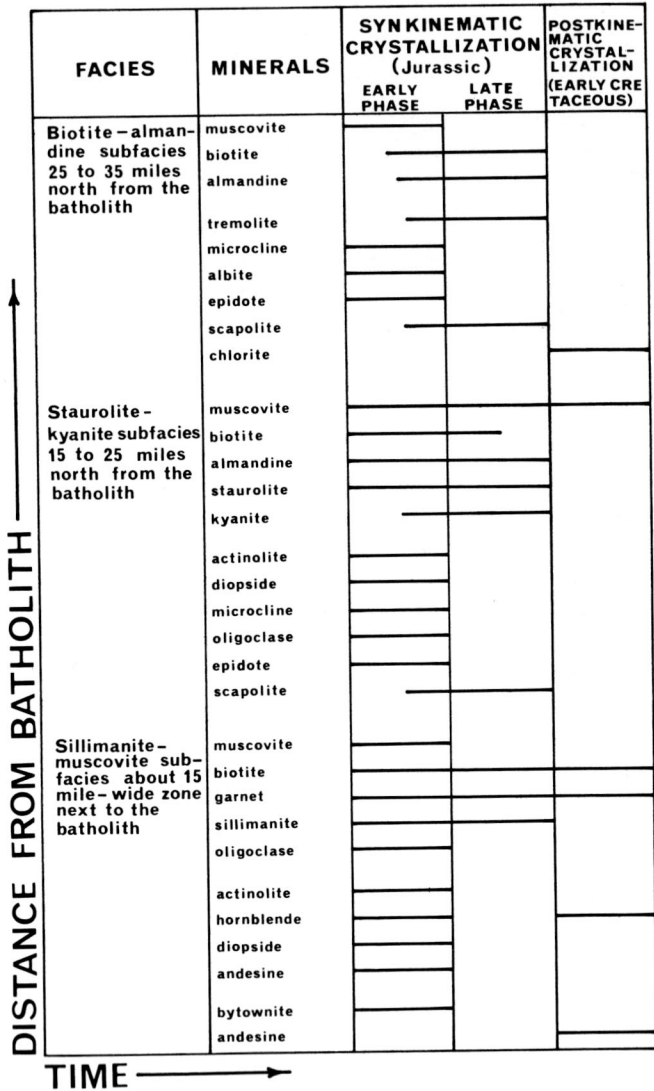

Figure 439. The sequence of crystallization events in various metamorphic facies in the area northwest of the Idaho Batholith, Idaho, U. S. A. (After Hietanen, 1961A, Fig. 345.2).

Reid (1957, 1963) studied the integration of the kinematic and metamorphic episodes in the pre-Beltian Pony and Cherry Creek Gneisses, Tobacco Root Mountains, Montana. Although these studies are in a more preliminary stage than those for the Pyrenees and the Scottish Highlands, Reid demonstrated that a long sequence of superposed deformations can be recognized with the aid of the distinctive facies mineral assemblages associated with the successive metamorphic episodes (Table 6).

Isoclinal F_1-folds (with axial-plane foliation) associated with concomitant high-grade regional metamorphism (M_a) represent the earliest episode that can be detected. These events were recognized on the basis

Table 6. – *Chronology of Events Affecting the Pony and Cherry Creek pre-Beltian Gneisses, Montana, U.S.A.*

Kinematic Episodes	Metamorphic Episodes	Mineral Assemblages		
		Amphibolitic Granulites	Mica Schists	Metabasalts
F_1 Precambrian	M_a High grade	Pigeonite Hypersthene Labradorite		
Intrusion of basic dykes				Pigeonite Labradorite
F_2 Precambrian	M_b Granulite facies	Diopsidic augite Garnet Olivine Plagioclase Rutile	Quartz Garnet Biotite Kyanite Plagioclase Orthoclase Rutile	Garnet Plagioclase Diopsidic augite Quartz
F_3 Precambrian ——— F_4 Precambrian	M_c Almandine amphibolite facies	Hornblende Plagioclase Quartz Cummingtonite Garnet	Quartz Garnet Sillimanite Microcline Plagioclase	Hornblende (brown to green) Biotite Plagioclase
F_5 Open folds Precambrian	M_d Transitional between almandine amphibolite and greenschist facies	Hornblende Biotite Oligoclase Green mica	Quartz Biotite Green mica	Hornblende (light to bluish green) Biotite Magnetite Oligoclase
F_6 Laramide	M_e Greenschist facies	Sericite Penninite Talc Antigorite Epidote Magnetite	Sericite Chlorite	Sericite Epidote

of relict kinematic and migmatitic patterns (i.e., gross structures and textures), although subsequent neocrystallization destroyed most of the original mineral assemblages and the individual grain textures. Specifically, shearing during the second kinematic episode (F_2) and the associated granulite facies metamorphism destroyed the earlier foliation and generated a new foliation (Table 6). A set of basic dykes provides the only basis for separating F_1 from the younger events. Renewed F_3-shearing

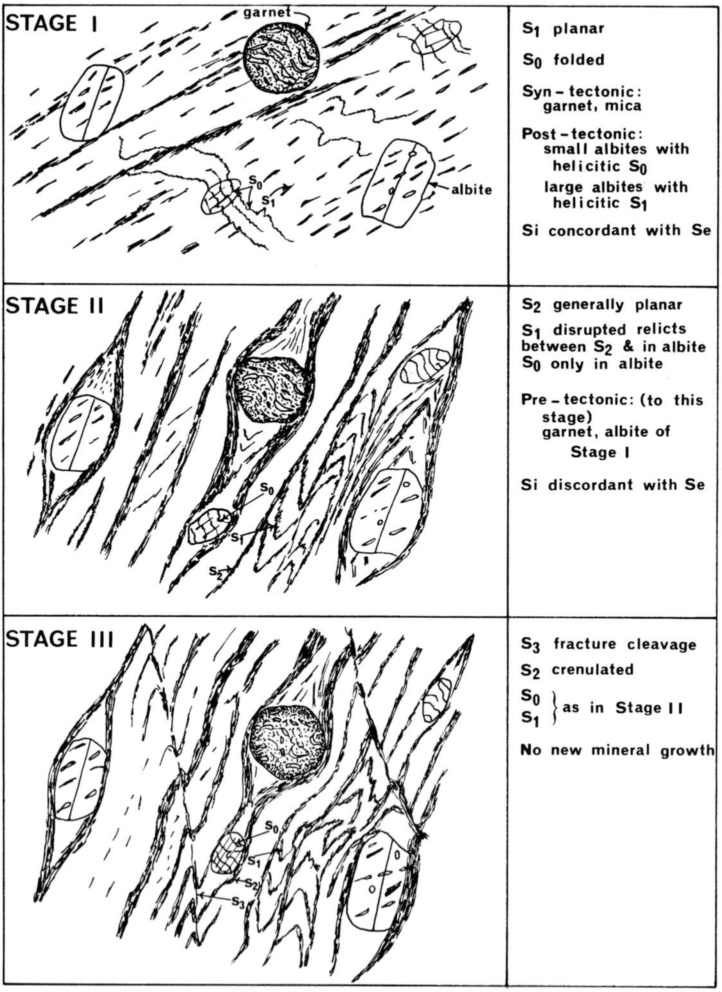

Figure 440. Schematic representation of the sequence of metamorphic and kinematic events seen in thin sections of Precambrian garnet-albite schists of the Franklin Group, Tasmania, Australia (After Spry, 1963B, Fig. 3).

(parallel to the F_1-axial planes), accompanied by almandine amphibolite facies metamorphism (M_c), caused new hornblende and biotite to crystallize parallel to the F_1-axial planes. The only bases for separating F_2 and F_3 are the distinctive mineral assemblages associated with the granulite and almandine amphibolite facies (M_b and M_c).

F_4 was associated with a late phase of the almandine amphibolite facies metamorphism (M_c), during which shear parallel to the axial planes of the isoclinal F_1-folds formed boudins and sheared-off fold closures; these boudins and fold closures were rotated and are now inclined to the foliation of the enclosing rocks. These events were followed by simple flexural-slip F_5-folding of the earlier foliation and retrograde metamorphism (Table 6). The kinematic and metamorphic model outlined in Table 6 is somewhat speculative, but additional work on this and neighboring areas should prove highly instructive.

Spry (1963B) made a detailed chronological analysis of the metamorphic and kinematic development of some Tasmanian Precambrian schists, phyllites, quartzites, and amphibolites. Figure 440 shows relationships in some garnet-albite schists of the Franklin Group in which two kinematic events (F_1 and F_2) occurred during the regional metamorphism of the Precambrian Frenchman orogeny. Spry thought the F_3-movements, which were only associated with slight neocrystallization, may include events ranging from Precambrian to Devonian times.

CONCLUDING REMARKS

Apart from the interest that studies of microtexture have in unravelling the sequence of tectonic events, the results are also important for understanding the characteristic sequence of progressive metamorphic changes.

When four or five kinematic and/or metamorphic episodes are superposed, relicts of the earliest events are hard to find. Indeed, they may have been completely lost; this is most likely when metamorphism and deformation gradually build up to a peak of intensity and then decline during a later sequence of episodes. Although the declining activity may leave relicts of the highest intensity activity, the latter is likely to have obliterated completely the relicts of the earliest phases.

Intuitively, it would seem reasonable to assume that a particular sample of metasedimentary rock collected from the core of an orogene might have undergone diagenesis and then deformation during a succession of progressively higher-grade regional metamorphic events. Nevertheless, Zwart (1963A, p. 58) claimed that the most remarkable feature about the metamorphic history of the Pyrenees is:

Table 7. — *Tentative Correlation Table of Geological Events at Cabo Ortegal, NW Spain (After den Tex and Vogel, 1963)*

Sedimentation and Igneous Events	Metamorphic Events	Tectonic and Epeirogenic Events
Deposition of semipelite, graywacke, and basic tuff sequence. Emplacement of ultramafic masses, basic sills, etc.		Isoclinal folding on NNE plunging axes
T I M E →	(1) ECLOGITE FACIES METAMORPHISM affecting basic rocks only (a) *Pyroxene granulite subfacies* Diopside-plagioclase symplectites, and kelyphitic rims round omphacitic pyroxenes. (b) *Hornblende-clinozoisite granulite subfacies* Hornblende-plagioclase symplectites in eclogites and pyroxene-granulites.	Gradual decrease of temperature Revivals of tectonic pressure (cross folding?)
	(2) GRANULITE FACIES METAMORPHISM	
Development of pegmatitic and granitic melts through metatexis and anatexis in zones of strong penetrative movement →	(3) ALMANDINE-AMPHIBOLITE FACIES METAMORPHISM Development of biotite and quartz laminae in granulites. Formation of rims around garnets and of sphene rims around rutiles.	Blastomylonitization, boudinage and flattening
	(4) GREENSCHIST FACIES METAMORPHISM Pistacite (in part as rims around clinozoisite). Serpentinization of rocks rich in olivine.	Further uplift and denudation
Intrusion of basic dykes and sills	(5) HYDROTHERMAL PHASE Quartz, prehnite, adularia, etc. in tension gashes.	Local mylonitization of epizonal character Faulting, mainly normal and with E-W trend

that such a complete record is preserved of the increase in temperature, in contrast to what is known from many regions, where metamorphism seems to start at the highest temperature and only the decrease of temperature is shown by the lower grade mineral assemblages.

In a study of the interrelationship of metamorphism and deformation at Cabo Ortegal, NW Spain, den Tex and Vogel (1963) seemed to imply that eclogite facies regional metamorphism was the first metamorphic event to affect the sedimentary and igneous rocks. Subsequent events in the granulite, almandine amphibolite, and greenschist facies were associated with a sequence of kinematic stages (Table 7).

It is important to emphasize that studies of the type outlined in this chapter identify the first *recognizable* kinematic event (see Reid, 1963, p. 297), and that existing methods commonly do not provide a means of deciphering the earliest history of those rocks that have passed through phases of highest-grade metamorphism and deformation.

Chapter 13

Sedimentary Characteristics Preserved in Folded and Metamorphosed Rocks

Shrock (1948) gave an excellent review of sequence in layered rocks, and in this book it has been assumed that the reader is familiar with those common features which permit the direction of facing of units to be determined. All clues to the direction of facing are extremely valuable in geometrical analyses of folded sequences. However, descriptions in the preceding chapters emphasize that penetrative deformation tends to obliterate the original sedimentary features — particularly those of the less competent and less coherent units. Bedding and foliations commonly become transposed during intense deformation and the formation of new foliations. Even with relatively open folding, considerable internal slip occurs; for example, the angle between the cross-bedding and the regional stratification in the Lake Superior Precambrian quartzites, Michigan, U.S.A., is greater than the angle of repose for dry sand, so that the structures must have been distorted during folding (Pettijohn, 1957B).

It was the recognition of sedimentary bedding in metamorphic rocks that originally led Hutton and Lyell to recognize these rocks as a distinct group. Read (1958, p. 92) recalled that:

> Alternations of layers of different composition, colour and thickness in the metamorphic rocks were so like those of ordinary sediments as to "leave scarcely any reasonable doubt that they owe this part of their texture to similar causes." (Lyell, 1833, *3*, 10).

In much of his writing Harker (e.g., 1932) emphasized that the finest details of sedimentary structure are frequently preserved in high-grade, regionally-metamorphosed rocks. This has encouraged many geologists to ascribe all compositional lamination in schistose rocks to original bed-

514

ding; however, this approach has often resulted in erroneous conclusions. During neocrystallization accompanying diagenesis and lithification of sediments, and during the lowest grades of regional metamorphism (zeolite facies) bedding and bedding planes commonly seem to become more distinct as a result of mimetic crystallization in or along them. Similar phenomena occur under higher-grade regional metamorphic conditions, although during kinematic activity sedimentary features are commonly obliterated and overprinted.

In consequence, many geologists contend, contrary to Harker's view, that bedding phenomena are rarely seen in metamorphic rocks. Read (1958, p. 92) summarized the situation thus:

> A century ago, Charles Darwin himself (1846) considered that much of the so-called stratification of the crystalline schists was not true bedding but was due to a process of segregation or concretion; he admitted, of course, that layers of marble and quartzite were true sedimentary beds. Lehmann (1884), the great Saxon metamorphic geologist, likewise held that banding and striped structures in metamorphic rocks were usually due to shearing. He described examples of banding that simulates the bedding of sediments but which is undoubtedly produced by intense movement along thrust-planes. We may here recall that Lapworth held that in the Moine Series of Scotland there were no planes of deposition preserved but only planes of shearing and cleavage — these produced a pseudo-bedded appearance. The opinion of these great masters has been reiterated by many workers. Balk & Barth (1936) have restated Lehmann's position and have concluded that "statements that alternating layers of different colour, each characterised by different mineral composition, are due to original stratification of the rock should be looked upon with suspicion." Similar views are held by many workers among the crystalline schists.

Misch (1949, pp. 693-4), when discussing static crystallization and granitization, wrote:

> Previously unmetamorphosed sediments, if thoroughly recrystallized, develop hornfelsic fabrics, with mineral associations varying according to temperature. Bedding is frequently preserved, in the form of differently colored layers differing in composition, and sometimes of division planes. Contrary to some statements in the literature, I have not seen a single case in which bedding in statically recrystallized sediments has become schistosity. The fundamental contrast between a hornfels and a schist illustrates this point. Crystalline schists are not made without differential rock deformation.

The precise method by which fine lamination or pseudostratification develops in recrystallized rocks is uncertain, although it is clearly associated with metamorphic differentiation (Stillwell, 1918; Eskola, 1932) and concomitant mechanical shearing (Schmidt, 1932; Wenk, 1937; Turner and Verhoogen, 1960). Neocrystallization accompanying transposition of bedding and other S-surfaces tends to promote strongly-developed pseudostratification that can be readily mistaken for original bedding because thin compositional layers develop parallel to the new S. The excellent example described by Turner (1941) from the Otago Schists, New Zealand, was referred to earlier. Such laminations may occur in most lithologies susceptible to plastic deformation, and, although pelitic and semipelitic rocks show it most commonly, the structure is not uncommon in metavolcanic rocks, metamorphosed calcareous shales, amphibolite, and many other rock-types.

Because it is often difficult to differentiate between pseudobedding and true bedding in the field, the validity of any structure called bedding in metamorphic rocks must be closely scrutinized (cf. Read, 1958, p. 91). The surest proof of pseudobedding stems from the recognition of isoclinal fold closures whose flanks define the S-surfaces (see Chapter 7).

When the original sedimentary bedding is not clearly defined, a "fracture cleavage" that simulates bedding sometimes develops in essentially unmetamorphosed rocks subparallel to the fold axial plane. Cloos (Cloos and Murphy, 1958, p. 26) drew attention to a clear example in the Cambrian Antietam Sandstone on the east flank of the South Mountain structure, near Smithburg, Maryland, U.S.A. The "fracture cleavage" is well exposed and dips gently eastwards, although Cloos showed that the almost-obscured bedding faces west and dips steeply east; occasional disjointed relicts of thin competent lithologic units define the real bedding. Similar pseudobedding in the Cambrian Weverton Sandstone is exposed clearly near the overlook at High Knob (Cloos and Murphy, 1958, p. 32).

Although various types of pseudobedding and pseudostratification are common in folded rocks, this must not obscure the fact that, when shearing during deformation was not excessive, genuine primary sedimentary structures tend to be preserved, even after high-grade regional metamorphism. Almost undeformed Devonian brachiopods were found in a fold closure in sillimanite-bearing mica schist, South Strafford, Vermont, U.S.A. (Doll, 1943). Billings and Cleaves (1935) and Billings and Sharp (1937) also described two brachiopods from schists of the Littleton area, New Hampshire, U.S.A., in which the bedding and foliation were said to be coincident.[1] It is by no means necessary to assume that old and deformed rocks must have lost all original features. Dons (1962, p. 264) for example, suggested:

[1]Such finds are not unique. Boucot and Thompson (1963) described reasonably well-preserved Silurian brachiopods from staurolite and sillimanite zones of regional metamorphism of west central New Hampshire, U.S.A. Higgins (1964) described well-preserved echinoid and crinoid debris in strongly deformed Mesozoic schists containing garnet, staurolite, and kyanite porphyroblasts south of the St. Gotthard pass, Switzerland.

One would assume that the rocks of the Telemark area [Norway], their age being 900-700 M.Y.... should have lost all primary features. This is not the case. One may find beautiful ripplemarks for instance in quartzite on the very top of Mt. Gausta, undeformed conglomerates and agglomerates, mudcracks in limy quartz schists, small scale slumping features, graded bedding and a structure ... found some years ago and described as Telemarkites enigmaticus, probably a fossil.

The remainder of this chapter is devoted to a few selected sedimentary attributes that are often preserved in folded rocks and that are particularly useful in geometrical analyses.

CROSS-STRATIFICATION (CROSS-BEDDING) RELICTS

There appears to be no consistency in the basic terminology used to describe cross-stratification and cross-bedding structures, although systematization was suggested by McKee and Wier (1953)—see Figure 441 and Table 8—and Allen (1963).

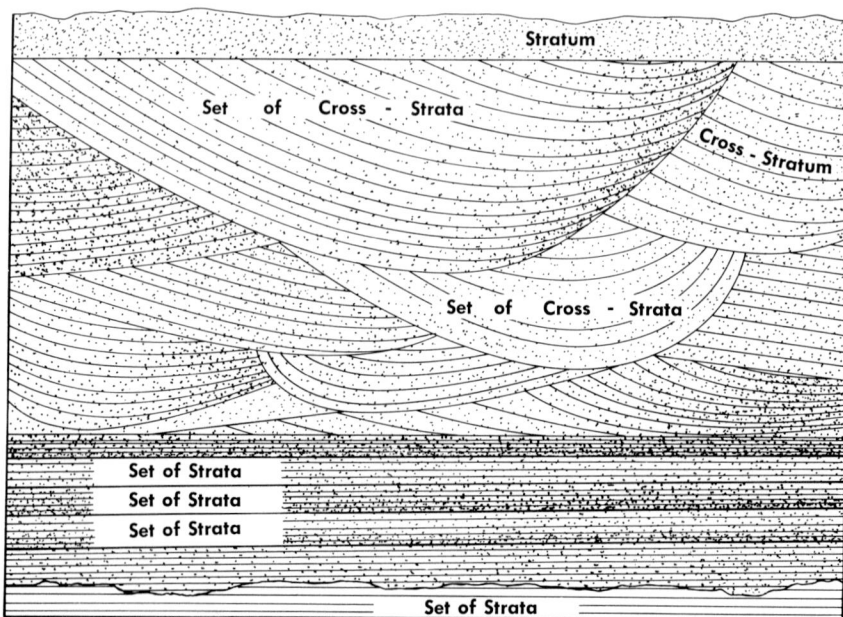

Figure 441. Terminology of stratified and cross-stratified sedimentary rock units (After McKee and Wier, 1953, Fig. 1).

Table 8.— Terms to Describe Cross-Stratification Structures (After McKee and Wier, 1953)

Terms to Describe Stratification		Terms to Describe Cross-Stratification		Thickness
Very thick-bedded	Beds	Very thickly cross-bedded	Cross-beds	Greater than 120 cm.
Thick-bedded		Thickly cross-bedded		Less than 120 cm.
Thin-bedded		Thinly cross-bedded		Less than 60 cm.
Very thin-bedded		Very thinly cross-bedded		Less than 5 cm.
Laminated	Laminae	Cross-laminated	Cross-laminae	Less than 1 cm.
Thinly laminated		Thinly cross-laminated		Less than 2 mm.

Cross-stratification is of considerable importance in structural geology because it provides one of the more important methods of determining the stratigraphic succession in a pile of metasedimentary rocks. The stratigraphic succession must first be established if the geometric structure of an area is to be solved satisfactorily (see Chapter 11). As Read (1958, p. 94) wrote: ". . . the rule must be: stratigraphical order first, geological structure second."

Initially cross-beds tend to form with asymptotic gradations of bedding at both top and bottom (Fig. 442A), but scouring action preceding deposition of the next younger unit can cause truncation prior to further sedimentation. It is this truncation of the cross-beds (Fig. 442B) that can provide unequivocal evidence of succession. The principle was described in detail at an early date by Kelly (1864) in his *Notes upon the Errors of Geology Illustrated by Reference to Facts Observed in Ireland.* Kelly implied that the technique was reasonably well known in 1864. The first descriptions in the U.S.A. were by Leith (1913) and Cox and Dake (1916), although the method had apparently been used previously in field work for some years. Twenhofel (1926) carefully described cross-stratification, but did not comment on its usefulness in structural geology until the 1932 edition of his book.

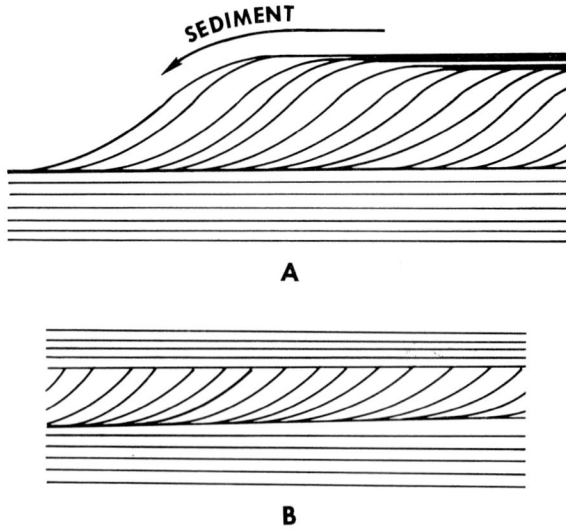

Figure 442. Diagrammatic sketch to show the development of cross beds.

Many fundamental concepts in geology were introduced at an early date but went unnoticed for many years. The usefulness of cross-stratification in structural geology is a case in point. Despite Kelly's (1864) work in Ireland, and Green's (1924) having used cross-bedding to determine the orientation of a quartzite in Islay, Scotland, the principle was not used in most of the earlier detailed work in Scotland; this is the more surprising because in 1934 Professor O. T. Jones (in Bailey, 1934, p. 524) commented that many British geologists had, where necessary, made use of criteria similar to cross-stratification, although they had not considered it worth writing about. However, interest in the phenomenon suddenly exploded in 1930.[2] This marked a major turning point in Scottish geology, and, on the basis of Bailey's earlier excellent lithological mapping, it became possible to prove that the Ballachulish succession is upside down for a distance of at least 10 kilometers measured across the strike (Bailey, 1930).

Cross-stratification is commonly preserved in competent quartzitic members of a succession even when repeated folding and severe metamorphism have occurred. Thus, perfectly-preserved ripple marks and cross-stratification can be studied in the thoroughly-recrystallized and vertically-dipping Precambrian Baraboo Quartzite, Wisconsin, U.S.A. (cf. Brett, 1955). Similarly, well-exposed cross-bedding can be seen in the

---

[2]The related papers by Vogt (1930), Tanton (1930), and Bailey (1930) are of historical interest in this regard; the "new" tool was rapidly acquired by others working in the Highlands and used in structural interpretation (e.g., Bailey, 1934; Anderson, 1935; Read, 1936, p. 468).

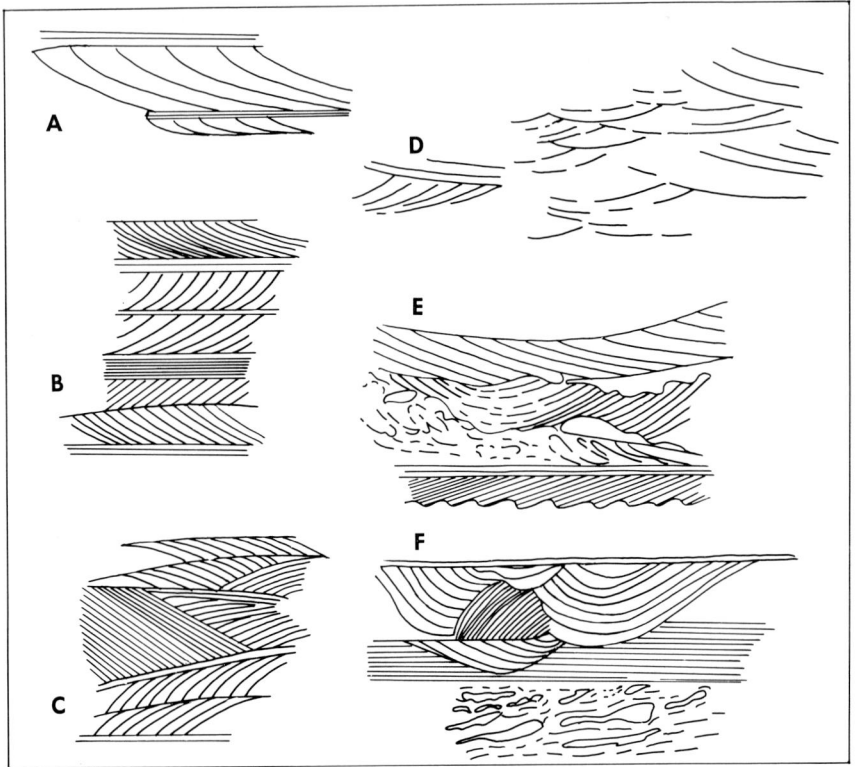

Figure 443. Sedimentary structures in Moine granulites, Fannich Forest, Northern Highlands, Scotland. A, B, C, and D show current (or cross) bedding, C being inverted. E and F appear to be slumped beds (After Sutton and Watson, 1954, Fig. 2).

new road cuts through the Karelian quartzites west of Kaltimo, SE Finland. Ripple marks and cross-bedding are preserved within some massive Tasmanian quartzites (Australia), although the contacts with the associated metasedimentary rocks have been completely transposed during the Frenchman orogeny, and the laminations of the neighboring schists are S_2 (Spry, 1963B).

Working at Morar, Richey and Kennedy (1939) were the first to use cross-stratification to establish the correct stratigraphic succession in the Moine Series of Scotland. Their detailed illustrated descriptions of sedimentary structures included information about cross-bedding, ripple marks, slump structures, sand-filled cracks, minor contemporaneous erosion surfaces, etc. Sutton and Watson (1954) also described cross-bedding and slump structures from the Moine tectonites of Fannich Forest (Fig. 443).

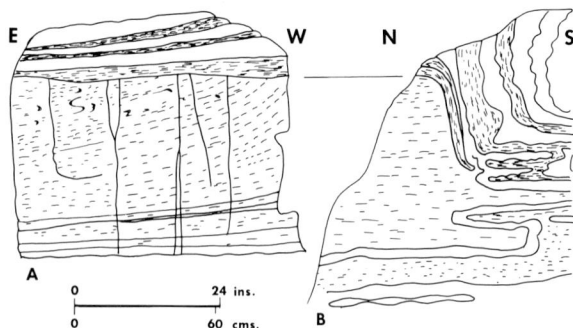

Figure 444. Tectonic structure simulating current bedding in Loch Tay Limestone, Ardchyle Burn, Glen Dochart; Central Highlands, Scotland. A: North-facing exposure. B: West-facing exposure (After Bowes and Jones, 1958, Fig. 2b).

However, structures in some tectonites are very deceptive. Bowes and Jones (1958) published a well-illustrated account of pseudo-cross-bedding structures produced by transposition and small slides during kinematic deformation; they pointed out that under such conditions it is doubtful whether recognizable sedimentary structures can remain. Figure 444 illustrates pseudo-cross-stratification in a marble (the Loch Tay Limestone), Perthshire, Central Scottish Highlands. Relationships of this type are difficult to spot in the field. In consequence, it may be unwise to follow Wilson (1952) in using relict cross-stratification, which only shows on suitably weathered surfaces of isoclinally folded acid charnockites (Musgrave Ranges, Australia), to help prove the sedimentary origin of the rocks; however, the ultimate sedimentary origin of these particular charnockites need not be questioned.

In addition to determining the direction in which a folded sequence faces, cross-stratification has often been used to analyze the original current directions. Despite the emerging evidence for widespread super-posed folding, Wilson, *et al.* (1953) concluded that the majority of cross-stratification in the Moine Series over a large area of NW Scotland originally dipped northwards. Similarly, Brett (1955) and Pettijohn (1957B) studied the prefolding orientation of cross-stratification in the Baraboo Quartzite, Wisconsin, and in seven other Precambrian quartzites in the Lake Superior area, U.S.A. Ross (1962A, 1962B) attempted to use stereographic techniques to determine the original current direction implied by cross-stratification in the Yellowknife Group at Mesa Lake, Northwest Territories of Canada. Such techniques are fraught with difficulties, especially when multiple folding is involved (see below). Ross has invoked three periods of folding at Mesa Lake; Ramsay (1963) strongly criticized the methodology used by Ross (1962A).

Two major problems are involved in attempting to reconstruct the prefolding nature and orientation of cross-stratification:

(a) Considerable distortion occurs in sedimentary units (e.g., cross-stratified quartzites) during the folding (Brett, 1955; Pettijohn, 1957B). The maximum possible inclination of cross-strata in sandy sediments is about 32° to 34°, whereas in folded rocks this angle is commonly exceeded. For example, the angle ranges up to 54° at Baraboo (Brett, 1955).

(b) The sedimentary elements can be rotated back to horizontal in several different ways depending on certain assumptions. Ten Haaf (1959) and Ramsay (1961) pointed out that simple flexural-slip folding about horizontal axes should not be assumed automatically. Commonly fold axes plunge, and errors are involved if allowance is not made for this in the stereographic manipulation. More serious is the fact that slip folding commonly plays a significant role in the total deformation, even when slip planes are not reflected by a visible foliation or schistosity.

In the following sections the constructions for unrolling sedimentary structures that have been affected by flexural-slip and slip folding are described; the methods are largely based on the work of ten Haaf (1959) and Ramsay (1961). Although Ramsay's paper is essentially theoretical, Craig and Walton (1962) used his methods to evaluate paleocurrent data for the strongly-folded Silurian metasedimentary rocks of Kirkcudbrightshire, SW Scotland. They used stereographic unrolling techniques to investigate the implications of assuming that slip and flexural-slip folding occurred in the region; within each area the plunge of the nearest observed fold axis was used for unrolling purposes. Craig and Walton recognized that any conclusions regarding paleocurrent directions are inadmissible while the mechanism of folding remains obscure.

FLEXURAL-SLIP FOLDING OF CROSS-STRATIFICATION

Figure 445 is a block diagram of a cross-stratification structure. Prior to flexural-slip folding the regional bedding may dip, but, for simplicity, it is assumed that the regional bedding is horizontal. In Figure 446 the primitive circle represents the regional bedding (S_0) and a cross-stratum (S_1) dips at 225 (30), although, in most natural cases, the angle of dip will vary along the cross-stratum. If flexural-slip folding occurs about a horizontal axis trending at 202, flexure of S_0 develops a family of rotated planes, S_0', and their poles ($\pi_{S_0'}$) lie on the great circle normal to

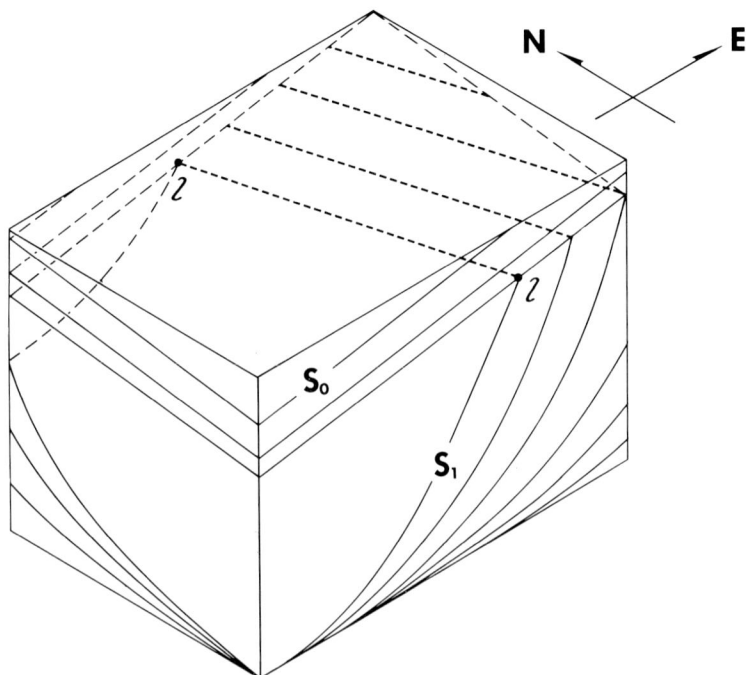

Figure 445. Block diagram of a cross-stratified unit. S_0 are the regional bedding, S_1 are the cross-strata, and $l—l$ is the intersection of one pair of S_0- and S_1-planes.

the fold axis F (Fig. 446). If l is the intersection of S_0 and S_1, l rotates during folding to follow a small circle about F and it assumes the positions l' on each S_0'. Similarly, poles to S_1 follow small circles about F and assume positions $\pi_{S_1'}$. Since the angle between the cross- and regional-strata was 30°, the angle between π_{S_0} and π_{S_1} is 30°, and Figure 446 shows that each pair of $\pi_{S_0'}$ and $\pi_{S_1'}$ is always 30° apart (i.e., the angle between cross- and regional-strata remains constant). In Figure 447 the folded S_0 are shown as broken lines, and the rotated cross-strata (S_1') as solid lines.

Suppose that the orientation of cross- and regional-strata are measured in a folded sequence of sedimentary rocks. In Figure 448 regional bedding (S_o) dips at 315 (40) and cross-strata (S_1) at 292 (50), while the observed fold axis (F) plunges at 252 (20). Commonly it is difficult (or impossible) to be certain of the original orientation of beds folded about plunging axes. Although it does *not* necessarily produce a correct picture of the original orientations, it is often convenient to unroll the folded unit until both limbs assume the same minimum dip (i.e., the same dip value as

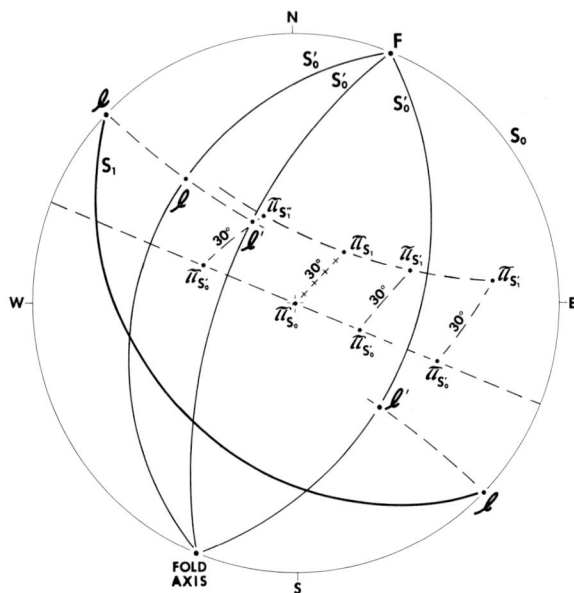

Figure 446. The geometry of cross-stratification affected by flexural-slip folding. S_0' are the orientations of folded regional bedding (S_0); the initial cross-stratification (S_1) intersects S_0 at *l*. See text for explanation.

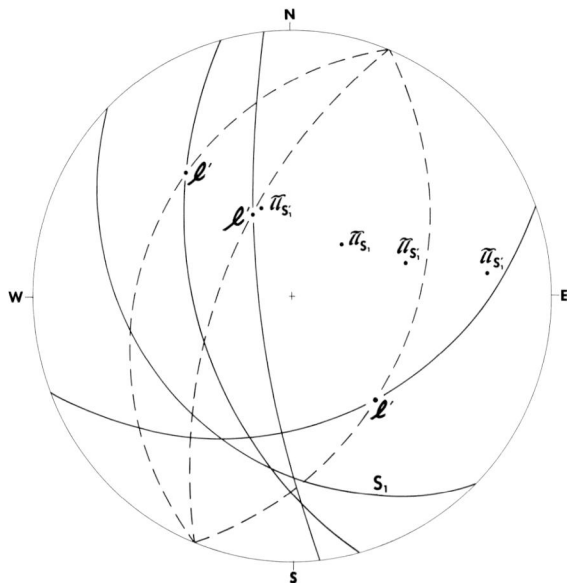

Figure 447. β-diagram to illustrate the folded rocks depicted in Figure 446. The broken lines are folded regional beds (S_0'); solid lines are folded cross-beds (S_1') whose poles are $\pi_{S_1'}$. S_1 and π_{S_1} refer to the original cross bed. *l'* are intersections of S_0' and S_1'.

the plunge of the fold axis), and then to rotate this plane to a horizontal position. Without guidance from external evidence, which is commonly lacking, such procedures are rather arbitrary; in complex areas some variation of the arbitrary assumptions may be more appropriate.

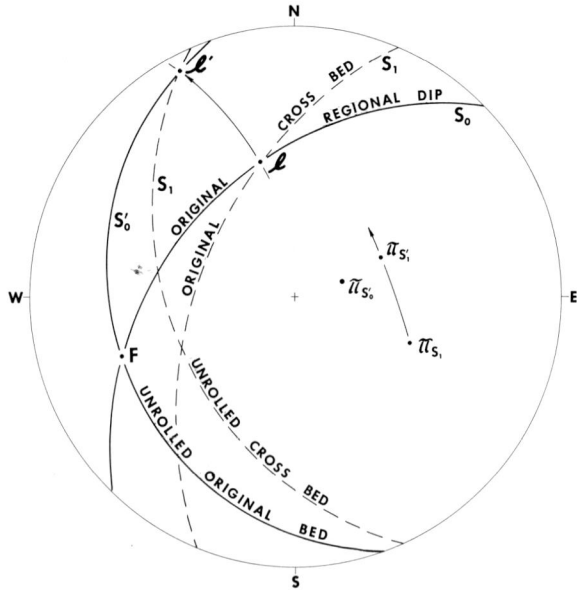

Figure 448. The geometry of cross-stratified rocks affected by flexural-slip folding about a plunging fold axis (F). See text for explanation.

However, by following the rules outlined, the regional bedding (S_0) is rotated to S_0' (Fig. 448), so that the dip of S_0' equals the plunge of F. Accompanying this rotation the intersection l and the pole to the cross-bedding (π_{S_1}) are rotated equal angular distances about F along their respective small circles. Using the rotated pole π_{Sl}, the trace of the rotated cross-bedding plane (S_1') can be drawn (Fig. 448) to pass through

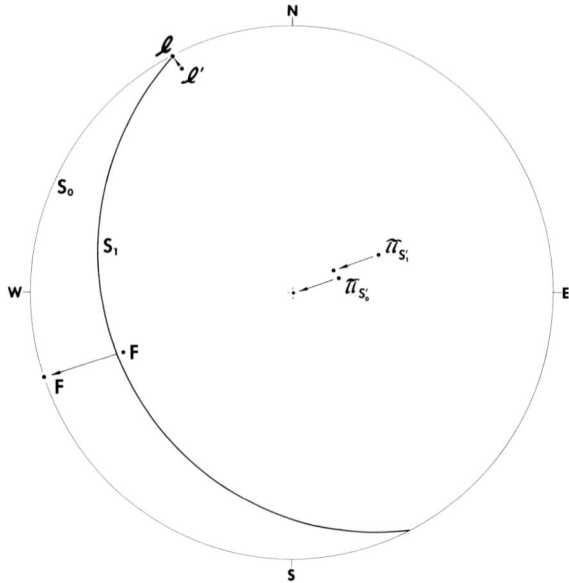

Figure 449. Rotation of the structural elements of Figure 448 to estimate the initial orientation of the cross beds S_1. See text for explanation.

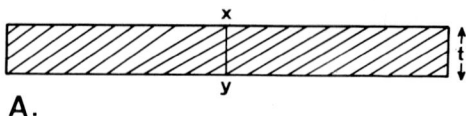

A.

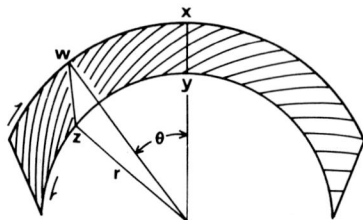

B.

Figure 450. The effect of flexural-slip folding on cross-bedding. A: Foreset beds uniformly inclined at 30° to the regional bedding before folding. B: The effect of folding; see text for explanation (After Ramsay, 1961, Fig. 14).

the rotated intersection (l'). The unrolled cross-bed now dips at approximately 248 (39).

If the dip of S_0' is to be eliminated, the bed can be rotated 20° about its strike, to bring F to horizontal (Fig. 449). For this rotation both π_{St} and l' rotate 20° along their respective small circles. The completely unrolled cross-bed can then be drawn through the new position of l by using the pole to the bed (rotated π_{St}); its dip is about 249 (19) in Figure 449.

Because simple rotation of the whole rock about an axis has been considered, the angle between regional- and cross-beds remains constant. However, internal slip during flexural-slip folding results in movement parallel to S_0 and, as Ramsay (1961) showed, consequent distortion of cross-bedding structures (Fig. 450); Wunderlich (1962) demonstrated that

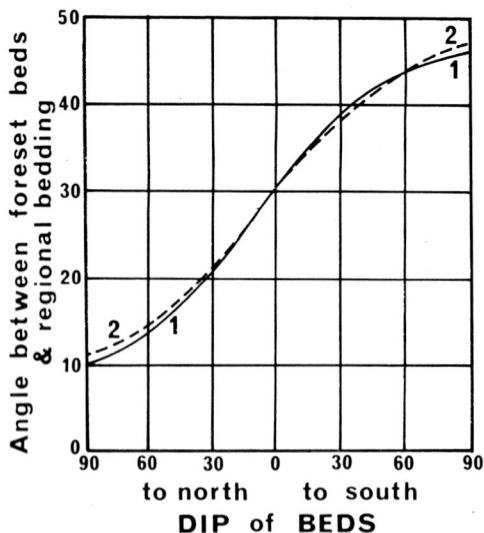

Figure 451. Distortion of the original 30° angle between foreset beds and regional bedding accompanying flexural-slip folding (cf. Fig. 450). Curve 1: $l \wedge B = 60°$. Curve 2: $l \wedge B = 30°$ (After Ramsay, 1961, Fig. 15).

flow sometimes occurs parallel to *B* during folding (see Chapter 9), and this would involve more complex distortion than suggested by Ramsay. With the planar distortion considered by Ramsay, and the fold axis sub-parallel to the *l*-direction, the angle between the cross- and the regional-stratification increases on one limb and decreases on the other. Where the bed has thickness *t*, and the fold has a radius of curvature *r*, *t* changes to *wz* (Fig. 450) during flexural-slip folding. Without indicating the precise method of computation, Ramsay (1961, p. 95) "accurately calculated" the amount of distortion of an original 30° angle between cross- and regional-stratification (Fig. 451).

SLIP FOLDING OF CROSS-STRATIFICATION

When cross-stratification is deformed by slip folding the resulting pattern is completely different from that associated with flexural-slip folding. If the cross-bedding illustrated in Figure 445 is affected by slip folding with the direction of slip (*a*) vertical, *b* and *c* are horizontal and the structure becomes like Figure 452. This pattern is shown stereograph-ically in Figure 453. The plane of slip is *ab* (here trending E-W). If the

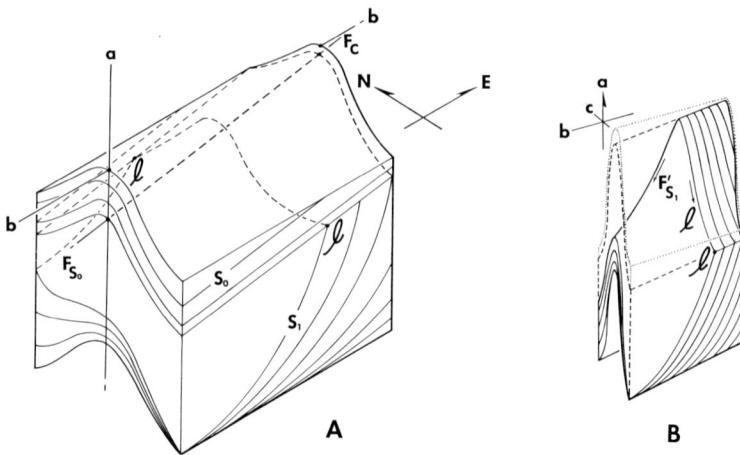

Figure 452. Block diagram to illustrate the influence of slip folding on the cross-stratified rocks shown in Figure 445. The *a*-kinematic axis is vertical, and, because S_0 had an initial dip to the southwest, F_{S_0} is not parallel to the *b*-kinematic axis. A: Simple slip parallel to *a*. B: Slip parallel to *a* accompanied by "compression" involving shortening parallel to *c*, elongation parallel to *a*, and no change parallel to *b*. F'_{S_0} is the fold axis defined by the cross-strata S_1 (Fig. 452B is based on Ramsay, 1961, Fig. 6).

bedding had been horizontal originally, the fold axis would coincide with *b*. However, since the regional-stratification was originally inclined to the southwest (Fig. 445), the fold axis ($F_{S_0'}$) coincides with the intersection of S_0 and *ab*; a representative set of rotated bedding planes is indicated as S_0' (Fig. 453). During slip folding the intersection of cross- and regional-stratification (*l* in Fig. 453) moves along the great circle passing through *a* and *l*, and its rotated positions, *l'*, help to define the orientations of the rotated cross-strata (S_1') for each position of S_0'.

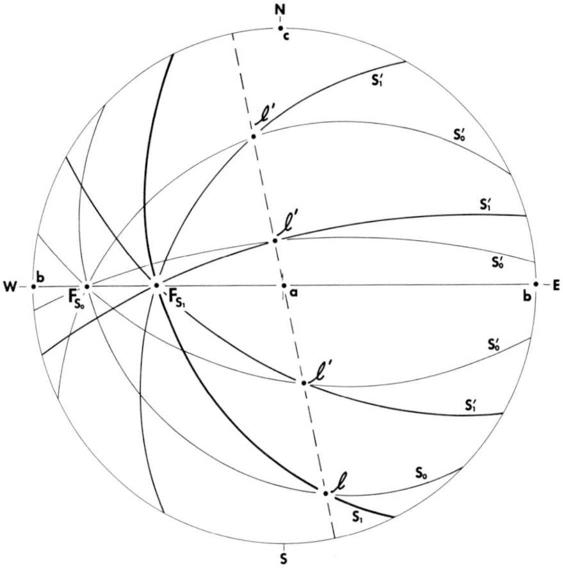

Figure 453. β-diagram illustrating the geometry of Figure 452A. Note that the regional bedding (S_0') and cross-bedding (S_1') define different fold axes: F_{S_0} and F_{S_1}, respectively.

When cross-stratification is observed in rocks affected by slip folding an independent estimate of the orientation of the slip direction, *a*, must be obtained before unrolling can be undertaken. In Figure 454 it is supposed that regional- and cross-stratification (S_0' and S_1', respectively) were measured in the field, and it is *assumed* that *a* is known to be vertical. The angular distance between the measured stratification planes is 39° ($\pi_{S_0'}$ to $\pi_{S_1'}$). Next, some decision has to be made about the orientation to which S_0' (regional bedding) is to be rotated in order to simulate the prefolding attitude. In Figure 453 the original bedding plane had an initial tilt before slip folding, but, in the absence of external evidence for an original primary tilt, it is customary to rotate the regional bedding until it has a minimum dip (i.e., a dip equal to the plunge of the fold axis, F_{S_0}). Using this assumption, S_0 has been constructed in Figure 454, and *l* is plotted to lie on the great circle passing through *a* and *l'*. With *l* plotted, S_1 can be constructed to pass through F_{S_1}. The poles to S_0 and S_1 are 34° apart

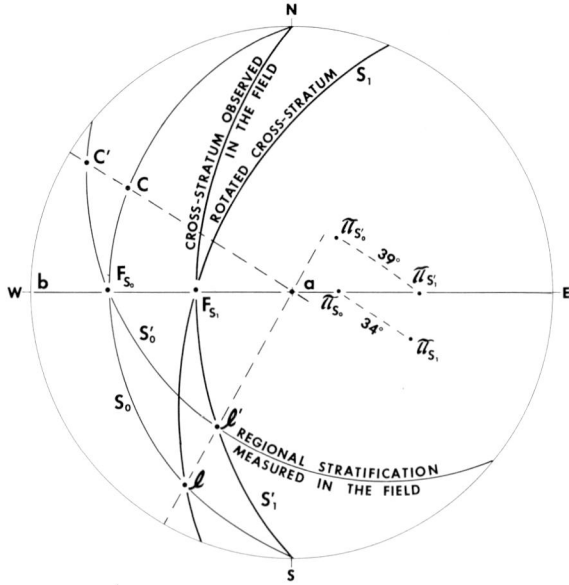

Figure 454. Stereogram to illustrate the method of unrolling slip folding that has affected a cross-bedded unit when the kinematic a-axis is assumed to be vertical. For complete explanation see text.

following rotation, instead of the observed 39°. As in the case of flexural-slip folding, S_0 can be rotated to horizontal. However, unrolling such structures is liable to serious error without external evidence for the exact orientation of both the a-kinematic axis and the regional bedding (S_0) before the folding. Unfortunately, in actual field studies, the required information is extremely difficult to obtain.

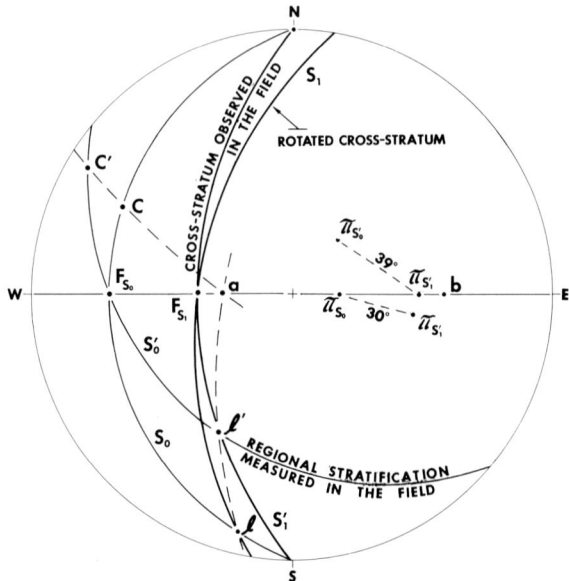

Figure 455. Stereogram to illustrate the method of unrolling slip folding that has affected a cross-bedded unit when the kinematic a-axis is not vertical. Note that the trend of the current direction observed in the field (C') changes when the effects of slip are eliminated (cf. Fig. 454).

The local current direction is usually considered to have been down the dip of the cross-strata (i.e., normal to the intersection of the cross- and regional-stratification). C indicates the current direction in Figure 454, and, being a direction, the locus of C during folding is the great circle passing through both C and a. The trends of both C and l remain constant throughout slip folding when the a-kinematic axis is vertical, although the plunges of both change. In the general case in which a is inclined, both plunge and trend of C and l change during the folding. This is demonstrated in Figure 455, which shows the effect of unrolling the observed data used in Figure 454 about an inclined a-axis.

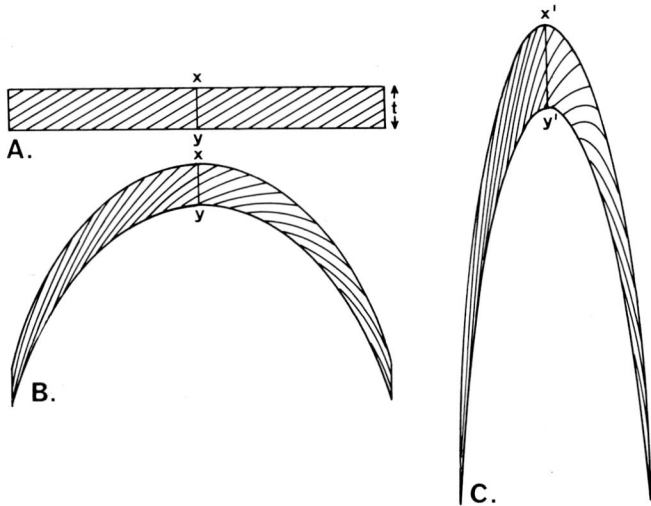

Figure 456. The effect of slip folding on cross-bedding. A: Foreset beds uniformly inclined at 30° to regional bedding before folding. B: The effect of folding (cf. Fig. 450). C: The effect of slip folding accompanied by 50 per cent compression (After Ramsay, 1961, Fig. 16).

Distortion due to internal slip parallel to a during slip folding results in abnormally low inclinations of cross-strata in the flanks of folds (Fig. 456). Figure 457 shows Ramsay's (1961) estimate of the distortion of an original 30° angle between cross- and regional-stratification. Figures 451 and 457 should be compared and contrasted.

With both flexural-slip and slip folding, dissimilar patterns develop as the orientation of the fold axis is changed with respect to dip direction of the cross-stratification (cf. Wilson, et al., 1953). Figure 458 is a diagrammatic sketch of some relationships developed in a cross-stratified unit affected by flexural-slip folding.

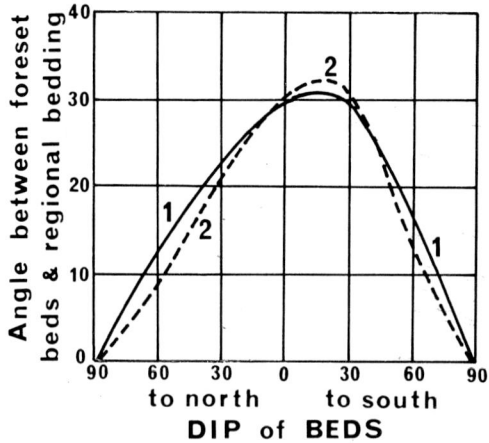

Figure 457. The distortion of the original 30° angle between foreset and regional bedding during slip folding (cf. Fig. 456). Curve 1: $l \wedge B = 60°$. Curve 2: $l \wedge B = 30°$ (After Ramsay, 1961, Fig. 17).

In practice it is often unrealistic to consider flexural-slip and slip folds as discrete systems. Combinations of the two kinematic models are well known, and then it is acutely difficult to unravel the distortion and rotation of cross-stratification and other sedimentary characteristics. Another difficult problem concerns the shortening perpendicular to the *ab*-plane that Ramsay (1961) called "compression." In the case of slip folds Ramsay considered shortening of *c*, with consequent elongation of *a*; *b* was assumed to remain constant. Ramsay also discussed analogous "compression" in flexural-slip folds and with, say, 50 per cent shortening parallel to *c*, gross distortion of internal structures (e.g., cross-stratification) occurs. Figures 452 and 456 illustrate the type of structures that can occur.

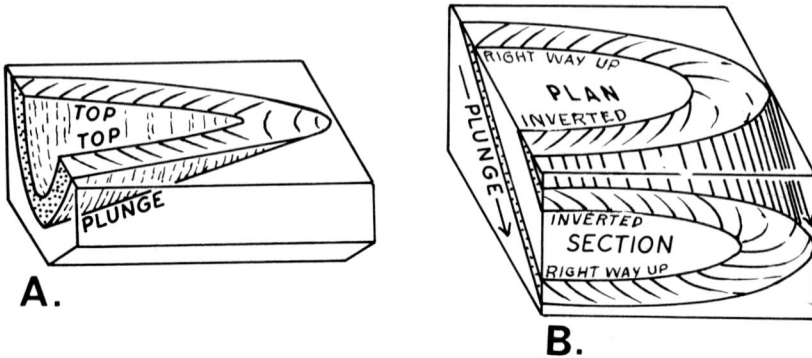

Figure 458. Diagrammatic block diagrams to show geometries that can develop with flexural-slip folding of cross-bedded units. A: Fold axis approximately parallel to dip of cross-beds. B: Fold axis approximately perpendicular to dip of cross-beds (After Wilson, *et al.*, 1953, Figs. 2 b and c).

Evidence is slowly accumulating to suggest that such "compression" is common in all types of folding, but, unfortunately, the magnitude of the shortening is most difficult to measure. It seems unnecessary to assume with Ramsay (1961) that b would remain undistorted (cf. Flinn, 1962; Wunderlich, 1962); in fact, "compression" of the c-geometric axis might be expected to result in elongation parallel to both the a- and b-geometric axes. Studies of this type of deformation might lead to evaluation of the strain pattern within the total rock, although the magnitude of any compression appears to be indeterminable without an accurate picture of the total strain within the rock mass involved.

Cross-stratified sedimentary rocks rarely preserve features suitable for determination of the total strain, although fossils (crinoid ossicles, brachiopods, etc.), ooliths, and certain sedimentary structures (sole marks, ripple marks, etc.) can give valuable information about local strain. However, it was emphasized in Chapter 9 that the strain indicated by rock components (e.g., fossils, pebbles) does not necessarily reflect accurately the total rock strain.

"Compression," or shortening, might be either homogeneous or heterogeneous in each structure of different size. Ramsay (1961) considered "compression" homogeneous in objects the size of the individual folds in Figures 452 and 456 where a normal to the axial plane remains parallel throughout the "compression." It will be noticed that in the models illustrated in Figures 452 and 456 the "compression" or shortening is postfolding, whereas the final geometry would be different if folding and "compression" were concomitant processes.

Hence, there are many hazards associated with unrolling folds and attempting to develop an accurate picture of the predeformation planar, linear, and directional properties of the original sedimentary rocks.

RELICTS OF GRADED BEDDING

Graded bedding has been used extensively by structural geologists for establishing the stratigraphic sequence in deformed rocks; the structure is equally important as cross-stratification for this purpose. In a typical graded sandstone or siltstone bed there is a transition in grain-size of the clastic particles from coarser at the bottom to finer at the top. Commonly, when one graded bed has been found, many similar beds can be located in the neighboring stratigraphic section. Individual graded beds range from about 4.5 meters to a fraction of a centimeter in thickness.

Although graded beds and cross-stratification often develop under dissimilar environmental conditions (cf. Bailey, 1930, 1936), many examples of their close association have been recorded. They occur together in the Ordovician Ardwell Flags, Ayrshire, Scotland (Henderson, 1935), in the Dalradian Fahan Slate-Grit Group of Inishowen, NW Ireland (McCallien, 1935), and in the Upper Dalradian rocks of Banffshire, NE Scotland (Sutton and Watson, 1955). Again, repetitive graded sandstone-shale sequences are a conspicuous feature of the Lower Mississippian Horton Series, Cape Breton Island, Canada. Whitehead and Shrock (Shrock, 1948, pp. 31-33) measured seven of these cycles along Southwest Mabou River, where each cycle commenced with (a) a massive sandstone, with large fore-sets and cross-stratification indicative of aeolian deposition, that grades upwards into (b) ripple-marked and cross-stratified silty and fine-grained sandstone, and ends with (c) sandy shale and shaly sandstone.

Leith (1913) referred to the value of graded bedding in structural geology and according to Cox and Dake (1916) the criterion was first used by W. O. Hotchkiss for the slates in the Florence district, Wisconsin, U.S.A., some years prior to 1916. Graded bedding was not generally used for structural interpretation in Scotland until after 1930 (see Vogt, 1930), although Bailey (1930, p. 85) wrote that he:

> ... first recognized graded bedding—and its significance as a criterion of succession... at Kilmory Bay on the shore of Argyll. The occasion was a sunrise visit in the Spring of 1906, during a 60 or 70 mile tramp from Inverary to Tayvallich. I have never lost sight of this observation ([in Peach, *et al.*], 1911, p. 64; 1913, p. 295; 1922, p. 97), but, meanwhile, have missed many comparable examples of graded bedding in other districts.[3]

Examples of unequivocal graded bedding in rocks affected by severe folding and metamorphism are now numerous in the literature. Allison (1933) specifically set out to ascertain the Dalradian succession in Islay and Jura, Inner Hebrides, Scotland, and made convincing use of graded bedding. Later, in the Loch Awe area, Allison (1941) equally convincingly demonstrated that the Tayvallich Limestone Group is younger than the Crinan Grit and Quartzite Groups; in doing this he vindicated Bailey's (1922) reading of the structure and settled a long-standing debate in Highland geology. When Allison read his paper in 1936 there was some reluctance to accept the criterion of cross-bedding (e.g., by J. F. N. Green).

Whereas cross-stratification often presents equivocal relationships in the field due to the lack of clear-cut asymptotic bottoms or clearly truncated tops, graded bedding tends to be a very reliable guide when used

---

[3]Watson (in Peach, *et al.,* 1911, pp. 63-64) described Bailey's observation at Kilmory Bay, and this was apparently the first description in Scotland.

with discrimination. However, despite this general reliability, special care is required for several reasons:

1. Sometimes grain-size may increase upwards as a normal feature of a sedimentary unit. For example, in the closely-folded Cambrian Caldwell Quartzites, Thetford district, Quebec, Canada, gradations occur from fine-grained at the base to coarse-grained at the top of the units (Cooke, 1931). Cooke correctly suggested that graded bedding should only be used to determine the succession when (a) several layers show concordant textural change—a view supported by Boswell (in Allison, 1941, p. 144) and Walton (1956, p. 263), or when (b) confirmation can be obtained on the basis of several independent criteria. Walton (1956, p. 270) found that: "An additional and surer guide to the sequence of rocks is provided by sharp and eroded contacts, 'flame structures', and irregularities at the bases of the coarse-grained beds, and by erosion surfaces in ripple marks and current bedding."

Other examples of "reversed" graded bedding were described by (a) Barrell (1917, pp. 803-4) from rhythmically-banded "ribbon slates" of Pennsylvania, U.S.A., in which soft black shale below grades up into clear gray sand above, (b) Ksiazkiewicz (1952) from the Carpathian Flysch, (c) Walton (1956) from the Glenkiln graywackes at Currarie Port, Ayrshire, Scotland, and (d) J. L. Knill and D. C. Knill (1961) from the Fahan slates and grits of the Upper Dalradian succession near Culdaff, NW Ireland.

2. Metamorphism can introduce difficulties. The correct interpretation of the direction of facing depends upon identification of the grain-size of the original clastic grains, but metamorphism can cause reversal of the original grain-size sequence. The finer-grained portion of a graded bed is commonly much richer in clay (aluminous) components, which, during metamorphism, readily recrystallize to coarse-grained aluminum silicates (e.g., andalusite). The coarser-grained base of each original graded unit, being more sandy, gritty, or pebbly, recrystallizes to a quartzitic rock which is finer-grained than the aluminous top.

Excellent examples were described by Read (1936, p. 473) from the Boyndie Bay Group, Upper Dalradian, Banffshire, NE Scotland. The base of each unit comprises pebbly grit recrystallized to a medium-grained association of biotite and quartz, abundant ore-grains, small staurolites, and rare tiny garnet and tourmaline crystals. Upwards in each bed andalusite begins to appear, and the top comprises the same assemblage as that at the bottom, except that in the ground-mass are set large porphyroblasts (up to 2.5 cm.) of andalusite with a few patches of cordierite. Such metamorphically-reversed graded bedding ". . . is constantly in use in such terrains as the Donegal Dalradian" of NW Ireland, according to Read (1958, p. 96).

3. Bowes and Jones (1958) described a fascinating example of pseudo-inverted graded bedding from the Dalradian rocks south of Balquhidder, Perthshire, Scotland (Fig. 459); in this case the pseudopebbles are quartz masses which comprise small rootless folds that were isolated by attenuation of minor fold limbs during isoclinal folding.

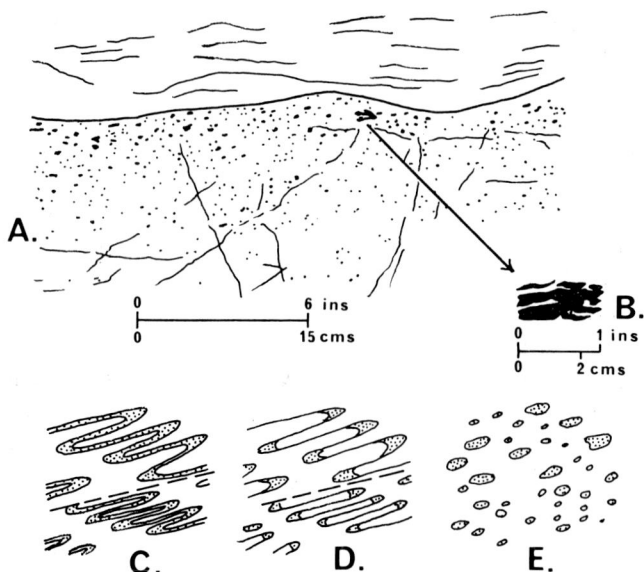

Figure 459. Pseudo-inverted graded bedding in siliceous band of Dalradian rocks, Calair Burn, Balquhidder, Central Highlands, Scotland. A: Sketch of apparent inverted graded bedding. B: Enlargement of part of A to show relicts of tight folds. C-E: Diagrammatic sketches of three stages in development of structures shown in A and B (After Bowes and Jones, 1958, Figs. 3 and 4 b and c).

4. In complex areas tectonic discontinuities can make difficulties in the interpretation of the facing direction. Professor O. T. Jones (in Allison, 1941) emphasized that in such areas (a) planes of dislocation may intervene between the graded bed and the lithologic unit contact, or (b) small-scale folding of the graded beds can lead to deceptive observations. Even when there are consistent indications of facing in one direction, repetition by folding accompanied by repeated thinning and shearing of one fold limb, can lead to misleading conclusions (King and Rast, 1956A). The vigorous debate between Shackleton and Anderson (in Shackleton, 1958) over dissimilar readings of the graded bedding in the Highland Border area, Scotland, points up some of the practical problems involved.

RELICTS OF OTHER SEDIMENTARY FEATURES

Special emphasis has been given to cross-stratification and graded bedding because these are the sedimentary phenomena most consistently used to determine the stratigraphic succession in deformed rocks. Virtually all structures that can aid in the determination of the tops of sedimentary and volcanic units have been used in structural work. Shrock's (1948) book on sequence in layered rocks and Read's (1958) centenary lecture to the Geologists' Association on stratigraphy in metamorphism provide complete reviews of these attributes. This is not the place to enumerate all the structures available to the structural geologist, but a few illustrative examples are cited.

In many folded rocks, and even in severely recrystallized terranes there is enough paleontological evidence to aid in unravelling the structure (cf. Bucher, 1953). Peach and Horne (in Peach, *et al.*, 1907, p. 482) cited the classical argument between Murchison and Nicol concerning the Cambrian quartzite near Loch Eireboll, NW Scotland. Murchison assumed the existence of two distinct quartzites, while Nicol showed that the upper of the two is an inverted equivalent of the lower unit. At Camas Bay, where the "upper" quartzite (pipe-rock) dips below Lewisian Gneiss, the trumpet-shaped openings of the annelid-tubes and the ripple marks are "on the lower faces, showing that there has been a complete reversal of the strata" (Nicol, 1861). Lapworth's (1889) classic paleontological demonstration of the isoclinal folds in the Southern Uplands of Scotland is also often quoted in connection with unravelling complex structures, although Walton's (1961; in Johnson and Stewart, 1963) remapping in the Southern Uplands has necessitated some revision of the overall geometric plan.

Pillow lavas are common components of Precambrian (e.g., Peach, 1904; MacGregor *et al.*, 1963, etc.) and more recent deformed unfossiliferous rocks. They commonly provide clear evidence for the facing direction of the sequence, because succeeding pillows tend to sag into the trough between adjacent pillows of the underlying horizon. In basic lava flows, pipe amygdules can sometimes be useful too. By finding pipe amygdules terminating at the present base of the Tayvallich Lava, Peach (in Peach, *et al.*, 1911, pp. 60 ff.) concluded that the Dalradian succession on the Knapdale peninsula, Loch Awe area, SW Scottish Highlands, faces upwards. Caution is necessary in using isolated examples of such way-up criteria, unless they are confirmed by evidence from other structures.

Bowes and Wright (1962, p. 54) illustrated well-preserved washout structures within the Dalradian Appin Quartzite, southwest of Ballachulish, Argyllshire, W Scotland. Apparently:

> At one horizon washouts filled with massive quartzite cut a thick bed of laminated quartzite. Along this horizon, fifteen large washouts occur in a distance of 200 feet. They are about 6 feet in breadth, spaced at intervals of about 6 feet and are 6 to 9 inches deep. Washouts are also common in the 6 feet of strata above this band.

Walls (1937A) and Sutton and Watson (1955) described well-preserved slump structures (Fig. 460) in the Upper Dalradian succession of the Banffshire coast, NE Scotland. Many other sedimentary structures are also clearly preserved in these metamorphosed rocks which suffered extraordinarily little internal deformation; Sutton and Watson even claimed that the clastic grains have retained their original angular shapes in one area. Along this coast section the more gritty rocks are often unfoliated, but the argillaceous rocks commonly possess moderately good foliation and after strong folding those in the Whitehills and Boyndie Bay groups were subjected to regional metamorphism with development of andalusite, cordierite, and staurolite.

Ripple marks and various sole marks on bedding planes are commonly very useful for determining the direction of facing. Such structures are beautifully preserved in the strongly folded Silurian rocks of Kirkcudbrightshire, SW Scotland, but, in describing the structures of these rocks, Craig and Walton (1962, p. 102) sounded a note of caution because:

> ... in certain areas the intersection of the cleavage with the bedding has produced pseudo-sedimentary structures. Longitudinal ridges, transverse ripples, groove moulds and nodule trails all have their tectonic counterparts. The form of these pseudo-sedimentary structures depends on the cleavage interval and cleavage strength, both of which are related to grain size. Rock flowage has also been a factor. The cleavage is generally developed in the mudstone and stops at the base of the succeeding greywacke. It is at this interface that the pseudo-sedimentary structures are found.
>
> We admit that we were at first misled by some of these pseudo-structures, but by rigorously excluding all doubtful cases we are confident that no tectonic structures are present in the data considered.

Dr. J. L. Talbot (personal communication) has mapped well-preserved sole-marks in folded schists of the biotite grade near Adelaide, Australia.

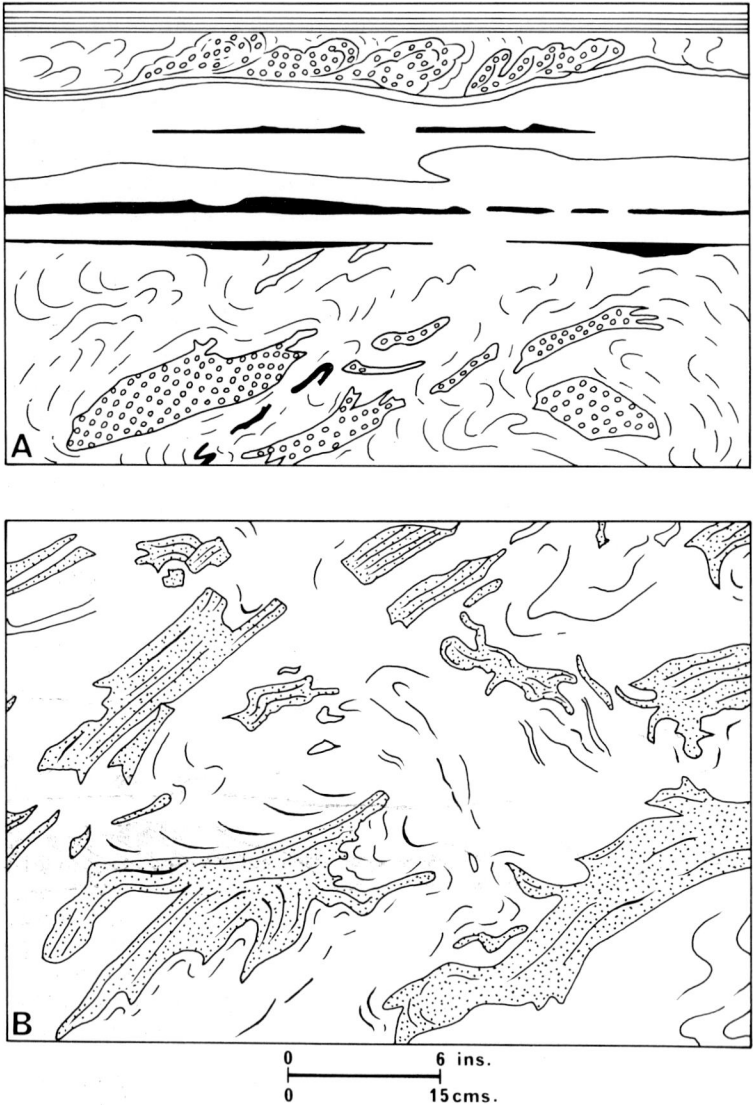

Figure 460. Slump structures in Upper Dalradian metasedimentary rocks, Banffshire, NE Scotland. A: Slip folding and slumping, Boyndie Bay; calcareous bands (black) are pulled apart; andalusitic pelites (circles) form slump balls in semipelite. B: Disrupted siliceous and calcareous fragments in argillaceous matrix which has lost most of the bedding structure, Elf Kirk, Scotstown (After Sutton and Watson, 1955, Figs. 6 and 7).

At the opposite extreme, strongly-marked major sedimentary features are commonly totally obliterated. In Chapter 11 considerable attention was given to the manner in which the unconformable relationships of the Moine rocks lying on the Lewisian basement at Glenelg, W Scotland, were almost completely obliterated by isoclinal folding (cf. Ramsay and Spring, 1963). Bederke (1963) also described the way in which an unconformity between infrastructure and younger superstructural rocks can be totally effaced by penetrative kinematic activity. Neither concordance, homotactic fabrics, nor common metamorphism can necessarily be used to prove a common history in rocks subjected to intense neocrystallization. Another classic example of effaced stratigraphic relationships involves the boulder conglomerate (tillite?) that lies unconformably on the granite migmatite at Mount Fitton, South Australia; because of subsequent neocrystallization, the unconformable contact now simulates a zone of progressive metasomatism or granitization (Bowes, 1954, 1956; Campana, 1955; Chinner, *et al.,* 1956A, 1956B).

Small crenulations in schistose rocks have sometimes been confused with genuine sedimentary ripple marks. Ingerson (1940, p. 569) showed that, as would be anticipated intuitively, (a) unmetamorphosed ripple marks show little or no preferred orientation of quartz grains, and (b) the microscopic petrofabric pattern of metamorphosed ripple-marked sandstone is related to the deformation accompanying the metamorphism, rather than to the ripple-mark axes.[4] Ingerson established that alleged ripple marks in Precambrian schists of the Grand Canyon, Arizona, U.S.A., are really small fold crenulations. Unlike ripple marks, such crenulations are not surface phenomena affecting a single S-plane, but are penetrative features affecting a whole sucession of S-surfaces (cf. Spry, 1963C).

Although the distinctions drawn by Ingerson may appear obvious, sedimentary features in regionally-metamorphosed rocks can be quite equivocal. A good example may be drawn from the structures at Broken Hill, Australia. Because Gustafson, *et al.* (1950) reported that no sedimentary features exist to indicate the stratigraphic sequence in the gneisses and schists, Condon (1959) specifically searched for relict structures and described abundant cross-stratification, ripple marks, and scour-and-fill structures; some of these structures were found in granite-gneisses and aplitized schists. Williams (in Condon, 1959) wrote that, with one possible exception, he could find no sedimentary structures in the outcrops described by Condon. Williams claimed that the current bedding

---

[4]Compare with Fairbairn's (1954) petrofabric analyses of cross-stratified quartzites from two Scottish and two Irish localities.

and scour and fill structures described by Condon are small parasitic folds that indicate significant internal flowage within the lithologic units (cf. Fig. 444). Clearly these exposures require careful and detailed reappraisal, but the dissimilar interpretation of the same minor structures by Condon and Williams is not atypical. Commonly very detailed study is necessary before the significance of structures can be correctly determined. Read's (1958, p. 90) comment about cross-bedding is equally valid in this connection: "A few good reliable observations are of the greatest significance; a multitude of doubtful, possible or likely observations is of no value whatsoever."

STRATIGRAPHIC RELATIONSHIPS

Finally, the importance of recognizable key stratigraphic horizons for understanding structurally-complex areas should be reemphasized. If attention is limited to mapping structural elements (e.g., fold axes, lineations, foliations, etc.) within a region, only a limited part of the total picture can possibly emerge. Unfortunately, where monotonous series of pelitic and semipelitic schists have been affected by several kinematic episodes, it is exceedingly difficult to isolate marker horizons and to elucidate the fold pattern. Recently, the recognition of a general stratigraphic sequence within the Moine Series of the Scottish Highlands has played an important role in deciphering the structural pattern within these stratigraphically-monotonous rocks (Chapter 11).

Recognition of mappable stratigraphic units and marker horizons often requires very intimate knowledge of the lithologies involved. A fine example comes from the detailed mapping of the Grenville Marble in the Adirondack Mountains, New York, U.S.A., by Brown and Engel (1956). By careful work fifteen units were mapped within a 600-meter succession of carbonates. Many of these units are extremely similar in lithology and they could only be separated by mapping distinctive interstratified members (e.g., talcose and pyritic units). Recognition of the units permitted Brown and Engel to complete a detailed structural study of the Balmat-Edwards district (Fig. 168).

Similarly, by mapping previously-unrecognized distinctive lithologic members within the Cambrian Conococheague Limestone in the Lebanon and Richland Quadrangles, Pennsylvania, U.S.A., Gray, *et al.* (1958) and Geyer, *et al.* (1958) established an unexpected structural pattern involving multiple thrusts and isoclinal folds (Fig. 461). Without recognition of the component members of the Conococheague Limestone, it is unlikely

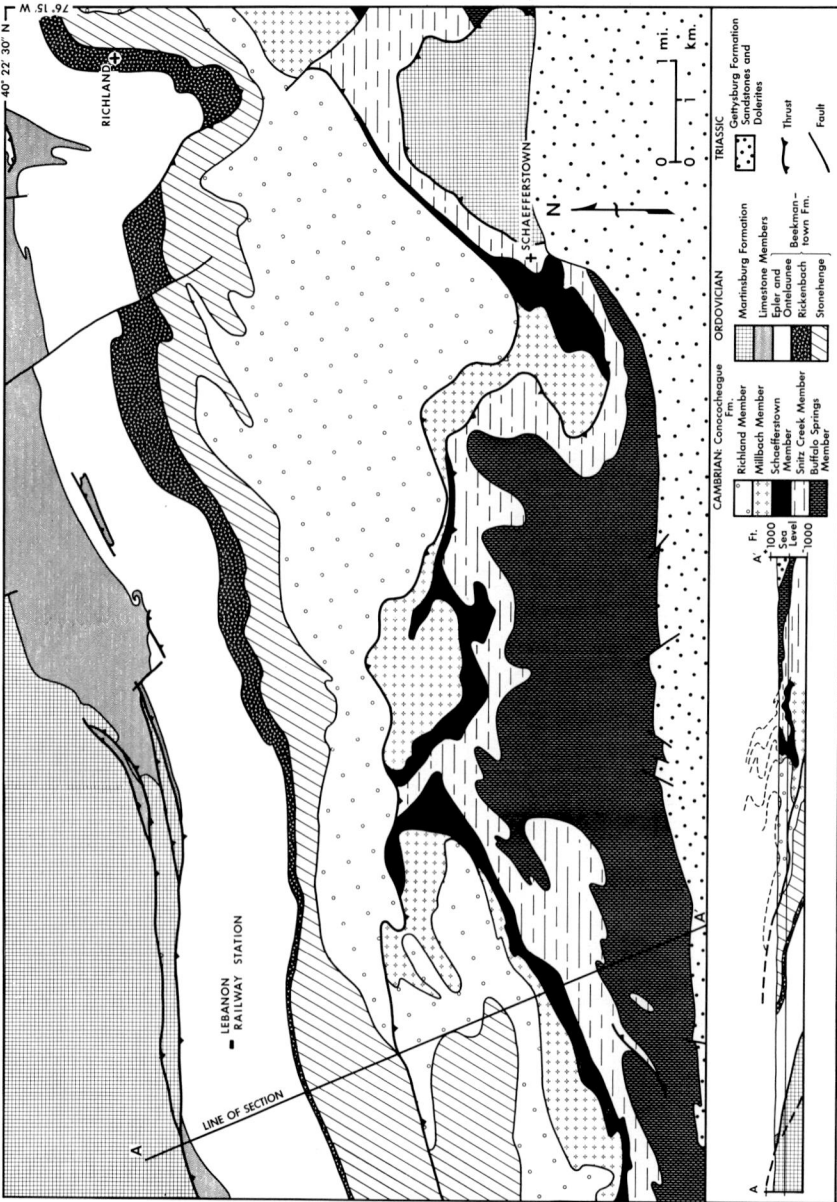

Figure 461. Thrust and isoclinal fold structures within the Cambro-Ordovician carbonate rocks of the Lebanon-Richland area, Pennsylvania, U. S. A. (Based on Gray, *et al.,* 1958, and Geyer, *et al.,* 1958).

Devonian system
  Chemung fm. Dch
  Brallier fm. Db
  "black shale" Dbs
  Huntersville chert $\Big\}$ Dho
  Ridgeley ss. $\Big\}$ Dho
  Helderberg group $\Big\}$ Doh
Silurian system
  Cayugan group Scy
  Clinch ss.   Ss
Ordovician system
  Juniata fm. Oj
  Martinsburg fm. Omb
  Bays ss. Oby
  Liberty Hall fm. Olh
  Effna ls.
  Lincolnshire ls. $\Big\}$ Ols
  New Market ls.
  Beekmantown group Ob
  Stonehenge ls. Os
Cambrian system
  Conococheague fm. €c
  Elbrook fm. €el
  Rome fm. €r
  Erwin fm.
  Hampton fm. $\Big\}$ €b
  Unicoi fm.

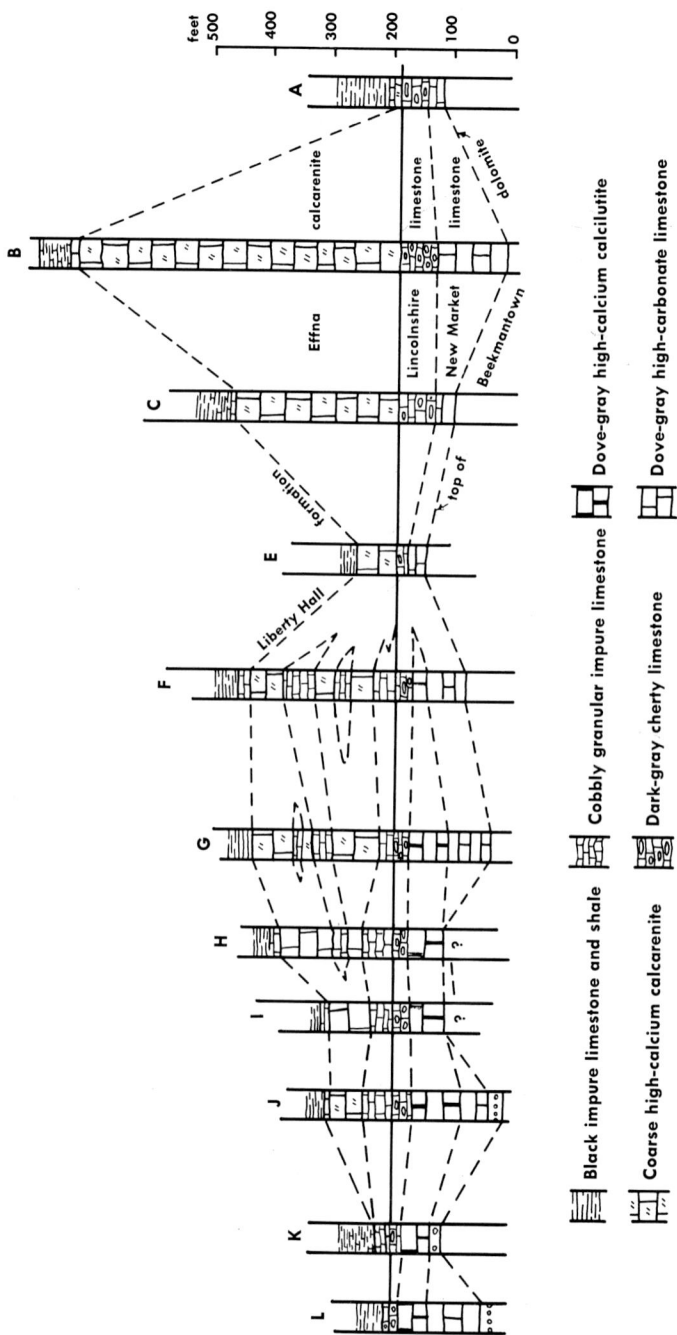

Figure 462. Relationships of the lenticular Effna Limestone in the Catawba syncline, immediately north of Roanoke, Virginia, U. S. A.; the measured sections are located by letter on the map (After Cooper, 1960, Figs. 3 and 5).

that any mapping of structural elements alone would have revealed this complicated geometry. Mapping is proceeding in adjacent areas of the Conococheague Limestone belt, and equally complex structures are being found (e.g., Wise, 1958; Freedman, *et al.,* 1964). However, the areal persistence of the particular stratigraphic members mapped in the Lebanon and Richland Quadrangles has not yet been established.

Careful mapping and much more experience are required before any predictions can be made about the type of three-dimensional variability that is common in different lithic units of a stratigraphical sequence in a geosynclinal belt. Some units persist for great distances with little lithological or thickness change. A good example appears to be the Dalradian Loch Tay Limestone. This Precambrian carbonate (marble) unit can be mapped for well over 300 kilometers from Banffshire (E Scotland) westwards into NW Ireland. The marble seems to have had a more-or-less constant thickness over this distance. By contrast, some of the limestone units in the central Appalachian fold belt of North America originally had very erratic thickness and lateral variations. This can be illustrated by the Ordovician Effna Limestone that was studied by Cooper (1960) north of Roanoke, Virginia, U.S.A. This calcarenite comprises a gigantic lens in the trough of the Catawba syncline, and the stratigraphic variability is illustrated in Figure 462. Cooper accumulated considerable evidence to support his contention that most of the units within the geosynclinal belt of Virginia possess very rapid original thickness changes. Clearly, if there is any likelihood that such stratigraphic variability is present, structural analyses must be preceded by detailed litho-stratigraphic mapping. One must agree with Pettijohn (1960, p. 447) that:

> ...no map of complex, non-fossiliferous terranes—most certainly Precambrian terranes—can any longer be regarded as adequate unless the primary structure of both sedimentary and volcanic strata utilized to determine stratigraphic order and structure of the area mapped are shown by appropriate symbols at the appropriate places.

Chapter 14

Models and New Methods for Quantitative Fold Description

Although one may not wholly concur with Griffiths (in Milner, 1962, p. 565) when he suggested that: "Progress in scientific investigation in any specialized field is generally measured by the degree to which the subject is pervaded by mathematics . . ." there is much truth in his assertion. Within the physical sciences it is common to test the validity and accuracy of conceptual response models with newly acquired data. Although qualitative, descriptive, and ordinal data permit some analyses, rigorous tests are more practicable on the basis of quantitative data collected according to a specified statistical sampling scheme.

PROCESS-RESPONSE MODELS

Conceptual response models are utilized extensively in the physical sciences; they are also being used more widely in the earth sciences (cf, Krumbein, 1962B; Potter and Pettijohn, 1963; Whitten, 1964A; Chorley, 1964). Because process-response models have only recently been introduced into structural geology, a brief review of the concepts and logic involved is desirable. Krumbein and Graybill's (1965) most significant volume—*An Introduction to Statistical Models in Geology*—was published after this book was in proof; although reference to this new work was not possible here, their book should be consulted for additional background.

Krumbein and Sloss (1963, p. 501) suggested that in the search for ". . . generalizing principles it is a useful philosophical device to recognize *models*—actual or conceptual frameworks to which observations are referred as an aid in identification and as a basis for prediction."

A model can take many forms, but it is essentially a framework for relating or "structuring" the data relevant to a problem. A geological **process model** is a framework for the processes that operate within a specified geographic domain. A geological **response model,** by contrast, is concerned with the attributes of a population of objects. Commonly it is convenient to consider a **process-response model** as a framework within which a specific set of process factors is linked to a specific rock formation.

When approaching a geological problem the rocks and processes exist (or existed) as real entities but the models are concepts. To express the nature and variability of the population of objects comprising a rock formation, a response model can be constructed on the basis of the attributes observed. Initially this may be a **conceptual model** in that many facets are inferred. As more data are collected, a **statistical model** can be considered, in which the relationships among the simultaneously-varying attributes can be analyzed; the contribution of each attribute to the total picture can be evaluated. The original conceptual model may eventually evolve into a **deterministic model** in which all relevant response characteristics are incorporated in a unifying mathematical equation that can be used for accurate description and prediction about the attributes of samples not actually observed.

Similarly an initial hypothesis may be structured into a **conceptual process-response model.** For example, the broad outlines of the interrelationships of process factors (winds, waves, etc.) and the observed nature of a beach may be embraced in a single conceptual model. Later all facets of the conceptual model might be treated in a statistical manner. Finally, the whole system may be expressed in a rigorous mathematical model in which an equation permits predictions to be made; then, either observations of the actual attributes of a beach permit predictions about the processes that were involved, or observations of the regime of processes permit predictions to be made about the type of beach structures that will evolve (cf. Krumbein, 1963A).

In many geological process-response models, time is an important factor. For example, as a front of progressive regional metamorphism passes through a prism of sedimentary rocks, a succession of process factors affects each individual unit of rock. For each successive event a process-response model can be erected, but, in a complete model incorporating time, the whole sequence can be incorporated. Similarly, for polyphase folding a conceptual model can be erected for each phase of folding, but, in a conceptual model for the complete orogenic sequence of events, the process-response models with respect to F_1, F_2, ... F_n could be linked in a single model.

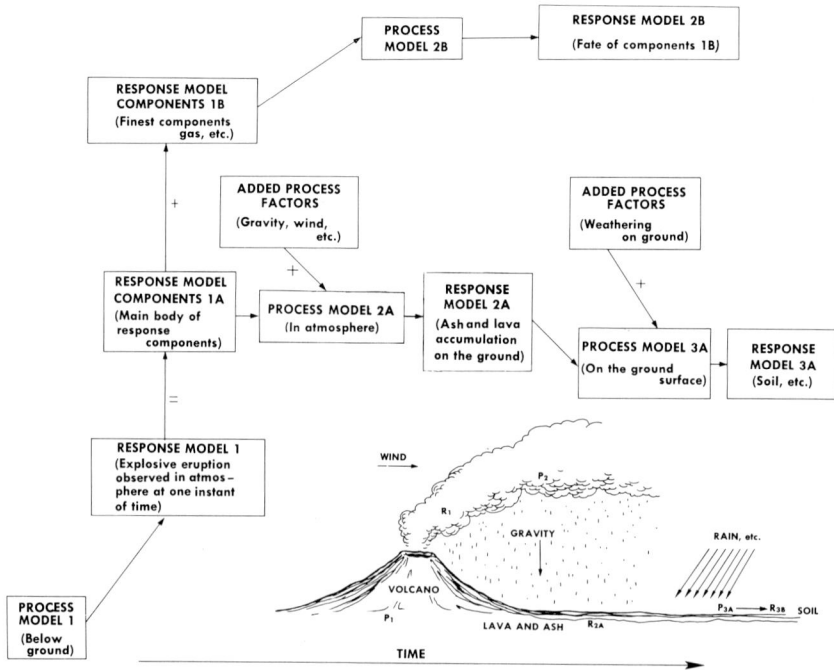

Figure 463. Sequential process-response model for some of the factors involved in a volcanic eruption (Adapted from Whitten, 1964A, Table 3).

In certain circumstances a geological event, such as the eruption of a volcano, can be considered as a geological response controlled by factors incorporated in a process model (Fig. 463); in such a case, the response model becomes all or part of the process model in the next phase of a continuing cycle of process-response phenomena (Whitten, 1964A).

Because, within the earth sciences generally, a very large range of factors enters into the evolution of a particular response, a clearly-defined conceptual framework can be of decisive value. It is unfortunate that commonly the specific objectives of a geological study are not clearly defined, and a field study is considered complete when all of the traditional facets have been described qualitatively. However, in most studies the optimum sampling plan is dependent on the objective of the study; hence, it is important to define objectives early. Also, as Chayes and Suzuki (1963) suggested, greater clarity is frequently brought to a problem by close definition of the objectives of a study (cf. Whitten, 1963C). Commonly, vast numbers of observations are recorded, the adequate planning of the data-collection process is overlooked, and a systematic attempt to maximize the information abstracted from the expensive raw data is not made. In many geological studies any vagueness about the objectives can be eliminated if the processes involved, and the responses to those processes, are expressed in the context of a unifying model.

The large volume of quantitative geological data that can now be made available and that can be analyzed with modern computing equipment, means that geological process and response models can be erected and tested to advantage. Several examples of the use of models for different geological problems have been described. For example, Olson and Miller (1951, 1958) used a model that provided a mathematical basis for operations involving biological theory and interpretation, and that enabled them to utilize both quantitative and qualitative considerations. In this chapter particular emphasis is placed on quantitative variables. Other published examples include the work of Melton (1958A, 1958B) on drainage systems and their controlling agents, of Hurley, *et al.* (1962) on a radiogenic strontium-87 model of continent formation, of Wyllie (1962) on a petrogenetic grid, of Krumbein (1963A) on a process-response model for analyzing beach phenomena, and of Whitten and Boyer (1964) on a model for the distribution of accessory minerals in a granite in Colorado, U.S.A. One purpose of such models is to provide a conceptual framework within which all the significant factors involved in a problem can be marshalled prior to incorporation in a definite hypothesis susceptible to quantitative test and modification as new observational data require (cf. Snow and Sironko, 1962). For example, Ringwood (1962A) proposed a model for the Earth's upper mantle, but subsequently (Ringwood, 1962B) modified it slightly to maintain consistency with new experimental data.

Models involving irregularly-spaced areally or three-dimensionally distributed observations have demonstrated value in sedimentation and stratigraphy (e.g., Miller and Ziegler, 1958; Krumbein, 1962B; Sloss, 1962; Krumbein and Sloss, 1963). Whitten (1963B, 1964A, 1964B) and Whitten and Boyer (1964) showed that analogous models are equally important in igneous and metamorphic petrology, although few have been developed explicitly.

In studies of present-day sedimentation both processes and responses can be analyzed. However, in most geological studies, the responses (e.g., an assemblage of rocks) can be examined, but the processes involved in their development can only be inferred. Often the uniformitarian principle appears to be valuable, and suggests that modern processes can be correlated with ancient responses. For example, modern beach processes may be invoked in making inferences about the process model appropriate to, say, a Cambrian sandstone (response model). An intensive study of the Cambrian sedimentary rock, coupled with use of uniformitarian principles, should permit realistic hypotheses to be erected about the appropriate paleogeographic process model.

In many geological problems, and specifically in most structural problems, the uniformitarian tool is of little value. The situation is easily illustrated by reference to axial-plane foliation, for which the actual processes that were involved cannot be observed and measured. Because of the radically dissimilar hypotheses about conditions under the Earth's surface, and the numerous rival genetic hypotheses for foliations, widely dissimilar inferences can be made about the processes involved. Hence, several different conceptual process-response models can be constructed for any particular foliation. In most cases, it is logical to conclude that one, and only one, set of processes was actually involved for a particular set of structures; but which set must, of necessity, be deduced from critical quantitative examination of the rocks and structures accessible for study by direct or indirect methods. When the response characteristics have been closely defined in terms of observed populations of attributes, it should eventually be possible to discriminate between appropriate and inappropriate process models. Experimental work in the laboratory provides considerable assistance. However, the paucity of quantitative information about the character and variability of foliation in nature, makes it difficult to determine whether an experimental product truly simulates the natural structures.

Directional attributes are of particular importance in structural geology but, because nondirectional attributes can be visualized more easily, the latter are used in the following elaboration of model concepts.

WHEN PROCESSES CAN BE OBSERVED

The simple example of development of a modern beach can be considered in terms of a process-response model (cf. Krumbein, 1963A). First, a few factors contributing to development of a beach can be measured in the field. These factors are components of the process model, and some examples are listed in Table 9. Intuitively some measurable phenomena appear relevant and others irrelevant to the process model, but it is necessary to test objectively and quantitatively the geological significance of all possible process factors (cf. Krumbein, 1959B). Hence, every possible factor is defined and measured in an attempt to embrace those processes relevant to the particular model.

Table 9. — Process-Response Model for a Modern Beach

Process model	*Response model*
mean wave height	beach slope
storm wave direction(s) ⟶	mineral composition
mean wind direction	organic content
rate of supply of detritus	penetrability
rate of scour of detritus	grain-size distribution
grain size of detritus	heavy mineral content
mean rainfall	ripple marks
wind velocity distribution	etc.
phase of moon	
latitude	
longitude	
salinity of sea	
faunal content	
floral content	
etc.	

Feedback loop
beach slope
beach wetness
grain size
etc.

Krumbein (1960B, p. 86) used an implicit function

$$f\,[q,\ G,\ P,\ (U,\ V,\ W),\ T\,] = 0 \qquad\qquad \text{(i)}$$

to define some aspects of sedimentation phenomena, where $q_1, q_2, q_3, \ldots$ represent an unspecified number of physical, chemical, and mineralogical properties of the sediments; $G_1, G_2, G_3, \ldots$ represent a number of geometrical and other properties of the deposit as a whole (such as slope of a beach foreshore, the width of an offshore bar); $P_1, P_2, \ldots$ represent individual elements of geological processes, such as wave characteristics on a beach, current velocity in a stream, and so on; U, V, and W represent geographic coordinates including elevation W; T is a time or stratigraphic factor. Krumbein noted that additional factors, such as biological controls on sedimentation and the fossil content of the deposits, can be included. All of the variables in this implicit function are factors of the process model (Table 9).

Second, numerous characteristics of the response (i.e., the resultant beach) can be identified and measured quantitatively. Examples are given in Table 9. Included in the response model are some attributes that effect "feedback" and that affect the nature of some process-model factors. For example, beach slope (a response-model variable) affects some process factors such as the rate of scour. An appropriate feedback loop (cf. Melton, 1958B; Krumbein, 1963A) is included in Table 9.

The process-response model is not complete until all the significant process factors and response characteristics have been identified. Each of the many response-model characteristics $(r_1, r_2, r_3, \ldots r_n)$ is a function of the process-model factors $(p_1, p_2, p_3, \ldots p_m)$. That is, the assemblage of factors $(p_1, p_2, \ldots)$ governs the response characteristics of the sampled beach. This can be expressed mathematically, thus:

$$r_1 = f_1 (p_1, p_2, p_3, \ldots, p_m) \qquad \text{(ii)}$$
$$r_2 = f_2 (p_1, p_2, p_3, \ldots, p_m) \qquad \text{(iiA)}$$
$$.$$
$$.$$
$$.$$
$$r_n = f_n (p_1, p_2, p_3, \ldots, p_m) \qquad \text{(iiB),}$$

where $f_1, f_2, \ldots, f_n$ are different functions. If quantitative measurements of all the significant contributing phenomena can be made, the role of each process factor and of each response characteristic can be evaluated. The geological effect of modifying one factor involved in the process-response model could then be determined.

It is apparent that both the process model (Table 9) and Krumbein's implicit function (equation i) include both (a) overall regional controlling factors and (b) local factors peculiar to the particular target population studied. Complete analyses involve separation of these two groups. In practice, minor process models subsidiary to the main process model can be defined. For example, rates of supply and scour of detritus (local factors) are functions of numerous other process factors, such as mean wave height, wind velocity distribution, gravity, etc. (regional factors). Hence, several relationships exist such as

$$p_1 = f (p_2, p_3, p_4, \ldots) \qquad \text{(iii);}$$

in consequence, only a proportion of the process factors is independent of the others.

Finally, the total beach can be represented by the sum of all the response characteristics defined by the equations (ii); this total response model may be designated R. It should be clear that R can be considered a function of the component characteristics, so that:

$$R = F (r_1, r_2, r_3, \ldots, r_n) \qquad \text{(iv).}$$

WHEN ACTUAL PROCESSES CANNOT BE OBSERVED

A small simple granite complex provides a good example. Although considerable experimental work has been done on the silicate systems relevant to granites, man has never observed the formation of a granite mass; hence, it is only possible to make inferences about the precise nature and quantitative values appropriate to the process factors. The vigor of the controversy about the genesis of granitic rocks reflects the fact that the processes cannot be established unequivocally, and that the principle of uniformitarianism is of little assistance because granites forming at the present time cannot be observed. However, on the basis of an extensive three-dimensional exploration program, it would be possible to prepare a complete response model (equation iv) for a particular granite complex. Since such a response model (R) is a function of an array of response characteristics $(r_1, r_2, r_3, \ldots)$, and each characteristic is a different function of the process factors $(p_1, p_2, p_3, \ldots)$, it should be possible to deduce the specific processes responsible (i.e., the process model, P) for a particular granite mass. Because the processes cannot be observed, several dissimilar conceptual process models $(P_a, P_b, P_c, \ldots)$ can be erected, and it may be difficult to determine which is the most appropriate.

Consider two rival conceptual models for a particular granite pluton. Let intrusion of magma be involved in the first model (P_a), in which many process factors $(p_{a_1}, p_{a_2}, p_{a_3}, \ldots p_{a_n})$ are involved (see Table 10). Let a granitizational process resulting from ionic migration be involved in the second model (P_b), where the factors are $(p_{b_1}, p_{b_2}, p_{b_3}, \ldots p_{b_n})$. Probably additional conceptual models $(P_c, P_d, P_e, \ldots, P_k)$ could also be developed (Whitten and Boyer, 1964).

For each process model $(P_a, P_b, \ldots)$ there is a corresponding response model $(R_a, R_b, \ldots$ respectively), and an array of predicted response characteristics:

$$
\begin{array}{llll}
r_{a_1} & r_{a_2} & r_{a_3} & r_{a_4} \ldots r_{a_n} \text{ corresponding to } P_a \\
r_{b_1} & r_{b_2} & r_{b_3} & r_{b_4} \ldots r_{b_n} \text{ corresponding to } P_b \\
r_{c_1} & r_{c_2} & r_{c_3} & r_{c_4} \ldots r_{c_n} \text{ corresponding to } P_c \\
\cdot & \cdot & \cdot & \cdot \\
\cdot & \cdot & \cdot & \cdot \\
\cdot & \cdot & \cdot & \cdot \\
r_{k_1} & r_{k_2} & r_{k_3} & r_{k_4} \ldots r_{k_n} \text{ corresponding to } P_k
\end{array}
$$

It is probable that, for a particular pluton, only one response model is wholly valid. However, some characteristics of each response model may be identical; for example, it might be found that $r_{a_2} = r_{b_2} = r_{c_2} = \ldots = r_{k_2}$. Then response characteristic r_2 is of no value in discriminating between the appropriate and inappropriate process models $(P_a, P_b, P_c, \ldots)$. As a specific example, each process model might predict exactly the same quantitative pattern of three-dimensional composition and variability for quartz percentage. However, some other response characteristics $(r_2, r_3, \ldots)$ will be dissimilar in the conceptual response models

Table 10.—*Two Possible Process-Response Models (With Feedback Loops) for a Granite Pluton*

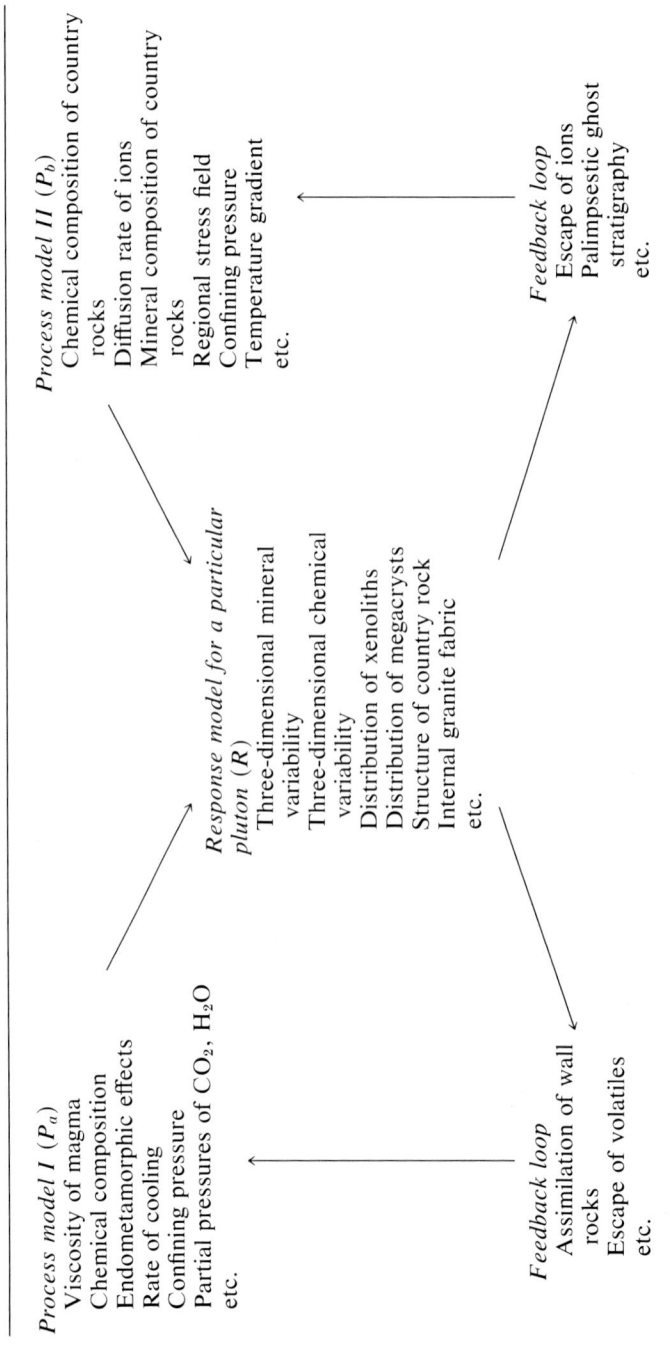

Process model I (P_a)
Viscosity of magma
Chemical composition
Endometamorphic effects
Rate of cooling
Confining pressure
Partial pressures of CO_2, H_2O
etc.

Process model II (P_b)
Chemical composition of country
    rocks
Diffusion rate of ions
Mineral composition of country
    rocks
Regional stress field
Confining pressure
Temperature gradient
etc.

Response model for a particular pluton (R)
Three-dimensional mineral
    variability
Three-dimensional chemical
    variability
Distribution of xenoliths
Distribution of megacrysts
Structure of country rock
Internal granite fabric
etc.

Feedback loop
Assimilation of wall
    rocks
Escape of volatiles
etc.

Feedback loop
Escape of ions
Palimpsestic ghost
    stratigraphy
etc.

N.B. Two representative process models are illustrated out of the *n* possible models.

corresponding to P_a, P_b, P_c, ... ; quantitative estimates of such variables (response characters) for a particular pluton could permit identification of the correct process-response model. The general lack of knowledge about the three-dimensional variability of attributes in granitic plutons, makes it difficult to forecast which response characteristics are likely to be critical in identifying the correct model.

ATTRIBUTES AND SAMPLES FROM STRUCTURAL COMPLEXES

To the present most structural field data have been partly qualitative, and most attributes have been expressed in terms that involve subjective judgments or intuitive estimates. Although an S-surface or fold axis may be measured carefully in the field, the choice of which structural element is actually measured is commonly partly controlled by erosion and partly by the subjective or intuitive choice of the geologist: where several elements are available for measurement there is a tendency to eliminate the "unusual" (cf. Flinn, 1962). Virtually no attention has been given to adequate sampling or statistical evaluation of the three-dimensional variability of sampled populations of structural attributes of rocks. In the preceding chapters many attributes of actual deformed rocks have been described in qualitative terms. An attempt has been made to clarify the descriptions of common structural elements because, before conceptual models can be developed realistically, rock structures must be precisely defined and measured. There is, for example, relatively little point in developing an elaborate hypothesis about the stress and strain relationships responsible for structure A, if the nature and variability of A are not known accurately.

Sampling problems associated with the collection of directional data (e.g., dip readings, lineation plunges, etc.) from complex structural provinces are acute. Analysis on equal-area projections has been used extensively in structural work within the past two decades, but, although it has specific valuable uses, the method actually discards significant information, namely the U,V,W-locations of the sample points (cf. Agterberg, 1961). In the case of a granitic complex, to plot observed model data on, say, a ternary diagram masks any spatial U,V,W-variability possessed by the variables; such a diagram divorces the attributes from their essential U,V,W-locations. Similarly, a synoptic equal-area projection for an attribute aids in identifying the mean value (if the population has been adequately sampled), but it can easily mask local or regional U,V,W-variability within the rocks investigated. To understand the structural geometry of a sequence it is most important to recognize the three-dimensional variability of the measured attributes. However, there is no satisfactory method of analyzing the three-dimensional variability in *general use* at present. The closest approach involves subdivision of a region into subareas, and then plotting the individual observations for each subarea on a projection, but this still divorces data from their U,V,W-locations.

In Chapter 3 the desirability of differentiating between the populations of objects and the populations of attributes was emphasized. Modern techniques now permit a geologist to acquire an astounding volume of data about rocks observed in the field. It is appropriate to pause to consider how many attributes can be measured for a particular sample. Consider first the nondirectional properties that might reflect the three-dimensional variability within, say, a geosynclinal graywacke unit. How many attributes can be studied? Ten? Twenty? Two hundred? Houser and Poole (1959) and Izett (1960) recorded quantitative data for over 120 attributes of granite samples from the Climax Stock, Nevada, U.S.A. Without having to devise any unusual or new variables, several hundred attributes could be measured for a graywacke sample (Table 11). Intuitively, it is reasonable to assume that most of these

Table 11.—Attributes That Can Be Measured in a Graywacke Unit and Whose Variability Could Be Mapped

Mean grain-size of sample (may vary for samples of different size)
Maximum and minimum grain-size
Skewness and kurtosis of grain-size
Weight percentage of 10 major oxides in bulk sample
Weight percentage of each trace element in bulk sample
Weight percentage of 10 major oxides in each size fraction of rock
Weight percentage of each trace element in each size fraction of rock
Volume percentage of each major mineral phase in bulk sample and/or in each individual size fraction
Volume percentage of each accessory mineral phase in bulk sample and/or in each individual size fraction
Mean (and variance) of chemical composition (major and minor elements) in each major and minor mineral in rock
Specific gravity of rock and of each mineral phase
Electrical resistivity and radioactivity of bulk rock
Mean and variance of refractive index of each mineral phase
Mean and variance of 2V, pleochroism, birefringence, etc. of each mineral phase.
Triclinicity of alkali feldspars
Mean thickness and variance of bedding units (and, if more than one bedding unit present in sample, the range and variance of all above attributes in each unit)
etc.

attributes would show some variability within a granite stock or a gray-wacke unit. Any petrogenetic hypothesis about the genesis of a lithic unit should be capable of accounting for the spatial U,V,W-variability of the observed attributes.

If a series of folded schists is studied, rather than a relatively undeformed graywacke unit or a granite mass, both directional and non-directional attributes are involved. Most of the attributes in Table 11 can be measured for samples of the folded schist unit, although the field-sampling plan required to obtain a realistic estimate of the three-dimensional variability of the schists is likely to be dissimilar. In addition, directional properties can be measured. The nature of measurable directional properties depends on the size of the objects studied, but, for objects of a specified size, they have considerable three-dimensional variability within a fold complex.

Most geological processes (including the development of fold structures) are very complex. The processes characteristically involve (a) interactions between a large number of attributes, and (b) simultaneous variation among all or most of the many variables involved. Krumbein (1960B, p. 83) pointed out that the problems of applying statistical methods in such circumstances:

> ... revolve in part around at least three considerations; severe sampling restrictions in some geological studies, the multiplicity of variables in even the seemingly simplest geological situation, and the high "noise level" of some geological data.

However, despite certain restrictions imposed by the nature of structural geology problems, most are susceptible to quantitative analysis.

APPLICATION OF PROCESS-RESPONSE MODELS TO STRUCTURAL GEOLOGY

The majority of structural geological work has been either descriptive or experimental, although a resurgence of interest within the past decade has been centered on semiquantitative geometric analyses. The widespread use of equal-area projections probably typifies this phase of structural geology.

The closest approaches to process-response models have possibly been in the field of microscopic petrofabric analysis. Particularly in the study of carbonate fabrics Turner and others (e.g., Turner and Ch'ih, 1951; Turner, *et al.,* 1954, 1956) amassed abundant quantitative data concerning response characteristics, and this has led to the development of specific kinematic and kinetic models. Experimental work under controlled kinetic conditions produced microfabrics that could be measured and specifically equated with certain natural carbonate rock fabrics.

The literature abounds with descriptions of experimental scale models in which various structural elements are said to have been simulated. Apart from the inherent problems associated with scale models and, in particular, of knowing whether every process factor (including time) has been scaled correctly, it is difficult to determine whether or not the experimental models simulate actual rock systems precisely. To decide whether an experimental model—for example, the series of fold structures produced by Ramberg (1963A, 1963B)—corresponds to observed natural deformed rock systems is commonly a matter of subjective qualitative judgment.

Many models have been proposed for the development of major orogenic fold systems, and the whole subject has long been one of vigorous controversy. It has often been contended that fold mountains reflect compression associated with shortening of the Earth's crust, but the immense shortening apparently required by such models has been criticized. Space does not permit listing all the rival models that have been proposed, although, by way of illustration, the model advocated by Beloussov (1961) is summarized in Figure 464. Beloussov believed that folding phenomena represent the reactions of layered strata to the differential vertical movements of separate blocks of the crust, and that fold phenomena yield no evidence for general compressional forces in the Earth's crust. According to Beloussov, horizontal compression only existed locally in narrow belts within geosynclines, whereas vertical movements of separate blocks of the crust are primary phenomena that can give rise to gravity tectonics during orogeny (Fig. 464). The model proposed by den Tex (1963) may be cited as a second illustration; this model (Fig. 436) concerns the interrelationships of metamorphism and the kinematic events in an orogenic belt (see Chapter 12).

Observed three-dimensional quantitative data are required in order to test these and other conceptual and experimental process models. It seems certain that the next decade will see model testing developed in a formal manner. At present, the immediate hurdles concern (a) identification of existing and new attributes of deformed rocks that can be expressed quantitatively, and (b) methods for describing the three-dimensional nature and variability of attributes that characterize deformed rocks. These topics are discussed more fully in a later section of this chapter.

The foregoing remarks show that models have significance in a wide range of structural problems. Four additional concepts are briefly outlined in the context of specific models, in order to indicate the variety of models that is useful in structural geology.

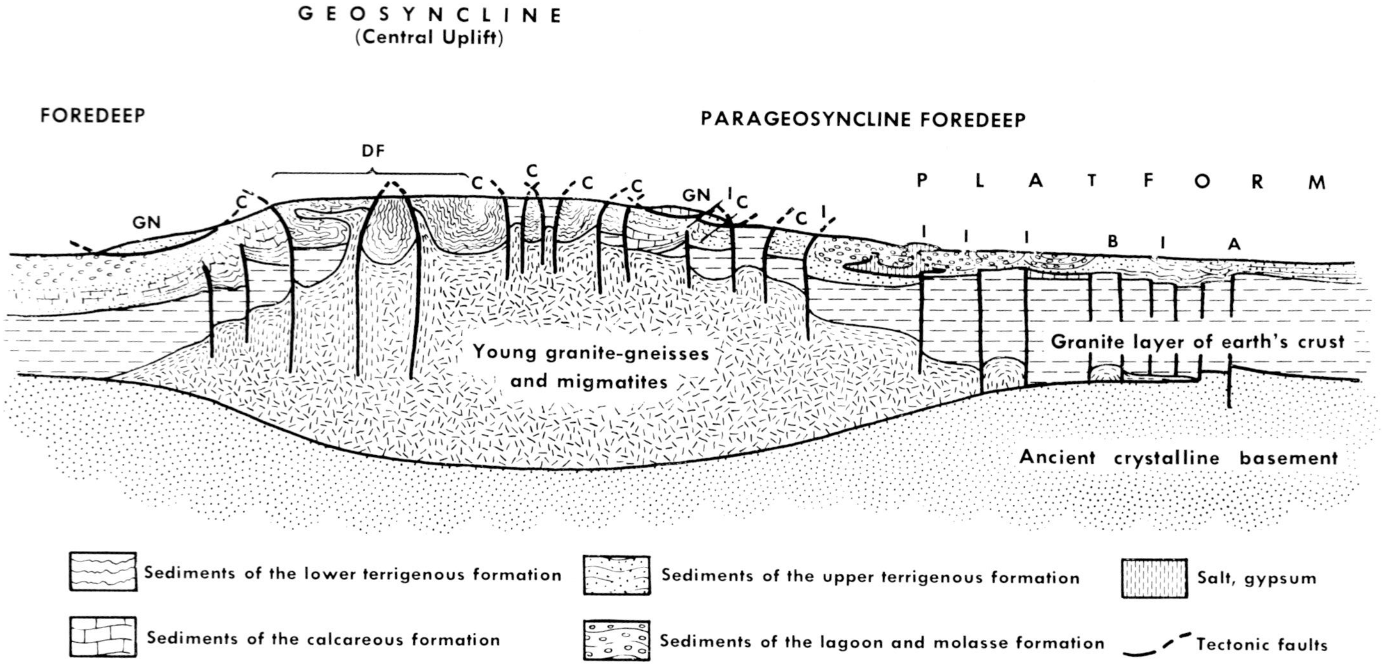

DF, Deep folding. C, The folding of general crumpling. GN, Gravitational tectonic nappe. I, Injection folds. B, Block fold. A, A welt of block origin

Figure 464. A generalized model for the various types of folding in the Earth's crust (After Beloussov, 1961, Fig. 14).

1. A SIZE MODEL

A conceptual model relating the dimensions of time and size in geotectonic phenomena was proposed by Carey (1962B). He pointed out that the range from global geology to astrophysics involves only three orders of magnitude, whereas within the field of rock deformation, structural geology, and geotectonics, the linear dimension varies through sixteen orders of magnitude (Fig. 465). Recognizing that time has a comparable span of magnitudes, Carey erected the model illustrated in Figure 466. He reasoned that mistakes are likely if empirical concepts about the states of matter are applied far beyond the range of their empirical foundations; such concepts have long been held, although

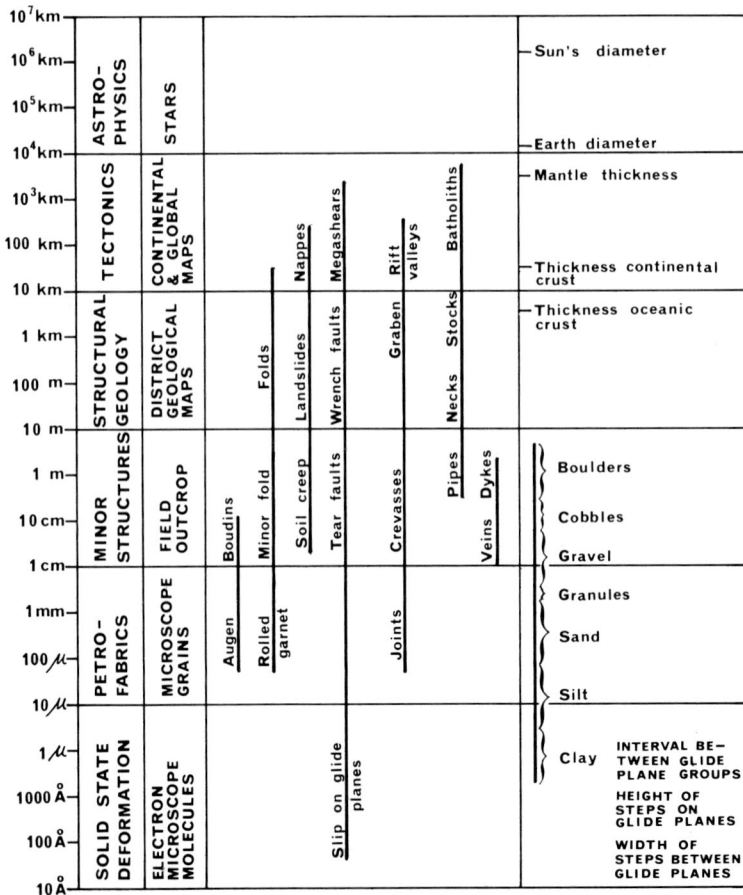

Figure 465. The size of geotectonic phenomena (After Carey, 1962B, Fig. 2).

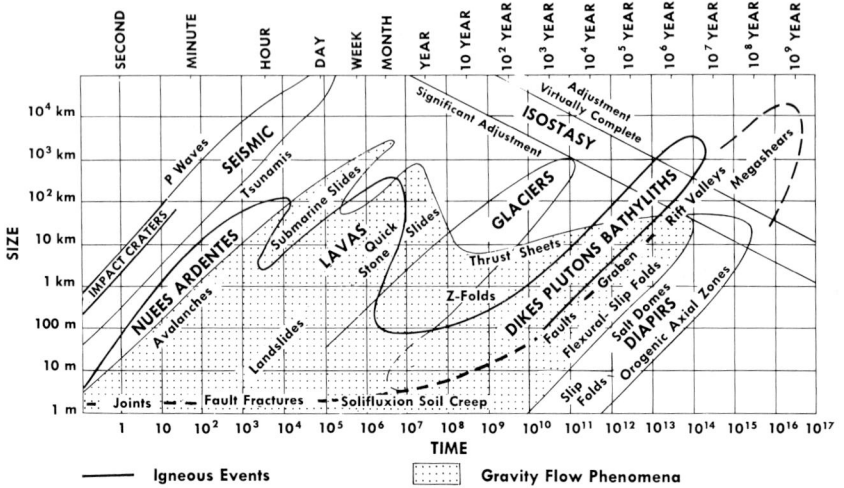

Figure 466. A size-time model for geotectonic phenomena (After Carey, 1962B, Fig. 2).

not expressed formally in a generalized model. Thus, for example, Professor Sollas (in Bailey, 1910, p. 619) said:

> In considering the flow of solid rocks, it was important to bear in mind not only the element of time, but also that of space. It was impossible to argue, from the physical constants of a rock examined in a hand-specimen, to those of the same rock buried deep within the crust and forming part of a flake 12 miles long or more.

Carey's model suggested that most geological phenomena occupy space-time fields that slope upwards to the right at 45° in Figure 466, and that these fields reflect rates of deformation largely controlled by the viscosities of natural media within the Earth's gravity field. Certain parts of this model might be criticized, but Figure 466 provides a useful conceptual framework against which new data can be tested. If existing or new data violate a model, that model requires modification; similarly, until a model is proved untenable, it can be used to test hypotheses about geological structures. This process is illustrated by the following quotation from Carey (1962B, pp. 103-4):

> The time scale for impact craters and explosion craters is very short, but not instantaneous. The shock propagates with the velocity of compressional waves. But even for a crater several kilometers in diameter the impulsive blow is over in seconds. For loads of this duration all rocks are brittle. Deformation is

by fracture only. Since the advent of crater-forming nuclear explosions, it has become the fashion to attribute many ancient circular structures to meteorite impact. Dietz (1961) has suggested that the South African Vredefort dome,[1] which has a central boss of granite with a halo of upturned and outwardly overturned sediments some 15,000 meters thick and nearly 100 km. in diameter, may have been produced by such an impact. Dietz calculates that a meteorite 2.3 km. in diameter would have sufficient energy. But this interpretation is improbable in the extreme. For structures of the size of the Vredefort dome, impact craters fall ten orders or more on the time scale from any rock deformation structures. Dietz's problem is not to get sufficient energy to do such work, but to apply it sufficiently slowly to cause these rocks to deform rather than to shatter. Folding as commonly understood is quite improbable on so short a time scale.

Carey's model suggested that gravity plays a very significant role in deformations of major size, so that, if gravity were decreased in magnitude many phenomena (e.g., isostasy, gravity sliding, etc.) would move to the right on Figure 466, and thus take longer to reach completion.

2. AN EMPIRICAL MODEL FOR THE AMPLITUDE OF FOLDS

Currie, *et al.* (1962) found that field observations and laboratory experiments indicate that the stratification of sedimentary rocks is significant in determining the response to deformation within a basin. The physical properties and thickness of a dominant member in a sedimentary sequence apparently control the fold wave length that develops during the early stages of deformation. On the basis of a limited number of measurements of the attributes indicated in Figure 467, Currie, *et al.* found that a linear relationship exists between fold wave length and dominant-member thickness (Fig. 468). This linear model (Fig. 468) is approximately valid for structures whose spacing ranges from 0.3 meters to several kilometers; the thickness of the dominant member that controls the fold wave length ranges from less than a centimeter to about 550 meters. However, this elementary model does not take into account all of the factors that contribute to the total strain pattern. A more complete model should probably take into account the progressive sequence

[1]The Vredefort structure is shown in Figure 470. Recent cratering experiments with explosives in basaltic and other rocks at the U.S. Atomic Energy Commission's test site, near Mercury, Nevada, U.S.A., have shown that small isoclinal recumbent folds are sometimes produced within the debris cone surrounding a crater.

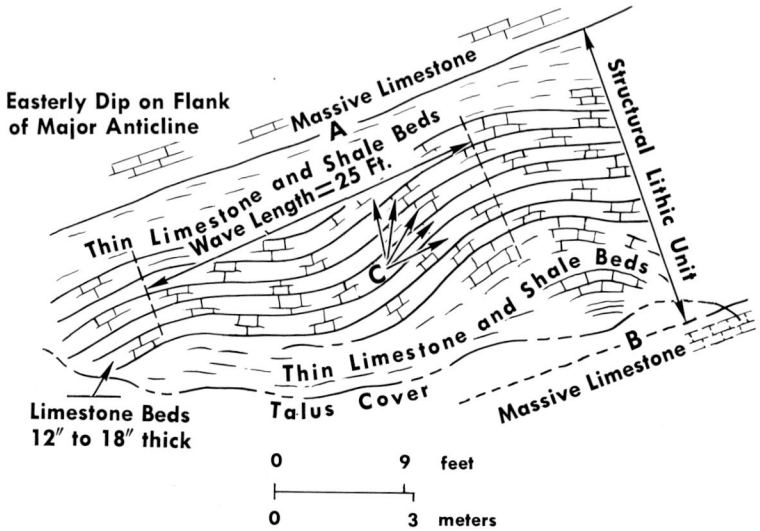

Figure 467. Sketch to indicate operational definitions used in the model shown in Figure 468. Minor folds are shown that occur on the flanks of a larger flexure in Lower Devonian limestone strata of the northern Appalachian Basin near Catskill, New York, U.S.A. A and B represent the boundaries of the lithic units, and the dominant members that control the period of the minor folds are presumably the limestone beds labelled C (After Currie, *et al.*, 1962, Fig. 9A).

of events in the development of fold systems; Currie, *et al.* (1962, pp. 672-3) recognized that:

> These events may influence the nature of structural lithic units, the effectiveness of a dominant member, and the significance of incompetent beds to the structural process. For example, two competent members of nearly equal thickness may act together within a single structural unit in the early stages of deformation, but at a later stage they may begin to act independently and modify the original unit toward a major element that contains a minor lithic unit within it. In other cases, a dominant member that controls deformation may fold initially, but in later stages it may serve as the locus of displacement on a fault. Also, a sequence of structural events that places an increasing restriction on incompetent beds at the core of a fold in the dominant member may finally curtail the folding process and require that further relief be obtained by fault displacement.

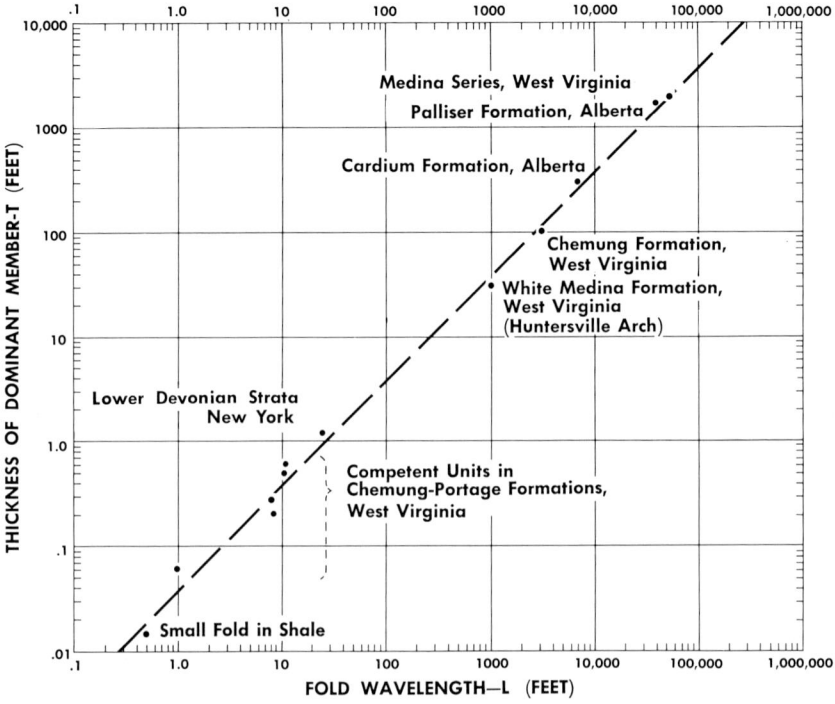

Figure 468. Relationship between wave length (L) of a fold and the thickness (T) of the dominant member within the fold (Based on Currie, *et al.,* 1962, Fig. 6).

3. AN EXPERIMENTAL MODEL OF GRAVITY TECTONICS

Ramberg has made many experimental studies of tectonic structures. He (Ramberg, 1963C) recently studied gravity tectonics with the aid of a large-capacity centrifuge (to simulate the force of gravity), and produced structures similar to submarine ridges, rift valley systems, and a number of plutonic bodies such as stocks, batholiths, sills, etc. These structural forms were accompanied by distortion of the layered structure of the host "rocks" (Fig. 469). Such experiments are useful if they are considered as models to be tested with actual geological quantitative and qualitative field data, and to be corrected as necessary.

It is interesting that Ramberg (1963C, p. 38), like Carey (1962B), considered the Vredefort granite dome, South Africa. Ramberg's model suggested that domes piercing the surface of the layered overburden commonly fold the surrounding strata back on themselves so that they dip towards the funnel-shaped intrusion (Fig. 469). Ramberg suggested

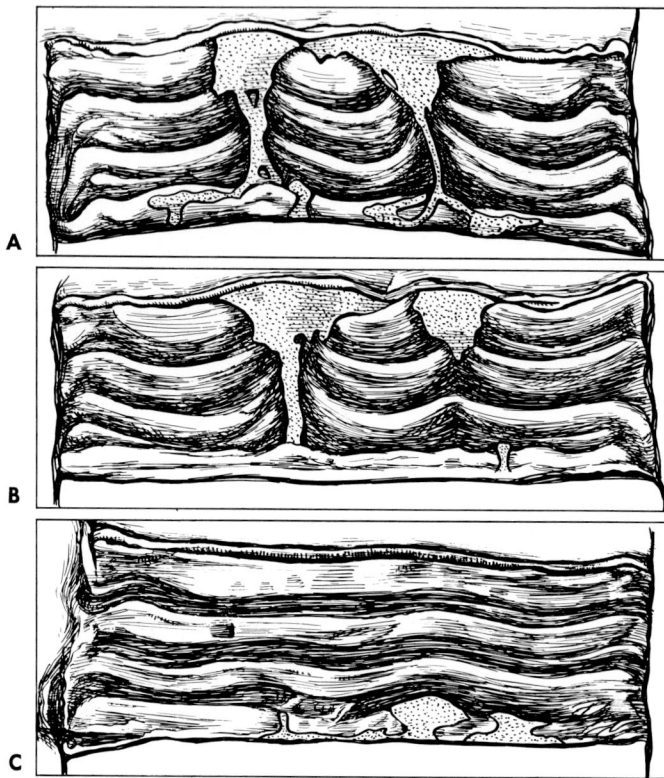

Figure 469. Simulated granite piercement domes produced in a large-capacity centrifuge; thick layers of the "host rock" are painter's putty of higher specific gravity than the "diapiric plutons" (bouncing putty). The extreme top layer capping the whole model is modelling clay (After photographs from Ramberg, 1963C, Fig. 3).

that this structure is a good analog of the Vredefort dome (Fig. 470) as mapped by du Toit (1954). According to Ramberg's model, such structures cannot develop if there is a very large viscosity contrast between the country rocks and the intruded material. In consequence, he concluded that, since the Witwatersrand and Ventersdorp rocks adjacent to the Vredefort massif were unquestionably crystalline when the dome rose, the intruded material must have been either (a) completely crystalline or (b) a mush with so little liquid material that its strength and viscosity approached that of the surrounding solid rocks.

On the basis of more field work at Vredefort, and at many other areas, the validity of this model might be confirmed; alternatively, new field data may indicate that the model requires drastic modification. Actually, numerous other conceptual process models have been pro-

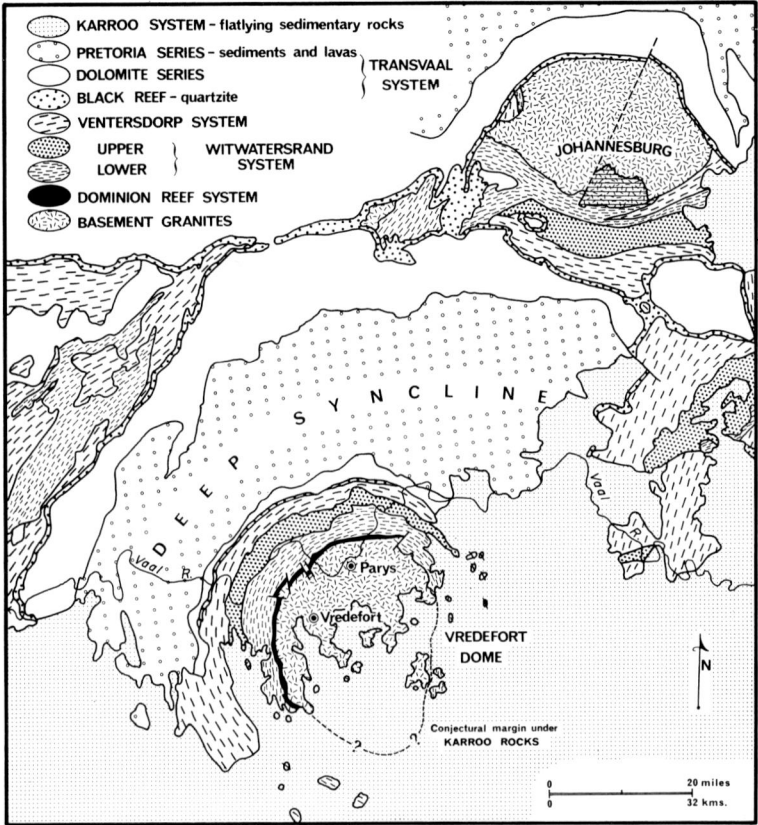

Figure 470. Geological map of the Johannesburg-Vredefort area, Republic of South Africa (Based on Borchers, 1961, Diags. 1 and 2).

posed for the Vredefort dome; Eskola (1949) referred to several of those that have been published when he suggested that a close analogy exists between Vredefort and the Karelian mantled gneiss domes of Finland.

4. CONCEPTUAL MODELS FOR SLIDES AND FAULTS

Although little attention has been given to faults in the preceding chapters, informal conceptual models have tended to dominate ideas about the genesis of slides and faults.

The slide concept has played an important role in recent work on disharmonic folds within the Caledonides (see Chapters 7 and 11) and other areas in which Alpine-type structures occur. The term **slide** was introduced into the literature by Bailey (1910, p. 593), although the word was suggested by Lapworth (see Bailey, 1934, p. 467; 1938). Bailey demonstrated that the limbs of the enormous recumbent folds in the Dalradian metasedimentary rocks of the Scottish Highlands commonly appear to be replaced by fold-faults, or slides, that are interpreted as having formed during folding. Thus, Bailey supposed that between Spean Bridge and Onich (*ca.* 30 kilometers) the lower limb of the Appin fold is largely replaced by the Fort William slide (Figs. 193 and 194). The conceptual model was that slides give the freedom for development of recumbent folds — that recumbent folds have been squeezed forward like intrusive masses, so that sliding is not confined to the lower limbs of recumbent anticlines but can occur along either flank.

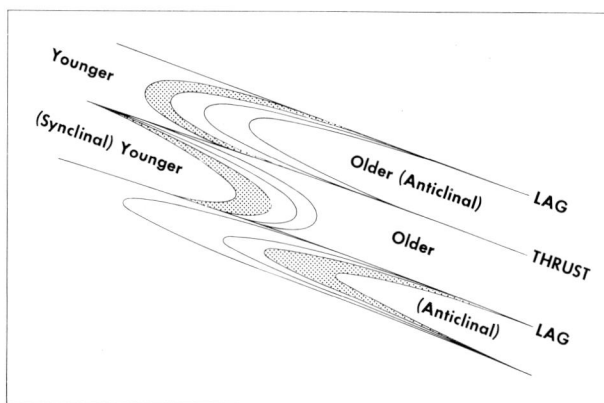

Figure 471. Diagram illustrating slide terminology.

Today "slide" is a useful and widely used term for a fault "... formed in close connection with folding, which is broadly conformable with a major geometric feature (either fold limb or axial surface) of the structure, and which is accompanied by thinning and/or excision of members of the rock-succession affected by the folding" (Fleuty, 1964B, p. 454). When the direction of facing can be determined, Bailey preferred to use (a) **thrust** — for a slide that replaces the inverted lower limb, actual or ideal, of a recumbent anticline (Bailey, 1934, p. 467),[2] and (b) **lag** — for a slide which (Bailey, 1938, p. 609) "... more or less completely replaces the unreversed limb of an overturned anticline (real or imaginary)."

---

[2]This is a restricted use of thrust; the term is commonly used for any low angle fault plane along which one mass of rocks has been moved over another block (cf. Fig. 471 and Anderson, 1951). Contrary to the implication of Patterson and Storey (1963, p. 291), either younger or older rocks may have been thrust over the structurally-lower block. Bailey's terms lag and thrust correspond to Lovering's (1932) underthrust fault and overthrust fault, respectively. Bailey based the term lag on the subhorizontal "lag" faults described by Marr (1900, p. 461).

Slide tectonics, although invoked by Bailey to explain major stratigraphic anomalies, are characteristic of structures of all sizes on which observations are possible (see Chapter 7). In units of small size, the most notable evidence for slides stems from the systematic attenuation of the layers in corresponding limbs of isoclinal folds (cf. King and Rast, 1956A, p. 251). Hence, the general conceptual slide response model is apparently applicable to structures of many dissimilar sizes. However, the model is neither inviolate nor necessarily applicable to every fold terrane; continuing tests are necessary if the model seems relevant to a particular study or to a particular area. Two examples may be cited:

(a) To establish the existence of a slide it is commonly necessary to assume that the lithostratigraphic units were areally persistent; facies changes without sliding could provide similar field relationships and map patterns. A significant feature of the slides mapped by Bailey (e.g., 1910) was that portions of the stratigraphic succession are cut out without apparent discordance or visible signs of mechanical disturbance between the units (Fig. 193). For one locality southeast of Ballachulish, Hardie (1952) suggested that a major unexposed sedimentary transition, or facies change, between the Eilde Quartzite and the Eilde Schists accounts for the rapid variations in thickness of these units better than the slide invoked by Bailey's (1934) model. Hardie's (1952) evidence was not unequivocal and was vigorously debated (Bailey, 1953, Hardie, 1953); more recent field work showed that the complex structures do not involve facies change, and that slides probably represent a correct conceptual model for this particular area (Hardie, 1955).

(b) Voll (1960, p. 562) expressed doubt about the slide model for the Dalradian rocks in the western Central Scottish Highlands. He pointed out that (1) "... the Iltay slide at Loch Creran or the Fort William slide in Allt Ionndrain have not produced any thinning or other deformation in the adjacent sediments ... this although miles of sediment are supposed to be missing at these contacts" and (2) S_0 in the supposed-nappe structures shows no evidence of distortion prior to development of a foliation (S_1) that passes uninterrupted across the "slides." Because this foliation is related to folds that flex the "nappe" boundaries, Voll concluded that the nappes must be premetamorphic. This evidence might have been considered detrimental to the slide model, but recent work (see Chapter 11) has shown that all the major nappe structures of the Dalradian and Moine Series either were premetamorphic or were associated with zeolite facies, or with very low greenschist facies, regional metamorphism; several workers have associated these structures with gravity slide tectonics. Thus, Voll's observations appear to embellish and amplify the slide model, rather than to contradict it. On the basis of new

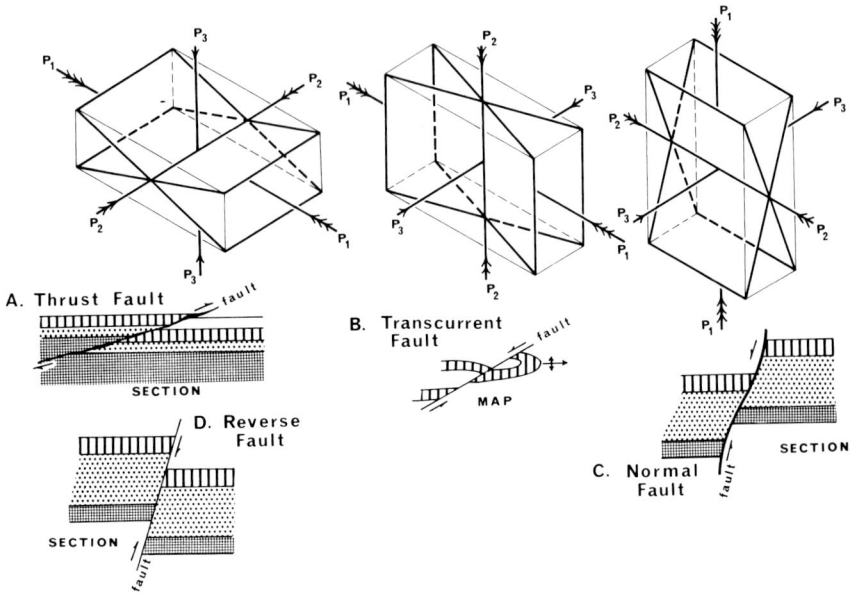

Figure 472. Possible types of fault according to the model proposed by Anderson (1951). The stress system is resolved about three orthogonal axes ($P_1 > P_2 > P_3$) and the fault movement occurs along planes between P_1 and P_3. The model includes thrust (A), transcurrent (B), and normal (C) faults, but not high-angle reverse faults (D).

work Voll (1965) recently suggested that these slides do not exist; this additional evidence will have to be evaluated against the existing tectonic models.

Anderson (1951) proposed one of the most comprehensive process-response models to embrace all types of faults; this model has been widely accepted in the geological literature (e.g., de Sitter, 1956B; Harland and Bayly, 1958). The model involved resolving the stress system about three orthogonal axes (Fig. 472). According to this simple and attractive model, one axis is always approximately vertical and three varieties of faults can occur: (a) normal, (b) thrust, and (c) transcurrent, according to whether the vertical stress of gravity is maximum, minimum, or intermediate, respectively, in relation to the two subhorizontal components of the stress system. Several more detailed analyses have amplified Anderson's original model. Some of the more notable include (a) the series of papers dealing with the role of fluid pressure in the mechanics of thrust faulting (Rubey and Hubbert, 1959; Birch, 1961; Hubbert and Rubey, 1961A, 1961B; Moore, 1961; Raleigh and Griggs,

1963), (b) the work reported in a Canadian symposium on the mechanics of faulting with special reference to fault-plane work (Hodgson, 1959), and (c) the detailed study of varied displacements associated with the San Andreas transcurrent fault system that spans several hundred kilometers of western California, U.S.A. (Crowell, 1962).

Anderson's fault model did not include reverse faults (Fig. 472). According to his model, reverse faults cannot occur because, with horizontal compression, the component of force parallel to a potential fault plane steeper than 45° is insufficient to promote movement. Anderson assumed that apparent reverse faults seen in small outcrops are really sinuous sections of normal faults (Fig. 472) or local non-vertical sections of transcurrent fault planes. Folding subsequent to faulting can alter the apparent geometry of some normal faults so that they simulate reverse faults (e.g., van Bemmelen, 1960). Only a small proportion of the numerous mapped reverse faults can be accounted for by such special circumstances. Diapiric forces can supply large upward forces capable or producing reverse faults (e.g., Beloussov, 1961), but it is not intended to enter further into the genesis of particular types of faults. The point of interest is that field observations clearly show that Anderson's model is incomplete; therefore, it requires revision and extension.

RESPONSE CHARACTERISTICS AND DIRECTIONAL ATTRIBUTES FOR QUANTITATIVE ANALYSES

Harland (1956) and Harland and Bayly (1958) outlined conceptual frameworks that are not too dissimilar to process-response models. Harland (1956) considered the five variables **tectonic facies, style, orientation, sequence,** and **date.** He suggested that a proper description of tectonic facies includes at least three related attributes of **form, size,** and **composition.** Harland considered that the dissimilar sizes of the similar forms found in slump, glacial, salt, and nappe structures reflect different conditions of formation, and therefore different tectonic facies. Tectonic style was considered to be a product of tectonic facies and orientation, although Harland recognized that so many variables are compounded in these concepts that it is almost impossible to isolate, define, and account for each.

Harland and Bayly (1958) analyzed **tectonic regime,** that is, the nature and orientation of the bulk strain during tectonic deformation. Tectonic regime designates the movement pattern common to all fold

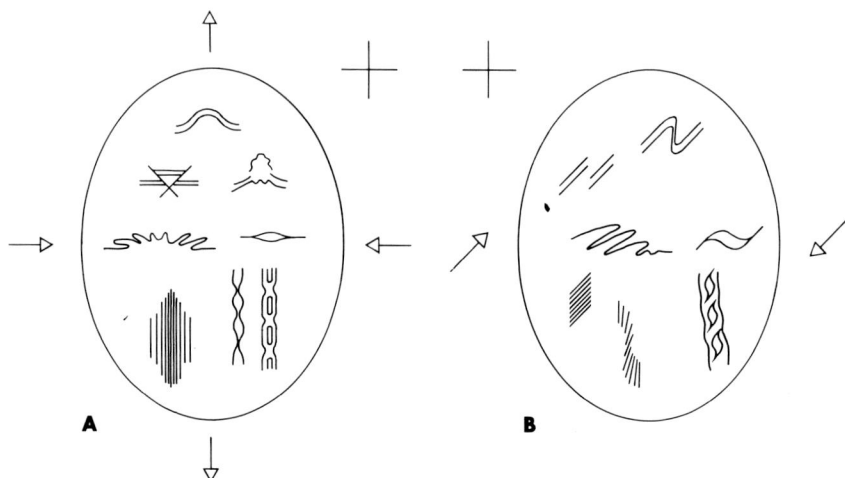

Figure 473. The variety of small structural elements that can be accommodated within the same bulk movement pattern. A: Symmetrical structures. B: Asymmetrical or predominantly shear structures (After Harland and Bayly, 1958, Fig. 3).

structures within a unit of deformed rocks. On the assumption that rocks are characteristically compositionally heterogeneous, Harland and Bayly showed that a single simple large-scale movement can generate a variety of structures; Figure 473 shows small structural elements, and, as a large example, an orogenic welt is shown diagrammatically in Figure 474.

Figure 474 illustrates a conceptual process-response model within which different regimes may be in operation at different small localities (considered at any one moment of time during the orogeny); at any one place, a series of regimes may have operated sequentially (Harland and Bayly, 1958). Harland (1956) thought that the marked similarities of style that commonly occur in different orogenic sequences, despite the apparently-endless series of variables involved, suggests that correlation of crustal movement patterns with specific tectonic styles may eventually be possible.

Studies of the three-dimensional variability of structural fabric elements involve practical difficulties in connection with the definition of populations of objects and populations of attributes. For example, folds developed during two discrete phases of folding may possess very similar styles, and thus be difficult to differentiate in the field. In other words, it is sometimes most difficult to segregate the members of two populations of objects that are mixed together in one area. Before evaluating a set

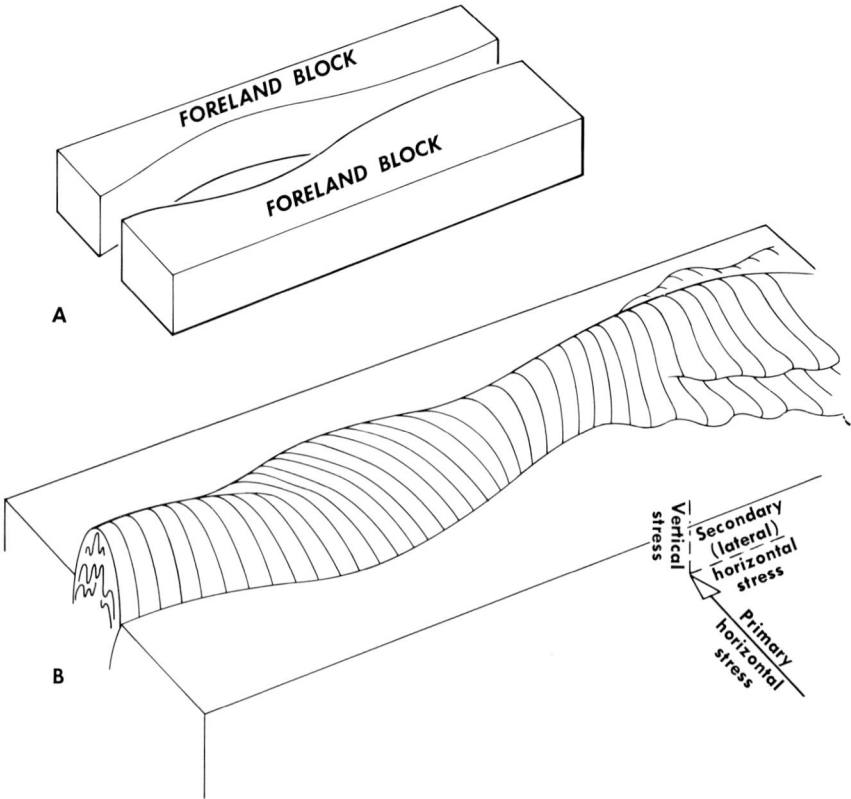

Figure 474. Diagrammatic sketch of a conceptual model of an orogenic welt. A: The "jaws" of the irregular foreland blocks shown without the welt. B: The deformed welt showing the major structural elements (After Harland and Bayly, 1958, Fig. 7).

of structural observations, however, it is important to make sure that they are drawn from a single population. Agterberg (1961) suggested that if the attributes are plotted on an equal-area projection, it is only when a single maximum appears that the objects represent a single population of objects. This is a necessary but not a sufficient condition to establish one population. Superposed discrete fabrics may involve subparallel sets of elements that might not be segregated by plotting on an equal-area projection (cf. D'Amico, 1964).

If the principles outlined in Chapter 3 are followed, precise operational definitions can be erected for each directional attribute. Bearing in mind the size of the objects to be studied, the populations of objects can be defined in relation to the target populations of interest. Commonly only two or three attributes have been measured in the past. However, in planning a field program, it is recommended that a complete list be

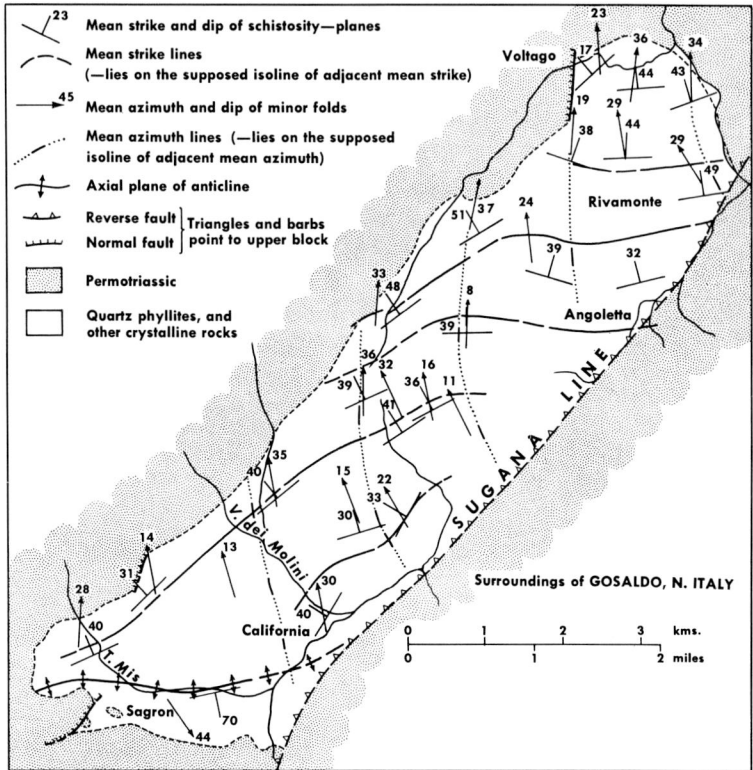

Figure 475. Structure map of the area near Gosaldo, Dolomites, Italy (After Agterberg, 1961, Map Sheet VII).

prepared of all attributes that could be measured for the population of objects. As field and laboratory work proceeds, additional attributes will probably be recognized, and the initial list can be augmented. At present little is known about the nature and variability of the available directional attributes. Probably some attributes will prove to be strongly correlated, so that measurement of them all would give rise to data redundancy. Also, some attributes will undoubtedly be more useful than others for discriminating between rival conceptual process-response models. This is a current area of active research; in many fields of geology it is clear that attributes that have been traditional subjects of enquiry are not necessarily those most suitable for quantitative analyses. Only by study of a large variety of attributes will it be possible to discern which are the most significant.

Agterberg (1961) made a useful contribution to quantitative structural geology when he prepared isopleth maps for (a) mean strike of schistosity, (b) mean trend of minor folds, and (c) mean plunge of minor folds. His maps[3] (e.g., Figs. 475 and 476) are of the crystalline basement rocks in the Dolomite area, northern Italy, where the tectonites were affected by two Hercynian and one Alpine phase of folding. Construction of these free-hand contour maps involved determination of the mean directional attributes for each small subarea. The mean value was plotted at the map location defined by the mean map coordinates for the individual sample stations (correct to 50 meters), in order to show (a) the geographic position of the mean attribute values and (b) the size of the sampled subareas.

In this work Agterberg subdivided each major region into small subareas within which the fabric elements are empirically found to be scattered around a distinct mean[4]; then, for each attribute within each subarea, he determined the mean value, thus:

$$\overline{X} = \sum_{i=1}^{N} x_i \Big/ N \qquad \text{(v)}$$

and the sample dispersion, thus

$$S = \sqrt{\sum_{i=1}^{N} (\overline{X} - x_i)^2 \Big/ N} \qquad \text{(vi)},$$

where x_i is a single observed value of an attribute, and N observations were made. Care must be exercised in using equations (v) and (vi) when directional attributes are involved, because two angular measurements are required to define a linear structure; for example, a fold axis has a trend and a plunge, both of which must be known before the fold axis is adequately defined. The mean of a series of fold axes, for example, involves both of these variables, and is best considered in terms of direction cosines (see below).

Because rocks like quartzites tend to be better exposed than, say phyllites, more measurements tend to arise from the former; as a result, sample variances computed for observations combined from all rock types may not be representative of the target population defined for all rocks within a domain.

[3]At this point concern is with useful methodology, but some of the attribute definitions used by Agterberg (1961, 1964A) were probably wrong, so that some of his maps seem to require revision (D'Amico, 1964).

[4]Agterberg actually isolated domains within which attributes are homogeneous; he apparently assumed that all attributes are homogeneous within each domain, but from Chapter 3 it is clear that some attributes are heterogeneous within a domain defined in this manner.

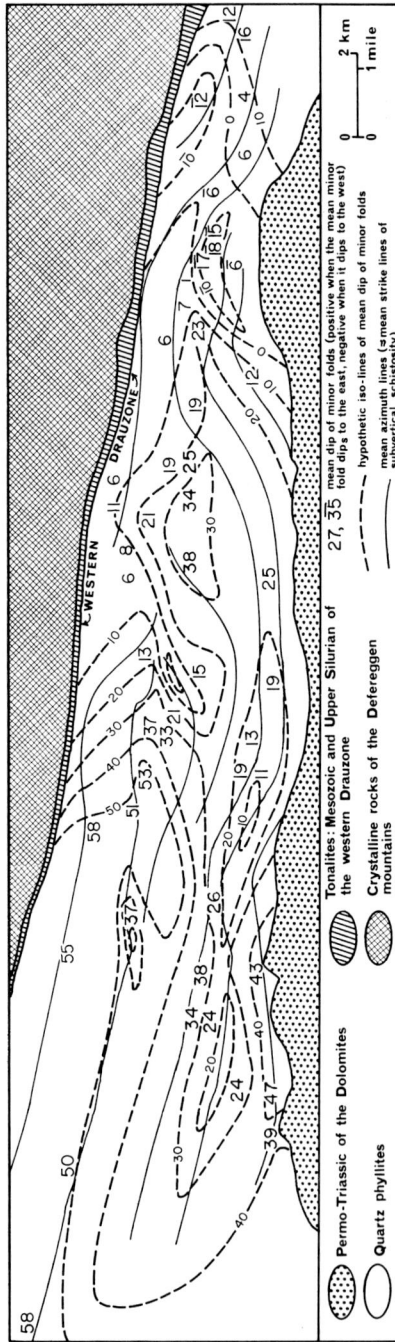

Figure 476. Structure map of the quartz phyllites between Brunico and Prato alla Drava, Dolomites, N Italy (After Agterberg, 1961, Plate IIIa).

OPERATIONS INVOLVING THE USE OF DIRECTION COSINES

Although Agterberg (1961),[5] Agterberg and Briggs (1963), and Potter and Pettijohn (1963) all made the assumption that the arithmetic mean values of directional properties are adequate provided that the range of azimuth variation does not exceed 104°, direction cosines have much more general utility in quantitative analyses. The nature of direction cosines can be illustrated by means of an example.

A fold axis can be considered as a unit vector, that is, as having both direction in three-dimensional space and also unit magnitude. Such a vector can be uniquely represented by three components parallel to the orthogonal reference axes, which for convenience, can be the south (U), east (V), and vertical (W) directions (Fig. 477). If the angles between the

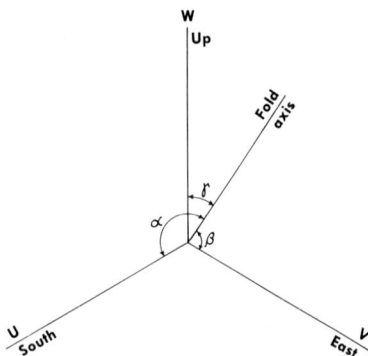

Figure 477. A fold axis referred to orthogonal U, V, W-axes.

reference axes (U, V, and W) and a fold axis are α, β, and γ, the three components of the fold axis are $1 \cdot \cos \alpha$, $1 \cdot \cos \beta$, and $1 \cdot \cos \gamma$, respectively; each cosine is multiplied by unity because the fold axis is considered to be a unit vector. These components are the direction cosines. Now, if a series of N homogeneous fold axes (unit vectors) has been measured within a region, each can be expressed in terms of direction cosines u_i, v_i, and w_i with respect to U, V, and W, respectively. The best estimate of the true mean orientation of a set of fold axes is given by $\bar{u}$, $\bar{v}$, and $\bar{w}$, the arithmetic mean of the individual cosines (Fisher, 1953), thus:

$$\bar{u} = \sum_{i=1}^{N} u_i \bigg/ N, \qquad \bar{v} = \sum_{i=1}^{N} v_i \bigg/ N, \text{ and} \qquad \bar{w} = \sum_{i=1}^{N} w_i \bigg/ N \quad \text{(vii)}$$

[5]Agterberg (1961) gave a geometric proof that the trend and plunge of the mean fabric element are sufficiently approximated by the arithmetic mean trend (where the range of directions is less than 104°) and the arithmetic mean plunge of the individual elements.

Table 12. – Steps in Calculation of Mean Fold Axis (α', β', γ') for Figure 478 Using Direction Cosines

Fold Axis	Trend	Plunge	α^*	β^*	γ^*	cos α u_1	cos β v_1	cos γ w_1
1	020	20	152	71.5	110	-0.8829	0.3173	-0.3420
2	030	20	144.5	62	110	-0.8141	0.4695	-0.3420
3	030	30	138	64	120	-0.7431	0.4384	-0.5000
4	035	30	135.5	61	120	-0.7133	0.4848	-0.5000
5	050	15	129	42.5	105	-0.6293	0.7373	-0.2588
6	100	20	81	22	110	0.1564	0.9272	-0.3420
						$\Sigma u_1 = -3.6263$	$\Sigma v_1 = 3.3745$	$\Sigma w_1 = -2.2848$
						$\bar{u} = -0.6044$	$\bar{v} = 0.5624$	$\bar{w} = -0.3808$

$$\sqrt{(\bar{u}^2 + \bar{v}^2 + \bar{w}^2)} = \sqrt{(0.3653 + 0.3163 + 0.1450)} = \sqrt{0.8266} = 0.9091$$

cos $\alpha' = -0.6044/0.9091 = -0.6648$, cos $\beta' = 0.5624/0.9091 = 0.6186$, cos $\gamma' = -0.3808/0.9091 = -0.4189$

so mean fold axis given by (α', β', γ') = (131°40', 51°47', 114°46')†

*can be calculated trigonometrically, or, alternatively, can be read from stereogram (Fig. 478).

†plotted as ▲ on Fig. 478.

N.B. The most suitable method for determining the statistical significance of an estimate of a vector mean, and some of the underlying sampling problems, were described by Fisher (1953), Watson and Irving (1957), and Watson (1966).

By simple trigonometry, the angles α', β', and γ' between the reference axes U, V, and W and the mean orientation of the fold axes (the vector represented by $\bar{u}$, $\bar{v}$, and $\bar{w}$) are given by:

$$\cos \alpha' = \frac{\bar{u}}{\sqrt{(\bar{u}^2 + \bar{v}^2 + \bar{w}^2)}}, \qquad \cos \beta' = \frac{\bar{v}}{\sqrt{(\bar{u}^2 + \bar{v}^2 + \bar{w}^2)}}, \text{ and}$$

$$\cos \gamma' = \frac{\bar{w}}{\sqrt{(\bar{u}^2 + \bar{v}^2 + \bar{w}^2)}} \qquad \text{(viii)}$$

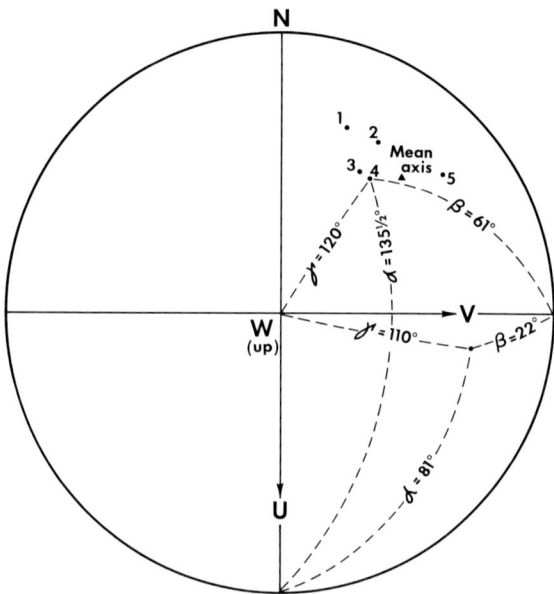

Figure 478. Stereogram showing mean fold axis based on direction cosines of individual fold axes. See text and Table 8 for full explanation.

The steps involved in an actual calculation are shown in Table 12 and Figure 478.

Loudon (1963, 1964) showed that the standard methods of matrix algebra can be applied to fabric elements such as S-surfaces and fold axes if they are expressed in terms of direction cosines. These techniques represent a most exciting advance in structural geology. The use of matrix algebra means that digital computers can be utilized to handle large volumes of structural measurements, and Loudon (1964) developed a FORTRAN program to handle most of the operations described below.

The observed field data have sample geographic locations defined by U,V,W-coordinates. Linear and planar structural elements can be

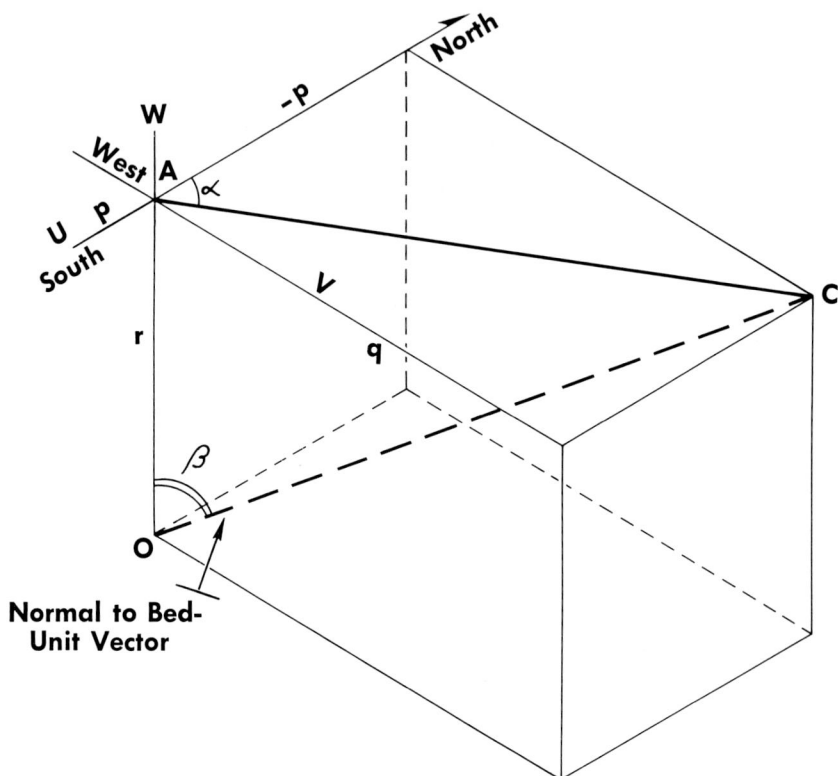

Figure 479. The direction cosines (p, q, r) of the normal to a bedding plane (S_0) that dips α degrees east of north at β degrees. Since $\sin \beta = AC/1$, or $AC = \sin \beta$, $\cos \alpha = -p/AC$ so $p = -\cos \alpha \cdot \sin \beta$
$\sin \alpha = q/AC$ so $q = \sin \alpha \cdot \sin \beta$
and $r = \cos \beta$.

expressed as vectors represented by direction cosines related to the same orthogonal U,V,W-axes; linear structures are represented by a unit vector along the element, and planar structures by a unit vector normal to the plane.

In the following discussion it is assumed that a set of poles to S_0-planes has been measured in a small area. Suppose that the direction of dip and the angle of dip of a S_0-plane are α degrees east of north and β degrees, respectively (Fig. 479); then the direction cosines p, q, and r of the normal to S_0 (taken in the way-up direction of S_0, so that $\beta > 90°$ for inverted beds) are shown by elementary trigonometry to be:

$$p = -\cos \alpha \cdot \sin \beta, \qquad q = \sin \alpha \cdot \sin \beta, \qquad \text{and } r = \cos \beta. \quad \text{(ix)}$$

When the orientations of S_0-planes have been measured for N samples the individual direction cosines can be built into a matrix of the following form (Loudon, 1964):

$$\begin{bmatrix} \dfrac{\Sigma\ p^2}{N} & \dfrac{\Sigma\ pq}{N} & \dfrac{\Sigma\ pr}{N} \\[2mm] \dfrac{\Sigma\ qp}{N} & \dfrac{\Sigma\ q^2}{N} & \dfrac{\Sigma\ qr}{N} \\[2mm] \dfrac{\Sigma\ rp}{N} & \dfrac{\Sigma\ rq}{N} & \dfrac{\Sigma\ r^2}{N} \end{bmatrix} = \mathbf{A} \qquad \text{(x)}$$

Matrix $\mathbf{A}$ can be considered a dispersion matrix which records the amount of folding, or the variability in orientation, of the normals to S_0 in different directions. The initial observed data are referred to geographical U, V, and W coordinate axes, but it is more convenient for subsequent analyses if their variability is described in terms of the symmetry or geometric axes of the folds. Such axes can be found geometrically or algebraically by techniques similar to those used in factor analysis for finding the **principal axes** (Harman, 1960; Kendall, 1961).

EIGENVALUES AND EIGENVECTORS

The principal axes are such that, if a set of normals to S_0 are referred to them, and a dispersion matrix $\mathbf{A}$ is prepared, all the cross-product terms (i.e., those involving pq, pr, and qr) are zero. Let the dispersion matrix $\mathbf{A}$, when referred to the principal axes, be

$$\begin{bmatrix} d_{11} & 0 & 0 \\ 0 & d_{22} & 0 \\ 0 & 0 & d_{33} \end{bmatrix} = \mathbf{D} \qquad \text{(xi)}$$

The elements of this diagonal matrix $\mathbf{D}$ are the **eigenvalues,** which are the second moments of the distribution of normals to S_0 referred to the principal axes. Let

$\mathbf{R}^\mathrm{T}$ be the matrix $\begin{bmatrix} b_{11} & b_{12} & b_{13} \\ b_{21} & b_{22} & b_{23} \\ b_{31} & b_{32} & b_{33} \end{bmatrix}$ where $\mathbf{R}^\mathrm{T}$ is the transpose of matrix $\mathbf{R}$,

such that $\qquad \mathbf{R\,A\,R}^\mathrm{T} = \mathbf{D}$ \qquad\qquad\qquad\qquad (xii)

Then $\mathbf{R}^\mathrm{T}$ is the matrix of **eigenvectors** of the matrix $\mathbf{A}$.

Loudon (1964) drew attention to the fact that the eigenvalues and eigenvectors express important characteristics about the distribution of normal to S_0. It is convenient to arrange for $d_{11} \geqslant d_{22} \geqslant d_{33}$ in $\mathbf{D}$ (equation xi). Then:

(1) $\mathbf{R}^\mathrm{T}$ gives the eigenvectors of $\mathbf{A}$; the columns of $\mathbf{R}^\mathrm{T}$ are the direction cosines of the principal axes, thus:

(a) $b_{11}\ b_{21}\ b_{31}$ are direction cosines of the A_I-axis, which is the mean orientation of the normals to S_0;[6] the corresponding eigenvalue d_{11} is a measure of the amount of folding (i.e., of the variability of the poles to S_0) parallel to the A_I-axis.
(b) $b_{13}\ b_{23}\ b_{33}$ are the direction cosines of the B-geometric fold axis,[7] the A_III-axis, and the smallest eigenvalue d_{33} is a measure of the folding parallel to the B-axis.
(c) $b_{12}\ b_{22}\ b_{32}$ are the direction cosines of the A_II-axis, which is perpendicular to the B-axis and to the vector mean; the eigenvalue d_{22} is a measure of the tightness of folding in the plane perpendicular to B (specifically, parallel to A_II).

(2) The matrix of eigenvectors $\mathbf{R}^\mathrm{T}$ can be used as a rotation matrix to transform direction cosines of vectors and geographical U,V,W-coordinates referred to the old axes into the framework of the new principal axes. This permits profiles of the folds to be studied in the three principal planes, and it also transforms the data into a convenient form for computing the higher statistical moments (see below).

(a) For any particular S_0, represented by its normal with direction cosines p, q, and r, the direction cosines referred to the new principal axes are p', q', and r', where:

$$[p\ q\ r]\ \mathbf{R}^\mathrm{T} = [p'\ q'\ r'] \tag{xiii}$$

(b) The U,V,W-coordinates of an individual sample location become U',V',W'-coordinates when referred to the new principal axes, where:

$$[U\ V\ W]\ \mathbf{R}^T = [U'\ V'\ W'] \tag{xiv}$$

In these operations it is naturally assumed that the original matrix $\mathbf{A}$ is based on either a random or a grid sample of S_0 observations of adequate size. As the actual observations depart from such a sample, less significance attaches to the resulting conclusions.

OBTAINING THE EIGENVECTORS
AND EIGENVALUES

There are several methods of determining the eigenvectors and eigenvalues of the matrix $\mathbf{A}$; although more direct methods are available, the iterative Jacobi method of successive approximation (cf. Harman,

---

[6]As folds approach isoclinal form, the vector mean corresponds to d_{22} instead of d_{11} and the eigenvectors are $b_{12}\ b_{22}\ b_{32}$.

[7]When an appropriate sample has been collected A_III is the B-geometric fold axis (Fig. 482), but several situations can arise in which the available sample defines a principal axis not quite coincident with B (e.g., data from one part of the fold only).

1960, p. 179; Clenshaw, *et al.,* 1961, p. 30) is used here because it is easily handled by digital computer (Loudon, 1964). In the geometric analog of the Jacobi method, the matrix **D** is obtained from **A** by successive rotations of **A** about each of the original axes until all the cross-product terms are eliminated.

A rotation of θ degrees around the vertical axis is achieved if **A** is premultiplied by matrix **T** and postmultiplied by $\mathbf{T}^\mathrm{T}$ (the transpose of **T**), thus $\mathbf{TAT}^\mathrm{T}$, where

$$\mathbf{T} = \begin{bmatrix} \cos\theta & \sin\theta & 0 \\ -\sin\theta & \cos\theta & 0 \\ 0 & 0 & 1 \end{bmatrix}, \text{ so that } \mathbf{T}^\mathrm{T} = \begin{bmatrix} \cos\theta & -\sin\theta & 0 \\ \sin\theta & \cos\theta & 0 \\ 0 & 0 & 1 \end{bmatrix}$$

Rotation is effected about the vertical axis because 1 is in the a_{33} position of the **T** matrix, and the cross-product term eliminated is that between the other two axes (i.e., the a_{12} term). To reduce a_{12} to zero the value of θ given by the equation

$$\tan 2\theta = \frac{2a_{12}}{(a_{11} - a_{22})} \tag{xv}$$

is used.

To rotate about the north-south and the east-west axes, the orthogonal matrices

$$\begin{bmatrix} 1 & 0 & 0 \\ 0 & \cos\theta & \sin\theta \\ 0 & -\sin\theta & \cos\theta \end{bmatrix} \text{ and } \begin{bmatrix} \cos\theta & 0 & \sin\theta \\ 0 & 1 & 0 \\ -\sin\theta & 0 & \cos\theta \end{bmatrix}$$

are used, respectively, instead of **T**, and values of θ are obtained from the following equation by rotating subscripts and using values (a_{ij}, etc.) from the matrix produced by the last previous multiplication. The largest off-diagonal element, a_{ij}, of a 3×3 matrix becomes zero if θ is given by the equation

$$\tan 2\theta = \frac{2a_{ij}}{(a_{ii} - a_{jj})} \tag{xvi}$$

where i and j $(i \neq j)$ are 1, 2, and 3; this applies to **A**, $\mathbf{TAT}^\mathrm{T}$, or the matrices resulting from subsequent rotations. Thus, a series of matrices similar to **T** is chosen and used successively to premultiply by **T** and to postmultiply by $\mathbf{T}^\mathrm{T}$; each **T** matrix is chosen to make zero the largest pair of off-diagonal elements in the matrix remaining at that stage. Successive rotations make the cross-product terms smaller and smaller, and ultimately a close approach to a diagonal matrix is obtained; thus, after k multiplications

$$\mathbf{T}_k\mathbf{T}_{k-1}\ldots\mathbf{T}_1\mathbf{A}\mathbf{T}_1^T\mathbf{T}_2^T\ldots\mathbf{T}_k^T = \mathbf{D} \qquad \text{(xviiA)}$$

which can be rewritten as

$$\mathbf{R}\mathbf{A}\mathbf{R}^T = \mathbf{D} \qquad \text{(xviiB)}$$

which provides the matrices $\mathbf{R}$, $\mathbf{R}^T$, and $\mathbf{D}$ for equation (xii) and thus the required eigenvectors and eigenvalues.

Loudon (1964) drew attention to the fact that, in addition to rotation, the geometric operations of stretching and flattening are easily accomplished by matrix multiplication as was described by, for example, Sawyer (1955). The geometric effects of a specified strain upon the fabric elements can be approximated by these methods.

OBTAINING THE FOLD AXIS OF
CYLINDROIDAL AND CONICAL FOLDS

The operations described above can be used for any set of poles to S-structures. It is possible to use these methods to find the fold axis and to recognize which type of fold geometry (cylindroidal or conical folds, domes, etc.) is present.

Loudon (1964) showed how to differentiate cylindroidal and conical folds. Suppose that the normal to the ith S_0-plane can be expressed by the direction cosines p_i, q_i, and r_i related to the original U,V,W-axes, or as p_i', q_i', and r_i' related to the new principal axes. Now suppose that the normal to each S_0-plane, which can be considered a unit vector, is plotted on the surface of a sphere (as if for a preliminary step in stereographic projection). The points plotted on the sphere lie on a great circle if S_0 is cylindroidally folded, and on a small circle if conically folded (Fig. 480). With the axes origin at the center of the sphere, the ith vector (the normal to the ith S_0) has direction cosines p_i, q_i, and r_i. The normal to the circle plane is the fold axis of the S_0-planes; let its direction cosines be x, y, and z. The circle plane is then given by the equation (see, for example, Cohn, 1961, p. 22):

$$p_i x + q_i y + r_i z = \cos\theta \qquad \text{(xviii)}$$

but $\cos\theta = k$, the distance of the plane from the U,V,W-origin (Fig. 481). For a cylindroidal fold, which gives a great circle, $k = 0$, but for a conical fold $\theta < 90°$ and $k > 0$; from Figure 481 it is seen that the apical angle of the cone of S_0-planes is $(180 - 2\theta)$ degrees.

When numerous orientations of S_0 have been measured, equation (xviii) can be solved by the standard method of least squares as follows. By rewriting equation (xviii)

$$p_i = \frac{k}{x} - \frac{y}{x}q_i - \frac{z}{x}r_i \qquad \text{(xix)}$$

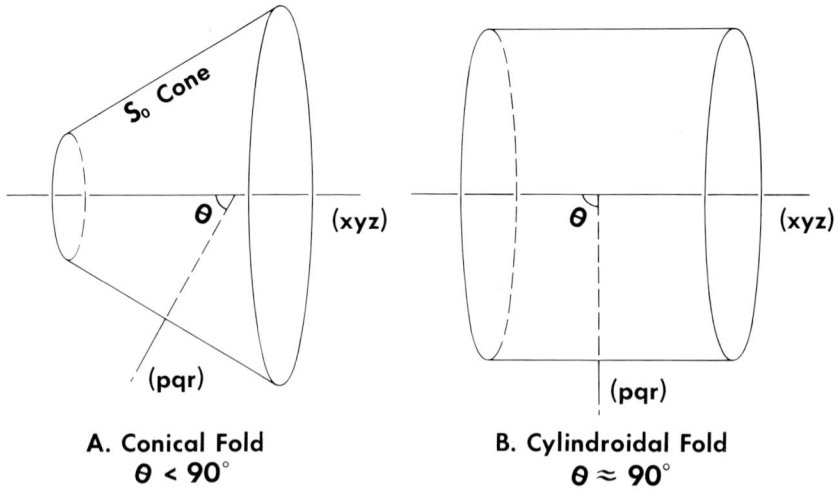

A. Conical Fold
$\theta < 90°$

B. Cylindroidal Fold
$\theta \approx 90°$

Figure 480. Diagrammatic conical and cylindroidal folds. $[p, q, r]$ are the direction cosines of a normal to S_0 which is inclined at θ degrees to the fold axis $[x, y, z]$.

in which p_i can be considered an observed value and the right-hand side of the equation a calculated value. Then the sum of squares of the differences between the observed p_i and its calculated value is

$$\sum_{i=1}^{N} (p_{i_{\text{observed}}} - p_{i_{\text{calculated}}})^2 = \sum_{i=1}^{N} \left(p_i - \frac{k}{x} + \frac{y}{x}q_i + \frac{z}{x}r_i\right)^2 \qquad (\text{xx})$$

To obtain the least squares solution the sum of squares of the differences must be minimized. It can be shown (e.g., Hoel, 1947, p. 90) that these

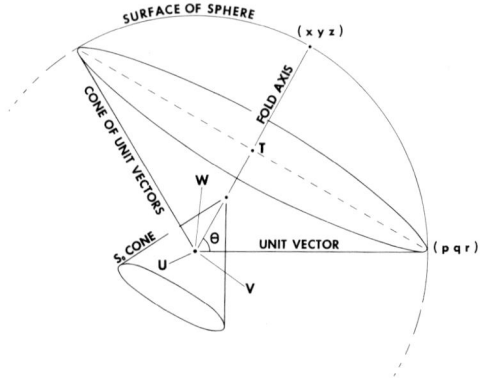

Figure 481. A conical fold (S_0 cone) related to the orthogonal axes U,V,W. Normals to S_0 are considered unit vectors with direction cosines p_i, q_i, and r_i. The fold axis to the S_0 cone has direction cosines x, y, and z, and it intersects the circle plane defined by the unit vectors at T.

sums of squares are a function of $\frac{k}{x}, \frac{y}{x}$, and $\frac{z}{x}$ only, so that equation (xx)

can be expressed as $F\left(\frac{k}{x}, \frac{y}{x}, \frac{z}{x}\right)$; to minimize this function, it is necessary that

$$\frac{\partial F}{\partial \frac{k}{x}} = \frac{\partial F}{\partial \frac{y}{x}} = \frac{\partial F}{\partial \frac{z}{x}} = 0.$$

These partial derivatives yield three normal equations, thus:

$$
\left.
\begin{aligned}
\frac{\partial F}{\partial \frac{k}{x}} &= \sum_{i=1}^{N} 2\left(p_i - \frac{k}{x} + \frac{y}{x} q_i + \frac{z}{x} r_i\right) \cdot (-1) = 0 \\[2mm]
\frac{\partial F}{\partial \frac{y}{x}} &= \sum_{i=1}^{N} 2\left(p_i - \frac{k}{x} + \frac{y}{x} q_i + \frac{z}{x} r_i\right) \cdot (q_i) = 0 \\[2mm]
\frac{\partial F}{\partial \frac{z}{x}} &= \sum_{i=1}^{N} 2\left(p_i - \frac{k}{x} + \frac{y}{x} q_i + \frac{z}{x} r_i\right) \cdot (r_i) = 0
\end{aligned}
\right\}
\qquad \text{(xxi)}
$$

These equations can be simplified as:

$$
\begin{bmatrix}
N & \Sigma q_i & \Sigma r_i \\
\Sigma q_i & \Sigma q_i^2 & \Sigma q_i r_i \\
\Sigma r_i & \Sigma q_i r_i & \Sigma r_i^2
\end{bmatrix}
\begin{bmatrix}
\frac{k}{x} \\
-\frac{y}{x} \\
\frac{z}{x}
\end{bmatrix}
=
\begin{bmatrix}
\Sigma p_i \\
\Sigma p_i q_i \\
\Sigma p_i r_i
\end{bmatrix}
\qquad \text{(xxii)}
$$

Since the terms involving p, q, and r are known, equation (xxii) can be solved by multiplying both sides by the inverse of the 3×3 matrix; if the 3×3 matrix is called $\mathbf{Q}$, the solution is:

$$
\begin{bmatrix}
\frac{k}{x} \\
-\frac{y}{x} \\
-\frac{z}{x}
\end{bmatrix}
= \mathbf{Q}^{-1} \cdot
\begin{bmatrix}
\Sigma p_i \\
\Sigma p_i q_i \\
\Sigma p_i r_i
\end{bmatrix}
\qquad \text{(xxiii)}
$$

It also follows from the Pythagoras theorem that

$$x^2 + y^2 + z^2 = \qquad \text{(xxiv)}$$

Equations (xxiii) and (xxiv) provide solutions for the four unknowns, x, y, z, and k; x, y, and z are the direction cosines of the fold axis. If k is zero, the folds are cylindroidal, but when $k > 0$, the folds are conical and since $k = \cos\theta$ the apical angle of the cone $(180 - 2\theta)$ is given.

QUANTITATIVE DESCRIPTION OF FOLD GEOMETRY

Objective numerical descriptors have not been developed for the fold attributes used by Harland (1956). The same is true for attributes used by Turner and Weiss (1963, pp. 110-12) for the classification of folds; their classification was based on the:

(a) geometric relationship between the fold axis or the hinge line and the axial surface;
(b) style, or general form, as seen in the profile;
(c) orientation or attitude of the linear and planar fabric elements with respect to the geographic coordinates; and
(d) symmetry of the fabric elements (without regard to the orientation in space of the structure).

While (a) and (c) are susceptible to objective measurement, until now it has only been possible to express (b) qualitatively; (d) is not entirely independent of (a), (b), and (c).

The necessary and sufficient attributes required to describe the geometry of a fold include:

(a) orientation of fold axis
(b) orientation of axial surface
(c) shape of the fold (particularly of the profile)
(d) size of the fold

Both (a) and (b) can be quantitatively described by two angular measurements each, and they can be expressed as scalars by using direction cosines; (c) is commonly qualitatively and subjectively described, but (d) can be precisely measured (although commonly described in terms of an ordinal scale). The lack of a quantitative method for describing all necessary fold attributes has been a serious handicap. However, Loudon (1963, 1964) showed how fold shapes can be described by scalars, and he erected operational definitions for six attributes which he termed **attitude, tightness, asymmetry, shape, skewness,** and **kurtosis.**

It was mentioned that the matrix of eigenvectors $\mathbf{R}^T$ is important for rotating S-structures into the framework of the principal axes when the statistical moments are to be computed. Suppose Figure 482 represents the profiles of two cylindroidal folds referred to principal axes (defined as above), and that at eleven equally-spaced points along A_{II} the in-

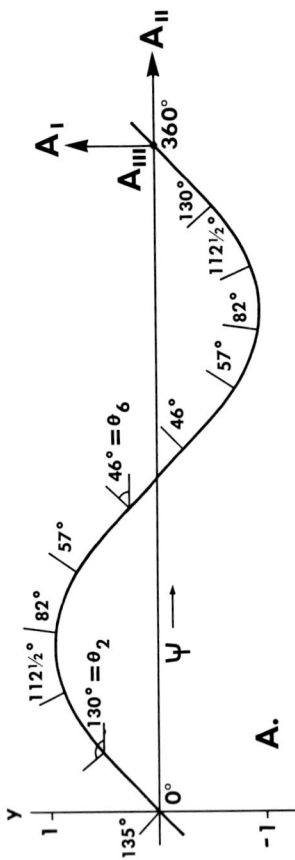

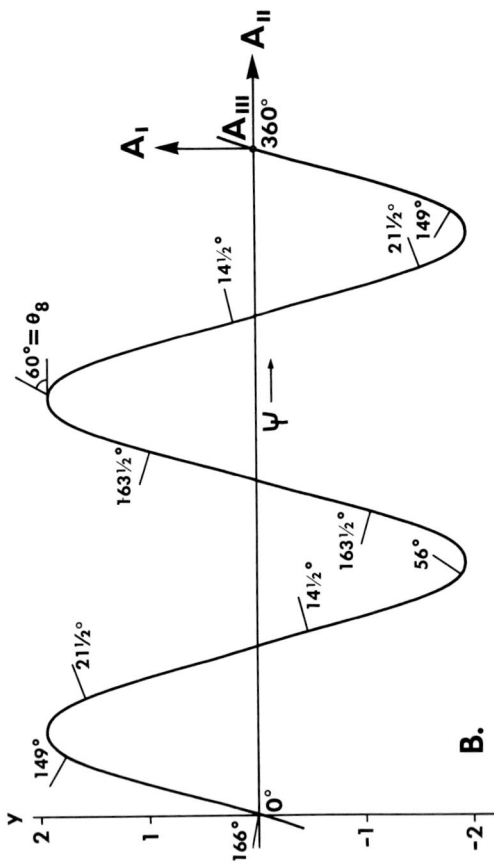

Figure 482. Profiles of two cylindroidal folds used to compute statistical moments with respect to their A_{II}-axes, as follows:

A: Fold is $y = \dfrac{180}{\pi} \sin \psi_i$, so $\dfrac{dy}{d\psi} = \cos \psi$ and $\theta_i = [90 + \tan^{-1}(\cos \psi_i)]$ degrees

ψ_i	0°	32°44'	65°28'	98°12'	130°56'	163°40'	196°24'	229°8'	261°52'	294°36'	327°20'	
θ_i	135°	130°4'	112°33'	81°53'	56°46'	46°11'	46°12'	56°48'	81°57'	112°36'	130°5'	
$\cos \theta_i$	−0.7071	−0.6437	−0.3835	0.1412	0.5481	0.6923	0.6921	0.5476	0.1400	−0.3843	−0.6439	$\dfrac{\Sigma \cos \theta}{N} = 0.0642 = m_1$
$\cos^2 \theta_i$	0.4999	0.4143	0.1471	0.0199	0.3004	0.4793	0.4790	0.2999	0.0196	0.1477	0.4146	$\dfrac{\Sigma \cos^2 \theta}{N} = 0.2929 = m_2$
$\cos^3 \theta_i$	−0.3535	−0.2667	−0.0564	0.0028	0.1647	0.3318	0.3315	0.1642	0.0027	−0.5675	−0.2669	$\dfrac{\Sigma \cos^3 \theta}{N} = -0.0002 = m_3$
$\cos^4 \theta_i$	0.2499	0.1716	0.0216	0.0004	0.0902	0.2297	0.2294	0.0899	0.0004	0.0218	0.1719	$\dfrac{\Sigma \cos^4 \theta}{N} = 0.1161 = m_4$
$\sin \theta_i$	0.7071	0.7653	0.9235	0.9900	0.8366	0.7216	0.7218	0.8368	0.9901	0.9232	0.7651	$\dfrac{\Sigma \sin \theta}{N} = 0.8346 = M_1$

Skewness $= m_3/m_2^{3/2} = 0.0000$
Kurtosis $= m_4/m_2^2 = 1.3533$

B: Fold is $\dfrac{y}{2} = \dfrac{180}{\pi} \sin 2\psi$, so $\dfrac{dy}{d\psi} = 4 \cos 2\psi$ and $\theta_i = [90 + \tan^{-1}(4 \cos 2\psi)]$ degrees

ψ_i	0°	32°44'	65°28'	98°12'	130°56'	163°40'	196°24'	229°8'	261°52'	294°36'	327°20'	
θ_i	165°58'	148°57'	20°54'	14°36'	56°20'	163°28'	163°26'	60°6'	14°36'	21°42'	149°5'	
$\cos \theta_i$	−0.9702	−0.8567	0.9342	0.9677	0.5543	−0.9586	−0.9585	0.4986	0.9677	0.9291	−0.8580	$\dfrac{\Sigma \cos \theta}{N} = 0.0227 = m_1$
$\cos^2 \theta_i$	0.9413	0.7339	0.8728	0.9365	0.3072	0.9189	0.9188	0.2486	0.9365	0.8632	0.7362	$\dfrac{\Sigma \cos^2 \theta}{N} = 0.7649 = m_2$
$\cos^3 \theta_i$	−0.9133	−0.6287	0.8153	0.9020	0.1713	−0.8808	−0.8806	0.1239	0.9062	0.8021	−0.6316	$\dfrac{\Sigma \cos^3 \theta}{N} = -0.0191 = m_3$
$\cos^4 \theta_i$	0.8861	0.5386	0.7618	0.8770	0.0944	0.8443	0.8442	0.0618	0.8770	0.7451	0.5420	$\dfrac{\Sigma \cos^4 \theta}{N} = 0.6429 = m_4$
$\sin \theta_i$	0.2425	0.5157	0.3567	0.2521	0.8323	0.2846	0.2851	0.8669	0.2521	0.3697	0.5138	$\dfrac{\Sigma \sin \theta}{N} = 0.4338 = M_1$

Skewness $= m_3/m_2^{3/2} = -0.0286$
Kurtosis $= m_4/m_2^2 = 1.0988$

clination of the normal to S (from the A_{II}-axis) is measured and expressed as a direction cosine. With N observations of θ_i (Fig. 482) the first four statistical moments are:[8]

$$m_1 = \frac{\sum\limits_{i=1}^{N} \cos \theta_i}{N}$$ A measure of the average **attitude** of S

$$m_2 = \frac{\sum\limits_{i=1}^{N} \cos^2 \theta_i}{N}$$ A measure of the **tightness** of the folds

$$m_3 = \frac{\sum\limits_{i=1}^{N} \cos^3 \theta_i}{N}$$ A measure of the amount and direction of **asymmetry** of the folds

$$m_4 = \frac{\sum\limits_{i=1}^{N} \cos^4 \theta_i}{N}$$ A measure of fold **shape** (i.e., whether closure is angular or rounded)

The tightness (m_2) expresses the variability of the S-structure about the A_{II}-principal axis, and this is given by the d_{22} eigenvalue, so that $m_2 = d_{22}$.

Loudon suggested that $M_1 = \dfrac{\sum\limits_{i=1}^{N} \sin \theta_i}{N}$ can be considered an alternative to the second moment (m_2); M_1 is a measure of the average **slope** of the limbs of a fold. He also suggested that **skewness** is a better measure of asymmetry than m_3; skewness is given by $\dfrac{m_3}{m_2^{3/2}}$, which is zero when the fold profile is perfectly symmetrical, positive when the steeper limb faces the positive end of the A_{II}-principal axis (from which θ_i is measured), and negative when the steeper limb faces the negative end of the A_{II}-axis (Fig. 483).

[8]For an actual fold these moments $m_1, m_2, \ldots$ can be calculated with respect to each of the three principal axes. In the following discussion moments with respect to the A_{II}-axis are considered, although, in an actual project, moments about the other axes may be significant. For example, curvature of the B-geometric fold axis would involve interest in the moments about the A_{III}-axis.

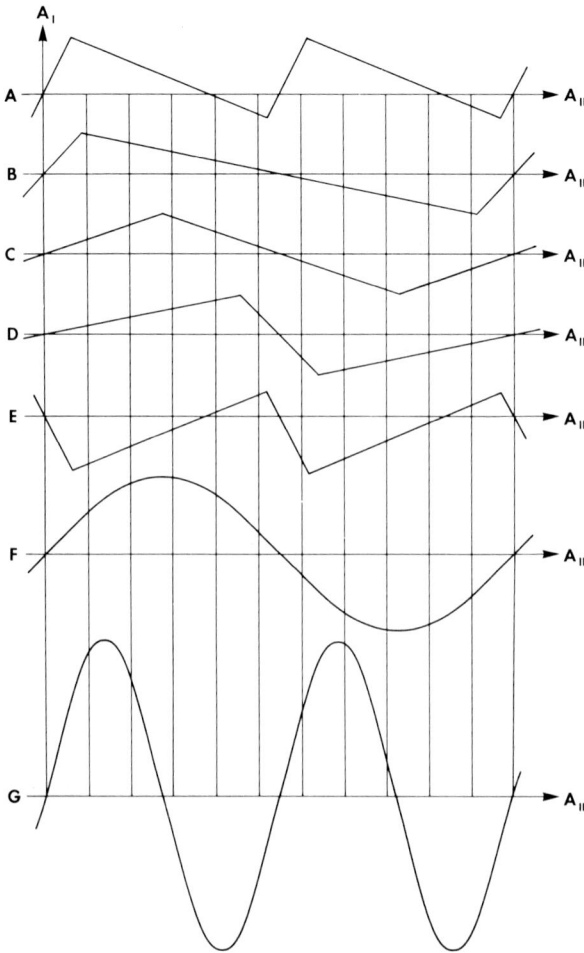

Figure 483. Fold profiles to illustrate some quantitative attributes used for their description. The inclination of the normals to each of the folded surfaces was measured (with respect to A_{II}) at each point intersected by the vertical lines; these measurements yield the following values:

Attribute	Attitude m_1	Tightness m_2	Asymmetry m_3	Shape m_4	Slope M_1	Skewness	Kurtosis
Fold A	0.1445	0.2592	−0.0856	0.1307	0.8412	−0.6449	1.9455
Fold B	0.1092	0.0785	−0.0258	0.0239	0.9566	−1.1727	3.8790
Fold C	0.0205	0.1003	0.0001	0.0101	0.9486	0.0026	1.0040
Fold D	−0.0275	0.1207	0.0586	0.0465	0.9317	1.3966	3.1893
Fold E	−0.1460	0.2566	0.0821	0.1266	0.8440	0.6315	1.9226
Fold F	0.0642	0.2929	−0.0002	0.1161	0.8346	0.0000	1.3533
Fold G	0.0227	0.7649	−0.0191	0.6429	0.4338	0.0286	1.0988

In addition to m_4, **kurtosis** is useful as a measure of fold shape. Kurtosis, $\frac{m_4}{m_2^2}$, is actually a measure of the frequency of particularly high values, so that if folds have straight limbs and angular closures the kurtosis is small, whereas with rounded crests the value is larger (Fig. 483).

All of these attributes based on the statistical moments are independent of fold size; hence, average values for each attribute can be based on measurements made for a number of folds. However, these quantitative attributes provide a basis for determining whether folds of different size have similar or dissimilar geometry; information of this type will be useful for determining whether observations for minor folds can be used for extrapolation to the geometry and style of major folds. Again, at the present time, essentially nothing is known about the statistical variance of fold attributes within regions of different size; these variables could form a basis for calculating variances. Such information would not only be of scientific interest; knowledge of the variance of fold attributes would also be highly significant in economic geology problems (e.g., petroleum and mineral exploration).

These attributes also clear the way for a start in comparing the characteristics of folds in different lithologies within the same region of folded rocks. Also, quantitative variables of this type can be used to study the three-dimensional variability of the folds in a specified lithology within an area (see below).

SOME NEW USES OF QUANTITATIVE ATTRIBUTES OF STRUCTURAL COMPLEXES

Quantitative measurements of the type advocated by Loudon (1963, 1964) have several immediate uses; several examples follow:

CORRELATION OF FOLD GEOMETRY WITH LITHOLOGY AND SIZE OF LITHIC UNITS

The dissimilar geometric behavior of different lithologies and different unit-thicknesses deformed within the same tectonic framework can be expressed quantitatively. This could lead to a correlation of tectonic environments by the recognition of similar fold structures, and would be analogous to the recognition of metamorphic facies on the basis of the assemblage of minerals present in a rock (cf. Fyfe, *et al.,* 1958).

REGIONAL VARIABILITY
OF FOLD GEOMETRY

The three-dimensional variability of fold attributes in each lithologic type can be described separately. In addition to the variables mentioned above, numerous additional numerical attributes can be developed and used in the study of the total three-dimensional variability of a structural complex. With such quantitative measurements to characterize fold structures and tectonic characteristics, it should be possible to map the areal and three-dimensional variability within orogenic provinces. In fact, such variables could be studied in a four-dimensional time-space framework if a sequence of structures within an orogenic belt can be elucidated by methods like those outlined in Chapter 12. Analyses of this type would lend a quantitative element to process-response models for orogenic belts, like those represented diagrammatically in Figures 436 and 464. When structural attributes can be expressed numerically, specific objective tests of elementary models can be designed. Two possibilities seem to be immediately applicable:

(a) To use contour methods to express areal and three-dimensional variability

Agterberg (1961) demonstrated that subjective free-hand contours can be drawn for average data for three directional variables measured in a series of subareas. Working with averages has certain disadvantages and tends to mask the inherent variance of the data. Although this technique represents a significant advance, it is well known that such maps tend to create a false impression of quantitative objectivity, whereas their construction is essentially subjective (cf. Chayes, 1961; Dawson and Whitten, 1962; Whitten, 1963A). The subjectivity can be demonstrated readily in the classroom by comparing maps drawn by different students for the same data. Unless the data points are very evenly distributed and are also very numerous, the maps are likely to be remarkably dissimilar. Bishop (1960, p. 45) discussed three methods of contouring maps — mechanical, equal spacing, and interpretative. With the last, without violation of the data, a geologist can readily infuse a contour map with subjectivity which reflects impressions gained in the field or any other preconceived ideas. Krumbein (1960C) suggested that this interpretative method is to be preferred for most maps because it permits inclusion of the "geologist's judgment of the 'grain' of the maps."

Again, contours of this type need not reflect real differences or gradients in the samples. Where the data have been collected according to a geographical grid, isopleths become more realistic, or significant, as the between-sample-unit (area) variability increases and the within-sample-unit (area) variability decreases (Whitten, 1957B); that is, if the range of values within individual grid-squares is large compared with the range of the mean values for each grid-square in the total area, the significance of the isopleths tends to be small.

A number of objective methods of analyzing nondirectional mapped data have been developed, and these can be used on a routine basis for the study of directional attributes expressed as direction cosines and their statistical moments.

Of several possible methods of map analysis, **trend surface analysis** is of particular value. The method has been used very extensively in sedimentary, stratigraphic, economic, and igneous geology. The advantages of trend surface maps over hand-drawn contour maps include complete objectivity and the fact that statistical confidence limits can be associated with each computed surface (cf. Mandelbaum, 1963; Krumbein, 1963B; Agterberg, 1964B). In addition, trend surface analysis permits stratification of the spatial variability into three components, namely (i) the regional trend or gradient which reflects the gross pattern of variability within the region, (ii) the local deviations from the regional trend which characteristically have geological significance, and (iii) local sporadic and chance "noise" that has little or no significance in the regional pattern of variability.

If quantitative measurements (e.g., skewness of folds) are made at numerous localities within a map-area, the sample locations can be defined by U,V-map coordinates (initially ignoring W); in a model, ordinates with heights proportional to the value of the measured attribute can be erected at each U,V-location. In this way points representing the dependent variable X (e.g., skewness) are located in three dimensions and the variability of X can be approximated by a mathematical surface, $X = f(U,V)$, as shown in Figure 484. If several attributes $X_1, X_2, X_3, \ldots X_n$ of the folded rocks have been measured, a family of surfaces representing each of these variables can be constructed.

Various types of function can be used to approximate the best mathematical surface, but polynomials such as

$$X_n = a_0 + a_1 U + a_2 V + a_3 U^2 + a_4 UV + a_5 V^2 + a_6 U^3 + a_7 U^2 V$$
$$+ a_8 UV^2 + \ldots \tag{xxv}$$

have been used extensively and successfully. To obtain the coefficients $(a_0, a_1, a_2, \ldots)$ of the surface most closely approximating the observed data, the conventional method of least squares is employed (Krumbein,

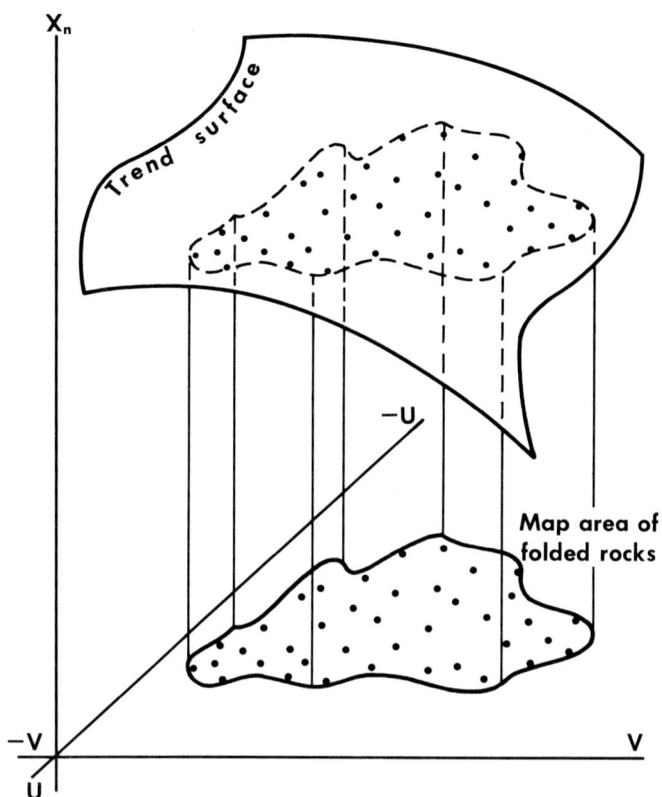

Figure 484. Diagram showing a hypothetical area of folded rocks projected onto a trend surface, $X_n = f(U,V)$. Dots on the outcrop area indicate sampled localities for which values of X_n were determined.

1959A; Whitten, 1963B). In practice, surfaces of successively higher degree are computed; the first-degree surface

$$X_n = b_0 + b_1 U + b_2 V \tag{xxvi},$$

then the second-degree surface

$$X_n = c_0 + c_1 U + c_2 V + c_3 U^2 + c_4 UV + c_5 V^2 \tag{xxvii},$$

and so on. Successive surfaces of higher degree account for larger proportions of the total sum of squares of the observed attribute (X_n). The proportion of the total sum of squares accounted for by a surface provides some measure of the "closeness" or "goodness" of fit to the observed attribute data.

Commonly, it is convenient to use the polynomial expression to compute contours on the trend surface within the whole map-area. Naturally, the distinction between these isopleths and those drawn manually for the observed data should be kept in mind clearly. The computed (trend surface) isopleths provide a useful method of assaying

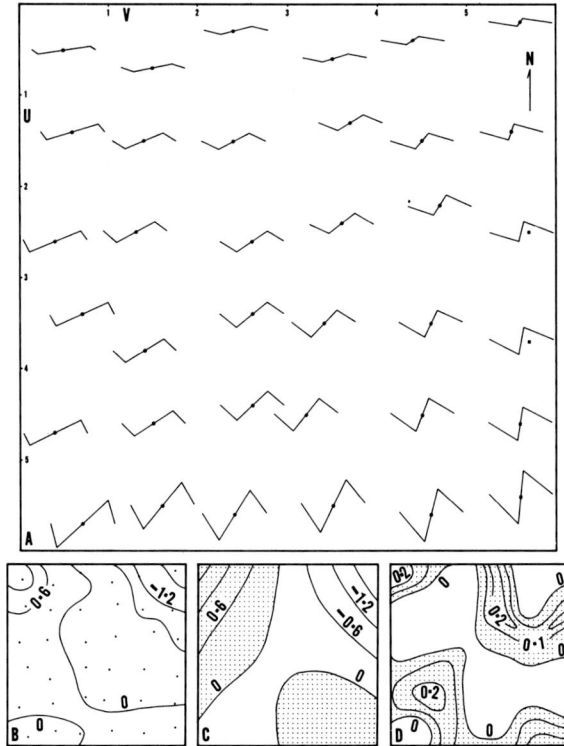

Figure 485. Hypothetical map showing the variation of fold profiles. A: Map showing fold profiles at their actual *U,V*-geographical locations (precise points located by dots); for convenience in this example the wavelength of each fold has been made the same and the A_{II}-axis is assumed to be a horizontal line on the profile. B: Manually-contoured map of skewness values computed for the profiles in Fig. 485A; the values, given in the same relative position as on the map, are:

1.5928	0.5304	0.0000	−0.1910	−1.2000	−2.1557
0.6387	0.3778	−0.1072	−0.2029	−0.8102	−1.1905
0.3683	0.1084	−0.0661	−0.1639	−0.3707	−0.7311
0.2552	0.1676	−0.0816	−0.1344	−0.0785	−0.2918
0.1187	0.1824	−0.0532	−0.0823	−0.1651	−0.0026
−0.4677	−0.0746	−0.0412	0.0880	0.5320	0.4543

C: Skewness degree 3 trend surface map which accounts for 94.7 per cent of the total sum of squares. D: Contours for positive deviations (departures) of observed values (Fig. 485B) from computed values (Fig. 485C) which comprise only 5.3 per cent of the total sum of squares in this example.

the regional trend inherent in the data. Commonly, such trends are masked by the local variability of the observed data (Whitten, 1959B, 1963C).

When the regional variability has been approximated, valuable geological information can usually be obtained by making a separate map of the departures of individual original observations from the computed surface (Fig. 485D).

The theoretical bases for such surfaces, the technical details of the computations, the nature of the confidence levels associated with surfaces, and the development of the concepts associated with trend surface analysis, have been described in a series of papers, notably those by Oldham and Sutherland (1955), R. L. Miller (1956), Grant (1957), Krumbein (1959A, 1963B), Whitten (1959B, 1963A, 1963B), and Mandelbaum (1963). The actual arithmetic involved in calculating the trend surfaces and the deviation maps is tedious, but both surfaces can be computed by digital computer on a routine basis. Published computer programs in languages suitable for different computers include those by McIntyre (1963B), Harbaugh (1963), Whitten (1963B), and Whitten, *et al.* (1965); Peikert (1962, 1963) also prepared a trend surface computer program suitable for expressing X_n as a function of three independent variables such as U, V, and W. These programs should be valuable tools in analyzing the three-dimensional variability of the directional attributes of folded rocks.

(b) To use the methods of numerical biological taxonomy to relate structural attributes to environment within an orogenic belt

Within the past decade numerical taxonomy has advanced rapidly, and Sokal and Sneath (1963, p. 282) drew attention to the fact that: "...numerical taxonomy methods may prove useful in investigating the degree to which the environment affects the phenotype,[9] both as regards overall size (or its equivalent) and shape." Now that more numerous quantitative attributes are becoming available, the methods advocated by Sokal and Sneath can be adapted to structural geology. For example, if pressure, temperature, stress, and metasomatic effects varied within an orogenic belt, it seems reasonable to assume that the tectonites developed will reflect the three-dimensional and temporal variability of these process factors. When numerous attributes of the response characteristics (the tectonites) can be expressed in quantitative terms, the numerical taxonomic methods and factor analysis can be utilized to sort out these relationships.

[9]An organism, or all of the individuals within a group, distinguished by visible characters rather than by hereditary or genetic traits.

CLASSIFICATION

A more realistic nomenclature and classification of fold S and related structures can be developed. Chapter 15 shows that the nomenclature of folds is very confused, and that few objective criteria have been used in naming and classifying folds. This subject deserves urgent reappraisal, and the quantitative methods introduced into numerical taxonomy (e.g., Michener and Sokal, 1957, Sokal and Michener, 1958, Sokal, 1961, Sokal and Sneath, 1963) are of direct application (cf. Whitten, 1963A, pp. 83-84). The relevance of these methods may be seen from the work of Sneath and Sokal (1962, p. 855) who wrote that numerical taxonomy is:

> ... 'the numerical evaluation of the affinity or similarity between taxonomic units and the ordering of these units into taxa on the basis of their affinities.' The primary aims of the method are repeatability and objectivity, on both of which taxonomy to-day is open to strong criticism. It is the aim of numerical taxonomy to develop methods by means of which different scientists, working quite independently, will and must arrive at identical estimates of the affinity between two organisms given the same characters on which to base their judgements. We believe that these methods will lead to stable classifications which will not need extensive revision as new knowledge becomes available.

*   *   *

These are a few of the exciting new fields of exploration in structural geology; it would seem that this type of thinking can provide a breakthrough from qualitative to quantitative work. Data will have to be specifically collected for such analyses, and a number of students of structural geology are engaged on this task at the present time. With such analyses it will be possible to test specific process-response models. In particular, any conceptual models that relate to the regional variability of style and type of folding could be tested by the use of quantitative attributes.

Chapter 15

Review of Terminology Used for Folds

An enormous number of terms has been proposed to designate types of folds. Terms are accumulating at an alarming rate, and in many cases without the use, or even the consideration, of any systematic concepts or operational definitions. Some terms are defined with respect to purely geometric characters, some with respect to genetic concepts, and others have names based on mixtures of geometric and genetic properties. Commonly, little attention has been paid to the magnitude of the difference between existing and proposed new fold types. Many of the names seem to be completely unnecessary, and commonly they are maintained solely because they have been used in the literature; their continued use tends to lend confusion, rather than clarity, to the subject. However, a large proportion of these unnecessary terms has to be known if current structural literature is to be followed and read.

Classification should be a guide to thought, and in order to erect a classification it is essential that operational definitions be evolved and names allocated to identify the classes defined. To illustrate the nomenclatorial confusion, it is interesting to review the terminology used to describe the main types of folds and folding by authors of four modern books on structural geology. Additional lists could be included. Turner (1948, pp. 172-4) recognized:

1. slip folding
2. flexure folding
3. flexural-slip folding

Billings (1954, pp. 88-92) subsequently recognized:

1. flexure folding (= true folding; this category includes flexure and flexural-slip folding)
2. flow folding (= incompetent folding)
3. shear folding (= slip folding)
4. folding due to vertical movements

De Sitter[1] (1956B, p. 183) two years later used:

1. concentric folding (= parallel = distance-true folding)
2. similar folding: (a) slaty-cleavage folding; (b) fracture-cleavage folding; (c) accordion folding; (d) chevron folding (=oblique-shear folding).
3. flow (folding)

Badgley (1960, p. 25) stated that there are three main types of folding, namely:

1. flexure or competent folding
2. flow or incompetent folding
3. shear folding

The two main types of fold have long been recognized, although they have been variously named; thus, Van Hise (1896, p. 599) distinguished between concentric or parallel folds on the one hand, and similar folds on the other. Van Hise (1896) and de Sitter (1956B) used concentric and parallel folds as synonyms, although, Stočes and White (1935) used these terms for two major dissimilar categories of flexure; such problems compound the nomenclatorial confusion. Table 13 shows how the categories recognized by Billings and de Sitter appear to equate with the terminology advocated in Chapter 6.

No attempt is made to present a complete glossary of the terms which have been used for folds. Terms used in some of the well-known textbooks (e.g., Decker, 1920; Leith, 1923; Willis and Willis, 1929; Stočes and White, 1935; Nevin, 1949; Hills, 1953, 1963A; Billings, 1954; Russell, 1955; de Sitter, 1956B, 1964; Badgley, 1960; Turner and Weiss, 1963) are reviewed below, together with a few additional ones which appear to be significant. Many of the terms used are in one or two of the books, but not in the others; the absence of a reference to one of these books usually implies that the term was not specifically defined in it. A very small proportion of the terms quoted below is used elsewhere in this book. It is strongly urged that quantitative objective descriptions be utilized in future work (cf. Chapter 14) and that general descriptive terms

---

[1]In the second edition of his book, de Sitter (1964, pp. 170-2) gave a revised classification as follows:

(1) concentric folding (=parallel = distance-true = concentric cleavage folding).

(2) axial-plane cleavage folding (= similar = schistosity folding). This includes: (a) oblique-cleavage folding; (b) accordion folding; (c) axial-plane cleavage folding; (d) flow folding = irregular folding.

Table 13.—Comparative Fold Terminology

This Book	Billings (1954)	de Sitter (1956B)
Flexural-slip folds	Flexure folds Folds due to vertical movements	Concentric folds
Slip folds	Shear folds	Similar folds (a) slaty-cleavage folds (b) fracture-cleavage folds
Cross between flexural-slip and slip folds		Similar folds (c) accordion folds (d) chevron folds
	Flow folds	Flow folds
Flexure folds		

be kept to a minimum. Both genetic and purely geometric terms are included below; particular care is required in the use of the former.

In order to emphasize the relative importance of the terms listed below the most useful names are set in boldface capitals, and the useful but less important names are capitalized but set in ordinary type. A term in italics should be suppressed and not used in new descriptions of folds. An asterisk implies that the name is particularly misleading or unnecessary in modern work. Hence, in terms of decreasing usefulness, the following convention is employed:

ANTIFORM
CHEVRON FOLD
Allochthonous fold
*Arcuate fold

It should be emphasized that the remainder of this Chapter is a glossary of terms used by other writers; in numerous cases, the definitions listed are at variance with those adopted in this book and with those used by other authors. It is hoped that this catalog will be a useful reference and guide when reading other geology books and the journal literature.

TYPES OF FOLD AND FOLDING

*Abnormal anticlinorium: An anticlinorium in which the axial planes of minor folds converge upwards (Leith, 1923, p. 165, after Van Hise, 1896).

*Abnormal synclinorium: A synclinorium in which the axial planes of minor folds diverge upwards (Leith, 1923, p. 165, after Van Hise, 1896).

Accordion fold: Similar folds with planar limbs of constant thickness and with thickening of the units in the hinge zone. Foliation is accentuated in the hinge zone (Fig. 486), but slip during the folding occurred along bedding and not the foliation planes (de Sitter, 1956B, p. 216; Badgley, 1960, p. 27). Hills (1963A, p. 238) used a more geometric description: straight or nearly straight limbs and sharply curved, or even pointed, crests (= *zig-zag, chevron,* and *mitre* folds). De Sitter (1964, pp. 66 and 294) wrote that foliation is restricted to the hinge zone in accordion folds and considered that accordion folding is intermediate between cleavage folding and concentric folding; he equated with *concertina fold* and *chevron fold.*

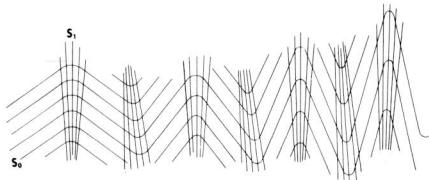

Figure 486. Accordion fold (After de Sitter, 1956B, Fig. 155).

Allochthonous fold: Where "strata no longer lie on the beds on which they were deposited, but have been torn free from them by folding and piled upon other rocks," (Stočes and White, 1935, p. 140, who stated that term is equivalent to *displaced fold*).

AMEBOID FOLD: A fold with low dips, no prevailing trend, and no definite shape; found in relatively undeformed areas of low regional dip (Russell, 1955, p. 78).

ANGULAR FOLD: A fold with straight or nearly straight limbs and sharply curved or even pointed crests and troughs (Hills, 1963A, p. 238, who used as equivalent to *zig-zag, chevron, accordion,* and *mitre* folds).

ANTECLISE: Regional up-arched structural form, commonly extended along the strike of an ancient fold of the cratonic basement platform, and possessing relatively thin mantles of platform-type sedimentary rocks. Term used by Shatski and Bogdanov (1957) and Potapov (1960) who cited the Belorussian and Volga-Urals anteclises as examples.

ANTICLINE: A fold in which the flanks diverge downwards and older rocks occur on the concave side of the fold. Hills (1953, p. 76; 1963A, p. 213) did not insist on younger beds being in the core of the fold.

ANTICLINORIUM: A large anticlinal structure comprising, or having within it, many smaller folds whose axes are subparallel to the main fold (cf. Decker, 1920, p. 14; Willis and Willis, 1929, p. 41; Hills, 1953, p. 77; 1963A, p. 215).

To Leith (1923, p. 164) a composite or complex arch, and to Nevin (1949, p. 39) a composite anticline in which the "superposed folds are minor, not only because they are smaller but also because their deformation has largely been controlled by the major structure."

ANTIFORM: A fold whose flanks diverge downwards and in which either older or younger rocks occur in the core (i.e., the concave side) of the fold (Hills, 1963A, p. 214); a useful term introduced by Bailey and McCallien (1937).

*Arcuate fold: Badgley (1960, Fig. 283, p. 216) illustrated an arcuate fold following Cloos (1946, Figs. 7 and 12). The type was not defined specifically, but it appears to be a fold in which the *B*-geometric axis is warped in the axial surface during a second phase of movement.

ASYMMETRICAL FOLD: This term is defined in dissimilar ways by different authorities. The most logical use is for a plane fold whose profile is not bilaterally symmetric about the axial plane (Turner and Weiss, 1963, p. 122; cf. de Sitter, 1964, p. 272). Stočes and White (1935, p. 116) used a somewhat similar definition, and stated that the limbs are unequally inclined to the axial plane or are of unequal length.

Less acceptable definitions were used by Hills (1953, p. 81; 1963A, p. 214): the dips of the two limbs are different; and by Billings (1954, p. 40): the axial plane is inclined and the two limbs dip in opposite directions but at different angles. Nevin (1949), Russell (1955), and de Sitter (1956B) used concepts similar to those of Hills and Billings, which do not require a lack of bilateral symmetry about the axial plane; Matthews (1958) defined three lengths measured perpendicular to the fold axis which can be used to define the form and size of such asymmetrical folds.

Autochthonous fold: Folds which lie on their original basement (Stočes and White, 1935, p. 410); most folds, except possibly parallel folds, have been more or less displaced.

*Axial-plane folding: "A special case of double folding...where the second folds are chiefly in a horizontal plane, so that the axial planes of the first folds are bent" (Hills, 1963A, p. 276).

Axial-plane stress-cleavage fold: The simplest variety of cleavage fold with axial-plane cleavage (de Sitter, 1964, p. 273).

BASIN: A periclinal flexure in which dips are towards the center of the structure at all points (Hills, 1953, p. 76); a synclinal depression with no distinct trend (Billings, 1954, p. 50), which is one type of non-cylindrical fold (Mertie, 1959, p. 107) and of syneclise (Potapov, 1960). This term is usually employed for structures measured in kilometers.

Bending fold: This misleading term implies a fold "...caused by vertically acting forces of different intensity in different places [which] does not involve lateral shortening of the section of folded rocks" (Hills, 1953, p. 90, who stated the term includes *supratenuous* and *Plains-type folds*). These are the "*folds due to vertical movements*" of Billings (1954).

   Ramberg (1963B, p. 1) used bending and buckling folds as the two main genetic subdivisions of folds; basing his terminology on that of strength-of-materials studies, he used bending fold for cases where the "periodic or heterogeneous transversal motion represents heterogeneous strain not generated by compression" along the lithological layers. This usage includes shear folds and supratenuous folds as varieties of bending fold.

Box fold: Stočes and White (1935, p. 157) and Hills (1953, p. 58; 1963A, p. 215) described these as folds which are subrectangular in cross section, and in which the widely-spaced anticlines are separated by undisturbed synclinal troughs. According to de Sitter (1956B, p. 199) a box fold is a flat-topped broad anticline with steep flanks.

*Brachyanticline: An anticline having great breadth in comparison with its length (Stočes and White, 1935, p. 124). A very short anticline whose axis plunges at both ends (Goguel, 1962, p. 130).

*Brachysyncline: A syncline having great breadth in comparison with its length (Stočes and White, 1935, p. 124). A very short syncline whose axis plunges at both ends (Goguel, 1962, p. 130; equated with *cuvette*).

BUCKLE FOLDING: One type of flexural-slip folding in which an individual competent layer in a heterogeneous sequence develops flexure without significant change in thickness (Turner and Weiss, 1963, p. 473; Hills, 1953, p. 89). Ramberg (1963B, p. 1) used buckling and bending folds as the two main genetic subdivisions of folds; to him buckling folds result when the "periodically varying transversal displacement is a secondary effect of compression [shortening] parallel to layering" which displays unlike mechanical properties.

Cardinate fold: A single isoclinal fold (i.e., having limbs parallel) within a sequence of little-folded rocks (Decker, 1920, p. 14; Willis and Willis, 1929, p. 35).

Cascade fold: One of a group of subsidiary folds developed in limestone by collapse and sliding down a dip slope under gravity at the Earth's surface (Hills, 1953, p. 19; 1963A, p. 342; and de Sitter, 1956B, p. 274; 1964, p. 240; following Harrison and Falcon, 1934).

CHEVRON FOLD: Slightly dissimilar definitions have been employed by different authorities. Billings (1954, p. 42), Russell (1955, p. 77), and Turner and Weiss (1963, p. 114) simply referred to sharp and angular bends at the hinge lines, but Hills (1963A, p. 238) also included straight or nearly-straight limbs in his definition. Turner and Weiss pointed out that chevron folds are commonly asymmetric (according to their definition).

De Sitter (1956B, p. 216, followed by Badgley, 1960, p. 27) thought that in chevron folding "one flank is sheared obliquely to the principal stress and parallel to the axis; the other flank is simply tilted with slip along the bedding-planes" (Fig. 487); such genetically-defined structures would comprise part of the geometric group defined by Billings, Russell, and Turner and Weiss. However, in 1964 de Sitter (1964, p. 294) claimed chevron fold is synonymous with *accordion fold* and *concertina fold*.

Hills (1953, p. 84) and Russell (1955, p. 77) equated chevron fold with *zig-zag fold,* Hills (1963A, p. 238) with *accordion* and *mitre folds* too; Turner and Weiss (1963, p. 114) equated chevron fold with *kink fold*.

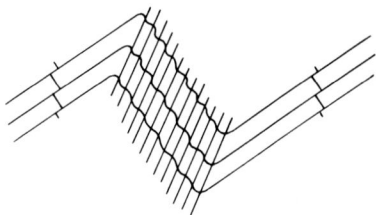

Figure 487. Chevron fold (After de Sitter, 1956B, Fig. 152b).

Cleavage fold: de Sitter (1956B, p. 182) and Turner and Weiss (1963, p. 113) used the descriptive term for folds of similar type that have secondary foliations symmetrically related to their axial surfaces. De Sitter (1964, p. 168) claimed that the most important feature is that the movement plane is planar and at an angle to bedding.

*Closed fold: Decker (1920, p. 9) and Stočes and White (1935, p. 121) used this term for folds whose limbs are parallel, and thus folded back on one another. Willis and Willis (1929, p. 38) referred to folds whose limbs "cannot bend closer without distorting the bedding" (cf. *squeezed fold*).

To Billings (1954, p. 45) closed folds were those in which "deformation has been sufficiently intense to cause flowage of the more mobile beds so that these beds thicken and thin": equated with *tight fold*. Since the term seems to be used by Billings for strongly folded units, this definition probably comes closest to current usage.

Compaction fold: A fold formed by compaction of sedimentary rock over buried hills, still-active faults and folds, and over interbedded lenses of sand, reef-rock, or other little-compactable rocks (Hills, 1963A, pp. 57-58; equivalent to *supratenuous fold*).

*Competent fold: Willis and Willis (1929, p. 246) and Stočes and White (1935, p. 122) used this for folds produced by compression in which the anticlines lifted material above, and sinking synclines displaced the underlying material; this genetic term is wholly unsatisfactory.

Busk (1929, p. 9) equated competent fold with *parallel fold*, while Turner and Weiss (1963, p. 112) equated it with *concentric* and *parallel folds*. Badgley (1960, p. 26) used competent folding interchangeably with *flexure folding* for *parallel folding*; to Badgley the lithic units have not undergone "thinning" in a competent fold, or in the competent part of a fold.

*Complex fold: A fold in which the axis has been flexed by another fold (Leith, 1923, p. 164; Stočes and White, 1935, pp. 121-2).

Composite fold: Leith (1923, p. 163) used this for a major fold with minor superimposed crenulations whose axes are more or less parallel to the axis of the major fold; Willis and Willis (1929, p. 39) used it for folds in which there are "minor folds involved in their limbs"; Stočes and White (1935, p. 121) for "a larger simple flexure of crenulated beds."

Compound folds: Hills (1953, p. 94) used this term for structures in which a combination of fold mechanisms operated.

CONCENTRIC FOLD: Folded layers maintain a more or less constant thickness within one fold, so that the radius of curvature decreases towards the fold core; such folds cannot persist far without dying out (Turner and Weiss, 1963, p. 112; they equated with *parallel* and *competent folds*). This is a generally accepted definition.

Leith (1923, p. 172) and de Sitter (1956B, p. 182) equated with parallel fold, but preferred *concentric fold*; Hills (1953, p. 81; 1963A, p. 229), Billings (1954, p. 58), and Russell (1955, p. 77) preferred the equivalent term *parallel fold*. Stočes and White (1935, p. 127) used a similar definition which is different from that for their parallel fold. De Sitter (1956B, p. 182) also equated with *distance-true fold* and later (1964, p. 168) with *flexure fold*.

*Concertina fold: Equivalent to *angular, zig-zag, chevron,* and *mitre fold* (Hills, 1963A, p. 238), and equivalent to *accordion* and *chevron fold* (de Sitter, 1964, p. 294).

Congruous minor fold (and *congruous drag fold*): Minor or drag fold whose axis is parallel to those of the major folds (Hills, 1953, p. 98; Billings, 1954, p. 83).

CONICAL FOLD: A noncylindroidal fold generated by a straight line passing through a fixed point (Turner and Weiss, 1963, p. 108, following Dahlstrom, 1954, and Haman, 1961; Badgley, 1960, p. 222, following Stockwell, 1950).

CONJUGATE FOLDS: Coined by Johnson (1956A) for closely-associated folds of similar style, which occur in pairs with mutually inclined and approximately conjugate axial surfaces (Turner and Weiss, 1963, p. 114).

CONVOLUTE FOLD: Turner and Weiss (1963, p. 115) defined convolute folds as "resembling disharmonic, conjugate, or polyclinal folds, but having axial surfaces that are curved, smoothly branching, or whorled, and hinges that are complex and convolute.... The individual folds are generally cylindrical in extension and resemble folds which have been called refolded folds" by Clifford, *et al.* (1957).

CROSS FOLD: Fold with its axis perpendicular to, or oblique to, the main folds (Stočes and White, 1935, p. 126, who equated with *transverse fold*; Russell, 1955, p. 88; Badgley, 1960, p. 217; Hills, 1963A, p. 275; and Turner and Weiss, 1963, p. 121, who equated with *oblique fold*). When the cross fold is younger than the main folds it becomes a *superposed* or *superimposed* fold (Turner and Weiss, 1963, p. 121; de Sitter, 1964, p. 303), or a *folded fold* (Hills, 1963A, p. 275).

*Curvilinear fold: A fold with smooth curves in cross section in contrast to one having straight or nearly straight limbs (Hills, 1963A, p. 214).

*Cuspate fold: In cuspate folds the limbs are smoothly curved in sectors, but adjoining sectors meet in sharp cusps (Hills, 1963A, p. 214).

CYLINDRICAL FOLD: A fold with rectilinear hinges that can be generated geometrically by a line parallel to the hinge moving with constant orientation (Turner and Weiss, 1963, p. 107, who equated with *cylindroidal fold* which they considered less desirable; Badgley, 1960, p. 52). The axial surface may be, but is not necessarily, a plane.

CYLINDRICAL PLANE FOLD: A cylindrical fold having an axial plane (as opposed to an axial surface) (Turner and Weiss, 1963, p. 108).

CYLINDROIDAL FOLD: A fold with rectilinear hinges that can be generated geometrically by a line parallel to the hinge moving with constant orientation (Hills, 1963A, p. 258). Used by Clark and McIntyre (1951), and is equivalent to *cylindrical fold* of Turner and Weiss (1963, p. 107).

*Derma fold: Fold in metamorphic rocks developed in deeper structures of the Basement—the *"plis de fond"* of Argand (1916) (de Sitter, 1956B, p. 116, attributed to Escher).

DIAPIR FOLD: Anticline in which a mobile intrusive core of sedimentary rocks breaks through more brittle overlying rocks, as in a salt dome (Billings, 1954, p. 59, equivalent to the preferred *piercing fold*; Russell, 1955, p. 76). The *diapiric fold* of Hills (1953, p. 87; 1963A, p. 275).

DIAPIRIC FOLD: Where "mobile beds are actually injected through overlying strata at anticlinal axes" (Hills, 1953, p. 87, and Badgley, 1960, p. 27, who both equated with *piercement fold*; Hills also considered it equivalent to *plis diapirs* and *Injektivfalten*). The *diapir fold* of Billings (1954) and Russell (1955).

*Dipping fold: A fold in which the "apex of the fold is depressed below the axis" (Stočes and White, 1935, p. 116).

Discordant fold: There are two different uses. (1) Stočes and White (1935, p. 149) equated with *unconformable fold* and seem to have used it for *disharmonic folds*. (2) Turner and Weiss (1963, p. 121) used it for varieties of oblique folds with axes inclined to those of the longitudinal folds in the same area.

DISHARMONIC FOLD: In disharmonic folds an abrupt change in profile occurs in passing from one surface or lithic unit to another within the lithologic section (Hills, 1953, p. 86; 1963A, p. 282; Billings, 1954, p. 58; Russell, 1955, p. 72; de Sitter, 1956B, p. 229; Badgley, 1960, p. 27; Turner and Weiss, 1963, p. 114). Stočes and White (1935, p. 149) used *discordant* or *unconformable fold* for this type.

DISJUNCTIVE FOLD: If "relatively brittle beds are interbedded with plastic rocks that are strongly deformed by flowage, the brittle beds may fracture and the parts become separated, while retaining an overall fold pattern." (Hills, 1953, p. 86; 1963A, p. 170).

*Displaced fold: See *allochthonous fold* (Stočes and White, 1935, p. 140).

*Distance-true fold: See *concentric fold* (de Sitter, 1956B, p. 182; equivalent to *parallel fold*).

DOUBLY-PLUNGING FOLD: A fold that "reverses its direction of plunge within the limits of the area under discussion. Most folds, if followed far enough, are doubly plunging" (Billings, 1954, p. 49). Russell (1955, p. 69) suggested using a different term because there is ambiguity as to whether a saddle or a closed area is involved.

DOWNFOLD: A fold that closes downwards, i.e., has the form of a nonrecumbent syncline (Bailey, 1934, p. 467).

DRAG FOLD: Because of the widespread use of this term in the literature several definitions are quoted in full.

    1. Leith (1923, p. 167) thought drag folds "are folds formed by the drag of one bed over another, as an incident of either major folding or faulting. Most folds are probably drag folds on a larger or smaller scale."

    2. Willis and Willis (1929, p. 268) said that drag folds are minor folds occurring in connection with major folding and "produced by the shearing stresses set up within a fold by the relative movements of beds along bedding planes, or by the crushing of incompetent beds under the arch of the competent bed." The axial-planes of the drag folds are roughly parallel to that of the major fold, and the plunges of the axes correspond too.

    3. Stočes and White (1935, p. 151): "Deep-seated rocks in the zone of flowage readily readjust themselves to the changes required by folding because they are in a state of high plasticity. Under these conditions pressure cannot be transmitted for long distances, and minute and complex folds, e.g., the so-called drag folds, are formed."

    4. Hills (1953, p. 97): In parallel folding where "competent and incompetent strata are interbedded, the movements of the former relatively to each other subject the incompetent strata to shearing stress, which may cause small folds, called *drag folds,* to develop in them." Hills (1953) claimed to be following Van Hise and Leith (1911); Hills (1963A, p. 284) later suggested "drag fold" lacks precision, and he mentioned three dissimilar types of structure to which the term has been applied.

    5. Billings (1954, pp. 45 and 82) wrote that drag folds "form when a competent ('strong') bed slides past an incompetent ('weak') bed.... Such minor folds may form on the limb of larger folds because of the slipping of beds past each other, or they may develop beneath overthrust blocks.... In many regions all the strata have essentially the same competency. Drag folds, in the

narrowest sense of the word, do not form. Nevertheless, large masses of rock move past one another, and the strata are thrown into *minor folds*. The minor folds usually bear the same relation to the major folds as do drag folds, and they are even called drag folds by many geologists."

6. Russell (1955, p. 82) wrote that drag folds are formed in incompetent beds contained between competent beds; as the competent beds slide over one another friction causes the axial planes of the drag folds to be systematically oriented.

7. De Sitter (1956B, p. 226) described drag folds as "small asymmetrical folds in a cleaved bed" produced by irregular slip along foliation planes. "The name drag-fold is not very satisfactory as it again introduces the concept of drag by the competent beds on the incompetent bed . . . nor are drag-folds influenced by this friction." De Sitter (1958) suggested *parasitic fold* is more appropriate than drag fold.

8. Turner and Weiss (1963, p. 117) said that although drag fold is sometimes applied to *intrafolial folds* and *parasitic folds,* the term should not be used because of its genetic connotation.

9. Ramberg (1963A, p. 97) included among drag folds those folds "with quasi-monoclinic symmetry usually, but not exclusively, developed in relatively thin competent layers in the flanks of larger folds. . . the component of compressive strain and stress parallel to layering causes slack in the competent layer which is accommodated by buckling whereas the component of shear gives the buckles their tilted monoclinic symmetry so characteristic for drag folds."

Drape fold: This term is "normally applied for open folds of the Plains type" rather than for close or overturned folds as implied by the French equivalent of *plis de revêtement* (Hills, 1963A, p. 328).

EN ÉCHELON FOLDS: Folds arranged in an *en échelon* map pattern (Billings, 1954, p. 54; Turner and Weiss, 1963, p. 121; de Sitter, 1964, p. 261). Hills (1963A, p. 262) used a more specialized definition: a parallel series of short folds whose long axes are oblique to some major structure such as folds of a lower order, or the trend of strike-slip faulting or regional cleavage.

Epidermic fold: Superficial folds developed at relatively shallow depths in sedimentary rocks—the "*plis de couverture*" of Argand (1916) (de Sitter, 1956B, p. 116, attributed to Escher).

*Expansion fold: A fold formed near the Earth's surface due to expansion of clay and shale on hydration during weathering (Hills, 1963A, p. 68).

*Fan fold and Fan-shaped fold: Billings (1954, p. 42) wrote that *"fan fold* is one in which both limbs are overturned. . . . In an anticlinal fan fold, the two limbs dip towards one another; in the synclinal fan fold, the two limbs dip away from each other. Fan folds are not as common as was formerly supposed; fifty years ago they were thought to be abundant in the Alps." Willis and Willis (1929, p. 38), Stočes and White (1935, p. 157), Russell (1955, p. 75), and Hills (1963A, p. 215) defined the terms in a similar manner, while Leith (1923, p. 165) described the structure as a specialized variety of *normal anticlinorium.*

*Fault-fold: Fault-folds occur when there is a close association between folding and nearly vertical faulting; parallel fault systems cause little-disturbed sedimentary rocks to drape over horsts, and thicker sequences of strata to be crumpled in the synclines in graben (Hills, 1953, p. 68, ascribed to Stille, 1910, etc., who introduced the concept as *Bruchfaltung*; Hills, 1963A, pp. 208-10).

*Faulted overturned fold: "Nappes due to the reduction and breaking of the middle limb are faulted overturned folds" (Stočes and White, 1935, p. 135).

Flap structure: Part of a limestone sheet that has bent over backwards away from the crest of an anticlinal fold, without fracturing and in response to gravity-collapse at the Earth's surface (Hills, 1953, p. 19, and de Sitter, 1956B, p. 274, 1964, p. 240; both following Harrison and Falcon, 1934).

FLEXURAL-SLIP FOLD: A fold in which *S*-surfaces were kinematically active during folding (Turner and Weiss, 1963, p. 473). Hills (1963A, p. 227) used an analogous definition.

A less acceptable definition was given by Hills (1953, p. 94) who described flexural-slip folds as a variety of *compound fold* "involving flexing of competent strata in buckle-folds and slipping of inter-bedded incompetent strata along planes of false cleavage."

Billings (1954, p. 90) did not separate flexural-slip from *flexural folds,* but noted that some geologists (e.g., Turner, 1948) distinguished two discrete categories.

FLEXURE FOLD: During formation of flexure folds the lithic units maintained unaltered their original thickness, and, during bodily rotation of the fold limbs, the upper and lower surfaces of each unit remained parallel; slip parallel to *S* (bedding, etc.) did not occur. Commonly, because of their lamination, rocks deform by flexural-slip rather than by flexure folding (Turner, 1948, p. 172).

Billings (1954, p. 88) said a flexure fold may result from either compression or a couple, and that in sedimentary rocks a very important factor is sliding of beds past one another; he equated the term with *true fold*. De Sitter (1956B, p. 310) only used the term to describe the flexure fold of Schuld, Rheinisches Schiefergebirge, on the basis of Cloos' (1948) account; he noted that bedding-plane slip occurred throughout the structure. This usage of Billings and de Sitter is confusing and is not to be recommended; equally confusing is de Sitter's (1964, p. 168) suggested equivalence of flexure and *concentric folding*. Billings included both flexure and flexural-slip folds as flexure folds.

Badgley (1960, p. 26) equated flexure fold with *competent fold* and defined it like *parallel fold*.

FLOW FOLD: Folds somewhat irregular in pattern and without development of visible discrete slip surfaces; during folding the *S*-surfaces were passive and there is no obvious secondary foliation (Turner and Weiss, 1963, p. 481).

Less precise usage has been common. For example, Hills (1953, p. 92) wrote that in flow folding, beds offered so little resistance to deformation that they assumed any shape impressed on them by surrounding more rigid rocks or the general stress-pattern; *ptygmatic folds* and folds in salt deposits and Archean formations were cited as examples. However, Hills (1963A, p. 345-6) later used flowage fold for gravitational flowage phenomena in sedimentary basins (cf. Bain, 1931). Billings (1954, p. 91) thought that in many respects flow folds do not differ in appearance from flexure folds (in Billings' sense), but minor folds are more abundant; flow folds were considered characteristic of the central parts of some orogenic belts, where the whole mass was forced to move under compression, and exhibited behavior analogous to that of a viscous liquid. Both Billings and Badgley (1960, p. 25) equated flow folding with *incompetent folding* (citing Fairbairn, 1949). De Sitter (1956B, p. 182; 1964, p. 170) claimed that flow folding is a fundamental mode of folding, in which a fixed orientation of shear planes to stress direction is lost; to him it represented the only true plastic deformation of rocks and is characteristic of weak rocks such as salt.

Ramberg (1963B, p. 9) wrote: "... flow folds probably do not either represent a category of folds different from the bending- and buckling types. Flow folds may be of either class, the sole peculiarity about flow folds being that the rocks constituting such folds either were more plastic, or that the process of deformation was more slow than in the cases of more 'ordinary' bending- and buckling folds."

FOLD NAPPE: A very large recumbent fold whose reversed middle limb has been completely sheared out as a result of great (often several kilometers) horizontal translation (Hills, 1953, p. 55; 1963A, p. 246).

Folded fold: The product of one or more younger phases of folding superposed on an older fold (Hills, 1963A, p. 275).

Folds resulting from vertical movements: Billings (1954, p. 92) described sedimentary layers raised into domes by vertical forces without any associated fractures (a mechanism dissimilar to flexure folding). This term is similar to Hills' (1953, p. 67) *Plains type of folding.*

*Foresyncline: The syncline adjacent to an asymmetric anticline on its steep side (Busk, 1929, p. 29).

*Fracture-cleavage fold: One class of *similar fold* according to de Sitter (1956B, p. 183). Not rigorously defined but presumably a variety of slip fold (as used in this book).

Geanticline: An uplift of the Earth's crust covering thousands of square miles (Stočes and White, 1935, p. 124, attributed to Dana, 1873) or many miles (Nevin, 1949, p. 39). To Goguel (1962, p. 282) a term to be avoided but applied to cordilleras or zones of deposition characterized by numerous lacunae and signs of emergence. This usage presumably relates to Schuchert's (1923) definition of a geanticline as a positive area related to a geosyncline, or rising from a geosyncline.

GEOSYNCLINE: A regional depression of the Earth's crust covering thousands of square miles (Stočes and White, 1935, p. 124) or many miles according to Nevin (1949, p. 39), who cited the Mississippi embayment, U.S.A., as an example.

Dana (1873, p. 430) used geosynclinal for a long-continued subsidence which is not a true syncline since the rocks of the bending crust may have had in them many true or simple synclinals as well as anticlinals, and a consequent long-continued accumulation of sediments.

These definitions are dissimilar to the common present-day concept of a geosyncline which involves a particular variety of large sedimentary basin; the concept has been used in many varying senses and at present "geosyncline" is rarely used without a prefix to designate the particular variety of basin involved. Krumbein and Sloss (1963, pp. 395, ff.) gave an extensive review of the many different uses (cf. de Sitter, 1964, pp. 390 ff.); geosyncline no longer has a place in the geometric description of fold form (cf. Hills, 1953, p. 44; 1963A, p. 317).

Generative fold: Folds "that increase in amplitude in successive beds with the accompaniment of an increase in stratigraphic thickness towards the axial region." (Hills, 1953, p. 86).

Gleitbrett fold: Synonym of *shear fold* (Hills, 1963A, p. 234).

HARMONIC FOLD: Term used by Billings (1954, p. 66) without explanation, to imply that no abrupt change in profile occurs in passing from one surface or lithic unit to another through the lithologic section (cf. Disharmonic fold, Billings, 1954, p. 58).

HOMOAXIAL FOLD: An assemblage of folds in which all the major and minor folds, foliations, etc., are related genetically to one longitudinal axial direction (Hills, 1963A, p. 261). In this book geometric parallelism of axes, rather than genetic relationship, is used for the basis of definition.

HORIZONTAL FOLD: A plane or nonplane cylindroidal fold with a horizontal or subhorizontal fold axis (Turner and Weiss, 1963, p. 118); should be compared with *level fold* (Stočes and White, 1935, p. 126) and *nonplunging fold* (Billings, 1954, p. 46).

INCLINED FOLD: Common usage for folds in which the axial surface is inclined but the steeper limb is not overturned (Willis and Willis, 1929, p. 34; Stočes and White, 1935, p. 116; Nevin, 1949, p. 47; Hills, 1953, p. 81). A specific definition seemed unnecessary to Leith (1923, p. 164). By contrast, Turner and Weiss (1963, p. 119) departed from the definition commonly used and suggested that an inclined fold is a cylindroidal or noncylindroidal plane fold in which the axial plane dips at less than 90° (equated with *overturned fold*).

*Incompetent folding: Dissimilar definitions of this term were used by different geologists:

Willis and Willis (1929, p. 257): "Incompetent folding is the result of a deflecting pressure which is initially directed at right angles to the strata, or nearly so, when they are in a horizontal or gently inclined position, and which also is resolved into components at right angles to and parallel to the stratification.... Incompetent is a necessary preliminary to competent folding."

Busk (1929, p. 9) equated incompetent and *similar folding*.

Stočes and White (1935, p. 122) merely wrote that "an incompetent fold is weak, collapsing under the load."

Billings (1954, p. 90) and Badgley (1960, p. 26) equated incompetent and *flow folding*.

Incongruous minor fold and *incongruous drag fold:* Minor folds and drag folds whose fold axes, whether dependent or independent, are not parallel to the axis of the major fold (Hills, 1953, p. 98, followed by Billings, 1954, p. 83).

INCONSTANT FOLD: Associated folds, either plane noncylindroidal or nonplane noncylindroidal, varying in both axial trend and plunge (Turner and Weiss, 1963, p. 121, attributed to Clifford, *et al.,* 1957).

*Inequant fold: A fold in which one limb is considerably longer than the other (Hills, 1963A, p. 214).

INTRAFOLIAL FOLD: "Isolated folds with distinctive characters are commonly observed in otherwise unfolded tectonites. They are tightly appressed plane folds occurring sporadically as relatively slight distortions of dominantly planar foliation. . . . They commonly grow from an unfolded foliation surface only to die out again at a higher level. The axial planes either conform approximately or are significantly inclined to the average attitude of the surrounding unfolded foliation; and this usually curves gently around the local swelling caused by folding. . . . The presence of intrafolial folds is one of the criteria of transposition of s-surfaces. . . . Because of its genetic connotation the term 'drag fold,' sometimes applied to intrafolial folds, is not used" (Turner and Weiss, 1963, pp. 116-7).

*Intraformational fold: "While some of the folds in the region south of Lake Erie are limited to only a few feet in vertical extent, it is significant that they do not commonly show close, overturned, and recumbent types, which are very common in minor intra-formational folds." (Decker, 1920, p. 36).

Hills, (1963A, p. 60) used intraformational fold for contortions of sedimentary rocks that occurred prior to complete lithification (e.g., slumping) so that folded units lie between undeformed strata.

*Irregular fold: Synonymous with *flow fold* (de Sitter, 1964, p. 170).

ISOCLINAL FOLD: Ideally, a fold with parallel limbs. A concentric fold can be truly isoclinal, but a similar fold can only approach this form (Turner and Weiss, 1963, p. 114; Decker, 1920, p. 14; Leith, 1923, p. 164; Willis and Willis, 1929, p. 38; Stočes and White, 1935, p. 123; Nevin, 1949, p. 47; Hills, 1953, p. 81; 1963A, p. 214; Billings, 1954, p. 41; Russell, 1955, p. 74; Badgley, 1960, p. 29).

*Jura-type fold: Folds in superficial rocks which have been able to fold independently of Basement as a result of décollment in, for example, the Jura and the High Atlas Mountains; equivalent of *plis de couverture* of Argand (1916) (Hills, 1963A, p. 327; de Sitter, 1956B, p. 188, 1964, p. 171, used *"Jurassic" fold* and equated with *superficial fold*).

*Kink fold: See *chevron fold* (Turner and Weiss, 1963, p. 114).

*Knee fold: A variety of fold in surficial gravity-collapse structures. When the "crest of a sinusoidal fold in interbedded limestones and shales is broken through by erosion, and the limestone sheets move apart by slipping downwards over underlying shales, a *knee fold* is first formed in the limestones" (Hills, 1953, p. 19, 1963A, p. 342, following Harrison and Falcon, 1934, 1936). The fold has a profile similar to the shape of a partially-flexed human leg.

*Ladder-fold: Occurs where "successive anticlinal crests are found higher and higher proceeding in the direction of the movement" (Stočes and White, 1935, p. 158).

Level fold: Such folds have horizontal crest lines (Stočes and White, 1935, p. 126); compare with *horizontal fold* (Turner and Weiss, 1963, p. 118) and *nonplunging fold* (Billings, 1954, p. 46).

Longitudinal fold: Folds with axes trending parallel to the regional elongation of an orogenic body (Turner and Weiss, 1963, p. 121; Stočes and White, 1935, p. 126, who equated with *strike fold*).

Macrofold: A large fold, as contrasted with a microfold (de Sitter, 1956B, p. 68).

MAJOR FOLD: A large fold, which may have minor folds on its limbs (Leith, 1923, p. 164).

Marginal fold: The minor folds, developed in the foreground of folded areas, which die out gradually towards the unfolded region (Stočes and White, 1935, p. 158).

Microfold: A small fold, as contrasted with a macro- or large fold (de Sitter, 1956B, p. 68; 1964, p. 179).

MINOR FOLD: Decker (1920, p. 16), Leith (1923, p. 164), and de Sitter (1964, pp. 179 and 306) applied this term to small as opposed to large folds. Willis and Willis (1929, p. 268) said ". . . minor folding is very apt to occur in connection with major folding. It is produced by the shearing stresses set up within a fold by the relative movements of beds along bedding planes, or by the crushing of incompetent beds under the arch of the competent bed. The former is called *drag folding.*" Billings (1954, p. 82) said that in "many regions all the strata have essentially the same competency. Drag folds, in the narrowest sense of the word, do not form. Nevertheless, large masses of rock move past one another, and the strata are thrown into *minor folds*. The minor folds usually bear the same relation to the major folds as do drag folds, and they are even called drag folds by many geologists." Hills (1953, p. 97) also differentiated drag folds from the minor folds of an anticlinorial or synclinorial structure. Russell (1955, p. 80) made the surprising statement that minor folds are "only a few hundred feet across [and] are apparently produced by the same forces which produced the larger ones. . . . All local dips due to minor folding are worthless for working out the structure of the major anticlines which produce oil and gas."

Mitre folds: Angular folds with inclined kink planes (Hills, 1963A, p. 240).

*Mixed fold: Any fold in which dissimilar interlaminated lithologies have responded differently to the kinematic events (de Sitter, 1964, p. 290).

Monocline: A monocline involves a local steepening of an otherwise uniformly dipping or horizontal sequence of rocks, so that an anticlinal bend occurs above and a synclinal bend at a lower level; by contrast, in oilfield geology, the dip is more or less uniform in direction (Hills, 1953, p. 76; 1963A, p. 216; Decker 1920, p. 14, who attributed the term to Gilbert, 1876; Leith, 1923, p. 165; Stočes and White, 1935, p. 122; Nevin, 1949, p. 38; Billings, 1954, p. 42; Russell, 1955, p. 70).

NAPPE: "A *nappe* ... is a sheet of rocks, of large dimensions (of the order of miles), that has moved forward for a considerable distance (again of the order of miles) over the formations beneath and in front of it. ... A nappe may be either the hanging wall ... of a great low-angle overthrust (*thrust nappe, Überschiebungsdecke* ...) or a recumbent fold (*fold nappe, Überfaltungsdecke* ...), of which the reversed middle limb has been completely sheared out as a result of the great horizontal translation" (Hills, 1953, pp. 54-5; 1963A, p. 246). German equivalent is *Decke*.

Billings (1954, p. 189) noted that a "*nappe* is a large body of rock that has moved forward more than one mile from its original position, either by overthrusting or by recumbent folding. The term is thus not synonymous with either *over-thrust sheet* or *recumbent fold*. A large recumbent fold is a nappe, but a small one is not."

Note that all nappes do not comprise individual folds.

Nappes de recouvrement: "... overturned folds often of large dimensions thrust over another—and often younger—series of strata" (Stočes and White, 1935, p. 134, who used as equivalent to *recumbent folds*).

Neutral-surface fold: A term used by Hills (1963A, p. 220) for structures similar to *flexure folds*.

NONCYLINDRICAL FOLD: A fold which is not cylindroidal (Turner and Weiss, 1963, p. 108), which to Mertie (1959, p. 107) includes "quaquaversal folds, elongate domes, or canoe-shaped folds, and the plunging ends of cylindrical folds."

NONCYLINDRICAL PLANE FOLD: A fold which has an axial plane but which is not cylindroidal (Turner and Weiss, 1963, p. 108).

NONPLANE CYLINDRICAL FOLD: Cylindroidal fold which does not have a *plane* axial surface (Turner and Weiss, 1963, p. 108).

NONPLANE NONCYLINDRICAL FOLD: A fold which is not cylindroidal and which does not have a *plane* axial surface (Turner and Weiss, 1963, p. 108).

Nonplunging fold: A fold with a horizontal fold axis (Billings, 1954, p. 47), which equates with *level fold* of Stočes and White (1935, p. 126) and *horizontal fold* of Turner and Weiss (1963, p. 118).

*Normal fold: "Associations of persistent, long and relatively narrow folds may be termed *normal fold structures,* in contrast with the recumbent and highly complex folds of the *Decken.* . . . Seeing that the term *normal fold* is no longer in general use as a synonym for *symmetrical fold,* it might be used with advantage in the sense suggested above, without leading to confusion" (Hills, 1953, p. 57).

   Turner and Weiss (1963, p. 119) used normal fold for cylindroidal and noncylindroidal plane folds in which the axial plane is vertical, and equated the term with *upright fold.*

*Normal anticlinorium: An anticlinorium in which the axial planes of minor folds converge downwards (Leith, 1923, p. 164, following Van Hise, 1896).

*Normal synclinorium: A synclinorium in which the axial planes of associated minor folds converge upwards (Leith, 1923, p. 165).

NOSE: ". . . a rather short, plunging anticline without closure. Presumably an anticline of considerable length without closure would be referred to as a plunging anticline rather than as a nose. Usually a typical nose shows a flattening of the dip or of the plunge of its axis along a part of its course" (Russell, 1955, p. 70).

*Oblique cleavage fold: A variety of *similar fold* (de Sitter, 1964, p. 171).

Oblique fold: Synonym for *cross fold, superimposed fold,* and *superposed fold* (Turner and Weiss, 1963, p. 121).

OBLIQUE SHEAR-FOLD: According to de Sitter (1956B, p. 182, Fig. 125c) a variety of *cleavage fold,* which apparently corresponds with what he called *chevron fold* (*note:* de Sitter's chevron fold is dissimilar to that described by many geologists). De Sitter (1964, p. 302) only used the term for some conjugate folds associated with a conjugate set of shears.

OPEN FOLD: To Decker (1920, p. 9), a fold ". . . in which the strata spread widely from the axial plane" and to Willis and Willis (1929, p. 38) one "which may be bent closer without disturbing the parallelism of the strata on any limb." Stočes and White (1935, p. 119) defined an open fold as one in which "the limbs are inclined away from the axial plane."

   Billings (1954, p. 45) thought that in an open fold the more mobile beds did not flow in response to the deformation, and consequently these beds do not thicken and thin; he recognized a continuous gradation to *closed folds* (or *tight folds*) in which flowage took place.

*Overfold: A fold in which the axial plane is inclined and the steeper fold limb is overturned (Busk, 1929, p. 28; Hills, 1953, p. 81; 1963A, p. 214; Billings, 1954, p. 40, who preferred the term *overturned fold*).

*Overturned fold:[2] A fold in which the axial plane is inclined and the steeper fold limb is upside down (Decker, 1920, p. 7; Willis and Willis, 1929, p. 34; Stočes and White, 1935; Billings, 1954, p. 40 equated with *overfold;* Dahlstrom, 1954, p. 8; Russell, 1955, p. 73). Leith (1923, p. 164) found definition unnecessary.

A completely different structure was defined by Turner and Weiss (1963, p. 119), namely, a cylindroidal or noncylindroidal plane fold in which the axial plane dips at less than 90°; this was equated with *inclined fold.*

PARALLEL FOLD: Van Hise (1896) originally defined parallel or *concentric folds* in which the strata are bent into parallel curves and retain a constant thickness throughout; such folds resemble a bent pack of cards, the flexing of which involves considerable bedding plane slip (Leith, 1923, p. 172; Busk, 1929, p. 9, equated with *competent fold;* Hills, 1953, p. 81; Hills, 1963A, p. 229, thought it undesirable to use concentric fold for parallel fold; Billings, 1954, p. 58; Russell, 1955, p. 77; de Sitter, 1956B, p. 182, equated with *distance-true fold;* Turner and Weiss, 1963, p. 112, equated with *competent folds* of sedimentary rocks). Decker (1920, p. 37) referred to parallel folds as "Numerous small folds with axes parallel with the trend of valleys. . . . The folds generally are small, involve only a few feet of strata, and usually are limited entirely to the valley floor, affecting the walls of the valley in only a few instances."

Stočes and White (1935, pp. 126-7) also used a definition completely different from that of all other geologists, and, in fact, for parallel fold they defined and illustrated *similar folds* of other workers, although their concentric fold is equivalent to the standard concept of parallel fold defined here.

PARASITIC FOLD: Small folds on the limbs of larger folds which share the same geometric elements (Turner and Weiss, 1963, p. 121). Used informally by D. J. Shearman, and adopted by de Sitter (1958, p. 280) who applied the term to folds commonly called "drag folds" (and minor folds, de Sitter, 1964, p. 179).

PIERCING FOLD: An anticline in which a mobile intrusive core of sedimentary rock broke through more brittle overlying rocks, as with

---

[2]Although commonly used, this name is misleading because the fold is not upside down (overturned).

salt domes (Billings, 1954, p. 59, who equated with *diapir fold*). Synonymous with *piercement fold* of Hills (1953, p. 87).

PIERCEMENT FOLD: A structure in which "mobile beds are actually injected through overlying strata at anticlinal axes" (Hills, 1953, p. 87, and Badgley, 1960, p. 27, both of whom equated with diapiric fold).

PLAINS TYPE OF FOLDING: Anticlines and domes produced by local uplift in cratonic areas without corresponding depression. The folds are more pronounced at depth, and above the crests of the folds thinning of the strata occurs. Commonly such folds are asymmetrical with the crestal plane dipping towards the steeper limb. These folds are said to be commonly associated with normal faulting (Hills, 1953, p. 67, and 1963A, p. 251, following Clark, 1932). These structures appear to equate with Billings' (1954, p. 92) *folds resulting from vertical movements*.

PLANE FOLD: A cylindroidal or noncylindroidal fold with an axial plane (Turner and Weiss, 1963, p. 108).

Plis de couverture: See *superficial, epidermic,* and *Jura-type folds* (de Sitter, 1956B, p. 187, 1964, p. 171; Hills, 1953, p. 62, 1963A, p. 326). Attributed to Argand (1916).

Plis de fond: See *derma fold* (de Sitter, 1956B, p. 187). Used by Hills (1953, 1963A) and de Sitter (1964), and attributed to Argand (1916).

PLUNGING FOLD: A plane or nonplane cylindroidal fold with the fold axis inclined to the horizontal (Turner and Weiss, 1963, p. 118; Billings, 1954, p. 46). Russell (1955, p. 69) added that in petroleum geology terminology such a fold must not show closure within the area under consideration.

POLYCLINAL FOLD: A group of adjacent related folds in which the axial surfaces have approximately random orientation but a common line of intersection (the fold axis, *B*) (Turner and Weiss, 1963, p. 115, after Greenly, 1919).

PTYGMATIC FOLD: Complex fold patterns distinguished from convolute folds by their lobate form; commonly their limbs are attenuated and the hinges are essentially concentric in form (Badgley, 1960, p. 27; Turner and Weiss, 1963, p. 116). Hills (1953, p. 93) included this structure as part of his *flow fold*. Hills (1963A, p. 279) wrote that the "curious meandrine contortions of quartzo-felspathic veins typically seen in gneisses especially in injection complexes and transfused rocks are known as ptygmatic structures or ptygmatic folds." The term was originated by Sederholm (e.g., 1913).

RECLINED FOLD: A cylindroidal plane fold in which the strike of the axial plane is normal to the trend of the fold axis (Turner and Weiss,

1963, p. 119, following Sutton, 1960A; this geometric relationship
was referred to as *recumbent fold* by Dahlstrom, 1954, Fig. 3, and by
Naha, 1959).

RECUMBENT FOLD: A cylindroidal or noncylindroidal plane fold in
which the axial plane is horizontal (Decker, 1920, p. 9; Stočes and
White, 1935, p. 116; Nevin, 1949, p. 47; Billings, 1954, p. 41;
Russell, 1955, p. 74; Turner and Weiss, 1963, p. 119). Hills (1953, p.
81) used this definition, but later (1963A, p. 215) insisted that both
fold limbs must be horizontal or nearly so.

Willis and Willis (1929, pp. 36-7) required an axial plane dipping
at between 60° and horizontal; Leith (1923, p. 164) said that a
definition was unnecessary.

Naha (1959) defined a recumbent fold as one with an axial plane
that has a subhorizontal trace on the profile plane, and a fold axis
which plunges at any angle up to 80° (cf. analogous use by Dahlstrom,
1954, Fig. 3). This is a *reclined fold* of Sutton (1960A) and Turner
and Weiss (1963, p. 119).

RECURRENT FOLD: According to Russell (1955, p. 80) a very
common variety of fold "recognized by the thinning of formations
and intervals near their crests, by the disappearance of certain
formations near their crests, and by local angular unconformities
which are present only in the higher parts of anticlines." The structure is a result of periodic uplift.

REFOLDED FOLD: Turner and Weiss (1963, p. 115, following Clifford, *et al.*, 1957) wrote that "individual convolute folds resemble
refolded folds." The term can be used to describe a fold which has
been affected by a second phase of folding.

Stočes and White (1935, p. 359, Fig. 569) described as a refolded
fold a structure resulting from folding over of the upper of a pair of
recumbent isoclinal folds, so that it partially envelops the lower fold.

RHEID FOLDING: "Geometrically there is a close analogy between 'rheid folds' as described by Carey and our models of slip
folding" (Turner and Weiss, 1963, p. 525, in reference to Carey,
1954).

Rheomorphic fold: A term based on genetic interpretation of features
supposed to indicate great mobility (e.g., exaggerated thinning of fold
limbs, and convolutions due to folding of first-formed folds); "rheomorphic folds exhibit a combination of shear folding and bulk distortion or flow, with shortening normal to the shear direction and
elongation parallel to the shearing" (Hills, 1963A, p. 245).

*Roof structure: A special case of an abnormal anticlinorium (Leith,
1923, p. 165).

Rootless intrafolial fold: An isolated closure, or a pair of opposing closures, in a disrupted portion of a tectonite; such closures appear to be floating as tectonic inclusions in relatively unfolded foliated rock (Turner and Weiss, 1963, p. 117).

Sag: A depressed basin or down warp of regional extent (Hills, 1963A, pp. 215, 313).

*Schistosity fold: Synonymous with *similar fold* (de Sitter, 1964, p. 170).

Shear fold: Fold produced by differential movement along closely-spaced shear planes; commonly called *slip fold* rather than shear fold (Hills, 1953, p. 90; Billings, 1954, p. 91; Badgley, 1960, p. 27, Fig. 28; Turner and Weiss, 1963, p. 480). Hills (1963A, p. 234) equated shear fold with *Gleitbrett fold* but not slip fold. De Sitter (1956B, pp. 182, 214, and 219) appeared to treat shear-fold, *cleavage fold,* and *similar fold* as synonyms. Ramberg (1963B) considered shear fold to be a variety of *bending fold.*

De Sitter (1964, p. 278) used shear fold for a variety of isoclinal recumbent asymmetric cleavage folds, different from axial-plane stress-cleavage folds in that foliation is not steeply dipping (commonly horizontal) and is commonly parallel to any relics of bedding; such folds were claimed to be characteristic of high-grade metamorphic rocks.

*Shear-cleavage folding: A type of folding intermediate between cleavage folding and concentric folding (de Sitter, 1964, p. 66).

SIMILAR FOLD: The profiles of folded *S*-surfaces are approximately similar; the lithologic units do not retain their original thicknesses, the fold limbs are attenuated, and, theoretically, individual folds persist indefinitely normal to the fold hinge (Leith, 1923, p. 174; Busk, 1929, p. 9, who equated with, and preferred, *incompetent fold*; Willis and Willis, 1929, p. 35; Hills, 1953, p. 84; 1963A, p. 233; Billings, 1954, p. 56; Russell, 1955, p. 77; Turner and Weiss, 1963, p. 112). De Sitter (1956B, pp. 182 and 216) did not specifically define similar fold, but used as above and equated it with his *cleavage fold*; to de Sitter similar folding included *fracture-cleavage, slaty-cleavage, chevron,* and *accordion folding*; later de Sitter (1964, pp. 170-71) equated similar fold with *axial-plane stress-cleavage fold* and used these categories as including *oblique-cleavage fold, accordion fold, flow fold,* and *schistosity fold.*

SIMPLE FOLD: A simple flexure without minor folds within the beds (Leith, 1923, p. 163; Willis and Willis, 1929, p. 39; Stočes and White, 1935, p. 121).

*Slaty-cleavage fold: One variety of *similar fold* (de Sitter, 1956B, p. 183).

SLIP FOLD: A fold in which the *S*-surfaces were kinematically passive and were transformed by differential movement along closely-spaced planes of slip (Hills, 1953, p. 90; Billings, 1954, p. 91; Turner and Weiss, 1963, p. 480; all equated with *shear fold*).

Slump fold: Intraformational fold formed by slumping of unlithified sediments (Hills, 1963A, p. 65).

*Squeezed fold: "A squeezed fold is one in which the compression has gone so far as to alter materially the thickness and form of the strata or to shear them off" (Willis and Willis, 1929, p. 38; term stated to be after Heim and de Margerie).

*Stress-cleavage fold: Apparently synonymous with *axial-plane stress-cleavage fold* (de Sitter, 1964, p. 80).

*Strike fold: Folds that have "axes parallel to main axis of the folded mountain chain" (Stočes and White, 1935, p. 126, who equated with *longitudinal fold*).

SUPERFICIAL FOLD: Fold independent of the Basement as a result of décollment (de Sitter, 1956B, p. 188; 1964, p. 171, who equated with *"Jurassic" fold* and *plis de couverture;* de Sitter, 1956B, p. 188, also equated with *epidermic fold*).

SUPERIMPOSED FOLD: A cross or oblique fold younger than the longitudinal or main fold; synonymous with *superposed fold* (Turner and Weiss, 1963, p. 121; de Sitter, 1964, p. 303). Stočes and White (1935, pp. 121-2) used *complex fold* for a fold which flexes the axis of another fold.

SUPERPOSED FOLD: Synonymous with *superimposed fold* (Turner and Weiss, 1963, p. 121).

SUPRATENUOUS FOLD: An anticline or a dome formed by compaction over buried hills is a supratenuous fold if the formations are thinner at the crest of the arch than in neighboring trough (Hills, 1953, p. 85, and 1963A, p. 251, considered that supratenuous folds are analogous to *compaction folds* and that folds of *Plains type* are commonly supratenuous; Billings, 1954, p. 56; Russell, 1955, p. 72). Term attributed to Nevin (1931).

Swell: A regional upwarp thousands of square kilometers in area (Hills, 1963A, p. 183).

SYMMETRIC FOLD: Challinor (1945, p. 87) pointed out that there is a sharp difference of opinion concerning this term; usages include:

1. A fold in which the limbs have an equal angle of dip and the axial plane is vertical (Hills, 1953, p. 80; 1963A, p. 214; Decker, 1920, p. 4; Willis and Willis, 1929, p. 33, who equated with *upright fold*; Russell, 1955, p. 71). This was an *upright fold* to Stočes and White (1935, p. 116).

2. A plane fold whose profile is bilaterally symmetric across the plane (Turner and Weiss, 1963, p. 122; Stočes and White, 1935, p. 116). Hills (1953, p. 81) thought this contrary to modern usage.

3. Nevin (1931, pp. 36-8, and 1949, p. 40) only required the fold limbs to be equally inclined to the axial plane (which can be in any orientation). Hills (1953, p. 81) claimed this is contrary to modern usage.

SYNCLINE: A fold which is concave upwards with younger rocks in the flexure formed of older rocks (Willis and Willis, 1929, p. 28). Decker (1920, p. 13) attributed to Gilbert (1876). Hills (1953, p. 76; 1963A, p. 213) did not insist on younger beds being in the core of the fold.

SYNCLINORIUM: A large syncline composed of many smaller folds (Decker, 1920, p. 14; Leith, 1923, p. 164; Willis and Willis, 1929, p. 41; Nevin, 1949, p. 39; Hills, 1953, p. 77; 1963A, p. 215).

SYNECLISE: Monometric or elongated large depressed portions of cratonic platforms (basement) covered with thick series (commonly 3 to 4 kilometers but ranging up to 10 kilometers) of cratonic sedimentary rocks (e.g., Caspian Syneclise). Term used by Shatski and Bogdanov (1957) and also Potapov (1960) in reference to the Moscow Syneclise.

SYNFORM: A fold whose flanks converge downwards, and in which either younger or older rocks occur in the core, or concave side, of the structure (Bailey and McCallien, 1937; Hills, 1963A, p. 214).

*Tight fold: Equivalent of *closed fold* (Billings, 1954, p. 45).

Transverse fold: Equivalent of *cross fold* (Stočes and White, 1935, p. 126) and used for folds which are at right angles by Dale (1896, pp. 553-4). Decker (1920, p. 38) used the term for folds with "axes transverse to the valley, so they are exposed in the flood plains, terraces and valley walls."

*Trough: An elongate depressed or downwarped basin of regional extent (Hills, 1963A, p. 216, who also used the term for the lowest point of any syncline, p. 213).

*True fold: Equivalent of *flexure fold* (Billings, 1954, p. 88).

*Unconformable fold: Equivalent of *discordant fold* (Stočes and White, 1935, p. 149).

*Undulating fold: "Minor folds with rounded apexes." (Stočes and White, 1935, p. 132).

Unsymmetric fold: Fold with unequal dip of beds on each flank (Decker, 1920, p. 5; Willis and Willis, 1929, p. 33). Compare with an *asymmetric fold.*

UPFOLD: A fold that closes upwards, i.e., form of a nonrecumbent anticline (Bailey, 1934, p. 467).

UPRIGHT FOLD: A fold in which the axial plane is vertical (Willis and Willis, 1929, p. 34; Stočes and White, 1935, p. 116; Turner and Weiss, 1963, p. 119, who equated with *normal fold*). To Hills

(1963A, p. 214) a fold in which the bisecting plane is vertical. Such structures are *symmetric folds* to Hills (1953, 1963A), Decker (1920, p. 4), Willis and Willis (1929, p. 30), and to Russell (1955, p. 71).

VALLEY BULGE: An anticlinal structure formed by squeezing up of clay in a valley floor where the superincumbent beds have been removed (Hills, 1953, p. 22; 1963A, p. 71, following Hollingworth, *et al.,* 1944). Similar bulges result from squeezing up of other plastic units, such as bedded rock salt or anhydrite.

VERTICAL FOLD: A plane or nonplane cylindroidal fold with a vertical axis (Turner and Weiss, 1963, p. 119).

*Zig-zag fold: Equivalent of *chevron fold* (Stočes and White, 1935, p. 132; Hills, 1953, p. 84; Russell, 1955, p. 77) and of *angular fold, accordion fold, concertina fold,* and *mitre fold* (Hills, 1963A, p. 238).

While it must be admitted that descriptive terms are often useful in accounts of deformed areas, it will be obvious from the above catalog that a great deal of ambiguity can arise from the use of terms that are defined imprecisely and that, in some cases, are used in dissimilar senses by different authors. In the majority of cases an accurate record of the axial plunge, an orientated sketch of the fold profile, and a note of the size of the structure are to be preferred to descriptive terminology. Commonly profiles of neighboring folds are dissimilar, so that an area requires careful sampling according to a statistical plan; subjective selection of an individual fold as a "typical" example worthy of illustration can be extremely misleading. The quantitative attributes based on direction cosines of the normals to S-planes provide a much more desirable picture of a fold than a series of qualitative terms.

Chapter 16

References Cited

The following references are cited with the abbreviations used in the fourth edition of the *World List of Scientific Periodicals published in the years 1900-1960*. The three volumes of this *List* were edited by P. Brown and G. B. Stratton and were published in 1963, 1964, and 1965 by Butterworth, Inc., Washington, D.C., U.S.A.

ACKERMANN, VON E., & FORSTER, A., 1960, Grundzüge der Stratigraphie und Struktur des Irumiden-Orogens: *Int. geol. Congr. XXI Copenhagen*, Pt. 18, pp. 182-92.

ACKERMANN, K. J., NICHOLSON, R., & WALTON, B. J., 1960, Mineral development and deformation in metasedimentary rocks in the Glomfjord region, northern Norway: *Int. geol. Congr. XXI Copenhagen*, Pt. 19, pp. 54-63.

ADERCA, B. M., 1957, Un cas de "boudinage" à grande échelle; la mine de Rutongo au Ruanda: *Annls. Soc. géol. Belg.*, v. 80B, pp. 279-85.

AGRON, S. L., 1950, Structure and petrology of the Peach Bottom Slate, Pennsylvania and Maryland, and its environment: *Bull. geol. Soc. Am.*, v. 61, pp. 1265-306.

———, 1963A, in *New England Intercollegiate Geological Conference Guidebook* (55th Ann. Mtg.), Providence, Rhode Island, U.S.A., 55 pp.

———, 1963B, Tectonic boulders in Carboniferous Conglomerate, Rhode Island: *Spec. Pap. geol. Soc. Am.*, 76, p. 3.

AGTERBERG, F. P., 1959, On the measuring of strongly dispersed minor folds; *Geologie Mijnb.*, v. 21, pp. 133-7.

———, 1961, Tectonics of the crystalline basement of the Dolomites in North Italy: *Geologica ultraiect.*, No. 8, pp. 1-232.

———, 1964A, The method of statistical structural analysis (as applied to the crystalline basement of the Dolomites in North Italy): *Geologie Mijnb.*, v. 43, pp. 222-35.

———, 1964B, Methods of trend surface analysis: *Colo. Sch. Mines Q.*, v. 59, pp. 111-30.

———, & Briggs, G., 1963, Statistical analysis of ripple marks in Atokan and Desmoinesian rocks in the Arkoma Basin of east-central Oklahoma: *J. sedim. Petrol.*, v. 33, pp. 393-410.

AKAAD, M. K., 1956, The Ardara granitic diapir of County Donegal, Ireland: *Q. Jl. geol. Soc. Lond.*, v. 112, pp. 263-90.

ALLEN, J. R. L., 1963, The classification of cross-stratified units, with notes on their origin: *Sedimentology*, v. 2, pp. 93-114.

ALLISON, A., 1933, The Dalradian succession in Islay and Jura: *Q. Jl. geol. Soc. Lond.*, v. 89, pp. 125-44.

———, 1941, Loch Awe succession and tectonics: Kilmartin-Tayvallich-Danna: *Q. Jl. geol. Soc. Lond.*, v. 96 (for 1940), pp. 423-49.

ANDERSON, E. M., 1923, The geology of the schists of the Schichallion district (Perthshire): *Q. Jl. geol. Soc. Lond.*, v. 79, pp. 423-42.

———, 1948, On lineation and petrofabric structure and the shearing movement by which they have been produced: *Q. Jl. geol. Soc. Lond.*, v. 104, pp. 99-126.

———, 1951, *The dynamics of faulting and dyke formation with applications to Britain* (2nd edit.): Oliver and Boyd, Edinburgh, 206 pp.

ANDERSON, J. G. C., 1935, The Dalradian succession in the Pass of Brander District, Argyll: *Geol. Mag.*, v. 72, pp. 74-80.

———, 1954, The pre-Carboniferous rocks of the Slieve League promontory, Co. Donegal: *Q. Jl. geol. Soc. Lond.*, v. 109 (for 1953), pp. 399-422.

———, 1955, Superposition of Caledonoid folds on an older fold-system in the Dalradians of Malin Head: *Geol. Mag.*, v. 92, pp. 83-4.

———, 1956, The Moinian and Dalradian rocks between Glen Roy and the Monadhliath Mountains, Inverness-shire: *Trans. R. Soc. Edinb.*, v. 63, pp. 15-36.

ARGAND, E., 1912, Sur la segmentation tectonique des Alpes Occidentales: *Bull. Soc. vaud. Sci. nat.*, v. 48, pp. 345-56.

———, 1915, Extrait de la *Suisse liberale* du 9 juin 1915: *Soc. Neuchat. Sci. Nat.*, pp.1-3.

———, 1916, Sur l'arc des Alpes occidentales: *Eclog. geol. Helv.*, v. 14, pp. 146-204.

ATKINSON, D. J., 1960, Caledonian tectonics of Prins Karls Forland: *Int. geol. Congr. XXI Copenhagen*, Pt. 19, pp. 17-27.

AYRTON, W. G., 1964, *A structural study of the Chandler-Port Daniel area, Gaspé Peninsula, Quebec:* Ph. D. Thesis, Northwestern University, 219 pp.

BADGLEY, P., 1960, *Structural methods for the exploration geologist and a series of problems for structural geology students:* Harper and Brothers, New York, 280 pp.

BAILEY, E. B., 1909, Glencoe and Glen Nevis: *Mem. geol. Surv. Summ. Prog.* for 1908, p. 51.

———, 1910, Recumbent folds in the Schists of the Scottish Highlands: *Q. Jl. geol. Soc. Lond.*, v. 66, pp. 586-620.

———, 1913, The Loch Awe Syncline (Argyllshire): *Q. Jl. geol. Soc. Lond.*, v. 69, pp. 280-307.

———, 1922, The structure of the south-west Highlands of Scotland: *Q. Jl. geol. Soc. Lond.*, v. 78, pp. 82-131.

———, 1925, Perthshire tectonics: Loch Tummel, Blair Atholl and Glen Shee: *Trans. R. Soc. Edinb.*, v. 53, pp. 671-98.

———, 1930, New light on sedimentation and tectonics: *Geol. Mag.*, v. 67, pp. 77-92.

———, 1934, West Highland tectonics: Loch Leven to Glen Roy: *Q. Jl. geol. Soc. Lond.*, v. 90, pp. 462-525.

———, 1935, *Tectonic essays mainly Alpine:* Oxford University Press, London, 200 pp.

———, 1936, Sedimentation in relation to tectonics: *Bull. geol. Soc. Am.*, v. 47, pp. 1713-26.

——, 1938, Eddies in mountain structure: *Q. Jl. geol. Soc. Lond.,* v. 94, pp. 607-25.

——, 1950, The structural history of Scotland: *Int. geol. Congr. XVIII London,* 1948, Pt. 1, pp. 230-55.

——, 1953, Facies changes versus sliding: Loch Leven, Argyll: *Geol. Mag.,* v. 90, pp. 111-13.

——, 1955, Moine tectonics and metamorphism in Skye: *Trans. Edinb. geol. Soc.,* v. 16, pp. 93-166.

——, 1959, Structural geometry of Dalradian rocks at Loch Leven, Scottish Highlands: a discussion: *J. Geol.,* v. 67, pp. 246-7.

——, & LAWRIE, T. R. M., 1960, The geology of Ben Nevis and Glen Coe and the surrounding country (explanation of Sheet 53): *Mem. geol. Surv. U. K.,* pp. 1-307.

——, & MACKIN, J. H., 1937, Recumbent folding in the Pennsylvania piedmont —preliminary statement: *Am. J. Sci.,* v. 33 (5th ser.), pp. 187-90.

——, & MCCALLIEN, W. J., 1934, The metamorphic rocks of north-east Antrim: *Trans. R. Soc. Edinb.,* v. 58, pp. 163-77.

——, & ——, 1937, Perthshire tectonics: Schichallion to Glen Lyon: *Trans. R. Soc. Edinb.,* v. 59, pp. 79-117.

——, & TILLEY, C. E., 1952, Rocks claimed as conglomerate at the Moinian-Lewisian junction: *Int. geol. Congr. XVIII London,* 1948, Pt. 13, p. 272.

BAIN, G. W., 1931, Flowage folding: *Am. J. Sci.,* v. 22, pp. 503-30.

BAIRD, A. K., 1962, Superposed deformations in the Central Sierra Nevada foothills east of the Mother Lode: *Univ. Calif. Publs. Bull. Dep. Geol.,* v. 42, pp. 1-69.

——, MCINTYRE, D. B., WELDAY, E. E., & MADLEM, K. W., 1964, Chemical variations in a granitic pluton and its surrounding rocks: *Science, N.Y.,* v. 146, pp. 258-9.

BALK, R., 1927, Die primäre Struktur des Noritmassivs von Peekskill am Hudson, nördlich New York (auch bekannt als "Cortlandt Norit"): *Neues Jb. Miner. Geol. Paläont. BeilBd.,* v. 57, Abt. B, pp. 249-303.

——, 1936, Structural and petrologic studies in Dutchess County, New York: *Bull. geol. Soc. Am.,* v. 47, pp. 685-774.

——, 1949, Structure of the Grand Saline Salt Dome, Van Zandt County, Texas: *Bull. Am. Ass. Petrol. Geol.,* v. 33, pp. 1791-829.

BARRELL, J., 1917, Rhythms and the measurements of geologic time: *Bull. geol. Soc. Am.,* v. 28, pp. 745-904.

BARROW, G., HINXMAN, L. W., & CRAIG, E. H. C., 1913, The geology of upper Strathspey, Gaick, and the Forest of Atholl (explanation of sheet 64): *Mem. geol. Surv. U.K.,* pp. 1-116.

BARTH, T. F. W., 1936, Structural and petrologic studies in Dutchess County, New York, Part II. Petrology and metamorphism of the Paleozoic rocks: *Bull. geol. Soc. Am.,* v. 47, pp. 775-850.

——, CORRENS, C. W., & ESKOLA, P. E., 1939, *Die Enstehung der Gesteine: Ein Lehrbuch der Petrogenese:* Springer, Berlin, 422 pp.

——, & DONS, J. A., 1960, Precambrian of Southern Norway: *Norg. geol. Unders.,* No. 208, pp. 6-67.

BATES, T. F., 1947, Investigation of the micaceous minerals in slate: *Am. Miner.,* v. 32, pp. 625-36.

BEAVIS, F. C., 1964, Strain slip cleavage in the Ordovician sediments of Central Victoria: *Geol. Mag.,* v. 101, pp. 504-11.

BECKER, G. F., 1882, Geology of the Comstock Lode and the Washoe District: *Monogr. U.S. geol. Surv.*, No. 3, pp. 1-422.

———, 1893, Finite homogeneous strain, flow and rupture of rocks: *Bull. geol. Soc. Am.*, v. 4, pp. 13-90.

———, 1896, Schistosity and slaty cleavage: *J. Geol.*, v. 4, pp. 429-48.

———, 1904, Experiments on schistosity and slaty cleavage: *Bull. U.S. geol. Surv.*, 241, pp. 1-34.

BEDERKE, E., 1963, Altersgliederung und Dichteverteilung im kristallinen Grundebirge: *Geol. Rdsch.*, v. 52 (for 1962), pp. 1-12.

BEHRE, C. H., JR., 1933, Slate in Pennsylvania: *Bull. Pa. geol. Surv.*, M16, pp. 1-400.

BELOUSSOV, V. V., 1961, The origin of folding in the earth's crust: *J. geophys. Res.*, v. 66, pp. 2241-54.

———, 1962, *Basic problems in geotectonics:* McGraw-Hill Book Co., Inc., New York, 816 pp.

BELOV, A. A., DOLGINOV, Y. A., KROPACHEV, S. M., ORLOV, R. Y., & SOKOLOV, B. A., 1960, The Cherkessk-Kelasuri transverse dislocation in the structure of the Greater Caucasus: *Izv. Acad. Sci. USSR geol. Ser.*, 1961, No. 6, pp. 17-24 (English translation of *Izv. Akad. Nauk. SSSR*, 1960).

BEMMELEN, R. W. VAN, 1960, Zur Mechanik der ostalpinen Deckenbildung: *Geol. Rdsch.*, v. 50, pp. 474-99.

BENTZ, A., 1949, Ergebnisse der erdölgeologischen Erforschung Nordwest-deutschlands 1932-1947, ein Überblick (p. 7-18) in *Erdöl und Tektonik in Nordwestdeutschland:* Bodenforschung, Hannover-Celle, 387 pp.

BERTHELSEN, A., 1957, The structural evolution of an ultra- and polymetamor-phic gneiss-complex, west Greenland: *Geol. Rdsch.*, v. 46, pp. 173-85.

———, 1960A, An example of a structural approach to the migmatite problem: *Int. geol. Congr. XXI Copenhagen*, Pt. 14, pp. 149-57.

———, 1960B, Structural studies in the Pre-Cambrian of Western Greenland, II Geology of Tovqussap Nunâ: *Meddr. Grønland*, v. 123, pp. 1-226.

———, 1960C, Structural contour-maps applied in the analysis of double fold structures: *Geol. Rdsch.*, v. 49, pp. 459-66.

BEST, M. G., 1963, Petrology and structural analysis of metamorphic rocks in the southwestern Sierra Nevada Foothills, California: *Univ. Calif. Publs. Bull. Dep. Geol.*, v. 42, pp. 111-58.

BHATTACHARJI, S., 1958, Theoretical and experimental investigations on cross-folding: *J. Geol.*, v. 66, pp. 625-67.

———, 1959, Errata to "Theoretical and experimental investigations on cross-folding": *J. Geol.*, v. 67, pp. 125-7.

BIEMESDERFER, G. K., 1949, A new tool for plotting points on the equal area net: *Proc. Pa. Acad. Sci.*, v. 23, pp. 72-5.

BILLINGS, M. P., 1937, Regional metamorphism of the Littleton-Moosilauke area, New Hampshire: *Bull. geol. Soc. Am.*, v. 48, pp. 463-566.

———, 1942, *Structural Geology* (1st edit.): Prentice-Hall, Inc., New York, 473 pp.

———, 1954, *Structural Geology* (2nd edit.): Prentice-Hall, Inc., Englewood Cliffs, 514 pp.

———, & CLEAVES, A. B., 1935, Brachiopods from mica schist, Mt. Clough, New Hampshire: *Am. J. Sci.*, v. 30, pp. 530-6.

———, & SHARP, R. P., 1937, Petrofabric study of a fossiliferous schist, Mt.

Clough, New Hampshire: *Am. J. Sci.*, v. 34, pp. 277-92.

BIRCH, F., 1961, Role of fluid pressure in mechanics of overthrust faulting — Discussion: *Bull. geol. Soc. Am.*, v. 72, pp. 1441-3.

BISHOP, M. S., 1960, *Subsurface mapping:* John Wiley and Sons, Inc., New York, 198 pp.

BONNEY, T. G., 1886, Anniversary address of the President: *Proc. geol. Soc.*, v. 42, pp. 38-115.

———, 1919, Foliation and metamorphism in rocks: *Geol. Mag.*, v. 56, pp. 196-203, and 246-50.

BORCHERS, R., 1961, Exploration of the Witwatersrand System and its extensions: *Proc. geol. Soc. S. Afr.*, v. 64, pp. lxvii-xcviii.

BORN, A., 1929, Ueber Druckschieferung im Varistìschen Gebirgskörper: *Fortschr. Geol. Palaeont.*, v. 7, pp. 329-427.

BOSCHMA, D., 1963, Successive Hercynian structures in some areas of the Central Pyrenees: *Leid. geol. Meded.*, v. 28, pp. 103-76.

BOSWELL, P. G. H., 1949, *The Middle Silurian rocks of North Wales :* Edward Arnold and Co., London, 448 pp.

BOUCOT, A. J., & THOMPSON, J. B., JR., 1963, Metamorphosed Silurian brachiopods from New Hampshire: *Bull. geol. Soc. Am.*, v. 74, pp. 1313-33.

BOWES, D. R., 1954, The transformation of tillite by migmatization at Mount Fitton, South Australia: *Q. Jl. geol. Soc. Lond.*, v. 109 (for 1953), pp. 455-81.

———, 1956, Tillite-granite transformations: *Geol. Mag.*, v. 93, pp. 181-2.

———, & JONES, K. A., 1958, Sedimentary features and tectonics in the Dalradian of western Perthshire: *Trans. Edinb. geol. Soc.*, v. 17, pp. 133-40.

———, & WRIGHT, A. E., 1962, Washout structures in the Dalradian near Kentallen, Argyll: *Geol. Mag.*, v. 99, pp. 53-6.

BRACE, W. F., 1953, The geology of the Rutland area, Vermont: *Bull. Vt. geol. Surv.*, 6, pp. 1-124.

BRAITSCH, O., 1957, Zur Petrographie und Tektonik des Biotitgneises im Südlichen Vorspessart: *Abh. hess. Landesamt. Bodenforsch.*, v. 18, pp. 73-99.

BREDDIN, H., 1931, Über das Wesen der Druckschieferung im Rheinischen Schiefergebirge: *Zentbl. Miner. Geol. Paläont.*, Abt. B., pp. 206-16.

———, 1956A, Die tektonische Deformation der Fossilien im Rheinischen Schiefergebirge: *Z. dt. geol. Ges.*, v. 106 (for 1954), pp. 227-305.

———, 1956B, Tektonische Gesteinsdeformation im Karbongürtel Westdeutschlands und Süd-Limburgs: *Z. dt. geol. Ges.*, v. 107 (for 1955), pp. 232-60.

———, 1957, Tektonische Fossil- und Gesteinsdeformation im Gebiet von St. Goarshausen (Rheinisches Schiefergebirge): *Decheniana*, v. 110, pp. 289-350.

———, 1958A, Tektonische Gesteinsdeformation im kaum gefalteten Karbon des Erkelenzer Steinkohlenrevieres: *Neues Jb. Geol. Paläont. Mh.*, 1958, pp. 172-88.

———, 1958B, Tektonisch deformierte Fossilien von der Zeche Mathias Stinnes im der Emscher-Mulde und ilre Bedeutung für Tektonik und Paläontologie des Ruhrkarbons: *Glückauf*, v. 94, pp. 1095-101.

———, 1958C, Die regionale tektonische Fossil- und Gesteinsdeformation in der Molasse der Ost- und Mittelschweiz: *Eclog. geol. Helv.*, v. 51, pp. 378-9.

———, 1962, Zur geometrischen Tektonik des altdevonischen Grundgebirges im

Siegerland (Rheinisches Schiefergebirge): *Geol. Mitt. Aachen,* v. 2, pp. 227-82.

BRETT, G. W., 1955, Cross-bedding in the Baraboo quartzite of Wisconsin: *J. Geol.,* v. 63, pp. 143-8.

BRINKMANN, R., GIESEL, W., & HOEPPENER, R., 1961, Über Versuche zur Bestimmung der Gesteinsanisotropie: *Neues Jb. Geol. Paläont. Mh.,* Abt. B., 1961, pp. 22-33.

BROBST, D. A., 1962, Geology of the Spruce Pine District, Avery, Mitchell, and Yancey Counties, North Carolina: *Bull. U.S. geol. Surv.,* 1122-A, pp. 1-26.

BROUGHTON, J. G., 1946, An example of the development of cleavages: *J. Geol.,* v. 54, pp. 1-18.

BROWN, J. S., & ENGEL, A. E. J., 1956, Revision of the Grenville stratigraphy and structure in the Balmat-Edwards district, northwest Adirondacks, New York: *Bull. geol. Soc. Am.,* v. 67, pp. 1599-622.

BROWN, P. E., MILLER, J., SOPER, N. J., & YORK, D., 1965, The potassium-argon age pattern of the British Caledonides: *Proc. Yorks. geol. Soc.,* v. 35, pp. 103-38.

BRYAN, W. H., & JONES, O. A., 1955, Radiolaria as critical indicators of deformation: *Pap. Dep. Geol. Univ. Qd.,* v. 4, pp. 1-5.

BRYHNI, I., 1962, Structural analysis of the Grøneheia area, Eikefjord, western Norway: *Norsk. geol. Tidsskr.,* v. 42, pp. 331-69.

BUCHER, W. H., 1944, The stereographic projection, a handy tool for the practical geologist: *J. Geol.,* v. 52, pp. 191-209.

———, 1953, Fossils in metamorphic rocks: a review: *Bull. geol. Soc. Am.,* v. 64, pp. 275-300.

BUDDINGTON, A. F., 1929, Granite phacoliths and their contact zones in the northwest Adirondacks: *Bull. N.Y. St. Mus.,* 281, pp. 51-107.

———, 1934, Geology and mineral resources of the Hammond, Antwerp, and Lowville quadrangles: *Bull. N.Y. St. Mus.,* 296, pp. 7-246.

———, 1939, Adirondack igneous rocks and their metamorphism: *Mem. geol. Soc. Am.,* 7, pp. 1-354.

———, & CHAPIN, T., 1929, Geology and mineral deposits of southeastern Alaska: *Bull. U.S. geol. Surv.,* 800, pp. 1-398.

BUSK, H. G., 1929, *Earth Flexures, their geometry and their representation and analysis in geological sections with special reference to the problem of oil finding:* Cambridge Univ. Press, Cambridge, 106 pp.

CADIGAN, R. A., 1962, A method for determining the randomness of regionally distributed quantitative geologic data: *J. sedim. Petrol.,* v. 32, pp. 813-8.

CAMPANA, B., 1955, The stratigraphy of the northern Flinders Ranges and the alleged granitization of tillite in the Mt. Fitton area: *Aust. J. Sci.,* v. 18, pp. 75-7.

CAMPBELL, J. D., 1958, *En Echelon* folding: *Econ. Geol.,* v. 53, pp. 448-72.

CAREY, S. W., 1954, The rheid concept in geotectonics: *J. geol. Soc. Aust.,* v. 1 (for 1953), pp. 67-117.

———, 1962A, Folding: *J. Alberta Soc. Petrol. Geol.,* v. 10, pp. 95-144.

———, 1962B, Scale of geotectonic phenomena: *J. geol. Soc. India,* v. 3, pp. 97-105.

CHALLINOR, J., 1945, The primary and secondary elements of a fold: *Proc. Geol. Ass.*, v. 56, pp. 82-8.

CHAYES, F., 1960, On correlation between variables of constant sum: *J. geophys. Res.*, v. 65, pp. 4185-93.

———, 1961, Numerical petrography: *Yb. Carnegie, Instn. Wash.*, v. 60, pp. 158-65.

———, & SUZUKI, Y., 1963, Geological contours and trend surfaces (A discussion): *J. Petrology*, v. 4, pp. 307-12.

CHINNER, G. A., 1961, The origin of sillimanite in Glen Cova, Angus: *J. Petrology*, v. 2, pp. 312-23.

———, SANDO, M., & WHITE, A. J. R., 1956A, On the supposed transformation of tillite to granite at Mount Fitton, South Australia: *Geol. Mag.*, v. 93, pp. 18-24.

———, ———, & ———, 1956B, Tillite-granite transformations: *Geol. Mag.*, v. 93, p. 263.

CHORLEY, R. J., 1964, Geography and analogue theory: *Ann. Ass. Am. Geogr.*, v. 54, pp. 127-37.

CHRISTENSEN, M. N., 1959, Cleavage and foliation in the Mineral King area, California: *Bull. geol. Soc. Am.*, v. 70, p. 1580.

———, 1963, Structure of metamorphic rocks at Mineral King, California: *Univ. Calif. Publs. Bull. Dep. Geol.*, v. 42, pp. 159-98.

CHRISTIE, A. M., 1953, Goldfields-Martin Lake map-area, Saskatchewan: *Mem. geol. Surv. Brch. Can.*, 269, pp. 1-126.

CHRISTIE, J. M., 1956, Tectonic phenomena associated with the Post-Cambrian movements in Assynt: *Advmt. Sci., Lond.*, v. 12, pp. 572-3, 577.

———, 1963, The Moine Thrust Zone in the Assynt region Northwest Scotland: *Univ. Calif. Publs. Bull. Dep. Geol.*, v. 40, pp. 345-440.

CLABAUGH, P. S., & MUELHBERGER, W. R., 1962, Orientation of Salt Crystals in Grand Saline Salt Dome, Van Zandt County, Texas: *Spec. Pap. geol. Soc. Am.*, 68, p. 149.

CLARK, R. H., & McINTYRE, D. B., 1951, The use of the terms pitch and plunge: *Am. J. Sci.*, v. 249, pp. 591-9.

CLARK, S. K., 1932, The mechanics of the Plains-Type Folds of the Mid-Continent Area: *J. Geol.*, v. 40, pp. 46-61.

CLENSHAW, C. W., GOODWIN, E. T., MARTIN, D. W., MILLER, G. F., OLVER, F. W. J., & WILKINSON, J. H., 1961, *Modern computing methods* (2nd edit.): Philosophical Library, Inc., New York, 170 pp.

CLIFFORD, P., 1960, The geological structure of the Loch Luichart area, Ross-shire: *Q. Jl. geol. Soc. Lond.*, v. 115 (for 1959), pp. 365-88.

———, FLEUTY, M. J., RAMSAY, J. G., SUTTON, J., & WATSON, J., 1957, The development of lineation in complex fold systems: *Geol. Mag.*, v. 94, pp. 1-24.

CLIFFORD, T. N., 1957, The stratigraphy and structure of part of the Kintail district of southern Ross-shire: its relation to the Northern Highlands: *Q. Jl. geol. Soc. Lond.*, v. 113, pp. 57-92.

CLOOS, E., 1941, Flowage and cleavage in Appalachian folding: *Trans. N.Y. Acad. Sci.*, v. 3 (2nd ser.), pp. 185-90.

———, 1943, Method of measuring changes of stratigraphic thicknesses due to flowage and folding: *Trans. Am. geophys. Un.*, v. 24, pp. 273-80.

———, 1945, Correlation of lineation with rock-movement: *Trans. Am. geophys. Un.*, v. 25 (for 1944), pp. 660-2.

————, 1946, Lineation a critical review and annotated bibliography: *Mem. geol. Soc. Am.*, 18, pp. 1-122.

————, 1947A, Oölite deformation in the South Mountain fold, Maryland: *Bull. geol. Soc. Am.*, v. 58, pp. 843-918.

————, 1947B, Boudinage: *Trans. Am. geophys. Un.*, v. 28, pp. 626-32.

————, 1953A, Lineation Review of literature 1942-1952: *Mem. geol. Soc. Am. (Supp.)*, 18, pp. 1-14.

————, 1953B, Appalachenprofil in Maryland: *Geol. Rdsch.*, v. 41, pp. 145-60.

————, 1958, Lineation und Bewegung, eine Diskussionsbemerkung: *Geologie*, v. 7, pp. 307-11.

————, 1961, Bedding slips, wedges, and folding in layered sequences: *Bull. Commn géol. Finl.*, No. 196, pp. 105-22.

————, & HIETANEN, A., 1941, Geology of the "Martic Overthrust" and the Glenarm series in Pennsylvania and Maryland: *Spec. Pap. geol. Soc. Am.*, 35, pp. 1-207.

————, & MURPHY, T. D., 1958, Guidebooks 4 & 5 Structural Geology of South Mountain and Appalachians in Maryland: *Johns Hopkins Univ. Stud. Geol.*, No. 17, pp. 1-85.

CLOOS, H., 1936, *Einführung in die Geologie; Ein Lehrbuch der inneren Dynamik:* Gebrüder Borntraeger, Berlin, 356 pp.

————, 1948, Gang und Gehwerk einer Falte: *Z. dt. geol. Ges.*, v. 100, pp. 290-303.

CLOUGH, C. T., 1898, *Mem. geol. Surv. Summ. Prog.* for 1897, pp. 36-8.

————, 1900, *Mem. geol. Surv. Summ. Prog.* for 1899, pp. 14-8.

————, 1901, *Mem. geol. Surv. Summ. Prog.* for 1900, pp. 8-10.

COCHRAN, W. G., MOSTELLER, F., TUKEY, J. W., & JENKINS, W. O., 1954, *Statistical problems of the Kinsey Report on sexual behavior in the human male:* American Statistical Assoc., Washington, D.C., 338 pp.

COE, K., 1959, Boudinage structure in West Cork, Ireland: *Geol. Mag.*, v. 96, pp. 191-200.

————, 1962, *Some aspects of the Variscan fold belt:* Manchester Univ. Press, Manchester, 163 pp.

COHN, P. M., 1961, *Solid Geometry:* Routledge and Kegan Paul, London, 72 pp.

COLLOMB, P., 1960, La linéation dans les roches: *Bull. trimest. Serv. Inf. géol. Bur. Rech. géol. géophys.*, No. 48, pp. 1-11.

CONDON, M. A., 1959, Sedimentary structures in the metamorphic rocks and ore-bodies, Broken Hill: *Proc. Australas. Inst. Min. Metall.*, No. 189, pp. 47-79.

COOKE, H. C., 1931, Anomalous grain relationship in the Caldwell quartzites of Thetford District, Quebec: *Proc. Trans. R. Soc. Can.*, v. 25, Sect. 4, pp. 71-4.

————, 1948, Regional structure of the Lake Huron-Sudbury area: *in Structural Geology of Canadian ore deposits: Canadian Institute of Mining and Metallurgy, Geological Division*, pp. 580-9.

COOPER, B. N., 1960, The geology of the region between Roanoke and Winchester in the Appalachian Valley of Western Virginia: *Johns Hopkins Univ. Stud. Geol.*, No. 18, pp. 1-84.

CORIN, F., 1932, A propos du boudinage en Ardenne: *Bull. Soc. belge Géol. Paléont. Hydrol.*, v. 42, pp. 101-14.

COTE, L. J., DAVIS, J. O., MARKS, W., MCGOUGH, R. J., MEHR, E., PIERSON, W. J., ROPSEK, J. F., STEPHENSON, G., & VETTER, R. C., 1960, The directional spectrum of a wind generated sea as determined from data obtained by the stereo wave observation project: *Met. Pap. N.Y. Univ.*, v. 2, pp. 1-88.

COWIE, J. W., & ADAMS, P. J., 1957, The geology of the Cambro-Ordovician rocks of central East Greenland; Part I—Stratigraphy and structure: *Meddr. Grønland*, v. 153, No. 1, pp. 1-193.

COX, G. H., & DAKE, C. L., 1916, Geological criteria for determining the structural position of sedimentary beds: *Bull. Mo. Sch. Mines tech. Ser.*, v. 2, pp. 1-59.

CRADDOCK, J. C., 1957, Stratigraphy and structure of the Kinderhook Quadrangle, New York, and the "Taconic Klippe": *Bull. geol. Soc. Am.*, v. 68, pp. 675-723.

CRAIG, G. Y., & WALTON, E. K., 1962, Sedimentary structures and palaeocurrent directions from the Silurian rocks of Kirkcudbrightshire: *Trans. Edinb. geol. Soc.*, v. 19, pp. 100-19.

CROOK, K. A. W., 1964, Cleavage in weakly deformed mudstones: *Am. J. Sci.*, v. 262, pp. 523-31.

CROWELL, J. C., 1962, Displacement along the San Andreas Fault: *Spec. Pap. geol. Soc. Am.*, 71, pp. 1-61.

CUMMINS, W. A., & SHACKLETON, R. M., 1955, The Ben Lui recumbent syncline (S. W. Highlands): *Geol. Mag.*, v. 92, pp. 353-63.

CURIE, P., 1894, Sur la symétrie dans les phénomènes physiques, symétrie d'un champ électrique et d'un champ magnétique: *J. Phys. théor. appl.*, v. 3, pp. 393-415.

CURRIE, J. B., PATNODE, H. W., & TRUMP, R. P., 1962, Development of folds in sedimentary strata: *Bull. geol. Soc. Am.*, v. 73, pp. 655-74.

CUTHBERT, F. L., 1946, Differential thermal analyses of New Jersey clays: *Bull. geol. Surv. New Jers.*, 60, pp. 5-20.

DAHLSTROM, C. D. A., 1954, Statistical analysis of cylindrical folds: *Bull. Can. Inst. Min. Metall.*, v. 57, pp. 140-5.

DALE, T. N., 1896, Structural details in the Green Mountain region and in eastern New York: *Rep. U.S. geol. Surv.*, 16, pp. 543-70.

———, 1899, The slate belt of eastern New York and western Vermont: *Rep. U.S. geol. Surv.*, 19, pp. 153-300.

———, et al., 1914, Slate in the United States: *Bull. U.S. geol. Surv.*, 586, pp.1-220.

DALY, R. A., 1912, Reconnaissance of the Shuswap Lakes and vicinity (South-central British Columbia): *Summ. Rep. geol. Surv. Brch. Can.*, 1911, pp. 165-74.

———, 1915, A geological reconnaissance between Golden and Kamloops, B. C., along the Canadian Pacific Railway: *Mem. geol. Surv. Brch. Can.*, 68, pp. 1-260.

———, 1917, Metamorphism and its phases, *Bull. geol. Soc. Am.*, v. 28, pp. 375-418.

D'AMICO, C., 1964, Petrography and tectonics in the Agordo-Cereda Region (Crystalline of the Southern Alps): *Geologie Mijnb.*, v. 43, pp. 236-44.

DANA, J. D., 1873, On some results of the Earth's contraction from cooling, including a discussion of the Origin of Mountains, and the nature of the Earth's Interior-Part I: *Am. J. Sci.*, v. 5 (3rd ser.), pp. 423-43.

DARWIN, C., 1846, *Geological Observations on South America. Being the third part of the geology of the Voyage of the Beagle, under the command of Capt. Fitzroy, R. N. during the years 1832 to 1836:* Smith, Elder and Co., London, 279 pp.

DAWSON, K. R., 1954, Structural features of the Preissac-Lacorne batholith, Abitibi County, Quebec: *Geol. Surv. Pap. Can.,* 53-4, pp. 1-22.

———, & WHITTEN, E. H. T., 1962, The quantitative mineralogical composition and variation of the Lacorne, La Motte, and Preissac granite complex, Quebec, Canada: *J. Petrology,* v. 3, pp. 1-37.

DAWSON-GROVE, G. E., 1955, Analysis of minor structures near Ardmore, County Waterford, Eire: *Q. Jl. geol. Soc. Lond.,* v. 111, pp. 1-21.

DEARMAN, W. R., SHIELLS, K. A. G., & LARWOOD, G. P., 1962, Refolded folds in the Silurian rocks of Eyemouth, Berwickshire: *Proc. Yorks. geol. Soc.,* v. 33, pp. 273-85.

DEARNLEY, R., 1963, The Lewisian complex of South Harris with some observations of the metamorphosed basic intrusions of the Outer Hebrides, Scotland: *Q. Jl. geol. Soc. Lond.,* v. 119, pp. 243-312.

DECKER, C. E., 1920, *Studies in minor folds:* Univ. Chicago Press, Chicago, pp. 89.

DERRY, D. R., 1939, Some examples of detailed structure in early Pre-Cambrian rocks of Canada: *Q. Jl. geol. Soc. Lond.,* v. 95, pp. 109-34.

———, HOPPER, C. H., & McGOWAN, H. S., 1948, Matachewan Consolidated Mine: *in Structural Geology of Canada ore deposits: Canadian Institute of Mining and Metallurgy, Geological Division,* pp. 638-43.

DEWEY, J. F., & McMANUS, J., 1964, Superposed folding in the Silurian rocks of Co. Mayo, Eire: *Lpool. Manchr. geol. J.,* v. 4, pp. 61-76.

———, & PHILLIPS, W. E. A., 1963, A tectonic profile across the Caledonides of South Mayo: *Lpool. Manchr. geol. J.,* v. 3, pp. 237-46.

DIETZ, R. S., 1961, Vredefort ring structure: Meteorite impact Scar?: *J. Geol.,* v. 69, pp. 499-516.

DIETZEL, G. F. L., 1960, Geology and Permian paleomagnetism of the Merano Region, Province of Bolzano, N. Italy: *Geologica ultraiect.,* No. 4, pp. 1-58.

DOLL, C. G., 1943, A brachiopod from mica schist, South Strafford, Vermont: *Am. J. Sci.,* v. 241, pp. 676-9.

DONATH, F. A., & PARKER, R. B., 1964, Folds and folding: *Bull. geol. Soc. Am.,* v: 75, pp. 45-62.

DONS, J. A., 1962, The Precambrian Telemark area in south central Norway: *Geol. Rdsch.,* v. 52, pp. 261-8.

DREVER, H. I., 1964, An experiment in the use of light drilling techniques in petrological research: *Proc. geol. Soc.,* No. 1615, pp. 50-2.

DUNBAR, W. R., 1948, Structural relations of the Porcupine ore deposits: *in Structural Geology of Canadian ore deposits: Canadian Institute of Mining and Metallurgy, Geological Division,* pp. 442-56.

ENGEL, A. E. J., 1949, Studies of cleavage in the metasedimentary rocks of the northwest Adirondack Mountains, New York: *Trans. Am. geophys. Un.,* v. 30, pp. 767-84.

ENGELS, B., 1955, Zur Tektonik und Stratigraphie des Unterdevons zwischen Loreley und Lorchhausen am Rhein (Rheinisches Schiefergebirge): *Abh. hess. Landesamt. Bodenforsch.,* v. 14, 1-96 pp.

———, 1959, Die kleintektonische Arbeitweise unter besonderer Berücksichtigung ihrer Anwendung im deutschen Paläozoikum: *Geotekt. Forsch.,* v. 13, pp. 1-129.

ESKOLA, P., 1932, On the principles of metamorphic differentiation: *Bull. Commn géol. Finl.,* 97, pp. 68-77.

———, 1949, The problem of mantled gneiss domes: *Q. Jl. geol. Soc. Lond.,* v. 104 (for 1948), pp. 461-76.

EVANS, A. M., 1963, Conical folding and oblique structures in Charnwood Forest, Leicestershire: *Proc. Yorks. geol. Soc.,* v. 34, pp. 67-80.

FAIRBAIRN, H. W., 1935, Notes on the mechanics of rock foliation: *J. Geol.,* v. 43, pp. 591-608.

———, 1942, *Structural petrology of deformed rocks:* Addison-Wesley Publishing Co., Inc., Cambridge, 143 pp.

———, 1949, *Structural petrology of deformed rocks:* Addison-Wesley Publishing Co., Inc., Cambridge, 344 pp.

———, 1954, The stress-sensitivity of quartz in tectonites: *Mineralog. petrogr. Mitt.,* v. 4, pp. 75-80.

FALCON, N. L., 1961, Discussion on Memoir No. 2: Geological results of petroleum exploration in Britain 1945-57: *Proc. geol. Soc.,* No. 1583, p. 15.

———, & KENT, P. E., 1960, Geological results of petroleum exploration in Britain 1945-57: *Mem. geol. Soc. Lond.,* 2, pp. 1-56.

FELLOWS, R. E., 1943A, Flowage and recrystallization in Paleozoic quartzites: *Trans. Am. geophys. Un.,* v. 24, p. 271.

———, 1943B, Recrystallization and flowage in Appalachian quartzites: *Bull. geol. Soc. Am.,* v. 54, pp. 1399-432.

FERMOR, L. L., 1909, The manganese-ore deposits of India: *Mem. geol. Surv. India,* 37, pp. 1-1294.

———, 1924, The pitch of rock rolds: *Econ. Geol.,* v. 19, pp. 559-62.

FISHER, O., 1884A, On faulting, jointing, and cleavage: *Geol. Mag.,* v. 1 (Dec. III), pp. 266-76.

———, 1884B, On cleavage and distortion: *Geol. Mag.,* v. 1 (Dec. III), pp. 396-406.

FISHER, R. A., 1953, Dispersion on a sphere: *Proc. R. Soc.,* Ser. A, v. 217, pp. 295-305.

FLETT, J. S., 1906, On the petrographical characters of the inliers of Lewisian rocks among the Moine gneisses of the North of Scotland: *Mem. geol. Surv. Summ. Prog.* for 1905, pp. 155-67.

FLEUTY, M. J., 1961, The three fold-systems in the metamorphic rocks of Upper Glen Orrin, Ross-shire and Inverness-shire: *Q. Jl. geol. Soc. Lond.,* v. 117, pp. 447-79.

———, 1964A, The description of folds: *Proc. Geol. Ass.,* v. 75, pp. 461-92.

———, 1964B, Tectonic slides: *Geol. Mag.,* v. 101, pp. 452-6.

FLINN, D., 1952, A tectonic analysis of the Muness Phyllite Block of Unst and Uyea, Shetland: *Geol. Mag.,* v. 89, pp. 263-72.

———, 1956, On the deformation of the Funzie conglomerate, Fetler, Shetland: *J. Geol.,* v. 64, pp. 480-505.

———, 1962, On folding during three-dimensional progressive deformation: *Q. Jl. geol. Soc. Lond.,* v. 118, pp. 385-433.

FORSTER, A., 1963, Altersbeziehungen zwischen den Metamorphosezyklen und den orogenen Bewegungen im östlichen Teil des nordrhodesischen Grundgebirges: *Geol. Rdsch.,* v. 52 (for 1962), pp. 280-92.

FOSLIE, S., 1941, Tysfjords geologi; beskrivelse til det geologiske gradteigskart Tysfjord−geology of the Tysfjord map area: *Norg. geol. Unders.*, No. 149, pp. 1-298.

FOURMARIER, P., 1932, Observations sur le développement de la schistosité dans les séries plissées: *Bull. Acad. r. Belg. Cl. Sci.*, v. 18, pp. 1048-53.

———, 1951, Schistosité, foliation, et microplissement: *Archs Sci., Genève*, v. 4, pp. 5-23.

———, 1952, Aperçu sur les déformations intimes des roches en terrains plissés: *Annls Soc. géol. Belg.*, v. 75B, pp. 181-93.

———, 1953A, L'allure du front supérieur de schistosité dans le Paléozoique de l'Ardenne: *Bull. Acad. r. Belg. Cl. Sci.*, v. 39, pp. 838-45.

———, 1953B, Schistosité et grande tectonique: *Annls Soc. géol. Belg.*, v. 76B, pp. 275-301.

———, 1953C, Schistosité et phénomènes connexes dans les séries plissées: *Int. geol. Congr. XIX, Algiers, 1952*, Pt. 3, pp. 117-31.

———, 1956, Remarques au sujet de la schistosité en général avec application aux terrains paléozoiques de l'Ardenne et du massif schisteux Rhénan: *Geologie Mijnb.*, v. 18, pp. 47-56.

FRANCIS, G. H., 1956, Facies boundaries in pelites at the middle grades of regional metamorphism: *Geol. Mag.*, v. 93, pp. 353-68.

FRÄNKL, E., 1953, Geologische Untersuchungen in Ost-Andrées Land (NE-Grønland): *Meddr. Grønland*, v. 113, No. 4, pp. 1-160.

FREEDMAN, J., BENTLEY, R. D., & WISE, D. U., 1962, Pattern of folded folds across the Appalachian Piedmont near the Susquehanna River: *Spec. Pap. geol. Soc. Am.*, 68, p. 179.

———, WISE, D. U., & BENTLEY, R. D., 1964, Pattern of folded folds in the Appalachian Piedmont along Susquehanna River: *Bull. geol. Soc. Am.*, v. 75, pp. 621-38.

FULLER, M. D., 1964, On the magnetic fabrics of certain rocks: *J. Geol.*, v. 72, pp. 368-76.

FURTAK, H., 1963, Die "Brechung" der Schiefrigkeit: *ForschBer. Landes NRhein-Westf.*, No. 1200, pp. 1-33.

———, & HELLERMANN, E., 1963, Die tektonische Verformung von pflanzlichen Fossilien des Karbons: *ForschBer. Landes NRhein-Westf.*, No. 1199, pp. 1-37.

FYFE, W. S., TURNER, F. J., & VERHOOGEN, J., 1958, Metamorphic reactions and metamorphic facies: *Mem. geol. Soc. Am.*, 73, pp. 1-259.

FYSON, W. K., 1962, Tectonic structures in the Devonian rocks near Plymouth, Devon: *Geol. Mag.*, v. 99, pp. 208-26.

———, 1964, Folds in the Carboniferous rocks near Walton, Nova Scotia: *Am. J. Sci.*, v. 262, pp. 513-22.

GAIR, J. E., 1950, Some effects of deformation in the central Appalachians: *Bull. geol. Soc. Am.*, v. 61, pp. 857-76.

GALWEY, A. K., & JONES, K. A., 1962, Inclusions in garnets: *Nature, Lond.*, v. 193, pp. 471-2.

GANGOPADHYAY, P. K., & JOHNSON, M. R. W., 1962, A study of quartz orientation and its relation to movement in shear folds: *Geol. Mag.*, v. 99, pp. 69-84.

GARRETTY, M. D., 1943, Use of geology at North Broken Hill Mine: *Chem. Engng. Min. Rev.*, v. 35, pp. 333-45.

GASTIL, G., & KNOWLES, D. M., 1960, Geology of the Wabush Lake area, southwestern Labrador and eastern Quebec, Canada: *Bull. geol. Soc. Am.*, v. 71, pp. 1243-54.

GAULT, H. R., 1945, Petrography, structures, and petrofabrics of the Pinckney-ville quartz diorite, Alabama: *Bull. geol. Soc. Am.*, v. 56, pp. 181-246.

GEALEY, W. K., 1951, Geology of the Healdsburg quadrangle, California: *Bull. Div. Mines Calif.*, 161, pp. 1-50.

GEYER, A. R., GRAY, C., McLAUGHLIN, D. B., & MOSELEY, J. R., 1958, Geology of Lebanon quadrangle: *Atlas Pa. geol. Surv.*, 167C.

GILBERT, G. K., 1876, The Colorado Plateau Province as a field for geological study: *Am. J. Sci.*, v. 12 (3rd ser.), pp. 16-24.

GILETTI, B. J., MOORBATH, S., & LAMBERT, R. St. J., 1961, A geochronologi-cal study of the metamorphic complexes of the Scottish Highlands: *Q. Jl. geol. Soc. Lond.*, v. 117, pp. 233-72.

GILLOTT, J. E., 1956, Structural geology of the Manx Slates: *Geol. Mag.*, v. 93, pp. 301-13.

GILLULY, J., 1934, Mineral orientation in some rocks of the Shuswap terrane as a clue to their metamorphism: *Am. J. Sci.*, v. 228, pp. 182-201.

————, 1963, The tectonic evolution of the western United States: *Q. Jl. geol. Soc. Lond.*, v. 119, pp. 133-74.

GILMOUR, P., & McINTYRE, D. B., 1954, The geometry of the Ben Lui fold (S.W. Highlands): *Geol. Mag.*, v. 91, pp. 161-6.

GINDY, A. R., 1953, The plutonic history of the district around Trawenagh Bay, Co. Donegal: *Q. Jl. geol. Soc. Lond.*, v. 108 (for 1952), pp. 377-411.

GOGUEL, J., 1936, Description tectonique de la bordure des Alpes de la Bléone au Var: *Mém. Serv. Carte géol. dét. Fr.*, pp. 1-360.

————, 1962, *Tectonics:* W. H. Freeman & Co., San Francisco, 384 pp.

GOLDRING, D. C., 1956, The structural petrology of the Dalradian schists of N.E. Antrim: *Advmt. Sci. Lond.*, v. 12, pp. 576-7.

————, 1961, The relationship of the micro-fabric to the small-scale structures of the Dalradian rocks of north-eastern County Antrim: *Proc. R. Ir. Acad.*, v. 61B, pp. 345-68.

GOLDSMITH, R., 1961, Axial plane folding in southeastern Connecticut: *Prof. Pap. U.S. geol. Surv.*, 424C, pp. 54-7.

GONZALEZ-BONORINO, F., 1960, The mechanical factor in the formation of schistosity: *Int. geol. Congr. XXI Copenhagen*, Pt. 18, pp. 303-16.

GOSSELET, J., 1888, L'Ardenne: *Mém. Serv. Carte géol. dét. Fr.*, pp. 1-881.

GRANT, F., 1957, A problem in the analysis of geophysical data: *Geophysics*, v. 22, pp. 309-44.

GRAY, C., GEYER, A. R., & McLAUGHLIN, D. B., 1958, Geologic map of the Richland quadrangle, Pennsylvania: *Atlas Pa. geol. Surv.*, 167D.

GREEN, J. F. N., 1924, The structure of the Bowmore-Portaskaig district of Islay: *Q. Jl. geol. Soc. Lond.*, v. 80, pp. 72-105.

GREEN, R., 1964, Available methods for the analysis of vectoral data: *J. sedim. Petrol.*, v. 34, pp. 440-2.

GREENLY, E., 1919, The geology of Anglesey (vol. 1): *Mem. geol. Surv. U.K.*, pp. 1-388.

————, 1930, Foliation and its relations to folding in the Mona Complex at Rhoscolyn (Anglesey): *Q. Jl. geol. Soc. Lond.*, v. 86, pp. 169-90.

GREENSMITH, J. T., 1957, The status and nomenclature of stratified evaporites: *Am. J. Sci.*, v. 255, pp. 593-5.

———, 1958, Reply to Professor F. H. Stewart: *Am. J. Sci.,* v. 256, pp. 219-20.

GREENWOOD, R., 1960, Sedimentary boudinage in Cretaceous limestones of Zimapan, Mexico: *Int. geol. Congr. XXI Copenhagen,* Pt. 18, pp. 389-98.

GREGORY, H. E., 1914, The Rodadero (Cuzco, Peru), -A fault plane of unusual aspect: *Am. J. Sci.,* v. 37 (4th ser.), pp. 289-98.

GRIGGS, D., & HANDIN, J. W., 1960, Rock deformation (a symposium): *Mem. geol. Soc. Am.,* 79, pp. 1-382.

———, & MILLER, W. B., 1951, Deformation of Yule marble; Part I — Compression and extension on dry Yule Marble at 10,000 atmospheres confining pressure, room temperature: *Bull. geol. Soc. Am.,* v. 62, pp. 853-62.

———, TURNER, F. J., BORG, I., & SOSOKA, J., 1951, Deformation of Yule marble; Part IV — Effects at 150° C: *Bull. geol. Soc. Am.,* v. 62, pp. 1385-406.

———, ———, ———, & ———, 1953, Deformation of Yule marble; Part V — Effects at 300° C: *Bull. geol. Soc. Am.,* v. 64, pp. 1327-42.

GUITARD, G., 1955, Sur l'évolution des gneiss des Pyrénées: *Bull. Soc. géol. Fr.,* v. 5 (6th Ser.), pp. 441-68.

GUNN, W., CLOUGH, C. T., & HILL, J. B., 1897, The geology of Cowal including the part of Argyllshire between the Clyde and Loch Fine: *Mem. geol. Surv. U.K.,* pp. 1-333.

GUNNING, H. C., & AMBROSE, J. W., 1940, Malarctic area, Quebec: *Mem. geol. Surv. Brch. Can.,* 222, pp. 1-142.

GUSTAFSON, J. K., BURRELL, H. C., & GARRETTY, M. D., 1950, Geology of the Broken Hill ore deposit, Broken Hill, N.S.W., Australia: *Bull. geol. Soc. Am.,* v. 61, pp. 1369-437.

HAAF, E. TEN, 1959, *Graded bedding of the Northern Apennines:* Doctoral Thesis, Univ. Groningen, 102 pp.

HABICHT, K., 1945, Geologische Untersuchungen im südlichen Sanktgallisch-appenzellischen Molassegebeit: *Beitr. geol. Karte Schwiez.,* v. 83, pp. 1-166.

HALLER, J., 1955, Der "Zentrale Metamorphe Komplex" von NE-Grönland; Teil I — Die Geologische Karte von Suess Land, Gletscherland und Goodenoughs Land: *Meddr. Grønland,* v. 73, I Afd., No. 3, pp. 1-174.

———, 1956A, Geologie der Nunatakker Region von Zentral-Ostgrönland zwischen 72°30′ und 74°10′ N.BR.: *Meddr. Grønland,* v. 154, No. 1, pp. 1-172.

———, 1956B, Die Strukturelemente Ostgrönlands zwischen 74° und 78° N.: *Meddr. Grønland,* v. 154, No. 2, pp. 1-27.

———, 1956C, Probleme der Tiefentektonik. Bauformen im Migmatitstockwerk der ostgrönländischen Kaledoniden: *Geol. Rdsch.,* v. 45, pp. 159-67.

———, 1957, Gekreuzte Faltensysteme in Orogenzonen: *Schweiz. miner. petrogr. Mitt.,* v. 37, pp. 11-29.

HAMAN, P. J., 1961, *Manual of the stereographic projection of geology and related sciences:* West Canadian Research Publications, Calgary, 67 pp.

HANDIN, J. W., & GRIGGS, D., 1951, Deformation of Yule marble; Part II — Predicted fabric changes: *Bull. geol. Soc. Am.,* v. 62, pp. 863-86.

HARBAUGH, J. W., 1963, Balgol program for trend-surface mapping using an IBM 7090 Computer: *Spec. distribution Kansas Geol. Survey,* 3, pp. 1-17.

HARDIE, W. G., 1952, The Lochaber Series south of Loch Leven, Argyllshire: *Geol. Mag.,* v. 89, pp. 273-85.

———, 1953, Facies changes versus sliding — a reply: *Geol. Mag.,* v. 90, pp. 114-6.

————, 1955, The problem of facies changes and sliding, south of Loch Leven, Argyllshire: *Geol. Mag.*, v. 92, pp. 506-11.

HARKER, A., 1885A, The cause of slaty cleavage: compression *v.* shearing: *Geol. Mag.*, v. 2 (Dec. III), pp. 15-7.

————, 1885B, On the successive stages of slaty cleavage: *Geol. Mag.*, v. 2 (Dec. III), pp. 266-8.

————, 1886, On slaty cleavage and allied rock-structures, with special reference to the mechanical theories of their origin: *Rep. Br. Ass. Advmt Sci.*, 1885 (55th meeting), pp. 813-52.

————, 1889, On the local thickening of dykes and beds by folding: *Geol. Mag.*, v. 6 (Dec. III), pp. 69-70.

————, 1932, *Metamorphism:* Methuen & Co. Ltd., London, 360 pp.

HARLAND, W. B., 1956, Tectonic facies, orientation, sequence, style, and date: *Geol. Mag.*, v. 93, pp. 111-20.

————, & BAYLY, M. B., 1958, Tectonic regimes: *Geol. Mag.*, v. 95, pp. 89-104.

HARMAN, H. H., 1960, *Modern factor analysis:* Univ. of Chicago Press, Chicago, 469 pp.

HARRIS, A. L., & RAST, N., 1960A, Oriented quartz inclusions in garnets: *Nature, Lond.*, v. 185, pp. 448-9.

————, & ————, 1960B, The evolution of quartz fabrics in the metamorphic rocks of central Perthshire: *Trans. Edinb. geol. Soc.*, v. 18, pp. 51-78.

HARRIS, R. L., JR., 1959, Geologic evolution of the Beartooth Mountains, Montana and Wyoming. Part 3. Gardner Lake area, Wyoming: *Bull, geol. Soc. Am.*, v. 70, pp. 1185-216.

HARRISON, J. E., & WELLS, J. D., 1959, Geology and ore deposits of the Chicago Creek area, Clear Creek County, Colorado: *Prof. Pap. U.S. geol. Surv.*, 319, pp. 1-92.

HARRISON, J. V., & FALCON, N. L., 1934, Collapse structures: *Geol. Mag.*, v. 71, pp. 529-39.

————, & ————, 1936, Gravity collapse structures and mountain ranges, as exemplified in south-western Iran: *Q. Jl. geol. Soc. Lond.*, v. 92, pp. 91-102.

HARRISON, P. W., 1957, New technique for three-dimensional fabric analysis of till and englacial debris containing particles from 3 to 40 mm. in size: *J. Geol.*, v. 65, pp. 98-105.

HAUGHTON, S., 1856, On slaty cleavage, and the distortion of fossils: *Phil. Mag.*, v. 12 (4th ser.), pp. 409-21.

HEIM, A., 1878, *Untersuchungen über den Mechanismus der Gebirgsbildung* (vol. 2): B. Schwabe, Basel, 246 pp.

————, 1900, Gneissfältelung in alpinem Centralmassiv, ein Beitrag zur Kenntnis der Stauungsmetamorphose: *Vjschr. naturf. Ges. Zürich*, v. 45, pp. 205-26.

HELLMERS, J. H., 1955, Krinoidenstielglieder als Indikatoren der Gesteinsdeformation: *Geol. Rdsch.*, v. 44, pp. 87-92.

HELM, D. G., ROBERTS, B., & SIMPSON, A., 1963, Polyphase folding in the Caledonides south of the Scottish Highlands: *Nature, Lond.*, v. 200, pp. 1060-2.

HENDERSON, J. F., 1948, Structural control of ore deposits in the Canadian Shield between Great Slave and Great Bear Lakes, Northwest Territories: *in Structural Geology of Canadian ore deposits: Canadian Institute of Mining and Metallurgy, Geological Division*, pp. 238-43.

HENDERSON, S. M. K., 1935, Ordovician submarine disturbances in the Girvan district: *Trans. R. Soc. Edinb.*, v. 58, pp. 487-509.

HEWETT, D. F., CALLAGHAN, E., MOORE, B. N., NOLAN, T. B., RUBEY, W. W., & SCHALLER, W. T., 1936, Mineral resources of the region around Boulder Dam: *Bull. U.S. geol. Surv.*, 871, pp. 1-197.

HIETANEN, A., 1961A, Relation between deformation, metamorphism, metasomatism, and intrusion along the northwest border zone of the Idaho batholith, Idaho: *Prof. Pap. U.S. geol. Surv.*, 424D, pp. 161-4.

———, 1961B, Metamorphic facies and style of folding in the Belt Series northwest of the Idaho batholith: *Bull. Commn géol. Finl.*, 196, pp. 73-103.

———, 1961C, Superposed deformations northwest of the Idaho Batholith: *Int. geol. Congr. XXI Copenhagen 1960*, Pt. 26, pp. 87-102.

———, 1962, Metasomatic metamorphism in western Clearwater County, Idaho: *Prof. Pap. U.S. geol. Surv.*, 344-A, pp. 1-116.

HIGGINS, A. K., 1964, Fossil remains in staurolite-kyanite schists of the Bedretto-Mulde Bündnerschiefer: *Eclog. geol. Helv.*, v. 57, pp. 151-6.

HILLS, E. S., 1945, Examples of the interpretation of folding: *J. Geol.*, v. 53, pp. 47-57.

———, 1946, Some aspects of the tectonics of Australia: *J. Proc. R. Soc. N.S.W.*, v. 79 (for 1945), pp. 67-91.

———, 1953, *Outlines of structural geology* (3rd edit.): John Wiley & Sons, Inc., New York, 182 pp.

———, 1963A, *Elements of structural geology:* John Wiley & Sons, Inc., New York, 483 pp.

———, 1963B, Conjugate folds, kinks and drag: *Geol. Mag.*, v. 100, pp. 467-8.

———, & THOMAS, D. E., 1954, Turbidity currents and the graptolitic facies in Victoria: *J. geol. Soc. Aust.*, v. 1 (for 1953), pp. 119-33.

HINXMAN, L. W., ANDERSON, E. M., et al., 1915, The geology of mid-Strathspey and Strathdearn including the country between Kingussie and Grantown (explanation of sheet 74): *Mem. geol. Surv. U.K.*, pp. 1-97.

HITCHCOCK, E., HAGER, A. D., & HITCHCOCK, C. H., 1861, *Report on the geology of Vermont*, 2 vols: A. D. Hager, Claremont, N.H., 988 pp.

HODGSON, J. H., 1959, The mechanics of faulting, with special reference to fault-plane work (a symposium): *Publs. Dom. Obs.*, v. 20, pp. 251-418.

HOEL, P. G., 1947, *Introduction to mathematical statistics:* John Wiley & Sons, Inc., New York, 258 pp.

HOEPPENER, R., 1955, Tektonik im Schiefergebirge: eine Einführung: *Geol. Rdsch.*, v. 44, pp. 26-58.

———, 1956, Zum Problem der Bruchbildung, Schieferung und Faltung: *Geol. Rdsch.*, v. 45, pp. 247-83.

———, 1960, Ein Beispiel für die zeitliche Abfolge tektonischer Bewegungen aus dem rheinischen Schiefergebirge: *Geologie Mijnb.*, v. 39, pp. 181-8.

———, 1961, Grundlagen einer Systematik tektonischer Gefüge: *Geol. Rdsch.*, v. 50 (for 1960), pp. 77-83.

HOFFMANN, K., 1961, Die Tektonik am Südrand des Hohen Venns: *Neues Jb. Geol. Paläont. Abh.*, v. 111, pp. 341-67.

HOLLINGWORTH, S. E., TAYLOR, J. H., & KELLAWAY, G. A., 1944, Large-scale superficial structures in the Northampton Ironstone Field: *Q. Jl. geol. Soc. Lond.*, v. 100, pp. 1-44.

———, WELLS, M. K., & BRADSHAW, R., 1960, Geology and structure of the

Glomfjord Region, northern Norway: *Int. geol. Congr. XXI Copenhagen*, Pt. 19, pp. 33-42.

HOLMES, A., 1928, *The nomenclature of petrology* (2nd edit.): Thos. Murby & Co., London, 284 pp.

HOLMQUIST, P. J., 1931, On the relations of the "boudinage-structure": *Geol. För. Stockh. Förh.*, v. 53, pp. 193-208.

HOOTS, H. W., BEAR, T ᵀ & KLEINPELL, W. D., 1954, Geological summary of the San Joaquin Valley, California: *Bull. Div. Mines Calif.*, v. 170, pp. 113-29.

HORNE, J., 1898, *Mem. geol. Surv. Summ. Prog.* for 1897, 176 pp.

———, HINXMAN, L. W., PEACH, B. N., & CRAIG, E. H. C., 1914, The geology of the country around Beauly and Inverness: including a part of the Black Isle (explanation of sheet 83): *Mem. geol. Surv. U.K.*, pp. 1-108.

HORTON, C. W., HEMPKINS, W. B., & HOFFMAN, A. A. J., 1964, A statistical analysis of some aeromagnetic maps from the northwestern Canadian Shield: *Geophysics*, v. 29, pp. 582-601.

———, HOFFMAN, A. A. J., & HEMPKINS, W. B., 1962, Mathematical analysis of the microstructure of an area of the bottom of Lake Travis: *Tex. J. Sci.*, v. 14, pp. 131-42.

HOUSER, F. N., & POOLE, F. G., 1959, "Granite" exploration hole, Area 15, Nevada Test Site, Nye County, Nevada—Interim report, Part A, Structural, petrographic, and chemical data: *Open file Rept. U.S. geol. Surv.*, TEM 836A, pp. 1-58.

HOUSTON, R. S., & PARKER, R. B., 1963, Structural analysis of a folded quartzite, Medicine Bow Mountains, Wyoming: *Bull. geol. Soc. Am.*, v. 74, pp. 197-202.

HOWELL, J. V., 1957, *Glossary of geology and related sciences:* American Geological Institute, Washington, D.C., 325 pp.

HOWKINS, J. B., 1961, Helicitic textures in garnets from the Moine rocks of Moidart: *Trans. Edinb. geol. Soc.*, v. 18, pp. 315-24.

HUBBERT, M. K., & RUBEY, W. R., 1961A, Role of fluid pressure in mechanics of overthrust faulting: I. Mechanics of fluid-filled porous solids and its application to overthrust faulting: Reply to discussion by Francis Birch: *Bull. geol. Soc. Am.*, v. 72, pp. 1445-52.

———, & ———, 1961B, Role of fluid pressure in mechanics of overthrust faulting: A reply to discussion by Walter L. Moore: *Bull. geol. Soc. Am.*, v. 72, pp. 1587-94.

HULL, E., KINAHAN, G. H., NOLAN, J., CRUISE, R. J., EGAN, F. W., KILROE, J. R., MITCHELL, W. F., M'HENRY, A., & HYLAND, J. S., 1891, Explanatory memoir to accompany Sheets 3, 4, 5 (in part), 9, 10, 11 (in part), 15 and 16 of the maps of the Geological Survey of Ireland, comprising north-west and central Donegal: *Mem. Geol. Surv. U.K.*, pp. 1-174.

HURLEY, P. M., HUGHES, H., FAURE, G., FAIRBAIRN, H. W., & PINSON, W. H., 1962, Radiogenic strontium-87 model of continent formation: *J. geophys. Res.*, v. 67, pp. 5315-34.

INGERSON, E., 1940, Fabric criteria for distinguishing pseudo ripple marks from ripple marks: *Bull. geol. Soc. Am.*, v. 51, pp. 557-70.

IVES, R. L., 1939, Measurements in block diagrams: *Econ. Geol.*, v. 34, pp. 561-72.

IZETT, G. A., 1960, "Granite" exploration hole, Area 15, Nevada Test Site, Nye County, Nevada—Interim report, part C. Physical properties trace elements memorandum report 836-C: *Open file Rep. U.S. Geol. Surv.*, TEM 836-C, pp. 1-37.

JANNETTAZ, M. E., 1884, Mémoire Sur les Clivages des Roches (schistosité, longrain), et sur leur reproduction: *Bull. Soc. géol. Fr.*, v. 12 (3rd ser.), pp. 216-36.

JOHNSON, F. J., 1961, The geology of the Astrolabe Lake area (west half) Saskatchewan: *Rep. Sask. Dep. Miner. Resour. (Min. Brch. Geol. Div.)*, 59, pp. 1-26.

JOHNSON, M. R. W., 1956A, Conjugate fold systems in the Moine Thrust Zone in the Lochcarron and Coulin Forest areas of Wester Ross: *Geol. Mag.*, v. 93, pp. 345-50.

———, 1956B, Time sequences of folding, thrusting and recrystallization in the Kishorn and Moine Nappes in the Lochcarron and Coulin Forest areas of Western Ross: *Advmt Sci.*, v. 12, pp. 575-8.

———, 1957, The structural geology of the Moine Thrust Zone in Coulin Forest, Wester Ross: *Q. Jl. geol. Soc. Lond.*, v. 113, pp. 241-70.

———, 1960, The strucutral history of the Moine Thrust Zone at Lochcarron, Wester Ross: *Trans. R. Soc. Edinb.*, v. 64, pp. 139-68.

———, 1961, Polymetamorphism in movement zones in the Caledonian Thrust Belt of northwest Scotland: *J. Geol.*, v. 69, pp. 417-32.

———, 1963, Some time relations of movement and metamorphism in the Scottish Highlands: *Geologie Mijnb.*, v. 42, pp. 121-42.

———, & HARRIS, A. L., 1965, Is the Tay Nappe post-Arenig? *Scottish J. Geol.*, v. 1, pp. 217-9.

———, & Stewart, F. H., 1960, On Dalradian structures in north-east Scotland: *Trans. Edinb. geol. Soc.*, v. 18, pp. 94-103.

———, & ———, 1963, *The British Caledonides:* Oliver and Boyd, Ltd., Edinburgh, 280 pp.

JOHNSTON, W. D., & NOLAN, T. B., 1937, Isometric block diagrams in mining geology: *Econ. Geol.*, v. 32, pp. 550-69.

JONES, A. G., 1959, Vernon map-area British Columbia: *Mem. geol. Surv. Brch. Can.*, 296, pp. 1-186.

JONES, D. L., 1959, Stratigraphy of Upper Cretaceous rocks in the Yreka-Hornbrook area, northern California: *Bull. geol. Soc. Am.*, v. 70, pp. 1726-7.

JONES, K. A., 1961, Origin of albite porphyroblasts in rocks of the Ben More-Am-Binnein area, western Perthshire, Scotland: *Geol. Mag.*, v. 98, pp. 41-55.

JONES, O. T., 1961, The Broad Haven monocline: *Lpool. Manchr. geol. J.*, v. 2, pp. 276-9.

KALLIOKOSKI, J., 1953, Interpretations of the structural geology of the Sherridon-Flin Flon Region, Manitoba: *Bull. geol. Surv. Can.*, 25, pp. 1-18.

KALSBEEK, F., 1962, Petrology and structural geology of the Berlanche-Valloire area (Belledonne Massif, France): Ph.D. Thesis, Univ. Leiden, 136 pp.

KELLY, J., 1864, *Notes upon the errors of geology illustrated by reference to facts observed in Ireland:* Longmans, Roberts, and Green, London, 158 pp.

KELVIN, LORD (THOMSON, W.), & TAIT, P. G., 1883, *Treatise on natural philosophy; Part II* (2nd edit.): Cambridge Univ. Press, Cambridge, 527 pp.

KENDALL, M. G., 1961, *A course in multivariate analysis:* Charles Griffin & Co. Ltd., London, 185 pp.

KENNEDY, W. Q., 1948, On the significance of thermal structure in the Scottish

Highlands: *Geol. Mag.,* v. 85, pp. 229-34.

————, 1955, The tectonics of the Morar anticline and the problem of the northwest Caledonian front: *Q. Jl. geol. Soc. Lond.,* v. 110 (for 1954), pp. 357-90.

KIENOW, S., 1934, Die Innere Tektonik des Unterdevons zwischen Rhein, Mosel und Nahe: *Jb. preuss. geol. Landesanst. Berg Akad.,* v. 54 (for 1933), pp. 58-95.

KING, B. C., 1955, The tectonic pattern of the Lewisian around Clashnessie Bay near Stoer, Sutherland: *Geol. Mag.,* v. 92, pp. 69-80.

————, 1956, Structural geology; Part III: *Sci. Prog., Lond.,* v. 44, pp. 88-104.

————, & RAST, N., 1956A, Tectonic styles in the Dalradians and Moines of parts of the Central Highlands of Scotland: *Proc. Geol. Ass.,* v. 66 (for 1955), pp. 243-69.

————, & ————, 1956B, The small-scale structures of south-eastern Cowal, Argyllshire: *Geol. Mag.,* v. 93, pp. 185-95.

————, & ————, 1959, Structural geometry of Dalradian rocks at Loch Leven, Scottish Highlands: a discussion: *J. Geol.,* v. 67, pp. 244-6.

KLEINSMIEDE, W. F. J., 1960, Geology of the Valle de Arán (Central Pyrenees): *Leid. geol. Meded.,* v. 25, pp. 127-244.

KNILL, D. C., & KNILL, J. L., 1961, Time relations between folding, metamorphism and the emplacement of granite in Rosguill, County Donegal: *Q. Jl. geol. Soc. Lond.,* v. 117, pp. 273-306.

KNILL, J. L., 1960A, The tectonic pattern in the Dalradian of the Craignish-Kilmelfort district, Argyllshire: *Q. Jl. geol. Soc. Lond.,* v. 115 (for 1959), pp. 339-64.

————, 1960B, A classification of cleavages, with special references to the Craignish district of the Scottish Highlands: *Int. geol. Congr. XXI Copenhagen,* Pt. 18, pp. 317-25.

————, 1961, Joint-drags in Mid-Argyllshire: *Proc. Geol. Ass.,* v. 72, pp. 13-9.

————, & KNILL, D. C., 1958, Some discordant fold structures from the Dalradian of Craignish, Argyll, and Rosguill, Co. Donegal: *Geol. Mag.,* v. 95, pp. 497-510.

————, & ————, 1961, Reversed graded beds from the Dalradian of Inishowen, Co. Donegal: *Geol. Mag.,* v. 98, pp. 458-63.

KNOPF, A., 1941, Petrology: *in Geology 1888-1938 Fiftieth Anniv. Volume Geol. Soc. America,* pp. 333-63.

KNOPF, E. B., 1931, Retrogressive metamorphism and phyllonitization: *Am. J. Sci.,* v. 21 (5th ser.), pp. 1-27.

————, & INGERSON, E., 1938, Structural petrology: *Mem. geol. Soc. Am.,* 6, pp. 1-270.

KNOWLES, D. M., MacPHERSON, W. A., & BLAKEMAN, W. B., 1962, Example of superposed folds in Labrador: *Spec. Pap. geol. Soc. Am.,* 68, p. 212.

KNOWLES, L., & MIDDLEMISS, F. A., 1958, The Lower Greensand in the Hindhead Area of Surrey and Hampshire: *Proc. Geol. Ass.,* v. 69, pp. 205-38.

KOARK, H. J., 1951, Zur tektonisch-petrographischen Analyse der Kalkbrüche bei Burträsk in Västerbotten: *Geol. För. Stockh. Förh.,* v. 73, pp. 261-99.

————, 1952, Über Querfaltung, Bewegung ∥B und Erzlagerung mit Beispielen aus Malmberget/Gällivare: *Bull. geol. Instn. Univ. Upsala,* v. 34, pp. 251-78.

KOBER, L., 1923, *Bau und Entstehung der Alpen:* Gebrüder Borntraeger, Berlin, 283 pp.

KRANCK, E. H., 1960, On lineation in gneisses and schists: *Bull. Commn. géol. Finl.*, 188, pp. 11-22.

KRUMBEIN, W. C., 1959A, Trend surface analysis on contour-type maps with irregular control-point spacing: *J. geophys. Res.*, v. 64, pp. 823-34.

——, 1959B, The "sorting out" of geological variables illustrated by regression analysis of factors controlling beach firmness: *J. sedim. Petrol.*, v. 29, pp. 575-87.

——, 1960A, The "geological population" as a framework for analysing numerical data in geology: *Lpool. Manchr. geol. J.*, v. 2, pp. 341-68.

——, 1960B, Some problems in applying statistics to geology: *Appl. Statist.*, v. 9, pp. 82-91.

——, 1960C, Stratigraphic maps from data observed at outcrop: *Proc. Yorks. geol. Soc.*, v. 32, pp. 353-66.

——, 1962A, Open and closed number systems in stratigraphic mapping: *Bull. Am. Ass. Petrol. Geol.*, v. 46, pp. 2229-45.

——, 1962B, The computer in geology: *Science, N.Y.*, v. 136, pp. 1087-92.

——, 1963A, A geological process-response model for analysis of beach phenomena: *A. Bull. Beach Eros. Bd.*, v. 17, pp. 1-15.

——, 1963B, Confidence intervals on low-order polynomial trend surfaces: *J. geophys. Res.*, v. 68, pp. 5869-78.

——, & GRAYBILL, F. A., 1965, *An introduction to statistical models in geology:* McGraw-Hill Book Co., Inc., New York, 475 pp.

——, & SLOSS, L. L., 1963, *Stratigraphy and Sedimentation* (2nd edit.): W. H. Freeman & Co., San Francisco, 660 pp.

KSIAZKIEWICZ, M., 1954, Graded and laminated bedding in the Carpathian Flysch: *Annu. Soc. géol. Pologne,* v. 22 (for 1952), pp. 399-449.

KULP, J. L., LONG, L. E., GIFFIN, C. E., MILLS, A. A., LAMBERT, R. ST. J., GILETTI, B. J., & WEBSTER, R. K., 1960, Potassium-argon and rubidium-strontium ages of some granites from Britain and Eire: *Nature, Lond.,* v. 185, pp. 495-7.

KÜRSTEN, M., 1959, Die Geologie der Schottischen Hochlande: *Geol. Rdsch.,* v. 46 (for 1957), pp. 602-11.

KURTMAN, F., 1960, Fossildeformation und Tektonik im nördlichen Rheinischen Schiefergebirge: *Geol. Rdsch.,* v. 49, pp. 439-59.

KVALE, A., 1945, Petrofabric analysis of a quartzite from the Bergsdalen quadrangle, western Norway: *Norsk. geol. Tidsskr.,* v. 25, pp. 193-215.

——, 1948, Petrologic and structural studies in the Bergsdalen Quadrangle, western Norway: Part II – Structural geology: *Bergens Mus. Årb.,* 1946-7, Naturv. rekke 1, pp. 1-255.

——, 1957, Gefügestudien im Gotthardmassiv und den angrenzenden Gebieten (Vorläüfige Mitteilung): *Schweiz. miner. petrogr. Mitt.,* v. 37, pp. 397-434.

LAHEE, F. H., 1941, *Field Geology* (4th edit.): McGraw-Hill Book Co., New York, 853 pp.

LAMBERT, J. L. M., 1959, Cross-folding in the Gramscatho Beds at Helford River, Cornwall: *Geol. Mag.,* v. 94, pp. 489-96.

LAMBERT, R. ST. J., 1958, A metamorphic boundary in the Moine Schists of the Morar and Knoydart districts of Inverness-shire: *Geol. Mag.,* v. 95, pp. 177-94.

——, 1959, The mineralogy and metamorphism of the Moine schists of the Morar and Knoydart districts of Inverness-shire: *Trans. R. Soc. Edinb.,* v. 63, pp. 553-88.

————, & POOLE, A. B., 1964, The relationship of the Moine Series and Lewisian Gneisses near Mallaigmore, Inverness-shire: *Proc. Geol. Ass.*, v. 75, pp. 1-14.

LAPWORTH, C., 1889, On the Ballantrae rocks of South Scotland and their place in the Upland Sequence: Part II—The sequence in the Southern Uplands: *Geol. Mag.*, v. 26, pp. 59-69.

LEGGO, P. J., COMPSTON, W., & LEAKE, B. E., 1965, Geochronology of the Connemara granites and its bearing on the antiquity of the Dalradian Series: *Circul. geol. Soc. Lond.*, 126, p. 1.

LEHMANN, J. G., 1884, *Untersuchungen über die enstehung der altkrystallinischen schiefergesteine, mit besonderer bezugnahme auf das sächsische granulitgebirge, erzgebirge, fichtelgebirge und bairische-böhmische grenzgebirge:* M. Hochgürtel, Bonn, 278 pp.

LEITH, C. K., 1905, Rock cleavage: *Bull. U.S. geol. Surv.*, 239, pp. 1-216.

————, 1913, *Structural Geology:* Henry Holt & Co., New York, 169 pp.

————, 1923, *Structural Geology* (2nd edit.): Henry Holt & Co., New York, 390 pp.

LINDSTRÖM, M., 1955A, Structural geology of a small area in the Caledonides of Arctic Sweden: *Acta Univ. lund.*, v. 51, Ard. 2, No. 15, pp. 1-31.

————, 1955B, A tectonic study of Mt. Nuolja, Swedish Lapland: *Geol. För. Stockh. Förh.*, v. 77, pp. 557-66.

————, 1957, Tectonics of the area between Mt. Keron and Lake Allesjaure in the Caledonides of Swedish Lapland: *Acta Univ. lund.*, v. 53, No. 11, pp. 1-33.

LINK, R. F., & KOCH, G. S., JR., 1962, Quantitative areal modal analysis of granitic complexes: discussion: *Bull. geol. Soc. Am.*, v. 73, pp. 411-14.

LOHEST, M., 1910, De l'origine des veines et des géodes des terrains primaires de Belgique. Troisième note: *Annls. Soc. géol. Belg.*, v. 36 (*Bull.* for 1908-9), pp. 275-82.

————, STAINIER, X., & FOURMARIER, P., 1908, Compte rendu de la Session Extraordinaire de la Société Géologique de Belgique tenue à Eupen et á Bastogne les 29, 30 et 31 août et 1er, 2 et 3 septembre, 1908: *Bull. Soc. belge Géol. Paléont. Hydrol.*, v. 22, pp. 453-512.

————, ————, & ————, 1909, Compte rendu de la Session Extraordinaire de la Société Géologique de Belgique tenue à Eupen et á Bastogne les 29, 30 et 31 août et 1er, 2, et 3 septembre 1908: *Annls. Soc. géol. Belg.*, v. 35 (*Bull.* for 1907-8), pp. 351-414.

LORETZ, H., 1882, Ueber Transversalschieferung und verwandte Erscheinungen im thüringischen Schiefergebirge: *Jb. preuss. geol. Landesanst. Berg-Akad.*, 1881, pp. 258-306.

LOUDON, T. V., 1963, *The sedimentation and structure in the Macduff district of North Banffshire and Aberdeenshire:* Ph.D. Thesis, Univ. Edinburgh, 107 pp.

————, 1964, Computer analysis of orientation data in structural geology: *Tech. Report Geog. Branch Off. Naval Res.*, O.N.R. Task No. 389-135, *Contr. No.* 1228 (26), No. 13, pp. 1-130.

LOVERING, T. S., 1932, Field evidence to distinguish over-thrusting from under-thrusting: *J. Geol.*, v. 40, pp. 651-63.

LÖWL, F., 1906, *Geologie:* F. Deuticke, Leipzig and Vienna, 332 pp.

LYELL, C., 1833, *Principles of geology, being an attempt to explain the former changes of the earth's surface, by reference to causes now in operation,* Vol. 3 (1st edit.): J. Murray, London.

MacCulloch, J., 1816, A geological description of Glen Tilt: *Trans. geol. Soc. Lond.*, v. 3, pp. 259-337.

MacGregor, A. G., 1948, Resemblances between Moine and "sub-Moine" metamorphic sediments in the western Highlands of Scotland: *Geol. Mag.*, v. 85, pp. 265-75.

———, Sheila, M. A., & Roberts, J. L., 1963, Dalradian pillow lavas, Ardwell Bridge, Banffshire: Geol. Mag., v. 100, pp. 17-23.

Mackin, J. H., 1950, The down-structure method of viewing geologic maps: *J. Geol.*, v. 58, pp. 55-72.

———, 1962, Structure of the Glenarm Series in Chester County, Pennsylvania: *Bull. geol. Soc. Am.*, v. 73, pp. 403-10.

Mandelbaum, H., 1963, Statistical and geological implications of trend mapping with nonorthogonal polynomials: *J. geophys. Res.*, v. 68, pp. 505-19.

Marmo, V., 1962, Kallioperämme ikä: *Terra*, v. 74, pp. 114-23.

Marr, J. E., 1900, Notes on the geology of the English Lake District: *Proc. Geol. Ass.*, v. 16, pp. 449-83.

Martin, J. C., 1916, The Precambrian rocks of the Canton quadrangle: *Bull. N.Y. St. Mus.*, 185, pp. 1-112.

Matthews, D. H., 1958, Dimensions of asymmetrical folds: *Geol. Mag.*, v. 95, pp. 511-3.

Maxwell, J. C., 1962, Origin of slaty and fracture cleavage in the Delaware Water Gap Area, New Jersey and Pennsylvania: *Mem. geol. Soc. Am.* (Buddington volume), pp. 281-311.

Mayo, E. B., 1941, Deformation in the interval Mt. Lyell-Mt. Whitney, California: *Bull. geol. Soc. Am.*, v. 52, pp. 1001-84.

McCall, G. J. H., 1954, The Dalradian geology of the Creeslough area, Co. Donegal: *Q. Jl. geol. Soc. Lond.*, v. 110, pp. 153-75.

McCallien, W. J., 1935, The metamorphic rocks of Inishowen, Co. Donegal: *Proc. R. Ir. Acad.*, v. 42B, pp. 407-42.

McCrossan, R. G., 1958, Sedimentary "boudinage" structures in the Upper Devonian Ireton Formation of Alberta: *J. sedim. Petrol.*, v. 28, pp. 316-20.

McIntyre, D. B., 1950A, Note on two lineated tectonites from Strathavon, Banffshire: *Geol. Mag.*, v. 87, pp. 331-6.

———, 1950B, Note on lineation, boudinage, and recumbent-folds in the Struan flags (Moine), near Dalnacardoch, Perthshire: *Geol. Mag.*, v. 87, pp. 427-32.

———, 1951A, Note on the tectonic style of the Ord Ban quartzites, Mid-Strathspey: *Geol. Mag.*, v. 88, pp. 50-4.

———, 1951B, The tectonics of the area between Grantown and Tomintoul (mid-Strathspey): *Q. Jl. geol. Soc. Lond.*, v. 107, pp. 1-22.

———, 1952, The tectonics of the Beinn Dronaig area, Attadale: *Trans. Edinb. geol. Soc.*, v. 15, pp. 258-64.

———, 1954, The Moine Thrust—its discovery, age, and tectonic significance: *Proc. Geol. Ass.*, v. 65, pp. 203-23.

———, 1963A, Precision and resolution in geochronology (pp. 112-34) *in The Fabric of Geology:* Addison-Wesley Publishing Co., Inc., Reading, Mass., 372 pp.

———, 1963B, Program for computation of trend surfaces and residuals of degree 1 through 8: *Tech. Rept. Seaver Geol. Lab. Pomona College, Claremont, California*, No. 4, pp. 1-24.

———, & Turner, F. J., 1953, Petrofabric analysis of marbles from mid-Strathspey and Strathavon: *Geol. Mag.*, v. 90, 225-40.

————, & WEISS, L. E., 1956, Construction of block diagrams to scale in orthographic projection: *Proc. Geol. Ass.*, v. 67, pp. 142-55.

MCKEE, E. D., & WIER, G. W., 1953, Terminology for stratification and cross-stratification in sedimentary rocks: *Bull. geol. Soc. Am.*, v. 64, pp. 381-90.

MCKINSTRY, H. E., 1961, Structure of the Glenarm Series in Chester County, Pennsylvania: *Bull. geol. Soc. Am.*, v. 72, pp. 557-78.

————, & MIKKOLA, A. K., 1954, The Elizabeth copper mine, Vermont: *Econ. Geol.*, v. 49, pp. 1-30.

MCLACHLAN, G. R., 1953, The bearing of rolled garnets on the concept of *b*-lineation in Moine rocks: *Geol. Mag.*, v. 90, pp. 172-6.

MCTAGGART, K. C., 1960, The geology of Keno and Galena Hills, Yukon Territory (105 M): *Bull. geol. Surv. Can.*, 58, pp. 1-37.

MEAD, W. J., 1940, Folding, rock flowage, and foliate structures: *J. Geol.*, v. 48, pp. 1007-21.

MEADE, R. H., 1964, Removal of water and rearrangement of particles during the compaction of clayey sediments — Review: *Prof. Pap. U.S. geol. Surv.*, 497-B, pp. 1-23.

MEANS, W. D., 1962, Structure and stratigraphy in the central Toiyabe Range, Nevada: *Univ. Calif. Publs. Bull. Dep. Geol.*, v. 42, pp. 71-110.

————, 1963, Mesoscopic structures and multiple deformation in the Otago Schist: *N.Z. Jl. Geol. Geophys.*, v. 6, pp. 801-16.

MEHNERT, K. R., 1939, Die Meta-Konglomerate des Wiesenthaler Gneiszuges im sächsischen Erzgebirge: *Mineralog. petrogr. Mitt.*, v. 50, pp. 194-272.

MELTON, M. A., 1958A, Geometric properties of mature drainage systems and their representation in an E_4 phase space: *J. Geol.*, v. 66, pp. 35-56.

————, 1958B, Correlation structures of morphometric properties of drainage systems and their controlling agents: *J. Geol.*, v. 66, pp 442-60.

MENDELSOHN, F., 1959, Structure of the Roan Antelope deposit: *Trans. Instn Min. Metall., Lond.*, v. 68, pp. 229-63.

MERTIE, J. B., 1959, Classification, delineation, and measurement of nonparallel folds: *Prof. Pap. U.S. geol. Surv.*, 314-E, pp. 91-124.

METZGER, A. A. T., 1945, Zur geologie der Inseln Ålö und Kyrklandet in Pargas-Parainen, S. W. Finnland: *Acta Acad. åbo, (Math. Phys. XV)*, No. 27, pp. 1-103.

————, 1947, Zum Tectonischen stil von Palingen-granit und Marmor in den Svekofenniden in Finnland: *Bull. Commn géol. Finl.*, 140, pp. 183-92.

————, 1954, The deposit of crystalline limestone of Pargas-Parainen: *Geotek. Julk.*, v. 55, pp. 1-6.

————, 1959, Svekofenniden und Kareliden: eine kritische studie: *Acta Acad. åbo, (Math. Phys. XXI)*, No. 41, pp. 1-27.

MICHENER, C. D., & SOKAL, R. R., 1957, A quantitative approach to a problem in classification: *Evolution, Lancaster, Pa.*, v. 11, pp. 130-62.

MICHOT, P., 1957, Phénomènes géologiques dans la Catazone profonde: *Geol. Rdsch.*, v. 46, pp. 147-73.

MIDDLEMISS, F. A., 1961, Discussion on Memoir No. 2: Geological results of petroleum exploration in Britain 1945-57: *Proc. geol. Soc.*, No. 1583, p. 15.

MILLER, J. A., & BROWN, P. E., 1965, Potassium-argon age studies in Scotland: *Geol. Mag.*, v. 102, pp. 106-34.

MILLER, J. M., 1956, Geology of the Kitale-Cherangani Hills area: *Rep. geol. Surv. Kenya*, 35, pp. 1-34.

MILLER, R. L., 1956, Trend surfaces: their application to analysis and description of environments of sedimentation: *J. Geol.*, v. 64, pp. 425-46.

———, & ZIEGLER, J. M., 1958, A model relating dynamics and sediment pattern in equilibrium in the region of shoaling waves, breaker zone, and foreshore: *J. Geol.*, v. 66, pp. 417-41.

MILNER, H. B., 1962, *Sedimentary Petrography* (Vol. 1): Macmillan Co., New York, 643 pp.

MISCH, P., 1949, Metasomatic granitization of batholithic dimensions; Part III — Relationships of synkinematic and static granitization: *Am. J. Sci.*, v. 247, pp. 673-705.

MOENCH, R. H., HARRISON, J. E., & SIMS, P. K., 1962, Precambrian folding in the Idaho Springs-Central City area, Front Range, Colorado: *Bull. geol. Soc. Am.*, v. 73, pp. 35-58.

MOORE, G. E., 1949, Structure and metamorphism of the Keene-Brattleboro area, New Hampshire-Vermont: *Bull. geol. Soc. Am.*, v. 60, pp. 1613-70.

MOORE, W. L., 1961, Role of fluid pressure in overthrust faulting: a discussion: *Bull. geol. Soc. Am.*, v. 72, pp. 1581-6.

MORRIS, T. O., & FEARNSIDES, W. G., 1926, The stratigraphy and structure of the Cambrian slate-belt of Nantlle (Carnarvonshire): *Q. Jl. geol. Soc. Lond.*, v. 82, pp. 250-303.

MOSEBACH, R., 1951, Zur Petrographie der Dachschiefer des Hunsrück-Schiefers: *Z. dt. geol. Ges.*, v. 103, pp. 368-76.

MUELHBERGER, W. R., 1959, Internal structure of the Grand Saline salt dome, Van Zandt County, Texas: *Rep. Invest. Bur. econ. Geol. Univ. Tex.*, 38, pp. 1-18.

———, 1960, Internal structures and mode of uplift of the Grand Saline salt dome, Van Zandt County, Texas, United States of America: *Int. geol. Congr. XXI Copenhagen*, Pt. 18, pp. 28-33.

NADAI, A. L., 1951, *Theory of flow and fracture of solids:* McGraw-Hill Book Co., New York, 572 pp.

NAHA, K., 1956, Structural set-up and movement plan in parts of Dhalbhum, Bihar: *Sci. Cult.*, v. 22, pp. 43-5.

———, 1959, Steeply plunging recumbent folds: *Geol. Mag.*, v. 96, pp. 137-40.

NAUMANN, C. F., 1839, Ueber den Linear-Parallelismus oder die Streckung mancher Gebirgssteine: *Arch. f. Min. Geog., Bergbau und Hüttenk*, v. 12, pp. 23-38.

NEVIN, C. M., 1931, *Principles of structural geology:* John Wiley & Sons, Inc., New York, 303 pp.

———, 1949, *Principles of structural geology* (4th edit.): John Wiley & Sons, Inc., New York, 410 pp.

NICHOLSON, R., 1963, Eyed folds and interference patterns in the Sokumfjell Marble Group, northern Norway: *Geol. Mag.*, v. 100, pp. 59-68.

NICOL, J., 1861, On the structure of the North-Western Highlands and the relations of the Gneiss, Red Sandstone, and Quartzite of Sutherland and Ross-shire: *Q. Jl. geol. Soc. Lond.*, v. 17, pp. 85-113.

NIGGLI, E., 1960, Mineral-Zonen der alpinen Metamorphose in den Schweizer Alpen: *Int. geol. Congr. XXI Copenhagen*, Pt. 13, pp. 132-8.

NISSEN, H. U., 1962, Analysis of strained crinoid stems in sandstone from Lindlar near Cologne, Germany: *J. geophys. Res.*, v. 67, p. 1650.

———, 1964A, Dynamic and kinematic analysis of deformed crinoid stems in a quartz graywacke: *J. Geol.*, v. 72, pp. 346-60.

——, 1964B, Calcite fabric analysis of deformed oölites from the South Mountain Fold, Maryland: *Am. J. Sci.*, v. 262, pp. 892-903.

NOBLE, J. A., HARDER, J. O., & SLAUGHTER, A. L., 1949, Structure of a part of the northern Black Hills and the Homestake Mine, Lead, South Dakota: *Bull. geol. Soc. Am.*, v. 60, pp. 321-52.

NOBLE, D. C., & EBERLY, S. W., 1964, A digital computer procedure for preparing beta diagrams: *Am. J. Sci.*, v. 262, pp. 1124-9.

NOLAN, T. B., MERRIAM, C. W., & WILLIAMS, J. S., 1956, The stratigraphic section in the vicinity of Eureka, Nevada: *Prof. Pap. U.S. geol. Surv.*, 276, pp. 1-77.

NORMAN, G. W. H., 1962, Faults and folds across Cordilleran trends at the Headwaters of Leduc River, Northern British Columbia: *Mem. geol. Soc. Am.* (Buddington volume), pp. 313-26.

NUREKI, T., 1960, Structural investigation of the Ryôké metamorphic rocks of the area between Iwakuni and Yania, Southwestern Japan: *J. Sci. Hiroshima Univ., Ser. C.*, v. 3, pp. 69-141.

O'DRISCOLL, E. S., 1962, Experimental patterns in superposed similar folding: *J. Alberta Soc. Petrol. Geol.*, v. 10, pp. 145-67.

——, 1964, Cross fold deformation by simple shear: *Econ. Geol.*, v. 59, pp. 1061-93.

OKAMURA, Y., 1960, Structural and petrological studies on the Ryôké gneiss and granodiorite complex of the Yanai District, Southwest Japan: *J. Sci. Hiroshima Univ., Ser. C.*, v. 3, pp. 143-213.

OLDHAM, C. H. G., & SUTHERLAND, D. B., 1955, Orthogonal polynomials: their use in estimating the regional effect: *Geophysics*, v. 20, pp. 295-306.

OLSON, E. C., & MILLER, R. L., 1951, A mathematical model applied to a study of the evolution of species: *Evolution, Lancaster, Pa.*, v. 5, pp. 325-38.

——, & ——, 1958, *Morphological Integration:* Univ. Chicago Press, Chicago, 317 pp.

OULIANOFF, N., 1953, Superposition successive des chaînes de Montagnes: *Scientia, Bologna*, v. 88, pp. 323-7.

PAGE, B. M., 1963, Gravity tectonics near Passo della Cisa, Northern Appenines, Italy: *Bull. geol. Soc. Am.*, v. 74, pp. 655-72.

PARK, R. G., 1964, The structural history of the Lewisian rocks of Gairloch, Wester Ross, Scotland: *Q. Jl. geol. Soc. Lond.*, v. 120, pp. 397-433.

PARKER, R. B., 1961, Petrology and structural geometry of Pre-Granitic rocks in the Sierra Nevada, Alpine County, California: *Bull. geol. Soc. Am.*, v. 72, pp. 1789-806.

PATERSON, M. S., & WEISS, L. E., 1961, Symmetry concepts in the structural analysis of deformed rocks: *Bull. geol. Soc. Am.*, v. 72, pp. 841-82.

PATTERSON, J. R., & STOREY, T. P., 1963, Caledonian earth movements in the vicinity of End Mountain and South Fork Ghost River, Alberta, Canada: *J. Alberta Soc. Petrol. Geol.*, v. 11, pp. 288-98.

PAVLOVSKIY, Y. V., 1958, *Izv. Akad. Nauk SSSR*, No. 6, pp. 23-47 (English translation: Brief outline of Precambrian and Lower Paleozoic of the Scottish Highlands: *Izv. Acad. Sci. USSR geol. Ser.*, No. 6, pp. 19-39).

PEACEY, J. S., 1961, Rolled garnets from Morar, Inverness-shire: *Geol. Mag.*, v. 98, pp. 77-80.

PEACH, B. N., 1904, *Mem. geol. Surv. Summ. Prog.* for 1903, pp. 1-196.

———, GUNN, W., CLOUGH, C. T., HINXMAN, L. W., CRAMPTON, C. B., ANDERSON, E. M., & FLETT, J. S., 1912, The geology of Ben Wyvis, Carn Chuinneag, Inchbae and the surrounding country, including Garve, Evanton, Alness and Kincardine (explanation of sheet 93): *Mem. geol. Surv. U.K.,* pp. 1-189.

———, & HORNE, J., 1914, *Guide to the geological model of the Assynt Mountains:* His Majesty's Stationery Office, London, 32 pp.

———, & ———, 1930, *Chapters on the geology of Scotland:* Oxford Univ. Press, London, 232 pp.

———, ———, GUNN, W., CLOUGH, C. T., GREENLY, E., *et al.,* 1913, The geology of the Fannich Mountains and the country around upper Loch Maree and Strath Broom (explanation of sheet 92): *Mem. geol. Surv., U.K.,* pp. 1-127.

———, ———, ———, ———, HINXMAN, L. W., & TEALL, J. J. H., 1907, The geological structure of the North-West Highlands of Scotland: *Mem. geol. Surv. U.K.,* pp. 1-668.

———, ———, HINXMAN, L. W., CRAMPTON, C. B., ANDERSON, E. M., & CARRUTHERS, R. G., 1913, The geology of central Ross-shire (explanation of sheet 82): *Mem. geol. Surv. U.K.,* pp. 1-114.

———, ———, WOODWARD, H. B., CLOUGH, C. T., HARKER, A., & WEED, C. B., 1910, The geology of Glenelg, Lochalsh and southeast part of Skye (explanation of one-inch map 71): *Mem. geol. Surv. U.K.,* pp. 1-206.

———, KYNASTON, H., MUFF, H. B., *et al.,* 1909, The geology of the seaboard of mid Argyll, including the islands of Luing, Scarba, the Garvellachs, and the Lesser Isles, together with the northern part of Jura and a small portion of Mull (explanation of sheet 36): *Mem. geol. Surv. U.K.,* pp. 1-121.

———, WILSON, J. S. G., HILL, J. B., BAILEY, E. B., GRABHAM, G. W., *et al.,* 1911, The geology of Knapdale, Jura, and North Kintyre (explanation of sheet 28): *Mem. geol. Surv. U.K.,* pp. 1-149.

PEIKERT, E. W., 1962, Three-dimensional specific-gravity variation in the Glen Alpine Stock, Sierra Nevada, California: *Bull. geol. Soc. Am.,* v. 73, pp. 1437-42.

———, 1963, IBM 709 program for least-squares analysis of three-dimensional geological and geophysical observations: *Tech. Rept. Geog. Branch Off. Naval Res. O.N.R. Task No.* 389-135, *Contr. No.* 1228(26), No. 4, pp. 1-72.

PETTIJOHN, F. J., 1949, *Sedimentary Rocks,* Harper & Brothers, New York, 526 pp.

———, 1957A, *Sedimentary Rocks* (2nd edit.): Harper & Brothers, New York, 718 pp.

———, 1957B, Palaeocurrents of Lake Superior Precambrian quartzites: *Bull. geol. Soc. Am.,* v. 68, pp. 469-80.

———, 1960, Some contributions of sedimentology to tectonic analysis: *Int. geol. Congr. XXI Copenhagen,* Pt. 18, pp. 446-54.

PHEMISTER, J., 1948, *British Regional Geology Scotland: The Northern Highlands* (2nd edit.): His Majesty's Stationery Office, Edinburgh, 94 pp.

———, 1958, Summary of recent research on the Pre-Tertiary geology of the Northern Highlands: *Trans. geol. Soc. Glasg.,* v. 23, pp. 53-78.

———, 1960, *British Regional Geology Scotland: The Northern Highlands* (3rd edit.): Her Majesty's Stationery Office, Edinburgh, 104 pp.

PHILLIPS, F. C., 1937, A fabric study of some Moine Schists and associated rocks: *Q. Jl. geol. Soc. Lond.,* v. 93, pp. 581-620.

———, 1945, The micro-fabric of the Moine Schists: *Geol. Mag.,* v. 82, pp. 205-20.

———, 1950, The Lizard-Start problem: *Geol. Mag.,* v. 87, p. 71.

———, 1951, Apparent coincidences in the life-history of the Moine Schists: *Geol. Mag.,* v. 88, pp. 225-35.

———, 1960, *The use of stereographic projection in structural geology* (2nd edit.): Edward Arnold (Publishers) Ltd., London, 86 pp.

PHILLIPS, J., 1844, On certain movements in the parts of stratified rocks: *Rep. Br. Ass. Advmt Sci.,* 1843 (Cork), pp. 60-1.

PICHAMUTHU, C. S., 1962, Some observations on the structure, metamorphism, and geological evolution of Peninsular India: *J. geol. Soc. India,* v. 3, pp. 106-18.

PILGER, A., & SCHMIDT, W., 1957A, Definition des Begriffes "Mullion-Struktur" (mullion structure): *Neues Jb. Geol. Paläont. Mh.,* 1957, pp. 24-8.

———, & ———, 1957B, Die Mullion Strukturen in der Nord-Eifel: *Abh. hess. Landesamt. Bodenforsch.,* v. 20, pp. 3-53.

PITCHER, W. S., & READ, H. H., 1960, The aureole of the main Donegal granite: *Q. Jl. geol. Soc. Lond.,* v. 116, pp. 1-36.

———, ———, CHEESMAN, R. L., PANDE, I. C., & TOZER, C. F., 1959, The main Donegal granite: *Q. Jl. geol. Soc. Lond.,* v. 114 (for 1958), pp. 259-305.

PLAS, L. VAN DER, 1959, Petrology of the northern Adula region, Switzerland (with particular reference to the glaucophane-bearing rocks): *Leid. geol. Meded.,* v. 24, pp. 415-602.

POLL, J. J. K., & ZWART, H. J., 1964, On the tectonics of the Sulcis area, S. Sardinia: *Geologie Mijnb.,* v. 43, pp. 144-6.

POTAPOV, I. I., 1960, Skhema klassifikatsii tektoniche-skikh form: *Sovetsk. Geolog.,* No. 8, pp. 66-74. (English translation *Int. Geol. Rev.,* 1961, v. 3, pp. 1168-73.)

POTTER, P. E., & PETTIJOHN, F. J., 1963, *Paleocurrents and basin analysis :* Academic Press, Inc., New York, 296 pp.

———, & PRYOR, W. A., 1961, Dispersal centers of Paleozoic and later clastics of the upper Mississippi Valley and adjacent areas: *Bull. geol. Soc. Am.,* v. 72, pp. 1195-250.

POWELL, D., 1964, The stratigraphical succession of the Moine Series around Lochailort (Inverness-shire) and its regional significance: *Proc. Geol. Ass.,* v. 75, pp. 223-50.

QUIRKE, T. T., 1923, Boudinage, an unusual structural phenomenon: *Bull. geol. Soc. Am.,* v. 34, pp. 649-60.

———, & LACY, W. C., 1941, Deep-zone dome and basin structures: *J. Geol.,* v. 49, pp. 589-609.

RALEIGH, C. B., & GRIGGS, D. T., 1963, Effect of the toe in the mechanics of overthrust faulting: *Bull. geol. Soc. Am.,* v. 74, pp. 819-30.

RAMBERG, H., 1952, *The origin of metamorphic and metasomatic rocks:* Univ. Chicago Press, Chicago, 317 pp.

———, 1955, Natural and experimental boudinage and pinch-and-swell structures: *J. Geol.,* v. 63, pp. 512-26.

———, 1959, Evolution of ptygmatic folding: *Norsk. geol. Tidsskr.,* v. 39, pp. 99-152.

————, 1963A, Evolution of drag folds: *Geol. Mag.*, v. 100, pp. 97-106.

————, 1963B, Strain distribution and geometry of folds: *Bull. geol. Instn. Univ. Upsala*, v. 42, No. 4, pp. 1-20.

————, 1963C, Experimental study of gravity tectonics by means of centrifuged models: *Bull. geol. Instn. Univ. Upsala*, v. 42, No. 1, pp. 1-97.

RAMSAY, A. C., & SALTER, J. W., 1866, The geology of North Wales: *Mem. geol. Surv. U.K.*, pp. 1-381.

RAMSAY, D. M., 1962, Microfabric studies from the Dalradian rocks of Glen Lyon, Perthshire: *Trans. Edinb. geol. Soc.*, v. 19, pp. 166-200.

————, & STURT, B. A., 1963, A study of fold styles, their associations and symmetry relationships from Sørøy, Northern Norway: *Norsk. geol. Tidsskr.*, v. 43, pp. 411-30.

RAMSAY, J. G., 1958A, Superimposed folding at Loch Monar, Inverness-shire and Ross-shire: *Q. Jl. geol. Soc. Lond.*, v. 113 (for 1957), pp. 271-307.

————, 1958B, Moine-Lewisian relations at Glenelg, Inverness-shire: *Q. Jl. geol. Soc. Lond.*, v. 113 (for 1957), pp. 487-523.

————, 1960, The deformation of early linear structures in areas of repeated folding: *J. Geol.*, v. 68, pp. 75-93.

————, 1961, The effects of folding upon the orientation of sedimentation structures: *J. Geol.*, v. 69, pp. 84-100.

————, 1962A, The geometry and mechanics of formation of "similar" type folds: *J. Geol.*, v. 70, pp. 309-27.

————, 1962B, The geometry of conjugate fold systems: *Geol. Mag.*, v. 99, pp. 516-26.

————, 1962C, Interference patterns produced by the superposition of folds of similar type: *J. Geol.*, v. 70, pp. 466-81.

————, 1963, The folding of angular unconformable sequences: a discussion: *J. Geol.*, v. 71, pp. 397-400.

————, 1964, The uses and limitations of beta-diagrams and pi-diagrams in the geometrical analysis of folds: *Q. Jl. geol. Soc. Lond.*, v. 120, pp. 435-54.

————, & SPRING, J., 1963, Moine stratigraphy in the western Highlands of Scotland: *Proc. Geol. Ass.*, v. 73 (for 1962), pp. 295-326.

RAST, N., 1956, The origin and significance of boudinage: *Geol. Mag.*, v. 93, pp. 401-8.

————, 1958A, Tectonics of the Schichallion complex: *Q. Jl. geol. Soc. Lond.*, v. 114, pp. 25-46.

————, 1958B, Metamorphic history of the Schichallion complex (Perthshire): *Trans. R. Soc. Edinb.*, v. 63, pp. 413-31.

————, 1962, The relationship between tectonic deformation and regional metamorphism: *Proc. geol. Soc.*, No. 1594, pp. 25-36.

————, 1964, Morphology and interpretation of folds — a critical essay: *Lpool. Manchr. geol. J.*, v. 4, pp. 177-88.

————, & PLATT, J. I., 1957, Cross-folds: *Geol. Mag.*, v. 94, pp. 159-67.

————, & STURT, B. A., 1957, Crystallographic and geological factors in the growth of garnets from central Perthshire: *Nature, Lond.*, v. 179, p. 215.

————, ————, & HARRIS, A. L., 1962, Inclusions in garnet: *Nature, Lond.*, v. 195, pp. 274-5.

READ, H. H., 1934A, Age-problems of the Moine Series of Scotland: *Geol. Mag.*, v. 71, pp. 302-17.

————, 1934B, On the segregation of quartz-chlorite-pyrite masses in Shetland igneous rocks during dislocation-metamorphism, with a note on an occurrence of boudinage-structure: *Proc. Lpool. geol. Soc.*, v. 16, pp. 128-38.

———, 1936, The stratigraphical order of the Dalradian rocks of the Banffshire coast: *Geol. Mag.*, v. 73, pp. 468-76.

———, 1940, Metamorphism and igneous action: *Advmt. Sci., Lond.*, v. 1, pp. 223-51.

———, 1949, A contemplation of time in plutonism: *Q. Jl. geol. Soc. Lond.*, v. 105, pp. 101-56.

———, 1955, The Banff Nappe: an interpretation of the structure of the Dalradian rocks of north-east Scotland: *Proc. Geol. Ass.*, v. 66, pp. 1-29.

———, 1956, The last twenty years' work in the Moine Series of Scotland: *Verh. K. ned. geol.-mijnb. Genoot. (Geol. Ser.)*, v. 16, pp. 330-54.

———, 1958, A Centenary Lecture: Stratigraphy in metamorphism: *Proc. Geol. Ass.*, v. 69, pp. 83-102.

———, 1961, Aspects of Caledonian magmatism in Britain: *Lpool. Manchr. geol. J.*, v. 2, pp. 653-83.

———, & FARQUHAR, O. C., 1956, The Buchan anticline of the Banff nappe of Dalradian rocks in north-east Scotland: *Q. Jl. geol. Soc. Lond.*, v. 112, pp. 131-56.

REECE, A., 1960, The stratigraphy, structure and metamorphism of the Pre-Cambrian rocks of North-West Ankole, Uganda: *Q. Jl. geol. Soc. Lond.*, v. 115 (for 1959), pp. 389-420.

REID, R. R., 1957, Bedrock geology of the north end of the Tobacco Root Mountains, Madison County, Montana: *Mem. Mont. St. Bur. Mines Geol.*, 36, pp. 1-27.

———, 1959A, Kinematic analysis for metamorphic rocks in the upper South Fork of the Clearwater River area, Idaho: *Bull. geol. Soc. Am.*, v. 70, pp. 1785-6.

———, 1959B, Reconnaissance geology of the Elk City region, Idaho: *Pamph. Idaho Bur. Mines Geol.*, 120, pp. 1-74.

———, 1963, Metamorphic rocks of the Northern Tobacco Root Mountains, Madison County, Montana: *Bull. geol. Soc. Am.*, v. 74, pp. 293-306.

REITAN, P., 1959, Pegmatite veins and the surrounding rocks; III—Structural control of small pegmatites in amphibolite, Rytterholmen, Kragerøfjord, Norway: *Norsk. geol. Tidsskr.*, v. 39, pp. 175-95.

REUSCH, H., 1887, Geologische Beobachtungen in einem regionalmetamorphosirten Gebiet am Hardangerfjord in Norwegen: *Neues Jb. Miner. Geol. Paläont. BeilBd.*, v. 5, pp. 52-67.

———, 1888, *Bømmeløen og Karmøen:* P. F. Steensballes, Kristiania, 422 pp.

REYER, E., 1892, *Geologische und geographische experimente. Vol. I, Deformation und Gebirgsbildung:* Engelmann, Leipzig, 52 pp.

REYNOLDS, D. L., & HOLMES, A., 1954, The superposition of Caledonoid folds on an older fold-system in the Dalradians of Malin Head, Co. Donegal: *Geol. Mag.*, v. 91, pp. 417-44.

RICHEY, J. E., & KENNEDY, W. Q., 1939, The Moine and Sub-Moine Series of Morar, Inverness-shire: *Bull. geol. Surv. Gt. Br.*, 2, pp. 26-45.

RICHTER, D., 1963, Über Querfaltung in den spanischen Westpyrenäen: *Geol. Mitt. Aachen*, v. 3, pp. 185-96.

RICKARD, M. J., 1961, A note on cleavages in crenulated rocks: *Geol. Mag.*, v. 98, pp. 324-32.

———, 1962, The stratigraphy and structure of the Errigal area, County Donegal, Ireland: *Q. Jl. geol. Soc. Lond.*, v. 118, pp. 207-38.

———, 1963, Analysis of the strike swing at Crockator Mountain, Co. Donegal, Eire: *Geol. Mag.*, v. 100, pp. 401-19.

RINGWOOD, A. E., 1962A, A model for the upper mantle: *J. geophys. Res.*, v. 67, pp. 857-67.

———, 1962B, A model for the upper mantle, 2: *J. geophys. Res.*, v. 67, pp. 4473-7.

ROACH, R., 1960, *Proc. geol. Soc.*, No. 1583, p. 11.

ROBERTS, J. L., & TREAGUS, J. E., 1964, A re-interpretation of the Ben Lui fold: *Geol. Mag.*, v. 101, pp. 512-6.

ROBERTS, R. J., HOTZ, P. E., GILLULY, J., & FERGUSON, H. G., 1958, Paleozoic rocks of north-central Nevada: *Bull. Am. Ass. Petrol. Geol.*, v. 42, pp. 2813-57.

ROBERTSON, D. S., 1953, Batty Lake Map-area, Manitoba: *Mem. geol. Surv. Brch. Can.*, 271, pp. 1-55.

ROSENFELD, M. A., 1954, Petrographic variation in the Oriskany Sandstone: *Bull. geol. Soc. Am.*, v. 65, pp. 95-6.

ROSS, J. V., 1959, Geology, Mesa Lake, District of Mackenzie, Northwest Territories, sheet 86 $\frac{B}{14}$ west half: *Prel. map Geol. Surv. Canada*, 30-1959.

———, 1962A, The folding of angular unconformable sequences: *J. Geol.*, v. 70, pp. 294-308.

———, 1962B, Deposition and current direction within the Yellowknife Group at Mesa Lake, N. W. T., Canada: *Bull. geol. Soc. Am.*, v. 73, pp. 1159-62.

———, & McGLYNN, J. C., 1963, Concentric folding of cover and basement at Basler Lake, N. W. T., Canada: *J. Geol.*, v. 71, pp. 644-53.

RUBEY, W. W., & HUBBERT, M. K., 1959, Role of fluid pressure in mechanics of overthrust faulting; II — Overthrust belt in geosynclinal area of western Wyoming in light of fluid-pressure hypothesis: *Bull. geol. Soc. Am.*, v. 70, pp. 167-205.

RÜGER, L., 1933, Gefügekundliche Untersuchungen an den Geröllgneisen von Obermitweida (Erzgebirge): *Neues Jb. Miner. Geol. Paläont. BeilBd., Abt. A*, v. 66, pp. 275-93.

RUSNAK, G. A., 1957, A fabric and petrologic study of the Pleasantview sandstone: *J. sedim. Petrol.*, v. 27, pp. 41-55.

RUSSELL, W. L., 1955, *Structural geology for petroleum geologists:* McGraw-Hill Book Co. Inc., New York, 427 pp.

RUTLAND, R. W. R., 1959, Structural geology of the Sokumvatn area, north Norway: *Norsk. geol. Tidsskr.*, v. 39, pp. 287-338.

———, HOLMES, M., & JONES, M. A., 1960, Granites of the Glomfjord area, northern Norway: *Int. geol. Congr. XXI Copenhagen*, Pt. 19, pp. 43-53.

RUTLEDGE, H., 1952, The structure of the Fannich Forest area: *Trans. Edinb. geol. Soc.*, v. 15, pp. 317-21.

SAGGERSON, E. P., JOUBERT, P., McCALL, G. J. H., & WILLIAMS, L. A. J., 1960, Cross-folding and refolding in the Basement System of Kenya Colony: *Int. geol. Congr. XXI Copenhagen*, Pt. 18, pp. 335-46.

SAHA, A. K., 1959A, On some characteristics of the quartz fabric of granitic rocks: *Proc. natn. Inst. Sci. India*, v. 25A, pp. 281-90.

————, 1959B, Emplacement of three granitic plutons in southeastern Ontario, Canada: *Bull. geol. Soc. Am.*, v. 70, pp. 1293-326.

SANDER, B., 1911, Über Zusammenhänge zwischen Teilbewegung und Gefüge in Gesteinen: *Mineralog. petrogr. Mitt.*, v. 30, pp. 281-315.

————, 1926, Zur petrographisch-tektonischen Analyse; III Teil: *Jb. geol. Bundesanst., Wien*, v. 76, pp. 323-406.

————, 1930, *Gefügekunde der Gesteine:* Springer, Vienna, 352 pp.

————, 1934, Typisierung von deformierten Tonschiefern mit optischen und röntgenoptischen Mitteln: *Z. Kristallogr.*, v. 89A, pp. 97-124.

————, 1936, Beiträge zur Kenntnis der Anlagerungsgefüge (Rhythmische Kalke und Dolomite aus der Trias): *Mineralog. petrogr. Mitt.*, v. 48, pp. 27-139.

————, 1942, Über Flächen- und Achsengefüge (Westende der Hohen Tauern, III Bericht) (Geologie des Tauern-Westendes I): *Mitt. Reichsamts Bodenforsch. Zweigst. Wien*, v. 4, pp. 1-94.

————, 1948, 1950, *Einführung in die Gefügekunde der Geologischen Körper:* Springer, Vienna, Volume I, 1948, 215 pp., Part II, 1950, 409 pp.

SARKAR, S. N., 1957, Stratigraphy and tectonics of the Dongargarh System: A new system in the Pre-Cambrians of Bhandara-Drug-Balaghat area, Bombay and Madhya Pradesh: *J. Sci. Engng. Res.*, v. 1, pp. 237-68.

————, & MUKHERJEE, A., 1958, Studies on the Nischintapur fault and its relation to the cross-folding movement in East Singhbhum, Bihar: *Q. Jl. geol. Soc. India*, v. 30, pp. 147-51.

————, & SAHA, A. K., 1962, A revision of the pre-cambrian stratigraphy and tectonics of Singhbhum and adjacent regions: *Q. Jl. geol. Soc. India*, v. 34, pp. 97-136.

————, & ————, 1963, On the occurrence of two intersecting Pre-Cambrian Orogenic belts in Singhbhum and adjacent areas, India: *Geol. Mag.*, v. 100, pp. 69-92.

SAWYER, W. W., 1955, *Prelude to mathematics:* Penguin Books, Harmondsworth, 214 pp.

SCHENK, E., 1956, Gangspaltenbildung als Bebenursache?: *Z. dt. geol. Ges.*, v. 106 (for 1954), pp. 361-77.

SCHEUMANN, K. H., 1956, Boudinagen und Mikroboudinagen im Metagabbrischen Plagioklas-Amphibolit von Rosswein (Sächs, Granulitgebirge): *Abh. sächs. Akad. Wiss., Math.-Phys. Kl.*, v. 45, pp. 1-18.

SCHMIDT, W., 1918, Bewegungsspuren in Porphyroblasten kristalliner Schiefer: *Sber. Akad. Wiss., Wien, Abt.* I, v. 127, pp. 293-310.

————, 1925, Gefügestatistik: *Mineralog. petrogr. Mitt.*, v. 38, pp. 392-423.

————, 1932, *Tektonik und Verformungslehre:* Borntraeger, Berlin, 208 pp.

SCHMINCKE, H. U., 1961, Beitrag zum Kapitel "Mullion-Struktur": *Neues Jb. Geol. Paläont. Mh.*, 1961, pp. 225-35.

SCHOEMAN, J. J., 1949, Geology of the Sotik district: *Rep. geol. Surv. Kenya*, 16, pp. 1-39.

SCHOLTZ, H., 1930, Das varistische bewegungsbild, entwickelt aus der inneren tektonik eines profils von der Böhmischen masse bis zum massiv von Brabant: Fort Brabant: *Fortschr. Geol. Palaeont.*, v. 8, pp. 235-316.

————, 1931, Über das Alter der Schieferung und ihr Verhältnis zur Faltung: *Jb. preuss. geol. Landesanst. Berg Akad.*, v. 52, pp. 303-16.

————, 1932, Faltung und Schieferung im Ostsauerländer Hauptsattel: *Zentbl. Miner. Geol. Paläont., Abt. B*, pp. 321-35.

SCHREYER, W., 1959A, Über das Alter der Metamorphose im Moldanubikum des südlichen Bayerischen Waldes: *Geol. Rdsch.*, v. 46 (for 1957), pp. 306-17.

———, 1959B, Geologisch-petrographische Beobachtungen in den Schottischen Highlands: *Geol. Rdsch.*, v. 46 (for 1957), pp. 612-41.

SCHUCHERT, C., 1923, Sites and nature of the North American geosynclines: *Bull. geol. Soc. Am.*, v. 34, pp. 151-229.

SCHUILING, R. D., 1963, Some remarks concerning the scarcity of retrograde vs. progressive metamorphism: *Geologie Mijnb.*, v. 42, pp. 177-9.

SCHULMAN, N., 1959, Geology of Tornafort area, Central Pyrenees, Noguera de Pallaresa, Prov. de Lerida, Spain: *Leid. geol. Meded.*, v. 24, pp. 407-14.

SCHWARZACHER, W., 1963, Orientation of crinoids by current action: *J. sedim. Petrol.*, v. 33, pp. 580-6.

SCOTFORD, D. M., 1955, Axial-plane folding: *Bull. geol. Soc. Am.*, v. 66, p. 1614.

———, 1956, Metamorphism and axial-plane folding in the Poundridge area, New York: *Bull. geol. Soc. Am.*, v. 67, pp. 1155-98.

SCROPE, G. P., 1825, *Considerations on Volcanoes, the probable causes of their phenomena, the laws which determine their march, the disposition of their products, and their connexion with the present state and past history of the globe; leading to the establishment of a new theory of the earth:* W. Phillips, London, 270 pp.

———, 1862, *Volcanoes: The character of their phenomena, their share in the structure and composition of the surface of the globe, and their relation to its internal forces* (2nd edit.): Longmans, Green, Longman, and Roberts, London, 490 pp.

SECRIST, M. H., 1936, Perspective block diagrams: *Econ. Geol.*, v. 31, pp. 867-80.

SEDERHOLM, J. J., 1913, Über ptygmatische Faltungen: *Neues Jb. Miner. Geol. Paläont. BeilBd.*, v. 36, pp. 491-512.

SEDGWICK, A., 1835, Remarks on the structures of large mineral masses, and especially on the chemical changes produced in the aggregation of stratified rocks during different periods after their deposition: *Trans. geol. Soc. Lond.*, v. 3 (2nd ser.), pp. 461-86.

SHACKLETON, R. M., 1954, The structural evolution of North Wales: *Lpool. Manchr. geol. J.*, v. 1 (for 1953), pp. 261-96.

———, 1958, Downward-facing structures of the Highland Border: *Q. Jl. geol. Soc. Lond.*, v. 113 (for 1957), pp. 361-92.

SHARP, R. P., 1942, Periglacial involutions in northeastern Illinois: *J. Geol.*, v. 50, pp. 113-33.

SHARPE, D., 1847, On slaty cleavage: *Q. Jl. geol. Soc. Lond.*, v. 3, pp. 74-105.

———, 1849, On slaty cleavage (second communication): *Q. Jl. geol. Soc. Lond.*, v. 5, pp. 111-29.

SHATSKI, N. S., & BOGDANOV, A. A., 1959, Explanatory notes on the tectonic map of the U.S.S.R. and adjoining countries: *Internat. Geol. Review*, v. 1, pp. 1-49 (original Russian, 1957).

SHIELLS, K. A. G., & DEARMAN, W. R., 1963, Tectonics of the Coldingham Bay Area of Berwickshire, in the Southern Uplands of Scotland: *Proc. Yorks. geol. Soc.*, v. 34, pp. 209-34.

SHROCK, R. R., 1948, *Sequence in layered rocks*, McGraw-Hill Book Co., New York, 507 pp.

SIMPSON, A., 1963, The stratigraphy and tectonics of the Manx Slate Series, Isle of Man: *Q. Jl. geol. Soc. Lond.*, v. 119, pp. 367-400.

SIMPSON, S., 1940, Das Devon der Südost-Eifel zwischen Nelte und Alf; Strati-graphie und Tektonik mit einem Beitrag zur Hunsrückscheifer-Frage: *Abh. senckenb. naturforsch. Ges.*, no. 447, pp. 1-67.

SITTER, L. U. DE, 1952, Plissement croisé dans le Haut-Atlas: *Geologie Mijnb.*, v. 14, pp. 277-82.

————, 1954, Schistosity and shear in micro- and macrofolds: *Geologie Mijnb.*, v. 16, pp. 429-39.

————, 1956A, The strain of rock in mountain-building processes: *Am. J. Sci.*, v. 254, pp. 585-604.

————, 1956B, *Structural Geology:* McGraw-Hill Book Co., New York, 552 pp.

————, 1957, Cleavage folding in relation to sedimentary structure: *Int. Geol. Congr. XX Mexico,* 1956, Sect. 5, pp. 53-64.

————, 1958, Boudins and parasitic folds in relation to cleavage and folding: *Geologie Mijnb.*, v. 20, pp. 277-86.

————, 1959, The Rio Esla nappe in the zone of Leon of the Asturian Cantabric Mountain chain: *Notas Comun. Inst. geol. min. Esp.*, No. 56, pp. 1-23.

————, 1960A, Conclusions and conjectures on successive tectonic phases: *Geologie Mijnb.*, v. 39, pp. 195-7.

————, 1960B, Crossfolding in non-metamorphic of the Cantabrian mountains and in the Pyrenees: *Geologie Mijnb.*, v. 39, pp. 189-94.

————, 1962, The structure of the southern slope of the Cantabrian Mountains: explanation of a geological map with sections scale (1:100.000): *Leid. geol. Meded.*, v. 26, pp. 255-64.

————, 1964, *Structural Geology* (2nd edit.): McGraw-Hill Book Co., New York, 551 pp.

————, & ZWART, H. J., 1959, Geological map of the Paleozoic of the Central Pyrenees. Sheet 3, Ariège, France. 1:50,000: *Leid. geol. Meded.*, v. 22, pp. 351-418.

————, & ————, 1960, Tectonic development in supra- and infra-structures of a mountain chain: *Int. geol. Congr. XXI Copenhagen,* Pt. 18, pp. 248-56.

————, & ————, 1961, Excursion to the Central Pyrenees, September 1959: *Leid. geol. Meded.*, v. 26, pp. 1-49.

SLOSS, L. L., 1962, Stratigraphic models in exploration: *Bull. Am. Ass. Petrol. Geol.*, v. 46, pp. 1050-7.

SMITH, B., & GEORGE, T. N., 1948, *British Regional Geology North Wales* (2nd edit.): His Majesty's Stationery Office, London, 89 pp.

SNEATH, P. H. A., & SOKAL, R. R., 1962, Numerical taxonomy: *Nature, Lond.*, v. 193, pp. 855-60.

SNOW, J. P., & SIRONKO, P. T., 1962, Statistical analysis of mining system parameters: *Mineral Inds. J.*, v. 9, pp. 5-7.

SOKAL, R. R., 1961, Distance as a measure of taxonomic similarity: *Syst. Zool.*, v. 10, pp. 70-9.

————, & MICHENER, C. D., 1958, A statistical method for evaluating systematic relationships: *Kans. Univ. Sci. Bull.*, v. 38, pp. 1409-38.

————, & SNEATH, P. H. A., 1963, *Principles of numerical taxonomy:* W. H. Freeman, San Francisco, 359 pp.

SOMMER, M., 1957, Geologische Untersuchungen in den Praekambrischen Sedi-menten zwischen Grandjeans Fjord und Bessels Fjord (75-76° N. BR.) in NE-Grönland: *Meddr. Grønland*, v. 160, No. 2, pp. 1-56.

SOPER, N. J., 1964, Conditions in the metamorphic Caledonides during the period of late orogenic cooling: *Geol. Mag.*, v. 101, pp. 567-8.

SORBY, H. C., 1853, On the origin of slaty cleavage: *Edinb. New Phil. Journ.*, v. 55, pp. 137-48.

——, 1856, On slaty cleavage, as exhibited in the Devonian limestones of Devonshire: *Phil. Mag.*, v. 11 (4th ser.), pp. 20-37.

SPENCER, E. W., 1959, Geologic evolution of the Beartooth Mountains, Montana and Wyoming; Part 2 — Fracture Patterns: *Bull. geol. Soc. Am.*, v. 70, pp. 467-508.

SPRY, A., 1963A, The origin and significance of snowball structure in garnet: *J. Petrology*, v. 4, pp. 211-22.

——, 1963B, The chronological analysis of crystallization and deformation of some Tasmanian Precambrian rocks: *J. geol. Soc. Aust.*, v. 10, pp. 193-208.

——, 1963C, Ripple marks and pseudo-ripple marks in deformed quartzite: *Am. J. Sci.*, v. 261, pp. 756-66.

——, & GEE, D., 1964, Some effects of Palaeozoic folding on the Pre-Cambrian rocks of the Frenchmans Cap area, Tasmania: *Geol. Mag.*, v. 101, pp. 385-96.

STAINIER, X., 1907, Sur le mode de gisement et l'origine des roches métamorphiques de la région de Bastogne (Belgique): *Mém. Acad. r. Belg. Cl. Sci. 4°* (2nd ser.), v. 1, pp. 1-162.

STARK, J. T., & BARNES, F. F., 1932, The structure of the Sawatch Range: *Am. J. Sci.*, v. 224, pp. 471-80.

STAUB, R., 1924, Der Bau der Alpen Versuch einer Synthese: *Beitr. geol. Karte Schweiz.*, N. S. 52, pp. 1-272.

STEINMETZ, R., 1962, Analysis of vectoral data: *J. sedim. Petrol.*, v. 32, pp. 801-12.

——, 1964, Available methods for the analysis of vectoral data: a reply: *J. sedim. Petrol.*, v. 34, pp. 441-2.

STEWART, F. H., 1958, The nomenclature of evaporite textures: *Am. J. Sci.*, v. 256, p. 219.

——, & JOHNSON, M. R. W., 1960, The structural problem of the younger gabbros of north-east Scotland: *Trans. Edinb. geol. Soc.*, v. 18, pp. 104-12.

STILLE, H., 1910, Die mitteldeutsche Rahmenfaltung: *Jber. niedersächs. geol. Ver.*, v. 3, pp. 141-70.

STILLWELL, F. L., 1918, The metamorphic rocks of Adelie Land Section I: *Scient. Rep. Australas. Antarct. Exped.*, Ser. A, v. 3, Pt. 1, pp. 1-230.

STOČES, B., & WHITE, C. H., 1935, *Structural geology with special reference to economic deposits :* Macmillan & Co. Ltd., London, 460 pp.

STOCKWELL, C. H., 1950, The use of plunge in the construction of cross-sections of folds: *Proc. geol. Ass. Can.*, v. 3, pp. 97-121.

STRAND, T., 1945, Structural petrology of the Bygdin Conglomerate: *Norsk. geol. Tidsskr.*, v. 24, pp. 14-31.

——, 1951, The Sel and Vågå map areas; geology and petrology of a part of the Caledonides of central southern Norway: *Norg. geol. Unders.*, No. 178, pp. 1-117.

STURT, B. A., 1961A, Preferred orientation of nepheline in deformed nepheline syenite gneisses from Soroy, Northern Norway: *Geol. Mag.*, v. 98, pp. 464-6.

——, 1961B, The geological structure of the area south of Loch Tummel: *Q. Jl. geol. Soc. Lond.*, v. 117, pp. 131-56.

——, & HARRIS, A. L., 1961, The metamorphic history of the Loch Tummel area, Central Perthshire, Scotland: *Lpool. Manchr. geol. J.,* v. 2, pp. 689-711.

SUTTON, J., 1960A, Some crossfolds and related structures in Northern Scotland: *Geologie Mijnb.,* v. 39, pp. 149-62.

——, 1960B, Some structural problems in the Scottish Highlands: *Int. geol. Congr. XXI Copenhagen,* Pt. 18, pp. 371-83.

——, 1961, The Moine Series of Scotland; I – Stratigraphy, chronology and polymetamorphism: *Sci. Prog., Lond.,* v. 49, pp. 715-24.

——, 1962, The Moine Series of Scotland; II – Structure: *Sci. Prog., Lond.,* v. 50, pp. 76-86.

——, & WATSON, J., 1951, The pre-Torridonian metamorphic history of the Loch Torridon and Scourie areas in the North-West Highlands, and its bearing on the chronological classification of the Lewisian: *Q. Jl. geol. Soc. Lond.,* v. 106 (for 1950), pp. 241-307.

——, & ——, 1953, The supposed Lewisian inlier of Scardroy, Central Ross-shire, and its relations with the surrounding Moine rocks: *Q. Jl. geol. Soc. Lond.,* v. 108 (for 1952), pp. 99-126.

——, & ——, 1954, The structure and stratigraphical succession of the Moines of Fannich Forest and Strath Bran, Ross-shire: *Q. Jl. geol. Soc. Lond.,* v. 110, pp. 21-53.

——, & ——, 1955, The deposition of the Upper Dalradian rocks of the Banffshire coast: *Proc. Geol. Ass.,* v. 66, pp. 101-33.

——, & ——, 1956, The Boyndie syncline of the Dalradian of the Banffshire coast: *Q. Jl. geol. Soc. Lond.,* v. 112, pp. 103-30.

——, & ——, 1959, Structures in the Caledonides between Loch Duich and Glenelg, North-West Highlands: *Q. Jl. geol. Soc. Lond.,* v. 114 (for 1958), pp. 231-57.

——, & ——, 1962A, An interpretation of Moine-Lewisian relations in central Ross-shire: *Geol. Mag.,* v. 99, pp. 527-41.

——, & ——, 1962B, Further observations on the margin of the Laxfordian complex of the Lewisian near Loch Laxford, Sutherland: *Trans. R. Soc. Edinb.,* v. 65, pp. 89-106.

SUTTON, R. F., 1963, Involutions in surficial deposits, northwestern Ontario: *Bull. geol. Soc. Am.,* v. 74, pp. 789-94.

SUZUKI, T., 1963, A petrofabric study of shear micro-fold: *Geol. Rep. Hiroshima Univ.,* No. 12, pp. 445-61.

SWANSON, C. O., 1941, Flow cleavage in folded beds: *Bull. geol. Soc. Am.,* v. 52, pp. 1245-63.

TALIAFERRO, N. L., 1943, Geologic history and structure of the central Coast Ranges of California: *Bull. Div. Mines Calif.,* v. 118, pp. 119-63.

TANTON, T. L., 1930, Determination of age-relations in folded rocks: *Geol. Mag.,* v. 67, pp. 73-6.

TEICHMÜLLER, M., & TEICHMÜLLER, R., 1955, Zur microtektonischen Verformung der Kohle: *Geol. Landesanst. Geol. Jahr. Berlin :* v. 69 (for 1954), pp. 263-79.

TEICHMÜLLER, R., 1931, Zur Geologie des Tyrrhenisgebietes: Alte und junge Krustenbewegungen im südlichen Sardinien: *Abh. K. Ges. Wiss. Gottingen, Math.-Phys. Kl.,* Folge 3, pp. 1-94.

TEX, E. DEN, 1954, Stereographic distinction of linear and planar structures from apparent lineations in random exposure planes: *J. geol. Soc. Aust.*, v. 1 (for 1953), pp. 55-66.

————, 1956, Complex and imitation tectonites from the Kosciusko Pluton, Snowy Mountains, New South Wales: *J. geol. Soc. Aust.*, v. 3, pp. 33-54.

————, 1959, The geology of the Grey Mare Range in the Snowy Mountains of New South Wales: *Proc. R. Soc. Vict.*, v. 71, pp. 1-24.

————, 1963, A commentary on the correlation of metamorphism and deformation in space and time: *Geologie Mijnb.*, v. 42, pp. 170-6.

————, & VOGEL, D. E., 1963, A "Granulitgebirge" at Cabo Ortegal (N. W. Spain): *Geol. Rdsch.*, v. 52 (for 1962), pp. 95-112.

TISCHER, G., 1962, Über die Wealden-Ablagerung und die Tektonik des östlichen Sierra de los Cameros in den nordwestlichen Iberischen Ketten (Spanien): *Beih. geol. Jb.*, v. 44, pp. 122-63.

————, 1963, Über K-Achsen: *Geol. Rdsch.*, v. 52 (for 1962), pp. 426-47.

TOIT, A. L. DU, 1954, *The geology of South Africa*, Oliver & Boyd, Ltd., Edinburgh, 611 pp.

TREMLETT, W. E., 1963, Stress chronology of the mid-Paleozoic orogeny of the British Isles: *J. Geol.*, v. 71, pp. 793-800.

TURNER, F. J., 1941, The development of pseudo-stratification by metamorphic differentiation in the schists of Otago, New Zealand: *Am. J. Sci.*, v. 239, pp. 1-16.

————, 1948, Mineralogical and structural evolution of the metamorphic rocks: *Mem. geol. Soc. Am.*, 30, pp. 1-342.

————, & CH'IH, C. S., 1951, Deformation of Yule marble; Part III – Observed fabric changes due to deformation at 10,000 atmospheres confining pressure, room temperature, dry: *Bull. geol. Soc. Am.*, v. 62, pp. 887-906.

————, GRIGGS, D. T., CLARK, R. H., & DIXON, R. H., 1956, Deformation of Yule marble; Part VII – Development of oriented fabrics at 300°C-500°C: *Bull. geol. Soc. Am.*, v. 67, pp. 1259-94.

————, ————, & HEARD, H., 1954, Experimental deformation of calcite crystals: *Bull. geol. Soc. Am.*, v. 65, pp. 883-933.

————, & HUTTON, C. O., 1941, Some porphyroblastic albite schists from Waikouaiti River (south branch), Otago: *Trans. R. Soc. N.Z.*, v. 71, pp. 223-40.

————, & VERHOOGEN, J., 1960, *Igneous and metamorphic petrology:* Mc-Graw-Hill Book Co., New York, 694 pp.

————, & WEISS, L. E., 1963, *Structural analysis of metamorphic tectonites:* McGraw-Hill Book Co., New York, 545 pp.

TWENHOFEL, W. H., 1926, *Treatise on Sedimentation:* Williams & Wilkins Co., Baltimore, 661 pp.

————, 1932, *Treatise on Sedimentation* (2nd edit.): Williams & Wilkins Co., Baltimore, 926 pp.

VAN HISE, C. R., 1896, Principles of North American Pre-Cambrian Geology: *Rep. U.S. Geol. Surv.*, 16 (1894-5), Part I, pp. 571-843.

————, & LEITH, C. K., 1911, The geology of the Lake Superior region: *Monogr. U.S. geol. Surv.*, 52, pp. 1-641.

VARDABASSO, S., 1941, Osservazioni sulla tettonica dell'Iglesiente: *Rc. assoc. min. Sarda*, v. 46, no. 5, pp. 1-4.

————, 1956, La fase sarda dell'orogenesi caledonica in Sardegna: *in Geotektonisch. Symposium for Ehren von H. Stille*, pp. 120-7.

VOGT, M. C., 1954, *Tectonics and petrofabric structures in Ballachulish and Glen Coe, Argyllshire* : Ph.D. Thesis, Univ. Glasgow, 119 pp.

VOGT, T., 1930, On the chronological order of deposition of the Highland Schists: *Geol. Mag.*, v. 67, pp. 68-73.

VOLL, G., 1953, Zur Mechanik der Molasseverformung: *Geologica bav.*, v. 17, pp. 135-43.

———, 1960, New work on petrofabrics: *Lpool. Manchr. geol. J.*, v. 2, pp. 503-67.

———, 1965, Deckenbau und Fazies im Schottischen Dalradian: *Geol. Rdsch.*, v. 53 (for 1964), pp. 590-612.

WALLS, R., 1937A, *Andalusite-schists and associated rocks of north-east Scotland:* Ph.D. Thesis, Univ. Liverpool.

———, 1937B, A new record of boudinage-structure from Scotland: *Geol. Mag.*, v. 74, pp. 325-32.

WALTON, E. K., 1956, Limitations of graded bedding: and alternative criteria of upward sequence in the rocks of the Southern Uplands: *Trans. Edinb. geol. Soc.*, v. 16, pp. 262-71.

———, 1961, Some aspects of the succession and structure in the Lower Palaeozoic rocks of the Southern Uplands of Scotland: *Geol. Rdsch.*, v. 50, pp. 63-77.

WARD, R. F., 1959, Petrology and metamorphism of the Wilmington complex, Delaware, Pennsylvania, and Maryland: *Bull. geol. Soc. Am.*, v. 70, pp. 1425-58.

WATERS, A. C., & KRAUSKOPF, K., 1941, Protoclastic border of the Colville batholith: *Bull. geol. Soc. Am.*, v. 52, pp. 1355-418.

WATSON, G. S., 1966, The statistics of orientation data: *Tech. Report Logistics & Math. Stat. Branch Off. Naval Res. O.N.R. Task No. 042-232, Contr. 4010(09)*, No. 51, pp. 1-20.

———, & IRVING, E., 1957, Statistical methods in rock magnetism: *Mon. Not. R. astr. Soc. geophys. Suppl.*, v. 7, pp. 289-300.

WATSON, J., 1963, Some problems concerning the evolution of the Caledonides of the Scottish Highlands: *Proc. Geol. Ass.*, v. 74, pp. 213-58.

———, 1964, Conditions in the metamorphic Caledonides during the period of late-orogenic cooling: *Geol. Mag.*, v. 101, pp. 457-65.

WEGMANN, C. E., 1929, Beispiele tektonischer Analysen des Grundgebirges in Finnland: *Bull. Commn géol. Finl.*, 87, pp. 98-127.

———, 1932, Note sur le boudinage: *Bull. Soc. géol. Fr.*, (5th ser.), v. 2, pp. 477-91.

———, 1935A, Zur Deutung der Migmatite: *Geol. Rdsch.*, v. 26, pp. 305-50.

———, 1935B, Preliminary report on the Caledonian orogeny in Christian X's Land (North-east Greenland): *Meddr. Grønland*, v. 103, no. 3, pp. 1-59.

———, 1938, Geological investigations in southern Greenland: Part I – On the structural divisions of southern Greenland: *Meddr. Grønland*, v. 113, no. 2, pp. 1-148.

WEISS, J., 1949, Wissahickon schist at Philadelphia, Pennsylvania: *Bull. geol. Soc. Am.*, v. 60, pp. 1689-726.

WEISS, L. E., 1953, Tectonic features of the Hecla Hook Formation of the South of St. Jonsfjord, Vestspitsbergen: *Geol. Mag.*, v. 90, pp. 273-86.

———, 1954, A study of tectonic style: structural investigation of a marble-quartzite complex in southern California: *Univ. Calif. Publs. Bull. Dep. Geol.*, v. 30, pp. 1-102.

———, 1955, Fabric analysis of a triclinic tectonite and its bearing on the geometry of flow in rocks: *Am. J. Sci.*, v. 253, pp. 225-36.

————, 1956, *Structural analysis of rocks deformed by flow, with special reference to the concept of symmetry:* D. Sc. Thesis, Univ. Edinburgh.

————, 1959A, Structural analysis of the Basement System at Turoka, Kenya: *Overseas Geol. Miner. Resour.,* v. 7, pp. 3-35 & 123-53.

————, 1959B, Geometry of superposed folding: *Bull. geol. Soc. Am.,* v. 70, pp. 91-106.

————, & McINTYRE, D. B., 1957, Structural geometry of Dalradian rocks at Loch Leven, Scottish Highlands: *J. Geol.,* v. 65, pp. 575-602.

————, & ————, 1959, Structural geometry of Dalradian rocks at Loch Leven, Scottish Highlands: a reply: *J. Geol.,* v. 67, pp. 247-9.

————, ————, & KÜRSTEN, M., 1955, Contrasted styles of folding in the rocks of Ord Ban Mid-Strathspey: *Geol. Mag.,* v. 92, pp. 21-36.

WELLS, J. D., SHERIDAN, D. M., & ALBEE, A. L., 1961, Metamorphism and structural history of the Coal Creek area, Front Range, Colorado: *Prof. Pap. U.S. geol. Surv.,* 424-C, pp. 127-31.

WENK, E., 1937, Zur Genese der Bändergneise von Ornö Huvud: *Bull. geol. Instn. Univ. Upsala,* v. 26, pp. 53-89.

————, 1963, Das reaktivierte Grundgebirge der Zentralalpen: *Geol. Rdsch.,* v. 52 (for 1962), pp. 754-66.

WEST, I. M., 1964, Deformation of the incompetent beds in the Purbeck anticline: *Geol. Mag.,* v. 101, p. 373.

WETHERILL, G. W., KOUVO, O., TILTON, G. R., & GAST, P. W., 1962, Age measurements on rocks from the Finnish Precambrian: *J. Geol.,* v. 70, pp. 74-88.

WHITE, H. J. O., 1921, A short account of the geology of the Isle of Wight: *Mem. geol. Surv. U.K.,* pp. 1-219.

WHITE, W. S., 1949, Cleavage in east-central Vermont: *Trans. Am. geophys. Un.,* v. 30, pp. 587-94.

————, & BILLINGS, M. P., 1951, Geology of the Woodsville quadrangle, Vermont-New Hampshire: *Bull. geol. Soc. Am.,* v. 62, pp. 647-96.

————, & JAHNS, R. H., 1950, Structure of central and east-central Vermont: *J. Geol.,* v. 58, pp. 179-220.

WHITTEN, E. H. T., 1951, Cataclastic pegmatites and calc-silicate skarns near Bunbeg, Co. Donegal: *Mineralog. Mag.,* v. 29, pp. 737-56.

————, 1957A, A note on stereographic analysis of bedding planes with special reference to folding in the Isle of Wight, England: *J. Geol.,* v. 65, pp. 551-6.

————, 1957B, The Gola granite (Co. Donegal) and its regional setting: *Proc. R. Ir. Acad.,* v. 58B, pp. 245-92.

————, 1959A, A study of two directions of folding: the structural geology of the Monadhliath and Mid-Strathspey: *J. Geol.,* v. 67, pp. 14-47.

————, 1959B, Composition trends in a granite: modal variation and ghost-stratigraphy in part of the Donegal granite, Eire: *J. geophys. Res.,* v. 64, pp. 835-48.

————, 1961A, Comment on discussion on Memoir No. 2: Geological results of petroleum exploration in Britain 1945-57: *Proc. geol. Soc.,* No. 1588, p. 92.

————, 1961B, Quantitative areal modal analysis of granite complexes: *Bull. geol. Soc. Am.,* v. 72, pp. 1331-60.

————, 1962, Sampling and trend-surface analyses of granites: a reply: *Bull. geol. Soc. Am.,* v. 73, pp. 415-8.

————, 1963A, Application of quantitative methods in the geochemical study of granite massifs: *Spec. Public. Roy. Soc. Canada,* 6 (Studies in analytical geochemistry), pp. 76-123.

————, 1963B, A surface-fitting program suitable for testing geological models which involve areally-distributed data: *Tech. Report Geog. Branch Off. Naval Res., O.N.R. Task No.* 389-135, *Contr.* 1228 (26), No. 2, pp. 1-56. 1-56.

————, 1963C, A reply to Chayes and Suzuki: *J. Petrology*, v. 4, pp. 313-6.

————, 1964A, Process-response models in geology: *Bull. geol. Soc. Am.*, v. 75, pp. 455-64.

————, 1964B, Models in the geochemical study of rock units: *Colo. Sch. Mines Q.*, v. 59, No. 4, pp. 149-68.

————, & BOYER, R. E., 1964, Process-response models for the San Isabel granite, Colorado, based on heavy mineral content: *Bull. geol. Soc. Am.*, v. 75, pp. 841-62.

————, KRUMBEIN, W. C., WAYE, I., & BECKMAN, W. A., JR., 1965, A surface-fitting program for areally-distributed data from the earth sciences and remote sensing: NASA *contr. Rep.*, CR-318, pp. 1-146.

WHITTINGTON, H. B., & KINDLE, C. H., 1963, Middle Ordovician Table Head Formation, Western Newfoundland: *Bull. geol. Soc. Am.*, v. 74, pp. 745-58.

WILLDEN, C. R., 1958, Cretaceous and Tertiary orogeny in Jackson Mountains, Humboldt County, Nevada: *Bull. Am. Ass. Petrol. Geol.*, v. 42, pp. 2378-98.

WILLIAMS, A., 1959, A structural history of the Girvan District, S. W. Ayrshire: *Trans. R. Soc. Edinb.*, v. 63, pp. 629-67.

WILLIAMS, E., 1961, The deformation of confined, incompetent layers in folding: *Geol. Mag.*, v. 98, pp. 317-23.

WILLIS, B., & WILLIS, R., 1929, *Geologic structures* (2nd edit.): McGraw-Hill Book Co., Inc., New York, 518 pp.

WILSON, A. F., 1952, The charnockite problem in Australia: *Sir D. Mawson Anniv. Vol. Univ. of Adelaide*, pp. 203-24.

WILSON, G., 1946, The relationship of slaty cleavage and kindrid structures to tectonics: *Proc. Geol. Ass.*, v. 57, pp. 263-302.

————, 1951, The tectonics of the Tintagel area, North Cornwall: *Q. Jl. geol. Soc. Lond.*, v. 106, pp. 393-432.

————, 1953, Mullion and rodding structures in the Moine Series of Scotland: *Proc. Geol. Ass.*, v. 64, pp. 118-51.

————, 1961, The tectonic significance of small scale structures, and their importance to the geologist in the field: *Annls. Soc. géol. Belg.*, v. 84, pp. 423-548.

————, WATSON, J., & SUTTON, J., 1953, Current-bedding in the Moine Series of north-western Scotland: *Geol. Mag.*, v. 90, pp. 377-87.

WISE, D. U., 1958, An example of recumbent folding south of the Great Valley of Pennsylvania: *Proc. Pa. Acad. Sci.*, v. 32, pp. 172-6.

————, 1964, Microjointing in Basement, Middle Rocky Mountains of Montana and Wyoming: *Bull. geol. Soc. Am.*, v. 75, pp. 287-306.

WOODLAND, B. G., 1965, The geology of the Burke Quadrangle, Vermont: *Bull. Vt. geol. Surv.*, no. 28, pp. 1-151.

WUNDERLICH, H. G., 1959A, Erzeugung engständiger Scherflächen in plastischem Material: *Neues Jb. Geol. Paläont. Mh.*, 1959, pp. 34-44.

————, 1959B, Zur Entstehung von Boudins und Parasitärfalten: *Neues Jb. Geol. Paläont. Mh.*, 1959, pp. 132-7.

————, 1962, Faltenstereometrie und Gesteinsverformung: *Geol. Rdsch.*, v. 52, pp. 417-26.

————, 1963A, Neue Untersuchungen zur Struktur von Faltengebirgen: *Umschau*, v. 19, pp. 600-1.

————. 1963B, Faltenbau, Stratigraphie und fazielle Entwicklung Ostelbas: *Neues Jb. Geol. Paläont. Mh.*, 1963, pp. 161-81.

————, 1963C, Ablauf und Altersverhältnis der postvaristischen Tektonik und Metamorphose im Westalpenbogen: *Geologie Mijnb.*, v. 42, pp. 155-69.

WYLLIE, P. J., 1962, The petrogenetic model, an extension of Bowen's petrogenetic grid: *Geol. Mag.*, v. 99, pp. 558-69.

WYNNE-EDWARDS, H. R., 1957, Structure of the Westport concordant pluton in the Grenville, Ontario: *J. Geol.*, v. 65, pp. 639-49.

————, 1963, Flow folding: *Am. J. Sci.*, v. 261, pp. 793-814.

ZEN, E-AN, 1961, Stratigraphy and structure at the north end of the Taconic Range in West-Central Vermont: *Bull. geol. Soc. Am.*, v. 72, pp. 293-338.

ZIMMERLE, W., & BONHAM, L. C., 1962, Rapid methods for dimensional grain orientation measurements: *J. sedim. Petrol.*, v. 32, pp. 751-63.

ZWART, H. J., 1953, La géologie du massif du Saint-Barthélemy (Pyrénées, France): *Leid. geol. Meded.*, v. 18, pp. 1-228.

————, 1956, À propos des migmatites pyrénéennes: *Bull. Soc. géol. Fr.*, (6th ser.), v. 6, pp. 49-56.

————, 1959A, Metamorphic history of the central Pyrenees Part I; Arize, Trois Seigneurs and Saint-Barthelemy massifs (Sheet 3): *Leid. geol. Meded.*, v. 22, pp. 419-90.

————, 1959B, On the occurrence of chloritoid in the Pyrenees: *Geologie Mijnb.*, v. 21, pp. 119-22.

————, 1960A, Relations between folding and metamorphism in the central Pyrenees, and their chronological succession: *Geologie Mijnb.*, v. 39, pp. 163-80.

————, 1960B, The chronological succession of folding and metamorphism in the central Pyrenees: *Geol. Rdsch.*, v. 50, pp. 203-18.

————, 1963A, In the determination of polymetamorphic mineral associations, and its application to the Bosost Area (Central Pyrenees): *Geol. Rdsch.*, v. 52 (for 1962), pp. 38-65.

————, 1963B, Some examples of the relations between deformation and metamorphism from the Central Pyrenees: *Geologie Mijnb.*, v. 42, pp. 143-54.

Author Index

Ackermann, Von E., 470, 624
Ackermann, K. J., 430, 624
Adams, P. J., 433, 632
Aderca, B. M., 298, 624
Agron, S. L., 228, 253, 270, 274, 447, 624
Agterberg, F. P., 94-5, 444, 554, 571-5, 591-2, 624
Akaad, M. K., 311, 624
Albee, A. L., 452, 661
Allen, J. R. L., 515, 624
Allison, A., 531, 625
Ambrose, J. W., 468, 637
Anderson, E. M., 109, 266, 269, 325, 381, 392, 519, 566-9, 625, 639, 649
Anderson, J. G. C., 199, 201, 383, 418, 535, 625
Argand, E., 323, 606, 608, 613, 618, 625
Atkinson, D. J., 43-5, 625
Ayrton, W. G., 468, 625

Badgley, P., 598, 606, 610-3, 620, 625
Bailey, E. B., 53, 55, 66, 69, 199, 213-4, 223, 313, 358, 361, 371-2, 377, 383-9, 397, 401, 404, 408, 420, 423, 519, 532-3, 565-7, 601, 607, 622, 625-6, 649
Bain, G. W., 146, 610, 626
Baird, A. K., 73, 458-67, 626
Balk, R., 148, 173, 183, 186-8, 239, 244-5, 294, 515, 626
Barnes, F. F., 185, 657
Barrell, J., 534, 626
Barrow, G., 381, 626
Barth, T. F. W., 515, 626
Bates, T. F., 171, 226, 626
Bayly, M. B., 568-71, 638
Bear, T. L., 467, 640
Beavis, F. C., 236, 238, 254, 626
Becker, G. F., 80, 121, 133, 223, 627
Beckman, W. A., Jr., 595, 662
Bederke, E., 539, 627
Behre, C. H., 226, 243, 627

Beloussov, V. V., 50, 88, 557-8, 569, 627
Belov, A. A., 446, 627
Bemmelen, R. W. Van, 569, 627
Bentley, R. D., 427, 447-9, 544, 635
Bentz, A., 144, 627
Berthelsen, A., 298, 300, 434-8, 627
Best, M. G., 233, 271, 460, 627
Bhattacharji, S., 329, 627
Biemesderfer, G. K., 20, 627
Billings, M. P., 40, 105, 130, 167-8, 172-4, 217, 220, 230, 260-1, 264-6, 269, 315, 450, 516, 597-9, 601-20, 627, 661
Birch, F., 568, 628
Bishop, M. S., 591, 628
Blakeman, W. B., 468, 642
Bogdanov, A. A., 601, 622, 655
Bonham, L. C., 290, 663
Bonney, T. G., 219, 230, 628
Borchers, R., 565, 628
Borg, I., 101, 637
Born, A., 222, 248, 628
Boschma, D., 441, 443-4, 628
Boswell, P. G. H., 254-5, 258, 262-3, 534, 628
Boucot, A. J., 516, 628
Bowes, D. R., 521, 535, 537, 539, 628
Boyer, R. E., 548, 552, 662
Brace, W. F., 230, 243, 269, 628
Bradshaw, R., 430-2, 639
Braitsch, O., 330, 628
Breddin, H., 123-6, 161, 163-4, 224, 283, 446, 628-9
Brett, G. W., 519, 521-2, 629
Briggs, G., 94, 575, 624
Brinkmann, R., 127-9, 629
Brobst, D. A., 450, 629
Broughton, J. G., 244-5, 446, 629
Brown, J. S., 450, 540, 629
Brown, P. E., 366, 503, 629, 646
Bryan, W. H., 283, 629
Bryhni, I., 428-30, 629
Bucher, W. H., 10, 536, 629
Buddington, A. F., 185, 257, 450, 470, 629

Burrell, H. C., 144, 539, 637
Busk, H. G., 604, 611-2, 617, 620, 629

Cadigan, R. A., 94, 629
Callaghan, E., 467, 639
Campana, B., 539, 629
Campbell, J. D., 370, 629
Carey, S. W., 140-7, 164, 167, 173, 263, 351-6, 559-61, 563, 619, 629
Carruthers, R. G., 392, 649
Challinor, J., 40, 621, 630
Chapin, T., 470, 629
Chayes, F., 77, 547, 591, 630
Cheesman, R. L., 313, 650
Ch'ih, C. S., 101, 659
Chinner, G. A., 489-90, 539, 630
Chorley, R. J., 545, 630
Christensen, M. N., 191, 630
Christie, A. M., 298-9, 315, 630
Christie, J. M., 415, 630
Clabaugh, P. S., 173, 630
Clark, R. H., 38, 43, 101, 606, 630, 659
Clark, S. K., 618, 630
Cleaves, A. B., 516, 627
Clenshaw, C. W., 581, 630
Clifford, P., 63-4, 340, 382, 392, 397-401, 404, 413-4, 473, 605, 612, 630
Clifford, T. N., 401, 411, 630
Cloos, E., 123, 126, 162, 164, 177-8, 229, 264, 269, 272-83, 286, 292, 294, 308, 311, 319, 516, 601, 610, 630-1
Cloos, H., 427, 631
Clough, C. T., 133, 322, 366, 371, 391-2, 401, 404, 631, 637, 649
Cochran, W. G., 72-4, 631
Coe, K., 299, 303, 305, 309, 631
Cohn, P. M., 582, 631
Collomb, P., 318, 631
Compston, W., 503, 644
Condon, M. A., 539-40, 631
Cooke, H. C., 469, 534, 631
Cooper, B. N., 543-4, 631
Corin, F., 294, 631
Cote, L. J., 94, 632
Cowie, J. W., 433, 632
Cox, G. H., 518, 533, 632
Craddock, J. C., 361, 632
Craig, E. H. C., 381, 392, 626, 640
Craig, G. Y., 522, 537, 632
Crampton, C. B., 323, 392, 473, 649
Crook, K. A. W., 260, 632
Crowell, J. C., 569, 632
Cummins, W. A., 69, 361-2, 632
Curie, P., 112, 122, 632
Curie, J. B., 561-3, 632
Cuthbert, F. L., 171, 226, 632

Dahlstrom, C. D. A., 38, 48, 52, 56, 59, 63, 605, 617, 619, 632
Dake, C. L., 518, 533, 632
Dale, T. N., 46, 226-7, 230-1, 450, 622, 632
Daly, R. A., 224, 256-7, 632
D'Amico, C., 444, 571, 573, 632
Dana, J. D., 611, 632
Darwin, C., 217-8, 515, 633
Davis, J. O., 94, 632
Dawson, K. R., 100, 591, 633
Dawson-Grove, G. E., 260, 633
Dearman, W. R., 427-8, 633, 655
Dearnley, R., 418, 633
Decker, C. E., 598, 601-4, 613-9, 621-3, 633
Derry, D. R., 469, 633
Dewey, J. F., 421-2, 633
Dietz, R. S., 561, 633
Dietzel, G. F. L., 444, 633
Dixon, R. H., 101, 659
Dolginov, Y. A., 446, 627
Doll, C. G., 516, 633
Donath, F. A., 39, 131, 633
Dons, J. A., 430, 516, 626, 633
Drever, H. I., 73, 633
Dunbar, W. R., 469, 633

Eberly, S. W., 52, 648
Engel, A. E. J., 50, 185, 187-90, 193-6, 201, 232-6, 241, 450, 472, 540, 629, 633
Engels, B., 224, 248, 633
Eskola, P. E., 223, 515, 565, 626, 634
Evans, A. M., 64, 67, 151, 261, 634

Fairbairn, H. W., 99, 101, 114, 216-7, 221-4, 315, 539, 548, 610, 634, 640
Falcon, N. L., 60, 603, 609, 614, 634, 636
Farquhar, O. C., 489, 652
Faure, G., 548, 640
Fearnsides, W. G., 226, 647
Fellows, R. E., 286, 634
Ferguson, H. G., 455, 653
Fermor, L. L., 313, 634
Fisher, R. A., 575-6, 634
Fisher, O., 222, 634
Flett, J. S., 390, 392, 482, 634, 649
Fleuty, M. J., 178, 392, 395, 397, 399, 401, 404, 414, 566, 605, 612, 630, 634
Flinn, D., 121, 152, 248, 271, 275-6, 532, 554, 632
Forster, A., 470, 624, 634
Foslie, S., 472, 635
Fourmarier, P., 170-1, 222, 262, 635, 644
Francis, G. H., 488, 635
Fränkl, E., 433, 635

Freedman, J., 427, 447-9, 544, 635
Fuller, M. D., 128, 635
Furtak, H., 123, 221, 285, 635
Fyfe, W. S., 590, 635
Fyson, W. K., 151, 153, 261, 303, 467, 635

Gair, J. E., 259, 635
Galwey, A. K., 481, 635
Gangopadhyay, P. K., 286-7, 635
Garretty, M. D., 143-4, 539, 635, 637
Gast, P. W., 418, 661
Gastil, G., 468, 636
Gault, H. R., 311, 636
Gealey, W. K., 466, 636
Gee, D., 471, 657
George, T. N., 227, 656
Geyer, A. R., 448, 540-1, 636
Giesel, W., 127-9, 629
Giffin, C. E., 502, 643
Gilbert, G. K., 615, 622, 636
Giletti, B. J., 365-6, 502, 636, 643
Gill, W. D., 424-6
Gillott, J. E., 241, 636
Gilluly, J., 257-8, 455, 466-7, 501, 636, 653
Gilmour, P., 361, 636
Gindy, A. R., 305-6, 636
Goguel, J., 31-2, 46-7, 158, 174, 222, 602, 611, 636
Goldring, D. C., 423, 636
Goldsmith, R., 450, 452, 636
Gonzalez-Bonorino, F., 164, 636
Goodwin, E. T., 581, 630
Gosselet, J., 294, 636
Grabham, G. W., 649
Grant, F., 595, 636
Gray, C., 448, 540-1, 636
Graybill, F., 545, 643
Green, J. F. N., 519, 636
Green, R., 94, 636
Greenly, E., 191-2, 244, 391, 401, 425, 618, 636, 649
Greensmith, J. T., 100, 636-7
Greenwood, R., 312, 637
Gregory, H. E., 315, 637
Griffiths, J. C., 74, 93, 545
Griggs, D. T., 101, 219, 568, 637, 650, 659
Guitard, G., 493, 637
Gunn, W., 322, 391-2, 401, 637, 649
Gunning, H. C., 468, 637
Gustafson, J. K., 143-4, 539, 637

Haaf, E. ten, 175, 522, 637
Habicht, K., 63, 65, 637
Hager, A. D., 269, 639

Haller, J., 433-4, 637
Haman, P. J., 63, 605, 637
Handin, J. W., 101, 219, 637
Harbaugh, J. W., 595, 637
Harder, J. O., 453, 648
Hardie, W. G., 567, 637-8
Harker, A., 217-8, 222, 224, 253, 294, 391, 404, 484, 505, 514-5, 638, 649
Harland, W. B., 568-71, 585, 638
Harman, H. H., 579-80, 638
Harris, A. L., 266, 483-5, 503, 507, 638, 641, 651, 657
Harris, R. L., Jr., 53-4, 638
Harrison, J. E., 451, 638, 647
Harrison, J. V., 603, 614, 638
Harrison, P. W., 94, 638
Haughton, S., 222, 638
Heard, H., 101, 659
Heim, A., 182-3, 224, 229-30, 232, 638
Hellermann, E., 123, 285, 635
Hellmers, J. H., 283, 638
Helm, D. G., 427, 638
Hempkins, W. B., 95, 640
Henderson, J. F., 469, 638
Henderson, S. M. K., 533, 639
Hewett, D. F., 469, 639
Hietanen, A., 170, 259, 262, 311, 453-5, 506-8, 631, 639
Higgins, A. K., 516, 639
Hill, J. B., 391, 637, 647
Hills, E. S., 40, 81, 147-51, 174, 220, 224, 238-9, 476, 598, 600-13, 615-23, 637
Hinxman, L. W., 322, 381, 392, 402, 626, 639-40, 649
Hitchcock, C. H., 269, 639
Hitchcock, E., 269, 639
Hodgson, J. H., 569, 639
Hoel, P. G., 583, 639
Hoeppener, R., 113, 123, 127-9, 224, 232-3, 248-9, 446, 629, 639
Hoffman, A. A. J., 95, 640
Hoffmann, K., 446, 639
Hollingworth, S. E., 430-2, 623, 639
Holmes, A., 219, 314, 419, 640, 652
Holmes, M., 430, 653
Holmquist, P. J., 294, 640
Hoots, H. W., 467, 640
Hopper, C. H., 469, 633
Horne, J., 313, 322, 364, 391-2, 401, 404, 536, 640, 649
Horton, C. W., 95, 640
Hotz, P. E., 455, 653
Houser, F. N., 555, 638
Houston, R. S., 452, 640
Howell, J. V., 218, 640
Howkins, J. B., 488, 640

Hubbert, M. K., 219, 228, 568, 640, 653
Hughes, H., 548, 640
Hull, E., 640
Hurley, P. M., 548, 640
Hutton, J., 514
Hutton, C. O., 481, 659

Ingerson, E., 104, 112, 120, 224, 539, 640, 642
Irving, E., 576, 660
Ives, R. L., 31, 640
Izett, G. A., 555, 640

Jahns, R. H., 450, 661
Jannettaz, M. E., 127, 641
Jenkins, W. O., 72-4, 631
Johnson, F. J., 144-5, 641
Johnson, M. R. W., 151, 154, 286-7, 388, 404, 415, 482, 484, 486-91, 502-3, 605, 641, 655
Johnston, W. D., 1, 31, 641
Jones, A. G., 208-9, 224, 236, 243, 258, 264-6, 288, 297-8, 300-4, 315, 470, 641
Jones, D. L., 467, 641
Jones, K. A., 481, 483, 521, 535, 628, 635, 641
Jones, M. A., 430, 653
Jones, O. A., 283, 629
Jones, O. T., 262, 519, 535, 641
Joubert, P., 380, 470, 653

Kalliokoski, J., 469, 641
Kalsbeek, F., 311, 442, 641
Kellaway, G. A., 623, 639
Kelly, J., 518-9, 641
Kelvin, Lord, 80, 641
Kendall, M. G., 579, 641
Kennedy, W. Q., 401, 405-7, 488, 491, 520, 641-2, 652
Kent, P. E., 60, 632
Kienow, S., 148, 248, 642
Kilroe, J. R., 312, 640
Kindle, C. H., 175-6, 662
King, B. C., 191, 194-6, 201, 237, 244, 259, 373-4, 380, 383, 387, 420, 473-5, 535, 567, 642
Kleinpell, W. D., 467, 639
Kleinsmiede, W. F. J., 197, 199, 221, 237, 248, 493, 642
Knill, D. C., 156, 245-8, 259, 419, 534, 642
Knill, J. L., 150, 152, 156, 232, 245-54, 259, 262, 370, 419, 534, 642
Knopf, A., 216, 220, 642

Knopf, E. B., 80, 99, 103-4, 112-3, 119-20, 156-8, 165, 181-6, 191-2, 224, 230, 232, 642
Knowles, D. M., 336, 468, 636, 642
Knowles, L., 57, 642
Koark, H. J., 433, 472, 642
Kober, L., 323, 642
Koch, G. S., Jr., 73, 644
Kouvo, O., 418, 661
Kranck, E. H., 267, 643
Krauskopf, K., 300, 321, 660
Kropachev, S. M., 446, 627
Krumbein, W. C., 72-3, 76-7, 545-51, 556, 591-2, 595, 611, 643, 660
Ksiazkiewicz, M., 534, 643
Kulp, J. L., 502, 643
Kürsten, M., 380, 389, 643, 661
Kurtman, F., 124, 283-5, 643
Kvale, A., 220, 262, 269-70, 428, 441, 472, 643
Kynaston, H., 198, 222, 649

Lacy, W. C., 469, 650
Lahee, F. H., 193, 643
Lambert, J. L. M., 156, 475, 643
Lambert, R. St. J., 365, 405-8, 502, 643-4, 636
Lapworth, C., 536, 566, 644
Larwood, G. P., 427, 633
Lawrie, T. R. M., 214, 384-5, 626
Leake, B. E., 503, 644
Leggo, P. J., 503, 644
Lehmann, J. G., 515, 644
Leith, C. K., 164-7, 220-1, 224, 257, 260-1, 315, 323, 518, 533, 598, 600-1, 604-9, 613-22, 644, 659
Lindström, M., 324, 433, 472-3, 644
Link, R. F., 73, 644
Lohest, M., 294-5, 644
Long, L. E., 365, 502, 643
Loretz, H., 229, 644
Loudon, T. V., 577, 579-82, 585, 588-90, 644
Lovering, T. S., 566, 644
Löwl, F., 256, 644
Lyell, C., 514, 644

MacCulloch, J., 294, 645
MacGregor, A. G., 403, 405
MacGregor, S. M. A., 536, 645
Mackin, J. H., 53, 55-6, 626, 644
MacPherson, W. A., 468, 642
Madlem, K. W., 73, 626
Mandelbaum, H., 592, 595, 645
Marks, W., 94, 632

Marmo, V., 418, 645
Marr, J. E., 566, 645
Martin, D. W., 581, 630
Martin, J. C., 323, 450, 472, 645
Matthews, D. H., 601, 645
Maxwell, J. C., 171-2, 222, 225-8, 255, 263, 269, 286, 645
Mayo, E. B., 459, 645
McCall, G. J. H., 380, 419, 470, 645, 653
McCallien, W. J., 66, 313, 372, 397, 423, 533, 601, 622, 626, 645
McCrossan, R. G., 312, 645
McGlynn, J. C., 66, 653
McGough, R. J., 94, 632
McGowan, H. S., 469, 633
McIntyre, D. B., 31, 34-9, 43, 46, 73, 110-1, 191, 193-4, 201-2, 205-7, 264, 295, 318, 340, 361-3, 366-7, 380-3, 386-8, 412, 461, 595, 606, 626, 630, 636, 645-6, 661
McKee, E. D., 517-8, 646
McKinstry, H. E., 53, 68, 646
McLachlan, G. R., 287, 646
McLaughlin, D. B., 448, 540-1, 636
McManus, J., 421, 633
McTaggart, K. C., 137-40, 241, 646
Mead, W. J., 165, 167, 217, 220-1, 230, 646
Meade, R. H., 227-8, 646
Means, W. D., 454-7, 471, 646
Mehnert, K. R., 272, 646
Mehr, E., 94, 632
Melton, M. A., 548, 550, 646
Mendelsohn, F., 151, 646
Merriam, C. W., 467, 648
Mertie, J. B., 602, 615, 646
Metzger, A. A. T., 209-13, 417-8, 646
Michener, C. D., 596, 646, 656
Michot, P., 430, 646
Middlemiss, F. A., 57, 60, 642, 646
Mikkola, A. K., 68, 646
Miller, G. F., 581, 630
Miller, J. A., 366, 503, 627, 646
Miller, J. M., 470, 646
Miller, R. L., 548, 595, 647-8
Miller, W. B., 101, 637
Mills, A. A., 502, 643
Milner, H. B., 647
Misch, P., 515, 647
Moench, R. H., 451, 647
Moorbath, S., 365-6, 502, 636
Moore, B. N., 467, 639
Moore, G. E., 235, 450, 647
Moore, W. L., 568, 647
Morris, T. O., 226, 647
Mosebach, R., 224, 647
Moseley, J. R., 448, 540-1, 636

Mosteller, F., 72-4, 631
Muehlberger, W. R., 173, 630, 647
Muff, H. B., 198, 222, 249, 252-3, 649
Mukherjee, A., 471, 654
Murphy, T. D., 516, 631

Nadai, A. L., 300, 647
Naha, K., 68, 471, 619, 647
Naumann, C. F., 266, 647
Nevin, C. M., 329, 598, 601, 611-3, 615, 619-22, 647
Nicholson, R., 351, 430, 432-3, 624, 647
Nicol, J., 536, 647
Niggli, E., 504, 647
Nissen, H. U., 124, 278, 285, 647-8
Noble, D. C., 52, 648
Noble, J. A., 453, 648
Nolan, J., 312, 640
Nolan, T. B., 1, 31, 467, 639, 641, 648
Norman, G. W. H., 470, 648
Nureki, T., 100, 648

O'Driscoll, E. S., 345, 648
Okamura, Y., 99, 648
Oldham, C. H. G., 595, 648
Olson, E. C., 548, 648
Olver, F. W. J., 581, 630
Orlov, R. Y., 446, 627
Oulianoff, N., 418, 648

Page, B. M., 176, 648
Pande, I. C., 313, 648
Park, R. G., 418, 648
Parker, R. B., 39, 131, 452, 460, 633, 640, 648
Paterson, M. S., 80, 113-9, 122-3, 126-7, 648
Patnode, H. W., 561-3, 632
Patterson, J. R., 566, 648
Pavlovskiy, Y. V., 389, 648
Peacey, J. S., 287, 648
Peach, B. N., 198, 222, 313, 322, 364, 391-2, 401, 404, 536, 640, 648-9
Peikert, E. W., 595, 649
Pettijohn, F. J., 104, 115, 514, 521-2, 544-5, 575, 649-50
Phemister, J., 388-9, 649
Phillips, F. C., 10-1, 21-6, 31, 267, 291-3, 315, 321, 391-2, 473, 650
Phillips, J., 222, 650
Phillips, W. E. A., 421-2, 633
Pichamuthu, C. S., 416, 418, 650
Pierson, W. J., 94, 632

Pilger, A., 315-7, 650
Pinson, W. H., 548, 640
Pitcher, W. S., 313, 421, 650
Plas, L. van der, 492, 650
Platt, J. I., 324, 374-5, 651
Poll, J. J. K., 444, 446, 650
Poole, A. B., 405, 408, 644
Poole, F. G., 555, 640
Potapov, I. I., 601-2, 622, 650
Potter, P. E., 94, 545, 575, 650
Powell, D., 408, 650
Pryor, W. A., 94, 650

Quirke, T. T., 294, 469, 650

Raleigh, C. B., 568, 650
Ramberg, H., 129, 153, 168, 295-9, 301, 308-10, 507, 557, 563-4, 602, 608-10, 620, 650-1
Ramsay, A. C., 294, 651
Ramsay, D. M., 373, 377, 433, 651
Ramsay, J. G., 52, 133, 151, 154-5, 158-64, 267, 340, 346-51, 356, 366-70, 374, 392-8, 401-12, 415, 432, 442, 465, 471, 483, 490, 522, 526-7, 530-2, 539, 605, 612, 630, 651
Rast, N., 105, 112, 133, 191, 194-201, 300, 305, 311-2, 324, 361, 370-9, 382-3, 387, 420, 475, 483-9, 502, 507, 535, 567, 638, 642, 651
Read, H. H., 257, 288, 294, 313, 389, 391, 421, 489, 514-5, 518-9, 534-6, 540, 650-2
Reece, A., 470, 652
Reid, R. R., 303, 453, 455, 508, 513, 652
Reitan, P., 298, 652
Reusch, H., 269, 652
Reyer, E., 294, 652
Reynolds, D. L., 419, 652
Richey, J. E., 405, 518, 652
Richter, D., 418, 440, 652
Rickard, M. J., 88, 232, 236-8, 243, 253, 262, 419-21, 652-3
Ringwood, A. E., 548, 653
Roach, R., 329, 653
Roberts, B., 427, 638
Roberts, J. L., 361, 536, 645, 653
Roberts, R. J., 455, 653
Robertson, D. S., 469, 653
Ropsek, J. F., 94, 632
Rosenfeld, M. A., 72-4, 653
Ross, J. V., 66, 469-70, 521-2, 653
Rubey, W. W., 467, 568, 639-40, 653
Ruger, L. 269, 653

Rusnak, G. A., 94, 653
Russell, W. L., 598-609, 613-23, 653
Rutland, R. W. R., 403, 430-1, 653
Rutledge, H., 412, 653

Saggerson, E. P., 380, 470, 653
Saha, A. K., 418, 471, 653-4
Salter, J. W., 294, 651
Sander, B., 50, 96, 102-6, 112-5, 119, 122, 140-1, 156, 159, 183, 186, 223-4, 253, 259, 271, 289, 313, 318, 325, 340, 472, 477, 479, 654
Sando, M., 539, 630
Sarkar, S. N., 418, 471, 654
Sawyer, W. W., 121, 582, 654
Schaller, W. T., 467, 639
Schenk, E., 248, 654
Scheumann, K. H., 297-8, 654
Schmidt, W., 10, 137, 167, 315-7, 477, 482, 516, 650, 654
Schmincke, H. U., 315, 654
Schoeman, J. J., 470, 654
Scholtz, H., 148, 654
Schreyer, W., 389, 446, 655
Schuchert, C., 611, 655
Schuiling, R. D., 504, 655
Schulman, N., 440, 442, 655
Schwarzacher, W., 124, 655
Scotford, D. M., 448, 655
Scrope, G. P., 217-8, 322, 655
Secrist, M. H., 31, 655
Sederholm, J. J., 618, 655
Sedgwick, A., 222, 655
Shackleton, R. M., 69, 255-6, 361-2, 377, 535, 632, 655
Sharp, R. P., 178, 516, 627, 655
Sharpe, D., 222, 655
Shatski, N. S., 601, 622, 655
Shearman, D. J., 303, 617
Sheridan, D. M., 452, 661
Shiells, K. A. G., 427-8, 633, 655
Shrock, R. R., 69, 514, 533, 536, 655
Simpson, A., 241, 427, 638, 655
Simpson, S., 248, 656
Sims, P. K., 451, 647
Sironko, P. T., 548, 656
Sitter, L. U. De, 105, 132, 144-5, 148-52, 158-9, 162, 165-8, 171-4, 219, 222-5, 260-2, 299-302, 307-8, 437-42, 471, 492-5, 568, 598-621, 656
Slaughter, A. L., 453, 648
Sloss, L. L., 545, 548, 611, 643, 656
Smith, B., 227, 656
Sneath, P. H. A., 595-6, 656

INDEX

Snow, J. P., 548, 656
Sokal, R. R., 595-6, 646, 656
Sokolov, B. A., 446, 627
Sollas, W. J., 560
Sommer, M. 434, 656
Soper, N. J., 366, 503, 629, 656
Sorby, H. C., 223-4, 657
Sosoka, J., 101, 637
Spencer, E. W., 92, 657
Spring, J., 366-9, 401, 405-10, 539, 651
Spry, A., 471, 482, 510-1, 520, 539, 657
Stainier, X., 294, 644, 657
Stark, J. T., 185, 657
Staub, R., 323, 657
Steinmetz, R., 94, 657
Stephenson, G., 94, 632
Stewart, F. H., 100, 388, 487-9, 641, 657
Stille, H., 609, 657
Stillwell, F. L., 515, 657
Stočes, B., 168-9, 598, 600-23, 657
Stockwell, C. H., 38, 56-8, 63, 605, 657
Storey, T. P., 566, 648
Strand, T., 428, 472-3, 657
Sturt, B. A., 116-7, 266, 270, 291, 372-4,
    377-9, 383, 403, 433, 483-4, 489, 502,
    507, 651, 657
Sutherland, D. B., 595, 648
Sutton, J., 63, 69, 327, 359-60, 365-7, 389-
    92, 395-7, 401-12, 418, 473-5, 489,
    520-1, 530-3, 537-8, 605, 612, 619,
    630, 658, 662
Sutton, R. F., 177-8, 658
Suzuki, T., 244, 658
Suzuki, Y., 547, 630
Swanson, C. O., 217, 220-1, 224, 658

Tait, P. G., 80, 641
Talbot, J. L., 537
Taliaferro, N. L., 467, 658
Tanton, T. L., 519, 658
Taylor, J. H., 623, 639
Teall, J. J. H., 322, 649
Teichmüller, M., 248, 658
Teichmüller, R., 248, 444, 658
Tex, E. den, 26, 104, 476, 503-5, 512-3,
    557, 658-9
Thomas, D. E., 174, 639
Thompson, J. B., 516, 628
Tilley, C. E., 408, 626
Tilton, G. R., 418, 661
Tischer, G., 63-5, 659
Toit, A. L. du, 563, 659
Tozer, C. F., 313, 650
Treagus, J. E., 361, 653
Tremlett, W. E., 487, 503, 659

Trump, R. P., 561-3, 632
Tukey, J. W., 72-4, 630
Turner, F. J., 10, 22, 40, 50, 63, 80-1, 86,
    101, 105, 112-3, 121-2, 132, 141, 156-
    7, 169, 173-4, 199, 207, 219-20, 264,
    269, 290, 315, 342, 481, 516, 556, 585,
    588, 597-8, 601-23, 635, 637, 645, 659
Turner, J. S., 401
Twenhofel, W. H., 518, 659

Van Hise, C. R., 132-3, 164-7, 183, 193,
    217, 220, 236, 323, 598-600, 616-7,
    659
Vardabasso, S., 444, 659
Verhoogen, J., 112, 516, 590, 635, 659
Vetter, R. C., 94, 632
Vogel, D. E., 512-3, 659
Vogt, M. C., 213, 215, 383, 660
Vogt, T., 519, 533, 660
Voll, G., 124-6, 160-1, 164, 224, 254, 269,
    371, 377, 383, 403, 567-8, 660

Walls, R., 294, 309, 484, 537, 660
Walton, B. J., 430, 624
Walton, E. K., 522, 534-7, 632, 660
Ward, R. F., 308, 448, 660
Waters, A. C., 300, 321, 660
Watson, G. S., 533, 576, 660
Watson, J., 359-60, 365-7, 390-2, 397, 401-
    2, 407-12, 418, 473-5, 484, 489-90,
    502-3, 520-1, 530-3, 537-8, 605, 612,
    630, 658, 660, 662
Waye, I., 595, 662
Webster, R. K., 502, 643
Wegmann, C. E., 37, 48, 179-80, 207-8, 294,
    297-9, 303, 434, 481, 660
Weed, C. B., 391, 404, 649
Weiss, J., 230, 660
Weiss, L. E., 10, 22, 31, 34-5, 40, 43, 50, 63,
    80-1, 86-8, 96, 105-6, 111-23, 126-7,
    157, 173-4, 191-3, 199-207, 219-20,
    264, 269, 290, 315, 318, 326-9, 340-2,
    357, 380, 383, 386-8, 425, 433, 461,
    465, 470, 585, 598, 601-23, 646, 648,
    659-61
Welday, E. E., 73, 626
Wells, J. D., 451-2, 638, 661
Wells, M. K., 430-2, 639
Wenk, E., 504, 516, 661
West, I. M., 180, 661
Wetherill, G. W., 418, 661
White, A. J. R., 539, 630
White, C. H., 168-9, 598, 600-23, 657
White, H. J. O., 60-1, 661

White, W. S., 230, 236, 241-3, 450, 661
Whitehead, W. L., 533
Whitten, E. H. T., 57, 60, 73-6, 93, 103-4, 191, 194, 200, 204-7, 213-5, 289, 297-8, 327-9, 359, 380-2, 423, 478-80, 545-8, 552, 591-6, 633, 661-2
Whittington, H. B., 175-6, 662
Wier, G. W., 517-8, 646
Wilkinson, J. H., 581, 630
Willden, C. R., 467, 662
Williams, A., 49-50, 62, 662
Williams, E., 151, 165-7, 539, 662
Williams J. S., 467, 648
Williams, L. A. J., 380, 470, 653
Willis, B., 598, 601-4, 607, 609, 612-23, 662
Willis, R., 598, 601-4, 607, 609, 612-23, 662
Wilson, A. F., 521, 662
Wilson, G., 155, 220, 223, 230, 261-2, 282-3, 295-7, 302-5, 313-20, 405-6, 521, 530-1, 662

Wilson, J. S. G., 222, 649
Wise, D. U., 92, 427, 447-9, 544, 635, 662
Woodland, B. G., 235-6, 244, 450, 662
Woodward, H. B., 391, 404, 649
Wright, A. E., 537, 628
Wunderlich, H. G., 224, 300, 306-10, 444-5, 504, 526, 532, 662-3
Wyllie, P. J., 548, 663
Wynne-Edwards, H. R., 174-5, 469, 663

York, D., 366, 503, 629

Zen, E-An, 68, 243, 663
Ziegler, J. M., 548, 647
Zimmerle, W., 290, 663
Zwart, H. J., 171, 174, 262, 271, 440, 444-6, 482, 492-500, 507, 511, 650, 656, 661

General Index

Acadian Orogeny, 450, 469
Achsenverteilungsanalyse, 15
Adirondack Mountains, New York, 185, 195-6, 234-6, 241, 256-7, 323, 450, 472 540
Agglomerate fragments, elongated, 283
Alpine Orogeny, 437, 440-1, 444
Alps, 31-2, 323, 418, 440, 442, 492, 503-4, 572-4
Ammonoosuc volcanic conglomerate, Littleton-Moosilauke Area, U.S.A., 266, 269
Amygdule, elongated, 280, 283
Analysis, microtextural, 477-501
  petrofabric, 98, 289, 315, 477
  stereographic, 91
  trend surface, 592, 595
Anlagerungsgefüge, 104
Anteclise, 600
Anticline, 66, 68, 165, 601
Anticlinorium, 601
Antiform, 66, 68, 410, 601
Appalachian fold belt, 53, 126, 178, 229, 235-6, 242-4, 269, 277-8, 286, 311, 322, 359-60, 427, 447-52, 468-9, 516, 533, 540-4
Ardennes, Belgium, 294-5, 306-7
Assynt Area, NW Scotland, 292
Asymmetry, 585, 588-9
Attitude, 585, 588-9
Ausweichungsclivage, 182-3, 185, 224, 230, 261
Axes, fabric, 266
  geometric fold, 37, 106-7
  kinematic fold, 106-7, 109, 124, 131-8
Axis, principal, 579-80, 582, 585, 588

B, 50
Ballachulish, Argyllshire, Scotland, 116-7, 213-5, 371, 380, 383-5, 403, 405, 537, 567
Ballapel Foundation, 371, 380, 383

Banffshire, NE Scotland, 315, 371, 484, 489, 533-4, 537-8, 544
Baraboo Quartzite, Wisconsin, U.S.A., 519, 521, 522
Basin, 66, 602
Bedding, cross-, 514, 517-32, 540
  graded, 517, 532-4, 536
Beinn Dronaig, Northern Highlands, Scotland, 367, 412-3, 473
Belt Series, 506-7
Belteroporic texture, 258
Beta, 50
Bewegungsbild, 119
Block diagrams, 31-5
  isometric, 31
  orthographic, 31
  perspective, 31
Boudin, 293-312
Boudinage, 199, 293-312, 425, 478, 480, 512
  sedimentary, 312
Brachiopods, 123, 126, 283, 516, 532
Brachyanticline, 602
Brachysyncline, 602

Calamites, 286
Caledonian, 402, 433
Caledonide, 322, 363, 388, 425, 427-8, 430, 434, 472, 502-5, 565
Caledonoid, 383
Central Scottish Highlands, 155, 187, 191, 194-9, 201, 222, 232, 248-9, 253, 294-5, 359-88, 403, 475, 482-91, 501-2, 507, 519, 533, 536, 566-7
Characteristic, response, 551-2
Circle, great, 11, 49, 134
  primitive, 13-4
  small, 11, 18
Cleavage, 216, 219-20
  see foliation
  cataclastic, 230

close-joints, 261
false, 261
fault slip, 261
flow, 186, 220-1, 236, 255, 261
fracture, 152, 183, 186, 216, 219-21, 255, 260-2, 516
kinked, 248
reticulate, 260
secondary, 236
shear, 230, 261
slaty, 153, 186, 219, 221-2, 225, 262
slip, 183, 230, 236, 261
strain-slip, 183, 230, 249, 261-2
transposition, 230
Coast ranges, California, U.S.A., 466, 501
Orogeny, 466
Competent, 40, 174, 156-7
Compression, 531-2
orthorhombic, 271
Confidence band, 76, 80, 86, 88-9
Conglomerate, Bygdin, 472
Pennsylvanian, Newport, Rhode Island, 228
Conococheague Limestone deformation, 126, 277-8, 311, 540-1, 544
Craignish-Klimelfort District, Argyllshire, Scotland, 151, 156, 249-52, 259
Crinoid, 124, 284, 285

Date, 567
Decke, 615
Décollment, 249
Deformation, affine, 121, 136
penetrative, 111, 138, 268, 289, 380
Diablastic (sieve) texture, 484
Diagenesis, 129-30, 515
Diagram, beta, 50-1, 267
girdle, 99-100, 103
pi, 50-2, 58, 60, 94, 267
pole, 99, 103
synoptic, 332-3, 344
Differentiation, metamorphic, 199
Dip, average, 49-50
Direction cosine, 575-83, 585, 588, 592
Domain, 89, 91, 113
Domains, homogeneous, 85, 267
statistically homogeneous, 85
Dome, 66
Dome, salt, 144, 172-3
Donegal, Ireland, 22-3, 25, 232, 238, 243-5, 289-91, 297-8, 305-6, 312, 315, 318, 418-21, 478-80, 533-4
Dutchess County, New York, U.S.A., 186, 228, 239, 244
Dykes, sandstone, 225
Dynamics, 119

Eigenvalue, 579-80, 582, 588
Eigenvector, 579-80, 582, 588
Einengung, 271

Fabric, 97, 113-4, 269
apposition, 104
axes, 105, 266
belteroporic, 289
chemical precipitation, 104
data, 114
depositional, 104
element, 98, 102, 113
element, combined, 123
element, composite, 126
element, crystallographic, 114, 127
element, imposed, 123, 126
element, inherited, 123, 126
element, kinematically active, 126
element, kinematically passive, 123
element, noncrystallographic, 114, 123
final, 119
heterotactic, 119
homotactic, 119, 537
initial, 119
primary, 97
secondary, 97
Face, 69
Facies, tectonic, 569
Factor, process, 546, 551-2, 557
Fannich Forest Region, Ross-Shire, 390, 392, 397, 411, 473-5, 520
Fault, 565-6, 568-9
Highland Boundary, 377
Feedback, 550, 553
Fissility, 217, 261
Flattening, 140, 159, 162, 164, 356, 483
Flexure, level of, 50
Flow, gravitational, 175
laminar, 146
plastic, 175
rheid, 263
turbulent, 146
viscous, 175
Fold, Alpine, 441
asymmetrical, 41, 601
axial-plane, 448, 602
$B_1 \perp B_2$, 327, 472-6
$B_1 \wedge B_2$, 327-30, 358-472
bottoming, 68
Caledonoid, 373-4, 375, 383, 419-20
conical, 63, 582-3, 585, 605
conjugate, 151, 154-5, 404, 476, 499, 605
cross, 324, 329, 373-5, 420, 605
crossing, 329
cylindrical, 40, 605

cylindroidal, 36, 40, 174, 207, 582-3, 585, 606
diapiric, 66, 606
disharmonic, 40, 179-80, 186, 190, 201, 207, 606
drag, 151, 153, 164-8, 607-8
due to vertical movements, 172
eyed, 351, 427, 432, 442, 457
flexural-slip, 109, 130-4, 137-8, 147-52, 156-62, 165-74, 178-81, 221, 244, 252, 259, 265, 268-9, 324, 609
flexure, 130, 169, 324, 609-10
flow, 146, 173-4, 610
gleitbrett, 137, 139, 146, 499
helicitic, 496, 501
homoaxial, 45, 174, 612
isoclinal, 137, 175-6, 181, 185-6, 222, 241, 244, 405, 567, 613
minor, 133, 168, 174, 614
minor flexural-slip, 168, 614
minor slip, 168, 614
modal, 49
neutral, 397
noncylindroidal, 36, 63-6, 615
open, 222, 616
overturned, 137, 140, 245, 617
parallel, 109, 132, 133, 617
parasitic, 151, 164-8, 179, 189, 329, 395, 538, 617
reclined, 618-9
recumbent, 68, 137, 140, 201, 259, 383, 420, 448, 619
recumbent isoclinal, 204, 258
refolded isoclinal, 235
rheid, 142-4, 619
rootless, 199, 620
scar, 300, 302
similar, 132-3, 162, 620
size, 47
slip, 109, 130-48, 156-7, 161-4, 167, 169-74, 178, 221, 225, 259, 263, 269, 324, 499, 620-1
style, 37
superimposed, 329, 621
superposed, 83, 174, 237, 267, 322-476, 621
superposed flexural-slip, 329-40
superposed slip, 340-56
symmetrical, 41, 621
terminology, 597-623
topping, 68
transverse, 329, 622
wild, 173
Fold-fault, 566

Folding, $B_1 \perp B_2$, 325, 327
$B_1 \wedge B_2$, 325, 327-30, 332, 334
cleavage, 171, 601
concentric, 171, 604-5
penetrative slip, 341
true, 130, 622
Foliation, 216-63
axial-plane, 89, 153, 162, 186, 221, 225-8, 235, 238, 244, 249, 254-5, 259-60, 262-3, 267
axial-plane crenulation, 259
bedding, 156, 256, 258, 260, 221
crenulation, 167, 186, 221, 224, 230-44, 248, 251-6, 260-3, 267, 377
crenulation developed as conjugate sets, 238, 249, 251, 254
linear, 313
nonpenetrative, 239
penetrative, 236, 241, 244
penetrative crenulation, 263
Force, dynamic, 130
kinetic, 119, 130
of crystallization, 505
Foresyncline, 611
Form regel, 101
Fossil, strained, 162, 268, 283
Frenchman Orogeny, 511, 520
Front Range, Colorado, 108, 229, 238-9, 450-2

Garnet, rolled, 287
Garnets, snowball, 481
Gaspésian Orogeny, 468
Geanticline, 611
Gefüge, 113
Gefüge, funktionale, 113
gestaltliche, 113
Gefügeelemente, 102
Geosyncline, 611
Gitter regel, 101
Gleitbrett, 140, 167, 241
Glen Orrin, 390, 392, 395, 397-9, 404, 414
Glenelg, West Scotland, 340, 348-50, 359-60, 366-70, 390-1, 404-5, 407-9, 412-4, 539
Goniatites, 123, 283
Grampian Highlands, Scotland, *see* Central Scottish Highlands
Grantown Series, 380
Grantown-Tomintoul Area, Central Highlands, Scotland, 46, 110-1, 201-2, 367, 380, 382
Graticule, Wulff stereographic, 13

Greenland, 180, 207-8, 294-8, 300-1, 418, 433-8, 481
Grenville orogenic province, 468
  Series, Adirondack Mountains, 175, 187, 189-90, 232, 234, 540
Group, point, 115
  space, 115

Helicitic, 481-2, 484, 496, 500-1
Hercynian Orogeny, 418, 437-44, 492, 494
Homoaxial, 111, 207-8, 291, 329, 367, 380, 382, 392, 415, 470
Homogeneous, 80
  statistically, 80

Idaho Batholith, Idaho, U.S.A., 170, 259, 453, 506-8
Iltay Nappe, 371-2, 377-80, 485, 487, 489
Incompetent, 40, 156, 174
Inliers, Lewisian, 390-402
  supposed-Lewisian, 391-402
Interlock of variables, 77
Interngefüge, 481
Involution, 178
Ireland, 418-25, 502
Iron Ore Orogeny, 418, 471
Irumide Orogeny, 470
Isocline, 181, 187, 613

Joint, *ac-*, 267, 289
Joint-drag, 152, 248-9
Joints, analysis of, 92

Karelidic Orogeny, 416
Ketilide Orogeny, 207, 434
Kinematics, 119, 130
Kink zone, 249
Kintail Region, 401, 411-4
Knoydart Area, Scotland, 366, 368-9, 390, 405-7, 503
Kurtosis, 585, 589-90

Lag, 420, 566
Lamellibranchs, 123, 283
Lava bombs, elongated, 283
Lavas, pillow, 536
Laxfordian Orogeny, 365, 414, 418

Lepidodendron, 286
Linear element, nonpenetrative, 265
  penetrative, 265
  rotated, 287
Linear structures, 222, 264-321
Lineation, 264-321
Lithification, 515
Loch Leven, Scottish Highlands, 118-9, 191, 193, 213, 340, 383, 386-8, 403
Loch Luichart Area, Northern Highlands, Scotland, 382, 390, 392, 397, 399-401, 403-4, 407, 473
Loch Monar Area, Northern Highlands, Scotland, 348-9, 390, 392-5, 397, 403-4, 414
Loch Morar, Northern Highlands, Scotland, 287, 366, 390, 405-7, 410, 501, 518
Loch Tummel, Perthshire, Scotland, 270, 373-4, 377-80, 484-9, 502

Maps, 1-10
Martic overthrust, 262, 447
Martinsburg Slate, 171, 225-6, 244, 263, 269, 286, 446
Metamorphism, load, 224, 256-8
  static, 256-7
Microlithon, 162, 164, 167
Mid-Strathspey Area, Scotland, 34-5, 39-40, 43, 48, 103-4, 194, 200-2, 204-7, 213, 215, 266, 288, 359, 380-3, 423
Mimetic crystallization, 257, 289, 515
Modal limb, 49, 50
Model, conceptual, 546
  deterministic, 546
  experimental, 557
  kinematic, 556
  kinetic, 556
  process, 546-53
  process-response, 545-61, 553-4, 556, 569-70, 572, 591
  response, 545-53, 567
  scale, 557
  statistical, 545-6
Moine Thrust Zone, 117-8, 154, 286-7, 292, 364-5, 389-91, 404-5, 413, 415, 502
Moments, statistical, 580, 587-8
  affine, 119-20
  componental, 104, 130, 216
  direct componental, 102, 224
  indirect componental, 102, 224, 256
  nonaffine, 120-1
  penetrative, 286

Movement picture, 119, 121, 123, 130, 156
Mozambiquian Orogeny, 470
Mullion, 264, 293, 313-8, 421
  bedding, 315
  cleavage, 315
  fold, 315
  foliation, 314-7
  irregular, 317-8

Nappe, 154, 403, 438-41, 448, 491, 567, 569, 615
  Iltay, 371-2, 377-9, 380, 485, 487, 489
  Kishorn, 502
  Moine, 405
  Morar, 401, 405
Neocrystallization, 99-104, 114, 127, 130, 167, 222, 224, 228, 254, 258, 260, 263, 268, 289, 291, 403, 405, 480-1, 488, 500, 507, 510, 515-6, 539
  diagenetic, 324
Neomineralization, 99
Nevadan Orogeny, 457
Nonaffine, 135
Nonpenetrative, 114, 167, 239, 254, 256, 260
Nontectonites, 105
North Wales, 192, 226-7, 244, 253-5, 258, 263, 294, 425, 427
Northern Highlands, Scotland, 388-415, 490
Northwest Highlands, Scotland, 63-4, 151, 158, 286, 291, 319-20, 323, 391-2, 397, 401-4, 411-5, 473-5, 490, 503, 520, 536
Norway, 116-7, 153, 269-70, 291, 294-7, 351, 428-33, 472, 516
Nose, 616

Observations, typical, 71
Oolite, 273, 275, 277
Oolith, 123, 125, 268, 274, 278, 281, 532
  strained, 162, 164, 229, 277, 279
Ord Ban, mid-Strathspey, Scotland, 34-5, 39-40, 194, 205, 207, 380, 382
Orientation, 569
  habit, 101, 289, 291, 495, 500
  lattice, 101-2, 289, 291, 495
  nonpenetrative, 113
  penetrative, 113
  preferred, 98
Ossicle, crinoid, 283, 532
Otago Schists, New Zealand, 105, 199, 481, 516
Oykell Bridge, NW Scotland, 314-5, 317-8

Palimpsests, 477, 478, 492
Partition, 299
Pattern, subsystematic distribution, 82
  systematic, 75-6, 78-9, 86
  systematic areal, 82
  unsystematic, 75-6, 78-9
Peach Bottom Slate, U.S.A., 253, 447-8
Pebbles, 125, 271
  strained, 119, 264, 269, 275-6
Penetrative, 114, 167, 236, 241, 244, 257, 260, 263, 268
Pericline, 66
Perthshire, Scotland, 194, 197-9, 300, 311, 373-4, 377, 379, 481, 488, 505, 521, 535, 544
Petrofabrics, 97
Picture, kinematic, 130, 191, 196
Pitch, 26-7, 38
Plane, axial, 40
  bisecting, 7
  S-, 96-8, 102
Plattung, 140, 159
Plunge, 7, 16, 38
Pole, *S*-, 50
Population, available, 74
  existent, 74
  homogeneous attribute, 80
  hypothetical, 74
  of attributes, 71-2, 81-2, 85-6, 89, 555, 570
  of directional attributes, 87
  of objects, 71-3, 81-4, 88-92, 113, 546, 555, 570-1
  sampled, 74, 554
  subsystematic, 86
  target, 72, 74, 551, 573
Porphyroblast, 478-505, 534
  growth, 505
  rolled, 288
Profile, 42-3, 53, 56, 165
  coulisse, 203
  fold, 37, 135, 158
Projection, equal-area, 10-31, 90-3, 98
  isometric, 32-3
  Lambert, 10
  orthographic, 32, 34
  Schmidt, 10, 18-26
  stereographic, 10-31, 90
  Wulff, 10, 14, 18
Pseudo-bedding, 182, 186-7, 191-3, 197, 199, 202, 207, 234, 257-8, 516
Pseudostratification, 516
Pseudo-unconformities, 190
Pyrenees, 148, 152, 167, 171, 197, 199, 225, 240, 248, 437-40, 443, 478, 492-508, 511

Radiolaria, 283
Regime, tectonic, 569
Relicts, palimpsestic, 484, 492
Rheid, 141-2, 144, 146-7, 164, 167, 173, 175, 263
Rheidity, 141, 324
Rift, 261
Rigidity, 141
Ripple marks, 519, 532, 534, 536-7, 539
Rod, 264, 313, 319-21
Rodding, 293, 318-9
Rosguill, Donegal, Ireland, 156, 245-8, 259, 419
Rotation, penetrative, 287
Runnzelsclivage, 232

Sag, 620
Sample, probability, 92
   typical, 71
Sampling, 70
   probability, 73, 92
   upper semiprobability, 73-4
Scale, macroscopic, 86, 88
   mesoscopic, 86-7
   microscopic, 86
   submicroscopic, 86
Scardroy, Ross-Shire, 367, 390-2, 401-2, 405, 407, 412
Scherungstektonite, 103
Scherung-S, 138
Schichallion, 361, 372-8, 380, 383, 420, 475, 482, 484-8, 502
Schistosity, 216, 219
   axial-plane, 235
   bedding, 252, 259
   shear, 138
Schmelztektonite, 104
Scotland, 256, 363-415, 502
Scottish Highlands, 196, 327, 363-415, 418, 508
Scourian Orogeny, 365, 418, 473
S_e, 481, 483-501
S_e-fabric, 484
Sequence, 569
Shape, 585, 588-9
Shuswap Terrane, British Columbia, Canada, 208-9, 224, 236-7, 243, 256-8, 264, 266, 288, 298, 300-1, 303, 315, 470
S_i, 481, 483-501
Sierra Nevada Foothills, California, U.S.A., 233, 271, 460-9
S_i-fabric, 484
Sigillaria, 286

Singhbhum Orogeny, 418, 471
Size, 70, 80-2, 84-5, 88, 109, 113-4, 121, 559, 569, 590
Skewness, 585, 588-9, 594
Slickensides, 114, 265, 296
Slickensides-grooving, 314
Slide, 35, 201, 259, 395, 397, 401, 409-10, 420, 491, 565-8
   Ballachulish, 385
   Boundary, 373-4
   Fort William, 566-7
   Iltay, 565
   Morar, 405-6
   Sgùrr na Cairbe, 395, 397
   tectonic, 213, 245
Sliding, gravity, 176, 271
Slip, bedding, 178
   gravitational, 178
   nonaffine, 134, 142, 146
Slope, 586-7
Slump/slumping, 175, 517, 520, 537-8, 569
Sole marks, 532, 537
South Mountain, Maryland, U.S.A., 123, 126, 164, 229, 272, 275, 277-81, 283
Specimens, oriented, 35
Strain, 121, 141, 272, 281
   affine, 136
   elastic, 16, 140-1, 171
   elastico-viscous, 171
   ellipsoid, 121-3, 127-8
   finite, 121
   finite affine, 121
   firmoviscous, 140-1
   general, 121-2
   homogeneous irrotational, 223
   nonaffine viscous, 173
   plane, 121
   plastic, 140-1
   pure, 121-2
   viscous, 140, 164, 171
Stratification, cross-, 517-32
Stratum contour, 3-4, 6, 9
Stress, 121, 141
Striations, 265
Structure, corduroy, 313
   flame, 534
   flap, 609
   gleitbrett, 140, 167, 241
   linear, 222, 264-321
   mimetic, 257, 289, 515
   mullion, 312-8, 421
   penetrative linear, 266
   prelithification, 178
   S-, 97, 101
Style, 569-70
Subfabric, 114-7, 119

Subpopulations, 85-6, 90
  homogeneous, 85
Surface, axial, 40
  bisecting, 40
  flattening, 103
  S-, 96-8, 103
Susceptibility, magnetic, 128
Svecofennidic Orogeny, 211, 416-7
Swell, 621
Switzerland, 63, 65, 161-2, 182-3, 232, 283, 418, 444, 516
Symmetry, axial, 116, 258
  monoclinic, 106, 117, 151, 155, 258
  orthorhombic, 117, 151
  spherical, 115
  tetragonal, 119
  triclinic, 119, 151
Syncline, 66, 68, 165, 622
  stratigraphic, 68
  structural, 68
Synclinorium, 423, 622
Syneclise, 622
Synform, 66, 68, 398, 410, 450, 622

Tables, closed, 77
Tablettes de chocolat structure, 294, 303, 309
Taconic Orogeny, 450, 469
Taxonomy, numerical, 595
Tectonic inclusion, 196-9, 201, 203
Tectonics, gravity, 563, 567
Tectonite, 102-4, 130, 197, 199, 207, 269, 289, 293
  B-, 102-4
  fusion, 104
  R-, 102-3, 105

rotation, 103
S-, 102-3
slip, 103
Teilbewegung, 102
Thrust, 566
Tightness, 585, 588-9
Till, glacial, 76
Tintagel, Cornwall, England, 282-3, 295-6
Toiyabe Range, Nevada, U.S.A., 454-7
Transposition, 179, 181-5, 191, 198-9, 207, 232, 234, 342, 516
Trend, 16, 38, 49
  surface analysis, 94, 593-4
Trough, 622
Tumbide Orogeny, 470
Turoka Area, Kenya, 117-8, 187, 201, 203, 207, 380, 470

Überprägung, 119
Umfaltungsclivage, 183
Umprägung, 119
United States, 446-67

Variable, homogeneous, 91
Variscan Orogeny, 425
Vergence, 427
Vesicles (gas), elongated, 283
Viscosity, 141

Wachstumsgefüge, 104
Washout structure, 537
Wedge, 177-8

Young, 69